MOM
GENES

ALSO BY ABIGAIL TUCKER

The Lion in the Living Room

MOM GENES

INSIDE THE NEW SCIENCE
OF OUR ANCIENT
MATERNAL INSTINCT

ABIGAIL TUCKER

G
Gallery Books

New York   London   Toronto   Sydney   New Delhi

G

Gallery Books
An Imprint of Simon & Schuster, Inc.
1230 Avenue of the Americas
New York, NY 10020

First Gallery Books hardcover edition April 2021

GALLERY BOOKS and colophon are registered trademarks of
Simon & Schuster, Inc.

Some names and identifying details have been changed.

For information about special discounts for bulk purchases,
please contact Simon & Schuster Special Sales at 1-866-506-1949 or
business@simonandschuster.com.

The Simon & Schuster Speakers Bureau can bring authors to your
live event. For more information or to book an event, contact the
Simon & Schuster Speakers Bureau at 1-866-248-3049 or visit
our website at www.simonspeakers.com.

Manufactured in the United States of America

1 3 5 7 9 10 8 6 4 2

Library of Congress Cataloging-in-Publication Data

Names: Tucker, Abigail, author.
Title: Mom genes : inside the new science of our ancient maternal
instinct / Abigail Tucker.
Description: First Gallery Books hardcover edition. | New York :
Gallery Books, 2021.
Identifiers: LCCN 2020045858 (print) | LCCN 2020045859 (ebook) |
ISBN 9781501192852 (hardcover) | ISBN 9781501192869 (ebook)
Subjects: LCSH: Motherhood. | Mothers—Psychology.
Classification: LCC HQ759 .T928 2021 (print) | LCC HQ759 (ebook) |
DDC 306.874/3—dc23
LC record available at https://lccn.loc.gov/2020045858
LC ebook record available at https://lccn.loc.gov/2020045859

ISBN 978-1-5011-9285-2
ISBN 978-1-5011-9286-9 (ebook)

For Ross

Then all went on their knees, and holding out their
   arms cried, "O Wendy lady, be our mother."
"Ought I?" Wendy said, all shining. "Of course it's
   frightfully fascinating, but you see I'm only a little
   girl. I have no real experience."
"That doesn't matter," said Peter . . . though he was
   really the one who knew the least. "What we need
   is just a nice motherly person."
"Oh dear!" Wendy said. "You see, I feel that is exactly
   what I am."

— J. M. Barrie, *Peter Pan* (1911)

CONTENTS

MOM GENES

Introduction: Of Mice and Moms

"IT FEELS like I grew a new heart."

That's what my best friend told me the day her daughter was born. Back then, I rolled my eyes at her new-mom corniness. But ten years and three kids of my own later, Emily's words drift back to me as I ride a crammed elevator up to a laboratory in New York City's Mount Sinai Hospital, where cardiologists are probing the secrets of maternal hearts.

Every year, thousands of pregnant women and just-delivered mothers land in emergency rooms with a life-threatening type of heart failure. Symptoms include swollen neck veins and shortness of breath. Their hearts can hardly pump. The cause of this "peripartum cardiomyopathy" is unknown, but it's the kind of health disaster that for regular people ends in a prompt heart transplant, or oblivion.

Yet fate has a different design for the fledgling moms. About 50 percent spontaneously get better, exhibiting the highest rate of recovery for this type of illness. Some mom hearts are practically as good as new in as little as two weeks. Adult heart tissue doesn't rally easily, but new mothers may somehow be able to regrow heart cells the way salamanders sprout new tails.

At this Mount Sinai Hospital lab, a cardiologist named Hina Chaudhry thinks she's figured out why. After surgically injuring mother mice to simulate a heart attack, and then cutting out and dissecting their tiny tickers, she and her research team discovered just what they expected: heart cells with DNA that doesn't match the mother's own.

The mystery cells belong to unborn mice. During pregnancy the baby mouse's cells cross the placenta into the mother's body, joyriding around in her blood vessels until cardiac damage happens, at which point they sense inflammation and make a beeline for her wounded heart. It's a little like how my second daughter sprints at me with a Band-Aid when I scrape myself grating Parmesan for dinner.

"They just zoom in," says Chaudhry. "These cells home to the heart like heat-seeking missiles."

Multiplying in maternal chests, the fetal stem cells transform into blood vessel–like tubes and even something that looks an awful lot like the holy grail of cardiology: full-fledged heart muscle cells, which cardiologists have struggled for decades to re-create in a lab. The mother's crippled organ likely uses this fresh tissue to heal.

It feels like I grew a new heart.

On a nearby computer screen, Chaudhry pulls up highly magnified video footage of these fugitive baby mouse cells in a petri dish. Tagged with a green fluorescent protein, they look like fresh peas in a dish of gray gravy.

She hits play, and the peas begin to pulse, to twitch. *Ga-gung, ga-gung,* they seem to say, like Patrick Swayze in *Dirty Dancing.* I squint. Why on earth, I ask, are the fetal cells bopping around like that?

Chaudhry grins. "They're beating."

∾

It's not just hearts. A mother's body is like her living room, strewn with kid castoffs and debris. Scientists discover fetal cells in the darnedest places, like when I find somebody's shin guards stuffed behind the TV, or a tiara in the laundry basket. Our children colonize our lungs, spleens, kidneys, thyroids, skin. Their cells embed in our bone marrow and breasts.

Often they stay forever. Scientists find rogue fetal cells while dissecting the cadavers of old ladies, whose littlest babies are now middle-aged men. Long after giving birth, the bodies of surrogate mothers are scattered with the genes of strangers' progeny.

The phenomenon is called "fetal microchimerism"—"micro," because these are typically teeny numbers of cells, only a handful per millimeter of blood in pregnant women, and fewer in moms later in life.

A "chimera" is a type of awkward Greek monster remixed from various familiar creatures, until an entirely new organism arises.

On my computer screen I stare at statues of these ancient freaks cast in bronze: goat legs, lion heart, dragon wings and fire breath billowing out of one of three heads.

That's no monster, I think. *That's me most mornings. That's a mom.*

∽

Although fetal microchimerism is evolutionarily ancient and common in mammalian moms from cats to cows, modern researchers are just getting around to studying it. So it goes for much of the science surrounding the two billion or so human moms patrolling the planet today. Though in a sense, there are far more of us than that, since microchimerism also flows the other way, with mothers' stray cells trespassing into babies' bodies and living on through them. Thus, while one of my dearest friends died of cancer three years ago, a fraction of her cells are currently attending second grade.

Worldwide, more than 90 percent of all women become moms. But until pretty recently, few scientists, especially in cutting-edge fields like neuroscience, had been even vaguely curious about our inner happenings. Blame the historically macho scientific establishment, if you must: some thinkers trace this neglect all the way back to Charles Darwin, who grew up motherless and maybe couldn't bear to think about us too much, poor guy. It wasn't until 2014 that the National Institutes of Health confessed its "over-reliance on male animals and cells" in research studies and mandated the inclusion of female models, moms occasionally included.

Another long-standing weakness of mom science is that what little exists is often the disguised study of babies, which as models of the human condition are (we get it) way cuter and less encumbered by obnoxious variables like culture and personality, and can be compensated for their time in Ritz crackers. Compared to their rapidly developing infants, moms have a reputation for being dull and predictable, hardly hotbeds of sexy hypotheses. In nature, animals such as baby whales sometimes mistake ocean buoys and other large, inert orbs for their mothers; scientists may make similar assumptions.

But finally more scholars, many of them young women, are taking the time to actually investigate, sometimes by attaching spy cameras to babies' heads, or sewing microphones into their onesies. Their state-of-the-art experimental tools include the most ordinary materials in moms' lives—family photo albums, Froot Loops, Play-Doh. They're discovering that the moms themselves are not so ordinary. In fact, we may be more intriguing and complex than anybody imagined.

And that's what makes Chaudhry's heart work so eye-catching: it's indisputable proof that, if you peer closely enough, moms often look very different from the rest of humankind.

Scientists are still trying to understand why, and what it means for women. For even as Chaudhry and her team hope that their microchimerism research will ultimately lead to all sorts of promising heart therapies for a wide variety of people, at the moment nobody knows for sure what those kid cells are really doing inside their moms' bodies.

The hope is that they help us. "It's evolutionary biology," says Chaudhry, who published her first microchimerism paper in 2012.

"The fetus is designed to protect the mother," the organism most essential to its future survival. And the fetal cells do seem to mostly stick to the Goody Two-shoes act, as if they're shortly due an allowance payout. In addition to our hearts, they may mend moms' flesh wounds—fetal cells likely pack my C-section scars, for instance—and help shelter us from myriad dread diseases. One decade-long Dutch study tracked 190 women in their fifties and sixties, and those with detectable leftover baby cells were less likely to die of virtually everything. It's even been proposed that these swarming stem cells slow the aging process, no $300-per-ounce face cream required.

In a particularly famous case, doctors discovered that a son's lingering cells had rebuilt an entire lobe of one woman's ruined liver. (The case is notable mostly because the mother in question had no children. Her son had never been born but was living on, after an abortion, inside her.)

In some instances, though, our babies' cells may get up to mischief. Anyone who's watched little kids play dress-up realizes that it would be unwise to let them permanently refashion the human form. Greedy fetal cells—well, technically cells are just mindless entities, but even scientists tend to humanize them when they belong to people's kids—may collaborate with certain cancers, especially breast cancer, in a covert effort to max out moms' milk outputs. They may infest our thyroids, jacking up our body temperatures the better to warm themselves, causing various metabolic disorders while they're at it. Despite their sweet little Muppet voices, our children may actually puppeteer *us*, perhaps even bully us a bit. (Some evolutionary biologists also think that my three children's cells might be making war on each other inside my body, and honestly—I wouldn't put it past them.)

This sweet treachery, recognizable to any mom who's watched her kids lovingly snip up homemade confetti for her birthday and then turn around and booby-trap the dishwasher, explains why I did a double take upon learning that there's evidence for fetal microchimerism in mom brains.

Could undercover kid cells inside my own skull finally explain my baffling life for the last ten years: my sudden affinity for velvety cheeks, azure eyes, deep dimples, and daft smiles, and my persistent mental drift from best-laid plans, the eclipse of the old me by a different sort of self?

In fact, it turns out that what really happens inside the mom mind is so much weirder, and the story of this book.

∽

The first time I ever considered the hard science behind the tender maternal instinct was during a visit years ago to a famous vole laboratory at Atlanta's Emory University. Larry Young, the lead researcher, told me how prairie voles' unusual brain chemistry may enable them to form lifelong pair-bonds with their mates by recycling a much more basic and ancient mammalian system: the maternal circuitry that mobilizes when females become mothers. (In humans, similar cross-wiring of old mommy-brain parts may explain our somewhat creepy desire to call our lovers "baby.")

Though already expecting my second child at the time, I had always thought—or maybe willed myself to believe—that motherhood was an elective lifestyle rather than a biological predicament, a label not a state, and one hat among many that I sometimes chose to wear, as opposed to my entire head and all its expensively educated contents. But Young was describing motherhood as an unseen and poorly understood cellular-level revolution that rebuilds the female brain.

Okay, fine—it was true that I had been feeling more than a little out of sorts for these past few years as I muddled through two pregnancies while working full-time as a magazine writer. My mind seemed a little scattered, my thoughts quickly yanked out and discarded like so many baby wipes.

But surely I'd snap out of it as soon as I got a bit more sleep. My brain would bounce back, in much the same way my body would one day (I so innocently hoped) glide again into those pre-pregnancy jeans kept on the lowest shelf in my closet, within arm's length but so far out of reach. Until that day, in fact, I had fretted far more about my old jeans than my new brain.

This superficial focus is totally understandable. The visible changes of motherhood are downright distracting, even at moments when I'm not plastered in pirate stickers. Over three pregnancies I gained a total of well over a hundred pounds, losing . . . not quite all of it. (Still, it could be worse: blue whales gain a hundred thousand.) My sides remain riddled with stretch marks like lightning bolts.

In pregnancy our entire physical selves are in flux. Our moles may darken, our voices drop an octave (as a pregnant Kristen Bell's did while recording *Frozen*—it seems that the notorious soundtrack could have been still shriller in places). Our noses widen, our arches flatten, and our toenails fall off. Our hair can change color or gain curl. We may burp as if we've swallowed a bomb cyclone. Our livers may leak bile, causing us to itch like the dickens. And we become demonstrably more delicious to mosquitoes because of our increased body temperature and carbon dioxide output.

These sorts of whole-body revisions are nothing to sneeze at. They cause Serena Williams to fail to qualify for the French Open and Beyoncé to bail on Coachella, and they can last a long, long time— maybe forever. One scientific paper rather meanly documents the textbook Humpty-Dumpty mom bod, with "increased abdominal and reduced thigh girth." It turns out, too, that the old wives' saying "Gain a child, lose a tooth" rings sort of true—compared to childless peers, moms are more prone to forfeit teeth, whether the cause is our depleted calcium stores or all those skipped dentist appointments. Elderly moms also have more trouble walking. On the bright side, those who breastfed are less likely to have strokes.

And yet all of this turmoil pales in comparison to what's happening inside the maternal mind.

The writing is crayoned on the wall, if we stop to read it. Those toothless old moms may also have a different relationship with Alzheimer's disease, with one recent study of more than fourteen thousand women suggesting that those with three or more kids have a 12 percent lower risk of dementia.

Yet not all the neural news is good. Indeed, many dangerous and opaque mental problems hound moms, especially as they transition into the maternal mode. While more than half of new mothers weather the "baby blues," roughly one in five go on to develop full-blown postpartum depression. Scientists aren't really sure how or why. Moms are also at increased risk for depression not just around the time of birth, but for years after. Maternity may help resolve the conundrum of why women in general suffer from more than our share of mood disorders. In the first month of motherhood a woman is, for instance, twenty-three times more likely to develop bipolar disorder than she is at any other time in her life.

These are all heavy hints that what's transpiring within our brains is just as extreme as our somewhat unwelcome external makeover. As moms' neurons sop up the trippy chemicals of childbirth, the genes inside the cells turn off and on, causing change and brain growth. The upshot is that over the course of a few short months, our brains are abruptly demoed and renovated, HGTV style, causing us to reinterpret familiar stimuli—a stranger's face, or the color red, or the smell of a tiny T-shirt—in freaky new ways. Suddenly a child's smile is our alpha and omega. Our old systems of desire have been rewired.

So the most important change in motherhood isn't about how we look on the outside.

It's about how we see.

❧

It's no coincidence, of course, that this idea of being hijacked, hacked, overridden, reprogrammed, or otherwise assigned a new identity is the stuff of dystopian female fiction, from *The Stepford Wives* to *The Handmaid's Tale*.

But I've thought a lot about this idea of becoming a "new lady"— which is what my daughters christened me, after I complained about being called an "old lady"—while sitting at the dining room table many nights, drinking my "black wine" (their term for the opposite of white wine). I've come to the conclusion that it's actually rather refreshing.

Ever since I heard that first alien knock of the fetal heart at the doctor's office, I get a kind of mommy vertigo when I imagine my kids' six blinking eyeballs, or examine the X-ray of my daughter's anklebone after a bad ride down the slide at Chick-fil-A. *I made these people in my stomach.* It's one of the strangest conceivable thoughts; in a way it feels more normal to picture giving birth to myself.

Which is one striking way to imagine what mothers really do. In fact, the changes of motherhood are so singular and extreme that scientists are beginning to describe us in terms previously reserved for our great scientific rivals, the babies. Mothers are the opposite of dull and predictable. We're new beginnings, not dead ends. In psychologists' lingo, we're "developing."

∾

Is "maternal instinct" the right phrase to describe the senses and sensibilities formed by this rebirth? These days, *instinct* is a Giorgio Armani perfume, not a scientific buzzword. It's something that Jedi knights trust in, not scholars.

A century ago and even more recently, the *New York Times* and other newspapers used the term in scolding descriptions of disreputable women, like the hula dancer with poor fashion sense and "thick ankles" who stole somebody else's baby ("frustrated" maternal instinct) or the mom who skipped town on her husband and kids (a tsk-tsky "lack" of said instinct). It smacks of a time when women entered their babies like prize pigs in state fairs and tuned in to the US Department of Agriculture's housekeepers' chat.

But I still like it, and many researchers don't mind it, mostly because it's a "know it when I see it" term that women themselves continue to identify with and deploy. (Also, if scientists always had their linguistic way, I might not be talking about women at all, but rather "substrates with the capacity to become maternal.") It's satisfying to connect the latest scientific findings with Mindy Kaling's online gushing about her newfound "big maternal instinct" because women do indeed know what they are talking about. The maternal instinct is real and powerful, a spontaneously arising set of emotions and actions pertaining to the perception and care of babies.

But because it is a fraught term, to say the least, let me also mention what I *don't* mean by "instinct." It's common to hear childless women say that they don't have a maternal instinct as shorthand for saying they don't want to have kids. I'm (mostly!) not explaining why some women do or don't plan or aspire to have kids in the first place or whether that's good or bad. (Though, by the way, this was totally me. The whole bizarre maternal undertaking was, as we shall see, my husband's idea.) These are interesting but narrowly human and more or less modern questions—in general, female mammals don't want to have children. They want to have sex. Offspring simply happen. Besides, on this subject you can't always trust moms to reveal their true motives. A study last year suggested that many human moms are so bowled over by baby love that they don't accurately describe their past pregnancy intentions, tending to report accidental conceptions as planned.

What I'm more interested in is what happens to females once they become pregnant, for this is when mothers are manufactured, the maternal mindset unfurls, and the master plan, if there ever was one, goes out the window like a banana peel on the car ride to swimming lessons.

The other "instinctive" misperception that I want to dismiss at the outset is the idea that human moms magically know what they are doing. Much more on this to come, but quite obviously: we don't. The instinct I'm describing is a transformed mental state, a new repertoire of senses, feelings, impulses; it isn't a how-to guide for good mothering.

Instead, I'm drawn to two big questions about this mysterious new maternal repertoire. First, how are mothers different from other types of people, and similar to one another? For across the whole mammalian family, hamster moms and wallaby moms and human moms are all kindled by a common spark. And disconcerting as it sometimes seems, our close resemblance to our furry sisters is also lucky, since scientists are allowed to dissect them and not us, and animal models like sheep and mice have divulged much of what we know about ourselves.

My second question is, if we're all so similar to our distant mammalian cousins, why are we human moms also vividly different from one another? For like our birth canals, moms' stories have some serious twists. In Japan, hyper-involved "monster moms" reign; in Germany, "raven moms" care only for their careers. There are "late" mothers (to use a *très* French euphemism for "old") and "lone" mothers (that's sad British slang for "single"). "Murfers," or mom surfers, hang ten in Australia. And in America a million types claw for ascendancy: stay-at-homes, work-at-homes, and work-out-of-homes; free-range and helicopter; formula-feeders and breast-feeders; co-sleepers and cry-it-outers; clear Magna-Tile moms and solid-colored.

Some scientists have become convinced that the secrets of our differences can be found inside each mom's unique genome—if, fortune cookie–like, we can just find a way to crack inside. But we'll also see how every woman's maternal destiny is shaped by myriad and bizarre environmental factors—like whether you babysat, or had oboe lessons, or ate too much fast food, but also by who loved you.

My hope is not to momsplain, but for us to discover together what divides moms and what unites us. I want to witness—under a microscope, or inside a monkey corral—the forces that move us all. I want to know what rocks the hand that rocks the cradle.

෯

Now maybe mom biology just isn't your bag. Maybe you are like the twentysomething birth striker I heard on NPR the other day who knew all she needed to know about the maternal experience because somebody on her kickball team had a kid once. Maybe it's not so interesting to you that, in addition to furnishing the raw material for pair-bonding and mammalian social interaction in general, the maternal instinct likely fuels human phenomena as diverse as female friendship, religious experience, right-handedness, altruism, lesbianism, language, music, obsessive-compulsive disorder, and pet-keeping, and also may help explain why the fair sex beats the pants off men when it comes to surviving hardships such as potato famines and measles epidemics and, yes, Covid-19 and plagues in general. (Thank you, Great-great-great-great-grandma.)

But there are also plenty of practical, even Machiavellian reasons to understand this stuff. Tens of thousands of new moms are made worldwide each day. Many of these are in the developing world, like mom-rich Zimbabwe, where some labor and delivery wards apparently still charge women by the scream. But while the West's declining birth rate may suggest that moms are going out of vogue here, in truth we are still trending. We're having fewer kids, and waiting longer to have them, and yet a greater percentage of Americans are mothers today than a decade ago, with 86 percent of women still being reborn into momhood by their mid-forties. Even the millennials are joining up at the rate of a million moms per year.

That makes moms a force not just of nature, but also of economics. We comprise a staggering portion of the American labor market, with 70 percent of us working, the majority full-time, and we're the sole breadwinners for 40 percent of families. We are apparently pretty good at our jobs, since Goldman Sachs is attempting to retain new-mom employees by internationally airlifting our breast milk. Even MI6 actively tries to recruit mom spies—not for our honeypot appeal, alas, but for our "emotional intelligence."

Marketing companies are eager to figure out how our brains work, the better to sell us everything "from bras to booze," as the title of one recent seminar put it. According to the latest research, moms hit mobile consumer apps starting at 5 a.m., and reportedly shop 15 percent faster than other people. ("Remember the 'drudgery' " of motherhood, one analyst urged, advising businesses to ply harried moms with "easily digestible information.") Microsoft's eggheads have even developed a handy code that spots new mothers online based on our altered usage of impersonal pronouns and other linguistic tells.

Finally, we are a crucial voting bloc, since in recent elections women have checked more ballots than men, and the hidden changes of motherhood sometimes seem to track with political shifts—not just a no-brainer embrace of overtly mom-friendly policies, but also unseen effects, like potentially "warmer feelings toward the military." Yet these shifts are not globally consistent: there's a complex interplay between women and their political systems, and the maternal instinct can be harnessed by either side of the aisle. With nearly two dozen American congresswomen tending to minor children at home, a growing share of our politicians are knee-deep in diapers themselves.

∽

But although the prospect of channeling global mompower is beguiling, as you can probably tell I'm even more interested in what benefits *us*.

For understandable reasons, the more that motherhood is seen as a choice, one life path among many, the more women are inclined to wonder whether they are likely to be happy as their new selves. In fact, America's record share of older, educated moms means that many of us spent scores of perfectly contented years being somebody else. Maybe it's no wonder that today's moms-to-be have rates of depression 50 percent higher than our own mothers did. I can tell you straight off the bat that motherhood has made me the happiest and saddest that I've ever been in my life.

While *Will I be happy?* is a question somewhat outside the scope of science, biology reveals the forces that swing the pendulum. We are at the mercy of so many powers, infinitesimal and vast, from the stuff going on inside of our own cells to the prejudices of whole civilizations, not to mention the diseases that can suddenly drop on our societies, marooning us with our little darlings for months on end. There is no one maternal path, and every woman has the potential to morph into many different moms. In fact, you will soon see that I've been multiple mothers myself, and science has helped me to understand how each of these best and worst selves came to be.

This is the paradox and the wonder of the maternal instinct. It is both fixed and highly flexible, powerful and fragile, ancient and modern, universal and unique. As I learned from my dying friend who spent her last days monitoring her daughter's cupcake intake and curating a stylish wardrobe to last through middle school, death itself cannot destroy it. And yet, under the right circumstances, it can be undermined or undone.

It can also be repaired and nurtured. Scientists deep in this research look forward to new and better mom-specific medications, and to the day when brain scanners are as much a part of ob-gyn visits as blood pressure cuffs. But there are plenty of nonmedical measures that governments, communities, friends, and families can take to make a difference in moms' lives right now.

Do we really need help? Female *Homo sapiens*, after all, have been in the mom biz for some two hundred thousand years now. In some ways today's moms are better-equipped than ever, with our newfound ability to give birth how and when we want, even via strangers' transplanted uteruses if push comes to shove. We can pump breast milk while sleeping (aka "pump and slump") or running half marathons ("pump and thump," perhaps?). Pregnant women, whom society once tucked away "in confinement," can do anything: report the news from war zones, go for Olympic gold, summit various Alps, and preside as prime ministers and CEOs.

Yet despite our robo-strollers and fancy-pants baby monitors that let us sing lullabies from far-flung time zones while on business trips, we aren't always in charge here, nor are we precisely who we used to be. In the course of becoming mothers, we do not "change our minds" about the world. Our minds are simply changed.

In the present age of individualism and bespoke identities, this feels uncomfortable to say. And yet acknowledging our lack of agency, and understanding the aspects of motherhood that modify us without our consideration or consent, is the first step toward taking control.

One Princeton University–led study suggests that much of the world's maternal misery may stem from the simple mismatch between women's preconceptions about motherhood and the "information shock" of its lived reality, especially as it plays out in educational choices and the workplace. Pretending that we are just the same as we ever were—in other words, the same as everybody else—and that we have anything like the final say in the matter is deeply counterproductive and even dangerous.

Maybe some of us would rather stay sheltered from these truths, like the female hornbill moms who (using their own feces) seal themselves and their hatchlings inside tree cavities for the duration of young motherhood, allowing their mates to feed them the occasional ripe fig.

I'd rather face the music, even if it's Raffi on repeat again, and even if my mom brain does end up resembling a pile of scrambled eggs or pulled pork, as I secretly fear. Recognizing the fundamental shift in my center of gravity—both physical and mental—is the best way to move forward.

The other day, one daughter wrote my name on her leg in red marker, as is her habit. Standing over her this time, though, I noticed something new: the flip side of MOM is WOW.

conquering placenta and then by our own systems, change not only our bodies but also our minds.

In truth, I'm not sure I really want to know what has become of my brain, three pregnancies to the wind. Just thinking about it gives me an uncomfortable feeling, like peeking into the Tupperware drawer with all the mismatched and microwave-melted lids. Especially lately, I've been something of a mess.

And yet, as the daughter of a mother who—some four decades after my birth—will still tail me into wintery sheep barns, and as the mother of a daughter who presently aims to be the "first girl on Mars" but also to have twenty-two children, I have many questions about this shared female journey, and the places that women unknowingly go. Is the maternal instinct a real thing? Can we see and measure it? Do all mothers have it? Do only mothers have it? Are we permanently assigned to this new self?

Like the sheep farmer leaning over the barn railing, I wonder: Why? Why? And how do we know?

∽

Let's start with the obvious: To the extent that the term "maternal instinct" implies that human moms miraculously know what we are doing, clearly nothing could be further from the truth. "There is no maternal instinct" of that sort in human beings, says Jodi Pawluski, a neuroscientist who studies maternal behavior at France's University of Rennes 1. "Everybody has to learn to parent."

This is music to my ears. I've long ago given up on waiting for my inner supermom to show up.

The nagging worry that I hadn't the foggiest idea how to become or be a mom began nearly a decade ago, at the dawn of my first pregnancy. I was thirty years old. My high school babysitting days were a distant (and not particularly fond) memory, and I'd since spent only a handful of hours in the company of tiny children. It would be a stretch to say that I missed them. As Washington, DC, twentysomethings, my husband and I enjoyed a rather yummy existence, traveling the globe for journalism work and, while at home, patronizing the neighborhood's hip new Balkan restaurant or loping along local running paths at a ludicrously slow pace. My central complaint was that there were too many friends' weddings to attend on the weekends.

But now the jig was up. An invisible stowaway was biding its time inside of me. I was going to be a mom, though I had rarely imagined it. My mind was ominously devoid of motherly knowledge. I felt that I must take some type of preparatory action—but what? One day in my second trimester I wandered out to the mall. But instead of shopping for, say, a baby blanket, I spent a long while cruising the department store aisles for slippers and a matching robe, an ensemble that I had never previously possessed or desired, but which I felt could be just the thing for shuffling through the hallways of the labor and delivery ward among fellow well-appointed ladies, pausing now and then to elegantly wince.

And of course, as a lifelong grade-grubber, I had to sign up for a class. Nobody was sure if Lamaze even existed anymore, having been eclipsed by other birthing fads. But I figured that I was no slave to fashion and that my mom had breezed through this same course thirty years earlier, the "blow out the birthday candles"–style breathing techniques ultimately carrying her to maternal triumph.

The Lamaze teacher had coiffed gray hair and amazingly commodious hips. Those hips, she explained at the start of the class, had allowed her to squirt out her one and only kid in ten minutes flat, leaving virtually no time to deploy the Lamaze wisdom that she would be sharing with us that day.

A Methuselah's tally of extinguished birthday candles later, I graduated Lamaze with only one memorable takeaway. At the outset all of us moms-to-be had been pinned with very large and unusual round name tags made of construction paper. At some point, the teacher revealed that these bagel-sized circles were precisely ten centimeters wide, just like a fully dilated cervix. That image lodged in my head as more useful facts dispersed.

A decade and a trio of kids later, I'm not all that much wiser, having become a battle-tested matron who's alarmingly bereft of both timeworn wisdom and trendy tips pertaining to childbirth and children. I never figured out sleep regressions or which molar comes in when. I've been reduced to consulting experts that I once had no idea existed, to teach my kids how to go to sleep (sleep coach), how to eat (food coach), how to ride a bike (some poor soul at the bike store). I once took my daughter to the doctor to have a splinter yanked from her toe. For years I carried the business card of a professional lice picker.

Whenever I think I have a parenting move down pat, or feel a glimmer of what seems like motherly intuition, I'm quickly disabused of the notion. Take the time, not long ago, when I had to unexpectedly nurse a baby on the fly during a family hike and ended up shirtless—sports bras pose such a riddle in these situations—and surrounded by a crowd of camouflage-clad senior citizens armed with binoculars. ("This is a *hotspot* on the warbler migration," one disdainful birder reprimanded me.) Or the weekend when I waved away one of my kids' stomach flu symptoms so that we could embark on a long-planned family trip, a debacle that ended in epic communal hotel retching, a misplaced purse, pilfered keys, and ultimately the theft of our trusty family car. (The car was eventually recovered, its front end crushed, after a high-speed chase with police ended in a crash. "The stroller yours?" asked the policeman who canvassed the wreck for our belongings. "How about the brass knuckles?")

My husband and I have even coined a term for this special kind of snowballing domestic disaster: the "parental cascade."

∽

Thank heaven it's not just me. Study after study highlight human mothers' native incompetence. We are ignorant of the US Department of Agriculture's child nutrition guidelines. We haven't the faintest idea of how to treat a fever or how to stop kids from choking or how to put them to sleep safely. According to one headline, "Potty Training Is a Scientific Mystery" that moms are incapable of solving. (Indeed, as the average age of kid continence continues to rise—from two years in the 1950s to three and counting today—our meager maternal talents seem to be atrophying.) No wonder moms are lining up to join groups like Loom, a kind of country club for posh yet anxious Los Angeles women offering "judgment-free services" to "navigate the maze of contemporary childcare protocols." No wonder we download baby translators like ChatterBaby, an iPhone app that supposedly decodes what the heck your kid is crying about.

At first I guffawed upon reading about the invention of the Snoo, the $1,300 iPhone-enabled robotic bassinet bristling with microphones, speakers, and Wi-Fi switches that reads your baby's cues and cries and jiggles her automatically back to sleep.

Then, a few months later, I ordered one. (Good thing I only rented mine, because I couldn't ever quite figure out how to work it. The machine undoubtedly knew more than I did.)

Not every human mother is so clueless. But in many ways our capabilities lag well behind those of the busy ewes. While never completely predictable, other mammals boast far more of what scientists call "fixed action patterns"—innate and automatic mothering behaviors designed to get the job done.

After giving birth, a mother rat is pretty much on autopilot: she eats the placenta, cleans and retrieves and carries the pups, nurses and hovers over them, and engages in zestful anogenital licking. And that's about all there is to the job.

The mother rabbit has perhaps the most vivid and specific maternal routine. Precisely one day before birth, she starts madly plucking the fur from her thighs, which she uses to line her nest. If scientists shave her body to stop her from doing this, her other maternal habits will be derailed and her babies will likely die.

It's possible that human moms may have a touch of this "nesting instinct," with questionnaire responses suggesting that pregnant women are more likely to experience "an uncontrollable urge to reorganize and cleanse" their homes as their due dates draw near. (*"Organize hair elastics!"* I vehemently vowed on one pregnancy to-do list.) Yet, divested of her Windex, a woman will still care for her child.

Indeed, scientists have long struggled to find a human version of the "fixed action pattern," or any single behavior to which every *Homo sapiens* mother robotically defaults. One contender is "motherese," the high-pitched, cutesy speech patterns that moms use when addressing babies, which is broadly documented from America to Japan, with even deaf mothers seeming instinctively to adapt sign language along similar lines. Researchers can generally tell when mothers are speaking, not just by the absurd things we say during scientific observation ("Let's not eat the kitty cat," in one study) but by the timbre of our voices. Some go so far as to claim that the ancient duet between mother and child is the basis for all human language, and maybe music, too.

Yet even motherese is not a species-wide given—at least, not in the way that thigh-plucking is for rabbits, or the maternal bleat is for sheep. In some cultures mothers rarely speak to their infants at all, and seldom even look at them. (In Papua New Guinea, for instance, a baby disappears into a kind of droopy backpack suspended from his mother's forehead for most of the first two years of his life, and I've definitely heard worse ideas.) Singing to babies isn't universal either: a study of American moms in neonatal intensive care units showed that 40 percent did not spontaneously serenade their tiny sweethearts.

Even that definitive mammalian behavior, nursing, varies wildly among our kind. While mother rats nurse like clockwork for twenty-one days, human moms may breastfeed for five years, or not at all. If it were so natural, so deeply ingrained and instinctive, why would we need a four-hundred-page brick of a bestselling manual on the "womanly art of breastfeeding"? (Naturally, I enlisted a lactation consultant.)

Among *near*-universal mothering behaviors, perhaps the most striking is called "left-handed cradling bias." Something like 80 percent of right-handed women and, remarkably, almost as many *left-handed* women hold their babies automatically on the left. In most statues of the Virgin Mary, the baby Jesus is sitting on her left, and ordinary children often wind up in the same spot: though heavily right-handed, I cannot seem to manage to cradle a baby with my right arm. It just feels wrong. While this maternal tendency is most pronounced in the first three months, even today my school-aged kids still fight tooth and nail to be on my left side during story or movie time.

It turns out that left-leaning moms populate the animal kingdom. Researchers recently catalogued lefty preference across a braying array of mammal mothers, including Indian flying foxes and walruses, which like to keep their baby portside as they (respectively) hang upside down or float.

This global inclination probably has to do with the lopsided layout of the mammalian brain. Holding and observing a baby on the left helps transmit information to the mom's right hemisphere, where her emotions are processed. It likewise allows the infant to view the more expressive left side of the maternal face. Researchers who thumbed through family photo albums recently found that "more depressed and less empathetic" mothers tended to hold their babies on the right. Italian scientist Gianluca Malatesta, an expert in this area, pointed out to me that the depression-prone Princess Diana was given to right-cradling. (Or maybe, being a princess who quite literally never lifts a finger renders one unprepared to schlep stuff, babies included.) Some fascinating work suggests that babies cradled on their mothers' right sides grow up to have a diminished ability to read faces. Even little girls hold baby dolls on the left—although I wouldn't know firsthand, since I never played with baby dolls myself.

∽

But it's also possible that in humans, at least, left-handed baby lugging isn't a mom-only habit.

In a recent and rather adorable experiment, ninety-eight British kindergartners were given pillows to hold, which they did without favoring either arm. Then researchers painted the pillows with primitive faces. Suddenly many five-year-old girls *and* boys—none of them mothers, obviously—switched to cradling on the left. The lefty baby-snuggling bias isn't so pronounced in adult men, but it's still apparently present (although my own husband is a resolute right-cradler).

Which brings up the next challenge in defining the human maternal instinct. In most types of mammals, such as rats, males and females who are not themselves mothers will ignore babies or—worse—gobble them up. But humans are an alloparental species: super social, we have universal caregiving capacities, and babies occupy a special place in the hearts of all men and women, and in our neural circuitry as well.

So some of what we think of as the maternal instinct is common to the entire human race. A baby is one of the most arousing stimuli, regardless of a person's biological sex or parental status. Our body temperature rises when we look at babies, let alone cuddle them. Our brains generally tend to process baby faces differently than they do adult faces, and additional brain regions are involved. In one 2012 study, childless Italian adults viewed pictures of unfamiliar human babies, adults, and animals while fMRI machines mapped the blood flow in their brains. Baby faces activated distinctive swaths of gray matter. This "species-specific response," the researchers wrote, appears "to transcend the adult's biological relationship to the baby."

It transcends race and ethnicity as well: while adults tend to have varied neurological responses to people of different ethnicities, race is apparently irrelevant when it comes to baby faces, a comparison of Japanese and Italian subjects showed. The human brain goes gaga for them all.

It's much the same story for infant cries. In a study of British neurosurgery patients—chosen because they conveniently (for the researchers, at least) already happened to have electrodes planted deep inside their brains—scientists played a tape recording of a sobbing infant. A deep-down area called the "periaqueductal gray" fired up within 49 thousandths of a second at the sound of the baby's quavering wail. That's about twice as quickly as the brain responds to similarly structured noises, such as the distress calls of cats.

Primal infant signals seem to prepare people to act, as well as to look and listen. Adults who'd just heard a baby's whimper excelled, in one laboratory experiment, at fast-reflex games of Whac-A-Mole, compared to people who'd heard a more pleasant stimulus, like birds singing.

These studies and more suggest that all humans are hardwired to heed infants, to respond at least a little bit as a mother would even when they aren't one. If a man or woman spies an abandoned baby bawling in the gutter, the vast majority will fish the poor creature out. The average Jane or Joe might not vow to shelter and feed the kid forever, but will at least attempt to find help, and certainly won't treat the child as a tempting hors d'oeuvre. As bare-bones as this behavior sounds, it sets us apart from practically all other mammals.

But science also shows that, even among human beings, certain gifts are reserved for mothers alone.

∽

A few months after our lamb vigil, my mom and I again visit a maternity ward, this time my younger sister's, in a Pittsburgh hospital a plane ride away. I've left my own brood back in Connecticut, and, maternal instinct be damned, I'm enjoying the momcation: it's been ages since I've cleared an airport without decanting bottles of breast milk for security line explosives tests, or disassembling my stroller like a marine during rifle inspection. Well rested, recently showered, and laden with shelter magazines, I give a little beauty-pageant wave to my hollow-eyed brother-in-law, the new father waiting for us near the baggage claim.

"I held a leg," he says, and then keeps mum for the remainder of the long drive to the hospital. With deep satisfaction I note that he is suckling from a cup of Dunkin' Donuts coffee, a classic parental beverage that he and my sister had previously disparaged as "brown water." *No more home-roasted espresso beans for you guys!* I chortle to myself in the back seat. *No more hot yoga!* These unkind thoughts may be all too typical of what scientists would call a "multipara," or mom of multiple children, observing the ordeal of a first-timer, or "primipara"—my poor sister.

After a tense hallway standoff with a hospital robot delivering lunch trays, I enter my sister's recovery room and find her amid a jumble of Greek yogurt containers. My just-hatched nephew is out promenading with a nurse.

Only now that the baby is out of the room does my sister feel finally able to get out of bed. She's vowed never to stand up while holding him, for fear of keeling over. "It's the smell of his head," she explains. "It's like a drug. I feel like I'm going to pass out."

She's not crazy: babies not only smell especially *distinct* to moms, as those experiments with sheep perfume and fraudulent Baskin-Robbins containers have shown us, but they smell unaccountably *delicious* to us, too.

In another scent-based study, which involved whiffing cheese, spices, and babies' T-shirts, mothers of two-day-old infants gave the baby aromas higher "hedonic ratings" than non-mothers did. To new moms, saggy-diapered wretches are as aromatic as lilac trees, or chocolate chip cookies fresh out of the oven.

This deliciousness, as we shall see, is the rub and the wrinkle and the hidden twist. It's nature's secret weapon. The second birth of motherhood is a kind of neural renaissance that overhauls what women find rewarding. There's a paradigm shift in our experience of pleasure, a drug fiendish–like narrowing of desire. The hairless little life-form that nine months or so ago hacked into your immune system is all of a sudden your sun and stars, your new true north. It's not just that you will gamely liquefy your bones and your fat stores to breastfeed it. Your entire field of vision now has a (teeny) focal point.

Perhaps most remarkably, these thrilling feelings of pleasure and enjoyment come quite hard on the heels of profound fear and suffering. The adored little dreamer in a brand-new mother's arms has likely just put her through the worst hell of her life.

∽

Though wracked by two days of intensely painful labor, my sister still had what's called a "good birth."

My first was . . . not so good.

Everyone else seemed so confident that nature would take its course. At my initial ultrasound appointment, the doctor—who was supposed to be checking for things like genetic abnormalities, and twins—instead felt compelled to comment on my "enormous carrying capacity." Was he calling me chubby? Well, sort of. Upon interrogation, he explained his point that I was simply a big, sturdy girl built to handle the hard physical tasks of womanhood ahead, unlike certain poor slender waifs elsewhere in his practice. As my husband watched with large and frightened eyes, I managed to grit out a smile and refrain from decking the doctor.

I did indeed seem to have a certain genius for getting larger, but by the end of the third trimester nothing much else was happening. I felt like a time bomb, big and round and ticking, and hopelessly addicted to peanut butter–smeared apples.

Forty-one and then nearly forty-two weeks passed, and even though I'd now been pregnant for more than ten months as opposed to the falsely advertised nine, labor did not strike. Instead, on an appointed day—Super Bowl Sunday, as a matter of fact—I packed up my matching fuzzy slippers and robe and a lot of other random doodads and decamped to the hospital for induction.

That night I was dosed with a cervical softener, and the next morning with the birth stimulant Pitocin, and then still more Pitocin. The contractions started as ripples and quickly mounted to tidal waves. "Pit her," a nurse said matter-of-factly. The waves reared higher. I had heard somewhere that visualization techniques are effective in labor and so, dutifully, I attempted to imagine myself as a surfer, carving my way through these terrible swells. When that didn't work, I also tried the trick of fixating on a single object, but I had nothing symbolic to focus on, and so ended up staring at the red cap of a Coke bottle with all my earthly might.

My official goal of a painkiller-free birth was not guided by my parenting philosophy (I didn't have one) nor by concern for the unborn baby—whom, after all, I'd never even met—but rather by my lifelong fear of blood and needles and, most of all in recent days, C-sections. I knew that unmedicated birth was one strategy to avoid the scalpel and the primal terror of being disemboweled, if only on the most temporary and medically routine basis.

Alas, these induced contractions were getting the better of me. After a long morning of muffled, and then somewhat less muffled, screams, a nurse strode in to check my progression: four centimeters.

Four centimeters! I thought. *That's not even half a name tag!*

So maybe, after all, I wasn't a big ox of a gal ready to gut my way through nature's worst. I seemed to be failing at a task that I didn't even understand. The epidural—hey there, big needle—was somewhat nightmarishly inserted. Then a sweet, if brief, peace came.

Paralyzed from the waist down, I was unable to parade the halls in my fashionable birthing attire, so my husband and I binge-watched TV shows instead. It was a stern gray winter day. The blinds were tightly closed.

"Don't you guys want some sunlight in here?" a nurse asked in a disapproving tone.

Evening came. Nothing much appeared to be happening, but apparently something was, because at one point the doctor came in and said that I was now ten centimeters dilated and it was time to push.

The first push was a disaster. The baby's heart rate, previously a peppy rat-a-tat on the monitor, decreased suddenly, like a stone ricocheting off the sides of a deep, deep well, each thud later than the last. The doctor rushed back in. But the heart rate steadied itself. So I pushed again—and again, and again.

It seemed like progress was being made. The birth canal's stations of delivery range from −3 to +3, the last stop before daylight. A nurse reported that I had stormed from −3—through −2 to −1—all the way to zero. I had never been so happy to achieve zero. Zero was epic. Only three more stations to go!

But something was not quite right, it became clear perhaps twenty minutes later. Another check below deck, and yup—the previous nurse had measured incorrectly. I was stalled way back at −3. After several hours of me straining and huffing and puffing like the Big Bad Wolf himself, the baby hadn't budged a millimeter.

"I don't think you're going to be able to push this baby out," the new nurse said coolly, resting her chin on my quaking knee in the most nonchalant manner.

My husband, now frantic, took the opposite approach. "Come on!" he roared up the birth canal, urging a final charge of some valiant invisible cavalry.

How could I politely point out to these people that I was dying? I no longer cared enough to mention it. The once-vanquished pain came pouring back. My will, always formidable, ebbed away. A fever took hold and my temperature spiked. Everyone's faces began to shimmer and ooze. I took a long, steep step into a valley full of stars.

An operating room. A blue tarp rose up before my eyes, swift and almost festive, like a just-pitched circus tent, mercifully blocking my view. The surgeons chitchatted away—of all the emergency C-sections they'd seen, mine was hardly the most dramatic. They even let me stay awake, which was really not ideal from my perspective. There was lots of tugging and yanking in my belly, and at one point it felt like somebody jumped on my rib cage. I gathered that I was getting scooped out, like an enormous party melon. But at least there was no more pain, as I gazed at the twinkling surgical lights above. "The baby's out," somebody said. A long, ominous silence followed, and then finally, a tentative cry.

There was meconium—fetal poop—in the amniotic fluid, an

indicator that in labor the baby had been as panicked on the inside as I was on the outside, and might have inhaled some of the tarry black stuff. So the baby would need to go to the neonatal intensive care unit for at least twenty-four hours of observation.

I saw a baloney-colored blur, and then the creature was gone.

∽

Dawn. Although still adrift on many drugs, I had a growing awareness of the aching wound on my abdomen. My flesh was a yellow-green color from blood loss, and my ankles had begun to swell grotesquely from all the IV fluids. I still couldn't walk, but then again, I no longer wanted to. I didn't want to call my mom or talk to Emily. I didn't want to pencil in a newborn photo shoot or "make a game plan" with the lactation nurse who'd already popped in to say hi. I didn't want to think about yesterday or tomorrow or about the baby at all. What I wanted was to go back to sleep.

After a while—a few minutes or maybe hours later, I'm really not sure—my shell-shocked husband finally spoke up from his station on the hospital room's plastic-upholstered couch.

"Should we go see him?"

(We had, in fact, had a girl.)

I guessed so. I didn't care much one way or the other, but I did worry what the nurses would think if I said no.

My IV bag and I, slumped in a wheelchair, rolled down the hall as my husband (for once) pushed. We passed a procession of triumphant baby-flaunting new mothers, some of them wearing matching slippers and robes, just as I had always imagined. *So much for being a mom*, I thought.

The NICU was a small room. Several desolate little creatures lay in clear plastic "isolettes"—a word new to me, and one of the loneliest in the English language. The nurse pointed us toward a container in a far corner. My husband steered me over, and I looked down.

She was sprawled out, wearing only a diaper, and zigzagged with many, many wires and tubes, including one poking up her nose to deliver oxygen. But I didn't really see any of that.

I saw her. I saw her face. Her tiny mouth was bent in a frown. She had my husband's round ears and my pointy eyebrows.

"She has eyelashes on the bottom!" I breathed in wonder. "She's *so* cool."

She was more than cool. She was the most exquisite, vivid, and arresting thing I'd ever seen in my life. It felt like my eyes were branded with the sight, the way they'd felt when I watched the World Trade Center collapse on live TV, or when, in eighth grade, I saw the face of my father as he lay in his coffin. But somehow this was a happy cataclysm.

With shaking legs I stood up from my wheelchair for the first time, ready to take my baby in my arms. She looked enormous, so much bigger than the others. Part of this was because she was full term and then some, and so she really *was* bigger. At 8 pounds, 11 ounces, though, she also seemed much larger than life.

Between my IV tube and her IV tube, holding her was no easy feat. I could manage it for only about a minute.

But I beheld her.

∽

How can researchers re-create this primal epiphany in a laboratory? How can science prove that in that moment—or, more accurately, in ten months' worth of moments, and a zillion subtle genetic and neurochemical changes that one by one paved the way for that moment—my mental goalposts were uprooted and dragged beyond the normal field of human affection, far enough that I was now playing a different game entirely?

Curiously enough, on the exalted subject of motherly love, studies of lowly lab rats often supply the best answers.

Remember, before having her first litter, a virginal rat doesn't enjoy the presence of pesky rat babies in the slightest. Like my former self, a childless urbanite perhaps overly fond of a bottomless mimosa brunch, the pre-maternal rat will always choose eating snacks over hanging out with rat pups . . . and gluttonous rat maidens will happily nosh on the pups themselves, given half a chance.

This preference persists almost until the end of pregnancy. But just about three and a half hours before birth, something momentous happens inside the rat mom-to-be, and she starts preferring pups to food. (Likewise, while I felt out-of-the-blue clobbered by love upon meeting my daughter in the NICU, studies in humans suggest that my attitude toward babies likely began to subconsciously shift midway through pregnancy, as my brain chemistry gradually changed.)

How do we know that babies suddenly trump brunch?

In one early study, new rat moms were given the chance to press a bar to receive pups, which tumbled down a chute into a little cup. The moms hit the pup lever over and over, so frenetically that the end of the chute became clogged with "an accumulation of bodies"—a scene reminiscent of what happens at the bottom of playground slides, when human toddlers begin to pile up. Confronted with this spectacle, the human scientist decided that each mother rat would be allowed to keep only six pups in her cage, yet "this did not seem to slow down the determined behavior." One particularly demented mom hit the pup lever 684 times during the three-hour experiment. The scientist assumed that she would eventually exhaust herself and quit. In the end, though, only the scientist wearied, writing in his journal article that he "got tired" of stuffing the chute with fresh specimens.

The new rat moms didn't devour the pups once they were deposited in the cage. They just wanted the pleasure of the pups' company. And "pleasure" is the word: the rat mom will even choose quality time with an infant over a straight-up hit of cocaine, having become—like my sister, nodding off in her hospital bed—a type of baby addict. She will brave an electric grid to reach pups, which a virgin wouldn't risk even for the most lavish cornucopia. You can blind her, deafen her, muzzle her, amputate her nipples, disable her nose, even burn off certain bits of her brain. You can trap her babies in a glass bottle or try to stump her entirely by substituting newborn guinea pig impostors, or even little hunks of raw beef heart, for her pups—and, for better or for worse, scientists have done all of these things to rat moms. She won't waver in her devotion.

Clearly we can't study the habits of human moms by zapping them or ejecting human baby after human baby from a laboratory chute. But scientists have devised other clever ways of testing just how powerfully babies trigger us as mothers.

For instance, they've figured out how to peek into our skulls to see what's up when we inhale the fumes of those "hedonic" little baby heads. In a 2013 smell-based experiment, thirty women sniffed at a mystery item—a newborn's two-day-old pajamas—as scientists watched their brains react via an fMRI scanner. Only the mothers showed distinct activity in an area called the "thalamus," which regulates sensory signals and alertness.

Baby faces, too, are extra-stimulating to moms. One 2014 experiment, titled "Here's Looking at You, Kid," pitted the neural responses of twenty-nine first-time moms against thirty-seven non-mothers as they viewed the (somewhat eerie) pictures of disembodied heads of babies and adults floating against a black background. While both groups of women seemed to find the baby mug shots more stimulating than the adults' faces, the moms ogled the babies for measurably longer.

Perhaps most important, infant emotions move mothers more profoundly. Our pupils dilate rapidly when viewing distressed babies, and we are slower than others to look away. Our scalps register different electrical readings at the sound of baby screams.

Using a technique called "near-infrared spectroscopy," Japanese scientists tracked how the oxygen levels of moms' brains changed as they viewed emotional baby pictures—of happy babies, who had been playing with attractive toys; of enraged babies, from whom said toys were taken; and of fearful babies, who were being eyeballed by a strange male. The moms showed activation in a different area of the prefrontal cortex than women who had never been pregnant.

For non-moms, male and female, laughing babies are more stimulating than crying babies—which seems sensible enough. For mothers, though, our fMRI readings suggest crying initiates a more powerful cascade in our amygdalae—and we may even find cries strangely rewarding. This neural switcheroo likely helps explain why moms persist longer than others when experimentally tasked with soothing a blubbering infant mannequin, even when the doll has been programmed to be inconsolable (as so often seems to be the case in real life). Other people may avoid despondent children, but moms seem propelled to approach—and research suggests that we are especially vulnerable to cries of pain, as opposed to those of hunger.

All this underscores something already clear to veteran moms. Being a mom isn't as simple as riding high on baby fumes and vibing off their button noses. Just because we have a new source of jollies doesn't mean mom life is suddenly a picnic with sippy cups. As usual, pain accompanies pleasure.

Motherhood, as many of us well know, is often an intensely miserable business. Having been internally rejiggered to find babies immensely rewarding, we are also made intimately aware of all their cues, and rebuilt to perceive and read and interpret their states, which involve grocery store tantrums and night terrors at least as much as story time smiles and good-night kisses. This compulsive interest and monomaniacal focus, this constant obsession that first opened its gray eyes inside me that morning in the NICU, is essential to our transformation.

"Sensitization" is science's term for our experience. It's almost as though our nerves extend out of our bodies. I think sensitization explains why mothers have a hard time watching movies or even TV commercials involving suffering children. We feel it too deeply.

It's a little depressing to think of oneself as uniquely attuned to tears, but this perhaps explains why bawling babies on planes make me feel like I'm being boiled alive, a peeled tomato rolling across jagged pavement. That's maternal sensitivity for you. Of course, other mammal moms have it so much worse: deer hunters know that to attract does, you play the recorded bleats of fawns.

∽

And yet while all babies, including the screamer in seat 3F, hold some sway over mothers, a human mom's own infant rules supreme. Rat moms are equally attracted to all rat babies, since, living in private underground nests, they are unlikely to stumble on unrelated young and waste their precious milk and attention on them. (Besides, a mom rat can have more than a dozen babies at a pop, so maybe it pays not to play favorites.)

Sheep, on the other hand, typically give birth to just one or two little ones in the middle of a mob—and, as we've seen, are evolved to think that only their own offspring are the bee's knees.

Human moms fall somewhere in the middle. Like rats, we are extra-attuned to *all* babies. But our own child is also a special little lamb. A human mom's brain responds most intensely to her particular youngster, and these findings seem to hold across cultures: from Cameroon to South Korea, we are all more infatuated with our own infant than the rest.

"We don't really need neuroimaging to tell us this, but it does," says Linda Mayes, head of Yale University's Child Study Center.

Even if our one-and-only is draped in the exact same gray cloth as every other kid in the experiment, our brains react faster just for him, and additional reward areas catch fire at the sight. Looking back at my own three babies' early pictures, I can now acknowledge that newborns look a lot alike, especially for the first few weeks. Stripped of their fetching little outfits, almost all bear a certain regrettable resemblance to an uncooked Perdue chicken, plucked and ready for trussing. But in the heat of the new-mom moment, your own baby appears wildly distinct, a face full of unique character and panache and promise and fragile beauty.

She has eyelashes on the bottom!

Moms' minds process our own children uniquely, too. It's not only their scents that we home in on. Within a day of birth, we can pick them out visually from a lineup of seemingly identical angry, red-faced neonatal burritos. Studies suggest that we can recognize them simply by stroking the silken backs of their hands. Even their particular diapers smell dreamy to us—or at least, they don't smell terrible, according to a 2006 diaper-delving study rather bluntly titled "My Baby Doesn't Smell as Bad as Yours." (I would argue that my babies' poop doesn't smell like poop at all—more like carob.)

A woman's one particular munchkin amps up activity in parts of her brain involved with reward, emotion, empathy, social cognition, motor control, and other functions—practically the whole enchilada. Our heart rates accelerate for our own five-month-old's hollering, while they slow down below normal for an unknown five-month-old's, even if we aren't told which cry is which.

In fact, within forty-eight hours of birth, a new mom recognizes her own baby's cry so well that she would wake only for that sound, amid the strident screams of other same-aged babies in the hospital. (This finding comes not from some highly unethical experimental setup, but from real-life investigations of the midcentury multi-bed maternity wards into which our mothers and grandmothers were shoehorned for lengthy stays.)

In those first endless hospital days, my brand-new daughter's cry was so piercing, like the screech of a peregrine falcon, or possibly a pterodactyl. Every time she screamed, I felt like I was being zapped with a cattle prod. I screamed a bit myself. "She's spitting up!" I would holler, in a tone normally reserved for announcing an alien invasion.

She was out of the NICU after twenty-four hours, which was wonderful except now my husband and I had to figure out how to take care of her. Hunkered down in our hospital room, we held her like a swaddled grenade. We needed help with absolutely everything: diapering, burping, and especially breastfeeding.

But I would stop at nothing to learn how to do these things. I soon became the terror of the nurse's station, roving half-naked and (despite my wounds) quite nimbly through the halls at all hours in search of assistance.

Because all of a sudden my new baby was the most wonderful thing in the world, and exquisitely sensitive as I now was to all babies' emotions but most of all hers, I was extremely *motivated* to take action to protect and help her however I could.

These three things—infant-centric pleasure, heightened sensitivity to baby cues, and bullheaded motivation—make up the heart of the new mother's instinctive awakening.

Maybe I would never know when bedtime ought to be, or which model of pacifier to deploy, or what on earth to do at any given moment. Maybe my newly sensitized sister, a half-dazzled hostage in her hospital bed, wouldn't figure it out either. Maybe none of us ever truly will.

But we will *want* to know, in a way that an aunt or babysitter or friendly neighbor won't, even though they'll aid a distressed foundling. Motherhood isn't about knowledge. It's about desiring to do anything at all for your child, at every given moment, and to press on to the ends of the earth until something works. Moms are enthralled, in all senses of the word.

Roughly 90 percent of new mothers report being "in love" with their new babies, and neuroscience backs us up. Our brains burn for our special cupcakes in patterns that echo those activated by our lovers.

Except that this common analogy is backward, and the mimicry runs the other way. In our species' natural history, maternal love long predates candlelit dinners and likely explains their existence. Mother love is the planet's original romance.

❧

In Pittsburgh, my sister is at last discharged from the hospital, kid in tow, after the usual rigmarole of circumcision appointments and a blur of paperwork. The latest and greatest in Diaper Genies waits on her front doorstep, alongside a basket of blue flowers.

We order her favorite pumpkin curry for lunch, but she won't eat it because the spice might pollute her breast milk and give the baby gas, and she certainly won't touch the celebratory champagne, so Mom and I pick up the slack.

We discuss in detail the baby's cherubic thighs, the length of his fingers.

"Is he still breathing?" she asks every few minutes.

All of this is perfectly normal. Freshly fledged moms are explicitly thinking about their babies, on average, about fourteen hours every day. Scientists think baby mania might help explain the evolutionary basis of obsessive-compulsive disorder—and indeed, clinical-grade OCD symptoms manifest in about 11 percent of new mothers, compared to just 2 percent of the general population.

My sister used to listen to so much NPR that she joked that the baby would identify Ira Glass as his real dad. But now the house is as silent as a mausoleum, so that the little kumquat's every fart and warble reverberates. My sister, the elite and fearless athlete, is scared to carry eight pounds up the stairs. Her phone is off, her voice mail full, and, for the moment at least, she won't be returning any calls.

The fact is, my sister may never really come home.

I'm pretty sure I never did. And the latest brain scans support this hunch. Mom brains don't just act different. They are structurally unlike other people's brains as well. A Leiden University–led lab recently discovered stark differences and gray matter reductions in the brains of first-time moms versus the brains of childless women. Even more striking, the new moms' brains looked much different from their *own* old brains, revisited through scans taken before they got pregnant. These gray matter losses may total up to 7 percent in some mothers, another study found. Change of this magnitude is practically unheard of in mature humans, with the possible exception being survivors of traumatic brain injuries.

The Leiden lab even cooked up an algorithm that identifies mothers based on brain anatomy alone, with near-perfect accuracy. The maternal mind, it seems, is distinct to the point that it's actually diagnosable.

These mind-bending changes last for at least two years, and maybe for life. "Are you . . . European?" one nurse carefully asked during one of my subsequent maternity ward stays, observing my scantily clad condition as I plowed about my room.

In fact, I had spent most of my life as a prudish New Englander.

But now I was somebody else.

The term "maternal instinct" may undersell the extent of this invisible neurological revolution. An instinct sounds like one arrow among many, instead of the whole quiver. Robert Bridges of Tufts University prefers to describe a maternal "unmasking," the abrupt reveal of a hidden potential or latent identity that was lurking inside you all along.

I like his term a lot, both because it gives the grunt work of new motherhood the glamorous air of a costume ball, and because it implies that our maternal self is a legitimate entity, not some mushy-gushy, frizzy-haired pretender to the throne.

It's a little sad to part ways with your old identity. Every rebirth is also a goodbye.

But my new self and the organism formerly known as my little sister certainly have a lot to talk about.

Chapter 2

DAD GENES
How the Father Makes the Mother

What about my brother-in-law? Has he gone permanently missing as well? Or is he simply on yet another Dunkin' Donuts run, due back any minute?

A semipro martial arts fighter who also just happens to be a PhD candidate in child development and computer science, he is far better prepared in body and spirit than most first-time dads. He's read the latest high-tech baby books, bought the top gizmos, and crunched all kinds of kid data. And yet for much of our visit in Pittsburgh he looks a little green around the gills, chugging from various Styrofoam receptacles that almost certainly don't do justice to the coffee beans' complex flavor notes.

Might some kind of internal transformation be underway inside of him as well? Could new parenthood blow dads' minds just as much as moms'? Maybe we should be searching for a "parental instinct," not an exclusively maternal one.

As scientifically neglected as we moms have been, even we have snagged more research dollars than dads, leaving many fascinating questions about the paternal instinct on the table. But, so far at least, studies suggest that while the transition into fatherhood may be personally shocking and scientifically measurable, moms and dads experience parenthood in quite different ways.

Well . . . yeah. This fact was laid bare, along with my bladder and intestines, during my first C-section, and afterward as well. My vicariously traumatized husband was a trooper, to be sure. He spent that wretched first night of fatherhood atop a camping mattress the approximate thickness of a graham cracker and never complained once.

But something happened while he was snoozing. When they'd wheeled me up from surgery, nobody noticed that the call button on my hospital bed was broken. In the middle of the night, overtaken by agony, I needed more meds—but my increasingly frantic button-smashing summoned no one. I croaked to my husband for help, but try as I might, I couldn't wake him from his exhausted new-dad blackout. Finally an alert nurse gathered what was happening. The blinding beam of the repairman's flashlight bounced around the room and still my husband did not wake up. I told him all about it in the morning.

I realized that night that while my husband and I were legally full partners in this parenting enterprise, our fates had fundamentally diverged. I was stuck in my sawed-apart body, my stomach slowly deflating like a hot-air balloon come to ruin, and he was lounging over there, more or less intact. The physical differences between the two of us had been compounding for ten months. A journey of ten centimeters had just yielded a new world to me. And now, in birth's bizarre aftermath, our mental and emotional paths seemed to be dividing, too.

Maybe they had been separate all along.

This probably shouldn't have surprised me as much as it did. Just as men and women on average exhibit clear physical differences—pelvis shape, say, and body fat distribution—our neural anatomy also varies somewhat. Women have a bigger hippocampus and more wiring for language, while men have a larger amygdala. We use different parts of our brains to rotate shapes and to read John Grisham novels.

Even before we become parents, these sex-based differences guide how we process infant cues, or so experiments suggest. Yes, all people are primed to like babies, but on average women appear to be somewhat more sensitive to them from the get-go. Conveniently enough, reproductive-aged women—especially those in the fertile phase of their ovulation cycle—may be the non-parents most responsive to baby cues.

In one National Institutes of Health–led experiment, childless men and women listened to white noise interspersed with earsplitting baby screams. Women's brains perked right up at the screams, while "the men's brains remained in the resting state" for the duration as their minds continued to "wander."

Another experiment used facial temperature to measure emotional arousal in response to pictures of various faces. The tips of women's noses changed from a cool green to an excited yellow for baby pictures.

Meanwhile, the men's, ahem, tips warmed up more for women's pictures.

∽

Just how did women end up liking baby pictures better? Why are we emotionally entangled with our kids right off the bat in ways that men oftentimes are not?

The answer may stem from those age-old domestic essentials: eggs and milk.

Outside mammals' corner of the natural world, moms are not always the default caregivers. Among fish, when parental care is provided at all, males usually step—make that swim—up. Many human moms will be (perhaps endlessly) familiar with the dutiful daddy fish from *Finding Nemo*. In the Pixar spin, Nemo's mom is derelict because she has, most regrettably, been eaten. In real life, however, a mommy fish is more often the *eater*. Many fish enjoy a rather splendid adaptation called "unrestricted growth"—which means the female can get continuously larger throughout her life, gorging and growing and releasing ever more eggs as she expands. (These revelations have made me glare jealously into my daughters' goldfish tank, all of whose occupants are allegedly girls.) Because the female fish's fertilized eggs land on a sea-weed frond, or in some other likely spot outside her own body, the father may be thereafter deputized for egg- and larva-minding duty while the mom, finloose and fancy-free, shimmies through the world's oceans looking for the next opportunity to sow her wild eggs.

Female birds aren't quite so carefree, but about 90 percent of bird species equitably split chick-rearing duties with their mates, as though someone had stuck a chore chart on the refrigerator. Because, once again, the fertilized eggs mature outside the female, eggs and hatch-lings can benefit from both parties' steadfast efforts to guard and warm and provision them. Our backyard is home to a pair of hawks, and I often gaze up at their biparental nest in silent salute.

Mammalian females don't lay our eggs, however. We keep them tucked in our guts, even after fertilization. Internal gestation of em-bryos helps explain mammals' riotous global success: pregnancy keeps our youngsters warm, fed, and shielded from predators and lets us in-filtrate even the harshest environments.

But the same nifty adaptations that helped us to outlast the dino-saurs have also left females holding the diaper bag.

In our kind, the males are the gamete-spewers, who can theoreti-cally sire nearly infinite numbers of babies. Females are saddled with a cramped internal tenancy that's often months long. A mammalian mom has no choice but to invest heavily in the fetus(es) already jammed in-side her, and to postpone passing on her genes to more. Males, mean-while, may move on to fertilizing eggs in somebody else's basket.

Nine—make that ten, and then some—months of pregnancy (or twenty-two months, for the unfortunate she-elephant) is actually just the beginning of our maternal predicament. Milk further seals mam-mal moms' fate. Our "mammalian" identity flows straight from our mammaries, or breasts. A few other non-furry creatures—tiger sharks, garter snakes—have evolved internal gestation. But only mammal moms make milk.

Sometimes it feels almost liberating to tote around a ready-made food source for a child. I've been caught countless times without a spare onesie, a diaper, and even my stroller. Heck, I've nearly forgotten my kids. But a nursing mother never leaves behind her milk.

This 200-million-year-old convenience comes with consequences, though. In pre-Enfamil days, a built-in milk bar meant that moms alone could nourish newborns, a vast and intimate undertaking that often halts ovulation, intensifying mammal moms' investment in the hungry youngster at hand.

To complicate matters for human moms in particular, the type of milk that we make is unusually thin and watery. Other mammals spend far less time nursing their young. For wild rabbits, with their rich milk, it's about five minutes a day. Fur seals may nurse only once a week. Humans, though, can spend half the night trying to get the job done. And while other juvenile mammals are weaned within weeks, humans are built to nurse for years because childhood is so long.

In this respect, the by-products of our bodies—eggs and milk—help make up moms' minds. The way that mammals bear and nourish children means that the onus is on females to care for them, and in turn, that moms' brains are built to emotionally foster this paramount relationship.

As the maternal behavior scholar Laura Glynn noted in a TEDx Talk, it may be that "the burden has fallen on the female nervous system to protect our genetic legacy." In the overwhelming majority of mammalian species, a bundle of joy is exclusively Mom's to lug around. Feeling the joy helps her bear the weight.

From jaguars to giraffes, most mammal dads have nothing whatsoever to do with their offspring. As much as we human moms justifiably lament bad or absent fathers, maybe we ought to be thanking our lucky stars that we're part of the measly 5 percent of mammal species in which dads chip in any care at all.

∽

I confess that all of this chilly biological lingo sounds a bit outrageous to a privileged modern woman like myself, serving beside a patient and unflinching partner who is arguably the superior parent—a celebrated lunch-box chef whose diaper-changing arts border on origami. By nature, my husband is attuned to the plight of helpless creatures: he's always worrying, for instance, whether our kids' hamster, Clementine, has eaten enough fresh broccoli lately. Before we reproduced, I always took for granted that he and I were more or less evenly matched when it came to caregiving, and I wouldn't have been surprised if he even outdid me in one of those shrieking-baby "battle of the sexes" experiments. As a brand-new father, he was so jittery piloting our little trio home from the hospital that he got one of the only speeding tickets of his life. He shed tears at day-care graduation. He's the epitome of an involved dad.

He's far from alone. Modern fathers have clearly picked up plenty of the tricks of the parenting trade, with a record two million American stay-at-home dads today. There are more single fathers than ever, and dads are spending an average of more than seven hours per week with their kids, three times as much as in 1965. In the most competitive corners of our culture, elite paternal strategy seems to have shifted to a mom-like focus on showering resources onto just a couple of offspring.

If I can drag us back into the Stone Age, though, it's likely that even from the early days of *Homo sapiens* there were advantages to having a resident caveman around to keep an eye on the kids, not to mention bludgeon saber-toothed tigers, while you stirred that just-invented fire. Almost certainly those early dads did sometimes pitch in with the prolonged and vexatious task of raising a human child. Involved fatherhood actually caught on, and flourished, long before women began agitating for it—becoming hardwired into dad brains in measurable but distinct ways.

Obviously, mammalian fathers don't undergo pregnancy and labor. ("Why do men feel more attractive after childbirth?" one study mused. Hmm, go figure.) Nor do they lactate. (Well, a couple of kinds of weirdo daddy fruit bats do—although that's more likely due to something funky in their diets rather than to any sort of simmering Mr. Mom ambition.)

But a few special types of mammal dads, humans included, *may* experience hormonal changes upon the arrival of their young.

New fathers' testosterone levels, for instance, oftentimes plummet, the chemical drop possibly explaining "sympathetic pregnancy," or couvade syndrome, in which some unusually devoted dads gain their own little Buddha bellies. For somewhat less clear reasons, dads-to-be often grow beards, and in some South Pacific cultures they may even theatrically take to bed for the duration of the pregnancy, complaining of aching backs. New dads can experience a form of postpartum depression—though it's far less common than in mothers.

In baby-cue experiments, new dads are quantifiably more reactive to infant cues than childless men are, and they may find baby odors to be unusually beguiling. Some studies have spotted similar neural patterns in mothers and fathers interacting with their infants and—look away, dads!—men with the most robust activity had the lowest testicular volume to begin with. Resident dads can match even moms when it comes to recognizing a particular kid's cry.

All these documented changes, though, are still "nothing like what you would see in a woman," says Joe Lonstein, a maternal behavior researcher at Michigan State University. "The magnitude that you see in a mother is nothing like you see in any person at any other time."

More on the substance of those radical mom changes in just a minute. For now, let's simply note that the clever algorithm designed to pick out mom brains can't spot the dads. Fathers' neural anatomy isn't so trademarked. There aren't always the same predictable patterns in dads' revised brain structures, or the same reliable gray matter flux.

And while it's true that as parents my husband and I are a lot alike—sharing values and schedules and discipline strategies and a mutual inability to cut last year's Christmas tree twine off the minivan roof—I think we'd both agree that I *changed* more.

A famous pie chart from the 1980s illustrates this feeling. A year and a half after their kids were born, dads reported that fatherhood accounted for roughly 27 percent of their identity. For moms, though, it was 55 percent—such a big slice of the identity pie that it looked like a gaping Pac-Man mouth, ready to gobble up the "partner" slice and the "career" slice and the rest of the woman's former self.

You might think this is just cultural expectations at work. But before giving birth, the women themselves expected that motherhood would account for only 16 percent of their personal pies—a striking mismatch between anticipation and reality. Future dads estimated 17 percent, which was at least within hailing distance of the truth.

∾

Just as childless men and women react differently to babies' signals, moms' and dads' responses often diverge, too. Overall, mothers remain more sensitized to their child's emotions. When a child weeps, deeply situated brain zones related to pain and emotion light up in moms but not in dads. It takes somewhat more urgent-sounding cries to rouse a father, which may help explain why new moms get forty-five minutes less sleep per night than dads. (On average, that is: the gulf can yawn much wider on any given night, believe me.)

In some studies, researchers played the whimpers of babies hungry for food, and the howls of babies who had just been circumcised. Naturally, dads snapped to attention for the circumcision screams. But moms were overall better at telling the difference between the cry types and responding to both. As a rule, mothers react to infant emotions more viscerally, giving the most extreme ratings to intense unhappiness. We also have a better memory for babies' smiles, suggesting that they are extra rewarding to us.

Moms think about our babies twice as often as dads do. We speak to them far more, too. In fact, it's not entirely clear that "fatherese" exists. In one study, researchers plopped audio recorders in preschoolers' shirt pockets, then listened through more than 150 hours of their inane jabbering. When parents addressed the kids, the moms' pitch soared up 40 hertz: classic motherese. But the dads did not sound squeaky in the least.

These differences may lessen over time. Dads do seem to bond with their children more deeply later in childhood, with paternal interest mounting a year or so after birth when their kid is a tossable toddler as opposed to a still-furled waif. My husband was always a bit baffled by our newborns. "She looks like a rolled-up pair of socks," he said once, of a tiny swaddled daughter. Another time: "She looks likes a puppet from *Mister Rogers' Neighborhood.*"

But even with older children, mom-dad distinctions surface in the most mundane settings, like swim lessons, where mothers are apt to hold their one-year-olds facing them, so they can keep their eyes romantically locked together, while dads point the kids away. Moms who work outside the home are more likely than dads to take time off for kids' sick days. We tend to soothe and hold, while dads often bounce and throw, pedal legs and tickle bellies, issuing more distracting noises (I can confirm that one) and encouraging risk-taking, a tendency quantified by a rather harrowing-sounding study technique that involves putting toys at the top of a staircase and watching (perhaps from in between your own fingers) what happens next.

These amusing minor daily differences may hint at more extreme life-or-death divides in parents' priorities. Some economists have argued that giving aid money to mothers instead of fathers maximizes child welfare, presumably because moms are more disposed to care for children. In one experiment in rural Tanzania, in which parents were offered a choice of three gifts—children's shoes, cash, or sugar— diligent mothers more often scooped up the children's shoes, while dads preferred cash or sugar.

∽

Above all, the emergence of a paternal instinct—a core pro-baby motive akin to a maternal instinct, albeit buried more deeply and

slower to develop in males, and involving a somewhat different behavioral repertoire—seems to depend on the amount of exposure
that a man has to the mother of his child and, later, to the child itself. How this works is still something of a mystery, although smell
may have something to do with it. A recent study of ninety-one
men conducted by researchers at Newcastle University suggested
that after inhaling the body odors of pregnant women in an experimental setting, male subjects stared significantly longer at baby
pictures.

Simple practice is probably also in play. Like moms, human fathers
are flexible, smart, and motivated: what's more, their brains grow with
what they know. Just as London cabdrivers have enhanced mental
anatomy for navigation, and bird-watchers' brains have highly trained
circuits for facial recognition, dad brains and their behavior might
change if they were only allowed mom-like experience—which our
culture's expectations may sometimes deny them.

For instance, one study showed that those fathers who take substantial paternity leave and put in the early hours alongside new moms
are more likely to consistently undertake tasks like diapering and bathing nine months after the baby is born. (Let's not discount the power
of guilt trips in these dynamics, by the way. A rodent study revealed
that mice moms must ultrasonically squeak at mice dads in order to get
them to pitch in with childcare.)

But whatever the mechanism, a key difference between biological
moms and dads is this: new moms are hormonally primed to seek out
experience with infants, while new dads must have those experiences
in order to get their hormones rolling. Men don't metamorphose into
instinctive fathers after a one-night stand if they never have contact
with the woman again—the way the woman will (if impregnated)
automatically change into a mother. Fatherhood is a far more elective process. To become a father, the first thing a guy has to do is stick
around, and many don't.

The differences in human male behavior around the world, of course, are as varied as human culture itself. But still, almost everywhere, dads are poised to pivot depending on local circumstances, always with a finger to the wind, while ride-or-die moms are more prone to stay the course.

Several environmental variables can help predict a human dad's likelihood of lingering. They include the harshness of the setting: in rough conditions where it takes two to tan a deer hide (or perhaps to scrape together a down payment), dads are more likely to remain. On the other hand, dads are more prone to jump ship in areas with lots of infectious disease, perhaps because their presence isn't necessarily protective against germs, and if kids are sick or dying they would do well to plant more progeny elsewhere. (When reporters returned to check on Zika-affected mothers and their babies a year after the epidemic hit Brazil, almost all had been ditched by their sexual partners.)

A society's means of subsistence also shapes fathering patterns. In hunter-gatherer cultures where human fare is meat-heavy and big-game hunting is a must, paternal involvement may be more pronounced. Take the dad-friendly Inuit, whose traditional diet, nearly 100 percent animal flesh, requires the slaying of whales and other formidable creatures. By contrast, temperate farming cultures lift the upper limits on how many kids a man can sire, lowering his average fatherly involvement.

The certainty of paternity also matters—rates of cuckoldry hover at around just 3 percent in modern Western cultures, but have ranged to 10 percent or higher at other times and places. The current record belongs to Namibia's Himba tribe, where a 2020 DNA analysis found that nearly 50 percent of a man's supposed children are sired by somebody else.

One fascinating study showed that the rare area of infant-reading where dads on average outperform moms is "child facial resemblance detection." In other words, males are more skilled at noticing whether a child looks like them—and are presumably prepared to decamp if the chip is off some other old block.

∽

But even in ideal milieus for male involvement, parenthood rarely becomes truly equitable.

By now my husband and I have raised children in a variety of locales. My daughters were born in densely urban Washington, DC. Then we got the bright idea to bring them up on a farm, so we were living on a multi-acre spread in my rural Connecticut hometown when my son came along. After that plan went up in flames, we and our three pipsqueaks fled to a leafy neighborhood in the nearby college town of New Haven.

Our present street, I've come to believe, is a perfectly calibrated microenvironment for dads, with a high degree of paternity certainty, low rates of disease, a monogamy-oriented culture, and a harsh environment for offspring (at least when it comes to college admissions).

But you really never know.

I'm guessing that most of my neighbors, many of whom are professors, would profess that childcare burdens should be equal. Yet academics who (perhaps traitorously) study their own kind have discovered that only 12 percent of tenure-track men take paternity leave compared to 67 percent of tenure-track women. (Maybe this begins to explain why, throughout academia, married fathers have the best chance of getting tenure, while married mothers of kids six years old and younger have the worst.) In another study, academics with children under age two "said they believe that husbands and wives should share childcare equally, but almost none did so," the shocked-sounding authors disclosed. The female professors performed more of virtually all twenty-five inventoried childcare chores.

Again, it doesn't always happen this way. Preliminary work on gay dads is especially interesting: in Israeli research on dual-father households, the gay dads' brains and overall physiology more closely resembled maternal patterns than did the brains of heterosexual fathers. With no female caregiver around to hog the baby, perhaps these men are extra-steeped in infant stimuli, with minds to match.

At the same time, though, many men are clearly disinclined to seek out this mom-like intimacy, or sometimes any relationship at all. As father-friendly as some corners of our culture have become, there are more American children than ever in fatherless homes. Twenty-seven percent of kids live apart from their biological dads now, and more than half will reside at some point with an unmarried mother before the age of eighteen. America has the highest rate of single-mother-headed households in the world. Also, the oft-touted free time that American dads are now spending with their progeny sometimes reflects unemployment, disability, or reduced working hours rather than a burning desire to stay at home and play patty-cake.

Even the valiant single fathers, nearly unheard of before *Full House* hit the air, don't fare so well once scientists get to scrutinizing them. A recent study of Canadian single dads showed that they suffer from staggeringly high rates of heart attacks and cancer and are three times more likely to die than single moms in the same straits, suggesting that this unfamiliar role may be highly stressful.

"Men have so many other concerns that their children are often of secondary importance," writes anthropologist Wenda Trevathan, summing up the global anthropology literature. In "the vast majority" of human cultures, another scholar notes, moms and other females provide more care for infants than dads do. (I'm tempted to up the ante and say "all" cultures, but for a 2017 finding that Finnish dads spend eight minutes more with their kids per day. Those were school-aged kids, though, not needy babies.) This pattern holds true even for anthropologists' favorite pro-dad peoples like the Aka of Central Africa, where fathers hold kids for 57 minutes per day—which is indeed substantial, but nothing compared to mothers' 490 minutes.

Until very recently, some of my mom friends would have shot down the idea of this kind of "naturally" lopsided parenting setup, since our culture and clever career choices had always shielded us. But then Covid-19 came along, throwing moms and dads alike back into the home and into a kind of back-to-the-1950s time warp—and suddenly, whaddaya know, working mothers in the highest professional echelons were shouldering the new childcare burdens, and moms across the country were slashing their hours and forfeiting three times as many jobs as dads.

Among the primates, humanity's branch of close animal relatives, there are a couple of outlier species, most notably titi monkeys, where dads trump moms as central caregivers, holding the baby most of the time, to the extent that the infant actually prefers the father's arms. (For what it's worth, the titi dad still seems fairly indifferent to the infant, preferring in his spare time to entwine tails with his lifelong mate.)

But these exceptions evolved in a very specific environment: titi monkeys live precariously high in the jungle canopies of South America. Supplying gallons of breast milk and hurling herself from towering treetop to treetop with a baby in tow, a sky-dwelling titi mom would likely be doomed to starvation if abandoned by her mate.

I have wondered if my husband might have a few drops of titi blood in him. Sometimes my kids call out for him in the night, instead of me, and if we had tails, then on certain days maybe they'd rather twine theirs with his.

In reality, though, people are more closely related to ground-dwelling African primates like rhesus monkeys, who—in lockstep with most mammals—grow up with anonymous fathers. And as welcome as human paternal presence and activity may be, as indebted as we may feel for our mates' grilled-cheese slinging and stroller patrols, that's all gravy in the evolutionist's cold-eyed view, where what matters at the end of the day is who dies and who survives. For humans across cultures, according to scientists who've run the numbers, the presence of a nearby grandma seems to be more of a boon to a child's survival than a dad is. Some go further, contending that dads are essentially irrelevant to child survival.

Deep down, maybe even the most ostentatiously enlightened men suspect this. As a Christmas gift, I subscribed my husband to a magazine crafted for the eager-beaver race of flannel-clad ur-dads, who are adventurous yet devoted, heavily bearded yet soul-baring, and invariably play guitar. It was called *Kindling Quarterly*, and it featured articles on rock star dads and "Traveling Istanbul with a Baby."

It folded after six issues.

~

So as transformative as it can sometimes be, the paternal experience remains highly optional. And yet fickle and flighty human fathers do have one rock-steady part to play in parenthood.

They are essential to the manufacture of mothers.

Even if they're back on Bumble the day after conception, fathers are still driving the chemical process that induces a woman to surrender her body to his unborn child and then fall hook, line, and sinker for the kid. Dads, or at least dad genes, are a veiled force behind the maternal transformation. This weird and hidden drama plays out in a human organ that is not only often overlooked, but frequently tossed in the garbage: the placenta.

It's been a slithery downhill slide for the placenta since Egyptian days, when the pharaoh seems to have publicly paraded his beside him, brandished on a tall staff. The Egyptians believed that it was the seat of the soul, and it turns out that they weren't too far off.

The placenta isn't just a vital, if short-lived, fetal appendage through which the baby eats, poops, and breathes. It's also the sorcerer behind key maternal hormonal changes, supplying some of the main ingredients that make a mother.

Yet in all the new-mommy propaganda littering the waiting room of my own ob-gyn's office, this alchemist of pregnancy is conspicuously missing. Placenta science is a backwater even in the generally neglected field of mom studies, though for admittedly legit reasons: in animals the placenta is a maddeningly difficult organ to study. It's often consumed, frequently in the dead of night when mammals give birth, like the beastly equivalent of a bottle of delivery-room bubbly. One placenta biologist described how she obtains her samples from hungry new monkey mothers: she barters with marshmallows, which are apparently the only more delectable delicacy.

Physician and research scientist Harvey Kliman of Yale University has been on the cutting edge—and the goopy placenta does have edges, if you know where to look—of placenta research for nearly four decades. He's also an amateur photographer. Some of his favorite placenta portraits, mostly microscopic close-ups of the tissue that look very much like abstract art, decorate the walls of his New Haven office, just a couple of miles from my home. There's also a spiraling mobile of paper cranes in one corner—or perhaps, given his profession, they are storks.

On his desk sits a kind of steampunk sculpture consisting of several light bulbs, a microscope, and a plastic human skull riveted to a wooden box. This is Kliman's "ideas chest." When in need of creative inspiration, he twists the knob of the brightest Edison bulb all the way up.

"Men can't create life, so we have to create something," he says.

What men create are placentas.

Kliman shows me a family picture from the wedding of one of his identical twin daughters before deftly toggling on a computer screen to a snapshot of their identical twin placenta. "This is Rachael and Michelle right here." He beams.

While nobody paraded with a pharaoh's staff, Kliman did preserve his twins' placenta for three years, until someone in his lab accidentally cleaned out the storage cabinet. And why shouldn't he? It is his handiwork. Kliman's wife made the daughters, but the double-decker placenta is his own.

"Most people think it's the mom's placenta that she makes to feed the baby," he says. "Nothing could be more wrong."

A famous series of experiments from the 1980s helps explain why. Scientists long assumed that mother and father each contribute half of a baby's genes, going Dutch on every trait. Since the placenta is an external organ of the fetus, with identical DNA, the same rules ought to apply.

Yet they don't. Combining two sets of a mother's genes in a manipulated mouse egg, stunned geneticists watched the egg develop into a nearly complete fetus with a puny placenta.

Doubled *dad* genes, on the other hand, yielded a stunted fetus. The placenta, however, was extraordinarily large and in the rubbery pink of health. (This mix-up sometimes happens naturally in humans, if sperm multiply inside of an egg without a nucleus, resulting in a "molar pregnancy" that's just a huge placenta, resembling a ghastly crimson cluster of grapes.)

And so the unsung placenta first clued scientists in to the intriguing phenomenon of imprinted genes. Mom and Dad do indeed split most traits—like, say, earlobe shape—fifty-fifty, each chipping in one copy of a gene. In less than 1 percent of our genetic code, though, one parent's contribution is muted and the other parent "imprints," controlling the chemical messaging.

While imprinting does happen elsewhere in the body, animal models suggest that unusual numbers of imprinted genes are active in the placenta—and most of them are father-driven. The organ's short life might allow for such extreme genetic experiments. It has to survive for only nine or ten months, while the kidneys or the pancreas must endure roughly a hundred times longer.

Biologists see genomic imprinting in the placenta as an invisible shoving match between mother and father. Perhaps you think of your partner as a comrade-in-arms, the guy who cowers under the covers with you, bracing for the first fateful snuffle from the baby monitor. But as some evolutionary biologists conceive of conception, you two started out on opposite teams, locked in mortal combat in your belly.

This "womb as Thunderdome" idea caught on with me a little more quickly than you might imagine, given my husband's amicable nature. Way back in college, he showed up forty-five minutes early to pick me up for our first date, his roommates having tricked him by setting his watch ahead while he was in the shower. All that awkward extra time allowed me the chance to study this strange man from my English seminar more closely in the harsh overhead dorm-room light. I took a good, hard look at the face beneath his beard (which, for the record, he sprouted not in the course of sympathetic pregnancy, but in mourning for another woman in our class who'd spurned him).

I knew that face. "It's *you*," I said.

It turned out that we'd been paired up once years before, in a statewide high school debate tournament. I'd never forgotten the insufferable matching sweater vests that he and his debate partner wore, especially because they won—although I did beat my future spouse in speaking points, not that anyone is still keeping score. Our teams ultimately faced off a second time at the altar, where my husband's partner was his best man, and mine—my best friend Emily, of course—was my maid of honor.

As swimmingly as my husband and I later got along, this adversarial aspect of our relationship sometimes resurfaced, like when we debated how soon we ought to have children. I was on board with the idea in principle—kids were on my bucket list, for sure—but in practice I preferred to delay as long as possible, mostly because I enjoyed drinking beer with my Ethiopian food and eating mountains of raw cookie dough potentially full of fetus-frying bacteria, and because I'd handled enough diapers in my babysitting days not to take the matter lightly.

"Do you know what's in diapers?" I hissed during one skirmish in a Mexican restaurant, a few margaritas in. "*Poop* is in diapers!"

I plunged a fork into the dark brown mole sauce drowning my enchiladas.

Eventually he won—or, as I'd prefer to frame it, I gracefully conceded the point, in large part because I'd misjudged the nature of the maternal instinct. Other than sacrificing ten months' worth of cocktails, *of course* motherhood would be easy for me. Everybody did it, and I wasn't just anybody—heck, I'd even ended up the state debate champion, although my husband would argue that's because he jettisoned the debate team for the high school newspaper. How hard could it really be?

So I determined at last to humor my poor husband, to let him have his timeline and to give motherhood an early whirl. (Obviously I hadn't seen that pie chart yet.)

But, as placenta science suggests, the truly titanic clashes between us were just beginning.

∽

Since I've never actually seen this squishy architect of my destiny, all three of my husband's placentas having been whisked away behind the prim curtain of the C-section surgery room, Kliman arranges for me to observe a fresh one in the lab.

The still-warm placenta splayed on the countertop looks loosey-goosey and collapsed, like a scarlet omelet, or the stranded red jellyfish my kids poke at on the beach. But it is actually the most surprising structure of any mammalian organ, coming in an array of outlandish layouts.

For all their fur and other frills, mammals are mostly the same inside. A hippo's liver is much like a jumbo human liver, and a gerbil's stomach is basically ours in miniature. The placenta, though, is a shapeshifter. Some species' placentas look like rubber bands, and others like padded rooms, or threadbare sheets draped over lumpy furniture. In the uterus, ours has been said to resemble a yarmulke.

Some scientists think that, like imprinting itself, this extraordinary variety might be more evidence that the organ is a war zone, where mother and father are always striking back against each other, redrawing the map. Love is indeed a battlefield.

The lady—like, say, me—with the newly fertilized egg twirling around in her innards isn't a doting mother quite yet. Far from it. Her immune system is trying to protect her bodily reserves and may even be trying to put the kibosh on the whole pregnancy. Technically speaking, "having a pregnancy is a big, big problem for a woman," Kliman explains. It's risky and nutritionally expensive, and the female body's first answer is resistance: only a fraction of all pregnancies make it to term, in large part because the maternal immune system tries every trick in the book to attack the placenta in the first few weeks.

Meanwhile, the father—whether he's sitting beside you rewatching *Game of Thrones* for the fourth time, or already decamped for Kokomo—is fighting for the pregnancy via his imprinted avatar, the placenta.

It's messy but mighty, Kliman explains, expertly flipping the omelet, a blood clot dangling from his gloved thumb. The maternal side of the placenta, where it attaches to the uterine lining, is a hot mess, but the fetal side looks smooth, collected, and alien. "No matter how hard you twist this, you cannot compress the vessels," he says, yanking the umbilical cord. Thick as the docking rope on a sailboat, it disappears into the placenta's center.

I'd always thought of the cord as a lifeline that I'd flung out to my babies, but it's the exact opposite. The whole placenta is like a grappling hook swung overhead and cast into the body of the mother. It branches into smaller and smaller hooks, or blood vessels, all designed to draw nutrition from the mom into the fetus. A mature human placenta has more than thirty miles of surface area.

Though it's also called the "afterbirth," the placenta starts to form extremely early, at just five days' gestation, peeling off from the outside of the barely there fetal clump called the "blastocyst." And almost immediately, this dad-driven faction of cells embarks on a coup d'état.

Normally the pituitary gland, deep in a woman's brain, sends signals to her ovaries to produce progesterone. (When the progesterone drip stops every month, her period arrives.) But in the very first days of pregnancy, the placenta bypasses the woman's brain and takes the reins, messaging her ovaries directly for more and more progesterone, so that her pregnancy-preventing period never comes.

"The placenta says, 'You know what, let's do it my way,'" Kliman says, doing his best placental impersonation. "'We can remove your head and still be fine.'"

Other maternal body parts go on the chopping block, too. Even your ovaries become irrelevant after about nine weeks or so, since by now the placenta has taken matters into its own "fingers" (that's how Kliman describes the organ's weedy internal structure, reaching deep into the woman). The placenta starts brewing up bootlegged proges-terone *in its own tissue*, and components of estrogen, too, so that your ovaries can be surgically removed and the pregnancy will continue on its merry way.

"It's kind of like when the space shuttle goes up from Cape Ca-naveral and about ten seconds into the air everything is taken over by Houston," Kliman tells me. "The placenta is Houston. The pregnancy, through the placenta, takes over all operating systems of the mother."

Meanwhile, the placenta ratchets up a mom's appetite and thirst, even as placental hormones make it harder for her to access her own blood sugar. Three pregnancies' worth of pad thai had, unbeknownst to me, bypassed my own liver to feed my husband's placentas. Pla-cental hormones also ready the breasts for nursing, prepping for a future when the baby's nutrition will be delivered via milk instead of blood.

All of this baby-mama drama arises, in part, from one of those lin-gering difficulties with internal gestation, which is that a mammalian father never really knows for certain if a child is his. So, rather than provision the child from the outside at great personal expense, he has evolved ways to press his claims from within.

And because a man can't be confident that he'll ever have a shot at another kid with you, it's also in his biological best interests to ransack your carcass for all it's worth this time around and extract the biggest, healthiest brat possible. Despite his kind, twinkly eyes and habit of bringing home your favorite grain bowl unbidden, his genes want to take you to the cleaners.

Under a microscope, Kliman shows me a piece of a woman's uterine lining that, to the naked eye, looks like a slice of fine prosciutto. With a ghostly white arrow he shows how certain placental cells—"They're very aggressive," he says—actually leave the placenta proper and migrate into the tissue of the mother, where they attack her arteries like starved wolves.

Setting sail a few weeks into pregnancy, these invasive cell bodies, which look like tiny black polka dots in the pretty pink paisley of the mother's tissue, remind me of the thousand ships that the Greeks sent after Helen. There are far more than a thousand, though. Hundreds of millions of placental cells surge into the flesh of each pregnant mother. And they use military-style tactics that even Agamemnon might admire.

Once they've got the mother's juicy little artery surrounded, they assault its wall and—in a process that may sound all too familiar to mothers—turn its taut muscle into pink mush, a first step in commandeering the mom's blood supply.

The artery weakens, then widens: once almost invisible to the naked eye, it becomes a gusher as wide as a dime. What began as a tight little stream is now a roomy, and very much man-made, Panama Canal, carrying the mother's blood from the uterus through the placenta to the fetus.

Normally, about 5 percent of a woman's blood flows into the uterus. By late pregnancy, thanks to placental tampering, it's more like 25 percent—an ocean of blood to feed a developing child whose own total blood volume is less than a small bottle of Poland Springs.

Fascinatingly, humans may have the most aggressive placentas of any mammal. Our placentas launch greater numbers of invasive cells even than the placentas of other primates, with the possible exceptions being chimpanzees and gorillas, our closest great ape kin. This suggests that our vampiric placenta may be necessary to nourish humanity's other distinctive organ.

"The placenta has been hugely central in the development of the human brain," says Julienne Rutherford, a placenta researcher (the one who traffics in marshmallows) at the University of Illinois. "The brain is a very expensive, demanding organ, and where is the energy coming from? There has to be some kind of energetic shift, and a combination of placental invasiveness and placental surface area explains the energy that is coming in."

All the extra blood flow also likely explains why postpartum hemorrhages plague human women. Mostly unknown in other mammals, hemorrhage impacts roughly 10 percent of our births and is the leading cause of maternal death worldwide—still claiming 125,000 women per year (even as blood transfusions save many more in our medically advanced era), collateral damage in the dad's wholly unconscious yet ruthless campaign to make his offspring feel right at home.

Kliman shows me one last slide, on which another mother's rosy uterine tissue appears to disintegrate into chaos. I'm not sure what I'm seeing. This, he says, is an example of "placenta percreta," where the mother cannot bring her lover's siege to a stalemate. Left to its own devices, the placenta grows all the way through the uterus and sometimes even into neighboring organs, such as the bladder.

"I've seen a lot of these cases," says Kliman, growing uncharacteristically somber. Now I know that we are inspecting a fragment of a dead woman.

∽

The placenta doesn't just siphon away nutrients to grow big brawny brains for some random guy's spawn, co-opting our uteruses, breasts, and blood vessels. It also procures love for the father's helpless descendants.

It wants *our* brains, too.

The mechanics of how the placenta's hormonal tsunami prepares women's minds for childcare are complicated and still poorly understood, although progesterone, estrogen, and other hormones are well-established players in the process. Yet however the magic happens, scientists suspect that the imprinted dad genes are driven to maximize maternal care.

A British lab recently published an unnerving paper about how the placenta might reach its gooey fingers all the way up into our minds. Rosalind John of Cardiff University and her team focused on a gene called PHLDA2, which normally limits the amount of hormones the placenta can produce, acting as an important check on paternal power. Using genetically manipulated mouse offspring, the researchers wanted to see what would happen if they shut down the mom's copy of the gene so that the hormonal tariff was lifted and the dad gene's motives, so to speak, were exposed—letting the placenta churn out all the lovey-dovey chemicals it liked.

The extra placental hormones spewed into the mother mouse's system, suffusing her brain. And after the mouse babies were born, sure enough, the high-hormone moms spent extra time nursing and grooming their newborns. The researchers were even able to pinpoint changes in two brain regions key for maternal care. Just amping up placental hormones, and by extension paternal influence, enhanced moms' behavior.

John thinks these findings may be one key to the disorienting shift to intense maternal devotion.

"If someone woke you up at four o'clock in the morning, crying and screaming and covered in poo, you would not be happy under normal circumstances," John says. "But somehow a new mother says, 'Oh, look, you are awake, and I will take care of you.'"

In fairness, the new mother isn't precisely *happy*. But she is willing, for reasons that begin with the placenta, and the nutritional, immunological, and behavioral overthrow inside the woman that the father's genes help unleash.

In a sense, in fathering a baby, the dad also fathers the child's mom.

∽

It all sounds a bit like Stockholm syndrome. Having been immunologically hijacked and nutritionally abused, mothers are somehow browbeaten into adoring their new babies—and in certain cases, into demanding to have more.

For after the birth of our first, my husband found that the tables were permanently turned: in our discussions of future children, he was suddenly the one urging caution, which I was now more than ready to throw to the wind in pursuit of the next cutie-pie. If dads do indeed make moms, he had created a momster.

But paternal prisoner-taking isn't the only way that moms are made. The placenta and its hormones are evicted from our bodies at birth, so their temporary presence can't sufficiently explain the lifelong transformation we undergo, the maternal fallout that only seems to increase with time.

Our own systems are at work here. Moms also make ourselves.

Chapter 3

THE WHOLE SHEBANG
The Maternal Chemistry Kit

EVEN TODDLERS grasp that moms are distinct from other people. "You're not a maiden," my three-year-old son declares as we read fairy tales together on the couch. "You're a mama." My goofy smile back down at him is no doubt as damning as any brain scan.

Observing differences between moms and dads, and moms and pretty much everybody else, is relatively easy—on the outside, that is. (It's readily apparent whose hair has been styled with spit-up, for instance.) What's harder to grasp is how a female mammal—Sheep 513, my little sister, or me—transforms from maiden to mama *on the inside*, at a cellular level.

Hard to grasp, and—fair warning—hard to convey. I've recently bragged to you that before having kids I'd succeeded at most things I'd tried, but perhaps that's because I never tried to be a neuroscientist. This stuff is really challenging, even for those whose brains have not been impaired by successive childbirths. So what follows is just a glimpse of how scientists ask and answer questions about mom brains' inner workings, not a definitive account.

Still, this peek is worth moms' while, for what feels like a change of heart is actually a change of brain. Neurochemicals, and the genes that code for them, influence our actions at the playdate or the parent-teacher conference. They drive disastrous maternal health problems like postpartum depression and other perinatal mood disorders, which—as many of us have learned the hard way—doctors don't really comprehend on a chemical or personal level, making these conditions tough to treat even as they are on the rise. That's in large part because we don't yet fully fathom the healthy maternal brain and how it ought to function when everything is working right. To understand ourselves, it's vital that we learn the rough outlines of how, on a microscopic level, mom minds are made up.

There do seem to be a few secret ingredients involved. Since the 1970s, scientists have known how to concoct a mouse mother on short notice, without a father or his placenta or any of the other sloppy hassles of pregnancy. Give the virgin rat a simple injection in the lab, and she begins to act like a mom, snuggling babies instead of consuming them.

What is this special sauce, or fateful potion, inside the syringe?

The injections, it turns out, contained a very simple substance: another mother rat's blood. Yet decades later, scientists are still trying to understand the precise chemicals that bring a mom to life.

✖

It looks like the researchers at the NYU Grossman School of Medicine are contacting space aliens instead of studying mothers. Amid swirls of wires in one corner of the enormous lab, a beaker of crystal-clear liquid—a homemade version of cerebrospinal fluid—bubbles above a high-powered microscope. Tiny glass pipettes chill nearby in a purple ice bucket.

Postdoctoral fellow Soomin Song uses tweezers to lift what looks like a large white snowflake from a glass full of flurries.

It's a sliver of brain, drained of blood. About fifteen minutes ago the whole brain was inside a female mouse—though not a mother.

She was literally a sacrificial virgin.

"Her brain is actually still alive," Song says as he works hard to keep it that way, bathing it in the faux cerebrospinal fluid, and keeping it at a mousy temperature of about 95 degrees. He wants to make sure some neurons are still functional so that we can "spy" on them.

"We are essentially trying to listen in on the maternal brain," lead researcher Robert Froemke had informed me a few minutes earlier. "We break it down into parts to figure out how it works."

The part of the mouse brain that his researchers are zeroed in on is the auditory cortex, which processes sounds, and some very special sounds in particular: "We are trying to examine the low-level brain circuits that respond to the acoustic structures of baby cries."

Ultrasonic mouse pup distress calls, which mice babies most often make when they get cold, are like nails on a chalkboard to virgin mice. That's part of why they shun babies. But the exact same noise is a siren song for mouse mothers, luring them like ships onto rocks. The moms prefer the pups' sorrowful keening even to the sound of music. So what makes the mother hear so differently?

Song slides the virgin's brain sliver under the microscope. He zooms in on the part that contains the mouse's hearing zone, peering first at four times magnification, then at forty times. Seen this close, the landscape of the virgin's brain is reminiscent of a vast and rolling gray desert. Song soon spots one of the so-called pyramidal neurons he's seeking.

"There's our neuron right there," he says, manipulating the microscope's joystick with telling dexterity. ("I've spent many a quarter in the arcade," he admits.) Approaching the target brain cell with a near-invisible glass pipette, he prepares to perform a technique called "whole-cell recording." It's sometimes jokingly known as "kissing" the neuron. "I'm literally going to be smooching it," Song informs me. He puts his mouth on the end of a long tube and sucks, creating pressure until a dimple forms on the cell's surface beneath our probe. As he sucks, there are even a few audible smooching noises, of the type overheard as I plant a wet one (or ten) on my son's pillowy cheeks before he can wriggle free.

After accidentally exploding the first neuron, he swiftly gets a lock on the next, embedding the glass pipette just under the membrane so he can tap into what's happening within.

The goal is to measure this single brain cell's reaction when it is stimulated. Now Song zaps the entire brain slice with electricity. This stimulus mimics something very important: the real-life electrical impulses of a bereft pup's cry. The pipette suctioned to the cell functions as a tiny electrode, taking readings. On a nearby computer screen, we watch the lone virginal neuron respond on a tracing, like a heart monitor. It spits out a series of little spikes, its own electrical signals.

Song will repeat this process almost endlessly, comparing readings from dozens of virgin brain cells with readings from dozens of mombrain cells, looking for variations in the neurons' reaction to the baby-cry stimulus.

Typically, mom neurons fire while the virgins' fizzle. This means that the moms' cells are somehow more responsive, with spikier readings.

It's like the story of maternal sensitization on a single-cell level.

∽

One key sensitizer—or secret mom ingredient, if you will—that likely helps explain the difference between the neurons of a "maiden" and those of a "mommy" is the hormone oxytocin, which is made in a part of the woman's brain called the hypothalamus. "Oxytocin" means "swift birth" because it gushes into the bloodstream during labor and delivery—just as the placenta and its estrogen and progesterone are being evicted—and facilitates uterine contractions and milk letdown.

Scientists have recently become captivated by its impact on the brain as well. Sometimes called the "love hormone" or the "trust hormone" in humans (yes, men have oxytocin, too), it is associated with social and romantic bonding, and even with activities like charitable giving. Froemke and others suspect that oxytocin doesn't just prepare women's bodies for birth; doubling as a neurotransmitter, it also readies our brains for infant worship. It is likely one of the unseen substances in the rat mothers' blood that transformed the injected virgin rats in those intriguing laboratory studies from forty years ago.

Froemke's lab members wanted to see if they could use oxytocin to watch a rodent brain transitioning from virgin to mom in real time. They set up a series of experiments, published in 2015 and now considered classics.

The lead researcher, Bianca Jones Marlin, drew her high-tech tools from the new field of optogenetics. She picked mouse virgins whose DNA had been manipulated to include extra code for brain cells that produce light-sensitive reactions. In this case, a blue light from a laser shining into the mouse's skull would stimulate a natural oxytocin rush.

Marlin whisked these genetically engineered virgins off to the lab's own studio-grade sound booth and implanted brain probes to take readings from individual neurons. (It's like what I did with Soomin Song, only in Marlin's experiment the brains were still inside the squeaking, breathing mice.) She played pup distress calls, but the virgins remained unmoved. Their brains responded in a typically disinterested manner, with a jaded spike here or there.

Then she blasted the blue light.

Oxytocin flooded the auditory cortex, as it might in childbirth. When she played the pup distress calls, the virgin brains began to perk up, with more reactive spikes. Within three hours, the virgin readings matched those of the moms.

Exposure to oxytocin had somehow sensitized their neurons to the cries.

"That was a pretty amazing thing to see over three hours," Marlin says. "We replicated the birthing process in a single neuron. The first time it happened I got chills. I got tears in my eyes."

It turns out that female mice brains are built to sop up this oxytocin gush. Throughout their life spans, both male and female mice have low levels of oxytocin receptors throughout their bodies and brains. But Froemke's researchers have detected a unique surge in the number of receptors in the left auditory cortex of female mice who have passed through adolescence and are ready for mating (that is, mice that are about twenty days old). While this lab group is laser-focused on this sound-related region, the oxytocin receptor burst likely occurs elsewhere, too, perhaps in areas related to other senses. It seems that some of the special neural equipment to handle the chemical rush of childbirth gets automatically built in.

With live dissection of humans being a definite no-no, much less is known about the natural distribution of oxytocin receptors in our own brains. But the data we have suggests that oxytocin also modulates human maternal behavior, whether we're howling our way through labor and delivery or, ever so much more appealingly, getting paid to inhale the stuff in the lab.

In several experiments, when childless women snorted jolts of oxytocin into their noses, they, too, had enhanced responses to baby faces and infant cues like crying and laughing, compared to women who'd only sniffed placebos. And after the oxytocin snorting, fMRI scans showed that certain parts of the childless women's brains involved in feelings such as empathy likewise mirrored the neural activity of mothers' brains.

∽

But wait a sec. Before anybody gets too excited about this smoking gun—this "mother molecule," as one scientist described oxytocin to me—you should know that there is another well-regarded laboratory, just down the way and also part of NYU, that studies the transformative maternal effects of an entirely different neurotransmitter: the pleasure chemical dopamine, which, like oxytocin, is produced by the mother's own body.

Still other labs remain keen on the lingering behavioral impact of progesterone, estrogen, and other by-products of the father's meddlesome placenta, which throughout pregnancy seem to combine in precise ratios to prime the mother's mind for the climactic hormonal rush of birth.

And of course there's prolactin, the breastfeeding hormone, and stress hormones to boot.

Yes, artificial blasts of oxytocin can kick-start maternal behavior in virgin rodents—but so can injectable cocktails concocted from various other maternal molecules. Sure, oxytocin explains why, if a sheep farmer wants a ewe to mother an unrelated lamb, he must often manually perform vago-cervical stimulation on her, faking birth's "love hormone" surge. (This was another potential task I'd dreaded that night at the sheep barn.)

But of course, then there's me. I didn't deliver my kids the old-fashioned way, and I don't recall any kind of pleasurable oxytocin rush as the surgeons hacked me open. There was no handsy sheep farmer standing by. Yet I still love my kids plenty.

It would seem that rewiring a female rat, let alone the lady ommmmm-ing beside you in baby yoga class, is a pretty complex business. All those neurochemicals piggyback on one another in poorly understood ways. For instance, estrogen enhances oxytocin receptor expression in various parts of the brain. Meanwhile, "oxytocin and the dopamine reward system are separate systems, but they talk to each other," says Lane Strathearn, who studies maternal behavior at the University of Iowa. "In the brain there are direct connections between oxytocin-producing brain cells and reward centers."

And once all the neurochemicals together amp up brain activity and start various and sundry brain regions chattering, new connections form and old ones die, and various brain structures begin to physically metamorphose. This malleability is moms' signature feature. Like so much else in our lives—baby bottles, Barbie dolls—moms' brains are plastic, too.

These new connections help make canoodling with your scrunchy-faced baby seem just as delicious as the most decadent dessert—when in your previous life you hardly gave babies the time of day. The thoughts and sensations stemming from our newfound purpose persist far into the future, and probably forever, long after the hormonal bonanzas of birth and breastfeeding are distant memories. (If you can even recall them. More on the memory-deleting aspect of the mommy brain later, provided that I remember to tell you.)

∽

So which parts of our brains are changing here? Where, exactly, is the maternal anatomy?

When I first started talking to scientists, I thought that unlocking the secrets of the maternal brain would mean looking for something discrete and self-contained, and maybe even helpfully labeled, like the ladies' lounge in a department store. I thought finding it would be like the moment in late autumn when the last leaves drop off the backyard trees, and you can finally lay eyes on the exact branch where the bird's nest is cradled, the source of last summer's singing.

Of course, in reality the brain is just a football's worth of squashy tissue. It doesn't have rooms or alcoves, and to the extent that it resembles a forest, the crucial maternal structure is not sitting pretty on a single backyard branch. It's the whole raucous jungle.

"No behavior is controlled by a single brain region," warns neuroscientist Danielle Stolzenberg, of the University of California, Davis. This is not an "X marks the spot" situation. Different groups of scientists have spent careers studying various nooks and crannies.

Rodent-studying scientists, with their many extra tools, have scurried far ahead of the human-centric ones. To the extent that they've found a maternal locus of control, or a "central site" of mom behaviors, they often refer to a part of the hypothalamus, way down toward the brain's core. This is a sensible seat for such an ancient instinct, since the deepest parts of our brains are thought to be similar across the mammal class.

The hypothalamus is "really important for the four F's," Stolzenberg says. "Feeding, fleeing, fighting, and . . . uh, mating." And at the very front of the hypothalamus sits the "medial preoptic area," or the mPOA for short.

The mPOA can be stimulated to produce maternal behaviors. Surgically dismantling or anesthetizing it, on the other hand, makes maternal behaviors disappear in rats, to the extent that moms no longer collect their screaming pups. (The animals' mouths still work fine in these experiments, since the moms remain proficient at gathering up Charleston Chew cubes and other treats. Babies simply no longer seem sweeter than candy to them.)

The mPOA is not some coveted corner office in the brain, complete with brass nameplate. It's a barely visible cluster of cells—in a female rodent, smaller than the head of a pin. But it receives oodles of sensory inputs from the eyes, nose, and other sensory areas that gather baby-related data from outside the body. For instance, the auditory cortex—the oxytocin-rich mouse brain cranny that Froemke's lab probes—feeds information into the mPOA.

These many layers of input help explain why the maternal instinct is so robust, as demonstrated in those old experiments in which a mother rat's senses are eliminated one by one. Disable a mother rat's nose, and she can still see her babies. Blind her eyes, and she will smell them.

This zone is also rich in its own estrogen and oxytocin receptors that help it detect chemical signals, receptors that seem to naturally multiply—in female rodents, at least—in the forty-eight hours right before birth.

And within the brain itself, the mPOA tosses long nerve fibers, called axons, like lassos hither and yon to network with other important brain clumps. The most important axons tether it to the dopamine-rich ventral tegmental area, a reward center related to motivation. (This key synergy might help explain why we moms stay high on baby scents for months after the first hormonal lift of childbirth fades.) In fact, together these two synced-up areas—the mPOA and the ventral tegmental area—are sometimes called the "maternal circuit." This packages baby cues together with reward.

But the whole maternal shebang ultimately ropes in many other systems related to pleasure, stress, memory, and practically everything else—and the more these connections fire up, the stronger they grow. The maternal circuit also gets input from the amygdala, hot spot of fear and emotion. The pleasure-centric nucleus accumbens is also in the loop, of course. Nor is the striatum, involved in voluntary movement, left out in the cold. And who could forget the periaqueductal gray?

Although scientists favor gee-whiz circuits and flowcharts to describe the relationships and hierarchies inside our skulls, it's also something of a jumble in there, like the contents of a mother's handbag. ("Strangest thing in your mom purse?" one of my favorite mom websites asked recently. Answers included a tutu, an avocado, cat food, a Christmas ornament, and a plastic dragon.) The jumble gets even more complicated when you make the jump from rodent brains to people brains, with our supersized cortexes that can override primitive impulses.

Human mothers are not just giant hairless two-legged rats— although we may feel otherwise on our worst days, scrounging for the last scrap of cheese in the fridge. Not only are our mom brains larger and more complicated, but most inconvenient of all, from science's perspective, they are far less freely available to slice and dice and centrifuge. So while scientists suspect that a human mother's hypothalamus is very much in the limelight as well, we can't even really make out her mPOA on brain readings—it's too small to see on MRIs, and way too deep down for EEGs to detect. We won't be able to probe its true importance for maternal care until our tools improve. Meanwhile, human mom-centric scientists are hobbled in other ways, too—for instance, we can't just up and genetically engineer ourselves to get an oxytocin rush at the blink of a blue light. (Too bad, really.)

Many studies of human pregnant female and maternal brains are forced to focus on larger hunks of the brain, or on layers closer to the surface. And even here, there is a level of ambiguity that leaves the bright-est scientists scratching their heads. We don't *truly* know which brain bits grow and which shrink. Alongside the previously described and somewhat distressing evidence for the withering of moms' gray matter, with 7 percent losses in some women, other researchers have found that the maternal brain *grows* in similar areas. This contradiction is somewhat mystifying to researchers themselves, though it likely has to do with the different methods that various labs use to measure brain volume.

At the moment, scientists do not necessarily agree on the precise na-ture of human maternal metamorphosis and where it transpires. They are adamant only that change occurs—that mothers are molten creatures.

∽

Meanwhile, creating a mom rat, or the PTA president, doesn't just in-volve the *hardwiring* that female brains are born with or automatically develop as they mature, or the *hormonal priming* of pregnancy (hail, placenta!) and childbirth.

There's a third *H* in our alphabet soup recipe: *hands-on experience* with the baby, or I suppose paws-on, if you happen to be a rat.

Let's return to that fateful day, February 7, 2011, the Monday after Super Bowl Sunday, the agonizing space between when I was delivered of my first baby and when I truly met her. For nearly ten months, I'd been plied with various natural hormones and, in those final hours, some fake ones, too—*"Pit her"*—but was distinctly not reborn. My brain was bobbing about in a type of maternal limbo.

This unnatural, panicky pause between becoming a mother and meeting my baby is painful to recall but also, in retrospect, quite in-structive. For it shows me that hormones and neurotransmitters aren't all-powerful. Sure, they paved the way for the epiphany to come. But that tiny face grimacing up from the isolette? She was the clincher.

This lived revelation may also come as a surprise to scientist moth-ers, even Bianca Jones Marlin herself, who happened to give birth to her first child amid her oxytocin research.

"I thought I knew a lot about maternal behavior because I had a PhD in maternal behavior," she tells me. Oxytocin was her whole life. "Then I became a mom."

She had a natural birth, but shortly afterward was separated from her baby to receive medical treatment and had to give up on breast-feeding, which in nature keeps the oxytocin flowing. But of course her infant remained her everything, even without that particular hormonal flood.

Human motherhood is not just a scripted series of chemical explosions, choreographed like fireworks. It is also a chaotic reality dependent on interactions with an unpredictable other, the baby. At some point in every maternal life span, even for women who've given birth naturally and nursed for eons, the chemicals become secondary. A mother still fiercely loves the ten-year-old she hasn't breastfed since infancy. That's because our brain connections, the link between attention and reward, become strong enough that they function on their own, no longer dependent on hormonal refreshing. At a certain point mom brains are just, well, mom brains, responding to baby cues without constant chemical prompts and incentives.

And, under certain circumstances, helpless little babies themselves even seem to be a stimulus powerful enough to make moms without hormonal prompting.

We've seen how, in our brave stunt doubles the rats, virgins become mothers via special chemical injections that can induce maternal behavior. But scientists have also found that rat mothers can be engineered *without any chemical treatment at all*, by exposing virgins to rat pups for long periods.

You simply stick a virgin in a cage with a mother and her pups.

During the first seven days, nothing happens. (You have to be very, very careful that she doesn't eat anybody.) But after about a week in close quarters with the babies, the deep-down systems start to awaken, and the formerly cannibalistic virgin begins to act like a big softy around the pups.

While I'm at the NYU lab, Naomi López Caraballo, another one of Froemke's graduate students, shows me a virgin mouse whose maternal instinct has been "unmasked" this way.

With latex-gloved fingers, she expertly drops a couple of peanut-sized eight-day-old mouse pups into the cage with a tellingly svelte female who is not a biological mother, but who has been exposed to pups for a week. "Let's see if she's up to retrieving," López Caraballo says.

The deposited pups immediately open their mouths and begin to shudder with the force of their inaudible-to-human-ears distress cries.

Instead of fleeing, the valiant virgin inches closer. (Watching her adorable diligence, I feel a pang of regret about certain snap traps laid in certain kitchen pantries.) She runs her paws gently over the pups' tiny, shaking bodies, then hustles to fluff up the cotton fibers of her nest, much as how I plump my living room couch pillows prior to play-dates. These are maternal behaviors, and although the virgin doesn't retrieve the pups in my presence, she did in previous trials.

"We don't know what trains the virgins to retrieve," López Caraballo says.

One propellant may be the biological mother herself, who—in a seemingly desperate bid for adult help or company—will drag the reluctant virgin into her wretched nest again and again. (I sometimes try to do this with babysitters.)

But the virgins are also increasingly willing participants, their reluctance diminishing with time. The researchers use all kinds of cunning little cameras, ultrasonic microphones, and neural recording equipment to monitor the mice in an effort to isolate the factors shaping their learning process.

If a clear barrier is placed between the learning virgin mouse and the pups, in later trials she will still come to retrieve the pups like a pro. But if that same barrier is covered up with gray duct tape, she never catches on. This suggests that, as important as they are, baby cries alone can't trigger the onset of maternal behavior in virgins, and eyesight and the visual cortex are somehow involved in this mode of sensitization.

I inquire about an odd-looking L-shaped metal plate attached to the virgin's head, which I somewhat unscientifically refer to as "her hat."

"Oh, that's for the virtual reality trials," López Caraballo says. "It holds her head still." The researchers are trying to determine if they can spark the maternal metamorphosis in virgins simply by popping in some mouse parenting videos.

Of course, pup-sensitization studies in mice and rats are an experimental tool, not a simulation of real life. Wild mice don't get much screen time, for starters, and without a watchful scientist to referee, it's unlikely that a virgin rat would ever get access to unrelated pups for hours, let alone for a whole week. If she did, the fur—and the whiskers, and the tails, and the guts—would soon begin to fly.

Still, this work helps emphasize the nonhormonal components of the maternal conversion. In fact, maternal sensitization is possible even if the virgin's pituitary gland—her hormonal Hercules—is surgically removed. It's undeniable that the hormones of pregnancy, birth, and nursing kindle a sudden and startling change in female rodents. But experience with pups, acting on the same built-in brain systems inside all female mammals, is also a strong catalyst for maternal care, though the change happens much more slowly.

These multiple points of entry are further testament to the might of the maternal instinct. Virgin sensitization via pup experience suggests that there is a maternal kernel inside of females, even if they've never been pregnant, that with the right stimulus—hormonal or experiential or both—can expand into the maternal instinct in full.

For once they are exposed to pups for the long term, the brains of these initially reluctant rodent caregivers also start to change. Measurable physical alterations accompany the onset of maternal care in females who have never given birth. Pup-sensitized virgin rats gain more prolactin receptors in their brains, for instance, and new neurons may sprout in the animals' hippocampus in patterns similar to those found in biological mothers.

With enough cajoling, maternal care can even be experimentally induced in male rats, who have no contact with their pups in nature. But drawing it out is far more protracted and painstaking: "You can force males to respond positively to pups," says Joe Lonstein of Michigan State University, "but it is much, much harder." If administered via injection, far larger doses of hormones must be doled out over lengthier periods of time. Likewise, male caregiving in rats can be stimulated by exposure, but it takes longer rooming-in periods with pups than the week or so that virgin females require.

Still, this suggests that the maternal instinct is so essential to the mammalian makeup that males, too, have a maternal seed buried deep in their brains. It also suggests, though, that in most species it will never naturally sprout outside of the lab. "The female brain is more ready to be induced," Lonstein says. "The threshold for males is much higher."

∽

But what happens only in the lab with animals can be freely chosen by human beings as a way of life. Biological motherhood is like a riptide that sucks in even totally unwilling women. But other types of humans can choose to set sail on these same waters, and going through the motions of motherhood day in and day out may coax forth the latent maternal instinct. We've seen how certain dads can become mom-like organisms, even undergoing some hormonal changes, provided they're fully embedded in the baby experience. And of course there is the almost uniquely human phenomenon of adoptive parents.

Neuroscientific studies of human adoptive moms are rare, but as with the virgin rats, the evidence we have suggests that choosing to care deeply for a baby can awaken and physically mold the maternal brain. In one study, fourteen birth mothers and fourteen adoptive and foster mothers looked at the faces of their own children and of other people. "All mothers, regardless of type," showed extra neural excitement for their own kids.

In another experiment, foster mothers and their young infants were briefly separated, then allowed to cuddle for thirty minutes. Before-and-after analysis of the foster mothers' urine showed that oxytocin levels in their bodies rose via contact with their kids, and were associated with maternal "delight."

Interestingly, a bit like those slowly sensitized virgin mice who had no hormonal leg up on motherhood, the foster moms changed more gradually over time, with oxytocin levels mounting the more months they'd had custody of their child. (Likewise, hands-on experience remains key for biological moms, who may become less physiologically attuned to their new babies if they spend hardly any time with them.)

That said, the brains of adoptive and biological moms probably never become identical. The two groups seem to respond somewhat differently to baby cries as well as to visual cues. Which is not to say that one is better than the other: adoptive moms' behaviors may subtly differ in ways that favor adopted children—one study found that adoptive moms "nourished and caressed" their babies more than biological moms did.

Since human beings are naturally alloparental, with an unusual affinity for each other's offspring, it's quite possible that women have a lower threshold for baby sensitization than other female mammals. Adoption of non-related young, for instance, is more or less unheard of elsewhere in nature, except in accidental cases where group-living animals like seals and kangaroos get mixed up after major predation or weather events. Also, the rare animals that do "adopt" are in almost all cases already biological mothers themselves, who have been mentally sensitized the old-fashioned way.

But humans who have never been pregnant *can* take up the flag of motherhood—provided they're really determined to carry it. In fact, some evolutionary theorists think that humans' pan-maternal tendencies may help explain the mystery of why our aggressive placentas funnel away more resources than biological mothers can afford to give, risking maternal lives in a unique play that would seem counterproductive to a species' survival.

Maybe our systems take on the risk because maternal demise is not a death sentence for a human baby, the way it is for most mammals. A different calculus applies to our kind. For humans, it's not just you and your baby alone in the burrow, together against the cruel world. For us there's almost always an au pair (heaven forbid a pretty one) or an aunt or even an uncle waiting in the wings, ready and willing to tap into the maternal instinct lurking inside of all mammals, to be triggered into caring.

"Human communities have relaxed the strength of natural selection on how much the baby can demand from the mother," David Haig of Harvard University tells me. "After birth, her health becomes less necessary."

I give a sarcastic little snort when he says this, but in a way it's the scariest thing I've ever heard.

Chapter 4

MOMMY WEIRDEST

When Mom Brains Leave the Nursery

S PEAKING OF worst nightmares: over the past ten years or so, I've realized that moms dream differently than other people do, and that our dreams are often not sweet ones. We meet grizzly bears and great white sharks and wolves. (I'm still waiting to have my inevitable dream about a cannibal rat.) We endure landslides and obliterating blizzards. We wake up and check on children who seem to have suffocated in these nightmares but are in reality snug in their beds. We're the ones in danger because these weirdly vivid postpartum dreams dispose us to sleepwalking.

Scientists aren't sure exactly what disturbs maternal sleep, but it's clearly part of the fallout from embarking on an extreme makeover, brain edition. And if motherhood shifts our experience of dreamland, it alters our view of the waking world even more. Giving birth really does a number on you, and the behind-the-scenes neuro-voodoo of motherhood revamps far more than women's responses to tiny babies (although those responses are always at the root of the alteration). Pregnant women, for instance, mysteriously seem to get in more car crashes than other people—maybe because their minds are elsewhere.

Experimented-upon moms have been shown to be less grossed out by disgusting things like human fleas and dog poop. ("Reduced disgust sensitivity might facilitate handling . . . offspring," the scientists reasoned, quite reasonably.)

These peculiar distinctions are more important than ever today, as so many moms spend our daylight hours away from our kids, engaged with other types of people and tasks. Sleep researchers advise somnambulist moms to lock themselves up at night so nobody gets hurt. Yet all day we are free to wander.

Some of our new maternal edits seem random and harmless enough. Moms' taste for food may change. We might exhibit an increased appetite for salt—something evident in rat experiments involving an adorable-sounding tool called a "lickometer," and perhaps in my newfound Fritos lust. Mother rats are also hotter than their pre-maternal selves—alas, in the quite literal instead of the alluring sense. Their bodies are physically warmer, with a higher core temperature.

Some of these changes likely reflect tweaks to the mothers' metabolism. But other differences in the way moms perceive and interact with the world at large reflect our revised mommy agenda, our altered system of reward, and our sense of what's interesting and what's dangerous. Our inner landscapes recast, we look outward toward different horizons.

The most surprising change is that pregnant women and new mothers are calmer than other people in the face of environmental stress. *No way*, I thought when I first skimmed these papers, in between frantically marinating chicken thighs and checking my watch for soccer pickup.

But it's true. In experimental settings, mothers-to-be are less aroused by displays of deliberate rudeness and have lower heart rates during psychologically stressful events like mock job interviews.

We excel at enduring physical discomfort, such as withstanding a heat chamber. If science requires us to stick our hands in buckets full of ice water for a minute, moms' saliva has less of the stress hormone

cortisol than a childless woman's does. We are also relatively unruffled when viewing disturbing pictures of things like mutilated bodies,
guns, and angry dogs. The blunted stress response seems to strengthen
throughout pregnancy.

Some of the most interesting experiments on preternaturally calm
moms-to-be happen outside the laboratory. Women experiencing
"major life events" ranging from serious bodily injury to job loss to
deaths of close family members and friends report being much less
stressed if they're in the third trimester, compared to the first.

These life events can sometimes be literally earthshaking. In 1994,
a 6.8-magnitude earthquake rocked the area around Northridge, California. Parking garages pancaked, buildings crumbled, and along with
the direct casualties, a number of people died from stress-related heart
attacks. Afterward, maternal behavior researchers at the University of
California, Irvine, asked pregnant women—who lived on average just
fifty miles from the quake's spasming epicenter—to complete what's
called a "life inventory test," rating the quake from "not at all stressful" to "extremely stressful." Using a four-point scale, moms-to-be who
were *very early* in pregnancy rated the event close to the maximum it
logically deserved. Women in the third trimester, though, gave the seis-
mic cataclysm a "meh" rating of 2.38.

As luck would have it—well, at least for storytelling's sake—there
was an almost-unheard-of earthquake near my home in Washington,
DC, the very first time I screwed up the courage to leave my infant
daughter home with a babysitter. I was—where else?—in the changing room of Ann Taylor at the time, having desperately pillaged the
racks to find work clothes that fit my new body, when the mall building
began to sway like a baby swing. The quake was strong enough to crack
the Washington Monument and snap the National Cathedral's spire.
Since the shopping center was next door to the Pentagon, I assumed at
first that a bomb had gone off and that I was half-naked in the middle
of a terrorist attack.

Everyone around me was screaming and stampeding, but somehow

I found myself fully clothed and marching out Ann Taylor's doors in a no-nonsense manner. Although I hadn't slept in weeks and my mind had been teeming with a million back-to-work details just a few minutes earlier, suddenly it was stacked and orderly. I serenely decided to abandon my car—I couldn't risk getting caught in the parking garage so far from my baby—and strode outside the mall, hailing the first taxicab I saw, which happened to contain a well-dressed businessman visiting from Argentina who was almost as cool a customer as I was. (This was because, I soon learned, being in a moving car meant that he hadn't felt the earthquake. Watching people race willy-nilly out of the city's office buildings, he'd assumed it was simply lunchtime in America.)

So before paralysis even had time to set in around the city, I was back nursing my baby, none the worse for wear, although unfortunately still without anything to wear to work. The quake left the babysitter—a young woman who had no children of her own—in far sorrier shape. My husband had to pry our child from her arms.

My teenage self, once barred from giving blood because of my penchant for hysterics, would have shared her freak-out. But the new me—the mom-me—never broke a sweat.

Perhaps major earthquakes are small potatoes compared to the trials of new parenthood. But, in a broader sense, it probably pays for pregnant ladies and new mothers to stay poised, collected, and focused on themselves and their babies' safety when a severe threat comes along. And come they do, more often than you'd imagine—whether disaster takes the form of an earthquake or the pot of sterilized pacifiers that one neighbor mom accidentally left boiling on the stove in a brave but doomed effort to go once more out to brunch. (The subsequent blaze was known forever as the "paci-fire.") My strangely stable earthquake mind didn't save my daughter's life that day, but maybe it could have: in a terrible mudslide a few years ago, a California bank teller ended up trapped under her sofa, having somehow had the presence of mind to grab her infant son as her neighbor's house slalomed down upon them; they both survived.

As a bonus, Jennifer Hahn-Holbrook of the University of California, Merced, points out, feeling ultra chill even in more ordinary circumstances likely prepares moms for the combined boredom and reverie of breastfeeding, which involves hours upon hours of enforced "relaxing."

Some scientists have even proposed that this suppressed reactivity—this dulling of feeling, you might say—could explain the evolved drive behind the "baby blues" that more than 50 percent of new moms experience. Maybe mild forms of postpartum depression are adaptive—though scientists don't know what happens chemically to produce the dangerous, clinical-grade kind that strikes approximately one in five women, turning coolness under pressure into the coldness of despair.

❧

Paradoxically, even as moms and moms-to-be become emotionally insulated from our environments, we are also much more physically aware of our surroundings. The vigilance with which we monitor that shining center of the universe—our squalling newborn—seems to extend to its peripheries, which is to say, the fully potty-trained parts of society.

New moms' ears don't perk up just for human baby cries, and cries of young creatures in general. (Mammalian moms are, studies suggest, suckers for babies across the board, with deer moms charging to the rescue at the recorded yelps of kittens and sea lion pups, and so on.) Laboratory tests suggest that moms' brain waves are enhanced even in response to auditory cues that have little acoustically in common with baby cries, like neutral words and tones. Compared to non-mothers, human mothers of infants under fourteen months old in one experiment also "showed greater activity" in hearing-related areas of the brain upon hearing *adult* voices, not just kids'.

Similarly, moms' generally souped-up sense of smell can pick up all sorts of odors, not just those wafting from baby heads. (In my experience, not all of these olfactory encounters can be classified as pleasant.) We also seem to examine everything more closely. Policewomen who return to work after giving birth, for instance, report being unusually alert on their beats.

In one typical experiment comparing new moms' reactions to baby pictures with childless women's responses, researchers used images of houses as a control, assuming that all women would just glance quickly over these ultra-ordinary sights. They were shocked to see that the mothers seemed to scrutinize the houses more intensely, too. (Perhaps this excuses my egregious HGTV habit.) Another study showed that pregnant women have a "significantly enhanced ability to discriminate between colors." In a darkened room, mothers-to-be viewed eighty-five colored caps that spanned the visible spectrum, detecting stronger differences in hue.

Moms are also great face-readers, even when it isn't rug rats' pouts that we're decoding. Pregnant women may be unusually good at assessing "apparent health" in people's faces, which could serve as a safeguard against disease at a particularly high-stakes moment. Moreover, we are adept at noticing the emotions of others. In one experiment, mothers of toddlers were able to more accurately discern strangers' emotions when watching silent videos. We are skilled at recognizing other people's faces, even after very short exposures.

But we are particularly intent on scrutinizing a certain type of face. Usually women are best at interpreting the faces of other females: our friends and rivals. But mothers seem to switch to studying the faces of adult males. We excel at recognizing them after just a small period of exposure, and at reading their expressions, particularly negative expressions, like disgust. Mothers tend to rate the faces of potential criminals as more menacing. We tend to be warier of strangers.

Why would moms side-eye some innocent fellow on the subway, but "keep calm and carry on" (as those dreadful needlepoint pillows on some of our couches say) during tooth-rattling earthquakes? It probably has to do with the dangers we evolved to cope with. In our deep evolutionary background, strange men may have posed a far more routine threat to our babies than did earthly tremors. In many mammalian species, infanticide by unrelated males is very common. In our close relatives the chimpanzees, females with young babies avoid males in general. But that's not really possible in our world.

When you combine moms' increased environmental awareness and threat sensitivity with our grace under pressure, you get one of our most celebrated behaviors: maternal aggression. Moms are both extra tuned in to threats and decidedly unafraid to face them, which is quite literally a killer combination. And the one emotion that seems to be easily turned up amid our general chill-out is righteous maternal anger.

Everyone knows the stories: mother moose slay bears, and goat matriarchs head-butt wolves down mountainsides. We've read the headlines: "Protective Walrus Mom Sinks Russian Navy Boat in Arctic Sea." And we moms are even rather glad to hear it. Of course she did! What other choice did she have? It wasn't even a destroyer—more like a biggish rubber dinghy that I could probably make short work of myself, given a pair of thirty-inch tusks.

But you needn't dip a toe in the Arctic Sea to meet ready-to-rumble mothers. Some of nature's most murderous mommy aggressors lurk closer to home, apparently supplying many of my kids' go-to snack foods. Milk cows, it turns out, are far more dangerous to humans than bulls, and are the frequent source of "high-velocity trauma" sustained by farmers. There are some very satisfying online videos of dairy cows kicking the dirty diapers out of human men whom they suspect of menacing their calves.

When I was a child, my dad liked to tell a story about how, while backpacking in Yellowstone National Park in his carefree twenties, he hiked practically right on top of a female bear and her two cubs, and had to reverse-creep all the way back up the lonely wilderness trail. Even then my sister and I suspected that this was a tall tale. Who actually wanders between the proverbial mama bear and her young?

But many years after he died, my sister discovered a stack of old slides from Dad's ancient camera.

Lo and behold, there was the bear sow, captured in a hastily snapped souvenir suggesting we were both lucky to have been born.

Obviously such warrior moms are a little intimidating to observe up close, so biologists may choose to focus on less menacing critters—say, squirrels. In one fascinating study, researchers played the sounds of rattlesnakes to a bunch of California ground squirrels, knowing that snakes almost always go after the baby squirrels instead of the big ones (who can give the rattlers as good as they get by biting and kicking gravel). In the experiment, the mom squirrels responded most violently to the threat, aggressively shaking their tails at the fake snakes. Moms with the littlest pups shook their tails the most. And the dads? They barely glanced up—which is just as you'd expect for a species with sky-high paternal uncertainty.

Under the right circumstances, human moms may be the beastliest of all. We ram kidnappers' cars and wrest our kids from all kinds of slavering jaws. "I had a mom instinct, right?" one Canadian mom said in a thrillingly blasé interview about extracting her seven-year-old from a mountain lion's maw. "I just leaped on it and tried to pry its mouth open."

Strange men and mountain lions aren't our only targets. In a recent Modern Love column for the *New York Times*, a mom described hurling a playground ball at an older kid who played rough with her five-year-old in the swimming pool. When the stunned child turned to face his attacker, "I'm pretty sure he wasn't expecting to see a middle-aged woman in an unflattering swimsuit, but there I was," the vengeful mom recalled.

The violence may even be mom-on-mom: "Florida Mothers Slash Each Other with Broken Coffee Mug in Fight over Parenting Methods," one headline gasped. Moms have occasionally come to blows while wearing baby carriers. These maternal showdowns help explain why the most dangerous restaurant chain in America, according to one *Wall Street Journal* report, may well be Chuck E. Cheese—as if kid birthday party logistics aren't headache enough without the police showing up.

Chill yet vigilant, we mothers are a famously feisty bunch, forever forming clubs, sounding the alarm and marching "against" things— drunk driving, most famously, but also tackle football, video games, climate change, airport expansions, the Internet. I once heard the women's gym that I somewhat fruitlessly frequent described as "the Angry Moms' Club." I rather liked that. In wild-caught mother rats, the brain cells of the amygdala, associated with aggression, are bigger. Next to love, the most commonly cited maternal emotion is rage.

Unlike testosterone-fueled male aggression, this maternal rage seems to flow from a different (but by now familiar) set of neurochemicals. In one experiment, female rats were taught to associate a peppermint smell with a painful electrical shock. The rat moms quickly learned to freeze in fear whenever they smelled peppermint—but not if their pups were present. Then the mothers valiantly wheeled about and attacked the tube that was spraying the scent, or tried to stuff it up with bits of their bedding.

When the scientists disabled the oxytocin-receiving areas of the mothers' brains, however, the pugnacious behavior ceased.

Another experiment, this one on humans, employed what's called the enthusiastic stranger paradigm—with a premise familiar to any new mom who has dared set foot out of her own door with a toothsome newborn. The mothers and babies were placed in a "waiting room" outside the lab (which—spoiler alert!—really was the lab itself). A researcher posing as "an ebullient, socially intrusive" maintenance worker checking smoke alarms approached the mothers, exclaimed, "What a lovely baby!" and then tried to stroke the child's cheek. All the mothers reacted negatively, but those who'd recently ingested an extra dose of oxytocin repelled the stranger's advances with more vigor.

Then there is the nursing hormone prolactin, whose influence seems strong enough that maternal rage is sometimes specifically called "lactational aggression." Tellingly, prolactin fuels maternal cool as well, dampening anxiety. If you stick a lactating lady on a treadmill, for instance, she exudes only half the stress hormones of a typical woman.

Yet this Zen state dissolves quickly under threat. Even if their tots aren't present, lactating mothers are more aggressive than formula-feeders. In one experiment, human mothers competed in a computer test against a flagrantly rude, gum-snapping, cell phone–checking foe, who was really another sly researcher in disguise. Compared with the bottle-feeding mothers, the breastfeeders were twice as likely to take the opportunity to "punish" this obnoxious rival. (Even while exacting revenge, the breastfeeders had lower blood pressure than the other women, a fact suggestive of their kicked-backed state of mind.) The punishment was simple: the mothers could electronically blast the other person with a burst of a "punitive" sound, turned up as loud as they wanted.

Call it what it is: a roar. On a reporting trip to the Serengeti years ago, I watched a pack of hyenas stalk a lone female lion with her only cub. Rather than run, the lioness tucked her little one under a tree and walked slowly out to confront the hyenas as they came at her across the grassland. She was like a gunslinger out of the Old West. Something in her level, golden gaze finally made them skid to a halt and—after a brief stare-down and some nervous yips and giggles—turn and run away, despite the eight-to-one odds in their favor.

These days I spend my time on soccer field sidelines instead of on savannas. But lionesses roam there, too.

∽

I've never faced down a hyena pack myself, reserving my pent-up aggression for the luckless J.Crew salesman delegated to explain that my kid's Easter dress went missing in the mail. ("You don't understand—I have the cardigan," I shrieked, as my husband winced in the background. *I have the tights!*") I see myself looming huge and grouchy in a teenage camp counselor's eyes as I demand more fresh air for the children at lunchtime.

But am I really willing to go to war for my kids, like the brave mommy dino in *The Land Before Time* as she wheels to face the *Tyrannosaurus rex* amid a very unfortunately timed volcanic eruption? I wonder. I've been known to run away from dive-bombing bees, leaving my squealing daughters to fend for themselves. Sometimes they complain that I don't stand up for them enough. All three kids got mad at me recently when an old lady scolded them for horsing around at a lobster pound, and I was (in their words) "too busy devouring" my lobster to swoop in on their behalf. (Guilty as charged.)

The only moment I can recall embracing physical pain to protect my child was just after my third C-section. There had been some sort of extreme miscalculation about the amount of pain medicine I would require, and midway through the surgery the numbness wore off and I began to feel the cutting. It was the closest I've ever felt to the rake of a mountain lion's claws. My abdomen seemed crisscrossed with trenches of flame. The anesthesiologist administered more medication, but it was too late. I screamed, screamed some more, and threw up.

After my son was safely extracted, the nurses mercifully ordered an enormous shot of something—morphine, maybe?—to put me out of my misery for a nice long while.

But as I waited for relief with chattering teeth in the recovery room, I grew increasingly aware of unsettling things happening around my new child—who was bald as a cue ball, with pendulous cheeks. Strangers kept coming in to look at the baby. A second nurse had noticed an almost inaudible (and possibly imaginary) rasp in my son's breath that made her wonder if he'd inhaled fluid. She repeatedly leaned in close to listen. She was itching to take him to the NICU on a precautionary basis, I could just tell.

The saving shot of morphine soon materialized on what appeared to be a silver platter (although probably it was stainless steel).

I looked at it longingly for a moment. And then I remembered the

NICU tubes and wires slithering across the torso of my firstborn, like snakes.

Nobody would be removing this new baby "for observation" as I slipped off to la-la land.

"Take that back," I snapped at the nurse with the needle. "Things are happening with the baby, and I want to make sure I know what's going on."

This brusque and busy nurse had gone through some trouble to obtain the shot, and there was paperwork involved. It occurred to me that, having loudly shrieked for (and then having been officially pre-scribed) the medicine, and remaining in still-obvious distress, I would now be forced to take it. My body was still mostly paralyzed, but inside I was scraping together the last of my strength for a nasty fight when the nurse spoke.

"Do you know what I think?" she said, pursing her lips. "I think your baby is lucky to have you for a mom."

∞

Amid all the pain and internal refurbishment, all the love and rage and marathon diaper changes, maybe it's not surprising that maternal intel-lects sustain some collateral damage as well.

Though hotly debated, often exaggerated, and optimistically pooh-poohed, the concept of "mommy brain" describes an undeniable reality. Something like 80 percent of all new moms report cognitive problems, particularly related to memory, and scientists urge us to believe them. Much like our bodies, moms' heads may be thickened.

Some researchers have bleakly proposed that moms "cannibalize" our own brains to feed our little squirts, which would explain the (as we've seen, contested) evidence for brain-volume losses. Other indig-nant scientists, who sometimes are mothers themselves, retort that this apparent brain contraction might really just be an efficiency-enhancing "synaptic pruning," which does sound rather edifying, like a vigorous bit of garden work. (The idea is that the mom brain is "leaner and meaner." Am I meaner? You bet. Leaner? Sadly not.) A few labs have even argued that motherhood "makes us smarter."

Mother rats do seem extra proficient at certain tasks. Their spatial memories in particular are pretty solid. Mom rats rock, compared to virgins, at finding Froot Loops in a maze. (Interestingly, the "adoptive" virgins sensitized through pup exposure alone are also formidable Froot Loop detectives.) And thanks to their newfound fearlessness, mammalian moms may be better hunters, too. Rat researchers at the University of Richmond set up a gladiatorial "testing arena" for a mom-versus-virgin cricket match. (This involved the insect, not the inscrutable British sport.) Moms pounced on crickets three times faster, noted the researchers peeping through night vision goggles, and the moms were also less likely to unhand the cricket before the fateful "kill-crunch." Impressed, the scientists later wrote that these bug-munching rodents reminded them of "Artemis, the Greek goddess of childbirth and of the hunt."

The evidence for maternal hunting prowess outside of the laboratory and classical mythology is sadly somewhat skimpier. A study of leopards in the Kalahari Desert showed that mothers had a "higher-than-expected overall hunting success rate" than females without cubs, and even compared with males. Reading the fine print, however, it seems that the moms' impressive body count includes many more little black lizards than loping, meaty gazelles. This might be because hungry mothers—whose ability to travel is significantly restricted— are willing to eat literally anything that moves.

Alas, bravery—or desperation—in the hunt has its hazards: researchers have noted that, perhaps because prowling rat moms are so hormonally emboldened, they end up in traps more often than virgins.

In nature, hunting mothers face serious logistical hurdles that cramp their style. To achieve lift-off, milk-turgid mom bats must sometimes kidnap somebody else's baby to slurp away the extra weight. (Why didn't I think of that?) Postpartum elephant seal moms are too buoyant to dive for prey: like a human mother trying to shimmy back into a certain pencil skirt, they must drop hundreds of pounds of blubber to become svelte enough to hunt. Among *Homo sapiens*, hunting moms are documented but somewhat rare: the famous example is the Philippines' Agta tribe, whose nursing moms can net pigs and other prey.

In the end, any enhanced bacon-bringing abilities can't change the fact that moms likely have deficits as well. The findings are mixed and inevitably controversial, but a recent meta-analysis of twenty different studies hammers home the theme that moms' memories are quantifiably impaired. One interesting experiment showed that we especially struggled to remember to complete assigned small tasks outside of the laboratory, like mailing a letter. And we seem to have the most trouble recalling words, reciting lists backward, and, you know, yadda yadda.

Some of this momnesia might be a kind of short-term coping mechanism: studies have documented that mothers truly don't retain much about the experience of childbirth, and are justifiably horrified by watching highlight reels with researchers who force them to remember.

All the lost sleep—an estimated seven hundred hours per year for new mothers—takes a toll, too. And maybe that last recitation of *Guess How Much I Love You* has lulled us into a temporary coma.

But some of the memory loss has to do with that new brain that we inhabit. We aren't wired the way we used to be, and our new skills and interests come at a cost. "There's an economy of attention," says Linda Mayes of the Yale Child Study Center. "It's not that there is an atrophy. It's just that you are highly, highly focused on that thing. To a certain degree your biology is pulling you to be focused on that baby. And so certain other things have to go to the side."

If we are wholly intent on the texture of our child's bowel movements, we can't quite put our finger on the quadratic equation. If we are busy belting out the extra verses of "The Itsy Bitsy Spider" with everything we've got, it's harder to recite the poetry of old whatshisname. The brain has a "use it or lose it" policy. At present, the playroom-sprucing part of my brain is in good working order, as is the bit that scrubs my son's plump torso clean of what he calls his "war paint." The section in charge of the Spanish subjunctive, meanwhile, has crumbled into dust.

I do wonder whether these memory woes might also explain how mothers experience the passage of time. My days used to inch by, bedazzled with a million intrigues and moments to ponder later. Now the days seem over before they've even started, and I quite literally cannot remember yesterday. This makes life seem to sail past with alarming speed, like a seven-year-old on the overhead zip line at the Mall of America.

One possibility is that our most highly developed capabilities diminish in motherhood, while the ancient core of our mammalian brain prospers. Remodeling the maternal brain may benefit the older parts, anatomy that we roughly share with lab rats and rhinos and the rest of mammal-kind, at the expense of our trademarked civilization-building add-ons—language and verbal memory and all that jazz. Since these skills have less to do with babies—who, goodness knows, can't exactly pontificate—perhaps it's only logical that they would slip into at least temporary postpartum decline.

But this is still tough to swallow when those very skills were central to your sense of self. It's all well and good that researchers have proposed that we moms might excel at casing a supermarket— perhaps the human equivalent of snuffling for Froot Loops in a rat maze. But even if true (and as always, there's a debate), it's not nearly a fair trade if you aren't some notional everymom but a real person who met her husband on the debate team, thrilled to poems instead of baby pictures, and even once in a while risked her life for the sake of words. I used to tell tales for a living. Now it's a struggle to scratch down a few fragments in the depressing little pink journal somebody bought me, titled "Mom's One Line a Day." (Apparently, that's all we're allotted.)

This, of course, is a page from my own sob story. However, yours is likely quite different. While motherhood causes sweeping and predictable changes common across orcas and wombats and women, no two human moms are exactly alike, even though we all answer to the same name. And understanding what makes each one of us unique opens up a whole new can of organic gummy worms.

Chapter 5

MOTHER OF INVENTION

*The Diversity of Maternal Experiences
and Why They Matter*

S ubject 39 could be any mom: leggings, sneakers, hair elastic around the wrist. More than thirty-five weeks pregnant, she doesn't flinch when the research assistant at the Yale Child Study Center dots the midpoint of her scalp in blue Crayola marker. I'm betting this isn't her first child, since clearly she's been colored on before.

The skullcap of electrodes is something new, though. It looks a little like those netted sacks that onions come in, except this one can read minds.

The electroencephalogram, or EEG, is one of the only safe tools available to study female brains so late in the third trimester. Heavily pregnant women can't lie flat in an fMRI scanner for long periods because it compresses major blood vessels. (The preggos' constant need to pee is a scientific hurdle, too.)

The lab assistant asks Subject 39 to remove her earrings, and then drapes her in a salon cape.

"No electricity comes out of this thing," the researcher promises as she wiggles the close-fitting cap of wires from back to front onto the mom's head. "It's just measuring your own electricity."

Brain cells communicate via tiny electrical impulses, and when thousands fire together, their signals can be detected on the scalp's surface. The EEG records these brain wave patterns through the cap's dozens of electrodes. Scientists typically present the mother with a baby stimulus—a picture, say, or a recorded cry—and watch how her brain reacts.

Now it looks as though some suction-cup-happy sea creature has the mom in a death grip. She's even dripping a bit. To increase conductivity, the cap has been soaked in a mixture of salt water and baby shampoo, which trickles down into the darnedest places, I discover later when I model the contraption myself. Every damp electrode feels like the wet kiss of a frog.

"Oh my God," Subject 39 says to her iPhone reflection. "I feel like Dana Barrett in *Ghostbusters.*"

The researcher records the temperature and humidity in the room. It's time to begin.

The experiment's working title is "The Transformative Experience of Pregnancy," but unlike most of the scientists I've spoken with so far, these researchers are not looking for the differences between mothers and regular women.

Instead, "we are looking to see how signals vary mum by mum," says Helena Rutherford, the project's British-born lead investigator.

As much as society sometimes wants to lump all human mothers together and heave us onto a single sturdy pedestal, the latest research is revealing maternal diversity as well as maternal distinctiveness. Some mothers vibe more off baby smells, touch their babies extra often, and pay closer attention to infant shrieks. Some report greater levels of overall maternal satisfaction. Some neglect their kids entirely. But how and why these differences arise remains a major question of mom science.

"It's not just that there is a mum brain, or a mum response," Rutherford says. "We are trying to get a sense of individual differences amongst mothers. There is no one-size-fits-all approach to mum brains."

The idea is not to hand out Mother of the Year trophies, but rather to develop tools to support individual women, taking advantage of the once-in-a-lifetime neural plasticity that all new moms possess.

Often the ultimate goal is to improve children's outcomes. Fathers are known to display an even wider range of behaviors, but mothers' often subtler variations may matter more for children's well-being, since we play such a definitive role in childcare. I spoke to one pediatrician researcher who kept going on about the child's "environment," and it took a long while for me to realize that she meant *me*.

I find this style of mom science to be completely fascinating, but a little bit frightening, too. All moms know there are differences among mothers. What else would we gossip about at the bus stop? And in our heart of hearts, all of us are a teeny bit paranoid about finding chinks in our own maternal instincts. It's unnerving to put yourself under the microscope.

On a visit to a different lab, a researcher stuck sensors on the skin beneath my rib cage to see how my heart thudded during exposure to baby videos, and then strapped Velcro strips around the fingers of my left hand to measure my changing sweat levels.

I was definitely sweating. What on earth would the numbers say? Could the machine somehow sense my inability to sing lullabies, or my rank cowardice in the face of a bloodthirsty horsefly at the swimming pool last summer? Could my inability to sew a bioluminescent squid costume, or to construct a log cabin out of pretzel rods, somehow be electronically detected?

Fortunately, we're nowhere near that degree of diagnosis. But the Yale researchers are amassing one of the richest data sets about the striking differences in mothers' natures. Once the EEG readings are analyzed, and confounding factors like excessive blinking and occasional catnaps—hey, it's the third trimester!—are accounted for, they will likely show how differently dozens of pregnant women respond to the exact same set of infant stimuli.

The question is: why? The Yale scientists will combine the brain data with other seemingly random biographical tidbits. Is a given mom right-handed? How well can she memorize a series of squares? Was the pregnancy intended?

Now Subject 39 is instructed to stare at a white cross on a computer screen and to put on a pair of headphones. I don't have headphones; in fact, by this point in the experiment, I've been tucked behind what looks like a gigantic navy shower curtain so I don't mess anything up.

Luckily, I can hear this piercing sound without any headphones at all.

It starts out like a squeaky hinge in a horror movie, then becomes a parrot's insulted squawk. The baby's grief-stricken cry rises. On the other side of the blue curtain, each of the 128 electrodes stuck to the mom's scalp spits out a reading of brain waves: first foothills, then mountains.

Next, a parade of baby faces crosses the computer screens: some with sly smiles, others with aghast "I have just tangled with mushy peas" expressions.

Bump, bump, bump, go Subject 39's brain waves.

I can almost feel mine bumping, too.

❧

Human moms are by no means unique in our uniqueness. Other types of mammal mothers, from seals to sows, exhibit behavioral variation, too.

African elephants, red-necked wallabies, and eastern gray kangaroos all have multiple distinct maternal modes. A study at a Mississippi aquarium showed big-time fluctuations in bottlenose dolphins behavior, with some goody-goody dolphin moms hovering within a few feet of their calves at most times, while the bad girls sneaked off to play with pool toys.

Certain marmoset moms will press a button to shush a crying infant far more readily than others. Particular red squirrel moms appear extra motivated to retrieve their experimentally kidnapped young.

(I've previously implied that squirrel science is a mom biologist's easy way out, but maybe not in this scenario, which required shimmying up towering Canadian spruce trees.)

Heck, even guinea pig moms fall on a spectrum. Given identical conditions of light and dark, standard wood-chip bedding, and the exact same diet of apples and hay, some seek out tender maternal pastimes like baby "fur sniffing" and "nose-to-nose contact," while others remain aloof.

"Hence," a triumphant team of guinea pig researchers concluded after hours of nocturnal snooping, " 'mothering styles' exist in guinea pigs."

Mammalian moms of other species are, however, much easier to study and assess. To rate a rabbit mom's maternal instincts, you might simply measure how much cardboard she gnaws for her nest. For baboon moms, you could tabulate glance rate, or scratching behaviors in rhesus monkey moms. You can document how regularly a feral goat ditches her baby to frolic with the herd.

Alas, there is no standard "mother in box" test for human moms the way there is for German shepherd moms employed by the Swedish military, whose mothering habits are monitored via video surveillance while they share an enclosed space with their pups.

Human women *are* the box—the proverbial black one. Scientists are still trying to figure out how to look inside.

∽

Evolutionary psychologists have thus far found few reliable skin-deep signs of what might make a supermom—and believe them, they've checked. They've measured our hand-grip strength, the relative length of our ring and index fingers. A very small scattering of studies have suggested that women with more feminine faces, and shorter women, may be initially extra interested in having children. (If true, perhaps this explains my initial reluctance. I hit five foot eight in fifth grade.) And there is some evidence that, just as moose with thicker layers of rump fat have more surviving calves, human women with what's technically called "steatopygia"—that is, Kardashianesque junk in the trunk—may be built to better nourish babies. Science hails the ample contents of our mom jeans as a "privileged store of neurodevelopmental resources," and I guess I'll take the compliment.

However, at least when it comes to nurturing kids after they are actually born, it doesn't appear that the most beautiful or shapely women have any advantage over their homelier sisters.

To analyze human maternal performance, valiant scientists have attempted a variety of real-world observational tactics—tailing moms with their two-and-a-half-year-olds through grocery store aisles, scrutinizing our picture-book-reading techniques and our quixotic offers of green vegetables to preschoolers.

At sixteen frames per second, researchers analyze a mother's gestures, sounds, smiles, and silences as she interacts with her infant. They chart how her heart rate gallops at bath time (which it apparently does even when one kid isn't trying to drown the other).

But part of the problem with studying human moms as individuals, beyond the lingering inconvenience that it is illegal to vivisect us, is that we are far more complicated than guinea pigs: we have personalities, pasts, performance anxieties. We're each so unique it's hard to find a pattern, especially given the dearth of fixed behaviors. Isolating moms in sound booths, musically trained observers eavesdropped on them talking with their babies ("Mummy gonna change your bummy?" "Where did you get all this energy!?") and found that each woman even had what could be considered her own acoustically unique "signature tune."

Meanwhile, scientists sometimes struggle to define how the forms of maternal "sensitivity" they probe in the lab play out in the real world. "These behaviors are complex," one group of researchers wrote, involving "cross-situation adaptations." What counts as an appropriate perception of infant signals or an optimal response could mean something quite different in a woman at a backyard barbecue versus that very same woman at a piano recital—let alone for moms from different backgrounds, in different places, or during times of plenty versus plague.

Scientifically assessing moms is also, to a certain extent, a judgment call that can be easily complicated or confused by different values, contexts, and worldviews. Some anthropologists think that the very idea of being a "sensitive mother" is a by-product of mid-twentieth-century psychobabble, a privileged notion that verges on irrelevance in more hardscrabble times and places. Since various strategies may succeed in sundry environments, it's hard to tell which bits of moms' differences are innate or learned or contextual or culturally contrived.

Herein lies one perk—and there are precious few, let me tell you—of being a mother of many, or at least what passes for many these days. Within my own life, as my circumstances have shifted across pregnancies, I've glimpsed how much of my core maternal identity is actually up for grabs. I'm not just different from every other mom; with each new birth, the mother I become is different from the ones before. You'll be meeting a few of me.

∽

But if there are no easy answers for scientists studying these complexities, there are, at least, the reliable and measurable indicators that we keep encountering: all those maternal hormones. While human moms experience roughly the same sweeping hormonal changes through pregnancy and birth, they are by no means standardized, and the varying levels of hormones may help engineer our diverging maternal behavior.

One California laboratory recently determined that a woman's mid-pregnancy ratios of progesterone and estrogen predicted the quality of the care she would later deliver to her one-year-old child. Another experiment found that first-time mothers with higher levels of cortisol are cuddlier, more drawn to infant odors, and better at identifying the sound of their own baby's cries. Some women's do-pamine systems, meanwhile, are apparently less activated by infant face cues.

To the surprise of no one, oxytocin is an especially potent force. Very attentive human mothers, who gaze at their kids extra often and speak more motherese, seem to have higher levels of the stuff in their bodily fluids. These differences show up in the animal realm as well, with on-top-of-it gray seal mothers having increased levels of oxytocin in their systems—according to scientists who braved a Scottish beach to net them and sample their blood.

Just as our chemical makeup varies, the brain changes that we undergo are similar but not the same. Universal doesn't mean uni-form. Remember the lab that developed the algorithm to diagnose mom brains with 100 percent accuracy? (The researchers had per-formed MRI readings before and after pregnancy, using magnets to map the brain's anatomy instead of tracking its electrical impulses.) While mom brains had lots in common, researchers also confirmed that not everyone's gray matter shrank by the same amount. Even more fascinating, the degree of an individual woman's brain con-traction *seemed to foretell aspects of her future mothering.* The women with the biggest volume of lost gray matter on the MRI scans had stronger neural reactions to their babies two and a half months after giving birth.

Brain reactions as well as brain size have predictive power. One EEG study of forty mothers-to-be, similar to the one being conducted at Yale, showed that the women's neural responses to pictures of in-fant faces diverged over the course of the pregnancy, and the moth-ers who showed the strongest increase of activity went on to have more intimate bonds with their newborns. Different moms may also use different parts of their brains to perform the same maternal tasks. For instance, when flashed pictures of their own babies, more sensi-tive moms showed the strongest response in a brain region called the nucleus accumbens, while less sensitive moms lit up extra in the amygdala, the zone associated with aggression.

∾

So moms aren't monolithic. There are stark differences in our brains, bodies, and behaviors. But why do some women have Pinterest-worthy postpartum brain pix while others do not? Why does oxytocin surge more for some moms than for others? Are there forces outside us that drive our collective hormonal and anatomical differences? And how much difference can our own conscious choices make?

Most of us are desperate to be good mothers. Something like 90 percent of American moms read parenting books in our scant spare hours. And while scientists suggest that this studiousness is not a bad idea—nor is it wrong to enroll in those hospital parenting classes where you learn to properly swaddle a baby before tiptoeing off to buy the cheater Velcro kind—it also turns out that some of what we think boosts our maternal game actually might not.

For instance, one study titled "Does 'Wanting the Best' Create More Stress?" found that taking baby sign language classes, a hallmark of elite mothering, actually had potentially detrimental impacts on mothers' experience of their children, possibly because our babies only learned to gesture for "more, more, MORE" and it stressed us the heck out.

On the other hand, another group discovered that moms who have studied a musical instrument in the past may be more sensitive to infant cues, perhaps because they can better judge the pitch of screams.

So then should we scrap that sign language manual and start tickling the ivories in piano lessons? It's hard to know what to make of these highly specific and somewhat random-seeming findings. Can mom science reveal more general themes?

In fact, there *are* a few very basic life history factors—including a woman's previous childcare experience, her age at delivery, her method of delivery and feeding, and her number of previous kids—that scientists believe may help sculpt moms' individual physiology, brains, and behavior.

And some of these are in play long before the thought of having kids of our own even crosses our minds.

∽

I flamed out of piano lessons as a kid, so that's one strike against me. But I did spend much of my preteen and teenage years babysitting.

This had absolutely nothing to do with a precocious fondness for children. There was a Chinese restaurant within walking distance of my childhood home that let me and Emily pay for our General Tso's in IOUs. We also nursed a pretty serious Blockbuster Video habit.

So at the age of twelve, facing mounting debt, we founded our babysitting firm, the Brucker Brothers. Brucker was an amalgam of our last names, and although we were not brothers or even related, we liked the ring of it. We printed up official business cards in middle school shop class and started handing them out around town.

These days, few mothers in my circle would let a peculiar, pimply preteen—forget two of them—care for their precious-snowflake babies, but this was the early 1990s, and we had plenty of takers. Those Saturday nights went about as you might expect. Diapers were left to molder, sometimes still attached to toddlers. Our necks were often soaked with infant tears. Invariably we downed all the frozen mozzarella sticks in the freezer in addition to the pizza that was ordered for us. After our charges were asleep, we sometimes terrified ourselves with Alfred Hitchcock movies, hid beneath our employers' bed, and telephoned the police. (We never told anybody's parents—the children's or our own—about these midnight episodes, in which suburban officers showed up to inspect the premises. "Frightened Babysitter on Whipstick Road," the police log would report in the following week's town newspaper, which we sorely hoped nobody read.)

We were paid something like two dollars an hour. To justify charging three, we completed a professional babysitting course, practicing shuttle runs to salvage the maximum number of babies from imaginary fires. Dusty chalkboard erasers substituted for children in these drills. I distinctly remember thinking, *There is no way I'm doing this in real life.*

So it would seem that we were bottom-of-the-barrel babysitters—from the perspective of our clients and their disgruntled toddlers, at least. But science suggests that babysitting may have been good for *us*—or rather, for the mothers that we had no real inkling we would one day become.

I grew a new heart, Emily would tell a dumbfounded me nearly twenty years later. But her mom brain might have begun growing way back in middle school.

It turns out that the classic hustle of money-grubbing preteens is actually a bona fide biological leg up for mammalian females, with potentially lifelong consequences. In humans, first-time moms with in-depth childcare experience seem to have a knack for new motherhood.

I don't want to overstate the case here, as I cannot even handle my own kids alone at a pool club, let alone haul a six-week-old baby to remotest Russia on a whim, like another mom friend recently did. But the ex-babysitter's edge may transcend these practical matters. Our experience might not tell us exactly how to soothe a colicky baby's banshee screams, or how to tidy pigtails (my kids' salon charges by the tangle, I've learned). But it may protect us somewhat from postpartum depression and make us less averse to baby cries and more drawn to infant odors, inclining us to form snugglier bonds with our babies. We may even have easier pregnancies on average.

In one lab test asking adults to diagnose the mysterious cause of a baby's cries, women who were not yet mothers but who had significant childcare experience needed fewer clues to solve the riddle than the biological dads did—and they even gave biological moms a run for their money.

Biologists studying our close primate cousins know for a fact that previous exposure to infants is imperative for effective mothering. If a Japanese macaque has never seen a baby before, she may flee, terrified, once her own is born. In another experiment, scientists allowed a group of subadult chimpanzees exposure to youngsters while another gang capered around child-free. When all these chimps later gave birth themselves, the ex-babysitters got the hang of motherhood much more easily.

Indeed, for a few kinds of primates, babysitting seems to be an absolute parenting prerequisite. When marmoset and tamarin mothers have no prior infant exposure, the survival rate of their firstborns is pretty much zero.

As we've seen, many virgin mammals automatically avoid babies, or eat them. But female primates are a little different. At certain stages of life they seek out hands-on experience, with monkey tweens and teens acting particularly baby-crazed. In addition to borrowing their friends' and relatives' babies, juvenile female monkeys have also been known to snatch infants from other family groups and even other monkey species.

If real infants are scarce, pretend ones will do. To prime gorilla first-timers for motherhood, keepers at the Smithsonian National Zoo in Washington, DC, hand out little plush gorillas for practice. (They have electronics inside, for realistic baby gorilla babble.) And a recent and rather delightful study of wild chimps in Uganda's Kibale National Park found that, left to their own devices, juvenile female chimps seem to fashion their own baby dolls—toting around, patting, and cuddling certain sticks, behavior that ceased with their first birth.

Perhaps the power of these maternal prompts helps explain the spectacular flop of popular "baby simulator" programs in human high schools. Educators have tried saddling at-risk teenagers with high-tech baby dolls whose highly realistic emotional and gastrointestinal needs are supposed to discourage teenage pregnancy.

Unfortunately, as a large-scale study in Australia recently found, it turns out that girls enrolled in these programs are significantly *more* likely to get pregnant than peers without the baby dolls. Which is not all that surprising since, as a University of Chicago study showed, interest in babies in human women actually peaks in adolescence, just as it does in monkeys. (In human males, this curiosity stays at a low ebb throughout the life span. That said, childcare chops may yet be a boon to fathers, as a human dad's prior experience with small children helps predict how substantial his postnatal hormonal change will be. In fact, I credit my sister-in-law, nine years younger than my husband and a babysitting millstone throughout his early teens, with incubating his parenting skills.)

In part, babysitting experience might up mothers' games through traditional learning processes, since primates (we humans included) have unusually large brains and can pick up new skills through exposure. I do seem to be something of a slow maternal learner, though. I have two crescent moon–shaped scars on my forearm from folding my stroller wrong. I did it once, bled profusely—and *then I did it again.* I also once strategically moved a hotel coffee table to clear the way for midnight commutes to an unfamiliar crib, then forgot I did this, promptly rammed into it, and broke my toe. It's hard for me to believe that I consciously remember many skills from long-ago teenage babysitting.

Subtle hormonal shifts may be a more important factor. Virgin marmosets who tote new babies have different prolactin levels, and there is evidence from lowlier mammals that babysitting also triggers neurochemical changes—perhaps similar to some of the transformations that adoptive mothers experience. All of that little-kid exposure might approximate those sensitization experiments in which virgin rats' buried maternal seeds slowly germinate just from hanging out with rat pups for a week. Minding babies may be mind-altering.

∽

Teenage babysitters themselves, though, may not always be the best candidates to bear babies—not yet, anyway. In fact, the more time that elapses between a mother's mozzarella-stick-munching middle school days and the moment she gives birth herself, the more her eventual children may stand to gain.

A mother's age is one of the best-studied, most reliable determinants of her maternal capabilities. Old moms and young moms are materially different, and when it comes to various measures of their everyday parenting capacities, older is almost always better—which is a lucky thing for kids in the developed world, where moms are now practically geriatric.

The average age of first-time American moms is twenty-six, up from twenty-one in 1972, and college-educated women on both coasts wait even longer—delaying till an average age of 33.4 in San Francisco, they're practically at death's door. Reproducing shortly after my thirtieth birthday in Washington, DC, I was one of the youngest moms I knew (whereas my mom, who had me at thirty-two in the New York City area, felt almost unfathomably old for her era). As the national birth rate has cratered these last few years, the number of first-time moms over forty is still rising.

These later pregnancies carry extra risks for the mother and the baby, to say nothing of the challenges of conceiving in the first place. Yet by many maternal metrics older moms excel. (There's no hard-and-fast rule for what counts as an "old" mom, but by and large the early twenties might be a serviceable cutoff for "young," although there's a lot of variation in the studies.) Long-in-the-tooth moms are more physically affectionate, less abusive, more forthcoming with praise, better at structured play, less likely to spank, more vocal and interactive with infants, more supportive of children's imaginations, and more satisfied overall with the maternal yoke. In other mammals, notably elephant seals, older moms are also known to be extra aggressive in defense of their pups, chomping the bejeezus out of the other big girls on the beach.

The differences start early in pregnancy, when older human moms are more vigilant about following diets and getting prenatal care, and continue through birth, with the oldsters more likely to start breastfeeding and less likely to quickly stop. In the first year of baby life, older moms are more cautious about matters like safe sleep practices. Especially compared to moms in their teens, these women seem to be more affectionate overall, have distinct hormonal profiles, and respond more sharply to infant cues when researchers monitor their heart rates.

Meanwhile, the youngest mothers, especially teenage ones, have specific struggles of their own. These women are more likely to suffer from psychological problems associated with the rocky transition into new motherhood, especially postpartum depression: mental health woes hit teenage moms four times as often as mothers in their twenties, and another study found a pronounced drop-off in postpartum depression after age twenty-five. The most sobering evidence for a maternal-age advantage is the global infanticide rate. A brand-new baby's most likely murderer is his or her own biological mother—and across the world, mothers under age twenty are the most likely to kill their kids, and older moms are the least likely.

Of course, it's not all smooth sailing for us ancient ones either. Creaky-kneed, we may exert ourselves less in physical play. We can experience more parenting stress. We are more likely to have C-sections, and are at greater risk for bearing children with chromosomal abnormalities (although perhaps, as we'll see, for counterintuitive reasons). And it's unclear how many of our advantages are scripted by brain chemicals and neural architecture, as opposed to resting on a foundation of greater security and wealth. Obviously, the old-mom advantage could be less biological and more cultural, since socioeconomic factors have always been tough to disentangle from maternal age.

Until relatively recently, some scientists assumed that older moms might actually be worse moms because of their so-called "senescent" bodies. (Ouch!) But that was back when an over-forty mom was more likely to be an exhausted, resource-strapped woman on her sixth or sixteenth child. Today, she's stereotypically a Pilates-sculpted C-suite occupant who's finally ready to funnel several decades' worth of accumulated riches into her first kid. Teen and younger twentysomething moms, on the other hand, are now more common in poorer American communities.

Yet some scientists believe that moms with more than their share of stray grays still have advantages even when you control for socio-economic variables. For instance, in a study of Appalachian moms in a rural corner of West Virginia, the old still outdid the young in terms of sensitivity measures. A study in sub-Saharan Africa found that older women are more likely to seek out medical care for their children.

One compelling explanation for these differences is that maternal brain transformations interact with basic developmental changes. Just as the bodies of the youngest moms are still growing, their brains are blossoming, too. The mental upheaval of adolescence actually continues into a woman's mid-twenties. If the cognitive changes of motherhood rival those undergone in puberty, as scientists believe they do, then perhaps it's best not to embark on both neural odysseys at once. And while it's a common belief that humans are "meant" to have babies young, since American girls can start puberty at as early as age eight, this may be a modern illusion. Hunter-gatherers still inhabit the type of environment that we all evolved in way back when, and studies suggest that most girls exposed to these more natural conditions and (especially) diets don't even begin to menstruate until their mid to late teens.

To the extent that intellect matters for good mothering, old moms may get particular mileage out of their enhanced "executive functioning" skills, a set of cognitive capacities that teenagers and early twentysomethings notoriously lack. These include the ability to plan, remember, multitask, and manage time. Studies have shown that women who still excel at memory tasks even after taking the cognitive hit of pregnancy respond better to their children's cues. If maternal brain changes do indeed cost women something in this area, then being older when the mom metamorphosis kicks off might help.

Meanwhile, evolutionary biologists focused on our fellow female mammals offer a somewhat colder-eyed mode of thinking about why old moms outperform the young: perhaps we pour more resources and energy into late-in-life children because this is our last shot at passing on our precious genes, so we better make it count.

Younger moms, on the other hand, have an incentive to hold back, preserving calories and effort for future offspring.

In other words, not only mothering ability but mothering strategy may gradually shift as a woman ages and her reproductive time and options run short. "The forty-one-year-old mother who gives her life for her only child," the evolutionary biologist Sarah Blaffer Hrdy writes, "is not the same individual who decades earlier might well have aborted her first." It's a callous yet compelling explanation for why young women commit more infanticide.

As a very young child I remember once asking my own mom if she would die "for anyone"—meaning, I thought obviously, me—and receiving an uncomfortably long pause in reply.

In her defense, biologists envision a complicated back-of-the-napkin calculus involved in such an answer. Maybe Mom wouldn't have died for me at age thirty-two, when I was newly born and there was the possibility of several more babies yet to come. But by the time I was ten, with just one measly sibling, she might have changed her tune. Her willingness to martyr herself would have crystallized.

(Or so I tell myself.)

The distinctive dedication of older moms is described, catchily, in what's called the "terminal investment hypothesis." It's more a theory than a hard-and-fast rule, but stories from around the mammal clan, from red deer to rhesus monkeys, lend it some support.

Take killer whale moms, who are just like us—well, maybe not in the way that they crush the ribs of narwhals and devour them alive, but in the scope of their reproductive lives, which start around age ten and end shortly after forty, even as females potentially live on into their nineties. One study showed that the last calf of a killer whale mom is 10 percent more likely to survive than other babies in the pod, perhaps because its doddering mommy is extra motivated.

In humans, some scientists even suspect that terminal investment explains not only older mothers' more solicitous behavior toward their infants, but their bodies' treatment of the unborn. According to this theory, one reason that older mothers are more likely to bear more chromosomally abnormal babies is not simply that their aged eggs are somehow faultier. They may also have a naturally higher internal threshold for miscarrying, since perhaps a final baby with challenges beats no baby at all.

∽

Along with her age and prior childcare experience, a woman's mode of childbirth may also mold her maternal mind.

In mammals, remember, the physical stimulation of vaginal delivery—with its associated oxytocin flood—is often crucial to revving up maternal behavior. But C-sections bypass the vagina and cervix completely, which can have major behavioral consequences for some animals, including humankind's close relatives. When scientists performed C-sections on 211 monkeys, only seven accepted their babies.

In humans, C-sections now account for roughly a third of all American births—and I've done my part for the statistics, having become what obstetricians sometimes call a "three-peat." I'm not bellyaching about these abdominal surgeries, by the way. It was my call to scoff at what I then considered the ridiculously crunchy idea of hiring a doula for our first—*that's* so *not scientific*, I'd tittered to myself, as a friend described her doula-approved visualization method of hanging upside-down baby pictures around her apartment, to correct a breech presentation. (At the eleventh hour, her breech baby did indeed turn and she dodged the C-section.) Plus, I just happened to be born with non-birthing hips. Today I'm quite pleased to be alive, and as a bonus I can still jump rope without peeing. Fortunately, the power of the maternal instinct ensures that human moms love our infants no matter the delivery method, and sometimes even if we never give birth at all.

That said, there is some reason to believe the ugly pink sneer on my midsection is not the only indelible souvenir from these surgeries. C-sections may mark our brains as well.

One study of C-section mothers two to four weeks after delivery showed that they were less responsive to their infants' cries, although this difference seems to resolve after a few months. In another study of Israeli moms who had lived through a series of terrorist attacks, the moms who'd delivered via cesarean seemed less naturally buffered from the stress than other moms in those same conditions. And emergency C-sections, like my first one, are the most potentially damaging to the maternal bond. Women who have them are 15 percent more likely to suffer postpartum depression.

This may reflect the absence of the natural oxytocin deluge. Or it could have to do with the fact that many women—like myself—who end up on the operating table have also been dosed with Pitocin, the artificial form of oxytocin given to hasten labor, which some scientists think could potentially disrupt natural hormonal processes. Or it might be that moms who've just been minced by their doctors tend to have more underlying health issues and experience unusual lingering post-delivery pain, which impairs bonding no matter the baby's exit route.

∽

For reasons that likewise have to do with oxytocin, how you feed your kid in the weeks after birth also matters to the maternal makeover. Neural changes in breastfeeding women may well be somewhat different from those of formula-feeders.

Just as breasts grow during breastfeeding, so does the breast-centric part of the brain—or so scientists gather from studying other types of mammals. In nursing rats, the part of the cortex associated with the nipples and chest doubles in size, and there are other alterations as well. (Remember the menace of lactational aggression?) When mother rats are fitted with snug spandex jackets so that their chests can't be stimulated, their maternal care is somewhat disrupted.

Compared to their leak-free, formula-using counterparts, nursing human mothers have been found to be more sensitive to infant cries as well as less anxious and less tense. It could be all the extra skin-to-skin contact—another proven promoter of maternal sensitivity—that breastfeeding entails. Or maybe it's a self-selective effect, with cuddly-by-nature moms more game to breastfeed from the beginning.

But the chemical rush of nursing is also in the mix here. It definitely seems possible for both parties to become addicted. (One mom friend had to flee to Europe alone for two weeks in order to wean her son, who was pushing six.)

Some scholars think any differences between breastfeeding and bottle-feeding women are restricted to short hormonal surges during feeding sessions and disappear with the end of lactation. Other evidence, though, suggests longer-lasting neurological consequences of the different feeding styles.

Research on breastfeeding moms, from Boise State University, found that they remained somewhat more sensitive to their children's cues even by the time their kids had reached fifth grade. And a study of more than seven thousand Australian mothers of various backgrounds suggested that those who breastfed were roughly five times less likely to neglect their kids throughout childhood.

Nothing about these aggregates implies that women can't be excellent mothers however they nourish their children. The human maternal instinct is powerful and robust, not dependent on any single switch, and bottle-feeding has been going on for millennia—baby bottles encrusted with milk residue from extinct hoofed beasts are one of the oldest forms of human pottery. Also, in our modern world breastfeeding has clear drawbacks, with nursing mothers potentially experiencing more of the "motherhood penalty" at work, and perpetuating the childcare imbalance in modern marriages. Plus, it's literally draining. I sometimes feel like one of Dracula's victims levitating out of my bed for a midnight summons.

However, breastfeeding may be more within a given mom's control than other factors, like how you made your lo mein money in middle school or the dimensions of your pelvis. I ended up being particularly grateful for the option, despite heading into the hospital to deliver my first daughter having barely pondered breastfeeding. (In fact, I had the impression that nursing cover-ups were a type of special tent that moms pitched if they wanted their kids to nap on the go—hey, I just saw the tiny feet sticking out!) During my induction, I was initially far more focused on my hospital bed Netflix lineup—and very soon thereafter, when the delivery went south, on the matter of my physical survival.

As I lay shell-shocked and mostly immobilized in the maternity ward afterward, the lactation consultant looming over me with her plastic cone-shaped pumps like a bad '80s bra, I was tempted to squelch the whole mushy-gushy business. I had been bottle-fed myself, and I turned out just fine! The vague spiel about how breast milk helps baby immunity barely resonated, and nobody pitched breastfeeding as a boon to my own maternal mind, which, reeling after gory abdominal surgery, seemed to be having just a little trouble getting off the ground.

But the reality is that breastfeeding may be extra important for women like myself who've already suffered a series of setbacks, such as a nasty run-in with the surgeon's scalpel and a baby bound for the NICU, that threaten our potential bond. And learning how to breastfeed was the silver lining of my C-section experience. If I hadn't been stuck in the hospital for four extra nights, in the clutches of one Terminator-like lactation consultant in particular, I doubt that I would have figured it out.

Because on the day of my first daughter's birth, which was also the beginning of my own new life, I was swimming upstream against another unseen current.

I was a first-time mom.

∽

Well, duh. There's a first time for everything, as our own mothers have often lectured us. All moms are newbies at some point. It's a pretty unavoidable reality.

And yet this idea is a little less obvious than it initially seems.

Whether you're a rat, a river otter, or a human being, maybe the most important clue to what kind of mom you'll be is whether you've been a mom before.

How well I remember the smug smiles of the second-timers as I, having finally been discharged from the hospital, grunted and plunged with my firstborn toward the neighborhood playground, my unwisely chosen stroller with the subpar shock absorption system shuddering with every sidewalk crack.

In those early days I used to spend hours at Buy Buy Baby, roving in endless circles—were there nine, as in Dante's version of hell?—in search of the saving eco-plastic rattle or bendy giraffe that would magically calm our newborn.

Of course, the real missing equipment was inside me. The maternal change was still in progress in those first weeks and months; though I'd likely begun to transform during mid-pregnancy, I remained unfinished, still morphing into a mom.

It's not just that veteran moms know all the best spray fountains and have mastered the pacifier cram. The study of "parity," or the number of times a woman has given birth, covers more than the simple matter of learning the ropes of motherhood (or, perhaps more accurately, mastering the twisted straps of your infant carrier). Beneath these superficial differences in basic know-how, second-time mothers are deeply different creatures from first-timers.

For one thing, once a woman weathers the momentous transition into motherhood, she doesn't endure the whole unsettling experience all over again for her second or twelfth child. While disruptions can absolutely occur, her maternal brain is for the most part up and running, batteries included, no assembly required, needing only fine-tuning and refinement. She's already been sensitized, and her good-to-go mom mind makes her faster and more effective at processing her latest baby's cues compared to the new initiate the next hospital bed over.

When second-time moms report missing that giddy feeling remembered from their first birth, they aren't merely moping about the lack of extravagant baby showers. They're missing the sensation of transformation itself. A second-time mom isn't a woman becoming a mom; she is a mom having a second child. She's the same mom—only perhaps more so.

"You could say that her ticket has already been punched," Linda Mayes of the Yale Child Study Center told me. And motherhood is definitely a one-way trip.

This gives second-time moms an edge on many measurements of maternal capability. In general, repeat-customer moms find infant cries to be less aversive. It takes longer for beginner moms to derive meaning from their babies' particular wails and yodels, while experienced ones are often better at distinguishing the sounds of infant pain from other cry types. Multi-timers tend to touch infants more. They are typically less shaken by events like prolonged separation from an infant, a C-section, or a premature birth.

Throughout the animal kingdom, experienced moms kick first-timers' tails. Second-time cheetah moms select better dens. Practiced ewes lick their babies much sooner after birth. Sea lion veterans almost never nurse the wrong babies the way the neophytes do.

In lab tests, experienced rat moms catch crickets much faster than the first-time moms (who in turn outdo the virgins), and they also school the rookies in classic maze tests and at finding Froot Loops hidden in cunning little terra-cotta pots. Rat moms who've been around the block also dominate the first-timers in feats of maternal tenacity—traversing a slippery wooden rod, climbing a rope, and performing something called a "wire hang test." (All of these rat mom activities, by the way, are eerily reminiscent of my postpartum boot camp exercise classes.)

For infants, these maternal differences may be a matter of life and death. In our close relatives like gorillas and baboons, firstborns are twice as likely to die as subsequent babies. And it's not them—it's us. Certain types of mammalian moms, especially rodents, are notorious for eating their first brood of young. In humans, maternal cannibalism is happily quite rare, but maternal abandonment and neglect are more commonplace. Experienced moms are far less prone to both.

They're also more likely to display the maternal balance of coolness and readiness for action. Experienced rat moms are much swifter, for instance, to attack strange males, and they fight longer and harder for their pups. Seasoned human mothers are more collected than greenhorns in the face of infant distress—they perceive infant noises as less piercing to their ears. (In the first months listening to my colicky first daughter cry, I felt like Faye Dunaway in the final seconds of *Bonnie and Clyde*, each sob an invisible bullet—but that feeling didn't recur with her siblings.) At the same time, battle-tested moms are quicker to detect trouble, with faster heart accelerations at the sound of their own infant's distress.

Scientists are still trying to figure out the precise mechanisms behind this "not my first rodeo" maternal prowess. Some of it probably *is* just practice: human beings can get used to almost anything, baby war whoops included, and having basic familiarity with the ways of babies makes it easier to manage and interpret them.

But there is clearly something unseen going on in the brain as well—a way in which our minds might get even more mom-like in subsequent pregnancies, a "cumulative effect" that's particularly notable in an era when "one and done" mothers are on the rise.

Studies in rats and sheep suggest that multi-time moms have more estrogen, opiate, and oxytocin receptors in certain parts of their brains—including the all-important mPOA—which may indicate that they are already up and running when the new litter arrives. This means just a little dose of feel-good hormones goes further with these worldly females—kind of like how it takes only a smidgen of shampoo to suds up already clean hair.

But being a well-oiled mom machine isn't all sunshine and rainbows. Momnesia, for instance, is apparently compounded for mothers of more than one child. Human mothers of three have a much harder time with verbal recall tasks than first-timers do, which again suggests that there are mounting chemical and perhaps anatomical changes across successive pregnancies and that these changes are often, whatchamacallit, permanent.

Higher-order moms may also be more prone to hit the bottle: as a by-the-book first-timer, I remember gaping at an experienced mom one restaurant table over who swigged from a glass of wine while breastfeeding. Now I get it. We moms of many also stand accused of being somewhat lazier about playing with our later kids, and that's fair enough. We may not have as much time to, say, whip up a freshly baked Very Hungry Caterpillar birthday cake complete with home-rolled green fondant to fete a two-year-old. And so it's an open question whether later kids are better off: they get the benefits of our fine-tuned mom systems, our improved focus and response time, but they still may get less petting overall. Perhaps this trade-off explains the finding that firstborns often do shine brightest in school, while later kids are typically more laid-back and well-adjusted—quite possibly because their moms are, too.

As it happened, there were some very eerie similarities between my second birth and my first. The baby was another girl. Again, my due date passed and labor did not begin. Again, I was most unwillingly wheeled into the operating room for a C-section. In a truly *Groundhog Day*–like twist, it was *once again the Monday after Super Bowl Sunday.*

Yet somehow this was a completely different ball game.

There was none of the forged-in-fire stuff in the aftermath. All was calm. The baby's scream did not sound like a car alarm at close range. Nobody ran naked through the hospital halls. I had remembered to pack a cheater swaddle, and breastfeeding was a breeze. Even my wound didn't throb nearly as much. In fact, I felt chipper enough to take the once laughably unimaginable step of booking a shoot with the hospital's roving newborn photographer.

The baby behaved like a champ through her glamour-shot session. The sun even broke through the clouds at just the right moment, producing, if I do say so myself, some of the most beautiful newborn portraits I have ever seen. The bleached-white hospital sheets were basted in butter-yellow light as the baby basked in peace.

But now I wonder if the sun really shone so bright. It was early February, after all. Perhaps the clear skies were inside of me.

∞

My (rather dusty) babysitting résumé, my age, my surgical history, my nursing habits, and even the number of car seats cluttering up the back of my minivan—all of these factors chipped in to make me the mom that I've become, and I had at least some degree of control over most of them.

Others forces shaping motherhood, though, are entirely beyond our personal reach. Mom scientists have long noticed how maternal behaviors seem to run in families, with certain parenting patterns flowing across generations, almost as if they are inherited. And the hidden genetics of motherhood has become a hot topic in recent years, with researchers on the lookout for one gene or another that might set especially capable mothers apart.

But while you can increase your skin-to-skin baby cuddle time, reckoning with what's under your skin is a much trickier enterprise.

Chapter 6

IN SEARCH OF THE MOM GENE
What Your Mama
(and Your Great-Great-Great-Grandma) Gave You

THE FAMILY observation room at UNC Greensboro feels like the den at your neighbor's house, or any other plastic-festooned kid zone—except that several scientists and I are hunkered down behind its one-way mirror, secretly taking notes.

On the other side of the glass, a stout eighteen-month-old named Frederick sits with his twentysomething mother, whose outfit—a frilly pink romper—had briefly stumped the research assistant tasked with snaking heart-rate recording wires beneath her clothes.

But that's all sorted out now, and mother and child have been exploring a carefully curated selection of toddler toys, including a race-matched baby doll, a screeching electronic play phone, and a selection of board books.

And now, right on cue, in strolls the ogre. ("We couldn't use a clown," lead researcher Esther Leerkes whispers. "Too many phobias.")

"Hi, Frederick," the ogre says brightly from the doorway. "What are you doing? I'm an ogre."

Frederick looks up, thunderstruck.

The ogre wears a green cosmetology robe. Its voice is Southern-accented, female, and somewhat muffled behind a green plastic mask.

Frederick's eyes enlarge as the looming green visitor, occasionally consulting the script scribbled on the backs of its green plastic hands, continues its startling soliloquy:

"Do you know what an ogre is? I have a green face. I'm really tall. I have big green hands, don't I? I won't touch you. Have you ever seen an ogre before? What are you doing, Frederick? I'm interested in what you're doing. I see you there, Frederick. You're not an ogre. You're not green like me. You're a little child. Children like to play. I like to play, too. Do you know ogres like to play? I can't play with these big hands. They're too big! That's because I'm an ogre."

The ogre hums the tune of "This Old Man," bopping along to the music, then takes a sudden nap. Its delicate snores fill the room.

There is no telling how a given toddler will react to this part of the experiment. A few kids try for ogre high fives; others dissolve in tears. Frederick is somewhere in between, placing a plump hand nervously on his mom's knee, his eyes trained on the romping monster.

And all the while, Leerkes's eyes are trained on his mother.

Moms participating in this experiment are told that the researchers are interested in child temperament, which is technically true. What interests Leerkes far more, though, is the moms' performance in these highly unusual circumstances, which are calibrated to rattle kids but not adults.

Long before the ogre starts prattling—indeed, before the toy basket is dumped of its contents, and even during the experimental setup when a miniature backpack full of censors and wires is fitted on the child—the researchers are covertly measuring a mom's reactions, coding her behavior via video recording at some thirty frames per second. Did the mom smile reassuringly? Snap at her child? Straighten a barrette? Wipe a booger?

"If the child only cried for one second," Leerkes says, "I know exactly what the mother did and whether it was appropriate or not."

By "appropriate," she means whether the mom has responded to her child's cues in a sensitive manner—encouraging a curious kid, or comforting a scared one without scaring him further, and above all not tuning out entirely and thumbing through her text messages, as some of us may occasionally be wont to do.

Trying to determine what makes some mothers more ideally responsive, Leerkes has run this test on more than two hundred mother-toddler pairs over many years, starting shortly before 2010. In addition to the painstaking mom micro-coding, she uses the physiological info from all those hidden sensors, as well as mounds of questionnaire data on the moms' socioeconomic background and life history.

"Sometimes I think I'm nuts," the scientist and mother of three admits. And yet she always wants to look closer.

Around 2012, Leerkes began to wonder if there were clues to the mothers' friendly-ogre response that she and her team of eagle-eyed coders were missing, in large part because they are invisible.

Sure, her researchers noted each cheesy mom smile, and every moment of adoring eye contact or breezy avoidance. They knew how old each mom was, how many kids she had, and a million other details about her life.

But what if important stuff was happening inside the mom, entirely out of view? What if a woman's unique genome was a missing piece of each mom puzzle? Maybe the genetic lottery shelled out certain genes to some moms but not others, and these helped build our slightly different brains, accounting for some part of the startling rainbow of behaviors Leerkes sees from behind her mirrored window.

If the instructions for assembling your new crib left you with a migraine for two days, brace yourself. Building a mom is much harder. Our DNA is our own personal instruction manual, ensuring that our bodies construct the right proteins at the right time. The nucleus of every single cell in a mom's body houses identical genes, but different ones are active in her cornea as opposed to her colon. In motherhood, combinations of genes are uniquely expressed in our brain tissue, flipping on and off, commanding the raw materials that make our changing minds.

Hundreds, probably thousands, of genes are activated in the maternal transition, tweaking our bodies and brains in countless unseen ways. Scientists heartily agree that no one DNA snippet could possibly orchestrate moms', well, momming.

But *maybe* a few key genes could sway the *quality* of maternal behavior. Perhaps certain little blips of code could nudge a woman, even just a teeny-weeny bit, toward the more sensitive or less sensitive extremes.

While Leerkes's thoughts were turning in this direction, scientists working with human populations had already explored a handful of these single so-called candidate genes, which might potentially influence parental behavior. Meanwhile, the rodent literature suggested that brain systems related to social cognition were also key for maternal care, which made a lot of sense to Leerkes as an expert in this research area.

She wondered if she could translate the rodent work to human beings.

One of the biggest mysteries of social science, to say nothing of Freudian analysis, is how patterns of maternal behavior repeat themselves in families, sometimes across many generations. If you examined us closely, you'd likely see that my mom and I share more than our knobby knees and dislike of cannoli. We probably share mothering traits as well. If certain versions of key mom genes do in fact run in our bloodlines, passed on from mother to daughter, they could help explain the repetitive cycles of mothering behavior that shape whole families over time. Maybe genetic markers, or "risk alleles," could even help identify individual women who could use extra counseling, helping to break these destructive cycles.

And so, with funding from the National Institutes of Health,

Leerkes asked each of her two hundred–odd mom participants to deposit two millimeters of saliva into a vial.

Then she mailed the mom spit off to a lab in Colorado for genetic analysis.

∾

Once the ogre encounter (performed for my benefit and not scientific analysis) is safely over, Leerkes and I stroll back to her cavernous office. She's also one of the university's associate deans, and with her enviably long publications list, kindly sky-blue eyes, and desk-side bowl of fresh-picked strawberries, there are definite trappings of the supermom about her. So I'm mildly surprised to learn that, growing up on a farm in upstate New York, she had her first baby as a teenager. She toted her three-year-old daughter to college and then to graduate school in Vermont, where her studies into the varied nature of maternal behavior began.

Although she ponders motherhood from many angles, and suspects that there's a complex cocktail of factors involved, Leerkes doesn't think age is quite as important to mom behavior as other researchers believe. It strikes me that perhaps her own life story aligns more with the idea that maternal identity may be at least partially fixed within us before we are born, in our DNA.

We watch archived experimental footage on her office's big screen. It's quite astonishing how many ways a mom can react to a kid reacting to an ogre, or (in another version of the same test) to a remote-control truck disguised as a giant hairy spider that skitters about the room.

Some moms in the lab instinctively snuggle their spooked progeny, or try to distract them, or—for those who receive the highest sensitivity rating of 9—engage in a kind of tango in which the kid's emotions lead and the mom's choreography evolves in response.

Of course, there are varying views on how much sensitivity is necessary for good mothering, and what passes for basic parenting in suburban North Carolina might be seen as major overkill in, say, the

Amazonian rain forest, where giant hairy arachnids might be familiar pests. Still, a core awareness of a baby's cues helps mothers keep their kids alive, which all can agree is a good thing, and—for socially complex humans—responding to a child's emotions makes up much of the work of motherhood.

Leerkes points out one high-scoring mom who, sensing her child's growing interest in the spider, recasts the test into a fun game, lifting her feet to let the spider scurry underneath.

Other moms, though, ignore or even giggle at the children's fear, or make them pet the spider. "Touch it, boy!" one mom barks. A few, Leerkes says, never budge from the laboratory's love seat. (Some dark, deeply sleepy part of me wonders, *Would that mom be me?*) "We've seen moms on the phone and reading their magazines," Leerkes says. "I had one mom get out a nail file." (Phew. I don't even own a nail file.)

Could invisible genetic units with inscrutable names really explain the baffling behavioral diversity that we see in Leerkes's laboratory, not to mention near the Baby Gap sale rack or in the water park's interminable log flume line? I'd dearly love to get this fortune-teller's take on my own maternal destiny, reading my genes like tarot cards or tea leaves.

I would be surpassingly lucky to have snagged my mom's mom genes, if maternal behavior is indeed passed down this way. Mom always seemed omnipotent to me. Somehow she always managed to keep a stash of frozen PB&Js in the freezer, half on white bread and half on wheat (for nutrition), all the crusts neatly chopped off, so that even when she had to rush off to work early my sister and I could have a homemade lunch.

Unbenownst to me, though, Mom had nursed secret fears about what kind of mom I might become. "I was a little worried," she told me, tactfully waiting until a few years after my first child was born to share this.

Throughout my youth, she had watched me ominously ignore the neat rows of collectors' edition Madame Alexander dolls on my bedroom shelves, with their heavily lashed, downcast eyes. I preferred to catch beetles in the backyard and to peruse books about man-eating sharks. I must admit that I experience similar doubts about my own daughters, who have inherited, along with my DNA, my formerly pristine childhood dolls, now contorted in a naked tangle on the playroom floor that cannot possibly bode well for future grandchildren.

But maybe our mom genes will carry us all through.

∽

The search for the genetic keys to our mothering is a reasonable quest. After all, just as our noble modern mammaries gradually arose from the leaky sweat glands of some ancient opossum-like varmint, maternal behavior is an evolved trait, coordinated by genes that have emerged and diverged over a million generations, with new variations constantly surfacing and getting passed down or scrapped.

Today, basic mammalian similarities aside, hedgehog moms act pretty different from wolf moms. And even within species, humans very much included, it's only natural that genetic variation would continue.

Emily and I, soul mates since we were seven, may be similar types of moms, but probably not as similar as my sister and I—or so studies suggest. But is this because my sister and I grew up in the very same household, sharing a bedroom and sometimes (by accident) a toothbrush, or is it because we share roughly half of our genes?

To tease that apart, scientists compare regular sisters to identical twin sisters and also to adopted sisters. Identical twin sisters, who have 100 percent of their genes in common, are more similar as mothers than regular sisters are. Meanwhile, adopted sisters—who share our bedrooms and toothbrushes but not our blood—are on average less similar than biological sisters in their mothering styles.

In truth, you don't need access to a centrifuge to see that mother-daughter behaviors are recurrent and that familial histories repeat themselves. "Mirror, mirror, on the wall, I am my mother after all," says another silly needlepoint pillow that pops up on some of our couches. Parental resemblance—both physical and psychological—is a subject that transcends science. It's at the heart of much of human literature, not to mention the *Star Wars* franchise.

But perhaps it's best to begin a little further afield—like, in yonder pasture.

∽

Prior to writing this book, most of my knowledge of sheep derived from reading *Where Is the Green Sheep?* to toddlers over and over again. Yet on the subject of mom genes, those woolly wonders again emerge as model organisms.

For farmers, analyzing maternal behavior has little to do with self-reflection. Barnyard moms are quite literally cash cows for them. "That's what the maternal instinct does for us," one farmer told me. "Makes us money."

Among beasts of burden, attentive maternal behavior ensures off-spring survival and drives farms' bottom lines. In the livestock liter-ature, "fertility and maternal instinct" are female traits prized nearly as much as "a high-yielding carcass." (Finally, folks who appreciate a plump mom!) Farmers and shepherds are considered the world's ear-liest geneticists, skilled at cultivating desirable traits in animals even if they didn't know precisely which genes were involved. Today they are at pains to understand how maternal behavior is passed down in order to optimize their animals' bloodlines.

For years Cathy Dwyer, an animal behavioral scientist and sheep specialist at Scotland's Rural College, listened to local farmers' reports that some breeds of local sheep moms were simply better at lamb-rearing. Specifically, one regional breed, the Suffolk sheep, was "rub-bish" at motherhood, with sky-high lamb death rates, while another common type, the Blackface sheep, excelled.

She decided to put the rumors to rest through an arduous series of barnyard investigations. Sure enough, after many hours of observing,

Dwyer concluded that the Blackface sheep indeed licked their newborns more, were quicker to suckle them, rarely head-butted them, and uttered more maternal bleats. In a Y-shaped-maze test, the Blackfaces located their babies faster and stayed with them longer.

Meanwhile, the Suffolks, she said, were "just a little bit casual in their maternal behavior, and a little bit more interested in getting to the food bar." Even the second- and third-time Suffolk moms struggled, sometimes rejecting or attacking their lambs.

Could the profound difference in behavior possibly stem from the lambs and not the dams? Neonatal sheep, like most mammal babies (and human ones are no exception), are not the savviest of creatures: unable to recognize their mothers, they are initially attracted "to any large object." (Let's try not to take that personally.)

But Blackface lambs do have a reputation for being hardier and more with-it than their Suffolk counterparts, perhaps making their mothers look more adept.

To rule out this variable, Dwyer performed a series of embryo transfers. She made Blackface moms give birth to Suffolk babies, and vice versa. The split in mothering styles endured, regardless of the baby's breed.

After this experimental deep dive, "my feeling is that there is a pretty hefty genetic component to maternal behavior," Dwyer says. After all, these striking breed-by-breed mothering differences crop up in many types of domesticated beasts—between golden retrievers and German shepherds, for instance, as well as among rabbit breeds, and even in various strains of white mice and lab rats.

What drives these genetic divergences? In the cases she studies, Dwyer believes it's the intensity of human care. The Blackface is a hands-off highland species that gives birth unattended in the wilderness, so natural selection weeds out the lambs born to lazy moms and whatever genes incline their dams toward inattentiveness.

The Suffolk, on the other hand, is a lowland meat species that has been more intensively cultivated—giving birth inside cozy shelters, with humans on hand to help jump-start the bond between mother and young.

So the selection pressures on top-notch maternal behavior may have been relaxed by human coddling over the past seventy-five years or so. When I ask if human interference could really reshape a beast's genetic code in such a short time, Dwyer reminds me that not long ago it took forty weeks for farmers to raise a broiler chicken, and now they're fat enough to eat in six.

Pioneering geneticists though farmers may be, in the Suffolk case they have accidentally coaxed out the opposite of the desired trait, creating a kind of Frankenstein's momster: a pampered strain of meaty but mediocre mothers.

Aware of this danger, many modern shepherds try to prevent something similar from happening in their own flocks. Every ewe from my Connecticut barn slumber party, for instance, would later be rated on a maternal performance scale, the results figuring into the math behind who heads to the slaughterhouse that year. The good moms get to live.

Farmers' maternal performance scales are fairly crude, though, often using indirect measures like the weight of a mother's lambs. It would be far more reliable and convenient if livestock professionals could identify genetic factors connected to optimal maternal habits, administer a DNA test, and then flood their pastures and pigstys with premier moms.

This is much more challenging than separating the Suffolk from the Blackface. Several of the more ambitious genetic studies have been done in pigs, perhaps because bad pig mothers are especially notorious. Known in the pork industry as "crushers," they are prone to roll over and squish their piglets. Pig farmers have a strong incentive to figure out how to breed sows with lower rates of piglet loss.

But stamping out the crushers is easier said than done. In pursuit of what is sometimes called a "super sow," one German team tried to find genetic links behind a mother pig's reaction to piglet screams. Hiding high-definition speakers in pig crates, the researchers played thirty seconds of a distressed piglet's squeals. They then tried to compare each sow's reaction to this "scream test" with her response to thirty seconds of a similar-volume song: "Lovefool" by the Cardigans. *Love me, love me, saaaaaaay that you love me. . . .*

Unfortunately for the experiment's aims, but in a testament to the power of mid-'90s pop music, most of the pig mothers seemed much more aroused by "Lovefool" than by the piglets' despairing shrieks.

To make matters even sloppier, the forms of heritable maternal behavior that are easier to pin down in domestic animals are not always the ones we'd like to enhance. One of the steadier measures of maternal prowess is aggression. There is likely a genetic link between animals' willingness to defend their young and their babies' survival rates. Yet farmers tasked with personally kidnapping and killing said youngsters might not wish to coax forth this trait.

"When this aggression is directed at stock workers or members of the public entering grazing fields," one group of wary livestock scholars writes, "it clearly becomes much more problematic."

Meanwhile, all of these studies are still looking at general patterns of inheritance, not specific genes: a genetic connection doesn't tell you exactly which gene is responsible. This leaves ranchers and shepherds still doing things the ancient way, weeding out the best and worst animal moms using old-fashioned mix-and-match tricks, observing traits and attempting to maximize them, rather than administering, say, a simple "keep or dump" blood test.

∾

Over here in the human world, for a time scientists were more optimistic that they could identify a few smoking-gun mom genes.

After all, geneticists seemed to be rattling off genes that explained plenty of other human qualities and capabilities. In 2008, they identified what was later dubbed the "faithfulness gene," which coded for a vasopressin receptor variant that predisposed men to be sexually loyal (or not). Others found the "wanderlust gene," supposedly driving globetrotting impulses and human migration patterns. Perhaps most famous was the "warrior gene," which allegedly triggered aggression and risk-taking in certain people.

A good mom gene—call it the "I can tell a mile off that you have a fever and also scrape the last bits of barf out of your car-seat crevices" gene—would be in this same vein.

To date, there are perhaps two dozen papers on whether having (or lacking) one culprit gene type or another predisposes a given woman to be a more sensitive mom. The usual suspects are genes related to maternal neurochemicals like oxytocin, dopamine, vasopressin, estrogen, and serotonin. For instance, all moms have genes that code to make dopamine receptors, which suck up the pleasurable chemical dopamine. This is a component of the brain's reward system that is likely vital to maternal care, making our poopy-pants babies seem as sweet as peonies.

But there are five types of dopamine receptor, and the genes that code for them come in a number of different genetic variations, like flavors. Different women are born with different flavors. Some receptor types may be more efficient at siphoning up the pleasurable neurotransmitter, perhaps making women gifted with these genes feel more rewarded when interacting with their children, and by extension making them superlative mothers.

When a friendly green ogre isn't called for, the studies that try to establish this kind of link, between specific genes and real-world mom hijinks, tend to follow rather adorably pedestrian designs. These types of experiments are commonly staged in moms' own homes, or in labs resembling living rooms, with cameras "inconspicuously installed in the ceiling." (Since the ceiling is currently the only clean surface in my living room, I bet I would notice.) To probe our genetic mettle, scientists use common mom paraphernalia, like blocks and Play-Doh, as their analytical tools. How does a mom go about building a tower with her child? Does she help him solve a vexing puzzle? Now then, a drizzle of spit, please, and we'll see which genes in particular forecast her triumph or failure.

In one Israeli experiment, researchers had moms and their three-year-olds play with "a colorful set of Play-Doh and modeling tools," and then coupled their observations with the results of simple genetic tests related to the neurotransmitter vasopressin, involved in social behavior.

They concluded that moms with a certain variation within the gene for the vasopressin receptor were slightly more likely to provide structure and gentle guidance in play. That is to say, moms born with this gene type were better at helping the child set Play-Doh goals, handle any Play-Doh setbacks (which might or might not involve eating the experimental materials), and so on.

Another group asked moms in their own homes to read to their eighteen-month-olds from "a wordless picture-book" and to use "a peg-board," whatever that is, to make various shapes. Combining observations with genetic results, scientists calculated that women with two copies of a long version of the vasopressin gene were less sensitive to their children.

A gene-hunting team from the University of Chicago really put us moms through our paces, staging a kind of iron mom competition. After encouraging an unsuspecting mother and her kindergarten-age child to play together under video surveillance, a researcher would suddenly storm in, scattering "clothes, papers and empty containers." Then the unlucky mother "was given an Etch-a-Sketch, worksheets, a magazine, a pencil and written instructions to complete tasks with her child in order . . . : (1) return toys to shelves, (2) put clothes in the box, (3) put paper and containers in the wastebasket, (4) count geometric shapes, (5) copy a set of geometric designs on paper, (6) dust the table with a cloth, (7) draw a diagonal line on the Etch-a-Sketch, and (8) choose one toy and play quietly while the mother reads and takes a telephone call."

The moms were given just fifteen minutes to accomplish these tasks, which are more than I can handle most weeks. But the pay-off was supposedly worth it: combining a woman's under-pressure Etch A Sketch performance with her spit sample and other tests, the researchers deduced that several variants of the gene coding for the oxytocin receptor were associated with maternal capability.

Using tools like the friendly-ogre test, the spider test, and others, Leerkes's lab, too, found that variation in a woman's genes can—albeit to a very, very slight degree—help explain her real-world behavior toward her child. In 2017, Leerkes and her colleagues published a paper showing that having a long (and less efficient) version of the gene for a dopamine receptor, along with one other genetic variant, was linked indirectly with less-sensitive mothering. That is to say, mothers with these "risk alleles" in their DNA interpreted their babies' behavior more negatively, which in turn predicted less-sensitive mothering.

∽

But if all this seems a little bit reductive, given the dazzling complexities of human motherhood even before an ogre shows up—well, you won't be surprised to learn that these days many scientists, Leerkes and her team included, rather agree.

On a family trip out west, I stop by the University of Colorado, Boulder, to meet Leerkes's collaborator, geneticist Andrew Smolen. He walks me through his lab, stopping at one device to show me how he heated the North Carolina moms' genes to 95 degrees and copied them a billion times. Next he ran them through a squat box that apparently costs $300,000—"You could have this or a Ferrari, and sometimes I think I'd rather have the Ferrari," he jokes of this genetic analyzer— which sorted the DNA snippets according to length, allowing him to figure out which subjects had what genetic versions.

"In general, I think parenting—good and bad—is heritable," Smolen tells me. "The apple does not fall far from the tree."

We end up in a room lined with enormous freezers, which—by my quick calculations—would have plenty of room for even the 72-count Eggo waffle box. This array of ultracold units, kept at negative 112 degrees Fahrenheit, contains the human samples from past experiments.

He opens the door to one big freezer, billowing with smoky cold. The mom genes are kept in neat little trays similar to the type used to make ice cubes, waiting for the next whiz kid to crack their secrets.

But those secrets may not yield as easily as scientists once hoped. "These genes may be involved," Smolen says, of the candidates his research has identified. "They may actually be. But it's kind of hard to believe that a single gene is going to be responsible for such a complex behavior, even though it has consequences down the line. Our ability to look at specific genes involved in behavior is surprisingly not as robust as we once thought."

Indeed, the tempting notion that "there's a gene for that," just as there's an app, belongs to a paradigm that a growing number of genetic researchers have moved past. The more I talk to researchers, the more I realize that the presence of any one suboptimal mom gene variant in the nuclei of my cells probably wasn't going to crush me as a mother.

Rodent research now suggests that many of the hundreds of genes activated in mom-rat brains are in chromosomal locations where nobody thought to look. The idea that having any one "flavor" of a single receptor would matter much in the grand scheme of human moms seems less and less likely.

"I see many things working together, interacting with so many other things," says maternal behavior researcher Stephen Gammie of the University of Wisconsin–Madison. The transition into motherhood is not a one-note affair but a whole genetic symphony, "a score sheet with some things going up and others going down."

Meanwhile, human genetic studies have been moving away from headline-grabbing candidate genes, and toward the less sexy but more thorough genome-wide association approach. Instead of picking a couple of shot-in-the-dark candidate genes, these studies scan the entire DNA sequences of enormous populations of people, seeking multiple variations that correlate with a particular trait, like sexual orientation, or a particular condition or disease.

Unfortunately, using these genome-wide techniques to study moms is daunting. For starters, the gold-standard genome-wide studies require tens of thousands of participants. Studies of hard-to-corral moms, though, often depend on painstaking one-by-one recruitment at childbirth and baby yoga classes, and sometimes even extra kiss-upping via Mothers' Day cards, and despite all this still usually round up only a few hundred people.

Then there's the fact that, while many mom studies rely on hours of individual observation, genome-wide studies typically ask participants to answer just one simple survey question ("Have you ever had a same-sex experience?" "Do you suffer from asthma?") and provide a DNA sample.

Even if you could get a critical mass of moms to do this, it's pretty hard to imagine a lone yes-or-no question to identify flying-colors mothers. ("Have you ever subsisted on discarded sandwich crusts?" "Do you own a laminator?" "Do Christmas decor companies start mailing you coupons in mid-July, and are you okay with that?") So to get a plausible genome-wide study up and running, you might still need to run observational experiments on every last mom, the green ogres punching in for overtime for years.

Scientists are always coming up with fresh approaches, and emerging genome-wide research on the genetics of personality may yet shed light on parenting habits, too. When it comes to mothering, though, "I don't know if we'll ever get to the point when we know the specific genes involved," says Ariel Knafo-Noam, an author of the Play-Doh paper.

Canadian researcher Viara Mileva-Seitz agrees. She spent years of her career at one of the world's most influential maternal behavior labs, compiling a suite of well-publicized papers on candidate mom genes— how variation in serotonin transporters influences a woman's attitude toward her six-month-old, or how an oxytocin gene type can help explain her breastfeeding duration.

Yet although that foundational work is still cited in the latest scientific literature, she sometimes wonders whether it truly gets at the heart of the matter.

expressed or not. Every person has 37 trillion or so cells, far more cells than there are stars in the Milky Way. Each cell has the same DNA at its core. Yet some become liver cells, and others skin cells. And those that become a female's brain cells will likewise function quite differently during her infancy versus after pregnancy.

The genes themselves are more or less set in stone in all of these cases. The changes are typically epigenetic ones, evoked by the female's environment and life experiences, which somehow turn certain genes on and off. The major experience of pregnancy kicks off certain domino effects in the brain, but so might experiences that are earlier and subtler, like how the female was treated in infancy—in the rat's case, via her experience of her own mother's tongue. A process called "methylation," for instance, silences particular genes by chemically coating them, so their recipes can't be read.

In the less-licked rats, Champagne's team discovered that certain DNA regions related to maternal chemistry seemed to shut down. With more of their stress-hormone receptor genes silenced, the under-licked rats were inefficient at processing stress in ways that seemed to make them less engaged, less lick-happy, when they became mothers in their turn.

The well-licked baby rats, meanwhile, were more likely to express their genes for certain estrogen receptors, which made them more sensitive to the key maternal hormone when they went on to have pups of their own. They were also more likely to express genes for oxytocin receptors, and to grow more oxytocin neurons in their brains. Which means, in turn, that when they had daughters of their own, the licking behavior was passed down—not through some be-all and end-all "licking gene," and not through learned behavior, but through the complicated interaction between a soft rat tongue and the expression of her daughter's genes.

Licking studies don't apply directly to human mothers . . . well, not usually, anyhow, although I can imagine how the "baby as lollipop" concept might tempt a particularly zealous back-to-nature mom group.

(This would be, in the truest sense, a gross misinterpretation.) But some promising parallel research suggests that touching or stroking human babies may be our equivalent of licking.

"Infants need to be touched," explains Lane Strathearn of the University of Iowa. "If you don't touch your teenagers, they will thank you for it. If you don't touch your baby, that baby will die."

One striking study from the British Columbia Children's Hospital asked parents to keep what's been described as a "cuddle diary" of daily physical interactions with their newborns. DNA swabs collected years later, when the kids were four, showed that epigenetic differences arose between "high-contact" and "low-contact" kids, just as they did between the licked and unlicked rats.

Of course, the examined genes came from human cheek swabs, not brain biopsies, which are a lot harder to come by. However, one small 2009 study did biopsy the brains of human suicide victims. Hippocampus samples from the ones who'd suffered from childhood abuse showed more gene silencing, or methylation, in the brain tissue. Another small study of maltreated children showed that if their parents had undergone a professional intervention (and presumably improved their behavior), the kids' DNA methylation patterns were altered, too.

These epigenetic changes may help explain not just patterns of observed maternal behavior through families, but some of the physical differences between our brains as well. One Baylor College–led study of thirty first-time mothers showed, via fMRI scans, that women who reported healthy childhood relationships with their own mothers had more powerful reactions in the reward-focused areas of their brains when shown pictures of their own kids. They also had more oxytocin in their systems when they frolicked with their seven-month-old infants.

Another study, from Yale University, showed that young mothers who had better childhood memories of their own moms had more gray matter in brain regions related to emotion processing, and had more pronounced responses to infant cries.

Women treated poorly by their mothers, on the other hand, tend to pay less attention to baby faces even if they've gone on to have children of their own. They seem to be more upset by the sounds of babies' cries. One British research group identified eighteen-month-old children who had insecure relationships with their mothers, and then ran brain scans on the same kids more than twenty years later. Their adult brains looked different, with larger amygdalae—the part associated with fear and aggression.

Fascinatingly, the same types of findings seem to apply in foster families. Researchers wondered whether the best predictor of an infant's relationship with a foster parent was the age at which the baby was taken into the family. But instead, it turns out to be the quality of the foster parent's *own childhood relationship with his or her own past caretakers.* Even in the absence of blood relationships, family history repeats itself.

Whether or not epigenetics fully solves the mystery of how mothering self-perpetuates, it does feel right that every mother is generations in the making, with maternal warmth passed like a candle or a whispered recipe down from one woman to the next. After all, a female fetus already has her full complement of several million eggs when she's occupying your pelvis. Thus, in true *matryoshka* doll style, in pregnancy you carry your future granddaughters as well.

When I experienced my first quickening—the goldfish flop of a daughter deep in my belly—I was about sixteen weeks pregnant and on a trip with my mom and sister in rural Ireland, home of some of my mom's ancestors. We had just wandered the rocky ruins of a great-grandmother's thatch-roofed seaside hut. I felt that first fetal tumble in my stomach afterward in a Galway pub, as I spooned up a bowl of piping-hot potato leek soup.

It's a bit unnerving to think that what all those unnamed mothers did centuries ago—in a world of strained sunshine and wet gray stone, a subsistence economy of sheep (of course), potatoes, and seaweed, and nights brightened only by firelight—might have somehow trickled down to make the modern me. What lullabies did they sing, between reviving the fire and boiling seaweed pudding for dinner? Could their actions help steer mine today, as I prance with my expensive stroller to grab falafel for lunch? I imagine these anonymous ancestors' gnarled, ghostly hands kneading dough, scrubbing clothes in cold water, touching a child's face. They are sculptors' hands.

As overwhelming as their presence feels, it is also liberating to think that almost any person—a biological mother, an adoptive or foster mom, a single dad—who resolves to love and cherish just one baby girl today may shape generations of mothers yet to come.

∽

Of course, our own moms and grandmas aren't our sole creators, nor are we their carbon copies. (Luckily, my mom's "vacuuming in your underwear three minutes before Thanksgiving company arrives" gene seems to have skipped a generation.)

To see who made us the mothers that we are today, we must look backward, yes, but also straight down—to knee height or maybe even a little bit lower, at the level where your toddler is coloring the carpet or locking his big sister in a death grip. Oh, right, our kids—remember them? Needless to say, they are not the least bit interested in how we got made, so long as we remain at our posts ready to pour the chocolate milk. When I tried to introduce my first daughter to the mysteries of how mothers become, by letting her tag along on one of my prenatal appointments, she seemed deeply unimpressed by the fancy ultrasound machine and the baby's cantering heartbeat and the rest, but one thing stuck. Upon returning home from our doctor's visit, she promptly stole a plastic kitchen cup and peed in it, just like she'd seen Mommy do.

She is at the core of my story anyway. For as much as mothers create mothers, children are a volcanic force in our parental forging. Our daughters (and sons) are our authors, too.

Chapter 7

ARE YOU KIDDING ME?
Why the Child Is the Father of the Mom

THEN ONE day, all the old familiar feelings: the cue ball rolling in reverse up the back of my throat, the syrupy tiredness oozing behind my eyelids, the kazoo-like public gas events. Fumbling in the bathroom late on a summer afternoon, I watch with only mild surprise as the usual two pink lines darken in parallel, pointing the path forward. At thirty-nine years old, I'm—in my obstetrician's weary words—"going back to the well again" for Baby Number Four.

Yet the same old pregnancy story is, as always, brand-new. I fantasize about hot sauce and cheap iced tea, instead of the ricotta pizza I craved the last time. I am compelled to play on repeat certain long-lost, previously unmissed melodies, including "It's My Party and I'll Cry if I Want To" and anything by Tiffany. My normally comatose nighttime television viewing gets a little livelier. Never much for recognizing faces, I gradually develop an uncanny ability to spot minor actors who've made cameos in '80s movies or short-lived sitcoms, awing my film-buff spouse. The English language is slipping away, though. My mom's in her seventies, I'm in my thirties, and sometimes we stare at each other trying to remember the same words. *Cairn. Mint.*

Maybe all this happened without my noticing in the previous three pregnancies. After all, I've just been reading papers about how mothers are extra suspicious of strangers' faces, and about how we hemorrhage words. Or maybe, like the scientific literature says, my mom powers and deficits are simply compounding with every ride on the hormonal roller coaster.

But what if it's just something special about Baby Number Four?

Our kids—or mine, anyhow—can wreak havoc at any time, in any place, but they seem especially disruptive to mom science. For researchers who wish to parse maternal differences, a child and all his or her adorable quirks can become a confounding (not to mention Dorito-extorting) variable.

If your kid is screaming bloody murder but some other lady's kid is picking his nose during an experiment, you and she will respond differently, regardless of the researchers' aims. This is one reason why scientists like to show moms the same pictures of other people's kids, and why they have also developed hyperrealistic infant simulators. These large electronic dolls are marvelously persuasive, especially glimpsed in passing: researchers who tote them about occasionally field congratulations from startled colleagues.

Other labs, though, are intent on mining, rather than minimizing, so-called child effects, and exploring how children, no less than placenta-depositing dads and distant Irish great-grandmas, are indispensable to moms' manufacture.

Every kid is a wild card, and each shapes us differently. Our chances of getting postpartum depression, our sleep habits, the frequency with which we smile, and even our willingness to have more babies are all influenced by our present baby's temperament, health, and other characteristics. Scientists who've trained their night vision cameras on infant cribs have no doubt noticed that it's the babies running the bedtime show, or rather shows, since every baby is a unique disruptor from birth (and even before, as we'll see). These frail little people have a lot more power over us than you'd think.

This phenomenon, perhaps screamingly obvious to moms of the past, may be harder for us to detect today, since our sample sizes are shrinking. In 1976, 40 percent of American moms had four or more kids, but today it's only 14 percent, while the number of one-child families has doubled. It's hard to gauge how much of your parenting is a response to a particular child's characteristics if you have just one little ringmaster to contend with.

Mothers of two or more, though, may make discoveries even more profound than finding last year's wad of crusty kid tissues stuffed in our winter coat pocket. We may be uniquely positioned to appreciate the titanic power of children's natural differences over our behavior.

In 2018, researchers from the University of Minnesota, Twin Cities—who, like so many researchers, did not necessarily set out to study moms—polled a thousand people about the degree to which they believed a person's genes or the environment (or both) influence various human traits, from eye color to intelligence.

The researchers were surprised to discover that, across a diverse population of respondents, the savviest answers came from mothers— and not just any mothers, but moms of multiple non-adopted children.

There is nothing like the birth of the next child to underscore how little about parenting we personally control—and to reveal the degree to which our lives and our minds are but Play-Doh in pudgy little hands.

∽

I'm back on the tail of Yale's extraordinarily energetic Helena Rutherford, following the fuchsia blur of her sweater as she swerves through the halls of Yale New Haven Hospital's Saint Raphael Campus. At the maternal fetal medicine ward, she hangs a left for room six.

It's like a different world in here, far removed from the blare of a hallway television and the heavy tread of pregnant women plodding past with full bladders for ultrasounds.

This doesn't look like a hospital exam space. Instead of a sterile white bed, room six has a recliner invitingly draped in a plaid throw. The overhead fluorescent bulbs are switched off in favor of a golden pool of lamplight. A faux orchid distracts from the glare of linoleum flooring. Rutherford has designed this place to feel as homey as possible, so moms-to-be—and yes, their fetuses, too—will feel at ease. (We're told that even the ward's nurses have taken to occupying the experimental suite on their lunch breaks.)

The afternoon's subject arrives, sporting big winter boots and a thirty-seven-week baby bump.

Rutherford ushers the pregnant woman to the recliner.

"I'm just going to ask you a few quick questions," she says. "Did you eat anything in the past hour and a half?" The subject mulls this over for a bit, and then shakes her head. "The reason we ask you not to eat is that if you had a lot of sugar and you came in, the baby's going to be super active, so we try to get the babies in the same state. Did you drink anything?"

Tea, around 11 a.m.

"Did you sleep well last night?"

(Another lengthy pause.)

"I know that's a tough one—just try for a yes or a no." The mom-to-be settles on a mild "yeah."

Next Rutherford positions the fetus's heart rate monitor, by buttoning an elastic strap around the lunar dome of the mom's bare abdomen, and ripping open a foil packet of jelly to help glide the monitor across the vast navel. Like a pro she finds the jig of the fetus's heartbeat, which hovers around the standard 140 beats per minute. The mom is already wearing a heart rate monitor of her own beneath her clothes. Soon mother and child heart rates are scrolling out, the fetus's on a nearby laptop screen and the mother's on Rutherford's digital wristwatch, to be downloaded later. The experimental stage is now set, so long as the equipment stays in the right place.

"If anything moves," Rutherford says as we step outside, "just give us a shout." The idea is to let mothers and babies rest for about twenty minutes to establish the baseline of their cardiac activity. As usual, Rutherford has a million questions about the invisible upheaval happening inside this woman. But she's especially eager to explore the relationship between the mother's and baby's levels of physiological alertness or "arousal." She'll be measuring this via variations in the mother's and baby's tandem heart rates. (*I think we're alone now*, Tiffany croons in my mind's ear. *The beating of our hearts is the only sound.*)

We leave the pregnant woman happily buried in a back issue of *In Touch*, with the eternal cover story "Jen and Brad: We Are Having a Girl!"

∾

Scientists have long known that a fetus responds to big changes in its mother's physical and mental states.

A mom's arousal levels can be crudely raised with tools like noisy door knockers or, in some highly regrettable experiments from the 1960s, by a false report from devious (and no doubt male) researchers that the fetus is not getting enough oxygen. When moms get distressed, fetuses react sharply, their heart rates soaring as they squirm inside us.

But scientists are only now realizing that this story has a flip side.

In 2004, researcher Janet DiPietro of Johns Hopkins University was monitoring pregnant women and their late-term fetuses, recording the heart rates of both.

"I was thinking in only one direction—the babies react to the women," she recalls. But the statistician panning through her data reported that he was seeing just the opposite. Whenever a fetus moved, the mother's nervous system seemed to get jolted. The fetus stimulated the mother; the tail wagged the dog.

At first DiPietro assumed that her landmark discovery was some kind of mistake. To the statistician, she said: "I don't think so—maybe you should make sure you have your X and Y axes correct."

But a second look at the data confirmed the pattern: About two to three seconds after every fetal shimmy, the mom's body responded in kind, with her rate of skin conductance (sweaty hands are another measure of maternal arousal) rising.

Sensing that she was on to something, DiPietro next designed experimental scenarios in which she disabled two of a pregnant woman's senses, not by permanently blinding her as a rat scientist might be tempted to do, but by covering her eyes with a gel mask and deafening her with noise-canceling earmuffs.

Meanwhile, a stealthy researcher crept up on the oblivious mom, armed with a cardboard tube full of unpopped popcorn kernels. Holding the tube a few inches above the woman's enormous stomach, the researcher rattled it loudly three times.

Since the baby alone could hear the noise, the researchers were able to watch the fetus's startle response cascade through the mom's system.

Moms don't consciously feel most fetal movements—just the odd karate chop here and there—yet our bodies seem to register them all. And DiPietro thinks these baby-to-mom messages serve an important psychological purpose, shaping us.

"My feeling is that there has to be a signaling function, meaning that the fetus is preparing the mother to pay attention to them," she says. The baby beckons the mother inward, subtly distracting her from the outside world—which may even help explain a pregnant woman's characteristically dampened responses to her physical surroundings.

Types of fetal activity vary wildly. In her inter-uterine spying, DiPietro has observed the unborn busting some pretty funky moves, including licking moms' insides. What's more, the *amounts* of activity may vary even more than the type. At thirty-six weeks, the wiggliest fetuses move at least five times as much as the slugabeds.

"If you have a real active fetus, you are getting stimulated more and more," DiPietro says, "and perhaps that is differentially preparing women to take care of different types of babies"—not some generic infant, but your very own beloved.

Maybe this explains why some moms have such a "remarkably vivid" understanding of their babies' nature before they're born, as one study based on interviews with first-time moms in childbirth classes suggests. Depending on the action that's going on in our abdomens, we seem to grasp months in advance whether we're having a bump on a log or a hell-raiser.

Dubbed "maternal programming," the ways that unborn babies condition their moms in utero may extend far beyond differences in fetal activity. Fetuses puppeteer us through the placenta's unique hormonal secretions, which vary pregnancy by pregnancy (as well as dad by dad) and maybe also through microchimerism, the direct insertion of whole fetal cells into the maternal body—into our hearts, yes, but also into our brains.

But for now, tracking simpler measures like heart rate is the surest way to isolate the fetal mode of influence, which is why Rutherford had recently visited DiPietro's lab to learn and implement some of her techniques.

Twenty minutes are up. Rutherford and I peek around the curtain at the contented pregnant woman and her magazine. Things are about to get slightly less pleasant in here.

"I'm going to ask you to close your eyes," the scientist tells her. "And I'm going to play a five-minute audio recording of a baby crying. I'm just going to ask you to imagine the baby crying. All right?"

It's a terrible, thin, chin-trembler of a wail that rises and falls— "cyclical," as Rutherford describes it once we're out of the room. It was harvested from an actual kid by professional cry researchers, which apparently exist. "At three and a half minutes it goes quiet for about eight seconds," Rutherford whispers to me, the tiniest twinkle in her eye. "But then it starts again!"

She's trying to test, first and foremost, how long it takes mother and baby to recover their equilibrium once the crying is over.

But she's also looking at what might even be called "fetal personality" and its impact on mothers.

Some fetuses respond more strongly to cry sounds and take longer to settle. Perhaps women who have been, over the course of nine to ten months, more thoroughly stimulated by especially acrobatic and reactive babies in their bellies will in turn have more intense responses to those standard mom tests, when women listen to baby cries or look at baby faces while wearing the electrode cap. Would the moms of the extreme rockers and rollers exhibit different brain wave patterns? Rutherford wondered.

And which life-form is taking the lead here? If mother-baby interactions are a kind of dance, who cranks up the rhythm? Is the baby's heart thumping away on the monitor like the drummer, while the mom's heart merely marches along to the beat? Or is it the other way around?

Mulling some of Rutherford's ideas later on, with my own personal project in the works, I find a rare quiet household moment to lie down on my bed. Who's in here, anyhow? On this day, my baby is just about seven weeks along—officially still an embryo, and not quite to the "gummy bear phase," as the ultrasound tech calls it. *You don't even have fingers yet,* I think. *How can you be molding me?*

∞

Since it's challenging to figure out how fetuses, cloistered and inchoate, manage to push moms' buttons, you might think that this line of inquiry would get easier once the kid is born, becoming more immediately available for study.

But actually it gets even harder.

That's because of an epic chicken-or-the-egg problem. Moments after birth, mother and baby are already vibing off each other to such an extent that it's hard to observe them as individuals—they are a matched set, or what scientists call a dyad.

"It's a convenience of research to think of each of them as separate," says Linda Mayes of the Yale Child Study Center. "But they aren't. They really are an interactive unit. One is evoking the other."

Before you know it, new moms and newborns are synchronizing

their circadian rhythms, brain waves, even the tones of their coos. Who can say who's at the helm? The two create never-ending mutual feedback loops.

Take the ever-tangled case of postpartum depression. Grouchy babies seem to promote depression in their mothers, but it's hard to say where the causal loop really starts or ends. Depressed moms are less aroused by infant stimuli; likewise, because of the lack of maternal interaction, their babies may change, even at a genetic level, with stress-related cellular damage to their DNA. At just a few months of age these infants are less responsive to their mothers' faces, and less able to read other people's expressions. Thus the babies under-stimulate their already-depressed mothers, tightening the spiral.

This confusing symbiosis is just the tip of the iceberg in the scientific quest to figure out who is shaping whom. For not only do moms and babies live and breathe in lockstep, but they share a physical environment, which certainly shapes at least some of their collective behaviors.

And of course they also share genes.

To pick apart the individual child's role in explaining his mom's behaviors, scientists have tried those same single-gene techniques they used on us moms, checking to see if one blip in a baby's DNA can throw a wrench into mother-child interactions.

But many kid candidate gene studies—and there are a bunch of them—hit the same walls as the mom gene studies: not only are the findings hard to replicate, but a single gene type for a serotonin transporter or a dopamine receptor might not amount to a hill of beans in the context of complicated child behaviors.

Plus, the fact remains that half of the baby's genes match the mom's. So even if a given gene is up to no good, it can be hard to know which party it's working through, or if it's working through both.

To sidestep the snarl, some geneticists fall back on an old-fashioned but tried and true tool: twin studies. We've already seen the research showing that identical twin *moms* parent more similarly than regular sisters. This next set of studies, though, explores twin *babies*, both identical and fraternal.

The big question is whether moms mother identical twins (who share 100 percent of their DNA) more similarly than they mother fraternal twins (who share just 50 percent of their DNA, like regular siblings). If so, it's a heavy hint that the kids are holding the reins and steering their mothers' behaviors.

"If all parents treat all types of twins the same, it suggests it has nothing to do with the child," explains Lisabeth DiLalla, who runs one such twin lab at Southern Illinois University. "If the parent treats kids differently, and more differently if they are genetically less similar, it suggests there is something about the children that is evoking different behaviors in the parents."

Granted, twin studies have some major limitations. Anyone who's spent time with multiples knows that twins' parents have a tough row to hoe. Moms of twins can't simply breastfeed: they must master a formidable technique called the "double football hold." Any specific twin parenting move might be a product of soul-crushing exhaustion, rather than of the ins and outs of toddler genomes. Young twins are much more likely, for instance, to die accidental deaths—not because of any daredevil gene, but because their mothers have only one set of eyes. My neighbor's stout twins used to climb up the wooden blinds of her living room windows the instant she left the room. Locked out of the house by one twin, she'd keep frantic tabs on the other while dialing the fire department, the first twin watching all the while through the window with sparkling eyes. (Mothers of twins are apparently more likely to die young as well, possibly from all the stress.)

However, the extraordinary lifestyles of their moms notwithstanding, the data does indeed suggest that in some ways identical twins are mothered more similarly than fraternal twins and other sibling types.

Fascinatingly, this even holds true, says twin scholar Jenae Neiderhiser of Pennsylvania State University, in cases where true doubles were misidentified as fraternal, or vice versa, and the scientists discovered the truth during later testing.

This means that moms aren't just responding to the charming *idea* of identical twins; they are triggered by their children's extra-similar personalities and innate characteristics. Researchers have estimated that roughly a quarter of the differences in maternal behaviors can be chalked up to children's genetic attributes.

Adoption studies also highlight kids' indelible influence on moms. These mother-child pairs typically share no bloodline at all. Yet the echo of the adopted child's genetics is loud enough that over time the adoptive mother may come to behaviorally resemble somebody she's not related to in the least, and may never even have met: the child's biological mother.

Other evidence of kids' independent power over their moms shows up in pharmaceutical studies, where researchers reverse engineer mom behavior by chemically altering child behavior. In one early study, from 1979, some "hyperactive boys" were prescribed ADHD medication and others were not; their mothers didn't know who had received treatment. Yet among the participants whose sons were actually dosed and made calmer, the moms' behavior metamorphosed as well.

In a more recent set of long-term studies, based on the African island of Mauritius, local children were administered nutritionally fortified juice boxes that also included brain-building lipids called omega-3 fatty acids. For comparison's sake, "other kids got the regular juice box," without the omega-3 fatty acids, says researcher Jill Portnoy of the University of Massachusetts, Lowell, who collaborated on the study with University of Pennsylvania psychologist Adrian Raine. Children who downed the brain-building juice boxes daily for six months experienced reductions in problem behavior a year later, presumably from the drink's neural boost. But just as strikingly, "we also saw improvements in the *parents'* behavior," Portnoy says. The caregivers, who were nearly all mothers, became less antisocial themselves once their children's behavior improved. There was even a reduction in mom-initiated intimate partner aggression that scientists attributed to the kiddie cocktails—even though the women didn't have so much as a sip.

"That's such an exciting concept to me," Portnoy says. "With a juice box, you might be able to improve the entire family."

∽

But if I had such a magic juice box at my disposal, I guarantee that at least one kid would whine about the flavor and I'd be foiled again. The peanut gallery of my own household contains a whole spectrum of personalities and perspectives. Sure, as full siblings my kids are similar in certain respects: for instance, not one of them got the memo that *Cats* was the worst movie ever made, as they all remain enraptured. But even though I call all three of my children the same names—sweet pea, honey pie, babycakes—they are three radically different people. And even though all of them call me Mommy, I am also a radically different person when dealing with each one of them. If they play the trio of me like fiddles, and they do, then I have to navigate a three-part harmony.

Take the gulf between my two daughters, born two years apart down to the Super Bowl Sunday, but total originals from their first swoops in my womb. One is a night owl; the other is a morning person. One runs screaming at the first buzz of a distant bug, while the other will let a beetle crawl all the way up to her shoulder. One likes Nacho Cheese Doritos; the other is a Cool Ranch girl.

Daughter One—shall I call her Lily of the Valley, as she would sometimes prefer to be known?—is my hothead, with feelings that course just below her surface. She will flee the room during romantic Disney scenes, unable to handle the tension. Beneath her passionate emotions lies a sweet and loving nature: she once braved a booming thunderstorm to rescue a parsley plant. One day I asked her why she was rubbing off my kiss. "I'm not rubbing it *off*," she said. "I'm rubbing it in!" But if you cross her, she will strike. At one wedding, somebody scolded her for sipping out of other guests' water glasses. In retaliation, Daughter One, who was four at the time, pinched this seventy-year-old woman, hard. Then she gave one of her great-aunts' bottoms a rude little pat. (In Daughter One's defense, the bottom was right at her eye level.)

At home afterward, my mom tried to teach a lesson about being sweeter to strangers.

"You catch more flies with honey than vinegar," she informed Daughter One.

With a grade school teacher's pedagogical gleam in her eye, Mom proceeded to pour out two little shot glasses—one full of honey, the other of vinegar—for an illustrative taste test, while I beat a hasty retreat to the next room.

A minute later, Mom stormed out of the kitchen.

"She likes the vinegar," she sputtered.

Daughter Two, meanwhile, born bathed in that mysterious February glow, is sunny still. Even back when she looked like a plump little koala bumbling around on the toddler playground, her preschool teachers had her pegged for a nimble politician, and she's matured into a shrewd kindergarten diplomat with "a pied piper personality," as one teacher recently said, leading epic games, and always knowing the right thing to say to friends with hurt feelings.

If her big sister mows like a machine gunner in arguments with me, Daughter Two snipes from an unseen hillside, lethal and precise. She's a little bit like a scrumptious chocolate truffle with a ball bearing for filling. I first tasted the steel beneath the sweetness when she quit napping before age two. "I will never surrender," she told me. My husband sometimes compares her to a dictator on the rise. On the bright side, she potty-trained herself, matter-of-factly reporting one day, "I'm all done with diapers"—and she was. She never really needed training wheels on her bicycle either.

When I used to sprawl on Daughter Two's bed, exhausted after another failed effort to get her to go to sleep, she would stroke my hair and whisper, "I know you're a beautiful princess," as if she were the mother and I were the child.

∽

Now that I've shown off my two equally wonderful, equally beautiful daughters, I'm a little hesitant to introduce this next prong of kid-and-mom science, which is about maternal favoritism. While our children can back us moms into emotional corners and manipulate us like crazy, moms may also treat kids differently for our own selfish reasons, based on who we think is the more valuable child.

Sometimes scientists call kids' inherent traits, like beauty, health, and intelligence, their "endowments at birth." It pains me to even type these words. Even though I know I mother my three children differently, I believe from the bottom of my heart that I love them all the same. No scientific paper will ever make me think otherwise.

But then again, I suppose if moms could see the whole truth about ourselves, we wouldn't need tattletale researchers to study us. Certainly there is ample evidence of cruel maternal favoritism from our furry peers. Mother grizzlies have been known to tree one cub and lumber off with the other. All mother pigs, even the non-crushers, can make different amounts of milk per teat, allowing their strongest piglets to monopolize the geysers while the runts starve.

Well, I would never do anything like that, you may tell yourself, just as I do. Yet you may well have participated in a fatal favoritism already, in the way that the female body seems to cull certain embryos in the womb, via miscarriages we may not even know about. And that culling can be conscious as well: in Europe, where genetic screening is more commonplace, abortion rates for fetuses with Down syndrome are about 90 percent.

What surprised me most, though, especially given all the trouble that we go through to grow our mom brains, and our newfound willingness to duel with mountain lions and so on, is the fact that human moms may actually be *more* prone to abandon or otherwise betray our offspring than other mammals, *even after our babies are born.*

This shocking reality may reflect the unparalleled cost and stress of raising a competent human being. And because of the extravagant length of human childhood, human moms must juggle multiple small dependents at once, a feat that sets us apart from our primate peers. Each new addition entails more than a decade of hard time and some 10 million extra calories.

In today's materially ample America, 10 million extra calories doesn't even seem like that much to ask for. *That's just a few extra runs to Costco!* you may well be saying. But in the premodern landscape, when our present-day mom tendencies were shaped, it was a different and much sadder tale. Moms likely killed or abandoned babies in a somewhat routine manner. In American cities infanticide remained fairly common even in the early twentieth century. There are still impoverished places in the world—like the Brazilian slums where the anthropologist Nancy Scheper-Hughes did her devastating fieldwork—where women regularly practice passive infanticide, or "mortal selective neglect." Even today a week-old American baby's most likely killer is his or her own mom.

The murderous tendencies of mothers are a frequent subject of Sarah Blaffer Hrdy's *Mother Nature*, a breathtaking scientific opus, to be sure, but not a book I'd necessarily recommend as a bedtime read. The famous primatologist floats some fairly harrowing theories, though at times I found myself almost wanting to laugh out loud. Harm my children? The idea of them experiencing the vaguest twinge of pain practically makes me faint, which is why I blow on their scraped knees to lessen the slight sting of hydrogen peroxide. As mad as they sometimes make me, I'm not about to expose them in a blizzard or feed them to wild dogs.

But scientists like Hrdy and others say a tendency to forsake our helpless babies may be constitutionally ingrained—which offers still another theory on the mystery of why certain postpartum mood disorders came to be. Perhaps this extremely common emotional numbing just after a new baby's birth has a cold-eyed, ruthless purpose—"to neutralize any elation a new mother might feel," in one scientist's chilly words, "and thus permit a more objective evaluation of offspring quality." Maybe the baby blues—striking early, before anybody gets too attached—are a kind of considered pause before moms make the full maternal commitment.

In the olden days, we couldn't afford to lavish precious years and calories on a child who wasn't up to snuff. We had to zero in on the strongest and best. Moms, in this dark view, evolved to play favorites.

Do mere minivan chauffeurs the likes of you and me still have a secret evolutionary mandate to winner-pick? Infanticidal logic just doesn't square with the rush of tender devotion I feel when I smooth a blankie over my slumbering son, or when I learn that he thinks the forbidden "b-word" is actually "bingo." The *unconditional* side of motherhood can certainly triumph—under the right conditions, that is. In our comfortable society, mothers don't typically abandon children born ailing or disabled, often suffering terribly alongside them, and maybe even loving them more in a way that puts the rest of us to shame. A friend of mine started her own foundation to raise research funds for her son's rare genetic disorder. I watched another friend empty herself financially and emotionally for a dying child. Studies suggest that a modern human mother's happiness is tied to her child's for life, and moms of very sick kids are often so stressed, so literally heartbroken, that they are much more prone to die of cardiac disease. But are modern Western women really fundamentally "better" than mothers of other times and places who acted differently?

Some studies insist that moms' hearts of darkness are still there, if you plunge deep enough beneath culture's cloak. In an era of surplus calories, perhaps we express our evolved ruthlessness a little differently, "cutting off" certain children using other methods, like emotional neglect.

In some rather upsetting long-term studies of very prematurely born Italian infants, the moms had lower-quality interactions with these babies at three months old compared to mothers of full-term babies. Even when the babies grew to be toddlers, the scientists, watching closely at snack time, still noted subtle but lingering differences in the mothers' behavior that reflected what they harshly termed "higher levels of maternal negative affection." (Although it's always possible that something else—like, say, the stress of extra costs incurred by premature birth—could explain the difference.) Another study suggests that Swedish moms, perhaps without even realizing that they did it, withdrew resources from offspring potentially affected by Chernobyl's nuclear fallout.

Enough. My hackles are up. The lioness stirs. My mom brain won't let me imagine any more.

∾

Favoring is easier for me to contemplate than forsaking. Maternal favoritism as we practice it today lingers mostly in the details. Some 80 percent of us allegedly—allegedly, kids!—prefer one of our children to the others, and more than half of parents demonstrate so-called differential treatment toward various progeny.

And, true to theory, moms seem to bet big on the most promising kids. Assisting offspring most when they seem to struggle, plunking down funds for extra Kumon lessons or SAT prep, appears to be a luxury of privilege. A long-term study of the academic program Head Start found that when resources are constrained, parents invest most in their brainiest kids from the beginning—and this is especially the case in large families, with potentially tighter budgets that force choices.

In one fascinating 2019 study from Malawi, an economist observed the impact of what is, in the West, considered a standard and quite harmless practice: delivering children's report cards, which are apparently not common in this part of Africa. The parents used the new data on their kids' academic performance not to cheer on the underachievers, but to bulk up their investments in the smarty-pants, sometimes yanking the poor performers out of school entirely.

But the most striking predictors of maternal favoritism are also the most superficial: moms appear to dote on their cutest kids.

There's a debate raging in evolutionary biology today about the meaning, or meaninglessness, of physical beauty. Some scientists think attractiveness is an honest signal of a potential mate's health or "good genes," while others think showy looks—like the peacock's famously over-the-top tail—are arbitrary and as fickle as fashion.

For the record, I'm rather partial to the beauty-as-fashion argument, just because it makes men look so entirely ridiculous. If human males were really in the market for the most serviceable baby incubators, then they'd be ogling gals with "solid, sturdy limbs and ankles," "broad hips," and ample waists—women, in other words, who resemble Cub Scout den mothers far more than supermodels.

When it comes to infants, however, there seems to be a lot less leeway in how humans perceive beauty. Yes, there can be subtle social dimensions to child attractiveness, with some cultures shunning babies born with too much or too little hair. In African American families, researchers have reported a troubling favoritism toward lighter-skinned babies that no doubt reflects bleak cultural undercurrents.

But overall the components of infant attractiveness—a suite of childish features known to researchers as *Kindchenschema*—are rigid and globally constant. These cues, which include big eyes, a large forehead, a small chin, and chubby cheeks, seem to transcend cultural and racial lines, translating even across species. Nearly all baby mammals exhibit similar traits, exaggerated in cartoon characters like Bambi, and in flesh-and-blood concoctions like French bulldogs and Persian cats. Scientists contend that even non-mammalian animals that rely on parental care may flaunt Kindchenschema, too. Hatchling dwarf crocodiles (whose mothers cradle them in their toothy mouths after birth) are cuter-looking than, say, California alligator lizards, which reap no such maternal perks.

What exactly these highly stable features signify, beyond extreme youth and vulnerability, is a bit of a mystery. Studies comparing a person's baby pictures with high school yearbook photos suggest that cute kids don't necessarily grow up to be the foxiest adults. I once babysat for an actual Gerber baby—or at least, a baby whose perfect circle of a face had graced diaper ads and the cover of parenting magazines—and while she turned out just fine, she definitely peaked around nine months or so.

Still, this enigmatic suite of baby features has indisputable effects. For instance, there is strong evidence that a child's degree of cuteness influences the attitudes and actions of caretakers who are not biologically related to the child. Attractive babies in the neonatal intensive care unit have better outcomes, "presumably because they receive more nurturing" from the staff, one study found. Day-care workers tend to underestimate the skills and intelligence of unattractive children. Cuteness is also a major predictor of orphaned children's adoption outcomes. And back in the days when unwanted human babies were advertised in turn-of-the-century newspapers like excess kittens, most were free—some moms would even pay you to take them—but the very cutest specimens could cost a hundred bucks.

But shouldn't parents be exempt from these prejudices? What about "the face only a mother could love"?

In fact, it's the father who seems more forgiving, since he's laser-focused on just one key aspect of kids' looks: whether a kid resembles him. Drawing on reams of data from the large Fragile Families and Child Wellbeing survey, which follows American households with nonresident dads, a 2017 study recently zeroed in on answers to a single question: "Who does the baby look like?" If both parents described the baby as a carbon copy of pop, it turned out the dad spent additional time each month hanging out with his progeny. Thanks to his extra investment, these look-alike kids were healthier one year after birth than those who more closely mirrored their moms. A French study—which employed a panel of judges to double-check that the dads were right about the resemblance, and found that they were—showed that dads were emotionally closer to their little duplicates. In Senegal, another group found that kids grow up bigger and are better fed if they look and in fact *smell* more like their dads. Somewhat hilariously, dads apparently prefer "mini-me" looks *even in adoption scenarios.*

All these dad benefits up for grabs probably explain why moms and maternal relatives like to harp on how the new baby is the spitting image of his old man, especially when the dad's within earshot. Before finding this vein of research, I always thought that my mom's sisters discerned my husband's ears here and his chin there because my newborns looked a little funny and they wanted to distance themselves. In fact the great-aunts were probably trying to help me out.

For our part, though, moms—permanently assured of our own maternity—don't give a hoot who a kid looks like: our husband, the handsome plumber, or second cousin Martha.

We are connoisseurs of raw cuteness.

In the 1990s, psychologists camped out in an Austin, Texas, labor and delivery ward to watch more than a hundred moms interact with just-born babies. From the very start the moms of the comeliest ones were "more affectionate and playful," and these differences in maternal behavior endured as the children aged. (It makes me feel a tiny bit better to learn that babies are rather shallow, too, apparently preferring the faces of attractive women.)

A few years later, a University of Alberta researcher picked up on another unnerving finding while exploring the seemingly sedate topic of shopping cart safety. Observing kids and parents cruising the frozen-pizza section and the cereal aisle, this supermarket gumshoe subsequently noticed that moms were more than twice as likely to securely buckle cuter kids into the cart. (Naturally, he had somebody other than the kids' parents evaluate their relative cuteness.) Other work has found that, by the time infant twins are eight months old, mothers gravitate more toward the larger, healthier baby.

Perhaps the most haunting study involves police photographs of child abuse victims from the 1980s, which revealed that kids with atypical craniofacial ratios were more likely to be maltreated than were stereotypically cute kids. While child abusers are often males—especially unrelated males, like a mother's new boyfriend—the sad truth might also be that mothers put more effort into protecting beautiful children.

More recently, scientists have used studies of cleft lip, a slight natural distortion of typical Kindchenschema, to probe these maternal taboos. One 2017 analysis of moms wearing eye-tracking glasses found that they gazed at their own cleft-lipped infants less often, compared to the moms of physically typical babies. Another study discovered that a speedy surgical repair of the cleft not only fixed the aesthetic problem, but might also potentially help heal the child's relationship with his or her mother. The faster the stereotypical cuteness was restored, the more tender the mom ultimately became.

Again, I can read these studies all day long and still deny that their implications implicate *me*. Each of my kids, I'm quite happy to report, came out pretty cute, although there were a few worrisome details early on. One newborn had pointed hairy ears. One had crossed eyes. One seemed short a lower lip. There was more than a passing resemblance, in certain cases, to Yoda; in others, E.T. "Have you contacted your spaceship yet?" I would whisper to these staring little alien life-forms dropped in my arms. But of course, they were all beautiful to me.

And doesn't this have to be the way things really work? Whatever studies claim, mustn't each mom unconditionally adore her writhing wee bairn, to the point that we are stone blind to its flaws and craving the scent of its feces? Aren't we all like Mrs. Jumbo, the devoted mama elephant who embarks on a violent rampage when circus-goers mock her son's funny looks?

The catch here is, of course, that the infant Dumbo is actually Disney's Kindchenschema masterpiece: with the giant ears serving mostly to frame his baby face, he isn't a tot only a mom could love, but precisely the kind of baby that a mom might, without quite realizing it, bend over backward to defend and even favor.

∽

At just ten weeks pregnant, I'm dispatched, due to my advanced age, for a newish type of test that snatches bits of the fetus's placental

DNA from the mother's bloodstream to screen for major chromo-somal flaws.

The draw seems to take ages, and as the tube slowly fills with crim-son, my usual "bodily fluids belong within the body" queasiness min-gles with pregnancy nausea, as well as with a more cerebral pang that maybe I shouldn't be getting access to my kid's genetic secrets, espe-cially this early in the game. Noticing my deepening green, the nurse asks if I have other children, and I weakly nod: two girls and a boy.

Within a few days emails start popping up from the test company, which by now has shuffled through the baby's personal blueprint. The results are normal and I breathe a big sigh of relief. Oh, yes, and they've determined the baby's sex, too—a detail of early testing that's consid-ered more an aside than a vital data point.

While I opted for a pink or blue balloon surprise the first time around, very shortly thereafter I decided that I'd had quite enough ob-stetrical revelations in the delivery room, thank you very much, and vowed to find out the sex of the rest of my kids as soon as possible. But especially after Daughter Two followed hard on the heels of Daughter One, I also felt confident that I'd just stay on my baby-girl roll. I'm from one of those girl families. I have a sister only. My dad had a sister, who had only daughters. My mom had three sisters and no brothers. Her mom had four sisters and no brothers, and so on. On my mom's side of the family there is a lone boy cousin, who understandably makes himself scarce at family get-togethers. My poor grandfather's sodium intake was constantly policed by the many female relations in his life, whom he referred to as a collective "they"—"they won't let you have a frankfurter," I overheard him mutter glumly once. ("You" was him.)

"In our family, boys are rare as hen's teeth," my aunts still like to say, digging into the hors d'oeuvres.

So I was amazed to learn, back in the early days of my third preg-nancy, that I was having a boy. I shared my anxiety with my obstetrician at the time, a kindly woman with a young son of her own who, I soon gathered, was something of a favorite.

"Don't worry!" she cried. "You're going to *love* that little frank and beans!"

No statement could have horrified me more. I next consulted the aunt who is the mother of the boy cousin. "Oh, you know . . . just roll him a ball," she said vaguely. (And now it suddenly dawned on me that she'd had no idea what to do with a boy either.)

These directives did not prepare me for my son. From the very first moments, he felt different in my arms, more solid somehow, like a little sack of cement. He cut his Chiclet teeth on his big sisters' pink chew toys and sported their hand-me-down heart-print pajamas, and as a toddler spent his Friday nights (somewhat unwillingly) watching *Project Runway* reruns by their sides. ("That's so '80s," he muttered of one failed ensemble.) And yet despite all this environmental girliness, almost as soon as he could scoot he established himself as a violent marauder, hell-bent on piracy and the Dark Side, referring to his sisters' old stroller as his "death coach." He likes to slash a rubber sword through the air and flex his marshmallow physique in front of his bedroom mirror, and he does have, I must say, some big-league "baby-releasers," as Kindchenschema features are sometimes known. Old ladies like to stalk him in grocery stores, blowing kisses. Yet he tells me that I'm his "favorite woman."

Having a boy altered me in certain palpable ways: I now know the difference between Barbary corsairs and Caribbean buccaneers, and between broadswords and cutlasses. But mom science suggests that bearing a male has influenced me in an unseen manner as well. Mothers of men are measurably different from the get-go, and we seem to receive the gender-reveal memo long before our brains can process the results of even the most prophetic genetic tests.

That memo includes some bad news. Women carrying male fetuses are prone to a host of pregnancy complications, including miscarriage, prenatal diabetes, preterm birth, and cesarean delivery. Nobody is sure of the reason, but it may be because larger, slower-growing male fetuses are both more physically demanding and more delicate in utero.

Even though we think of men as bigger and stronger, and they typically are, "frail males" (as scientists sometimes dub boys) are actually more likely to die than females throughout life. When my son developed that supposed hint of a gurgle in his lungs after his birth, the nurse was right to be on high alert: newborn boys are almost always more at risk for setbacks. The Y chromosome may explain why: in the womb, a mother's X-based immune systems may see her son's Y chromosomes as targets, and the absence of a duplicate X chromosome to fall back on may also make boys more prone to genetic disorders.

#Boymoms—as we sometimes call ourselves online, to lots of blowback—may suffer psychological in addition to physical challenges. We are about 70 percent more likely to get pregnancy-related depression, perhaps due to the inflammation in our immune systems, one recent study from the University of Kent showed. We are also, in our first and second trimesters, measurably more sensitive to disgust—which experimenters gauged, in one rather creative study, via moms' reactions to cockroaches, "throats full of mucous," and "a human hand preserved in a jar." This may be the case because fragile male fetuses are especially vulnerable to environmental threats, so their moms are rigged to be extra sensitive to our surroundings.

On the bright side, #boymoms-to-be are spared the worst of morning sickness, which is significantly more common in expectant #girlmoms, just as the old wives claim. We also get to eat about 10 percent more calories without gaining extra weight (allegedly), and we may be spared some of the cognitive problems associated with pregnancies, outperforming #girlmoms on some tests of working memory and spatial ability.

The fetal movements that may subtly "program" moms also vary by gender, with some reports suggesting that females, ever verbal, make more mouthing gestures, while males flail their legs around—possible manspreading?—and fidget. Female fetuses respond extra sharply to speech and startle sounds, and have higher heart rates. Perhaps this explains why, by the third trimester, their moms' hearts are thumping faster, too.

Chapter 8

THERMOMETERS
How the Physical Environment Makes (or Breaks) a Mom

T HE RATS' room is steeped in a faint red radiance as the scientists and I file in. It's about noon, Texas time, but the artificial lighting is programmed so that the "sun" sets first thing in the morning here. Now the nocturnal rats think it's the middle of the night, their favorite and most festive time, particularly for new moms and their pups.

My eyes can't seem to adjust. Squinting, I lean down toward one bustling rat habitat and say in a hearty voice: "Those are some big babies you've got there!"

"Those are actually adult males," says postdoctoral researcher Hannah Lapp, gently.

She steers me over to another clear-sided container, and now at last I can discern a mother atop a turbulent pile of six-day-old pups. Their fuzzy skin is still translucent, and we can see white milk pooling in some pups' bellies, evidence that they've just feasted.

Speaking of feasting: it's time to demonstrate the current experimental protocol, which involves Nilla wafers. These cookies have been on my mind of late, since I've just learned that you can turn two of

them (plus a Thin Mint in the middle) into a tiny faux hamburger for a
bake sale sensation. Another mom friend crumbles them over peaches
for cobbler-like lunchbox delights.

But Lapp and Frances Champagne, lead investigator at this University of Texas at Austin lab, are using these standard kid snacks for
somewhat more urgent purposes. They lace the wafers with the chemical bisphenol A (BPA), a major ingredient of modern plastics.

BPA is ubiquitous throughout the world, manufactured at the
rate of roughly 6 billion tons per year. It's in our dental fillings and
food wrappings and store receipts. Previously linked to a host of
problems—elevated cancer rates, developmental problems in kids—
it's now suspected that even low doses of BPA can alter maternal care
in animals, causing rodents to nurse their babies less and otherwise neglect their duties. It's possible that the chemical affects human moms'
behavior, too.

Nobody's sure of the exact mechanism, but certain man-made plastics seem to somehow gum up the natural plasticity of mom brains.
BPA is an endocrine disruptor, which means that it may mimic or mask
a pregnant female's natural estrogen processes, vital to her maternal
transformation.

Now Champagne and Lapp are testing to see whether two other
closely related chemicals—bisphenol S (BPS) and bisphenol F (BPF),
which plastic manufacturers began substituting for BPA after a recent
public outcry—also impact rodent moms. The scientists dosed a second batch of cookies with these substances.

Over the course of the past three weeks, a quarter of a wafer at time,
the researchers have been feeding the two varieties of Nilla wafers to
two groups of pregnant rats. They also served plain cookies to a third
group, as a control. Injecting the chemicals would have perhaps been
more straightforward, but in designing experiments Champagne is always mindful of the animals' quality of life. (As a former pregnant lady
herself, she may have guessed that the expecting rats would appreciate
a carb binge.)

Lapp breaks a cookie apart to demonstrate. With little pink paws, a female rat stuffs it down.

Infrared cameras and credit-card-sized computers are trained on the rats' habitats. Using the rats' ears, tails, and other body parts as coordinates, the machines are learning to spot specific mom activities like nursing and grooming amid the general chaos. The computer will ultimately be able to assess the maternal behaviors according to cookie type, to see who is on top of her mom game, and who might be lying down on the job (and not merely to nurse).

In the lab's conference room, I take a long swig out of the only water source I could find that morning, a crumpled disposable water bottle, hastily refilled from my motel room's bathroom tap. It tastes like dead goldfish. Since it's obvious that I'm pregnant, I brace for a reprimand.

"I drink from plastic all the time," Champagne says calmly.

Teaming up with some Columbia University scientists, the lab is also investigating the chemicals' effects on human mothers. Using urine samples, the scientists will study whether the women's plastic contamination levels correspond to their neural responses on a battery of baby tests, to see if their mom minds have been modified.

Ultimately, Champagne and Lapp will examine the rat moms' brains after death, looking for changes in gene expression, especially related to estrogen and oxytocin. So far, plastic studies in rodents suggest that the medial preoptic area—that's the good old mPOA, the brain's lodestar of maternal responsiveness—is particularly changed.

Though the baby rats have already been born, the data from the pilot phase of the Nilla studies is just starting to come in. Lapp discusses next steps with Champagne, who, as the author of those epic epigenetic rat-licking studies, is perhaps the closest thing mom science has to a superstar. She's since become even more interested in how a whole slew of environmental factors might influence moms' gene expression, invisibly shaping our behaviors.

"I don't know if this is interesting or relevant," Lapp says shyly to Champagne, "but here is the weight gain." Lapp pulls up a graph on her computer, with three color-coded lines, one for each rat cookie group.

The rat moms gained the same amount of weight until day six, when the different chemicals entered their diets. Suddenly, the colored lines diverged, one rising, one sagging, one somewhere in the middle.

"Oh wow," Champagne says, gazing at the charts.

Since the experiment is still blind, nobody yet knows which plastic ingredient is having what effect. But it's already clear that these chemically steeped pregnancies defy the norm, in ways that will likely have cascading effects for the moms.

The sturdy maternal instinct has soldiered on for tens of millions of years, driving wildebeests and manatees and marmots alike, encoded to a degree even in mammalian males.

But while our motivation to care is deeply ingrained, we moms are also exquisitely sensitive to environmental conditions beyond the nest, the den, or the condominium door.

This plasticity is, for the most part, a gift. Though mom bodies are ponderous, our brains can corner on a dime. Moms share a mandate, not a memorized script. Motherhood is a compass, a star to steer by, rather than a single path.

The built-in flexibility of the maternal instinct helps us thrive in all types of environments. Our transformation is never quite complete. One type of mother may become another, as circumstances demand. This allows us to rise to any number of occasions, usually making do.

But the world can also warp us against our wills.

That's why the smug "this is who I am and always will be" identity politics of motherhood make no sense. Having sent one of my children to a Waldorf preschool where moms churned their own butter, and another to an urban co-op where some moms got by on expired grocery donations, I marvel at maternal adaptability. Sometimes cloth diapers, free-range carrots, and two hours of pink-cheeked outdoor play even in the dead of winter just aren't in the cards. *Please stop*, I want to say to my fellow mothers (and sometimes to myself) when the judgments fly. *You have no idea who you could become.* Simple circumstances rather than some kind of magical banana bread–baking gene are a big part of what separates the passive-infanticide practicing mothers of Brazilian slums from the mommy-blogger paragons of hippest Brooklyn.

It's not just over the course of a human lifetime: if you look closely, a woman may become multiple mothers just on *an average day*, as our surroundings shift. The sleuthing scientists shadowing us as we run afternoon errands have shown how our behaviors fluctuate at the park versus the supermarket, or while giving baths versus changing diapers (which, not surprisingly, is a species-wide nadir in maternal sensitivity, with moms across the boards demonstrating "less positive regard").

Moms are twice as likely to spank in the evening as in the morning, maybe because our circadian rhythms change. But other patterns are probably more bureaucratic than biological. One study showed that Floridian moms are especially tempted to hit their five-to-eleven-year-olds on particular Saturdays—namely, the Saturdays after the kids' report cards are distributed. (Report cards really do seem to be a dangerous idea.)

"Don't ask what a gene does," the neuroscientist Robert Sapolsky cautions in his book *Behave*. "Ask what it does in a particular context." A whole host of contextual variables shape moms' emotions and behaviors. The environment is constantly reprogramming us in hidden ways, sometimes shutting down certain genes or dialing them up.

Man-made plastics, with their poisonous effects on moms' natural plasticity, are one of the more straightforward examples of how the environment can tamper with our genes, and the story is similar for other seeping chemicals. (Meadow-jumping mouse moms exposed to certain insecticides are unusually keen to eat their own babies, for instance.) But even the stuff in moms' regular diets may screw up their chemistry. In rats, and perhaps in people, too, eating a fat-loaded diet can heighten maternal anxiety, likely because all the cholesterol enlarges the adrenal glands that normally shrink during lactation. On the bright side, dining on fish or other heaping helpings of omega-3 fatty acids may help stave off postpartum depression.

As striking as these specific chemical pitfalls and benefits may be, all pale in comparison to the major but also most mysterious and thinly understood mode of environmental influence: stress.

∞

Stress isn't visible under a microscope. You can't just inject it or serve it up on a Nilla wafer. It also varies wildly in form and degree, from a stack of unsent thank-you notes to an outbreak of bubonic plague, and is highly personal. What stresses me out might roll right off your back. One mom's trial is another's trifle.

Nonetheless, overwhelming stress of the right type can challenge moms' built-in fight-or-flight systems, changing our behavior— sometimes forever.

Environmental stress helps explain why a mammalian mom may abandon her babies even when there is absolutely nothing wrong with them. There is nothing necessarily "wrong" with the mom either, at least in the evolutionary biologist's book. She's doing what has to be done to pass down her genes into eternity, waiting for life to improve so she can try again with another litter.

Environmental threats take many forms, including nutritional shortages, predation and other modes of violence, or disease outbreaks, which can befall mammal mothers from the bottom of the food chain to the top. When the going gets tough, a black-tailed prairie dog mom simply gets going—in the opposite direction of her pups, betting that the local outlook will brighten in time for her next batch. Roughly one in ten litters are abandoned in this straightforward manner.

In the worst of times even the mighty lioness will simply walk away from her mewling cubs and never look back.

Many human moms reading these pages are lucky enough to be buffered from some of the physical challenges that our mammal cousins face. But humans have unique stressors to contend with, too. Not long ago, the Yale Child Study Center set out to find the biggest single source of stress for low-income Connecticut women—that is to say, the environmental factor most consistently correlated with postpartum depression.

The danger they identified wasn't existential or mortal. It was diapers. Lack of access to disposable diapers, which were invented only in 1948, most strongly predicted poor moms' degraded mental health— more even than anxiety about food.

This puzzled me at first. Haven't we seen that new-mom brains are specially set up to withstand stress? The automatic postpartum downshift of the maternal stress system is, I'd thought, a signature flourish of the maternal transformation. Moms may look a bit disheveled on the outside, but inside we stay cool as cucumbers while other people melt down. Ice water runs through our varicose veins. That's how we clutch car seats tight during tornados, beat back bears with baseball bats, and flag down the first cab in the middle of an earthquake.

Once, zoned out on the couch a week or so after I'd come home from the hospital with newborn Daughter Number Two, I calmly noted that our dining room chandelier was on fire. (Note to future self: Never buy light fixtures that are partially constructed from cardboard, even— perhaps especially—if they are on sale.) No hoarded scrap of wisdom from that long-ago professional babysitting course consciously came to mind. Somehow my body just hopped off the couch, flipped off the relevant light switch, found the baking soda to sprinkle on the blaze, and, as the smoke cleared, hustled outside to hail the fire department.

I recently saw a video clip of an Arizona mom in a house fire far worse than mine. It took me an eerie moment to realize that this woman's world was ending while I watched. Flames soared around her as she hurled her toddler to men waiting below her balcony. Her own body was already burning by that point, but she seemed to have more important things on her mind, charging back into the blaze for a daughter still possibly trapped inside. Of course news reports focused on the guy who caught the toddler, but her name was Rachel Long. She never came out.

Come hell or high water, we moms take disaster in stride. And yet we are undone by . . . lack of diapers?

The stresses that threaten moms the most, it turns out, are very often *not* events like fires and earthquakes. We're built to handle sudden catastrophe. What messes with moms are creeping, chronic, and often invisible problems. Poverty. Hunger. Diapers. Make no mistake: moms can be wonderful in all types of situations, and the threshold for a modern human woman to forsake a child is extremely, almost imaginably high. But even if most will stay the course, parts of us may yet go missing.

Mothering under constant duress can disrupt our most fundamental maternal habits. Recall that cradling our babies on the left side is a nearly instinctive tendency, perhaps the closest we humans have to an automatic mothering behavior.

But one study showed that stressed-out mothers are more likely to trade arms when cradling babies, switching over to the right.

∾

When a neuroscientist wants to stress out a rat mother, he may remove her bedding, or dangle her by her tail.

David Slattery of Germany's Goethe University prefers what's called the "restraint test." Here, a mother rat is taken from her pups and placed in a narrow Plexiglas cylinder, where she can see and breathe fine but not move much.

This doesn't just happen once. She is put back in the tube over and over again.

The mom might shake off the experience at first. But after a few weeks of undergoing this harmless but jarring psychological stress once a day, a mom isn't the same animal anymore.

For one thing, reunited with her pups, she nurses 30 to 40 percent *more* than unstressed moms, seemingly seeking to soothe her jangled nerves with an oxytocin rush.

For another, placed in a maze, the chronically stressed rat doesn't act like a brazen mother, who under normal circumstances will boldly go where no rat has gone before, taking advantage of her muted stress response to beeline into the brightest and most exposed arms of the labyrinth, where she might find untold delicious tidbits for herself and her wee ones.

Slattery's wigged-out mothers cower in the maze's dark corners instead. They aren't bold and brave. They act just as wimpy as females who have never given birth.

"This was very surprising to us," Slattery says. "We expected that the mothers would deal with it better, but this turns out not to be the case. If there is too much stress, the systems that are in place to protect the mother are overcome."

Combing through the stressed-out rat mothers' frazzled brains after death, the researchers found startling signs that the unpleasant stint in the Plexiglas tube had physically eroded their maternal anatomy.

In regular rat mothers, for instance, the hippocampus stops creating new brain cells during pregnancy. This temporary stunting of the memory powerhouse is a hallmark of mammalian motherhood that may help explain human moms' many embarrassing mental bloopers, like blanking on the name of your favorite new mom friend, or losing the grocery list en route to Stop & Shop. These moments probably reflect some kind of adaptive trade-off, as brain circuits in other zones— like, say, the olfactory bulb, responsible for processing those suddenly yummy baby smells—grow stronger instead.

In a nervous-wreck mom, though, "this normal physiological change is reversed," Slattery says. Her hippocampus looks downright virginal.

The researchers detected more abnormalities in the stressed moms' brain tissue, including changes in gene expression possibly linked to lowered oxytocin production—which may be why they were so frantic to nurse. Maybe stress-related neural changes help explain why wild mothers in captivity, such as snow leopards, have been known to neglect their young.

Some chronically overburdened rodent moms maintain their freaked-out behaviors long after the stress has passed—even if, say, they never see the inside of a Plexiglas tube again. The maternal circuit may permanently weaken, as stress systems strengthen. The mom brain doesn't mature as it ought to. "Their maternal behavior is going to be different," says Elizabeth Byrnes of Tufts University, "even when they have their next litter."

Rodents don't deal with mounting medical bills and past-due rent. But Danielle Stolzenberg recently stationed a cage of her cosseted lab-rat moms in a remote California nature preserve, to see how they fared when exposed to compounding natural stressors such as drought conditions and forest fire smoke and the intimidating presence of wild turkeys.

"Somebody ate the camera," she said, a bit darkly, when asked how her work on the wild side was progressing.

Sometimes it's easier to look outside the cage entirely.

∽

Perhaps the most obvious and age-old environmental stress for mothers in nature is food availability. Dozens of our mammalian kin, from roe deer to brown bears, will not even reproduce in the first place unless there is the right amount of food around. Their reproductive tracts feature a nifty safe-deposit-box–like structure called a "uterine crypt," where they can stash fertilized embryos indefinitely in a state of suspended animation, not progressing in pregnancy until the berries on a nearby bush ripen or the environment otherwise sweetens to meet their standards.

Human mothers don't have this handy-dandy adaptation. Still, we can't grow babies or make milk without extra daily calories or else a goodly stockpile of cellulite—this is why a girl's body fat rises by more than 200 percent right around the time she starts to menstruate, and why scientists insist that well-stuffed mom jeans are a nice asset indeed.

These biological realities ensure that human mothers remain highly sensitive to environmental cues, especially in parts of the world where they still live hand-to-mouth. During the Bolivian planting seasons, when heavy manual labor burns through women's energy stores, farmers' wives lose four times as many pregnancies. In Ethiopia, where the circumference of nursing mothers' upper arms shrinks as the time since the last harvest lengthens, a few extra available calories can precipitate a baby boom. Installing a new plumbing system in one rural corner of the country, for instance, meant that women no longer had to burn their bodies' fuel trudging to far-off wells, and they had more babies.

Well-nourished mamas behave better, too. Cheetah moms with full bellies spend more time teaching their cubs to hunt. Meanwhile, hangry mammalian moms shirk their duties, with ewes on a calorie-restricted diet suddenly more apt to neglect their lambs, straying farther from them in the field.

Hungry moms may even get lazy about defending their young against predators. One scientific paper introduced this phenomenon with the slacker-mom title "Maternal Defense in Columbian White-tail Deer: When Is It Worth It?" To find out, scientists used hunting dogs to track does and their fawns. One factor that determined what happened next, the researchers found, was the mom's bodily condition and the food supply that year. The well-fed deer moms tended to stand their ground over their fawns, sometimes attacking the researcher's dog outright, ears back and hooves flailing.

The hungry ones, though, just snorted indignantly from the bushes.

∾

I've already hinted that I'm something of a coward when it comes to the study of how something as preventable as hunger might obliterate the tender mom-baby relationship. It's a surefire sign of my privileged station in the world and my pantry stuffed with chocolate chip pancake mix, but I almost can't imagine human beings forced to discard their children in hard times. In truth, I don't really *want* to imagine it, let alone understand it.

Yet these horrors help explain mom behaviors even in less extreme straits. It's a fact of life that our mothers understandably skipped.

To confront what maternal betrayal might look like on the simplest level, I stop very, very briefly by a University of Connecticut animal behavior laboratory where the evolution of motherhood is viewed through a more comfortably distant lens. Steve Trumbo studies parental care in insects, perhaps motherhood's most basic animal models. Involved bug moms are pretty rare, occurring in about 1 percent of known species. But certain creepy-crawlies, including cockroaches and earwigs, are dedicated mothers with neurochemicals not unlike our own.

Trumbo focuses on burying-beetle moms, who finesse a dead mouse on the forest floor into a kind of slimy meatball, laying their eggs in the nearby soil and raising their babies in the carcass.

Trumbo's lab is full of mouse meatballs in various states of production and decay, encased in Tupperware containers. He opens one box, and a nauseating stench drifts out. Naturally he waits until now to tell me that he lost his own sense of smell years ago in a pickup basketball–related concussion. Numerous rides on the dung beetle at the Bronx Zoo Bug Carousel have clearly not prepared me for this moment. I take a big step backward.

We tour another nearly pitch-black room full of bugs, the chinks beneath the doors stuffed with blankets to keep out the light. If you've seen the last scenes of *Silence of the Lambs*, this spot might look a little bit familiar.

Maybe it's a fitting place to find mothers' hearts of darkness.

Trumbo and I try not to breathe much, because while the beetles relish dead-mouse stink, they can't stand human breath. Pale and roly-poly, the bug babies remind me of swaddled human newborns. We strain our ears to hear the beetle moms scraping their wings over their abdomens, making a kind of love song to summon their fat, smooth little ones. "It's really soft, like a lullaby," Trumbo whispers. He shows me how a burying-beetle mom feeds her larva, twirling her antennae ecstatically as she lifts each pale grub to her mouth to deliver what looks like a smooch. Really, she's vomiting liquid mouse carrion into its mouth.

But sometimes this mother's kiss turns deadly, and she will shove the whole wriggling baby into her own open jaws like a foot-long sandwich.

By now it's almost second nature for Trumbo to sense just when these ghastly bouts of cannibalism are coming.

"It's basically math," he says. "A mouse of x grams can support y babies." Moms who get stuck with skinny mice must cull their offspring.

"Probably she won't go on to care for all of these," Trumbo muses as we pass one Tupperware container, its beetle mother standing athwart her meatball, babies squirming all around. "Probably she's going to have to kill some."

I can't tell if it's lingering mouse odors or mere ideas that are turning my stomach.

∾

Some mammals, highly sensitive to resource scarcity and other signs of a less-than-rosy future, practice a similar cannibalism, like the hamster who gave birth in my childhood bedroom, and—apparently not finding the environment on the top of my dresser to be a particularly promising one—ate her babies one by one, recouping her investment of precious protein while sometimes leaving just the pastel skins behind. (This long-remembered event explains why, when it recently came time for, er, Santa Claus to procure Clementine for my own kids, I relentlessly grilled the pet-store manager about his hamster husbandry practices, demanding what amounted to a virginity warranty.)

Humans are a lot more complex than beetles and hamsters, and most maternal organisms reading these words are probably pretty well insulated from the vicissitudes of a calorie-restricted lifestyle, with meatballs aplenty, unless we happen to be on some kind of hellacious diet. Yet our mom biology still responds viscerally to the experience of feast or famine. In the developed world, that might mean an economic downturn or a financial blow, which may modulate maternal behavior somewhat like the failed crops of old. Even when starvation isn't in the cards, our reactions remain gut-level.

I've come to think of this sub-discipline as Freakomomics.

Economists have long understood that the birth rate and the economy are intimately linked—a $10,000 increase in average housing prices leads to a 2 percent drop in renters' birth rates, and so on. (Conversely, regional windfalls like fracking booms can goose birth rates.) This isn't just a matter of conscious planning and practicality: scientists now suspect that women under psychological stress, including financial stress, are less likely to conceive even if they're trying to get pregnant. And for women who are already pregnant when financial disaster strikes, there may be an unseen, automatic downgrade in their investment in their unborn child. A study of unemployment rates in Denmark from 1995 to 2009, for instance, showed that jumps in joblessness corresponded with a national uptick in miscarriage rates. Although there was a matching rise in the abortion rate, some women's bodies—sensing long-term hardship on the horizon—seemed to cut short their pregnancies without any outside intervention or conscious choice.

"This idea that your decisions and your biology are separate is a false distinction," says Tim Bruckner of the University of California, Irvine, a leading expert in this grim field.

The looming threat of hard times also affects babies who do get born. In 2005, the US government unveiled nationwide plans to close military bases, a move that raised local unemployment in some areas by 20 percent. All of a sudden babies in those places started being born earlier, a potential sign that maternal bodies were investing fewer resources in their young. (Somewhat more hopefully, the Covid-19 pandemic touched off the opposite phenomenon—an unprecedented *dearth* of premature babies, as women's bodies apparently tried to hold off as long as humanly possible until danger passed. The doctors were shocked, but we moms have probably been pulling similar stunts since ancient times.)

Amid maternal stress, babies start shrinking, too. One study calculated that the announcement of five hundred layoffs is associated with a nearly 20-gram drop in newborn birth weight in local hospitals. Likewise, during the recent subprime loan crisis, foreclosed-upon moms bore lighter-weight babies than average.

Bruckner has published several rather soul-shattering papers linking economic ills with fatal maternal actions. He believes that sudden infant death syndrome, or SIDS, the medically opaque phenomenon of "crib death," often has clear economic precipitants. During California's unemployment surges, he's noted, SIDS deaths increase above expected levels, as mothers perhaps neglect to remove pillows and other unsafe items from cribs and become more apt to place their sleeping babies on their stomachs, against doctors' advice.

In fact, when a city's economy crumbles, child deaths rise in all kinds of accidental ways, as moms may fail to monitor bath time and secure car seats. Bruckner calculates that a 1 percent drop in a California city's employment predicts an 8 percent increase in "infant mortality due to unintentional injury" that same month.

Note the "unintentional" part. Not even Bruckner believes that America's stressed-out moms are intentionally icing their infants. Rather, he supports what he calls "the distraction hypothesis," which he also thinks explains why women are less likely to detect early-stage breast cancer lumps during times of economic stress. Our minds are simply elsewhere, mentally grappling with interview questions or agonizing over the heating bill. Perhaps the mom has forgotten to tell the new babysitter how to tighten the high chair straps. Perhaps she's frantically uploading an overdue job application, unaware of the eerie silence in the nursery.

This kind of stress, as any mother knows, isn't just a simple matter of a satisfied or empty belly, or the presence or absence of specific financial resources. It's about the frightening experience of uncertainty—the psychological stress of not knowing where your next meal or paycheck will come from is enough to hobble maternal behavior, even if mom and baby both get enough calories in the end.

In the 1980s and '90s, scientists conducted a now-famous series of experiments on captive macaque mothers. Instead of receiving their usual straightforward monkey chow rations, the moms were presented with "foraging carts" full of wood chips, their lunch hidden inside. Reaching through holes cut into the sides of the wheeled carts, the moms had to hunt around for their food as they would in the wild.

There were two types of foraging cart. One type was bountifully laden, with monkey chow strewn about the bottom. The other was more sparsely filled, with its monkey chow well concealed under wood chips. The monkeys who got stuck with the latter had to work harder to find their meals than the others.

But, fascinatingly, it wasn't the hustling monkeys with the poorly stocked carts whose mothering nosedived. (Despite all the food hoopla, nobody ever starved.) It was the moms who got an unpredictable mix *of both kinds* of wheelbarrows, on a schedule alternating every two weeks. These monkey moms, experiencing a bonanza one day and a bust the next, were the ones who fell apart, experiencing a more than 25 percent increase in their stress hormones and swiftly disintegrating caregiving.

"Even though there was no caloric shortage, the perception that there might be was there," says Jeremy Coplan of the State University of New York Downstate, who continues to work on these experiments today.

While the hardest-working mothers could anticipate what their day would bring and could even build in time at the end of their labors to cuddle the infants they'd earlier had to ignore, the mixed-up wheelbarrow group found no such rhythm. These stressed-out mothers became obsessed with patrolling the wheelbarrows and were measurably less affectionate with their babies, to the extent that the babies sustained cellular damage. The ends of these infants' chromosomes, called telomeres, were shorter than they should have been, a sign of stress and premature aging.

This unnerving shift from living high on the hog to scrounging is called "variable foraging demand." And it suggests that what mammalian mothers should fear most is fear itself.

∾

The variable foraging demand experiments hit close to home for me. In fact, they take me all the way back to the kitchen table one morning in 1987. My dad is staring at the *Wall Street Journal.* A black line—the stock market, crashing—crawls down the front page. Whenever I see graphs of a certain plummeting shape, whether they're tracking rainfall or heart rate, I get flashbacks to that Black Monday, when the market lost today's equivalent of 5,000 points.

My dad, who'd spent his career on Wall Street, was never able to earn money again, although he tried various other jobs. Over the course of a few years, we went from owning a stately brick home on the edge of a golf course to renting in the poorest condominium complex in town. My baby doll collection, porcelain-skinned instead of plastic, became a relic of our vanished riches. One day there was a Mercedes in the driveway and a mink coat in the closet, and the next we were rationing paper towels.

My dad had wanted to be an architect, the guy who drew the lines instead of riding them all the way down. Depression had long been a problem for him, and for people on his side of my family, and his exit from the workforce was far more complicated than just one bad trading day. But to my childish eyes, his mental health mirrored the cratering market.

He passed away, suddenly, when I was in middle school, and he spent a lot of his final years in a kind of daze. He gained so much weight that it was hard to remember that he had once been the football team quarterback and the fastest guy in his high school, although—let the record unfairly show—all his recruitable-college-athlete genes totally bypassed me and went straight to my little sister.

Big as he was, Dad still had fluted slender ankles, like a buffalo's.

Once when the wind ripped a kite out of my hands on Block Island, during perhaps the last family vacation we ever took, he chased it.

It was startling to see a large man run so fast, almost as if he were flying away from us.

∾

Maternal stress can also manifest itself through withdrawal. The anthropologist Robert Quinlan uses, among other tools, a heartbreaking yardstick to measure this emotional retreat: the distance that sleeping mothers maintain from their babies when both go to bed for the night. In places with lots of rampant diseases, the moms who don't ditch their kids outright may slumber farther away from them, a physical gulf that is also, perhaps, a measure of detachment.

Like disease, wartime encourages psychological distancing: in one fascinating study of the brains of Israeli moms who'd lived for years near the Gaza border under the constant threat of rocket fire, scientists detected less activation in regions related to social interaction and empathy, both involved in mothering. Any crisis, really, that flips from acute to chronic could foster this diminishment. As we've seen, moms excel during earthquakes and other out-of-the-blue natural disasters. But we may pull or drift away from our kids in the aftermath, especially if order takes a long time to be restored.

More than a year after an 8.0 earthquake rocked China's Sichuan province, pregnant women suffered from unusually high levels of depression. In the wake of Japan's Fukushima nuclear power plant meltdown, young mothers without physical symptoms were particularly prone to severe mental health problems—and in fact, another analysis found that young mothers suffered more emotional anguish than anybody else in the disaster zone, except for workers tasked with cleaning up the radioactive waste. Moms in refugee camps frequently have trouble breastfeeding, and post–Hurricane Katrina, infant death rates swelled for months as overwhelmed mothers floundered in unaccustomed chaos.

The lingering impact of unprocessed trauma of many kinds can be enough to create unhealthy distance between a mother and her children. And this doesn't just happen after once-in-a-lifetime natural disasters or way off in distant lands. There's a large and often hidden population of American mothers who've previously experienced trauma as a part of routine life—neighborhood gunplay, sexual violence, a death in the family, domestic abuse, or chronic neglect.

"There's big-*T* and small-*t* trauma," says Sohye Kim of the University of Massachusetts Medical School. "It doesn't have to be a war or an assault. It can be much less dramatic—a subtle but repetitive and sustained interactional pattern of emotional instability with someone important, especially a caregiver," a category that includes lots of people besides your own mom. In adult life, Kim says, a woman can address these painful past experiences through therapy, thoughtful conversations with loved ones, or other forms of self-reflection that bring these deep-down feelings to light.

If, however, there is no such reckoning, the old wounds may open when a woman gives birth. Thanks to the genetic grab bag of history, the faces of the dead have a funny way of resurfacing in our children, in the encore of a great-grandmother's dimple, say, or a long-lost uncle's chin.

Today I can watch my father watching television. His eyes are shining on the couch, in Daughter One's face, and his mouth—Daughter Two's—is laughing from the rocking chair.

Childhood trauma can, decades hence, recast the way a mom's genes get expressed and reshape her brain architecture. When Kim and her colleagues studied dozens of seemingly run-of-the-mill, high-functioning, middle-class American moms, the ones who had buried trauma in their past had amygdalae that looked different in the fMRI scanner when they gazed at pictures of their own babies. The amygdala is important for processing emotions and detecting relevant environmental cues. It's "what signals us that this is important, you have to pay attention," Kim says. A mother's amygdala should sizzle at the sight of her own baby's sad face, as part of the instinctive maternal response.

"That's very significant, because babies need their mothers most when they are distressed."

But the brains of the traumatized moms didn't react. "That area of the brain," Kim says, "is blunted."

This dulled neural activity, she believes, is an evolved self-protection mechanism, buffering moms from emotions linked to hurtful memories. "Yes, it's problematic from the point of view of sensitive caregiving. But it's very conducive to maternal survival. . . . It's incredibly adaptive."

So it serves a purpose, even as it exacts a cost.

⮵

Lucky for me, my mom did not drift away. When my dad got sick, she found a way to stabilize our world before it vaporized. She'd stayed at home for seven years—in other words, my whole life—but now she tracked down a job as a schoolteacher. People assume that moms make good teachers because they enjoy the company of children so much—the maternal instinct, and so on—and yet anybody who ever heard my mother cheer for a freshly declared snow day would know that's not necessarily the case. Although she was devoted to her sixth graders, what she found amid all the bulletin-board decorating and preposition drills was a solution to the variable foraging problem: a highly predictable schedule and generous benefits.

In our case, a teacher's salary wasn't quite enough to cover all our losses, so Mom also took on a newspaper route, waking up at 4 a.m. to drop the *New York Times* in driveways around town before anybody else's parents were up.

She did these things out of love for my sister and me, and we weren't particularly appreciative. We took it as our due, and our absolute faith in her was probably a sign that we had, for our whole lives, been loved by both our parents. But now as an adult I see that there were other potential paths, and other possible selves, with far less stabilizing outcomes.

In the middle of the night once, when we still lived in the big brick house but things had already taken a turn for the worse, Mom heard footsteps and then a loud sneeze downstairs. Our fancy house had been burglarized before. She didn't cower under the covers until the police came to confront the home invader; instead, something that I now recognize as maternal instinct took over. No more harm would come to us, if she had anything to do with it.

She strode to the top of the staircase.

"GET OUT OF MY HOUSE!" she hollered over and over, smashing an empty laundry basket against the railing for emphasis.

Before you pity the poor burglar too much, know that it was actually me, sleepwalking. The pale halo of the policeman's flashlight is my earliest memory.

∽

In addition to her natural tenacity and problem-solving skills, of course, my mom's socioeconomic background and her college degree were also very much in play here. A poor mother doesn't typically have a debt-free teacher's credential in her back pocket. And almost all the risk factors for maternal struggle are heightened for impoverished moms. They give birth at younger ages (a "now or never" life strategy that may in itself be a response to environmental instability), have higher C-section and lower breastfeeding rates, and even have intensified exposures to endocrine-disrupting chemicals and other pollutants, which may leach from nearby landfills or from the plastic-packaged foods that crowd their diets.

Poor mothers are more likely to have been exposed to substandard care as children and to be living with past trauma. They may give birth to smaller babies who have associated medical problems. They are twice as likely to be treated for postpartum depression.

Even just living in overly crowded conditions can make moms less responsive to their kids. An environmental factor as simple as an infestation of cockroaches can boost a woman's depression risk threefold.

And then, to top it all off, low-income moms already under chronic stress are extra vulnerable to sudden and unexpected shake-ups of all descriptions. They are the ones who get stuck with the crummy mortgage, or the house in the flood zone. Their incomes are the first to falter in an economic lurch. They have the least say over tomorrow.

Not surprisingly, poverty and its stresses may take a toll on the maternal brain itself—or so suggests the work of the University of Denver's Pilyoung Kim, an expert on poverty's impact on maternal behavior. As with long-buried maternal trauma, there's atypical activity in the amygdala—but instead of being blunted, like those of the traumatized middle-class moms, the poor women's amygdalae often seem *overactive*, burning more than average on brain scans at the sight of upset infants, their stress systems cranked into hyperdrive.

Perhaps being extra-stimulated by kids' sadness is a type of emotional triage. "If my environment is not stable or predictable," Kim says, "maybe it makes sense for me to pay more attention to distress cues, like a cry versus a smile, as a way to protect my child."

Not every poor woman exhibits these changes, and the differences between poor and middle-class mom brains by no means suggest that those differences are innate, as the eugenicists of past ages might have assumed. Quite the reverse: these physical variations are clear examples of how our material circumstances shape maternal biology, with poverty becoming a mind-body menace for moms.

"Everybody has stress," Kim says. "You don't have to be living in poverty to have stress. But for people with higher socioeconomic status it's rarer to experience multiple things at the same time. Interviewing moms in poverty, I cannot imagine how they cope with the stress of a newborn."

All things considered, perhaps it's not really so surprising that an American mother's sanity can hinge on the state of her diaper supply.

There's another way to look at the behavior of a mom who cannot curb or escape her stressors. Instead of being subpar or insensitive, she may be strategic, preparing her offspring to play the difficult hand she and her child have both been dealt. The implicit judgments sometimes lurking in mom science, in the very idea of defining optimal maternal behavior, may not always take this possibility into account.

Perhaps the most fascinating example is the way that overtaxed moms "decide," unconsciously but still in a somewhat calculated manner, whether to have a boy or a girl.

Wait a minute, you say, *doesn't the father's sperm determine the baby's gender? I, too, have taken tenth-grade biology!* Indeed—all of a mother's eggs contribute an X chromosome to the baby, while the dad's swiftest X- or Y-bearing swimmers each have roughly 50-50 odds.

But that's not the end of the story, because the mom's body scraps about half of all pregnancies after fertilization. Fetal sex appears to figure into this hidden cull, with our bodies offering safer harbor to boys or girls *depending on environmental cues.*

When the outlook is rosy and the mom is stress-free and in shipshape condition, according to some evolutionary biologists, her body is primed to favor sons. Boys are bigger, feebler, and more taxing to gestate, but—in good times, at least—they can later pay evolutionary dividends if they grow up strong and strapping, woo widely, and sire a bumper crop of grand-offspring.

Baby girls, on the other hand, may be the smarter play if a mom's world is wobbling. There's less of a physical and energetic down payment up front, and while daughters probably won't procreate like a Casanova or a Jagger, they are more likely to cough up a couple of grandchildren even in difficult environmental circumstances.

Maybe being part of a multigenerational matriarchy, as I am, isn't always an accident.

The Trivers-Willard hypothesis, as this evolutionary cipher is called, is still somewhat controversial, and doesn't hold true on a mom-by-mom level: obviously many stressed women have boys, and many chilled-out women have girls, if you look at their individual stories. The world, after all, is populated by roughly half of each.

But when larger populations are screened, there are strong indications that the hypothesis is on to something. In one recent Columbia University study of two hundred new moms with varying degrees of stress, nearly 70 percent of the most emotionally and physically maxed-out women ended up bearing girls. A different analysis of 48 million recent American births, from another group at Columbia, found that married, better-educated women bear more boys. And there is interesting evidence that some exceptionally well-situated women, specifically billionaires' wives, can be counted on to produce male heirs about 60 percent of the time. These patterns can figure beyond birth far into a child's future, with poorer parents investing more in daughters and richer ones in sons, on measures ranging from the amount we spend on elementary-school backpacks—poor families apparently splurge on fancier models for their daughters—to advanced degrees, which well-to-do sons seem to collect more of compared with their sisters.

Of course, stress is highly subjective and humans tend to acclimate to baseline levels—so scientists also like to study large and small environmental *changes* that impact women across social classes. For instance, amid recessions, some boy–girl spending patterns seem to change, with families betting bigger on daughters by suddenly plunking down more cash than usual on certain items, like girls' clothing, even though money is tighter than before. Meanwhile, boys are more likely to be born by C-section during these times, suggesting higher levels of fetal distress.

Or not to be born at all. After 9/11, scientists noticing the subsequent baby-girl skew in Manhattan's births initially blamed the toxic dust cloud that enveloped much of the island, which certainly may have been a factor in picking off the frailer male fetuses.

However, it turned out that the baby-boy bust reached all the way to California, affecting moms who'd simply watched the horrific footage on television. Stress alone was poison enough. Perhaps for similar reasons, the fraught aftermath of the bloody Paris terror attacks in 2015 led to a baby-girl heyday in France. Further analysis suggests that, compared to their never-born peers, the male fetuses that do make it to birth during these stressful periods may score unusually well on health and cognitive tests—as if they somehow have what it takes to run the gauntlet.

Studies show that even subtly stressful factors, like living in a parasite-infested area or in a neighborhood with lots of air pollution, or getting pregnant in unseasonably hot or cold weather, can slightly reduce a woman's odds of having a boy. So, apparently, can skipping breakfast, perhaps since this acts as a false signal to your body that local resources are dwindling, turning the ancient screws of stress, when in reality you just didn't want to pack on seventy pounds again.

It gets weirder, though. Stressed moms don't just customize fetal sex to environmental cues. They can reach deeper into their bag of tricks to tailor-make their kid's temperament, too. I've said that we can't edit our children's personalities, try as we might, and yet we may alter their natures in unintentional and subconscious ways, customizing them for a kind or cruel world.

In rats, this kid-engineering happens after birth, as we've seen, through physical touch, and how much moms "lick and groom" their babies. We naturally feel a little sorry for the low-licked rat babies, who may not grow into mothers who approximate our modern sentimental ideals. But these babies are perhaps better adapted, via gene-level changes to their fight-or-flight systems, to thrive in an especially stressful environment.

For mammals, breast milk may be another stealth mommy mode of influence. Primate milk varies in nutritional and hormonal content, and like soda jerks working the nozzles, moms can make many different kinds according to environmental signals, influencing a child's growth patterns and personality through the contents of an every-three-hours dairy drink.

This is called "lactational programming." Funneling a kid extra stress hormones, like cortisol, in your breast milk may result in a "more nervous, less confident" baby, born braced for disaster. In monkeys, these high-cortisol babies grow unusually quickly, "prioritizing" growth instead of social exploration, perhaps to up their chances of clobbering their unfriendly neighbors.

Whether it works the same way in humans is less clear. But one human study has linked children's temperaments to mothers' milk–borne stress hormones. And some scientists, like Bruce Ellis of the University of Utah, who runs something called the Hidden Talents Lab, have argued that stress-adapted kids may sport dispositions suited to succeed in dicier corners of the planet, including rough-and-tumble neighborhoods in our own country. A mom who finds herself in a hard spot—whose brain, say, over- or underreacts to the sound of her wailing child, causing her to act harsh or distant, or whose breasts gush milk that's full of cortisol—may be doing her absolute loving best for that kid.

"The mom's behavior isn't necessarily the wrong behavior," explains Elizabeth Byrnes of Tufts. "It's the right behavior for the wrong environment."

Maybe a spoiled upper-middle-class "special snowflake," whose fawning mommy's brain passed every "responsiveness" scan with flying colors, wouldn't have a snowflake's chance in a poor child's version of reality.

Ever since Emily informed me that she grew a new heart, I've heard other mothers deploy the same metaphor: a child is a second heart that can crawl and then toddle and tricycle and rollerblade about outside your body. Sure, the pavement is always rough. But what if the world outside your skin is full of land mines, or sexual predators? Wouldn't you strive to harden that defenseless little fluttering organ however you could? A mother creates a child who reflects her life experience. And that child, in turn, continues to shape the mother, cementing the feedback loop.

Poor moms' environmental responses may not just be acceptable—they may be smart.

"The concepts of 'good parenting' and 'bad parenting,' independent of context, are illogical," Ellis and his colleagues have written. "Instead, high and low effort parenting strategies are conditional; that is, different strategies are adapted to different social and ecological conditions."

❧

Ellis slipped a new word in there: "social." For while the wider world, with its natural and man-made perils, indisputably shapes moms, the networks of fellow human beings surrounding us are perhaps the mother of all mom forces.

I didn't pick up on this in a scientific journal or a book. I learned the hard way.

Chapter 9

NO MOM IS AN ISLAND
Unmaking a Mom

FETUS NUMBER Four's first kicks are tiny but concussive, as if my uterus were one of the endlessly battered free balloons my kids collect from the shoe store. Every time I stand up or sit down, I issue an Old Faithful–like belch.

So this pregnancy is progressing just as it should be. I'm only about twenty-four weeks along, but thanks to the scheduled C-section, the likely birthday has been on the calendar for months, and as a repeat surgical customer, I don't have to worry about a lot of weird stuff, like rolling an enormous "birthing ball" into the minivan's becrumbed back end. Especially given what I now know about the material realities of the maternal instinct, I ought to feel confident that the doctors will do their jobs, that nature and hormone receptors will take care of the rest, and that once the new baby is snug in my arms everything will once again fall into place.

Only last time it didn't.

My third child was conceived during a personal high for my husband and me: our careers were going better than we'd hoped, and we'd just made the somewhat out-of-the-blue decision to trade our compact row house in Washington, DC, the value of which had miraculously increased even as the walls seemed to shrink around us and our two rampaging preschoolers, for an idyllic antique farmstead in the far-out suburbs of Connecticut, just a few miles from my childhood home. It was not really a mansion, because we were journalists and not financiers, but it was pretty close—and truth be told, it had been a lifelong goal of mine to get back in the big house again, to shower my own kids with every conceivable material comfort, and to raise them in conspicuous view of all who'd observed my family's downward arc.

I wanted this house so much, in fact, that no amount of long, meaningful stares and pregnant pauses from the home inspector made any difference to me as we discovered this rotting post and that one, and my husband followed him down a green tunnel in between the rearing bushes that threatened to engulf the sun porch. The previous owners— who actually *were* financiers, come to think of it—had obviously died a slow fiscal death here, as Mother Nature took her course and then some. But for us, I thought, it would be merely romantic.

Baby Number Three was the capstone of our perfect plan. And while we didn't know it yet, this latest addition was a first for our family, a boy. Science might see this male-in-the-making as my body's subtle vote of confidence in our mutual environment and the promise of the future.

Then, on the very morning I proudly displayed the still-damp pee stick, mere days after that shaky house inspection had sealed the deal, my husband discovered a small red lump on his neck.

This bump, diagnosed at our local walk-in urgent care facility as "just a boil," soon vanished, but the symptoms that followed didn't. Three months, waves of pain and insomnia and phantom heart attacks, and a dozen doctor and emergency-room visits later, nobody could figure out what was wrong with him. Cardiologists, gastroenterologists, neurologists, and rheumatologists shrugged their shoulders. The psychiatrists had some definite ideas, but their prescriptions had no effect. Instead, my once-jovial and perpetually optimistic partner—our household singer of lullabies and clipper of toenails, my old high school debate foe who'd turned out to be my life's companion—withered away before my eyes, losing forty pounds and gaining a new personality, frantic and pain-wracked and tearful.

As vulnerable pregnant ladies go, I was in a fairly strong position to withstand these trials. This was not my first birth, so my maternal machinery was mostly in place. I was financially insulated (although every day less so as the costs of our "farm" mounted and my husband's capacity for work dwindled), well educated, a seasoned former babysitter and a breastfeeding champ, raised by a good mother who had always loved me, in excellent health, and with two beautiful daughters. I was happily married and, at thirty-five, plenty old.

However, I also had a few hidden risk factors for peripartum mood disorders, like that impressive family history of mental health issues scrawled deep in my genes somewhere, some medium-T childhood trauma (especially my father's early death), and a pattern of delivery by C-section. This time, too, there would be the unexpected problem of stingy anesthesia and rampant postpartum pain to boot.

Although this research tidbit hadn't been published when I was carrying my son, the birth of a third child, in particular, seems to herald a slight uptick of maternal mental woes, especially if the new kid is a different gender from the first pair. And having a boy ratchets up a mother's depression risk regardless.

But the most immediate threats to my maternal behavior had to do with factors beyond my own brain and body. In my case, environmental upheaval arose not from earthquake, typhoon, or war, nor grinding poverty. Instead, my world was being rocked by the turbulence of another human being, and by drastic changes in my "social support," as it's formally known.

I had given birth to my daughters with a feeling of safety, among trusted longtime friends going through similar experiences, tended by a husband who could be dispatched for Ben & Jerry's at the drop of a hat.

But now, in my new house with old bones, perched on the side of a crumbling stone cliff, I felt totally alone.

∽

A mother mammal's behaviors are informed—and sometimes even defined—by her communal context: the other nearby members of her species. This is true especially of people, with our evolved need for communal care. Social-support deficits and perinatal depression are intimately linked.

Research from Columbia University suggests that the strength of a pregnant woman's support system is the primary predictor of her mental health, for a whole raft of reasons. New mothers depend on others for physical help (like the fragrant chicken pot pie a kind neighbor slips into your barren fridge), for practical guidance (such as my mom's sage prepartum observation that it might indeed be wise to purchase more than a single undershirt for my forthcoming first baby), and also for the more mysterious matter of emotional sustenance.

This mushy-gushy last one—let's call it love, for short—is the least understood but probably the most important. A deep bench of friends and family has physical oomph, lowering maternal blood pressure and optimizing placental functioning throughout pregnancy. During birth, supported mothers undergo easier labors and fewer C-sections; afterward, they have less fatigue and a better shot at successful breastfeeding.

Yet social support is equally key for adoptive mothers, who don't physically gestate children, but whose future maternal well-being and competence rest in part on the amount of encouragement they receive before their new babies arrive.

To an extent, the identity of our cheerleaders doesn't even matter all that much. Pregnant women who are visited at home even occasionally by paid strangers, like nurses, often fare better as mothers later on and are less likely to abuse their kids. Women attended in childbirth by professional doulas are "more alert and responsive" to their babies from the very beginning.

But, not surprisingly, it turns out that particular people in a pregnant woman's life—her romantic partner, parents, and close friends—play essential roles in her making as a mother.

As my third child dug in his heels inside me, the people I had counted on for the first two pregnancies were gone. First there was the de facto disappearance of my husband, who was physically present, roaming the still-empty rooms of our decrepit farmhouse, but mentally and emotionally absent. Maybe the hard-charging life we'd chosen had imposed too much stress and now he'd finally cracked, like a Lego underfoot. Maybe my big dreams and greedy demands on behalf of our ragtag band of babies had broken him. Or maybe it had nothing to do with me, and some kind of unseen genetic glitch inside him was just now kicking in. He seemed consumed by either a dire physical disease or some bizarre perturbation of the mind—which wasn't much better, as my own childhood experience had taught.

Watching him through the ancient wavy glass of one of my new home's many painted-shut windows, I tried not to think about how family histories repeat themselves.

In the week or so before our big move north, I'd also bid adieu, in one fell swoop, to my group of close friends in Washington, DC—the comrade moms I'd grab coffee or hubcap-sized chocolate chip cookies with on a weekly basis. Oh, and I'd also recently quit my longtime magazine job, giving up good-natured communal griping to strike out on my own and write books in the attic by myself.

Oddly enough, the depression that quickly enveloped me brought to mind a scene from a comedy: *The Magic Flute*. I'm no opera buff, but decades earlier my elementary school music teacher had left a video of the performance for the third grade to watch whenever she was absent, which was apparently quite often, since it's seared into my memory.

Dressed in deepest navy, the Queen of the Night would inch onstage, the train of her sweeping, star-sequined gown flowing, growing, my third-grader self waiting for it to end, until it dawned on me that her dress was in fact endless, and that this crazy lady was the midnight sky itself, caterwauling at the top of her lungs in terrifying German:

> *Oh, don't tremble, my dear son!*
> *You are innocent, wise, and pious;*
> *A youth like you is best able*
> *To console this deeply troubled mother's heart.*

As I wailed and stormed about my new house in the months before and after my third birth, or stared bleakly at the bedroom ceiling, I understood that this was how my two tiny little girls—and once he stayed awake long enough to notice, "my dear son"—must now perceive me, a kind of infinite night queen who could roll in at noon: a living void, spangled with tears instead of stars.

How can a scientific discipline based largely on dissected rat noggins even begin to plumb this kind of uniquely human agony, which springs not only from a mother's singular experience and biochemistry, but from her complex and shifting social milieu?

Actually, even lab-rat moms are quite sensitive to social cues, with mothers behaving better if they are allowed to raise babies alongside their own sisters instead of all alone.

But to tap into the complexities of just one troubled mother's heart, some researchers look to an animal model that's almost as gregarious as we are.

∽

I'm waiting at the dusty end of Simian Lane, decked out in a white lab coat and disposable shoe covers exactly like the kind my day care requires for visitors.

Erin Kinnally, a scientist here at the UC Davis California National Primate Research Center, opens the door of her sedan to reveal a suspiciously immaculate interior, including the Cheez-It-free car seat of her four-year-old.

"Obviously I cleaned it before you came," she says. "Get in!"

We're headed for the Outdoor Colony, where the rhesus monkeys are housed in naturalistic paddocks, each one holding up to 150 monkeys, many of them mothers and infants.

Our car crawls past more than a dozen half-acre corrals. Baby monkeys sail through the air as though blown by unseen breezes; one glides down a support pole with the bravado of an old-fashioned fireman as the grown-ups squabble and lip-smack and bark below.

The enclosures are full of plastic slides, seesaws, and other salvaged pieces of human playground equipment, as well as hacked-up tree trunks and even a geodesic dome. Literal barrels full of monkeys dangle from the ceiling.

Yet these highly social and intelligent animals "also need alone time," according to Kinnally, so each habitat includes protected corners where monkeys can take shelter from simmering in-group tensions.

"This really is a great model of the human condition," Kinnally says. "They're genetically closely related to us, and they have such nuanced social lives. Each group is like its own little world, and to understand them you have to take into account the personality of the players, the demographics, how much conflict you have, whether there's a jerk in charge."

With their rigid hierarchies and pecking orders, macaques are what's called a "despotic species" (a term that reflexively reminds me of human toddlers). But monkey queens can rise and fall, so the intra-monkey relationships are complex and dynamic. To map out the dizzying social webs within a single group, primate researchers here recently employed the help of three statistical physicists.

And all the social slippage and friction and fallout informs the behavior of each and every mother.

We park in front of one paddock, which is the primate center's version of the O.K. Corral—a place with a storied reputation for rowdiness. That makes it one of the researchers' favorite stops during the maternal-behavior surveys they complete several times per week, tracking the monkey moms' every move.

Though we keep ten feet back from the perimeter, several macaques gallop off at our approach, while a curious older mom called Tubby plods over. In addition to a nickname, each mother has an official five-digit number tattooed on her chest and inner thigh, and random patches of her fur are dyed in unique patterns, although these identification aids can wash away in the occasional California rains.

I study Tubby's elfin ears and ruddy face, which is the exact same shade as her rear end, searching for something of myself in her mom paunch and almond eyes.

Kinnally rattles off the factors that can shape primate moms, most of which also mold me: age, number of births, genetics, her own mother's rearing history, the baby's sex and other characteristics, access to food and shelter and sundry other environmental factors. In the wild, some monkey moms' behavior even varies with the height of the tree canopy: the farther off the ground a family is living, the more anxious Mom becomes.

But social chemistry is perhaps the most potent force. Rhesus monkeys who grew up without their own moms are prone to abuse their infants. Those who live in proximity to their own mothers, meanwhile, tend to be more competent and relaxed.

"Having the grandmother around makes such a big difference," University of Chicago primate researcher Dario Maestripieri tells me later. "Grandmothers provide support by grooming and protection. They watch the kids and are vigilant. If you are a monkey, you are living in a dangerous society, and it matters if you are surrounded by family."

Monkeys with an even bigger network of female relatives around are more relaxed still, forming "play groups" and letting their babies range far and wide, confident in the support of their moms and sisters and cousins if trouble starts.

In macaques, rank—which is in some ways akin to social class in humans—gets handed down along maternal lines. If a mom's big family also happens to be high-status, then she has it made in the shade—often literally, since top-ranking monkey moms usurp the shady spots on hot days and the dry spots on rainy ones.

Kinnally points out Grapefruit, the group's alpha female. Tart as her name suggests, this scruffy queen bee brushes off the baby hanging from her left thigh and saunters away. She can afford to behave in this manner, since her offspring are literally untouchable, in the Italian mobster sense. Nobody would dare mess with her kid.

We also meet another, lower-ranked female, whose name, somewhat tellingly, Kinnally can't immediately recall. Keeping a wary eye out for conflict in the group, this skinny younger monkey likes her baby within closer reach and is reluctant to let it explore—which mirrors what sometimes happens among socioeconomically disadvantaged human moms, who may be more prone to separation anxiety and authoritarian discipline. This female visibly stiffens as the group's ultra-aggressive alpha male, Karate Kid, stalks past, his tail crooked in a puffy question mark. (Macaque paternity is always a bit of an open question, and moms are the only caregivers.)

Life here in the California sunshine is generally good—the monkeys don't have to cope with the monsoons of their native Southeast Asia, and their diet of monkey biscuits is generously supplemented with farm-to-table-style local produce and unwanted Halloween gourds from a local outfit called Bobby Dazzler's Pumpkin Patch.

But monkeys can still be stressed in this simian Shangri-la, especially low-ranking moms with spindly social-support systems. These moms have weaker immune systems and other distinct traits. A study of seventy monkeys by Maestripieri and others found that, compared to the high-fliers, the lowest-ranking moms had four times the amount of stress hormones in their blood. These females at the very bottom of the pecking order may also provide a certain amount of nasty amusement for their superiors.

"Just like we'd see in the wild, we've all seen things out here that are difficult," Kinnally says. "I have cried. I have seen a highly ranked juvenile 'play' with the infant of a low-ranked female for way too long." This "kidnapping" sometimes involves flattening the pilfered infant on the ground. The baby's mother is powerless to stop it. When juveniles engage in this rough-and-tumble activity, the colony's human caretakers often intervene.

Low-ranking moms grasp that they have to be vigilant at all times. Fascinating studies have shown that these moms are much more likely to try to shush their infants' cries when higher-ranked animals are around, for fear that the fussing will draw unwanted attention and attacks.

Woo! Woo! the monkeys scream as a white van screeches up to the perimeter.

In captivity, the animals have traded their myriad natural predators, which range from sharks to tigers, for the lab vehicles that whisk away certain clan members for lab tests or medical treatment. Upon the missing monkey's return, researchers sometimes stand by to observe the subsequent social-chain reaction.

To distract the monkeys, white-clad lab personnel hurl sunflower seeds into the enclosure like rice at a wedding. Fixated on claiming her outsized share, Grapefruit pounds past on all fours, her infant clinging gamely to her front, BabyBjörn style. She double-fistedly stuffs her face, her cheek pouches straining.

And way over there in the corner, almost unnoticed—but oops, not quite!—a wild brown blur detonates into the paddock, released by the researchers as monkey shrieks mount.

This returnee is likely somebody's sister or mother or rival, who could potentially shift the outward alliances of these mothers—and their hidden neurochemistry as well.

∽

Human moms diverged from macaques some 25 million years ago, and generally we don't pitch poop or inflict bloody wounds on each other for want of a cucumber. However, as models of social stress and support, the macaques illuminate the concrete importance of female networks, starting with the critical role of one female in particular: the maternal grandmother.

This was the silver lining of my family's doomed move to the woods of Connecticut. Our new home was just a few miles from my own mom's condo—a distance uncannily akin to the "day's walk" between mother and daughters' huts that anthropologists cite in their studies of remote villages.

This was a lucky break indeed, for one of the few strikes against privileged older moms is that we live so far from our parents, who are often quite decrepit by the time we finally get around to having kids. Poor women, meanwhile, tend to live close by their (much younger) parents, within the penumbra of their love. Postpartum depression seems to be much rarer in cultures where young mothers coexist with close kin.

It wasn't all kismet, of course: my husband and I always had a master plan—not technically approved by my mom, who by this point in her life was in Viking River Cruise mode—that involved her helping with the grandkids. (Skullduggery like this, Dave Barry has written, helps explain the very existence of Florida.) But in the end, most unexpectedly, Mom ended up taking care of me. That's one reason she was there watching with me at the sheep barn that night beneath the endless black sky, the air so cold that even the stars seemed to shiver.

Few other mammals enjoy such instrumental grandmas. Most mammal mothers simply ditch their full-grown young in what must be, to boomer moms, a rather refreshingly direct manner. Prairie dogs apparently sprint away from their pups once they're weaned. Brown bear moms bail the minute they find a new boyfriend. Rodents stonily ignore their young after a few weeks. ("Mother Is Not like Mother" reads the haunting title of a scientific paper about guinea pigs coming of age.)

Even among our closest primate cousins, like rhesus monkeys, where females within a tribe continue to enjoy quite cordial and supportive relationships with their adult daughters, the most solicitous grandmas remain fertile and breed to the very end of their own lives, so at best they must divide time and loyalties between their adult daughters and their newer, more demanding offspring. (In some New World monkeys, like marmosets, jealous pregnant grandmas have even been known to murder their own grandkids.)

Yet for human mothers, amid all our varied lifestyles and here-again-gone-again male partners, maternal grandmothers are a global bulwark, a core support system as trusty as our left hands.

"In some places fathers do more, in others less, but the maternal grandmother is much more constant," says Brooke Scelza, a University of California, Los Angeles, anthropologist who has studied grandmothers across the globe.

A century ago a famous (male) anthropologist embedded with Australia's Tiwi hunter-gatherer people puzzled over the sheer existence of post-reproductive women, whom he thought of as freaks of nature, "a terrible nuisance," and "physically quite revolting"—and certainly not worthy of his study. (Perhaps outrageous prejudices like these explain why my still-stylish mom resolutely shuns the g-word, answering only to "Mamie" and, briefly, "Foxy.")

But now there's a whole scientific literature celebrating human grandmas. The nearly uniquely human trait of menopause (only orcas seem to share it) makes their existence highly adaptive. "Given that her body is deteriorating over time," writes Sarah Blaffer Hrdy, describing Grandma's evolutionary rationale, "when should she throw in the towel, quit producing, and care for her daughter's offspring instead?" (Disclaimer: I should probably have massaged this messaging a bit when pitching my mom on her new babysitting duties.) It's another clever way that human women have evolved to hurl our genes into perpetuity: running out of fresh eggs (and milk) of our own makes us unusually helpful to our own daughters, and extra available for alloparenting. Older women with more kids may even have unusually lengthy telomeres at the end of their chromosomes, suggesting that they may somehow be aging more slowly, the better to help their daughters for the long haul.

We've seen the way mothering styles may insert themselves in human families, via early childhood interactions potentially etched atop our genes. But a mom's mom can make a big difference in the here and now, too, rolling up her sleeves and pitching in. Especially compared to skittish teenage babysitting trainees, maternal grandmas are the ultimate mother's helpers. They are close relatives whose mom circuitry—sometimes called "maternal memory"—comes prefabricated, if in the need of a little dusting-off, and they have loads of practical experience, allowing them to undertake the most highly skilled baby-minding tasks, like soothing and bathing.

No wonder the data indicates that involved grandmas affect infant survival rates more than fathers do, from Germany to rural Ethiopia. (A new study of preindustrial Finnish birth records showed that young kids' survival chances rose a whopping 30 percent if their maternal grandma lived nearby and was within the still-spry age bracket of fifty to seventy-five.) In Great Britain, a grandparental presence ups women's chances of getting pregnant in the first place, and grandmas are universally associated with healthier pregnancies. A hormonal survey of 210 women found a link between women's levels of placental corticotrophin-releasing hormone and family support (which most reliably comes from their own mothers, not the babies' fathers). This hormone is like a stopwatch for the onset of labor, and women with close ties to their own mothers are more chemically protected against preterm birth.

Postpartum traditions vary widely, but grandmothers are almost always integral. Nigerian grandmas install their recovering daughters in rather dreamy-sounding "fattening rooms." Chinese grandmas braise pigs' feet with ginger, perhaps to restore calcium to their daughters' depleted bones, while Indonesian grandmas brew a special medicinal soup to speed lactation. My mom makes spaghetti and meatballs.

Absent maternal grandmas leave a moat around a mother. The death of a mother's mother is a toxic pregnancy stressor, especially for those women carrying (always slightly more taxing) male fetuses. In a study of Puerto Rican mothers-to-be, the women on poor terms with their own nearby moms showed the worst pregnancy outcomes of any group.

And of course the women in deepest need of grandmotherly support—the ones who were abused or neglected in their own early childhood—are the least likely to get it.

Paternal grandmas, true, can also be instrumental. In childhood, because my mom was working, my dad's mom was my sick-day custodian, feeding me platters of bacon as we watched *The Price Is Right* together on the couch and she browsed her latest bodice rippers. My own mother-in-law is a beloved caregiver for my kids, pressing wildflowers with them and staging epic Easter egg hunts. Many paternal grandmas are on shaky footing with their sons' families not because they don't want to be involved, but because their daughters-in-law deliberately exclude them as part of a cruel practice called "kinkeeping."

But anthropologists and biologists, perched on some crag out there in the wilderness with an endless view, nonetheless see paternal grandmas as a less crucial category of alloparent. "Of course that doesn't mean you can't have a relationship with your mother-in-law," Scelza says. But there are evolutionary reasons why paternal grandmothers may be "less incentivized to care." Paternal uncertainty, again, is the big one: paternal grandmas are never guaranteed a blood relationship with their grandkids, and this may explain their more meager investments and negligible effect on child survival.

A paternal grandma also has less of a stake in the well-being of her grandchildren's mother: if this haughty and obnoxious young lady ultimately kicks the bucket in childbirth, her son can always sire children elsewhere. (Indeed—let's get down to brass tacks here—he may do so even if she lives.) Some potentially Thanksgiving dinner–destroying findings suggest that women dwelling near their mothers-in-law, as opposed to their own mothers, have larger families but poorer overall health. In a study of Chinese moms, those who received emotional support primarily from their mothers-in-law were twice as likely to be depressed.

To swiftly change the subject before my mother-in-law or yours gets wind of this controversial subfield, let's turn to the science of grandfathers. As the evolutionary psychologist Harald A. Euler notes, involved grandfathers don't really even exist in the rest of the animal world, "with the possible exception of bottlenose dolphins." And paternal grandfathers are doubly improbable: not only are they unsure that their son fathered a given crop of grandkids, but they can't even know for a fact they fathered their own son.

These academic objections make human grandpas' real-life presence a true gift for human moms. Grandfathers' engagement varies greatly, but anthropologists see them as a potential "stabilizing influence" for young families, offering material and emotional support.

My own father died nearly a quarter of a century before my husband's health collapsed, but luckily my husband's dad lived just about twenty minutes down the road from our unhappy new home. As our life unraveled, Grandpa busied himself with unsticking all those stuck farmhouse windows, childproofing precipices with sturdy wooden fences, and tackling other seemingly hopeless tasks. Maybe he endured fewer Queen of the Night encounters than my own poor mother, but he did foot the bill for an otherwise completely out-of-the-question part-time nanny who, in addition to being a huge help with the laundry, became, at a very lonely moment, something far more important: a friend.

∽

It turns out that even female vampire bats have girlfriends, with whom they share their regurgitated blood meals. In the depths of my depression, worlds away from my old friends, I really just wanted somebody to get ramen with.

Beneath that simple desire, though, lurk eons of selective pressures. Female friendship likely began as a stockpiling of allomothers, an all-hands-on-deck necessity given the length and depth of human child-rearing endeavors, our large numbers of simultaneously dependent young, the dangers of our environments, and the general flakiness of resident males. Indeed, lesbianism may have similar origins, a kind of trading of maternal favors that led to a much deeper relationship down the evolutionary road.

In subsistence societies, babies with more sociable mothers are more likely to survive, for a variety of reasons. Friends and neighbors in some human cultures routinely co-nurse infants, as she-wolves and lionesses do. Maybe most new moms in contemporary Boston don't swap boobs in this manner, but a study showed that they do seek out the advice of other, more experienced moms an average of ten times a day. And the steadfast emotional support of friends may outweigh any given favor. A study of mothers of four-year-olds found that those who rate themselves as satisfied with their social networks exhibited "more optimal maternal behavior."

Too bad I'd just left my girl squad in the dust, and Emily and my sister both lived in distant cities.

Previously in adult life, whenever I'd wanted a new friend, I'd scrounged one up at the office. This is a typical strategy: jobs are social hubs for many young moms, with 70 percent of us working and some 40 percent serving as primary breadwinners, compared to just 11 percent in 1960.

All moms obviously work hard one way or another. (My favorite examples from the animal world are the seal moms who sometimes grow barnacles on their epic hunting trips.) But the question of whether working *outside the home* is psychologically healthy for young human moms gets bogglingly nuanced and easily politicized, and the ability to make friends while on the clock is just one facet. The scholarly gist, though, is that some forms of work are toxic and stressful for moms, while others are socially rewarding and therefore beneficial. Some working moms are miserable, but stay-at-home moms may be more depression-prone, particularly those who live in the isolation of the modern suburbs. (I can now understand why: with the multi-acre zoning at our country manse, we didn't receive a single trick-or-treater, and over the course of several years never met any of our next-door neighbors.)

The workplace is another area of modern mom life where monkey paradigms apply. Remember those feast-or-famine foraging carts? Erratic professional environments undo mammalian mothers while predictability stabilizes us. In workplaces where moms can craft our own schedules, take ample leave, access flex time, hide from our children—jeepers, how did that get in there?—and perform elective telework, we can reap both professional rewards and social benefits without compromising the care we dole out at home. On the other hand, working moms tyrannized by swing shifts, seasonal work, and inflexible hours, and moms who—for financial or, especially, health-insurance-related reasons—have no choice but to take whatever work is available, may not fare so well. It's not terribly surprising that heading back to the salt mines less than a month after birth is associated with increased maternal stress and depression.

The brutality of a postpartum command performance affects many categories of working women, including corporate lawyers—I knew of one forced back to trial just days after birth—and surgery residents, an estimated 40 percent of whom consider quitting during pregnancy. But low-income single moms clearly bear the brunt of it, with the fewest escape hatches. The most stressed American moms, as measured by their death rates, are those who become single at later ages and have little control over their work options and schedules.

My working life, on the other hand, had always been rather jolly. Though not particularly well paid, as a journalist I enjoyed "a sense of freedom and choice" in my actions, an important factor in one Belgian study's findings on optimal working-mom mental health. I had juicy maternity leaves, lax supervision, lots of ego trips, and eccentric yet entertaining coworkers. There's research on the dangers of long commutes during pregnancy—more than ninety minutes correlates with a small decrease in male births, perhaps because of the associated practice of breakfast-skipping—but when we lived in Washington I could walk (or waddle, as pregnancies required) to work at my leisure. And believe me, there was no breakfast-skipping: I spent a fair chunk of my day gabbing in line at the empanada truck that parked in front of the

office. Prior to our unwise leap to Connecticut, my major disillusion-
ment as a professional mom was the abrupt discovery that the office's
breast-milk-pumping lounge doubled as the Muslim prayer room.

Human chains of command can be just as powerful as monkey clans
in terms of their impacts on maternal well-being. "Being a subordinate
really has negative effects on your health," explains primatologist Toni
Ziegler of the Wisconsin National Primate Research Center. "Say you
work somewhere and you're pretty much the low-ranking person in
your office, and you have no power, no one listens to what you say and
you offer suggestions and you don't get anywhere because people don't
consider you important. You are under chronic stress. You may get inflammation and your metabolic hormones running amok."

But my metabolism was fine. My name was near the top of the magazine's masthead. Most everybody at my job had made me feel good
about myself. Many were close friends.

Except now, in Connecticut, those coworkers were a memory. My
"office" was a discarded kitchen table shoved against a dingy wall.

∽

Class and privilege, of course, are crucial pieces in my own mom puzzle, explaining everything from my fond work memories to my husband's ability to pick up and move on a whim while remaining gainfully
employed to the presence of a grandpa who could spot us a nanny on
short notice.

But my own peculiar and complex experience of class wasn't always
protective. It also helps explain why the sensation of reliving my childhood arc—with a husband lost in a dark wood while our household
finances tanked—threw such a wrench into my mom brain.

It's hard to appreciate the power of class until you've dropped like
a stone out of one into another. In childhood, my family's financial
downfall meant more than no longer having pony rides at my birthday
parties. It was more than the fact that we were never again able to afford a vacation, while other girls in my class came back every February

sun-bronzed with their hair braided in brightly colored strings, status markers that stayed in until they rotted. It was more than trading the Mercedes for a Honda Civic so stripped down that it had only one side mirror. Many of our family's connections realigned as well: our social network shrank, our status faltered. Old friends no longer called. And my sister and I became fair game for bullying in ways that other kids in town weren't.

When you live in a place where there are no truly poor people—they'd been edged out of our hometown in a million ways long ago—the difference between being upper and lower middle class feels vast indeed. At the top of our town's hierarchy were corporate vice presidents and minor Wall Street tycoons. At the bottom were the school bus drivers and their kids, who now politely invited me to smoke cigarettes in a nearby gazebo.

I'd like to tell you that the good mothers of this nice town grasped what had happened to us and treated me kindly, but by and large they didn't. I was big and clumsy and got my period before anybody else, which some evolutionary biologists would tell you is a physical manifestation of growing up in an unstable environment, but in my case—since I'm from a family of big, sturdy gals—might just be my genes.

My sister and I spent much of our youth scheming to restore our family's lost honor. We rummaged around in bags of barbecue charcoal (we'd heard that coal transforms into diamonds, given long enough), and dug in every place we could think of for rare dinosaur skeletons, which we figured would fetch a pretty penny once the American Museum of Natural History got word.

At some point it occurred to us that it might be easier just to earn good grades. I was not the smartest cookie in my high school, but I was bound and determined, and accustomed enough to stress of many kinds that the SATs and even the biggest nail-biter of a high school debate seemed like no big deal. Emily, who was the daughter of our very strict middle school chorus teacher and thus a fellow outcast, joined me in the quest to turn our tables and defy our stars. And it worked:

she got into MIT, I got into Harvard, and soon I scrambled back into the comfortable class.

In returning to my hometown, I was willing a different ending for my family. I imagined my daughters' giggling wedding processions across a manicured side lawn at the new house, and quickly calculated that we could squeeze a couple of hundred guests out by the pool, if the weather only held.

Except now my white-knuckled climb back up the status ladder had ended in the familiar feeling of free fall. Trapped in a house that increasingly resembled the Overlook Hotel in the off-season, I'd been betrayed by my own ambitions, and that lifelong sense that I truly belonged back at the top. Class may be biologically inscribed, lingering as a throb of entitlement: even though I had ended up at the bottom of the local pecking order, with the scratches and bruises to prove it, there was always the ghost of the spoiled brat about me. Now that little ghost, wearing her very own rabbit-fur coat, had led me down the primrose path once more. And how the landscaping bills for all those finicky rosebushes piled up on the kitchen counter! It was a classic act of hubris.

Yet, perversely, my sense of entitlement would yet turn out to be a saving grace.

In humans, there are plenty of measurable differences between poor and rich moms—everything from the number of frozen soy desserts we serve to the number of bottoms we smack. But it only became clear how social privilege can be a mother's ace in the hole on the day that I finally decided to contact my doctor's office for help.

It was about eight weeks after my son's birth. He was sleeping downstairs in our emergency nanny's arms. My daughters were watching Disney's *Frozen* for the millionth time, their little faces pressed up near the TV; we had rolled right off the back of the no-screen-time wagon as soon as my husband's illness hit.

The Queen of the Night was in her bedroom, staring at the cracked ceiling as tears dribbled from the far corners of her eyes. At the time I hadn't yet contemplated the ancient potential evolutionary purposes of these numbing feelings, and I couldn't have cared less whether their point was to prep me to walk through some impending firestorm for my kids or to bundle them in bark and plop them in a convenient river. I hadn't really noticed whether I'd switched to holding my baby on the right, or whether his cries suddenly sounded duller to my ears. Maybe oxytocin was in short supply in my system. Maybe my nucleus accumbens just wasn't what it used to be. But right then I didn't give a fig about what my aching mess of a mom brain might look like on a scientist's scanner, which neurochemicals didn't quite add up, or whether some bum gene was to blame.

I just wanted it to stop.

I had put off this call for weeks, on the assumption that contacting my ob-gyn's office for help with my depression was an unpleasant but perpetually available last-ditch option. I was someone who'd proudly scorned therapy all her life, and it was not a move that came easily. I stalled and stalled. But when I finally picked up the phone that afternoon, I fully expected to be summoned in that day for commiseration or medication and (as in *Frozen*) "warm hugs"—preferably all of the above.

A message was taken. An hour or so later, a harried-sounding doctor called me back.

It wasn't the nice "frank and beans" doc who'd delivered my baby boy. ("Oh, I love his cheeks!" I'd heard her shout in triumph amid my own botched-anesthesia screams.) I'd met this other doctor once during a monthly in-office visit, but he didn't remember me and didn't seem to have read my file, and I was crying too hard to make clever conversation or to transmit any other subtle signals about myself.

Unlike my DC ob-gyn office, which catered expressly to the city's professional women, this was a diverse practice in a small Connecticut city, with all income levels and social backgrounds in the patient mix. This doctor couldn't have known from his cursory (if that) glance at the paperwork that I was a well-heeled, highly educated, not-to-be-trifled-with white lady who lived (for the moment, anyhow) in a large (albeit laughably dilapidated) home on one of the better streets in a wealthy town. At the moment I was just a sad little voice on the phone. I could have been any mom in America.

He listened to my story, and then said coldly: "You've been depressed before, haven't you?"

Not "Do you feel depressed?" or "Have you ever been depressed?" It was a flat accusation.

In fact, I had not been depressed before. Sure, I'd always been a bit high-strung—something of a stress freak, if you will. Cheerfulness was maybe not my default state. But I had also witnessed real depression, and knew all too well what it was and what it could do, and that I'd never lived it myself until now. Nothing in my chart would have indicated that I had, if he'd bothered to read it.

But what was "before," anyhow? Even though my life had been pretty spectacular nine months earlier, I had a hard time remembering way back then—or anything, really, more distant than the previous night's brutal 3 a.m. baby feeding. Had I ever in my whole life truly been happy? And then in a flash I recalled that once in college sixteen years earlier—faced with the due date on a particularly onerous English paper—I'd had to go to the campus hospital to lie down for an afternoon, receiving no medication or further treatment, just a few words of encouragement. Did that count as depression?

"I guess so," I blubbered in reply.

Clearly this was the out this doctor needed, as my answer translated somehow into "not his problem." He rattled off the phone number of a therapist in a nearby city and hung up.

His office never contacted me again.

When I collected myself enough to call this other number, I learned, via an answering service, that there was a two-month wait for a first appointment.

Now two months is a long time by any standard, but it's eons for somebody who has had a baby only eight weeks earlier, and whose every midnight lasts a million years. Mothers under these conditions can abandon their families. They can also kill themselves. It's a sign of how little respect the modern world has for mothers that many countries don't even track maternal suicide statistics. In rapidly aging Japan, though, where new mothers have become somewhat scarily scarce of late, a study found that about 30 percent of pregnant women and new moms who died in a given year had committed suicide.

In no way do I think I could ever contemplate, let alone commit, these actions. But then again, what mom imagines that she would?

This is the moment when the vital relevance of my station in life became apparent. If privilege is an inner sense of entitlement and personal indignation unfairly bestowed upon the upper class, then here it was. Someone with bared fangs and puffed-up fur reared inside me—call her my inner Grapefruit. My children and I were in trouble. How dare this man dismiss me? In my world, dark as it seemed right then, medical doctors were still a dime a dozen. I had many friends and relatives and even stoner college roommates who were now celebrated doctors of various types; I interviewed famous scientists for a living. This guy didn't know who he was dealing with. A deep, angry voice—the kind most commonly heard at monster truck rallies—rumbled deep in my throat. If a laundry basket had been handy, I would have smashed it. I would have swung a baseball bat at a bear. But this threat, and my gut-level maternal response, was more abstract than a hapless home invader or a grizzly. It was about my social standing in relationship to another human being's, this doctor's power versus my power, and his view of my reality versus my own. I would not accept his version of me, nor believe in the limited options he presented.

In a chattering rage, I called Emily in Minnesota, who (female support, check) always answers, and who is now a physician herself (hello, privilege). She'd already been able to pull some strings and connect me with an old pal of hers, the best general practitioner in the area, even though he technically wasn't taking any new patients. Because I wasn't quite thinking straight, it never occurred to me that another type of doctor could help with a textbook obstetric malady. But now she commanded me to call this other doctor, so I did.

By the following afternoon, two orange cylinders of pills rattled reassuringly inside my purse, and I had a scheduled follow-up appointment a couple of weeks out. Easy-peasy-lemon-squeezy, as my kids say.

But in truth I don't think the chemistry of the medications themselves helped me as much as the feeling of restored control that they conferred. I downed only a few doses; four years later, the bottles are un-refilled and still in my purse, a talisman. The helpless feeling engendered by that first short phone call reverberates today, along with the hard realization that my outcome could have been quite different.

Back then I was in mommy mode and cared only about saving myself and my own kids. Today, pregnant with my fourth, I ponder that phone call every time I sit in the waiting room for my monthly prenatal appointment, at my new (obviously) but just as diverse ob-gyn practice, sneaking looks at other with-child women from all walks of life as we yank at the waistbands of our maternity jeans, which sag universally no matter the brand. It's now so much less mysterious to me why the most vulnerable women often don't receive treatment for postpartum depression—even though they suffer at much higher rates—and have such miserable dealings with the medical system in general, counting far more on family members than on doctors for vital support.

Once I believed that there were many types of women who quite naturally became many types of mothers. Now I know that any one woman has the potential to become many different mothers, depending on conspiring circumstances, support systems, and access to resources, including powerful strangers' sympathy and respect.

I have been many mothers myself.

∽

Yet I don't want to pretend that passing beyond the dark side of the mommy moon was a mere matter of popping some pills—one of them rather distressingly nicknamed "housewife heroin"—and chilling with my mom while the nanny held down the fort.

There was still the matter of my husband.

Dads have taken some hard knocks in these pages—whether they're depositing placentas with one foot out the door, or calling it quits in times of plague. This is all pretty sturdy science.

But especially when it comes to humans, the most social of all beings, the story is a little more complicated. Macaque moms may not know or care who their infant's father is or give a hoot what happens to him—but I sure did.

In most other mammals, the maternal instinct reigns to the exclusion of all other loves. "What we see in animals is that they have one attachment," mainly to their babies, says Karen Bales, who studies pair-bonding at UC Davis. "In humans, though, we can have these very strong, very selective relationships with multiple people, and they do very different things for us. Our larger brain capacity may explain our expanded cognitive abilities, but also our emotional abilities."

I've already mentioned the compelling theory that the maternal instinct is at the heart of human romantic attachment. Mammalian pair-bonding—which is quite rare, occurring in less than 5 percent of species—may be a kind of hot-wiring of far more ancient maternal circuitry. Maternal and romantic bonds revolve around the same types of body parts: the recycled-circuitry theory may help explain, for instance, the mysterious male fascination with breasts, organs typically only spellbinding to infants. Indeed, some of the very same neurochemicals, like oxytocin, seem to bond a woman to a lanky, bearded adult human who is demonstrably not her baby.

I am a true pair-bonder. This hardwired human preference was, for me, likely amplified by the fact that my husband and I had been basically on our own for a decade together as we chased careers as a team, raising daughters in a distant city far from our parents and without any kind of family help.

He was the one beside me in the delivery rooms, not some

proverbial crone or even my own mom, and the one up with me all night. And he had helped me in many hidden ways as well, chipping away at the fear and distrust instilled by a rocky childhood, emotions that likely would not have served me in very good stead when I became a mother, had they been permitted to smolder in my amygdala or anywhere else. Through romance, marriage, and all the madcap and frequently incompetent parenting escapades that followed, our companionship meant for me a kind of peace. I loved him desperately and didn't think he would ever abandon me.

We've already seen that dads, even when they stick around, don't impact child survival rates—although I might personally dispute that finding, since my husband once fished a penny out of Daughter Number Two's throat. But dads definitely *do* give maternal grandmas a run for their money in terms of the social support they provide to moms.

In situations of intergenerational abuse, where patterns of poor mothering flow from mother to daughter, the loving presence of a spouse is one of the few factors that can break the vicious cycle. Moms with a supportive partner tend to be less stressed, more sensitive, and happier. They fare better in the wake of disaster—earthquake survivorship studies show that moms with a "functional marital relationship" had steadier mental health during any tremors to come. But the same goes for ordinary life, when a little TLC from a partner can protect a woman's postpartum psychology. A study of a newly instituted paternal leave policy in Sweden showed that the extra thirty days dads spent at home tracked with a 26 percent decrease in the amount of anti-anxiety prescriptions that new moms obtained.

Absent dads, on the other hand, are linked with premature birth, maternal anemia, high blood pressure, and depression—particularly among women who are not single by choice.

The reasons for this negative cascade are practical as well as emotional, since single moms often simply have more chores to do and fewer hours to spend with their kids, more economic stress and narrower paths for advancement, and substantially reduced social networks (yes, those mothers-in-law really do come in handy sometimes).

Likewise, the logic behind why a given dad stays while another shoves off forever is a complex blend of variables, including not only broader cultural mores and expectations but also highly personal influences, such as his relationship with his own father.

For a time, it was widely believed that long-term paternal involvement was more likely following the birth of a firstborn boy—so typical, right?

Fascinatingly, though, some social scientists now think that's because women are fractionally more likely to have boys when they're *already in* healthy, supportive relationships—making those baby boys an indicator of present relationship quality and lower stress levels, rather than a tempting T-ball-bashing incentive for dads to stay.

On the other hand, some disturbing dad studies continue to suggest that men do fine-tune their behavior toward a mother based on a child's looks—specifically, as the reader may already expect, on how much her baby looks like him. A survey of abusive dads in New York's Adirondack region showed that weak paternal resemblance predicted that men would more savagely injure their children's mothers in domestic disputes, as assessed by the mom's bruises, broken bones, and surgeries.

Whether the damage is physical or psychological, scholars agree that an abusive partner is worse for mothers' outcomes than no partner at all.

Leah Hibel of UC Davis recently asked mothers and fathers of six-month-olds to interact for ten minutes in her lab, instructing a "conflict group" to discuss thorny relationship issues and another group to chat about more pleasant topics.

"There were real fights, with people criticizing each other and arguing," Hibel recalls. Afterward, mothers stacked plastic rings and played with other toys alongside their infants as scientists watched. Based on mom and kid cortisol readings taken before and after these interactions, it seems that the stress of the parental conflict spilled over to the way the mother interacted with her child.

What especially fascinated Hibel, though, was how the moms' hormones and behavior correlated with the couple's real-life communication style, as opposed to their arbitrarily assigned group. Some couples who are supposed to be in the "conflict group" ended up having lovely and supportive conversations. Some in the "supportive group" went for each other's throats regardless, and were more likely to withdraw from their children afterward. The dynamics of a marriage are so deeply entrenched that they may defy experimental manipulation.

Just like jobs, "partners can be a source of support *or* stress," Hibel explains. "It's not that having a partner is a uniformly good thing. When you see pathology in the romantic relationship, you see pathology in the parenting relationship." In unhealthy partnerships a mother's behavior often corrodes—a study from another lab showed that moms who reported lower levels of "marital love" sometimes had impoverished communication patterns with their two-year-olds.

So while marital status remains a predictor of mothering quality, moms who master their own fates and end up as single mothers *by their choice*—there's that word again—may fare just as well as or better than moms whose marriages offer more pain and heartache than support.

∽

That was not *my* choice, however. My husband was my emotional buttress and—ever since we'd started having kids, even though it pained me to admit it—my economic mainstay as well. Raised by a widowed mother, I understood all too well the impacts of paternal absence. And although my sickly, skeletal husband and I now fought all the time without any experimental prompting whatsoever, I didn't want to divorce him, or to watch him die or otherwise disappear, which was what I had begun to believe really might happen.

While people tend to have rather passionate ideas about whether it's "good" or "bad" to be a single mom, the biological fallout from nonelective single motherhood is something of a scientific blind spot.

"It is surprising," writes neuroscientist Oliver Bosch of Germany's University of Regensburg, "that almost no data are available on the neurobiological basis of emotional changes following partner separation in females."

The neurochemistry of "maternal abandonment," as Bosch calls it, is hard to study in large part because humanlike pair-bonds are so scarce throughout the animal kingdom, and especially in the love-'em-and-leave-'em rodent realm where we recruit our trustiest animal models.

But Bosch works with prairie voles, the rare rodent species that performs biparental care, with dads huddling and grooming and otherwise puttering around the nest. (Despite how cute this sounds, the voles are "little monsters" to work with in the lab, reports Bosch, to the extent that the tiny fur balls must be handled with bulletproof Kevlar gloves.) These pugnacious little garden pests act differently from other types of lab rodents from the outset of courtship, with a female only ovulating upon meeting her first sexual partner, who typically becomes her mate for life.

For decades scientists—it must be said, often male ones—have been much more intent on the outlier neurochemistry of these unusually devoted vole daddies. But the vole moms intrigued Bosch, too. How would they mother without their life partners?

His team made all the necessary introductions, putting virgin female voles in cages with males, and then leaving the fuzzy soul mates to canoodle for eighteen days—from here to eternity in rodent time.

Then, just as the first crop of babies was about to be born, a researcher plucked the male from each cage, like the Kevlar-clad hand of God himself.

To Bosch's amazement, the moms gave birth and carried on with business as usual, tending to their babies more or less as they normally would. Everybody stayed fed and warm and, for the most part, alive.

But there were obvious differences in the moms themselves as the team put the voles through the standard array of stress tests. They cowered in the exposed arms of the maze they ought to have eagerly explored. Bosch also carefully slid each vole mom in a beaker of water—not at all a mean thing to do, since healthy voles are excellent swimmers. But the abandoned females hardly swam during the "forced swim test." They just floated there, barely trying. It was almost as if they'd lost the will to live.

Two experimental epiphanies have stayed with Bosch. On the one hand, he remains stunned by how tough the single vole moms were.

"It was fascinating to see with my own eyes how robust the maternal brain is, this one drive to take care of the offspring no matter what, and I think this is mostly the same with humans," he says. "The mother is still fully a mother even though she has to cope with this situation alone."

But the second epiphany was glimpsing the extreme fragility that existed beneath the moms' fighting spirit. The lab animals exhibited, he believes, "altered emotionality"—something akin to human depression.

His findings likely hold up in the wild, where something like one in three vole moms are naturally widowed. (Once described to me as "the potato chips of the prairie," vole daddies are, alas, an extremely popular prey item.) Most of the solo voles never pair-bond again, although they often continue to mate with male passersby and to mother litters.

At least his experimental subjects weren't widowed in vain: in some females, Bosch's team was ultimately able to reverse engineer, or "rescue," a depressive mom vole's floating behavior and other symptoms by chemically blocking certain receptors in her brain. Grasping the biological particulars of down-in-the-dumps vole moms, he believes, could blaze a trail to increasingly effective medications for human moms who need them.

∞

In my case, though, rescue came mainly when my husband began to get better. It turned out that he had a brutal but poorly understood bacterial illness. That initial lump was the bite of an infected deer tick. Instead of defective genes, or the half-dreaded revelation that our new house had been built above a Native American burial ground, the culprit was an environmental stressor that he'd unwittingly brushed up against, maybe even during that giddy home inspection. And when he started to recover—slowly, over many months of treatment and frustrating trial and error—then just as slowly, so did I.

Our fantasy of rural life died, however. I had imagined that a stylishly renovated country house would be a fortress for my children as they grew up. Instead we sold the farmhouse but didn't stray far, moving to a small nearby city still close to our moms and his dad, into a house with a tiny backyard. We know our neighbors and get swarmed by trick-or-treaters. There's a ramen spot right around the corner.

Material circumstances can indeed make or break mothers, but the people around us matter even more. A mother is her children's truest fortress. Yet she can't stand alone.

needs attention and resources, and when, and what kind. The most vulnerable moms are least likely to get the care they need.

A larger issue lurks here, too. The mistaken notion that maternal behavior is entirely innate may lead to the false assumption that moms won't benefit from help. After all, humanity has roughed it for two hundred thousand years—and our mammalian kin for 200 million—without gratitude journals or jasmine-infused rubdowns. If we moms are shaped by millions of years of evolution, a multigenerational legacy of caring, and a miraculous cascade of personal hormones, what more can anybody possibly do for us? If a mother's nature is so instinctive, why not just let nature take its course? Praising the inherent power of maternal biology and waiting for it to simply "kick in" can become an easy excuse to leave moms high and dry.

But of course, the instinct, while quite real, is not rote or automatic. It is dynamic, plastic, with a power that can be amplified or hushed. It responds to volatile material circumstances and is exquisitely sensitive to subtle social cues. Mothers are not Stepford robots, programmed to always be good and sweet and elbow-deep in homemade sourdough. We can be terrifying, unpredictable, and even violent. The same maternal instinct that protects our children can also damage them forever.

With all of this in mind, the best way to help the most moms is through engineering our shared environment, reducing maternal stress, and increasing support for all mothers. Our fellow humans can make us feel secure and provided for, or exposed and alone—empowered or helpless. At the end of the day mom science isn't really about navel-gazing and congratulating ourselves on how very special and complicated all of us moms are. (Believe me, after four pregnancies, my own navel is the last thing I want to gaze at.) It's about watching how women navigate the most fraught of human experiences, and finding the best ways to provision them. Together, we can transform the transformation.

The most tantalizing evidence that it's possible to revise the experi-ence of motherhood is the fact that this standard female rite of pas-sage already varies dramatically worldwide. The mysteries of human culture—hanging over our brains and bodies like a foggy marine layer—can be a sticking point and something of a bummer for scien-tists, which helps explain why they love working with rats so much. But for humans, culture is an unavoidable variable.

In one society, your baby's horoscope might be a cute conversation starter—but in another, it might shape the entire parent-child rela-tionship, with maternal investment varying according to the luckiness of her offspring's zodiac year. The estimated 93 percent of American moms who read *What to Expect When You're Expecting* presumably enter motherhood having gotten a totally different memo than the French mothers who still rely on Rousseau's *Emile*—yes, even though the distinguished philosopher famously abandoned all five of his kids to orphanages. Something as basic as the expectation that a mom should play with her kids may be a Western invention, reflecting the decline of communal modes of living. Two of the American mother's most dreaded charges, the so-called toddler and the alleged teenager, do not conceptually exist in some corners of the world—and maybe all moms should buy tickets to those places.

Indeed, some of the best evidence that human mothering is cul-turally inflected comes from studies of mothers on the move. Take, for instance, the contrast that's frequently drawn between East Asian and Western parenting. East Asian cultures are often called "collectiv-ist," because millennia of rice farming required community-wide irri-gation and terracing efforts, leading to selection for personality traits like group loyalty and compliance—or so the story goes. But Europe-ans, and especially their American descendants, like to bill ourselves as cultural "individualists"—mavericks and pioneers intent on self-expression rather than filial piety.

These divides appear to color many aspects of modern mothering, including the way moms perceive and narrate picture books—with Euro-Americans preaching loudly about the main character's self-actualizing feats and emotions, while East Asians croon about the background details—to how we speak to our kids, with American moms asking more questions and labeling objects like crazy, while Japanese moms tend to murmur and soothe. Play and discipline styles split down the same lines. East Asian moms ("tiger mothers," to borrow from Amy Chua's famous book) are reportedly more protective, maintaining constant contact with their children, and in Japan sometimes even co-bathing and co-sleeping until the kids are fifteen or so.

Some scientists have speculated that the moms' collectivist-individualist division is genetically locked in, perhaps via the 7R variant of the dopamine receptor DRD4, which is less efficient than other versions at sopping up pleasure chemicals and linked with the character traits of extroversion and impulsivity. This genetic variant is twenty-three times more common in European-bred populations than in East Asian ones.

Yet it seems a stretch to imagine that uneven sprinklings of a single gene could explain such complex hemispheric behavior—especially since it's easy enough to tick off potential nonbiological explanations for East Asian moms' actions. Maybe Chinese moms hover over their children, instilling rock-solid etiquette, because for decades the government's one-child policy let each woman have only one kid. Maybe Japanese moms co-sleep with their teenagers because Tokyo's real estate prices force families into cramped one-bedroom apartments.

And sure enough, research shows that when East Asians move to America, the tiger-mother style rarely survives past the first generation, and (for better or for worse) their daughters and granddaughters swiftly take up our more individualistic ways. In a study of 118 women interacting with their five-and-a-half-month-olds, for instance,

Japanese American mothers by some measures resembled native-born American moms. (Other remarkable changes transpire, too, with ancient son preferences eventually fading away in this daughter-proud new homeland.)

Meanwhile, some forms of supposedly bred-in-the-bone American iconoclasm are very recent inventions, not a legacy of the frontier spirit. Take baby names, which I've been dwelling on lately, for obvious reasons. Some cultures recycle the same few names, or refer to a government-approved list—but not Americans. Our love of unique monikers, exemplified by certain Hollywood starlets, seems like a prime example of our essential individualism.

Except, historical records suggest, this obsession is only about a century old, and it didn't really come into its own until the 1980s. Any American mom who's strolled through an eighteenth-century graveyard sardined with Sarahs and Elizabeths knows that once we all used the same ten names, too. And as East Asia has modernized, its urban moms are now coming up with their own outside-the-box baby names, alleged collectivism notwithstanding.

So American moms were not always as we are. What seems normal now was quite recently not. Indeed, those same Puritan forefathers, mostly guys named John and Thomas, often frowned on the same intense mother-infant play that's currently in vogue; as recently as 1914, mothering manuals warned that it "ruined the baby's nerves."

But if some of what seems instinctive about Western mothering— baby-play marathons, the tyrannical reign of the toddler—have cultural rather than biological origins, how were those cultural habits born?

Let's wheel our strollers back through that dreary eighteenth-century graveyard.

It's full of babies' graves.

As late as 1900, 10 percent of American babies died within their first year, and most mothers (especially poor ones) could expect to lose a young child. Even in the 1940s, these tragedies were common. My dad had a sibling—a sister, naturally—who passed away in her crib. Who knows why? Such events were routine enough that my grandmother, surviving two of her three children, never got around to mentioning it, and I only learned about this lost baby aunt when I was about twenty-five.

Infant mortality remains a big part of maternal experience in much of the world. Ten percent of babies born in Afghanistan still don't reach their first birthdays. Two out of three mothers in sub-Saharan Africa can expect to lose a child, or more.

But in the early twentieth century, a combination of scientific breakthroughs like the invention of the neonatal incubator, poverty-fighting programs, and environmental improvements sent American infant death rates into a swan dive. Today just three in five hundred American newborns die (and if those odds still don't sound that good to you, know that they also reflect efforts to save micro-preemies considered "stillbirths" just a few decades ago). The vast majority of our children reach adulthood.

I do not feel that I am built to survive a child's death, and indeed, I may very well not be, since we've seen that modern American moms who lose children frequently do perish of something like a broken heart. But what made me this way is my culture; my DNA is not very much different from that of my grandmother, who weathered this terrible loss and quietly went on with her life.

So the near-assumption that individual babies will live is an idea patented in our own parents' lifetimes, a triumph of public policies and scientific achievements that changed mothers' worlds forever.

This is likely a big part of why middle-class American moms have fewer kids than our predecessors did, treating each one like a crown prince, and living by a strategy that anthropologists call "hyperinvestment." It helps explain why we line up for preschool ukulele lessons, pretending that toddlers are an exalted race instead of just small stupid humans to be weaned via hot pepper on the nipple and sent out to muck the stalls.

Even our use of motherese, which feels so ancient and earth-motherly and instinctive, may be greatly exaggerated by low-infant-mortality conditions created by science-minded public strategies. Mothers across the globe have always loved and treasured and empathized with and wept for our babies, but their investment of time and energy was inevitably circumscribed by the knowledge that little ones might not be long for this world.

All this means that many aspects of American motherhood reflect a breathtaking collective accomplishment, offering a lot of reasons for even the most exhausted, toddler-tormented moms to feel grateful. We've created a new version of nature for ourselves.

And yet we can do so much better.

For all the innumerable marvels women in the wealthiest countries enjoy, such as disposable diapers and the hope that we will watch our babies grow old, we Western, educated, industrialized, rich, and democratic mothers—WEIRD moms, as the anthropologists call us—are still frequently unhappier than our peers in far poorer places.

When Jennifer Hahn-Holbrook of the University of California, Merced, mapped postpartum depression patterns across fifty-six countries, she was surprised to note that national wealth *did not* predict mothers' mental health.

Instead, her meta-analysis showed that Nepal—where the infant death rate is still five times higher than America's, and where 25 percent of the population lives below the poverty line—enjoyed one of the globe's lowest postpartum depression rates, second only to Singapore's.

What might mothers lose as societies grow richer? We might lose community. A culture of big houses in sprawling suburbs—as I myself discovered in my own teetering-on-a-precipice dream house—is perhaps more elegant than past maternal ecosystems, but also more isolating. Indeed, American moms' impulse to play constantly with our kids may be a symptom of this unnatural loneliness. As the anthropologist

David Lancy points out, the Inuit are one of the few indigenous groups to partake in this pain-in-the-neck practice, as they, too, are "stuck indoors" with their kids for long periods—except they're trapped by Arctic darkness and treacherous ice sheets rather than by commodious homes and tasteful multi-acre zoning.

We might also lose a sense of solidarity. When Hahn-Holbrook combed through the global data, she found that the most depressed-mom countries have the *highest income inequality rates*. Universally poor Nepal and wealthy Singapore don't have tons in common, but Singapore does boast milder income inequality than many rich societies. "That gap between the rich and the poor" is the real problem for many moms, Hahn-Holbrook tells me. "It's moms' *feeling* like they don't have enough resources."

After all, core mom biology is geared not just toward baseline levels of resources, but toward detecting subtle fluctuations in those resources while we respond to our spot in the larger social hierarchy.

Finally, in wealthy go-getter societies where women, moms included, are doing all kinds of wonderful things, mothers may sometimes lose status: the collective sense that motherhood deserves special support and protection. The gloomiest motherlands, Hahn-Holbrook also found, have the highest numbers of women of childbearing age who are working *more than* forty hours a week. Like low infant mortality rates, the opportunity to work long hours outside the home is a hard-won modern achievement. But the expectation that mothers should work those kinds of hours when their kids are very young points a finger at a suite of underlying ills: the lack of governmental support for new mothers, a shaky kinship system (whether through modern family breakdown or other factors), a rigid workplace culture, and a tendency for aging societies with low birth rates to essentially forget what it actually takes to raise a child.

Note that Hahn-Holbrook is a young working mother of a preschooler and a newborn who, to keep pace in a highly competitive field, frequently burns the midnight oil herself.

Yet she does so with a professor's rarefied privileges, status, and work-life balance, for a job she loves, on her own terms. "My son's sleeping in the next room right now," she tells me over the phone, "and that's a big comfort to me."

∽

To think about how mothers in general might be made more comfortable so we can all thrive, it's illuminating to inspect, side by side, two mom populations from economically similar countries, something Maria Gartstein of Washington State University recently did with America and the Netherlands.

Gartstein leads an infant temperament lab, studying marked global differences in infant personalities and behaviors. Babies are supposed to be free from culture's chains, but the cuffs go on early, maybe in the womb. In 2015, Gartstein ran studies comparing Dutch babies to infants of the same age from Idaho and Washington State, and found that Dutch youngsters on average seemed sunnier by nature, acting cuddlier, smiling more often, and soothing faster. American six- and twelve-month-olds, meanwhile, had "higher overall negative emotionality, fear, frustration, and sadness."

This news did not come entirely out of the blue—a 2013 Unicef study had already deemed Dutch babies the happiest not just on the block, but on the whole planet, while US infants limped in at twenty-sixth place, just about neck and neck with the little Lithuanians and Romanians.

Yet Gartstein's finding on sad American babies touched a nerve.

"I got all kinds of calls," she said. "People said, 'Are we overstimulating our babies?'" It was particularly surprising because Americans don't think of ourselves as being so very different from the similarly WEIRD Dutch. (Few people seemed to reflexively blame genetic differences, for instance.)

To get to the heart of the matter, Gartstein ran a second study, this one comparing the babies' *moms*—that is, pregnant women from those same two necks of the woods. It turned out that the Dutch expectant mothers were a pretty blissful bunch, while American mothers were comparatively quite miserable, with high levels of unhappiness and worry.

"I think it's because we're running a stress incubator for mothers over here," Gartstein says. "They do not get enough support in their health care or workplace." She believes that the moms' sadness can be transmitted to their babies in the womb, in a sense "programming" the children in a vicious cycle.

What do Dutch moms have that we don't?

Well, for starters, a *kraamverzorgster.*

∽

But before I get on a kraamverzorgster roll here, let's first acknowledge that political and cultural change comes hard.

If mom researchers—who themselves could use a lot more funding, by the way—struggle to show how simple tricks like foot massages or biweekly video chats can budge stubborn maternal behavior, it's not because they're ignoring the obvious players at work in our minds. But these forces are often too big, like, say, chronic poverty, or too small, like invisible DNA methylation patterns potentially dating back generations, for a single understaffed lab to tackle.

Truly transforming the maternal transformation would involve taking on some of the most grinding and deadlocked political issues of our day: not only income inequality, but also health care, education, and other topics that have consistently stumped our government. It would also require fully confronting racism, a menace as old as our country, its dire impacts on maternal behavior only now being fully exposed. This most insidious of all social stressors seems to physically damage the maternal body, contributing to pregnant Black women's higher blood pressure and elevated risks of prenatal diabetes, preterm delivery, and death. The chromosomes in placental cells inside Black mothers have shorter telomeres than those within white women, suggesting that this key organ of childbirth has aged, or "weathered," prematurely in the

womb. Black moms are less likely to get treatment for postpartum depression, and are many times more likely to be offered formula in the
hospital, as opposed to being cheered on to breastfeed.

Alongside these old challenges are newer ones, like the opioid epidemic. Narcotics hijack users' dopamine pathways, exploiting systems
that are built for babies, not drugs. Drug users of both sexes exhibit
dampened reward responses to normally awww-inspiring baby pictures, and addict moms seem less hooked on their own baby's cues,
the foundation of maternal sensitivity. Today, more opioid addicts are
giving birth, and more brand-new moms are becoming addicts, which
is quite understandable, given our modern maternal straits and the
traumatic past, stressful present, and uncertain future that so many of
us face. Nearly 2 percent of new mothers get permanently hooked on
the very painkillers they receive in the hospital.

A less dire but even more widespread challenge is the growing
role of technology in mothers' lives. Long before she gets around to
worrying about her kids' screen time, a mother's own use of technology can distract her from the essential physicality of her baby—from
the first Instagrammable moment after the child is born, to every time
she breastfeeds while scrolling through her phone. Scrambling babies'
cues, so-called "technoference" can change everything from how moms
read to kids to how determined we are to coax them to eat artichoke
hearts. Moms like me often text from the helm of our strollers, instead
of chatting with our babies. These attention-stealing technologies are
another coping mechanism for lonely moms, yet the competing stimulus can also be plainly dangerous. Around 2007, emergency rooms
saw a marked rise in child traumatic brain injuries, which some in the
Freakomomics camp blamed on the Great Recession, and mothers'
subsequent stress. A Yale study, though, linked this worrisome spike
not to the recent economic downturn, but to the spread of mobile
phones, and mothers' subsequent distraction.

All of these problems, some ancient and others brand-new, are daunting and complex. There are no quick fixes. But, in ideas and programs from around the world, there are ways to begin—baby steps, if you will.

More money for mothers is the simplest solution by far. I'm by no means above taking the blatant mom bribes that several European countries now dangle before multiple-birthers like myself. In Italy, I might—according to the recently proposed (and rather charmingly forthright) "land for children" plan—be awarded a piece of government-owned farmland for my third child. In Hungary, the birth of my fourth would mean no more income tax; in Poland, I'd score a lifetime pension.

Some of these plans may be adopted out of genuine solicitude for moms, others more out of panic at plummeting fertility rates. But the motivation matters less than the impact: this type of promised material reward is not a mere mom perk but a biological prompt, making mothers feel more secure and potential mothers more optimistic.

Their American equivalent, the child tax credit, is less reassuring because it's less generous. But ideas to make it more effective, with a larger credit or a child allowance (a kind of guaranteed basic income for families), exist in outline in Washington, DC, meaning that a solid financial floor for American moms is—you'll have to pardon the birth imagery—just one legislative push away.

Next, let's turn to the hospital. Labor and delivery ward quality varies wildly even across our country, with C-section rates at particular hospitals ranging from 7 to nearly 70 percent; a mom's chances of undergoing major surgery rise or fall tenfold depending on which set of doors she walks through. Often the most scalpel-happy facilities cluster in the poorest neighborhoods. Yet the highly controlled, full-service environment of the hospital can still be a good birthplace for emerging mothers. It's the perfect launching pad for scientifically tested, pro-mom practices like breastfeeding, skin-to-skin contact, and proper postpartum diet.

That said, the average two days' stay is not nearly enough to truly establish even these simple habits, let alone to guide those shell-shocked first-timers to something like maternal competence. (Japan invites its new mothers to linger for a week or longer, which seems more like it.)

Nor are conditions within hospitals always ideal. The behavior of all kinds of mammalian moms, from sheep to gorillas, suffers in artificially cramped or crowded conditions, with some animals neglecting their babies. Humans are no different: studies suggest that new mothers in private rooms are less stressed and much more likely to breastfeed, and that these differences may last after they're discharged. Yet even fully insured moms delivering at leading American hospitals are shaken down for hundreds of dollars *per night* just for a tiny single recovery room and the right to basic privacy. By contrast, nations like Israel prize mothers' comfort, and private rooms are just the beginning. Amenities at "maternity hotels" may include hot tubs, juice bars, massages, "elegantly arranged Dead Sea toiletries," and (maybe best of all, to any new mom who's crinkled and crunched her way through an interminable hospital night) real down pillows. Several observant Jewish communities in America now offer similar resorts for "Kimpatorin," as postpartum women are known.

Who delivers new mothers as they, too, are being born matters. Needy new mothers are primed to think of labor and delivery nurses as guardian angels and our next best friends. But then the shift changes. These staffing cycles are based on union contracts, not on what's best for maternal health, and they mean that newborn mothers don't get continuous coverage from the same caregivers, which distorts our perceptions of social support. One researcher recommended to me that the hospitals be restructured so that every mom, or at the very least the high-risk ones, undergoes the whole admission-to-discharge journey with a small team that truly knows her story, even if the exact same nurse (she'll need to sleep at some point, even if we can't) isn't there every step of the way. As it is, maternity wards have a freaky déjà vu feel as you endlessly recite the same information to different people, yet a care should be guaranteed to the most vulnerable type of new mothers—the first-timer, especially one who is single, young, poor, a person of color, with a history of depression or trauma, who's had a C-section and is struggling to breastfeed, whose own mom is gone or somewhere far away, and who is due back at work in six weeks.

If nothing else, our government shouldn't be stripping resources from any women during this vulnerable, volatile period of maternal metamorphosis. For many poor American moms covered by Medicaid during pregnancy, the program stops at two months postpartum, often with predictable psychological consequences.

"There couldn't be worse timing," Catherine Monk says. "It feels to them like they are falling off a cliff."

∾

Of course, as I write this, the coronavirus has made even more mothers lose our footing, not to mention our minds, exacerbating every conceivable challenge. Single and working moms are in a state of despair, as bars and restaurants open but schools stay closed or go online-only, a contrast that lays bare the government's unconcern—I'm being charitable here—for overburdened mothers and how we function. Meanwhile, we are often cut off from grandparents and other forms of traditional support, more socially at sea than ever. Racial and class divides yawn wider, as well-off moms pay for private learning "pods" and tutors for their kids, while working-class moms risk infection every day just to keep paying for food, rent, and those infernal diapers. More than ever, it feels like nobody will stand up for us.

With rates of maternal depression and insomnia through the roof, and our finances everywhere imperiled, it's likely that the birth rate will dip further. Almost inevitably, an America with fewer moms will become even less friendly to motherhood—a Covid-19 catch-22 from which, even post-pandemic, there is no obvious escape. If so, we'll become just another data point of the global trend of swooning birth rates and women opting out forever.

I type these dire tidings with several bitten-to-the-quick finger-nails painted an absurd shade of Pepto-Bismol pink, and all the rest bare, because my older daughters—cooped up in the house for going on five months now—got bored in the middle of giving me a mani-cure and wandered off.

Yet, despite everything, the sight makes me smile.

⁊

It's also possible that I'm mildly delirious, because right smack in the middle of the international madness, of course I had to go and have that fourth baby I've been telling you about.

I'd seen the birth as a chance for redemption—a (very) big-girl ver-sion of a do-over, if you will. This time my husband could be reliably dispatched for ice cream instead of staggering off into nearby forests. Financial ruin would be miles back in the rearview mirror. I wouldn't end up bawling in my bedroom quite so often. Maybe I'd even finally make my own baby food.

Instead, after my husband returned home from a cross-country plane trip in my seventh month of pregnancy, he went to the emergency room with breathing difficulty. The kids and I were soon hacking away as well. In the beginning, nobody really knew what Covid-19 would do to our family, especially the unborn baby, which was terrifying. ("I'll come right over," my mom said, undaunted, as soon as she heard we were sick. But we were all put under quarantine before she could fire up her retirement hot rod and make it to our door.) After we recovered, I was barred from my prenatal appointments for weeks. School and life in general got canceled. Again our finances were on the rocks. And once more I was trapped in the attic, trying to finish a book (this one).

In what I feared was a truly terrible omen for my own maternal fate, just before the illness hit, little Clementine the hamster some-how produced a litter of eraser-pink babies and proceeded to eat all of them one by one, despite our best efforts to reassure her with carrots. This spectacle, I fretted, was particularly hideous for my kids because of my ongoing pregnancy. After receiving the tongue-lashing of a life-time from yours truly in fiercest mama bear mode for exposing my poor children to such depravity, the small-animals manager at the pet store all but begged us to surrender Clementine and select some other non-infanticidal female as our family mascot, free of charge.

My daughters, though, decided to keep the mother hamster they had.

"Nobody's perfect, you know," Daughter Two told me.

For some reason, my eyes filled with tears when she said this.

I reported for my fourth C-section at the very peak of the outbreak. The maternity ward looked grim as a morgue. There were no balloon bouquets, no overly ambitious newborn photographers. Nurses' masks shielded their expressions.

But I was okay—or at least, so much better than the last time. Writing this book helped: having canvassed the research, I was now suddenly able to spot the forces arrayed for and against me. I knew, for instance, that disease outbreaks can warp parental behavior, but as the days passed it became clear that this particular sickness spared most children: a true blessing for mothers. I could also see my own personal points of weakness, and appreciate my strengths. I was in bad shape, sure, but so were a lot of other people. As a hierarchical being, I'm built to assess the direness of my own situation against that of others, and this time the catastrophe felt shared. I also knew that family members and new friends I'd made, people who cared about me and my child-to-be, were looking out for us. I wouldn't necessarily be able to see these loved ones for a while, but I knew they were out there somewhere because they'd heaped food upon our doorstep.

My husband was able to be with me in the hospital, which at the time had been very much in doubt—until people pushed back against a proposed delivery-room-partner ban that would supposedly save on protective equipment. This small victory had lifted my spirits im-mensely, and not only because I now fully fathomed the power of a supportive partner's presence. Quite simply, it proved to me that others can notice moms' plights, listen to us, and change. As usual, fellow mothers are one another's best advocates.

Despite the fact that this was the height of social distancing, at the hospital it also became clear to me that I'd lucked into a new medical practice full of caring people. When my husband had to wait outside during my anesthesia, the practically sword-sized needle sliding into my spine as I trembled and shook, my doctor even volunteered to hold my hand.

And while I have a natural penchant for closing my eyes like a corpse on the operating table and pretending to be absolutely anywhere else, somehow or other this doctor, perhaps sensing that I'm a sucker for a big reveal, also talked me into test-driving a nifty new invention that science had debuted since my last delivery: a transparent C-section drape deployed at the critical moment. While gracefully concealing my viscera, this clear plastic window might let me for once catch a glimpse of the baby's emergence, and see what all the fuss was about.

Which is how—with four pregnancies, multiple failed field trips to sheep barns, and about a million maternal biology papers under my belt—I at long last got the chance to witness a mammalian birth. It was all it's cracked up to be.

The baby was born at 8 pounds, 12 ounces—my biggest yet. I named her after my mom.

❦

For this new arrival and her sisters, who I very much hope will become mothers some day, I would add a few final pointers, since we moms do dearly love to dish out advice, and because our lives are so often ruled by the actions of clueless governments, stressed-out great-great-grandmas, or even former versions of our own selves.

Babysit. Eat lots of fish. Invest in glass food-storage containers instead of plastic. Make fast friends, especially female ones, and keep them for life.

Build a master plan that anticipates and respects your biology. Get the best education you can, but beware the weight of college loans. Screen your partner very, very carefully, for his genes can mess with yours. Also, notice that even though romantic relationships and men in general are now officially unnecessary to motherhood, American moms with the most privileges and options still tend to get and stay married.

Wait to have kids but not too long. Exercise extreme caution when selecting your employer and your ob-gyn. Pile your breakfast plate with pancakes if your heart is set on a boy, but keep in mind that girls are just as lovely. Hire a doula. Figure out some way to snag that private room.

Hold your babies lots, because it will likely shape their brains forever, but don't feel compelled to recite Shakespeare for them (provided you can still recall a couplet or two), or to contort yourself on the playroom carpet. Go ahead and take as many weeks off from work as your situation allows. Avoid suburbs without sidewalks. Find a house near a playground. Find a house near me.

Above all, I would say this: do all you can for your children, but don't entirely turn your back on the wider world. It's really not good for you, and besides, there's so much left for us to do. This business of being rebuilt from the ground up is deeply inconvenient at times, but it can also be a gift. Moms see the world with fresh eyes. To mothers' key environmental must-haves of security and control, I would add an element of challenge.

Start a medical practice treating addicted mothers, as Emily ended up doing. Sign on as the principal of a school serving refugee moms, as another lifelong friend did. Become a neuroscientist—or a sheep farmer, or a photographer, or, what the heck, a kraamverzorgster. Write a book. Vote for mothers, or run for office yourself.

And always look out for other women, especially struggling ones. Because only we know what it's like to be reborn as somebody else.

Acknowledgments

I AM INDEBTED to the dozens of scientists—named in these pages, or participating behind the scenes—who generously shared their work and perspectives with me, and I'm especially grateful to those who welcomed me into their labs to see where the magic happens.

I am thankful for the many mothers in my life, first and foremost my own mom, Maureen Tucker, and my grandmothers, Iva Gwendolyn Tucker and Helen Patricia O'Neill.

Special thanks to my dear friend Amanda Bensen Fiegl, whose website, www.lifeupstaged.com, keeps her voice alive.

Thank you to the excellent and unflinching Karyn Marcus at Gallery Books, a believer in this project back when it was but a twinkle in our eyes. (Let's just say that both of us may have taken the book-as-baby metaphor a bit too far this time!) Thanks also to the wonderful Rebecca Strobel, who rolled up her sleeves in a midwife-like capacity, and to researcher Vicky Hallett for steadfast and timely assistance. (And thank you to Celeste for letting me borrow your mom.)

Thank you to Scott Waxman and Ashley Lopez at the Waxman Literary Agency for your years of support. Thanks to Terry Monmaney and the editors and staff at *Smithsonian* magazine for their kindness and understanding as I serially vanished to have children.

I'm grateful to my sister, Judith Tucker, for the care and feeding of my website, which is her brainchild, and to adorable Amon (her actual child), as well as to Steven Dang, father and uncle par excellence. Thank you to Emily Brunner, who's been with me every step and stumble of the way. Thank you to my late father, Harold Tucker, and to all who advised, consoled, and distracted me and otherwise pitched in: Jeanne Snow, Patricia Snow, Charles Douthat, Julie Leff, Annie Murphy Paul, Amy Sudmyer, Angie Pepin, Steve Kiehl, Hilary Nawrocki, Virginia Shiller, Rachel Horsting, Lyn Garrity, Laura Helmuth, Richard Prum, Heide Hendricks, Sharon and Satish Rege, and Sarah Mahurin (whose lovely nose provides flesh and blood proof that devoted moms need not have a functional sense of smell). Little-known fact: thanks to Sam Moyn and his superb confections, my mom brain now consists of precisely 87 percent cake.

Special thanks to all those who wrangled my kids, especially Debbie Whitney, Amy Zuniga, and Indrani Narine. Hats off to all our amazing elementary school teachers. Thank you to Christa Doran and the coaches at Tuff Girl Fitness—should some maximally stressed mom (or anybody else) wish to burn quads in good company, join in virtually at www.tuffgirlfitnessct.com. (Also, for a glimpse of what a tough mother really is, see Christa's personal website at www.lessonsfromlea.com.)

I am forever grateful to my little ones, loved at first sight, who continue to make and break me on a daily basis. Gwendolyn, Eleanor, Nicholas, and Rosemary Maureen: you guys are now figuring out how to read, so if some line or another should make you spit out your Cheerios, please know that your old mommy was only trying to pay the orthodontist's ransom, and that she adores you from the depths of her soul.

And to Ross, my impossibly kind and patient husband and the love of my life: thank you for sharing your luminous mind with me, along with your genes, your pizza crusts, your sniffles, your sleepless nights, and your joyful days, too.

Notes

Introduction: OF MICE AND MOMS

1   *About 50 percent spontaneously get better:* Gregory Lim, "Do fetal cells repair maternal hearts?," *Nature Reviews Cardiology* 9, no. 67 (Feb. 2012); G. M. Felker et al., "Underlying Causes and Long-Term Survival in Patients with Initially Unexplained Cardiomyopathy," *New England Journal of Medicine* 342, no. 15 (Apr. 13, 2000): 1077–84.

1   *Some mom hearts are practically as good as new:* Lili Barouch, "Peripartum Cardiomyopathy," Johns Hopkins Heart and Vascular Institute, accessed Oct. 21, 2020, https://www.hopkinsmedicine.org/health/conditions-and-diseases/peripartum-cardiomyopathy; Felker et al. "Underlying Causes."

3   *So it goes for much of the science surrounding the two billion or so human moms:* Save the Children, *State of the World's Mothers 2000*, May 2000, https://www.savethechildren.org/content/dam/usa/reports/advocacy/sowm/sowm-2000.pdf.

3   *Worldwide, more than 90 percent of all women become moms:* Laura Glynn, "Decoding the Maternal Brain," TEDx Talks video, July 3, 2014, https://www.youtube.com/watch?v=71LT-MnfMEY.

4   *It wasn't until 2014 that the National Institutes of Health confessed:* Janine A. Clayton and Francis S. Collins, "Policy: NIH to balance sex in cell and animal studies," *Nature* 509, no. 7500 (May 14, 2014): 282–83.

4   *But finally more scholars:* R. Lee et al., "Through babies' eyes: Practical and theoretical considerations of using wearable technology to measure parent-infant behaviour from the mothers' and infants' view points," *Infant Behavior and Development* 47 (May 2017): 62–71.

4   *sewing microphones into their onesies:* Laura Sanders, "Here's some slim science on temper tantrums," *ScienceNews*, Apr. 22, 2016.

4   *For even as Chaudhry and her team:* Sangeetha Vadakke-Madathil et al., "Multipotent fetal-derived Cdx2 cells from placenta regenerate the heart," *PNAS* 116, no. 24 (June 11, 2019): 11786–95; Amy M. Boddy et al., "Fetal microchimerism and maternal health: A review and evolutionary analysis of cooperation and conflict beyond the womb," *BioEssays* 37, no. 10 (Oct. 2015): 1106–18.

4   *"It's evolutionary biology," says Chaudhry:* Rina J. Kara et al., "Fetal Cells Traffic to Injured Maternal Myocardium and Undergo Cardiac Differentiation," *Circulation Research* 110, no. 11 (Nov. 14, 2011): 82–93.

5   *One decade-long Dutch study:* Mads Kamper-Jørgensen et al., "Male microchimerism and survival among women," *International Journal of Epidemiology* 43, no. 1 (Feb. 2014): 168–73.

5   *In a particularly famous case:* Kirby L. Johnson et al., "Significant fetal cell microchime-

269

rism in a nontransfused woman with hepatitis C: evidence of long-term survival and expansion," *Hepatology* 36, no. 5 (Nov. 2002): 1295–97.

5   *This sweet treachery, recognizable to any mom who's watched her kids:* William F. N. Chan et al., "Male Microchimerism in the Human Female Brain," *PLoS ONE* 7, no. 9 (Sept. 26, 2012): e45592.

7   *One scientific paper rather meanly documents the textbook:* Jonathan C. K. Wells, Lewis Griffin, and Philip Treleaven, "Independent changes in female body shape with parity and age: A life-history approach to female adiposity," *American Journal of Human Biology* 22, no. 4 (July–Aug. 2010): 456–62.

7   *It turns out, too, that the old wives' saying:* Stefanie L. Russell, Jeannette R. Ickovics, and Robert A. Yaffe, "Exploring Potential Pathways Between Parity and Tooth Loss Among American Women," *American Journal of Public Health* 98, no. 7 (July 2008): 1263–70; Frank Gabel et al, "Gain a child, lose a tooth? Using natural experiments to distinguish between fact and fiction," *Journal of Epidemiology and Community Health* 72, no. 6 (2018): 552–56.

7   *Those toothless old moms:* Tara Bahrampour, "Women's reproductive history may predict Alzheimer's risk," *Washington Post,* July 23, 2018.

8   *While more than half of new mothers:* Jeffery C. Mays, "1 in 5 Mothers Gets Post-Partum Depression. New York City Plans to Help," *New York Times,* Feb. 5, 2020; "Baby Blues After Pregnancy," March of Dimes, Feb. 2017.

8   *In the first month of motherhood:* Laura M. Glynn, Mariann A. Howland, and Molly Fox, "Maternal programming: Application of a developmental psychopathology perspective," *Development and Psychopathology* 30, no. 3 (Aug. 2018): 905–19; Esther Landhuis, "Why Women May Be More Susceptible to Mood Disorders," ScientificAmerican .com, Apr. 14, 2020, https://www.scientificamerican.com/article/why-women-may -be-more-susceptible-to-mood-disorders.

9   *A century ago and even more recently:* Ira Henry Freeman, "Kidnapper Seized; Baby Well, Happy," *New York Times,* May 21, 1953; "Kidnappings Laid to 'Mother Mania,'" *New York Times,* Aug. 26, 1923; "Wife Who Leaves Explains Herself," *New York Times,* July 30, 1912.

10 *It's satisfying to connect:* Jessica Wang, "Mindy Kaling's Mother's Day Post Is About Seeing the Holiday Through Her Daughter's Eyes," Bustle, May 10, 2020, https://www.bustle .com/p/mindy-kalings-mothers-day-post-includes-ode-to-her-daughter-22889571.

10 *A study last year suggested that many human moms are so bowled over:* Lauren J. Ralph, Diana Green Foster, and Corinne H. Rocca, "Comparing Prospective and Retrospective Reports of Pregnancy Intention in a Longitudinal Cohort of U.S. Women," *Perspectives on Sexual and Reproductive Health* 52, no. 1 (2020): 39–48.

11 *"Murfers," or mom surfers, hang ten:* Carina Chocano, "The Coast of Utopia," *Vanity Fair,* Aug. 2019.

12 *Tens of thousands of new moms:* Marc H. Bornstein, "Determinants of parenting," in *Developmental Psychopathology,* 3rd ed., vol. 4, *Risk, Resilience, and Intervention,* ed. Dante Cicchetti (Hoboken, NJ: John C. Wiley and Sons, 2016), 1.

12 *But while the West's declining birth rate may suggest:* Claire Cain Miller, "The U.S. Fertility Rate Is Down, Yet More Women Are Mothers," *New York Times,* Jan. 18, 2018.

12 *Even the millennials are joining up:* Gretchen Livingston, "More than a million Millennials are becoming moms each year," *Fact Tank,* Pew Research Center, May 4, 2018, https:// www.pewresearch.org/fact-tank/2018/05/04/more-than-a-million-millennials -are-becoming-moms-each-year/.

12 *We comprise a staggering portion of the American labor market:* Mark DeWolf, "12 Stats About Working Women," U.S. Department of Labor Blog, Mar. 6, 2017, https://www .ishn.com/articles/105943-stats-about-working-women.

12 *Marketing companies are eager to figure:* Joe Pinsker, "How Marketers Talk About Motherhood Behind Closed Doors," *Atlantic*, Oct. 10, 2018.

12 *According to the latest research, moms hit mobile consumer apps:* Connie Hwong, "4 Ways that New Babies Influence Consumer Behavior," vertoanalytics.com, Mar. 13, 2017, https://vertoanalytics.com/4-ways-babies-influence-consumer-behavior/; Bill Page et al., "Parents and children in supermarkets: Incidence and influence," *Journal of Retailing and Consumer Services* 40 (Jan. 2018): 31–39.

12 *"Remember the 'drudgery'":* Rebecca Brooks, "Mothers Have Higher Fear and Anxiety Than Fathers: What Does It Mean for Brands?," Forbes.com, Mar. 3, 2020, https://www.forbes.com/sites/forbesagencycouncil/2020/03/03/mothers-have-higher-fear-and-anxiety-than-fathers-what-does-it-mean-for-brands.

13 *Microsoft's eggheads:* Munmun De Choudhury, Scott Counts, and Eric Horvitz, "Major life changes and behavioral markers in social media: case of childbirth," *CSCW '13: Proceedings of the 2013 Conference on Computer Supported Cooperative Work* (Feb. 2013): 1431–42.

13 *Finally, we are a crucial voting bloc:* Jill S. Greenlee, *The Political Consequences of Motherhood* (Ann Arbor, MI: University of Michigan Press, 2014), 166.

13 *With nearly two dozen congresswomen tending:* Julia Marin Hellwege and Lisa A. Bryant, "Congress has a record number of mothers with children at home. This is why it matters," *Monkey Cage* (blog), *Washington Post*, Feb. 15, 2019.

13 *Maybe it's no wonder that today's moms-to-be have rates of depression:* Rebecca M. Pearson, et al., "Prevalence of Prenatal Depression Symptoms Among 2 Generations of Pregnant Mothers," *JAMA Network Open* 1, no. 3 (July 2018): e180725.

15 *One Princeton University–led study suggests:* The Avon Longitudinal Study of Parents and Children, Ilyana Kuziemko et al., "The Mommy Effect: Do Women Anticipate the Employment Effects of Motherhood?" (NBER Working Paper No. 24740, National Bureau of Economic Research, June 2018).

Chapter 1: MOMENTUM

18 *Since they are herd animals:* P. Mora Medina et al., "Sensory factors involved in mother-young bonding in sheep: A review," *Veterinární medicína* 61, no. 11 (Jan. 2018): 595–611.

20 *In one experiment, researchers placed a lamb:* Frédéric Lévy, Matthieu Keller, and Pascal Poindron, "Olfactory regulation of maternal behavior in mammals," *Hormones and Behavior* 46, no. 3 (Oct. 2004): 284–302; Frédéric Lévy, "Neural Substrates Involved in the Onset of Maternal Responsiveness and Selectivity in Sheep," in *Neurobiology of the Parental Brain*, ed. Robert S. Bridges (Burlington, MA: Elsevier 2008), 26.

20 *Moments after birth, ewes memorize their baby's specific scent:* Barend V. Burger et al., "Olfactory Cue Mediated Neonatal Recognition in Sheep, Ovis Aries," *Journal of Chemical Ecology* 37, no. 10 (Oct. 2011): 1150–63.

20 *In one experiment Canadian researchers presented new human moms with Baskin-Robbins ice cream tubs:* Alison S. Fleming, Meir Steiner, and Carl Corter, "Cortisol, Hedonics, and Maternal Responsiveness in Human Mothers," *Hormones and Behavior* 32, no. 2 (Oct. 1997): 85–98.

24 *According to one headline, "Potty Training Is a Scientific Mystery" that moms are incapable of solving:* Melissa Dahl, "Potty Training Is a Scientific Mystery," *New York*, Sept. 15, 2014.

24 *Indeed, as the average age of kid continence continues to rise:* Susan Davis, "Potty Training: Seven Surprising Facts," Grow (blog), WebMD, https://www.webmd.com/parenting/features/potty-training-seven-surprising-facts#1; Zachary Crockett, "The Evolution

of Potty Training," Priceonomics, Sept. 16, 2014, https://priceonomics.com/the-evolu
tion-of-potty-training/.

24   *No wonder moms are lining up to join groups like Loom:* Sheila Marikar, "A Club for New
     Parents in Los Angeles," *New Yorker,* Sept. 11, 2017.

24   *It's possible that human moms may have a touch of this "nesting instinct":* Marla V. Anderson
     and M. D. Rutherford, "Evidence of a nesting psychology during human pregnancy,"
     Evolution and Human Behavior 34, no. 6 (Nov. 2013): 390–97.

25   *Researchers can generally tell when mothers are speaking:* Elise A. Piazza, Marius Cătălin
     Iordan, and Casey Lew-Williams, "Mothers Consistently Alter Their Unique Vocal
     Fingerprints When Communicating with Infants," *Current Biology* 27, no. 20 (Oct. 23,
     2017): 3162–67.

25   *Singing to babies isn't universal either:* Helen Shoemark and Sarah Arnup, "A survey of
     how mothers think about and use voice with their hospitalized newborn infant," *Journal
     of Neonatal Nursing* 20, no. 3 (June 2014): 115–21.

26   *Something like 80 percent of right-handed women:* Sarah Blaffer Hrdy, *Mother Nature: A
     History of Mothers, Infants, and Natural Selection* (New York: Pantheon Books, 1999),
     105.

26   *Researchers recently catalogued lefty preference:* Andrey Giljov, Karina Karenina, and
     Yegor Malashichev, "Facing each other: mammal mothers and infants prefer the posi-
     tion favouring right hemisphere processing," *Biology Letters* 14, no. 1 (Jan. 10, 2018).

26   *Researchers who thumbed through family photo albums found:* Gianluca Malatesta, "The
     left-cradling bias and its relationship with empathy and depression," *Scientific Reports* 9,
     no. 6141 (Apr. 2019): 1–9.

26   *Some fascinating work suggests that babies cradled on their mothers' right sides grow up to
     have a diminished ability to read faces:* M. P. Vervloed, A. W. Hendriks, and E. van den
     Eijnde, "The effects of mothers' past infant-holding preferences on their adult chil-
     dren's face processing lateralisation," *Brain and Cognition* 75, no. 3 (Apr. 2011): 248–54;
     A. W. Hendriks, M. van Rijswijk, and D. Omtzigt, "Holding-side influences on infant's
     view of mother's face," *Laterality* 16, no. 6 (2011): 641–55.

27   *In a recent and rather adorable experiment:* Gillian S. Forrester et al., "The left cradling
     bias: An evolutionary facilitator of social cognition?," *Cortex* 118 (Sept. 2019): 116–31.

27   *In one 2012 study, childless Italian adults viewed pictures of unfamiliar human babies:* An-
     drea Caria et al., "Species-specific response to human infant faces in the premotor cor-
     tex," *NeuroImage* 60, no. 2 (Apr. 2, 2012): 884–93.

27   *It transcends race and ethnicity as well:* Gianluca Esposito et al., "Baby, You Light-Up My
     Face: Culture-General Physiological Responses to Infants and Culture-Specific Cogni-
     tive Judgements of Adults," *PLoS ONE* 9, no. 10 (Oct. 29, 2014): e106705.

27   *In a study of British neurosurgery patients:* Christine E. Parsons et al., "Ready for action: a
     role for the human midbrain in responding to infant vocalizations," *Social Cognitive and
     Affective Neuroscience* 9, no. 7 (July 2014): 977–84.

28   *Adults who'd just heard a baby's whimper:* Christine E. Parsons et al., "Listening to infant
     distress vocalizations enhances effortful motor performance," *Acta Paediatrica* 101, no.
     4 (2012): e189–e191.

29   *In another scent-based study, which involved whiffing cheese, spices, and babies' T-shirts:*
     Alison S. Fleming et al., "Postpartum factors related to mother's attraction to newborn
     infant odors," *Developmental Psychobiology* 26, no. 2 (Mar. 1993): 115–32.

35   *Likewise, while I felt out-of-the-blue clobbered by love upon meeting my daughter:* Carl M.
     Corter and Alison S. Fleming, "Psychobiology of Maternal Behavior in Human Be-
     ings," in *Handbook of Parenting,* 2nd ed., vol. 2, *Biology and Ecology of Parenting,* ed. Marc
     H. Bornstein (Mahwah, NJ: Lawrence Erlbaum 2002), 147.

35   *In one early study, new rat moms were given the chance to press a bar to receive:* W. E.

Wilsoncroft, "Babies by bar-press: Maternal behavior in the rat," *Behavior Research Methods & Instrumentation* 1 (1968): 229–30.

35 *You can blind her:* Frank A. Beach and Julian Jaynes, "Studies of Maternal Retrieving in Rats. III. Sensory Cues Involved in the Lactating Female's Response to Her Young," *Behaviour* 10, no. 1 (1956): 104–25; L. R. Herrenkohl, P. A. Rosenberg, "Exteroceptive stimulation of maternal behavior in the naive rat," *Physiology & Behavior* 8, no. 4 (Apr. 1972): 595–98.

36 *In a 2013 smell-based experiment, thirty women sniffed at a mystery item:* Johan N. Lundström et al., "Maternal status regulates cortical responses to the body odor of newborns," *Frontiers in Psychology* 4, no. 597 (Sept. 5, 2013).

36 *One 2014 experiment, titled:* Chloe Thompson-Booth et al., "Here's looking at you, kid: attention to infant emotional faces in mothers and non-mothers," *Developmental Science* 17, no. 1 (Jan. 2014): 35–46.

36 *Using a technique called "near-infrared spectroscopy," Japanese scientists:* Shota Nishitani et al., "Differential prefrontal response to infant facial emotions in mothers compared with non-mothers," *Neuroscience Research* 70, no. 2 (Feb. 2011): 183–8.

37 *This neural switcheroo likely helps explain why moms persist:* Linda Mayes, "The Neurobiology of Parenting and Attachment," Sigmund Freud Institut video, Mar. 15, 2012, https://www.youtube.com/watch?v=feUjK2PRwIM.

37 *Other people may avoid despondent children, but moms are propelled to approach:* Erich Seifritz et al., "Differential sex-independent amygdala response to infant crying and laughing in parents versus nonparents," *Biological Psychiatry* 54, no. 12 (Dec. 15, 2003): 1367–75.

38 *Even if our one-and-only is draped in the exact same gray cloth as every other kid in the experiment:* G. Esposito et al., "Immediate and selective maternal brain responses to own infant faces," *Behavioural Brain Research* 278 (Feb. 1, 2015): 40–43.

39 *Studies suggest that we can recognize them simply by stroking the silken backs of their hands:* Marsha Kaitz et al., "Infant recognition by tactile cues," *Infant Behavior and Development* 16, no. 3 (July–Sept. 1993): 333–41.

39 *Even their particular diapers smell dreamy to us:* Trevor I. Case, Betty M. Repacholi, and Richard J. Stevenson, "My baby doesn't smell as bad as yours: The plasticity of disgust," *Evolution and Human Behavior* 27, no. 5 (Sept. 2006): 357–65.

39 *Our heart rates accelerate:* Alan R Wiesenfeld, Carol Zander Malatesta, and Linda L. Deloach, "Differential parental response to familiar and unfamiliar infant distress signals," *Infant Behavior and Development* 4 (Mar. 1981): 281–95.

39 *In fact, within forty-eight hours of birth:* David Fornby, "Maternal Recognition of Infant's Cry," *Developmental Medicine & Child Neurology* 9, no. 3 (June 1967): 293–98.

40 *Roughly 90 percent of new mothers report being "in love":* Bornstein, "Determinants of Parenting," 2.

40 *Mother love is the planet's original romance:* Michael Numan and Larry J. Young, "Neural mechanisms of mother–infant bonding and pair bonding: Similarities, differences, and broader implications," *Hormones and Behavior* 77 (Jan. 2016): 98–112.

41 *Freshly fledged moms are explicitly thinking about their babies:* Paul Raeburn, *Do Fathers Matter? What Science Is Telling Us About the Parent We've Overlooked* (New York: Farrar Straus & Giroux, 2013), 130.

41 *Scientists think baby mania might help explain the evolutionary basis of obsessive-compulsive disorder:* Emily S. Miller et al., "Obsessive-Compulsive Symptoms During the Postpartum Period," *Journal of Reproductive Medicine* 58, nos. 3–4 (Mar.–Apr. 2013): 115–22.

41 *A Leiden University–led lab recently:* Elseline Hoekzema et al., "Pregnancy leads to long-lasting changes in human brain structure," *Nature Neuroscience* 20 (2017): 287–96.

41 *These gray matter losses may total up to 7 percent:* Angela Oatridge et al., "Change in

Brain Size during and after Pregnancy: Study in Healthy Women and Women with Pre-eclampsia," *American Journal of Neuroradiology* 23, no. 1 (Jan. 2002): 19–26.

Chapter 2: DAD GENES

44 *Even before we become parents, these sex-based differences guide:* Rodrigo A. Cárdenas, Lauren Julius Harris, and Mark W. Becker, "Sex differences in visual attention toward infant faces," *Evolution and Human Behavior* 34, no. 4 (July 2013): 280–87; Irene Messina et al., "Sex-Specific Automatic Responses to Infant Cries: TMS Reveals Greater Excitability in Females than Males in Motor Evoked Potentials," *Frontiers in Psychology* 6, no. 1909 (Jan. 7, 2016): Amanda C. Hahn et al., "Gender differences in the incentive salience of adult and infant faces," *Quarterly Journal of Experimental Psychology* 66, no. 1 (Jan. 2013): 200–208.

45 *Conveniently enough, reproductive-aged women:* R. Sprengelmeyer et al., "The Cutest Little Baby Face: A Hormonal Link to Sensitivity in Cuteness in Infant Faces," *Psychological Science* 20, no. 2 (Feb. 2009): 149–54; Janek S. Lobmaier et al., "Menstrual cycle phase affects discrimination of infant cuteness," *Hormones and Behavior* 70 (Apr. 2015): 1–6.

45 *In one National Institutes of Health–led experiment:* Nicola De Pisapia et al., "Sex Differences in Directional Brain Responses to Infant Hunger Cries," *NeuroReport* 24, no. 3 (Feb. 13, 2013): 142–46; "Women's, men's brains respond differently to hungry infant's cries," National Institutes of Health news release, May 6, 2013, https://www.nih.gov/news-events/news-releases/womens-mens-brains-respond-differently-hungry-infants-cries.

45 *Another experiment used facial temperature:* Gianluca Esposito et al., "Using infrared thermography to assess emotional responses to infants," *Early Child Development and Care* 185, no. 3 (2015): 438–47.

46 *Female birds aren't quite so carefree:* Douglas W. Mock, "The Evolution of Relationships in Nonhuman Families," in *Oxford Handbook of Evolutionary Family Psychology*, ed. Catherine Salmon and Todd Shackelford (New York: Oxford University Press, 2011), 59.

47 *As the maternal behavior scholar Laura Glynn noted in a TEDx Talk:* Glynn, "Decoding the Maternal Brain."

47 *As much as we human moms justifiably lament bad or absent fathers:* Mock, "The Evolution of Relationships," 59.

48 *Modern fathers have clearly picked up plenty:* Gretchen Livingston, "Growing Number of Dads Home with the Kids," *Social & Demographic Trends*, Pew Research Center, June 5, 2014.

48 *There are more single fathers than ever:* Stephanie Kramer, "U.S. has world's highest rate of children living in single-parent households," *Fact Tank*, Pew Research Center, Dec. 12, 2019, https://www.pewsocialtrends.org/2014/06/05/growing-number-of-dads-home-with-the-kids/; Raeburn, *Do Fathers Matter?*, 213.

48 *"Why do men feel more attractive after childbirth?" one study mused:* Alicia D. Cast, Susan D. Stewart, and Megan J. Erickson, "Why do men feel more attractive after childbirth?" *Journal of Gender Studies* 22, no. 3 (2013): 335–43.

49 *New fathers' testosterone levels, for instance, oftentimes plummet:* Lee T. Gettler et al., "Longitudinal evidence that fatherhood decreases testosterone in human males," *PNAS* 108, no. 39 (Sept. 27, 2011): 16194–99.

49 *In baby-cue experiments, new dads are quantifiably more reactive to infant cues:* Alison S. Fleming et al., "Testosterone and Prolactin Are Associated with Emotional Responses to Infant Cries in New Fathers," *Hormones and Behavior* 42, no. 4 (Dec. 2002): 399–413;

Ina Schicker, "For Fathers and Newborns, Natural Law and Odor," *Washington Post*, Feb. 26, 2001.

49   *Some studies have spotted similar neural patterns:* Jennifer S. Mascaro, Patrick D. Hackett, and James K. Rilling, "Testicular volume is inversely correlated with nurturing-related brain activity in human fathers," *PNAS* 110, no. 39 (Sept. 24, 2013): 15746–51.

49   *Resident dads can match even moms when it comes to recognizing a particular kid's cry:* Erik Gustafsson et al., "Fathers are just as good as mothers at recognizing the cries of their baby," *Nature Communications* 4 (2013): 1698.

50   *A year and a half after their kids were born:* Caroline Pape Cowan et al., "Transitions to Parenthood: His, Hers, and Theirs," *Journal of Family Issues* 6, no. 4 (Dec. 1985): 451–81.

50   *Overall, mothers remain more sensitized to their child's emotions:* Christine E. Parsons et al., "Interpreting infant emotional expressions: Parenthood has differential effects on men and women," *Quarterly Journal of Experimental Psychology* 70, no. 3 (Mar. 2017): 554–64.

50   *When a child weeps:* Raeburn, *Do Fathers Matter?*, 133.

50   *It takes somewhat more urgent-sounding cries:* David Richter et al., "Long-term effects of pregnancy and childbirth on sleep satisfaction and duration of first-time and experienced mothers and fathers," *Sleep* 42, no. 4 (Apr. 2019): zsz015.

50   *In some studies, researchers played the whimpers of babies hungry for food:* Corter and Fleming, "Psychobiology," 167.

50   *We also have a better memory for babies' smiles:* ibid., 157.

50   *Moms think about our babies twice as often as dads do:* Raeburn, *Do Fathers Matter?*, 130.

50   *We speak to them far more:* Katharine Johnson et al., "Gender Differences in Adult-Infant Communication in the First Months of Life," *Pediatrics* 134, no. 6 (Dec. 2014): e1603–10.

50   *In one study, researchers plopped audio recorders:* Kevin Dudley, "Who Uses Baby Talk More—Moms or Dads?" WSU Health Sciences Spokane Extra, May 19, 2015, https://spokane.wsu.edu/extra/2015/05/19/who-uses-baby-talk-more-moms-or-dads/.

51   *But even with older children:* Daniel Paquette, "Theorizing the Father-Child Relationship: Mechanisms and Developmental Outcomes," *Human Development* 47, no. 4 (Aug. 2004): 193–219.

51   *Moms who work outside the home:* Adele Eskes Gottfried, Allen W. Gottfried, and Kay Bathurst, "Maternal and Dual-Earner Employment Status and Parenting," in Bornstein, *Handbook of Parenting*, 2:214.

51   *We tend to soothe and hold:* Daniel Paquette and Marc Bigras, "The risky situation: A procedure for assessing the father–child activation relationship," *Early Child Development and Care* 180, nos. 1–2 (2010): 33–50.

51   *In one experiment in rural Tanzania:* Jana Vyrastekova et al., "Mothers More Altruistic than Fathers, but Only When Bearing Responsibility Alone: Evidence from Parental Choice Experiments in Tanzania," *PLoS ONE* 9, no. 6 (June 25, 2014): e99952.

52   *A recent study of ninety-one men:* C. Allen et al., "Preparation for fatherhood: A role for olfactory communication during human pregnancy?," *Physiology & Behavior* 206 (July 1, 2019): 175–80.

52   *For instance, one study showed that those fathers:* Lenna Nepomnyaschy and Jane Waldfogel, "Paternity leave and fathers' involvement with their young children," *Community, Work & Family* 10, no. 4 (2007): 427–53.

52   *A rodent study revealed that mice moms:* Hong-Xiang Lu et al., "Displays of paternal mouse pup retrieval following communicative interaction with maternal mates," *Nature Communications* 4 (2013): 1346.

53   *But still, almost everywhere, dads are poised:* Fabian Probst et al., "Do women tend while men fight or flee? Differential emotive reactions of stressed men and women while viewing newborn infants," *Psychoneuroendocrinology* 75 (Jan. 2017): 213–21.

53 *They include the harshness of the setting:* David F. Bjorklund and Ashley C. Jordan, "Human Parenting from an Evolutionary Perspective," in *Gender and Parenthood: Biological and Social Scientific Perspectives,* ed. W. Bradford Wilcox and Kathleen Kovner Kline (New York: Columbia University Press, 2013), 74–76.

53 *On the other hand, dads are more prone to jump ship:* Hrdy, *Mother Nature,* 233.

53 *When reporters returned to check on Zika-affected mothers:* Ueslei Marcelino, "Mothers of babies afflicted by Zika fight poverty, despair," Reuters, Oct. 17, 2018, https://www .reuters.com/article/us-health-zika-brazil-widerimage/mothers-of-babies-afflicted -by-zika-fight-poverty-despair-idUSKCN1MR0F9.

53 *The certainty of paternity also matters:* Harald Euler, "Grandparents and Extended Kin," in Salmon and Shackelford, *Oxford Handbook,* 183.

53 *The current record belongs to Namibia's Himba tribe:* B. A. Scelza et al., "High rate of extra-pair paternity in a human population demonstrates diversity in human reproductive strategies," *Science Advances* 6, no. 8 (Feb. 19, 2020): eaay 6195.

53 *One fascinating study showed that the rare area of infant-reading:* Haiyan Wu et al., "The male advantage in child facial resemblance detection: Behavioral and ERP evidence," *Social Neuroscience* 8, no. 6 (2013): 555–67.

54 *Yet academics who (perhaps traitorously):* Robin Wilson and Piper Fogg, "On Parental Leave, Men Have It Easier," Chronicle of Higher Education, Jan. 7, 2005, https://www .chronicle.com/article/on-parental-leave-men-have-it-easier/.

54 *Maybe this begins to explain why, throughout academia, married fathers:* Katherine Ellison, *The Mommy Brain: How Motherhood Makes Us Smarter* (New York: Basic Books, 2005), 163.

54 *In another study, academics with children under age two:* Steven E. Rhoads and Christopher H. Rhoads, "Gender roles and infant/toddler care: Male and female professors on the tenure track," *Journal of Social, Evolutionary, and Cultural Psychology* 6, no. 1 (2012): 13–31.

54 *Preliminary work on gay dads is especially interesting:* Eyal Abraham et al., "Father's brain is sensitive to childcare experiences," *PNAS* 111, no. 27 (July 8, 2014): 9792–97.

55 *Twenty-seven percent of kids live apart from their biological dads now:* Raeburn, *Do Fathers Matter?,* 220; Sara McLanahan and Christopher Jencks, "Was Moynihan Right?," Education Next, Spring 2015, https://www.educationnext.org/was-moynihan-right/.

55 *America has the highest rate of single-mother-headed households in the world:* Kramer, "U.S. has world's highest rate of children living in single-parent households."

55 *A recent study of Canadian single dads:* Maria Chiu et al., "Mortality in single fathers compared with single mothers and partnered parents: a population-based cohort study," *Lancet Public Health* 3, no. 3 (Mar. 1, 2018): e115–e123.

55 *"Men have so many other concerns . . .":* Wenda Trevathan, *Ancient Bodies, Modern Lives* (New York: Oxford University Press, 2010), 16.

55 *In "the vast majority" . . . moms and other females provide more care for infants than dads do:* Marc H. Bornstein, "Parenting × Gender × Culture × Time," in Wilcox and Kline, *Gender and Parenthood,* 98–99.

55 *I'm tempted to up the ante:* Alexandra Topping, "Finland: The only country where fathers spend more time with kids than mothers," *Guardian,* Dec. 5, 2017.

55 *This pattern holds true:* Bornstein, "Parenting × Gender × Culture × Time," 98.

56 *But then Covid-19 came along:* Tim Henderson, "Mothers are 3 times more likely than fathers to have lost jobs in the COVID-19 pandemic," *Chicago Tribune,* Sept. 30, 2020.

56 *For humans across cultures, according to scientists who've run the numbers:* Rebecca Sear and Ruth Mace, "Who keeps children alive? A review of the effects of kin on child survival," *Evolution and Human Behavior* 29, no. 1 (January 2008): 1–18.

57 *It's been a slithery downhill slide for the placenta:* Roberto Romero, "Images of the human

placenta," *American Journal of Obstetrics and Gynecology* 213, no. 4, suppl. (Oct. 1, 2015): S1–S2.

59 *A famous series of experiments from the 1980s:* Anne C. Ferguson-Smith and Deborah Bourc'his, "The discovery and importance of genomic imprinting," *eLife* 7 (Oct. 22, 2018): e42368; Raeburn, *Do Fathers Matter?*, 46–66.

59 *In less than 1 percent:* Marco Del Giudice and Jay Belsky, "Parent-Child Relationships," in Salmon and Shackelford, *Oxford Handbook*, 74–76.

59 *While imprinting does happen elsewhere:* Xu Wang et al., "Paternally expressed genes predominate in the placenta," *PNAS* 110, no. 26 (June 25, 2013): 10705–10; Courtney W. Hanna, "Placental imprinting: Emerging mechanisms and functions," *PLoS Genetics* 16, no. 4 (Apr. 23, 2020): e1008709.

59 *Biologists see genomic imprinting in the placenta:* T. Moore, "Review: Parent–offspring conflict and the control of placental function," *Placenta* 33, suppl. (Feb. 2012): S33–S36; David Haig, "Maternal–fetal conflict, genomic imprinting and mammalian vulnerabilities to cancer," *Philosophical Transactions of the Royal Society* 370, no. 1673 (July 19, 2015): 20140178.

62 *A mature human placenta has more than thirty miles of surface area:* Hrdy, *Mother Nature*, 434.

65 *Mostly unknown in other mammals, hemorrhage impacts roughly 10 percent of our births:* Elizabeth Abrams and Julienne Rutherford, "Framing postpartum hemorrhage as a consequence of human placental biology: an evolutionary and comparative perspective," *American Anthropologist* 113, no. 3 (Sept. 2011): 417–30.

65 *A British lab recently published an unnerving paper:* H. D. J. Creeth et al., "Maternal care boosted by paternal imprinting in mammals," *PLoS Biology* 16, no. 7 (July 31, 2018): e2006599.

Chapter 3: THE WHOLE SHEBANG

73 *They set up a series of experiments:* Bianca J. Marlin et al., "Oxytocin enables maternal behaviour by balancing cortical inhibition," *Nature* 520 (2015): 499–504.

74 *In serveral experiments when childless women snorted jolts of oxytocin:* Sarah K. C. Holtfrerich et al., "Endogenous testosterone and exogenous oxytocin influence the response to baby schema in the female brain," *Scientific Reports* 8, no. 7672 (May 16, 2018); Madelon M. E. Riem et al., "Oxytocin Modulates Amygdala, Insula, and Inferior Frontal Gyrus Responses to Infant Crying: A Randomized Controlled Trial," *Biological Psychiatry* 70, no. 3 (Aug. 1, 2011): 291–97; Helena J. V. Rutherford et al., "Intranasal oxytocin and the neural correlates of infant face processing in non-parent women," *Biological Psychology* 129 (Oct. 2017): 45–48.

77 *The animals' mouths still work fine in these experiments:* Michael Numan and Keith P. Corodimas, "The effects of paraventricular hypothalamic lesions on maternal behavior in rats," *Physiology & Behavior* 35, no. 3 (Sept. 1985): 417–25.

77 *In fact, together these two synced-up areas:* Yi-Ya Fang et al., "A Hypothalamic Midbrain Pathway Essential for Driving Maternal Behaviors," *Neuron* 98, no. 1 (Apr. 4, 2018): 192–207.

78 *"Strangest thing in your mom purse?":* "Strangest thing in your mom purse?," DC Urban Moms (and Dads), Dec. 10, 2019, https://www.dcurbanmom.com/jforum/posts/list/15/844578.page.

79 *Alongside the previously described and somewhat distressing evidence for the withering:* Pilyoung Kim et al., "The plasticity of human maternal brain: Longitudinal changes in brain anatomy during the early postpartum period," *Behavioural Neuroscience* 124, no. 5 (Oct. 2010): 695–700.

83   *In one study, fourteen birth mothers . . . showed extra neural excitement for their own kids:* Damion J. Grasso et al., "ERP correlates of attention allocation in mothers processing faces of their children," *Biological Psychology* 81, no. 2 (May 2009): 95–102.

83   *In another experiment, foster mothers and their young infants:* Johanna Bick et al., "Foster Mother–Infant Bonding: Associations Between Foster Mothers' Oxytocin Production, Electrophysiological Brain Activity, Feelings of Commitment, and Caregiving Quality," *Child Development* 84, no. 3 (May–June 2013): 826–40.

84   *The two groups seem to respond somewhat differently:* M. Pérez-Hernández et al., "Listening to a baby crying induces higher electroencephalographic synchronization among prefrontal, temporal and parietal cortices in adoptive mothers," *Infant Behavior and Development* 47 (May 2017): 1–12; M. Hernández-González et al., "Observing videos of a baby crying or smiling induces similar, but not identical, electroencephalographic responses in biological and adoptive mothers," *Infant Behavior and Development* 42 (Feb. 2016): 1–10.

84   *Which is not to say that one is better:* Joan T. D. Suwalsky, Charlene Hendricks, and Marc H. Bornstein, "Families by Adoption and Birth: I. Mother-Infant Socioemotional Interactions," *Adoption Quarterly* 11, no. 2 (Oct. 2008): 101–25.

Chapter 4: MOMMY WEIRDEST

87   *Scientists aren't sure exactly what disturbs maternal sleep:* Leigh Ann Henion, "Do New Moms Dream Differently After Giving Birth?" *New York Times*, Apr. 16, 2020.

87   *Pregnant women, for instance, mysteriously:* Donald A. Redelmeier et al., "Pregnancy and the risk of a traffic crash," *Canadian Medical Association Journal* 186, no. 10 (July 8, 2014): 742–50.

88   *Experimented-upon moms have been shown to be less grossed out:* Pavol Prokop and Jana Fančovičová, "Mothers are less disgust sensitive than childless females," *Personality and Individual Differences* 96 (2016): 65–69.

88   *We might exhibit an increased appetite for salt:* Robert S. Bridges, "Long-Term Alterations in Neural and Endocrine Processes Induced by Motherhood," *Hormones and Behavior* 77 (Jan. 2016): 193–203.

88   *Mother rats are also hotter than their pre-maternal selves:* Alison Fleming and Ming Li, "Psychobiology of Maternal Behavior and Its Early Determinants in Nonhuman Mammals," in Bornstein, *Handbook of Parenting*, 69.

88   *stressful events like mock job interviews:* A. V. Klinkenberg et al., "Heart rate variability changes in pregnant and non-pregnant women during standardized psychosocial stress," *Acta Obstetrica et Gynecologica Scandinavica* 88, no. 1 (2009): 77–82.

88   *We excel at enduring physical discomfort:* Kalevi Vähä-Eskeli et al., "Effect of thermal stress on serum prolactin, cortisol and plasma arginine vasopressin concentration in the pregnant and non-pregnant state," *European Journal of Obstetrics & Gynecology and Reproductive Biology* 42, no. 1 (Nov. 3, 1991): 1–8.

88   *If science requires us to stick our hands in buckets full of ice water for a minute:* Martin Kammerer et al., "Pregnant women become insensitive to cold stress," *BMC Pregnancy and Childbirth* 2, no. 8 (2002).

89   *We are also relatively unruffled:* Heather A. Rupp et al., "Amygdala response to negative images in postpartum vs nulliparous women and intranasal oxytocin," *Social Cognitive and Affective Neuroscience* 9, no. 1 (Jan. 2014): 48–54.

89   *Women experiencing "major life events":* Laura M. Glynn et al., "Pregnancy affects appraisal of negative life events," *Journal of Psychosomatic Research* 56, no. 1 (Jan. 2004): 47–52.

89   *Afterward, maternal behavior researchers at the University of California, Irvine:* Laura M. Glynn et al., "When stress happens matters: Effects of earthquake timing on stress responsivity in pregnancy," *American Journal of Obstetrics and Gynecology* 184, no. 4 (Mar. 2001): 637–42.

90   *My strangely stable earthquake mind . . . a California bank teller:* Kirk Johnson, "Amid 'Exploding' Houses and a Wave of Mud, a Maternal Instinct Flared," *New York Times,* Apr. 9, 2014.

91   *Mammalian moms are, studies suggest, suckers for babies across the board:* Susan Lingle and Tobias Riede, "Deer Mothers Are Sensitive to Infant Distress Vocalizations of Diverse Mammalian Species," *American Naturalist* 184, no. 4 (Oct. 2014): 510–22.

91   *Laboratory tests suggest that moms' brain waves are enhanced even:* M. Purhonen et al., "Effects of maternity on auditory event-related potentials to human sound," *NeuroReport* 12, no. 13 (Sept. 7, 2001): 2975–79.

91   *Compared to non-mothers, human mothers:* Christine E. Parsons et al., "Duration of motherhood has incremental effects on mothers' neural processing of infant vocal cues: a neuroimaging study of women," *Scientific Reports* 7, no. 1727 (2017).

92   *Policewomen who return to work after giving birth:* Bornstein, "Determinants of Parenting," 16.

92   *In one typical experiment:* Helena J. V. Rutherford, Angela N. Maupin, and Linda C. Mayes, "Parity and neural responses to social and non-social stimuli in pregnancy," *Social Neuroscience* 14, no. 5 (Oct. 2019): 545–48.

92   *Another study showed:* Levente L. Orbán and Farhad N. Bastur, "Shifts in color discrimination during early pregnancy," *Evolutionary Psychology* 10, no. 2 (May 2012): 238–52.

92   *Pregnant women may be unusually good at assessing "apparent health":* B. C. Jones et al., "Menstrual cycle, pregnancy and oral contraceptive use alter attraction to apparent health in faces," *Proceedings of the Royal Society B* 272, no. 1561 (Feb. 22, 2005): 20042962.

92   *Moreover, we are adept at detecting the emotions of others:* R. M. Pearson, S. L. Lightman, and J. Evans, "Emotional sensitivity for motherhood: Late pregnancy is associated with enhanced accuracy to encode emotional faces," *Hormones and Behavior* 56, no. 5 (Nov. 2009): 557–63.

92   *In one experiment, mothers of toddlers:* Ellison, *Mommy Brain,* 132.

92   *Usually women are best at interpreting the faces of other females:* Catharina Lewin and Agneta Herlitz, "Sex differences in face recognition—Women's faces make the difference," *Brain and Cognition* 50, no. 1(Oct. 2002): 121–28.

92   *But mothers seem to switch to studying the faces of adult males:* Marla V. Anderson and M. D. Rutherford, "Recognition of novel faces after single exposure IS enhanced during pregnancy," *Evolutionary Psychology* 9, no. 1 (Feb. 2011) 47–60.

92   *Mothers tend to rate the faces of potential criminals as more menacing:* Daniel M. T. Fessler et al., "Stranger danger: Parenthood increases the envisioned bodily formidability of menacing men," *Evolution and Human Behavior* 35, no. 2 (Mar. 2014): 109–17.

93   *Everyone knows the stories:* S. D. Coté, A. Peracino, and G. Simard, "Wolf, *Canis lupus,* Predation and Maternal Defensive Behavior in Mountain Goats, *Oreamnos americanus,*" *Canadian Field-Naturalist* 111 (1997): 389–92.

93   *We've read the headlines:* Danielle Wallace, "Protective walrus mom sinks Russian navy boat in Arctic Sea," *New York Post,* Sept. 24, 2019.

93   *Milk cows, it turns out, are far more dangerous:* Colin G. Murphy et al., "Cow-related trauma: A 10-year review of injuries admitted to a single institution," *Injury* 41, no. 5 (May 2010): 548–50; M. Sheehan and C. Deasy, "A Descriptive Study of the Burden of Animal-Related Trauma at Cork University Hospital," *Irish Medical Journal* 111, no. 1 (Jan. 10, 2018): 673. Jess Staufenberg, "Cows officially the most deadly large animals in Britain," *Independent,* Nov. 9, 2015.

93   *There are some very satisfying online videos of dairy cows kicking the dirty diapers:* "Getting run over by an angry momma cow," YouTube video, May 21, 2016, https://www.you tube.com/watch?v=U99j7WzMAYA.

94   *In one fascinating study, researchers played the sounds of rattlesnakes to a bunch of California ground squirrels:* Ronald R. Swaisgood, Matthew P. Rowe, and Donald H. Owings, "Antipredator responses of California ground squirrels to rattlesnakes and rattling sounds: the roles of sex, reproductive parity, and offspring age in assessment and decision-making rules," *Behavioral Ecology and Sociobiology* 55 (2003): 22–31.

94   *"I had a mom instinct, right?":* Jeff Lawrence, "Mom who pried cougar's jaws off son shares chilling story," CTV News, Apr. 2, 2019, https://vancouverisland.ctvnews.ca /mom-who-pried-cougar-s-jaws-off-son-shares-chilling-story-1.4363100?.

94   *In a recent Modern Love column for the* New York Times: Susan Perabo, "When Mothers Bully Back," *New York Times,* Mar. 10, 2017.

94   *The violence may even be mom-on-mom:* Jared Leone, "Florida mothers slash each other with broken coffee mug in fight over parenting methods," *Atlanta Journal-Constitution,* Nov. 10, 2018.

94   *These maternal showdowns help explain why the most dangerous restaurant:* Anna Prior, "Calling All Cars: Trouble at Chuck E. Cheese's, Again," *Wall Street Journal,* Dec. 9, 2008.

95   *In wild-caught mother rats, the brain cells of amygdala, associated with aggression, are bigger:* Kelly G. Lambert and Catherine L. Franssen, "The Dynamic Nature of the Parental Brain," in Wilcox and Kline, *Gender and Parenthood,* 28.

95   *Next to love, the most commonly cited maternal emotion is rage:* Jennifer Verdolin, *Raised by Animals: The Surprising New Science of Animal Family Dynamics* (New York: The Experiment, 2017), 135.

95   *In one experiment, female rats were taught to associate a peppermint smell:* Elizabeth Rickenbacher et al., "Freezing suppression by oxytocin in central amygdala allows alternate defensive behaviours and mother-pup interactions," *eLife* 6 (2017): e24080.

95   *Another experiment, this one on humans, employed what's called the enthusiastic stranger paradigm:* Beth L. Mah et al., "Oxytocin promotes protective behavior in depressed mothers: a pilot study with the enthusiastic stranger paradigm," *Depression and Anxiety* 32, no. 2 (Feb. 2015): 76–81.

96   *If you stick a lactating lady on a treadmill:* Ellison, *Mommy Brain,* 88.

96   *In one experiment, human mothers competed . . . against a flagrantly rude, gum-snapping:* Jennifer Hahn-Holbrook et al., "Maternal Defense: Breast Feeding Increases Aggression by Reducing Stress," *Psychological Science* 22, no. 10 (Aug. 26, 2011): 1288–95.

98   *Something like 80 percent of all new moms report cognitive problems:* Teal Burrell, "Making Sense of Mommy Brain," *Discover,* Jan. 19, 2019.

98   *Some researchers have bleakly proposed that moms "cannibalize" our own brains:* Robert Martin, *How We Do It: The Evolution and Future of Human Reproduction* (New York: Basic Books, 2013), 146.

98   *Other indignant scientists . . . an efficiency-enhancing "synaptic pruning":* Jenni Gritters, "This Is Your Brain on Motherhood," *New York Times,* May 5, 2020.

99   *A few labs have even argued that motherhood "makes us smarter":* Ellison, *Mommy Brain.*

99   *Mom rats rock . . . formidable Froot Loop detectives:* Lambert and Franssen, "Dynamic Nature," 21–40.

99   *Rat researchers at the University of Richmond set up a gladiatorial "testing arena":* Craig Howard Kinsley et al., "The mother as hunter: Significant reduction in foraging costs through enhancements of predation in maternal rats," *Hormones and Behavior* 66, no. 4 (Sept. 2014): 649–54.

99   *A study of leopards in the Kalahari Desert:* J. du P. Bothma and R. J. Coertze, "Mother-

hood Increases Hunting Success in Southern Kalahari Leopards," *Journal of Mammalogy* 85, no. 4 (Aug. 16, 2004): 756–60.

99   *To achieve lift-off, milk-turgid mom bats must sometimes kidnap:* Carin Bondar, *Wild Moms: Motherhood in the Animal Kingdom* (New York: Pegasus Books, 2018), 165.

99   *Postpartum elephant seal moms are too buoyant:* ibid., 166.

100   *Among Homo sapiens, hunting moms are documented but somewhat rare:* Madeleine J. Goodman et al., "The compatability of hunting and mothering among the Agta hunter-gatherers of the Philippines," *Sex Roles* 12 (1985): 1119–209.

100   *The findings are mixed and inevitably controversial:* Sasha J. Davies et al., "Cognitive impairment during pregnancy: a meta-analysis," *Medical Journal of Australia* 208, no. 1 (Jan. 15, 2018): 35–40.

100   *One interesting experiment showed ... small tasks outside of the laboratory:* Carrie Cuttler et al., "Everyday life memory deficits in pregnant women," *Canadian Journal of Experimental Psychology* 65, no. 1 (Mar. 2011): 27–37.

100   *And we seem to have the most trouble recalling words:* Laura M. Glynn, "Giving birth to a new brain: hormone exposures of pregnancy influence human memory," *Psychoneuroendocrinology* 35, no. 8 (Sept. 2010): 1148–55; Serge V. Onyper et al., "Executive functioning and general cognitive ability in pregnant women and matched controls," *Journal of Clinical and Experimental Neuropsychology* 32, no. 9 (Nov. 2010): 986–95; Jessica F. Henry and Barbara B. Sherwin, "Hormones and Cognitive Functioning During Late Pregnancy and Postpartum: A Longitudinal Study," *Behavioral Neuroscience* 126, no. 1 (Feb. 2012): 73–85.

100   *Some of this momnesia might be a kind of short-term coping mechanism:* S. McKay and T. L. Barrows, "Reliving birth: maternal responses to viewing videotape of their second stage labors," *Image: Journal of Nursing Scholarship* 24, no. 1 (Spring 1992): 27–31; Eman Elkadry et al., "Do mothers remember key events during labor?" *American Journal of Obstetrics & Gynecology* 189, no. 1 (July 2003): 195–200.

100   *seven hundred hours per year for new mothers:* Louann Brizendine, *The Female Brain* (New York: Harmony, 2007), 105.

Chapter 5: MOTHER OF INVENTION

106   *A study at a Mississippi aquarium showed big-time fluctuations:* Heather M. Hill et al., "All Mothers Are Not the Same: Maternal Styles in Bottlenose Dolphins (Tursiops truncatus)," *International Journal of Comparative Psychology* 20, no. 1 (2007): 35–54.

106   *Certain marmoset moms:* C. R. Pryce, M. Dōbeli, and R. D. Martin, "Effects of sex steroids on maternal motivation in the common marmoset (Callithrix jacchus): development and application of an operant system with maternal reinforcement," *Journal of Comparative Psychology* 107, no. 1 (Mar. 1993): 99–115.

106   *Particular red squirrel moms appear extra motivated:* Sarah E. Westrick et al., "Attentive red squirrel mothers have faster growing pups and higher lifetime reproductive success," *Behavioral Ecology and Sociobiology* 74, no. 72 (2020).

107   *"Hence," a triumphant team ... " 'mothering styles' exist in guinea pigs":* P. C. H. Albers, P. J. A. Timmermans, and J. M. H. Vossen, "Evidence for the Existence of Mothering Styles in Guinea Pigs (Cavia aperea f. porcellus)," *Behaviour* 136, no. 4 (May 1999): 469–79.

107   *To rate a rabbit mom's maternal instincts:* For example, Gabriela Gonzáles-Mariscal and Jay S. Rosenblatt, "Maternal Behavior in Rabbits: A Historical and Multidisciplinary Perspective," in *Parental Care: Evolution, Mechanisms, and Adaptive Significance*, ed. Jay S. Rosenblatt and Charles T. Snowdon (San Diego: Academic Press, 1996), 338–39.

107   *Alas, there is no standard "mother in box" test for human moms:* Pernilla Foyer, Erik

Wilsson, and Per Jensen, "Levels of maternal care in dogs affect adult offspring temperament," *Scientific Reports* 6, no. 19253 (Jan. 13, 2016).

107   *They've measured our hand-grip strength:* Jeremy Atkinson et al., "Voice and Handgrip Strength Predict Reproductive Success in a Group of Indigenous African Females," *PLoS ONE* 7, no. 8 (Aug. 3, 2012): e41811; Emily Sutcliffe Cleveland, "Digit ratio, emotional intelligence and parenting styles predict female aggression," *Personality and Individual Differences* 58 (Feb. 2014): 9–14.

107   *A very small scattering . . . with more feminine faces:* Miriam J. Law Smith et al., "Maternal tendencies in women are associated with estrogen levels and facial femininity," *Hormones and Behavior* 61, no. 1 (Jan. 2012): 12–16; Denis K. Deady and Miriam J. Law Smith, "Height in women predicts maternal tendencies and career orientation," *Personality and Individual Differences* 40, no. 1 (Jan. 2006): 17–25.

107   *And . . . moose with thicker layers of rump fat have more surviving calves:* Alastair J. Wilson and Marco Festa-Bianchet, "Maternal Effects in Wild Ungulates," in *Maternal Effects in Mammals,* ed. Dario Maestripieri and Jill M. Mateo (Chicago: University of Chicago Press, 2009), 89; Hrdy, *Mother Nature,* 126.

108   *Science hails the ample contents of our mom jeans:* Trevathan, *Ancient Bodies,* 31.

108   *However, at least when it comes to nurturing kids after they are actually born:* Benedict C. Jones et al., "No compelling evidence that more physically attractive young adult women have higher estradiol or progesterone," *Psychoneuroendocrinology* 98 (Dec. 2018): 1–5.

108   *To analyze human maternal performance:* George W. Holden, "Avoiding Conflict: Mothers as Tacticians in the Supermarket," *Child Development* 54, no. 1 (1983): 233–40; Adriana G. Bus and Marinus H. van Ijzendoorn, "Affective dimension of mother-infant picturebook reading," *Journal of School Psychology* 35, no. 1 (Spring 1997): 47–60; Clare E. Holley, Claire Farrow, and Emma Haycraft, "If at first you don't succeed: Assessing influences associated with mothers' reoffering of vegetables to preschool age children," *Appetite* 123 (Apr. 1, 2018): 249–55.

108   *At sixteen frames per second, researchers analyze:* George W. Holden, *Parents and the Dynamics of Child Rearing* (Boulder, CO: Westview Press, 1997), 33.

108   *Isolating moms in sound booths, musically trained observers eavesdropped:* Tonya R. Bergeson and Sandra E. Trehub, "Signature tunes in mothers' speech to infants," *Infant Behavior and Development* 30, no. 4 (Dec. 2007): 648–54.

108   *"These behaviors are complex," one group of researchers wrote:* Alison S. Fleming et al., "Plasticity in the Maternal Neural Circuit: Experience, Dopamine, and Mothering," in Bridges, *Neurobiology of the Parental Brain,* 524.

109   *One California laboratory . . . a woman's mid-pregnancy ratios:* Laura M. Glynn et al., "Gestational hormone profiles predict human maternal behavior at 1-year postpartum," *Hormones and Behavior* 85 (Sept. 2016): 19–25.

109   *Another experiment found . . . first-time mothers with higher levels of cortisol:* Corter and Fleming, "Psychobiology," 149.

110   *Some women's dopamine systems . . . less activated by infant face cues:* Lane Strathearn, "Maternal Neglect: Oxytocin, Dopamine and the Neurobiology of Attachment," *Journal of Neuroendocrinology* 23, no. 11 (Nov. 2011): 1054–65.

110   *To the surprise of no one, oxytocin is an especially potent force:* Ruth Feldman et al., "Natural variations in maternal and paternal care are associated with systematic changes in oxytocin following parent-infant conflict," *Psychoneuroendocrinology* 35, no. 8 (Sept. 11, 2010): 1133–41.

110   *Very attentive human mothers, who gaze at their kids extra often:* Sohye Kim et al., "Maternal oxytocin response predicts mother-to-infant gaze," *Brain Research* 1580 (Sept. 11, 2014): 133–42; Yael Apter-Levi, Orna Zagoory-Sharon, and Ruth Feldman, "Oxyto-

cin and vasopressin support distinct configurations of social synchrony," *Brain Research* 1580 (Sept. 11, 2014): 124–32; Ruth Feldman et al., "Evidence for a Neuroendocrinological Foundation of Human Affiliation: Plasma Oxytocin Levels Across Pregnancy and the Postpartum Period Predict Mother-Infant Bonding," *Psychological Science* 18, no. 11 (Nov. 2007).

110   *These differences show up . . . on-top-of-it gray seal mothers:* Kelly J. Robinson et al., "Maternal Oxytocin Is Linked to Close Mother-Infant Proximity in Grey Seals (*Halichoerus grypus*)," *PLoS ONE* 10, no. 12 (Dec. 13, 2015): e144577.

110   *The women with the biggest volume of lost gray matter on the MRI scans:* Hoekzema, *Nature Neuroscience*, 287–96.

110   *One EEG study of forty mothers-to-be:* Joanna Dudek et al., "Changes in Cortical Sensitivity to Infant Facial Cues From Pregnancy to Motherhood Predict Mother-Infant Bonding," *Child Development* 91, no. 1 (Dec. 2018): e198–e217.

110   *For instance, when flashed pictures of their own babies:* James E. Swain, "Parental Brain Determinants for the Flourishing Child : Evolution, Family, and Society," in *Contexts for Young Child Flourishing*, ed. Darcia Narvaez et al. (New York: Oxford University Press, 2016), 134.

111   *Something like 90 percent of American moms read parenting books:* Ellison, *Mommy Brain*, 182.

111   *one study titled "Does 'Wanting the Best' Create More Stress?" found:* Neil Howlett, Elizabeth Kirk, and Karen J. Pine, "Does 'Wanting the Best' create more stress? The link between baby sign classes and maternal anxiety," *Infant and Child Development* 20, no. 4 (July/Aug. 2011): 437–45.

111   *On the other hand . . . moms who have studied a musical instrument:* Christine E. Parsons et al., "Music training and empathy positively impact adults' sensitivity to infant distress," *Frontiers in Psychology* 5, no. 1440 (Dec. 19, 2014).

113   *Our experience might not tell us exactly how to soothe a colicky baby's banshee screams:* Corter and Fleming, "Psychobiology," 151 and 158–61.

113   *In one lab test asking adults to diagnose the mysterious cause of a baby's cries:* George W. Holden, "Adults' Thinking about a Child-rearing Problem: Effects of Experience, Parental Status, and Gender," *Child Development* 59, no. 6 (Dec. 1988): 1623–32.

114   *In another experiment, scientists allowed a group of subadult chimpanzees:* Kim A. Bard, "Primate Parenting," in Bornstein, *Handbook of Parenting*, 2:125.

114   *When marmoset and tamarin mothers have no prior infant exposure:* Charles T. Snowdon, "Family Life and Infant Care: Lessons from Cooperatively Breeding Primates," in Wilcox and Kline, *Gender and Parenthood*, 51.

114   *At certain stages of life they seek out hands-on experience:* Sarah Blaffer Hrdy, *Mothers and Others: The Evolutionary Origins of Mutual Understanding* (Cambridge, MA: Harvard University Press, 2009), 217–19.

114   *To prime gorilla first-timers for motherhood, keepers at the Smithsonian National Zoo:* "#GorillaStory: Maternal Training with Calaya," Smithsonian National Zoo, Animal News Archive, Feb. 9, 2018, https://nationalzoo.si.edu/animals/news/gorillastory -maternal-training-calaya.

114   *And a recent and rather delightful study of wild chimps in Uganda's Kibale National Park:* Sonya M. Kahlenberg and Richard W. Wrangham, "Sex differences in chimpanzees' use of sticks as play objects resemble those of children," *Current Biology* 20, no. 24 (Dec. 21, 2010): R1067–R1068.

114   *Unfortunately, as a large-scale study in Australia recently found:* Sally A. Brinkman et al., "Efficacy of infant simulator programmes to prevent teenage pregnancy: a school-based cluster randomised controlled trial in Western Australia," *Lancet* 388, no. 10057 (Nov. 5, 2016): 2264–71.

114 *Which is not all that surprising since, as a University of Chicago study showed:* Dario Mae-stripieri and Suzanne Pelka, "Sex differences in interest in infants across the lifespan: A biological adaptation for parenting?," *Human Nature* 13, no. 3 (Sept. 2002): 327–44.

115 *childcare chops may yet be a boon to fathers:* Hrdy, *Mothers and Others,* 171.

115 *Virgin marmosets who tote new babies have different prolactin levels:* Ellison, *Mommy Brain,* 153; Kaitlyn M. Harding and Joseph S. Lonstein, "Extensive juvenile 'babysitting' facilitates later adult maternal responsiveness, decreases anxiety, and increases dorsal raphe tryptophan hydroxylase–2 expression in female laboratory rats," *Developmental Psychobiology* 58, no. 4 (May 2016): 492–508.

116 *The average age of first-time American moms is twenty-six:* Quoctrung Bui and Claire Cain Miller, "The Age That Women Have Babies: How a Gap Divides America," *New York Times,* Aug. 4, 2018.

116 *As the national birth rate has cratered these last few years:* Olga Khazan, "The Rise of Older Mothers," *Atlantic,* May 17, 2018.

116 *Yet by many maternal metrics older moms excel:* Hrdy, *Mother Nature,* 276, 314.

116 *In other mammals, notably elephant seals:* Joanne Reiter, Kathy J. Panken, and Burney J. Le Boeuf, "Female competition and reproductive success in northern elephant seals," *Animal Behaviour* 29, no. 3 (Aug. 1981): 670–87; T. S. McCann, "Aggressive and maternal activities of female southern elephant seals (*Mirounga leonina*)," *Animal Behaviour* 30, no. 1 (Feb. 1982): 268–76; W. Don Bowen, "Maternal Effects on Offspring Size and Development in Pinnipeds," in Maestripieri and Mateo, *Maternal Effects in Mammals,* 111.

116 *The differences start early in pregnancy:* Monica Akinyi Magadi, Alfred O. Agwanda, and Francis O. Obare, "A comparative analysis of the use of maternal health services between teenagers and older mothers in sub-Saharan Africa: Evidence from Demographic and Health Surveys (DHS)," *Social Science & Medicine* 64, no. 6 (Mar. 2007):1311–25; Ban al-Sahab et al., "Prevalence and predictors of 6-month exclusive breastfeeding among Canadian women: A national survey," *BMC Pediatrics* 10, no. 20 (2010); Katherine Apostolakis-Kyrus, Christina Valentine, and Emily DeFranco, "Factors Associated with Breastfeeding Initiation in Adolescent Mothers," *Journal of Pediatrics* 163, no. 5 (Nov. 2013): 1489–94; Bondar, *Wild Moms,* 190; Trevathan, *Ancient Bodies,* 169.

116 *In the first year of baby life . . . matters like safe sleep practices:* Michelle Caraballo et al., "Knowledge, Attitudes, and Risk for Sudden Unexpected Infant Death in Children of Adolescent Mothers: A Qualitative study," *Journal of Pediatrics* 174 (July 1, 2016): 78–83.e2.

116 *Especially compared to moms in their teens:* Katherine M. Krpan et al., "Experiential and hormonal correlates of maternal behavior in teen and adult mothers," *Hormones and Behavior* 47, no. 1 (Jan. 2005): 112–22; Jennifer Giardino et al., "Effects of motherhood on physiological and subjective responses to infant cries in teenage mothers: A comparison with non-mothers and adult mothers," *Hormones and Behavior* 53, no. 1 (Jan. 2008): 149–58.

117 *mental health woes hit teenage moms four times:* Ryan J. Van Lieshout et al., "The Mental Health of Young Canadian Mothers," *Journal of Adolescent Health* 66, no. 4 (Apr. 1, 2020): 464–69; Dawn Kingston et al., "Comparison of Adolescent, Young Adult, and Adult Women's Maternity Experiences and Practices," *Pediatrics* 129, no. 5 (May 2012): e1228–e1237.

117 *A brand-new baby's most likely murderer:* Catherine A. Salmon and James Malcolm, "Parent-Offspring Conflict" and Virginia Periss and David F. Bjorklund, "Trials and Tribulations of Childhood: An Evolutionary Perspective," in Salmon and Shackelford, *Oxford Handbook,* 85 and 158.

118 *For instance, in a study of Appalachian moms:* Marc H. Bornstein, Diane L. Putnick, and

Joan T. D. Suwalsky, "A Longitudinal Process Analysis of Mother-Child Emotional Relationships in a Rural Appalachian European American Community," *American Journal of Community Psychology* 50, nos. 1–2 (Sept. 2012): 89–100.

118   *A study in sub-Saharan Africa:* Caroline Uggla and Ruth Mace, "Parental investment in child health in sub-Saharan Africa: A cross-national study of health-seeking behaviour," *Royal Society Open Science* 3, no. 2 (Feb. 2016): 150460.

118   *And while it's a common belief that humans are "meant" to have babies young:* Hrdy, *Mother Nature,* 191.

118   *To the extent that intellect matters for good mothering:* Bornstein, "Determinants of Parenting," 16.

118   *Studies have shown that women who still excel at memory tasks:* Helena J. V. Rutherford et al., "Executive Functioning Predicts Reflective Functioning in Mothers," *Journal of Child and Family Studies* 27 (2018): 944–52; Helena J. V. Rutherford et al., "Investigating the relationship between working memory and emotion regulation in mothers," *Journal of Cognitive Psychology* 28, no. 1 (2016): 52–59; Elsie Chico et al., "Executive function and mothering: challenges faced by teenage mothers," *Developmental Psychobiology* 56, no. 5 (July 2014): 1027–35.

119   *"The forty-one-year-old mother who gives her life for her only child":* Hrdy, *Mother Nature,* 94.

119   *The distinctive dedication of older moms is described:* Bondar, *Wild Moms,* 198–99; Wilson and Festa-Bianchet, "Wild Ungulates," 88.

119   *One study showed that the last calf of a killer whale mom:* Eric J. Ward et al., "The role of menopause and reproductive senescence in a long-lived social mammal," *Frontiers in Zoology* 6, no. 4 (Feb. 3, 2009).

119   *In humans, some scientists even suspect:* Trevathan, *Ancient Bodies,* 71.

119   *According to this theory:* Tim A. Bruckner et al., "Down syndrome among primiparae at older maternal age: A test of the relaxed filter hypothesis," *Birth Defects Research* 111, no. 20 (Dec. 2019): 1611–17.

120   *C-sections on 211 monkeys:* Jay S. Rosenblatt, "Hormonal Bases of Parenting in Mammals," in Bornstein, *Handbook of Parenting,* 2:50.

121   *C-section mothers two to four weeks after delivery:* James E. Swain et al., "Maternal brain response to own baby-cry is affected by cesarean section delivery," *Journal of Child Psychology and Psychiatry* 49, no. 10 (Oct. 2008): 1042–52; Swain, "Parental Brain Determinants," 132.

121   *In another study of Israeli moms:* Marsha Kaitz, Guy Stecklov, and Noa Devor, "Anxiety symptoms of new mothers during a period of recurrent, local terror," *Journal of Affective Disorders* 107, nos. 1–3 (Apr. 2008): 211–15.

121   *Women who have them are 15 percent more likely:* Valentina Tonei, "Mother's mental health after childbirth: Does the delivery method matter?," *Journal of Health Economics* 63 (Jan. 2019): 182–96.

121   *In nursing rats, the part of the cortex associated with the nipples:* Ellison, *Mommy Brain,* 161.

121   *Compared to their leak-free . . . nursing human mothers have been found to be more sensitive:* Corter and Fleming, "Psychobiology," 152–53; Pilyoung Kim et al., "Breastfeeding, brain activation to own infant cry, and maternal sensitivity," *Journal of Child Psychology and Psychiatry* 52, no. 8 (Aug. 2011): 907–15.

122   *Research on breastfeeding moms, from Boise:* Jennifer M. Weaver, Thomas J. Schofield, and Lauren M. Papp, "Breastfeeding duration predicts greater maternal sensitivity over the next decade," *Developmental Psychology* 54, no. 2 (2018): 220–27.

122   *And a study of more than seven thousand Australian mothers of various backgrounds:* Lane Strathearn et al., "Does Breastfeeding Protect Against Substantiated Child Abuse and Neglect? A 15-Year Cohort Study," *Pediatrics* 123, no. 2 (Feb. 2009): 483–93.

122   *The human maternal instinct is powerful and robust:* J. Dunne et al., "Milk of ruminants in ceramic baby bottles from prehistoric child graves," *Nature* 574 (2019): 246–48.

125   *This gives second-time moms an edge:* For an overview, see Corter and Fleming, "Psychobiology," 141–81; Angela N. Maupin et al., "Investigating the association between parity and the maternal neural response to infant cues," *Social Neuroscience* 14, no. 2 (Apr. 2019): 214–25.

125   *It takes longer for beginner moms:* Jane E. Drummond, Michelle L. McBride, and C. Faye Wiebe, "The Development of Mothers' Understanding of Infant Crying," *Clinical Nursing Research* 2, no. 4 (Nov. 1993): 396–410.

125   *Multi-timers tend to touch infants more:* Sari Goldstein Ferber, "The nature of touch in mothers experiencing maternity blues: the contribution of parity," *Early Human Development* 79, no. 1 (Aug. 2004): 65–75.

125   *Second-time cheetah moms select better dens:* Kay E. Holekamp and Stephanie M. Dloniak, "Maternal Effects in Fissiped Carnivores," in Maestripieri and Mateo, *Maternal Effects in Mammals*, 231.

125   *Practiced ewes lick their babies much sooner after birth:* Levy, "Neural Substrates," 31.

125   *Sea lion veterans almost never nurse:* Bondar, *Wild Moms*, 162.

125   *In lab tests, experienced rat moms catch crickets:* Lambert and Franssen, "Dynamic Nature," 31–32.

125   *In our close relatives:* Hrdy, *Mother Nature*, 155; Bondar, *Wild Moms*, 190; Bornstein, "Determinants," 29.

126   *Experienced rat moms are much swifter . . . at attacking strange males:* Robert S. Bridges, "Long-term alterations in neural and endocrine processes induced by motherhood in mammals," *Hormones and Behavior* 77 (Jan. 2016): 193–203.

126   *Studies in rats and sheep:* Frances A. Champagne and James P. Curley, "Plasticity of the Maternal Brain Across the Lifespan," in *Maternal Brain Plasticity: Preclinical and Human Research and Implications for Intervention*, ed. Helena J. V. Rutherford and Linda C. Mayes (San Francisco: Jossey-Bass, 2016), 16–17; Emis M. Akbari et al., "The Effects of Parity and Maternal Behavior on Gene Expression in the Medial Preoptic Area and the Medial Amygdala in Postpartum and Virgin Female Rats: A Microarray Study," *Behavioral Neuroscience* 127, no. 6 (Dec. 2013): 913–22.

126   *Human mothers of three have a much harder time with verbal recall tasks:* Laura M. Glynn, "Increasing parity is associated with cumulative effects on memory," *Journal of Women's Health* 21, no. 10 (Oct. 2012): 1038–45.

127   *Higher-order moms may also be more prone to hit the bottle:* Jee-Yeon K. Lehmann, Ana Nuevo-Chiquero, and Marian Vidal-Fernandez, "The Early Origins of Birth Order Differences in Children's Outcomes and Parental Behavior," *Journal of Human Resources* 53, no. 1 (Winter 2018): 123–56

Chatper 6: IN SEARCH OF THE MOM GENE

132   *One of the biggest mysteries of social science:* For an excellent overview, see A.M. Lomanowska et al., "Parenting begets parenting: A neurobiological perspective on early adversity and the transmission of parenting styles across generations," *Neuroscience* 342 (Feb. 7, 2017): 120–39.

135   *To tease that apart, scientists compare:* Sandra H. Losoya et al., "Origins of familial similarity in parenting: A study of twins and adoptive siblings," *Developmental Psychology* 33, no. 6 (Nov. 1997): 1012–23.

136   *In the livestock literature, "fertility and maternal instinct" . . . "a high-yielding carcass":* For instance, "Sheepvention 2018: Texel sheep breed at peak performance," *Weekly Times,*

June 25, 2018; Dustin McGuire, "Common Beef Breeds of Oregon," Oregon State University—Beef Cattle Library, Apr. 2013: BEEF105.

136 *She decided to put the rumors to rest through an arduous series:* Cathy M. Dwyer, "Genetic and physiological determinants of maternal behavior and lamb survival: Implications for low-input sheep management," *Journal of Animal Science* 86, no. 14, suppl. (Apr. 2008): E246–E258.

137 *To rule out this variable, Dwyer performed a series of embryo transfers:* Cathy M. Dwyer and A. B. Lawrence, "Ewe–ewe and ewe–lamb behaviour in a hill and a lowland breed of sheep: a study using embryo transfer," *Applied Animal Behaviour Science* 61, no. 4 (Jan. 1999): 319–34.

137 *After all, these striking breed-by-breed mothering differences crop up:* Holekamp and Dloniak, "Fissiped Carnivores," 231; M. X. Zarrow, V. H. Denenberg, and W. D. Kalberer, "Strain differences in the endocrine basis of maternal nest-building in the rabbit," *Reproduction* 10, no. 3 (Dec. 1965): 397–401; Anstiss H. McIver and Wendell E. Jeffrey, "Strain differences in maternal behavior in rats," *Behaviour* 28, nos. 1/2 (1967): 210–16.

138 *But stamping out the crushers is easier said than done:* Inger Lise Andersen, Synne Berg, and Knut Egil Bøe, "Crushing of piglets by the mother sow (*Sus scrofa*)—purely accidental or a poor mother?," *Applied Animal Behaviour Science* 93, nos. 3–4 (Sept. 2005): 229–43.

138 *In pursuit of what is sometimes called a "super sow":* Marko Ocepek et al., "Can a super sow be a robust sow? Consequences of litter investment in purebred and crossbred sows of different parities," *Journal of Animal Science* 94, no. 8 (Aug. 2016): 3550–60; B. Hellbrügge et al., "Genetic aspects regarding piglet losses and the maternal behaviour of sows. Part 2. Genetic relationship between maternal behaviour in sows and piglet mortality," *Animal* 2, no. 9 (Sept. 2008): 1281–88.

139 *There is likely a genetic link:* S. P. Turner and A. B. Lawrence, "Relationship between maternal defensive aggression, fear of handling and other maternal care traits in beef cows," *Livestock Science* 106, nos. 2–3 (Feb. 2007): 182–88.

139 *"When this aggression is directed":* Marie J. Haskell, Geoff Simm, and Simon P. Turner, "Genetic selection for temperament traits in dairy and beef cattle," *Frontiers in Genetics* 5, no. 368 (Oct. 21, 2014).

139 *In 2008, they identified:* Hasse Walum et al., "Genetic variation in the vasopressin receptor 1a gene (*AVPR1A*) associates with pair-bonding behavior in humans," *PNAS* 105, no. 37 (Sept. 16, 2008): 14153–56; Eva G. T. Green and Alain Clémence, "Discovery of the faithfulness gene: A model of transmission and transformation of scientific information," *British Journal of Social Psychology* 47, pt. 3 (Sept. 2008): 497–517.

140 *Others found the "wanderlust gene":* Mark Ellwood and Laura Dannen Redman, "The Science of Wanderlust," *Condé Nast Traveler*, June 12, 2017.

140 *Perhaps most famous was the "warrior gene":* John Horgan, "Code rage: The 'warrior gene' makes me mad! (Whether I have it or not)," *Cross Check* (blog), ScientificAmerican.com, Apr. 26, 2011, https://blogs.scientificamerican.com/cross-check/code-rage-the-warrior-gene-makes-me-mad-whether-i-have-it-or-not.

140 *These types of experiments . . . "inconspicuously installed in the ceiling":* Ashlea M. Klahr et al., "Evocative gene–environment correlation in the mother–child relationship: A twin study of interpersonal processes," *Developmental Psychopathology* 25, no. 1 (Feb. 2013): 105–18.

141 *In one Israeli experiment, researchers had moms and their three-year-olds play:* Reut Avinun, Richard P. Ebstein, and Ariel Knafo, "Human maternal behaviour is associated with arginine vasopressin receptor 1A gene," *Biology Letters* 8, no. 5 (Oct. 23, 2012): 2012.0492.

141 *Another group asked . . . "a wordless picture-book":* R. Bisceglia et al., "Arginine vasopressin 1a receptor gene and maternal behavior: evidence of association and moderation," *Genes, Brain and Behavior* 11, no. 3 (Apr. 2012): 262–68.

141   *A gene-hunting team from the University of Chicago really put us moms through our paces:* Kalina J. Michalska et al., "Genetic imaging of the association of oxytocin receptor gene (OXTR) polymorphisms with positive maternal parenting," *Frontiers in Behavioral Neuroscience* 8, no. 21 (Jan. 3, 2014).

142   *In 2017, Leerkes and her colleagues published a paper:* E. M. Leerkes et al., "Variation in mothers' arginine vasopressin receptor 1a and dopamine receptor D4 genes predicts maternal sensitivity via social cognition," *Genes, Brain and Behavior* 16, no. 2 (Feb. 2017): 233–40.

144   *Then there's the fact that:* For example, Andrea Ganna et al., "Large-scale GWAS reveals insights into the genetic architecture of same-sex sexual behavior," *Science* 365, no. 6456 (Aug. 30, 2019).

144   *She spent years of her career at one of the world's most influential maternal behavior labs:* V. Mileva-Seitz et al., "Serotonin transporter allelic variation in mothers predicts maternal sensitivity, behavior and attitudes toward 6-month-old infants," *Genes, Brain and Behavior* 10, no. 3 (Apr. 2011): 325–33; W. Jonas et al., "Genetic variation in oxytocin rs2740210 and early adversity associated with postpartum depression and breastfeeding duration," *Genes, Brain and Behavior* 12, no. 7 (Oct. 2013): 681–94; V. Mileva-Seitz et al., "Dopamine receptors D1 and D2 are related to observed maternal behavior," *Genes, Brain and Behavior* 11, no. 6 (Aug. 2012): 684–94; V. Mileva-Seitz et al., "Interaction between Oxytocin Genotypes and Early Experience Predicts Quality of Mothering and Postpartum Mood," *PLoS ONE* 8, no. 4 (Apr. 18, 2013): e61443.

145   *In a sort of Russian nesting doll effect:* R. Arocho and C. M. Kamp Dush, "Like mother, like child: Offspring marital timing desires and maternal marriage timing and stability," *Journal of Family Psychology* 31, no. 3 (Apr. 2017): 261–72; Holden, *Parents and the Dynamics of Child Rearing*, 63.

146   *And yet the fact remains that your feelings about your own mother are a major predictor of:* Peter Fonagy, Howard Steele, and Miriam Steele, "Maternal Representations of Attachment during Pregnancy Predict the Organization of Infant-Mother Attachment at One Year of Age," *Child Development* 62, no. 5 (Oct. 1991): 891–905.

146   *He was initially interested in the crisis's effect:* Rand D. Conger, Thomas J. Schofield, and Tricia K. Neppl, "Intergenerational Continuity and Discontinuity in Harsh Parenting," *Parenting* 12, nos. 2–3 (2012): 222–31.

146   *Similar studies have since been carried out everywhere from England to Indonesia:* Rahma et al., "Predictors of sensitive parenting in urban slums in Makassar, Indonesia," *Attachment & Human Development* (2018); N. M. Kovan, A. L. Chung, and L. A. Sroufe, "The intergenerational continuity of observed early parenting: A prospective, longitudinal study," *Developmental Psychology* 45, no. 5 (Sept. 2009): 1205–13; Vaishnavee Madden et al., "Intergenerational transmission of parenting: findings from a UK longitudinal study," *European Journal of Public Health* 25, no. 6 (Dec. 2015): 1030–35.

147   *A New Zealand study followed three-year-olds:* Jay Belsky et al., "Intergenerational Transmission of Warm-Sensitive-Stimulating Parenting: A Prospective Study of Mothers and Fathers of 3-Year-Olds," *Child Development* 76, no. 2 (Mar.–Apr. 2005): 384–96.

147   *Some of the most arresting studies of cyclical mothering really do involve monkey moms:* Stephen J. Suomi, "Early determinants of behaviour: evidence from primate studies," *British Medical Bulletin* 53, no. 1 (Jan. 1997): 170–84.

147   *Vervet monkeys, for instance, spend almost identical amounts of time:* Dario Maestripieri, "Maternal Influences on Offspring Growth, Reproduction, and Behavior in Primates," in Maestripieri and Mateo, *Maternal Effects in Mammals*, 280.

147   *And in rhesus monkeys, abusive mothering stretches back:* Dario Maestripieri, "The Role of the Brain Serotonergic System in the Origin and Transmission of Adaptive and Mal-

173   *One 2017 analysis of moms wearing eye-tracking glasses:* Holly Rayson et al., "Effects of Infant Cleft Lip on Adult Gaze and Perceptions of 'Cuteness,'" *Cleft Palate–Craniofacial Journal* 54, no. 5 (Sept. 2017): 562–70.

173   *Another study discovered that a speedy surgical repair of the cleft not only fixed the aesthetic problem:* Lynne Murray et al., "The effect of cleft lip and palate, and the timing of lip repair on mother-infant interactions and infant development," *Journal of Child Psychology and Psychiatry* 49, no. 2 (Feb. 2008): 115–23.

175   *Women carrying male fetuses:* "Why baby's sex may influence risk of pregnancy-related complications," Obs Gynae & Midwifery News, July 23, 2018, https://www.ogpnews .com/2018/07/why-babys-sex-may-influence-risk-of-pregnancy-related-complic ations/31725.

176   *We are about 70 percent more likely to get pregnancy-related depression:* Sarah Myers and Sarah E. Johns, "Male infants and birth complications are associated with increased incidence of postnatal depression," *Social Science & Medicine* 220 (Jan. 2019): 56–64.

176   *We are also, in our first and second trimesters, measurably more sensitive to disgust:* Agnieszka Żelaźniewicz and Bogusław Pawłowski, "Disgust in pregnancy and fetus sex— Longitudinal study," *Physiology & Behavior* 139 (Feb. 2015): 177–81.

176   *We also get to eat about 10 percent more calories:* Ellison, *Mommy Brain,* 147; Claire M. Vanston and Niel V. Watson, "Selective and persistent effect of foetal sex on cognition in pregnant women," *NeuroReport* 16, no. 7 (May 12, 2005): 779–82.

176   *The fetal movements that may subtly "program" moms also vary by gender:* J. A. DiPietro and K. M. Voegtline, "The gestational foundation of sex differences in development and vulnerability," *Neuroscience* 342 (Feb. 7, 2017): 4–20.

177   *The jury is still out on which group of moms grows bigger boobs:* Agnieszka Żelaźniewicz and Bogusław Pawłowski, "Breast size and asymmetry during pregnancy in dependence of a fetus's sex," *American Journal of Human Biology* 27, no. 5 (Sept.–Oct. 2015): 690– 96; Andrzej Galbarczyk, "Unexpected changes in maternal breast size during pregnancy in relation to infant sex: An evolutionary interpretation," *American Journal of Human Biology* 23, no. 4 (July–Aug. 2011): 560–62.

177   *A study of several dozen healthy new moms in Massachusetts:* Camilla E. Powe, Cheryl D. Knott, and Nancy Conklin-Brittain, "Infant sex predicts breast milk energy content," *American Journal of Human Biology* 22, no. 1 (Jan.–Feb. 2010): 50–54.

177   *Many mammals make higher-fat milk for male babies:* Bondar, *Wild Moms,* 153–54.

177   *One interesting exception is dairy cows:* Katie Hinde et al., "Holsteins Favor Heifers, Not Bulls: Biased Milk Production Programmed during Pregnancy as a Function of Fetal Sex," *PLoS ONE* 9, no. 2 (Feb. 3, 2014): e86169.

177   *Animal studies . . . breast milk's chemical contents vary by gender:* Katie Hinde et al., "Daughter dearest: Sex-biased calcium in mother's milk among rhesus macaques," *American Journal of Physical Anthropology* 151, no. 1 (May 2013): 144–50; Katie Hinde and Lauren A. Milligan, "Primate milk: proximate mechanisms and ultimate perspectives," *Evolutionary Anthropology* 20, no. 1 (Jan.–Feb. 2011): 9–23.

177   *Presented with a strange infant, mothers act quite differently:* Bornstein, "Parenting × Gender × Culture × Time," 92.

177   *And there are plenty of curious:* Shelly Lundberg and Elaina Rose, "Investments in Sons and Daughters: Evidence from the Consumer Expenditure Survey," Department of Economics, University of Washington (Feb. 2003); Vicki L. Bogan, "Household Investment Decisions and Offspring Gender: Parental Accounting," *Applied Economics* 45, no. 31 (2013): 4429–4442; Rose Eveleth, "Young Girls Are More Likely to Want Braces Than Boys," Smithsonianmag.com, Nov. 25, 2013, https://www.smithsonian mag.com/smart-news/young-girls-are-more-likely-to-want-braces-than-boys; KJ Dell' Antonia, "Mothers Talk Less to Young Daughters About Math," *Motherlode* (blog), *New*

York Times, Feb. 24, 2012, https://parenting.blogs.nytimes.com/2012/02/24/mothers-talk-less-to-young-daughters-about-math/.

177 *American mothers of girls are more likely to tilt to the right politically:* Byungkyu Lee and Dalton Conley, "Does the Gender of Offspring Affect Parental Political Orientation?," *Social Forces* 94, no. 3 (Mar. 2016): 1103–27.

177 *Meanwhile, boys' mothers spend more on housing and pay out extra allowance:* Claire Cain Miller, "A 'Generationally Perpetuated' Pattern: Daughters Do More Chores," The Upshot, *New York Times*, Aug. 8, 2018, https://www.nytimes.com/2018/08/08/upshot/chores-girls-research-social-science.html.

177 *In some cultures, they also tend to breastfeed longer:* Seema Jayachandran and Ilyana Kuziemko, "Why Do Mothers Breastfeed Girls Less Than Boys? Evidence and Implications for Child Health in India," *Quarterly Journal of Economics* 126, no. 3 (Aug. 2011): 1485–538.

177 *But I wonder: Would a scientist watching my every move:* Amanda M. Dettmer et al., "First-time rhesus monkey mothers, and mothers of sons, preferentially engage in face-to-face interactions with their infants," *American Journal of Primatology* 78, no. 2 (2016): 238–46.

178 *If an eighty-year-old orca mom dies:* Sindya N. Bhanoo, "Orca Mothers Coddle Adult Sons, Study Finds," *New York Times*, Sept. 17, 2012.

178 *Although these poor women may deserve their earthly reward:* Samuli Helle and Virpi Lummaa, "A trade-off between having many sons and shorter maternal post-reproductive survival in pre-industrial Finland," *Biology Letters* 9, no. 2 (Feb. 23, 2013): 20130034.

178 *"Women wanted sons," the anthropologist Margaret Mead:* Sarah Harkness and Charles M. Super, "Culture and Parenting," in Bornstein, *Handbook of Parenting*, 2:257.

178 *In modern India, mothers pregnant with boys:* Prashant Bharadwaj and Leah K. Lakdawala, "Discrimination Begins in the Womb: Evidence of Sex-Selective Prenatal Investments," *Journal of Human Resources* 48, no. 1 (Winter 2013): 71–113.

178 *After birth, moms in these countries spend more time on average with their boys:* Seema Jayachandran and Ilyana Kuziemko, "Why Do Mothers Breastfeed Girls Less Than Boys? Evidence and Implications for Child Health in India," *Quarterly Journal of Economics* 126, no. 3 (Aug. 2011): 1485–538; Silvia H. Barcellos, Leandro S. Carvalho, and Adriana Lleras-Muney, "Child Gender and Parental Investments in India: Are Boys and Girls Treated Differently?," *American Economic Journal: Applied Economics* 6, no. 1 (Jan. 2014): 157–89.

178 *In one dismal . . . only three of eight thousand terminated fetuses were male:* Hrdy, *Mother Nature*, 322; Lisa R. Roberts and Susanne B. Montgomery, "India's Distorted Sex Ratio: Dire Consequences for Girls," *Journal of Christian Nursing* 33, no. 1 (Jan.–Mar. 2016): E7–E15.

179 *In one tribe from Turkmenistan:* David Lancy, *Raising Children: Surprising Insights from Other Cultures* (Cambridge, UK: Cambridge University Press, 2017), 123.

179 *For centuries, American moms likely also nursed:* Nora Bohnert et al., "Offspring Sex Preference in Frontier America," *Journal of Interdisciplinary History* 42, no. 4 (2012): 519–41.

179 *Some version of son preference:* Associated Press, "Study Finds Boys Preferred as First-borns," *New York Times*, July 6, 1982.

179 *American mothers at present desire a mix of genders:* Francine Blau et al., "Declining son preference in the US," VoxEU/CEPR, Mar. 12, 2020, https://voxeu.org/article/declining-son-preference-us; Ashley Larsen Gibby and Kevin J. A. Thomas, "Adoption: A Strategy to Fulfill Sex Preference of U.S. Parents," *Journal of Marriage and Family* 81, no. 2 (Apr. 2019): 531–41. Jacques D. Marleau and Jean-François Saucier, "Preference for a First-Born Boy in Western Societies," *Journal of Biosocial Science* 34, no. 1 (Jan.

2002): 13–27; Robert Lynch, Helen Wasielewski, and Lee Cronk, "Sexual conflict and the Trivers-Willard hypothesis: Females prefer daughters and males prefer sons," *Scientific Reports* 8 (2018): 15463.

179 *And on average today, American moms may lavish more time:* Michael Baker and Kevin Milligan, "Boy-Girl Differences in Parental Time Investments: Evidence from Three Countries," *Journal of Human Capital* 10, no. 4 (Winter 2016): 399–441.

179 *In the 1970s, son-only households were investing more:* Sabino Kornrich and Frank Furstenberg, "Investing in Children: Changes in Parental Spending on Children, 1972–2007," *Demography* 50, no. 1 (2013): 1–23.

179 *But by 2007 . . . daughter-only households were the big spenders:* Lambrianos Nikiforidis et al., "Do Mothers Spend More on Daughters While Fathers Spend More on Sons?," *Journal of Consumer Psychology* 28, no. 1 (Jan. 2018): 149–56.

179 *There are striking exceptions:* Douglas Almond and Yi Cheng, "Perinatal Health among 1 Million Americans" (NBER Working Paper, Aug. 2019).

Chapter 8: THERMOMETERS

182 *BPA is ubiquitous throughout the world:* Hidekuni Inadera, "Neurological Effects of Bisphenol A and its Analogues," *International Journal of Medical Sciences* 12, no. 12 (Oct. 30, 2015): 926–36; Jerome Groopman, "The Plastic Panic," *New Yorker*, May 31, 2010.

182 *Previously linked to a host of problems:* For example, Danielle Della Seta et al., "Bisphenol-A exposure during pregnancy and lactation affects maternal behavior in rats," *Brain Research Bulletin* 65, no. 3 (Apr. 2005): 255–60; Sarah A. Johnson et al., "Disruption of Parenting Behaviors in California Mice, a Monogamous Rodent Species, by Endocrine Disrupting Chemicals," *PLoS One* 10, no. 6(June 3, 2015): e0126284; Sofiane Boudalia et al., "A multi-generational study on low-dose BPA exposure in Wistar rats: Effects on maternal behavior, flavor intake and development," *Neurotoxicology and Teratology* 41 (Jan.–Feb. 2014): 16–26.

183 *So far, plastic studies in rodents suggest:* Mary C. Catanese and Laura N. Vandenberg, "Bisphenol S (BPS) Alters Maternal Behavior and Brain in Mice Exposed During Pregnancy/Lactation and Their Daughters," *Endocrinology* 158, no. 3 (Mar. 1, 2017): 516–30.

185 *The sleuthing scientists shadowing us as we run afternoon errands:* Holden, "Avoiding Conflict," 1983; Holden, *Parents and the Dynamics of Child Rearing*, 73; Bornstein, "Determinants of Parenting," 34–35.

185 *Moms are twice as likely to spank in the evening:* Holden, *Parents and the Dynamics of Child Rearing*, 71–72.

185 *One study showed that Floridian moms:* Melissa A. Bright et al., "Association of Friday School Report Card Release With Saturday Incidence Rates of Agency-Verified Physical Child Abuse," *JAMA Pediatrics* 173, no. 2 (Feb. 2019): 176–82.

185 *"Don't ask what a gene does," the neuroscientist:* Robert Sapolsky, *Behave: The Biology of Humans at Our Best and Worst* (New York: Penguin Press, 2017), 256.

185 *Meadow-jumping mouse moms exposed:* Mary C. Catanese, Alexander Suvorov, and Laura N. Vandenberg, "Beyond a means of exposure: a new view of the mother in toxicology research," *Toxicology Research* 4, no. 3 (May 2015): 592–612.

185 *In rats, and perhaps in people, too:* Clara V. Perani et al., "High-fat diet prevents adaptive peripartum-associated adrenal gland plasticity and anxiolysis," *Scientific Reports* 5 (Oct. 2015): 14821.

186 *On the bright side, dining on fish:* Kei Hamazaki et al., "Dietary intake of fish and n-3 polyunsaturated fatty acids and risk of postpartum depression: a nationwide longitudinal

study—the Japan Environment and Children's Study (JECS)," *Psychological Medicine* 50, no. 14 (Sept.–Oct. 2020): 1–9.

186   *Roughly one in ten litters are abandoned in this straightforward manner:* Hrdy, *Mother Nature*, 93.

187   *Not long ago, the Yale Child Study Center:* Anna E. Austin and Megan V. Smith, "Examining Material Hardship in Mothers: Associations of Diaper Need and Food Insufficiency with Maternal Depressive Symptoms," *Health Equity* 1, no. 1 (Sept. 1, 2017): 127–33.

187   *I recently saw a video:* "Watch: Man makes life-saving catch as mother throws young son from burning balcony in Arizona," CTV News video, July 8, 2020, https://www.youtube.com/watch?v=NsTup6SukWc.

188   *But one study showed . . . mothers are more likely to trade arms:* Nadja Reissland et al., "Maternal stress and depression and the lateralisation of infant cradling," *Journal of Child Psychology and Psychiatry* 50, no. 3 (Mar. 2009): 263–69.

188   *David Slattery of Germany's . . . called the "restraint test":* Katherine M. Hillerer et al., "Lactation-induced reduction in hippocampal neurogenesis is reversed by repeated stress exposure," *Hippocampus* 24, no. 6 (June 2014): 673–83.

189   *Maybe stress-related . . . known to neglect their young:* Holekamp and Dloniak, "Fissiped Carnivores," 232.

190   *Still, we can't grow babies or make milk:* Hrdy, *Mother Nature*, 125.

191   *During the Bolivian planting seasons:* Trevathan, *Ancient Bodies*, 71.

191   *In Ethiopia, where the circumference of nursing mothers' upper arms shrinks:* Kedir Teji Roba et al., "Seasonal variation in nutritional status and anemia among lactating mothers in two agro-ecological zones of rural Ethiopia: A longitudinal study," *Nutrition* 31, no. 10 (Oct. 2015): 1213–18.

191   *Installing a new plumbing system:* Trevathan, *Ancient Bodies*, 45.

191   *Cheetah moms with full bellies:* Holekamp and Dloniak, "Fissiped Carnivores," 237.

191   *Meanwhile, hangry mammalian moms shirk their duties:* Cathy M. Dwyer, "Genetic and physiological determinants of maternal behavior and lamb survival: implications for low-input sheep management," *Journal of Animal Science* 86, supp. 14 (Apr. 2008): E246–58.

191   *One scientific paper introduced this phenomenon:* Winston Paul Smith, "Maternal Defense in Columbian White-Tailed Deer: When is it Worth It?," *American Naturalist* 130, no. 2 (Aug. 1987): 310–16.

194   *Economists have long understood that the birth rate and the economy:* Lisa J. Dettling and Melissa S. Kearney, "House prices and birth rates: The impact of the real estate market on the decision to have a baby," *Journal of Public Economics* 110 (Feb. 2014): 82–100.

194   *Conversely, regional windfalls like fracking booms:* Melissa S. Kearney and Riley Wilson, "The Family Formation Response to a Localized Economic Shock: Evidence from the Fracking Boom," Nov. 10, 2016, https://papers.ssrn.com/sol3/papers.cfm?abstract_id=2866663.

194   *A study of unemployment rates in Denmark:* Tim A. Bruckner, Laust H. Mortensen, and Ralph A. Catalano, "Spontaneous Pregnancy Loss in Denmark Following Economic Downturns," *American Journal of Epidemiology* 183, no. 8 (Apr. 15, 2016): 701–8.

194   *All of a sudden babies in those places started being born earlier:* Kyle Carlson, "Red Alert: Prenatal Stress and Plans to Close Military Bases," *American Journal of Health Economics* 4, no. 3 (Summer 2018): 287–320.

194   *Somewhat more hopefully, the Covid-19 pandemic:* Elizabeth Preston, "During Coronavirus Lockdown, Some Doctors Wondered: Where Are the Preemies?," *New York Times*, July 19, 2020. See also: Claire E. Margerison-Zilko et al., "Post-term birth as a response to environmental stress: The case of Sept. 11, 2001," *Evolution, Medicine, and Public Health* 2015, no. 1 (Jan. 16, 2015): 13–20.

195 *One study calculated that the announcement of five hundred layoffs:* Kyle Carlson, "Fear itself: The effects of distressing economic news on birth outcomes," *Journal of Health Economics* 41 (May 2015): 117–32.

195 *Likewise, during the recent subprime loan crisis, foreclosed-upon moms bore lighter-weight babies than average:* Janelle Downing and Tim Bruckner, "Subprime Babies: The Fore-closure Crisis and Initial Health Endowments," *RSF: The Russell Sage Foundation Journal of the Social Sciences: RSF* 5, no. 2 (Mar. 2019): 123–40.

195 *During California's unemployment surges, he's noted, SIDS deaths increase above expected levels:* Tim Bruckner and Ralph A. Catalano, "Economic Antecedents of Sudden Infant Death Syndrome," *Annals of Epidemiology* 16, no. 6 (June 2006): 415–22.

195 *Bruckner calculates that a 1 percent drop:* Tim Bruckner, "Metropolitan Economic De-cline and Infant Mortality due to Unintentional Injury," *Accident Analysis & Prevention* 40, no. 6 (Nov. 2008): 1797–803.

196 *These monkey moms, experiencing a bonanza one day and a bust the next:* Jeremy D. Cop-lan et al., "Synchronized Maternal-Infant Elevations of Primate CSF CRF Concentra-tions in Response to Variable Foraging Demand," *CNS Spectrums* 10, no. 7 (July 2005): 530–36.

196 *These stressed-out mothers became obsessed with patrolling:* Shariful Syed et al., "Glucagon like Peptide 1 as a Predictor of Telomere Length in Non Human Primate Exposed to Early Life Stress," *Biological Psychiatry* 81, no. 10, suppl. (May 15, 2017): S344.

198 *The anthropologist Robert Quinlan:* Robert J. Quinlan, "Human parental effort and environmental risk," *Proceedings of the Royal Society B* 274, no. 1606 (Jan. 7, 2007): 20063690.

198 *Like disease, wartime encourages psychological distancing:* Jonathan Levy et al., "Chronic trauma impairs the neural basis of empathy in mothers: Relations to parenting and children's empathic abilities," *Developmental Cognitive Neuroscience* 38 (Aug. 2019): 100658.

198 *More than a year after an 8.0 earthquake:* Zhiyong Qu et al., "The Impact of the cata-strophic earthquake in China's Sichuan province on the mental health of pregnant women," *Journal of Affective Disorders* 136, nos. 1–2 (Jan. 2012): 117–23.

198 *In the wake of Japan's Fukushima nuclear power plant meltdown:* Aya Goto et al., "Im-mediate effects of the Fukushima nuclear power plant disaster on depressive symp-toms among mothers with infants: a prefectural-wide cross-sectional study from the Fukushima Health Management Survey," *BMC Psychiatry* 15, no. 59 (2015); Evelyn J. Bromet, "Emotional Consequences of Nuclear Power Plant Disasters," *Health Physics* 106, no. 2 (Feb. 2014): 206–10.

198 *Moms in refugee camps . . . infant death rates swelled for months as overwhelmed mothers floundered:* Liz Ford, "Young lives hang by a thread as past haunts Rohingya mothers," *Guardian*, Mar. 29, 2018; Sammy Zahran et al., "Maternal exposure to hurricane de-struction and fetal mortality," *Journal of Epidemiology and Community Health* 68, no. 8 (Aug. 2014): 760–66.

199 *When Kim and her colleagues studied dozens of seemingly run-of-the-mill:* Sohye Kim et al., "Mothers' unresolved trauma blunts amygdala response to infant distress," *Social Neuro-science* 9, no. 4 (2014): 352–63.

201 *They give birth at younger ages:* For more on this strategy, see Jay Belsky, Laurence Stein-berg, and Patricia Draper, "Childhood Experience, Interpersonal Development, and Reproductive Strategy: An Evolutionary Theory of Socialization," *Child Development* 62, no. 4 (Aug. 1991): 647–70; Frances A. Champagne and James P. Curley, "The Trans-Generational Influence of Maternal Care on Offspring Gene Expression and Be-havior in Rodents," in Maestripieri and Mateo, *Maternal Effects in Mammals*, 195; Laeti-tia A. N'Dri et al., "The Invisible Threat: Bisphenol-A and Phthalates in Environmental

Justice Communities," *Environmental Justice* 8, no. 1 (Feb. 2015): 15–19; Ami R. Zota, Cassandra A. Phillips, and Susanna D. Mitro, "Recent Fast Food Consumption and Bisphenol A and Phthalates Exposures Among the U.S. Population in NHANES, 2003–2010," *Environmental Health Perspectives* 124, no. 10 (Oct. 2016): 1521–28.

201 *Poor mothers are more likely to have been exposed to substandard care:* Paul W. B. Bywaters et al., *The relationship between poverty, child abuse and neglect: an evidence review,* Joseph Rowntree Foundation, Mar. 3, 2016; Dawn E. Dailey et al., "An Exploration of Lifetime Trauma Exposure in Pregnant Low-income African American Women," *Maternal and Child Health Journal* 15, no. 3 (2011): 410–18.

201 *They may give birth to smaller babies who have associated medical problems:* James W. Collins Jr. et al., "Women's lifelong exposure to neighborhood poverty and low birth weight: a population-based study," *Maternal and Child Health Journal* 13, no. 3 (May 2009): 326–33.

201 *They are twice as likely to be treated for postpartum depression:* Linda H. Chaudron et al., "Accuracy of Depression Screening Tools for Identifying Postpartum Depression Among Urban Mothers," *Pediatrics* 125, no. 3 (Mar. 2010): e609–e617.

201 *Even just living in overly crowded conditions:* Robert H. Bradley, "Environment and Parenting," in Bornstein, *Handbook of Parenting,* 2:290.

201 *An environmental factor as simple as an infestation of cockroaches:* Snehal N. Shah et al., "Housing Quality and Mental Health: the Association between Pest Infestation and Depressive Symptoms among Public Housing Residents," *Journal of Urban Health* 95 (2018): 691–702.

202 *As with long-buried maternal trauma:* Pilyoung Kim et al., "Socioeconomic disadvantage, neural responses to infant emotions, and emotional availability among first-time new mothers," *Behavioural Brain Research* 325, pt. B (May 15, 2017): 188–96; Pilyoung Kim, Christian Capistrano, and Christina Congleton, "Socioeconomic disadvantages and neural sensitivity to infant cry: role of maternal distress," *Social Cognitive and Affective Neuroscience* 11, no. 10 (Oct. 2016): 1597–607.

204 *In one recent Columbia University study of two hundred new moms:* Kate Walsh et al., "Maternal prenatal stress phenotypes associate with fetal neurodevelopment and birth outcomes," *PNAS* 116, no. 48 (Nov. 26, 2019): 23996–4005.

204 *A different analysis of 48 million recent American births:* Douglas Almond and Lena Edlund, "Trivers–Willard at birth and one year: evidence from US natality data 1983–2001," *Proceedings of the Royal Society B* 274, no. 1624 (Oct. 7, 2007): 2491–96.

204 *And there is interesting evidence that some exceptionally well-situated women:* Elissa Z. Cameron and Fredrik Dalerum, "A Trivers-Willard Effect in Contemporary Humans: Male-Biased Sex Ratios among Billionaires," *PLoS ONE* 4, no. 1 (Jan. 14, 2009): e4195.

204 *These patterns can figure beyond birth:* Shige Song, "Spending patterns of Chinese parents on children's backpacks support the Trivers-Willard hypothesis: Results based on transaction data from China's largest online retailer," *Evolution and Human Behavior* 39, no. 3 (May 2018): 336–42; Rosemary L. Hopcroft and David O. Martin, "Parental Investments and Educational Outcomes: Trivers–Willard in the U.S.," *Frontiers in Sociology* 1, no. 3 (Mar. 31, 2016).

204 *For instance, amid recessions, some boy-girl spending patterns seem to change:* "Parents Spend More on Girls Than on Boys in a Recession," NBCNews.com, May 19, 2015, https://www.nbcnews.com/better/money/parents-spend-more-girls-boys-recession-n361441.

204 *Meanwhile, boys are more likely:* Tim A. Bruckner et al., "Economic downturns and male cesarean deliveries: A time-series test of the economic stress hypothesis," *BMC Pregnancy and Childbirth* 14, no. 198 (2014).

204 *However, it turned out that the baby-boy bust:* Tim A. Bruckner, Ralph Catalano, and Jen-

Notes to Pages 204–214 *299*

nifer Ahern, "Male fetal loss in the U.S. following the terrorist attacks of Sept. 11, 2001," *BMC Public Health* 10, no. 273 (2010).

205 *Perhaps for similar reasons, the fraught aftermath:* Tim A. Bruckner et al., "Preterm birth and selection *in utero* among males following the November 2015 Paris attacks," *International Journal of Epidemiology* 48, no. 5 (Oct. 2019): 1614–22.

205 *Further analysis suggests:* Tim A. Bruckner and Jenna Nobles, "Intrauterine stress and male cohort quality: The case of Sept. 11, 2001," *Social Science & Medicine* 76 (Jan. 2013): 107–14; Timothy A. Bruckner, et al., "Culled males, infant mortality and reproductive success in a pre-industrial Finnish population," *Proceedings of the Royal Society B* 282 (Jan. 22, 2015): 20140835; Ralph Catalano et al., "Selection against small males *in utero*: a test of the Wells hypothesis," *Human Reproduction* 27, no. 4 (Apr. 2012): 1202–8.

205 *Studies show that even subtly stressful factors . . . can slightly reduce a woman's odds of having a boy:* Madhukar Shivajirao Dama, "Parasite Stress Predicts Offspring Sex Ratio," *PLoS ONE* 7, no. 9 (Sept. 26, 2012): e46169; Nicholas J. Sanders and Charles F. Stoecker, "Where Have All the Young Men Gone? Using Gender Ratios to Measure Fetal Death Rates" (NBER Working Paper 17434, National Bureau of Economic Research, Sept. 2011); Ralph Catalano, Tim A. Bruckner, and Kirk R. Smith, "Ambient temperature predicts sex ratios and male longevity," *PNAS* 105, no. 6 (Feb. 12, 2008): 2244–47.

205 *So, apparently, can skipping breakfast:* Fiona Mathews, Paul J. Johnson, and Andrew Neil, "You are what your mother eats: evidence for maternal preconception diet influencing foetal sex in humans," *Proceedings of the Royal Society B* 275, no. 1643 (Apr. 22, 2008): 1661–68.

206 *In monkeys, these high-cortisol babies grow unusually quickly:* Katie Hinde et al., "Cortisol in mother's milk across lactation reflects maternal life history and predicts infant temperament," *Behavioral Ecology* 26, no. 1 (Jan.–Feb. 2015): 269–81.

206 *But one human study has linked children's temperaments to mothers' milk–borne stress hormones:* Katherine R. Grey et al., "Human milk cortisol is associated with infant temperament," *Psychoneuroendocrinology* 38, no. 7 (July 2013): 1178–85.

207 *"The concepts of 'good parenting' and 'bad parenting'":* Tomás Cabeza de Baca, Aurelio José Figueredo, and Bruce J. Ellis, "An Evolutionary Analysis of Variation in Parental Effort: Determinants and Assessment," *Parenting* 12, nos. 2–3 (2012): 94–104.

Chapter 9: NO MOM IS AN ISLAND

211 *However, I also had a few hidden risk factors:* American Society of Anesthesiologists, "Postpartum depression linked to mother's pain after childbirth," ScienceDaily, Oct. 14, 2018, https://www.sciencedaily.com/releases/2018/10/181014142700.htm.

211 *Although this research tidbit hadn't been published:* Sarah Wilson, "The Other Costs of Children: Motherhood, Substance Use, and Depression," (abstract), Nov. 2019, https://ssrn.com/abstract=3483569.

212 *Research from Columbia University suggests that the strength:* Walsh et al., "Maternal prenatal stress phenotypes."

212 *During birth, supported mothers undergo easier labors:* Brooke A. Scelza and Katie Hinde, "Crucial Contributions: A Biocultural Study of Grandmothering During the Perinatal Period," *Human Nature* 30 (Dec. 2019): 371–97.

212 *Pregnant women who are visited at home even occasionally:* Hrdy, *Mothers and Others*, 104.

212 *Women attended in childbirth by professional doulas:* Corter and Fleming, "Psychobiology," 150.

214 *Actually, even lab-rat moms are quite sensitive:* James P. Curley and Frances A. Cham-

pagne, "Influence of maternal care on the developing brain: mechanisms, temporal dynamics and sensitive periods," *Frontiers in Neuroendocrinology* 40 (Jan. 2016): 52–66.

217   *A study of seventy monkeys by Maestripieri and others:* Eric Michael Johnson, "A primatologist discovers the social factors responsible for maternal infanticide," ScientificAmerican .com, Nov. 22, 2010, https://blogs.scientificamerican.com/guest-blog/a-primatologist -discovers-the-social-factors-responsible-for-maternal-infanticide/.

218   *Fascinating studies have shown:* Stuart Semple, Melissa S. Gerald, and Dianne N. Suggs, "Bystanders affect the outcome of mother–infant interactions in rhesus macaques," *Proceedings of the Royal Society B* 276, no. 1665 (Mar. 11, 2009): 2257–62.

219   *Our new home was just a few miles:* Scelza and Hinde, "A Biocultural Study."

219   *Poor women, meanwhile, tend to live close by:* Quoctrung Bui and Claire Cain Miller, "The Typical American Lives Only 18 Miles From Mom," *New York Times,* Dec. 23, 2015.

219   *Postpartum depression seems to be much rarer:* B. Campos et al., "Familialism, social support, and stress: Positive implications for pregnant Latinas," *Cultural Diversity & Ethnic Minority Psychology* 14, no. 2 (2008): 155–62.

219   *Prairie dogs apparently sprint away:* Verdolin, *Raised by Animals,* 3.

219   *Brown bear moms bail:* Bjørn Dahle and Jon E. Swenson, "Family Breakup in Brown Bears: Are Young Forced to Leave?," *Journal of Mammalogy* 84, no. 2 (May 30, 2003): 536–40.

219   *"Mother Is Not like Mother":* Marc Naguib, Melanie Kober, and Fritz Trillmich, "Mother is not like mother: Concurrent pregnancy reduces lactating guinea pigs' responsiveness to pup calls," *Behavioural Processes* 83, no. 1 (Jan. 2010): 79–81.

219   *In some New World monkeys:* Sarah Blaffer Hrdy, "Meet the Alloparents," *Natural History Magazine,* Apr. 2009; Natalie Angier, "Weighing the Grandma Factor: In Some Societies, It's a Matter of Life and Death," *New York Times,* Nov. 5, 2002.

220   *But now there's a whole scientific literature:* Susan C. Alberts et al., "Reproductive aging patterns in primates reveal that humans are distinct," *PNAS* 110, no. 33 (Aug. 13, 2013): 13440–45; Lauren J. N. Brent et al., "Ecological Knowledge, Leadership, and the Evolution of Menopause in Killer Whales," *Current Biology* 25, no. 6 (Mar. 16, 2015): 746–50.

220   *"Given that her body is deteriorating":* Hrdy, *Mother Nature,* 47.

220   *Older women with more kids:* Cindy K. Barha et al., "Number of Children and Telomere Length in Women: A Prospective, Longitudinal Evaluation," *PLoS ONE* 11, no. 1 (Jan. 5, 2016): e0146424.

221   *A new study of preindustrial Finnish birth records showed:* Bridget Alex, "The Grandmother Hypothesis Could Explain Why Women Live So Long," *Discover,* Apr. 1, 2019, https://www.discovermagazine.com/planet-earth/the-grandmother-hypothesis -could-explain-why-women-live-so-long.

221   *In Great Britain . . . grandmas are universally associated with healthier pregnancies:* David Waynforth, "Grandparental investment and reproductive decisions in the longitudinal 1970 British cohort study," *Proceedings of the Royal Society B* 279 11, no. 1 (Sept. 14, 2012): 1155–60.

221   *A hormonal survey of 210 women:* Jennifer Hahn-Holbrook et al., "Placental Corticotropin-Releasing Hormone Mediates the Association Between Prenatal Social Support and Postpartum Depression," *Clinical Psychological Science* 1, no. 3 (July 1, 2013): 253–64.

221   *Nigerian grandmas:* Scelza and Hinde, "A Biocultural Study."

221   *Chinese grandmas braise pigs' feet:* ibid. Anne Noyes Saini, "Pigs' feet and roasted ginger made my traditional postpartum month off," *The World,* Dec. 17, 2014, https://www .pri.org/stories/2014-12-17/pigs-feet-and-roasted-ginger-made-my-traditional-post partum-month.

221   *In a study of Puerto Rican . . . showed the worst pregnancy outcomes:* Scelza and Hinde, "A Biocultural Study."

222   *Some potentially Thanksgiving dinner–destroying findings:* ibid.
222   *In a study of Chinese moms:* Ellen Y. Wan et al., "Postpartum depression and traditional postpartum care in China: Role of Zuoyuezi," *International Journal of Gynecology & Obstetrics* 104, no. 3 (Mar. 2009): 209–13.
223   *As the evolutionary psychologist Harald Euler notes:* Euler, "Grandparents and Extended Kin," in Salmon and Shackelford, *Oxford Handbook,* 187.
223   *Grandfathers' engagement varies greatly:* Hrdy, *Mothers and Others,* 266.
223   *It turns out that even female vampire bats:* Gerald G. Carter, Damien R. Farine, and Gerald S. Wilkinson, "Social bet-hedging in vampire bats," *Biology Letters* 13, no. 5 (May 24, 2017): 20170112.
223   *Female friendship likely began as a stockpiling of allomothers:* Hrdy, *Mothers and Others,* 272.
223   *Indeed, lesbianism:* Barry X. Kuhle and Sara Radtke, "Born Both Ways: The Alloparenting Hypothesis for Sexual Fluidity in Women," *Evolutionary Psychology* (Apr. 1, 2013).
224   *Maybe most new moms in contemporary Boston don't swap boobs:* Ellison, *Mommy Brain,* 95–96.
224   *A study of mothers of four-year-olds:* Kay Donahue Jennings, Vaughan Stagg, and Robin E. Connors, "Social Networks and Mothers' Interactions with Their Preschool Children," *Child Development* 62, no. 5 (Oct. 1991): 966–78.
224   *This is a typical strategy: jobs are social hubs:* DeWolf, "12 Stats About Working Women."
224   *The scholarly gist, though, is that some forms of work are toxic:* Ben Renner, "Cruelty Begets Cruelty: A Toxic Workplace Turns Women Into Worse Mothers, Study Finds," Study Finds.org, Oct. 12, 2019, https://www.studyfinds.org/cruelty-begets-cruelty-toxic-workplace-turns-women-into-worse-mothers/; Klaus Preisner et al., "Closing the Happiness Gap: The Decline of Gendered Parenthood Norms and the Increase in Parental Life Satisfaction," *Gender & Society* 34, no. 1 (Feb. 1, 2020): 31–55.
224   *Some working moms are miserable:* Elizabeth Mendes, Lydia Saad, and Kyley McGeeney, "Stay-at-Home Moms Report More Depression, Sadness, Anger," Gallup.com, May 18, 2012, https://news.gallup.com/poll/154685/stay-home-moms-report-depression-sadness-anger.aspx.
225   *It's not terribly surprising that heading back:* Richard J. Petts, "Time Off After Childbirth and Mothers' Risk of Depression, Parenting Stress, and Parenting Practices," *Journal of Family Issues* 39, no. 7 (May 1, 2018): 1827–54.
225   *The brutality of a postpartum command performance:* Emma Goldberg, "When the Surgeon Is a Mom," *New York Times,* Dec. 20, 2019.
225   *The most stressed American moms:* Erika L. Sabbath et al., "The long-term mortality impact of combined job strain and family circumstances: A life course analysis of working American mothers," *Social Science & Medicine* 146 (Dec. 2015): 111–19; Peter Hepburn, "Work Scheduling for American Mothers, 1990 and 2012," *Social Problems* 67, no. 4 (Nov. 2020): 741–62. Kerri M. Raissian and Lindsey Rose Bullinger, "Money matters: Does the minimum wage affect child maltreatment rates?," *Children and Youth Services Review* 72 (Jan. 2017): 60–70. Erin K. Kaplan, Courtney A. Collins, and Frances A. Tylavsky, "Cyclical unemployment and infant health," *Economics & Human Biology* 27, pt. A (Nov. 2017): 281–88.
225   *Though not particularly well paid:* "What makes a happy working mom?," ScienceDaily, Dec. 6, 2017, https://www.sciencedaily.com/releases/2017/12/171206122517.htm.
225   *There's research on the dangers of long commutes:* Bhashkar Mazumder and Zachary Seeskin, "Breakfast Skipping, Extreme Commutes, and the Sex Composition at Birth," *Biodemography and Social Biology* 61, no. 2 (2015): 187–208.
227   *I was big and clumsy:* Paula Sheppard, Justin R. Garcia, and Rebecca Sear, "A-Not-So-Grim Tale: How Childhood Family Structure Influences Reproductive and Risk-

Taking Outcomes in a Historical U.S. Population," *PLoS ONE* 9, no. 3 (Mar. 5, 2014): e89539.

228 *In humans, there are plenty . . . everything from the number of frozen soy desserts:* Xiaozhong Wen et al., "Sociodemographic differences and infant dietary patterns," *Pediatrics* 134, no. 5 (Nov. 2014): e1387–e1398.

231 *It's a sign of how little . . . many most countries don't even track maternal suicide statistics:* Kimberly Mangla et al., "Maternal self-harm deaths: an unrecognized and preventable outcome," *American Journal of Obstetrics and Gynecology* 221, no. 4 (Oct. 1, 2019): 295–303.

231 *In rapidly aging Japan, though:* "Giving life and dying of loneliness: many new mothers commit suicide in Japan," AsiaNews.it, Sept. 11, 2018, http://www.asianews.it /news-en/Giving-life-and-dying-of-loneliness:-many-new-mothers-commit-suicide -in-Japan-44902.html.

233 *Mammalian pair-bonding—which is quite rare, occurring in less than 5 percent:* Ker Than, "Wild Sex: Where Monogamy Is Rare," LiveScience, Nov. 20, 2006, https://www.live science.com/1135-wild-sex-monogamy-rare.html.

234 *But dads definitely do give maternal grandmas:* Mary J. Levitt, Ruth A. Weber, and M. Cherie Clark, "Social network relationships as sources of maternal support and well-being," *Developmental Psychology* 22, no. 3 (1986): 310–16.

234 *In situations of intergenerational abuse:* Rand D. Conger et al., "Disrupting intergenerational continuity in harsh and abusive parenting: The importance of a nurturing relationship with a romantic partner," *Journal of Adolescent Health* 53, no. 4, suppl. (Oct. 1, 2013): S11–S17.

234 *Moms with a supportive partner:* Bornstein, "Determinants of Parenting," 35, 37; Holden, *Parents and the Dynamics of Child-Rearing*, 77.

234 *They fare better in the wake of disaster—earthquake survivorship studies show:* Jian-Hua Ren et al., "Mental Disorders of Pregnant and Postpartum Women After Earthquakes: A Systematic Review," *Disaster Medicine and Public Health Preparedness* 8, no. 4 (Aug. 2014): 315–25.

234 *A study of a newly instituted paternal leave policy in Sweden:* Petra Persson and Maya Rossin-Slater, "When Dad Can Stay Home: Fathers' Workplace Flexibility and Maternal Health" (NBER Working Paper 25902, National Bureau of Economic Research, Oct. 2019).

234 *Absent dads, on the other hand:* Tomás Cabeza de Baca et al., "Lack of partner impacts newborn health through maternal depression: A pilot study of low-income immigrant Latina women," *Midwifery* 64 (Sept. 2018): 63–68; T. Colton, B. Lanzen, and W. Laverty, "Family structure, social capital, and mental health disparities among Canadian mothers," *Public Health* 129, no. 6 (June 2015): 639–47; Raeburn, *Do Fathers Matter?*, 78.

235 *Fascinatingly, though, some scientists now think:* Amar Hamoudi and Jenna Nobles, "Do Daughters Really Cause Divorce? Stress, Pregnancy, and Family Composition," *Demography* 51, no. 4 (Aug. 2014): 1423–49.

235 *A survey of abusive dads in New York's Adirondack region showed:* Rebecca L. Burch and Gordon G. Gallup Jr., "Perceptions of paternal resemblance predict family violence," *Evolution and Human Behavior* 21, no. 6 (Nov. 2000): 429–35.

235 *Leah Hibel of UC Davis recently asked mothers and fathers of six-month-olds to interact:* Leah C. Hibel et al., "Marital conflict sensitizes mothers to infant irritability: A randomized controlled experiment," *Infant and Child Development* 28, no. 3 (May/June 2019): e2127.

236 *In unhealthy partnerships a mother's behavior often:* Nadia Pancsofar et al., "Family relationships during infancy and later mother and father vocabulary use with young children," *Early Childhood Research Quarterly* 23, no. 4 (4th Quarter 2008): 493–503.

236 *So while marital status remains a predictor of mothering quality:* "Children in single-mother-by-choice families do just as well as those in two-parent families," ScienceDaily, July 5, 2017, https://www.sciencedaily.com/releases/2017/07/170705095332.htm.

238 *At least his experimental subjects weren't widowed in vain:* Oliver J. Bosch et al., "Abandoned prairie vole mothers show normal maternal care but altered emotionality: Potential influence of the brain corticotropin-releasing factor system," *Behavioural Brain Research* 341 (Apr. 2, 2018): 114–21.

Chapter 10: MOTHERLAND

243 *In recent years researchers have piloted:* Michal Bat Or, "Clay sculpting of mother and child figures encourages mentalization," *Arts in Psychotherapy* 37, no. 4 (Sept. 2010): 319–27; Deirdre Timlin and Ellen Elizabeth Anne Simpson, "A preliminary randomised control trial of the effects of Dru yoga on psychological well-being in Northern Irish first time mothers," *Midwifery* 46 (Mar. 2017): 29–36; Jamshid Tabeshpour et al., "A double-blind, randomized, placebo-controlled trial of saffron stigma (*Crocus sativus* L.) in mothers suffering from mild-to-moderate postpartum depression," *Phytomedicine* 36 (Dec. 1, 2017): 145–52; R. F. Slykerman et al., "Effect of *Lactobacillus rhamnosus* HN001 in Pregnancy on Postpartum Symptoms of Depression and Anxiety: A Randomised Double-blind Placebo-controlled Trial," *EBioMedicine* 24 (Oct. 2017): 159–65; Rachel Y. Moon et al., "Comparison of Text Messages Versus E-mail When Communicating and Querying with Mothers About Safe Infant Sleep," *Academic Pediatrics* 17, no. 8 (Nov.–Dec. 2017): 871–78; Erin M. Murphy et al., "Randomized Trial of Harp Therapy During In Vitro Fertilization–Embryo Transfer," *Journal of Evidence-Based Complementary and Alternative Medicine* 19, no. 2 (Apr. 2014): 93–8; Dan A. Oren et al., "An Open Trial of Morning Light Therapy for Treatment of Antepartum Depression," *American Journal of Psychiatry* 159, no. 4 (Apr. 2002): 666–69.

243 *But these don't seem to reliably improve mothering or mental health:* R. Ne'eman et al., "Intranasal administration of oxytocin increases human aggressive behavior," *Hormones and Behavior* 80, (Apr. 2016): 125–31; Ritu Bhandari et al., "Effects of intranasal oxytocin administration on memory for infant cues: Moderation by childhood emotional maltreatment," *Social Neuroscience* 9, no. 5 (June 2014): 536–47.

244 *Antidepressants are more effective:* Whitney P. Witt et al., "Access to Adequate Outpatient Depression Care for Mothers in the USA: A Nationally Representative Population-Based Study," *Journal of Behavioral Health Services & Research* 38, no. 2 (Apr. 2011): 191–204.

244 *In 2019, the FDA approved the first-ever medication:* "FDA approves first treatment for post-partum depression," FDA news release, Mar. 19, 2019, https://www.fda.gov/news-events/press-announcements/fda-approves-first-treatment-post-partum-depression.

244 *More traditional talk therapy shows some promise:* James E. Swain et al., "Parent-child intervention decreases stress and increases maternal brain activity and connectivity during own baby-cry: An exploratory study," *Development and Psychopathology* 29, no. 2 (May 2017): 535–53.

244 *When the state of New Jersey pledged:* Katy Backes Kozhimannil et al., "New Jersey's Efforts to Improve Postpartum Depression Care Did Not Change Treatment Patterns for Women on Medicaid," *Health Affairs* 30, no. 2 (Feb. 2011).

246 *In one society, your baby's horoscope:* Chih Ming Tan, Xiao Wang, and Xiaobo Zhang, "It's all in the stars: The Chinese zodiac and the effects of parental investments on offspring's

cognitive and noncognitive skill development" (IFPRI Discussion Paper 1708, International Food Policy Research Institute, 2018).

246 *The estimated 93 percent of American moms:* Traig, *Act Natural*, 104; Ruth Franklin, "No Book Will Fix What's Wrong With American Parenting," *New Republic*, Feb. 22, 2012, https://newrepublic.com/article/100955/druckerman-parenting-french-children -bebe-brooklyn.

246 *Two of the American . . . the so-called toddler and the alleged teenager:* Traig, *Act Natural*, 164–65, 173.

246 *But Europeans, and especially their American descendants:* Sapolsky, *Behave*, 276–79.

246 *These divides appear to color:* Stacey N. Doan and Qi Wang, "Maternal Discussions of Mental States and Behaviors: Relations to Emotion Situation Knowledge in European American and Immigrant Chinese Children," *Child Development* 81, no. 5 (Sept.–Oct. 2010): 1490–503; Sapolsky, *Behave*, 276; Suero Toda, Alan Fogel, and Masatoshi Kawai, "Maternal speech to three-month-old infants in the United States and Japan," *Journal of Child Language* 17, no. 2 (June 1990): 279–94; Anne Fernald and Hiromi Morikawa, "Common Themes and Cultural Variations in Japanese and American Mothers' Speech to Infants," *Child Development* 64, no. 3 (June 1993): 637–56.

247 *East Asian moms:* Hyun-Joo Lim and Tina Skinner, "Culture and motherhood: Findings from a qualitative study of East Asian mothers in Britain," *Families, Relationships, and Societies* 1, no. 3 (Nov. 2012): 327–43; Harkness and Super, "Culture and Parenting," 273; Meg Murphy, "Surprising number of Japanese kids still bathe with their parents up until high school," Japan Today, Jan. 25, 2016, https://japantoday.com/category/features /lifestyle/surprising-number-of-japanese-kids-still-bathe-with-their-parents-up-until -high-school.

247 *Some scientists have speculated that:* Sapolsky, *Behave*, 279–81.

247 *In a study of 118 women interacting:* Marc H. Bornstein et al., "Modalities of Infant-Mother Interaction in Japanese, Japanese American Immigrant, and European American Dyads," *Child Development* 83, no. 6 (Nov./Dec. 2012): 2073–88; Linda R. Cote et al., "The Acculturation of Parenting Cognitions: A Comparison of South Korean, Korean Immigrant, and European American Mothers," *Journal of Cross-Cultural Psychology* 46, no. 9 (Oct. 1, 2015): 1115–30.

248 *Other remarkable changes transpire:* Ben Ost and Eva Dziadula, "Gender preference and age at arrival among Asian immigrant mothers in the US," *Economics Letters* 145 (Aug. 2016), 286–90.

248 *Except, historical records suggest:* Jean M. Twenge, Emodish M. Abebe, and W. Keith Campbell, "Fitting In or Standing Out: Trends in American Parents' Choices for Children's Names, 1880–2007," *Social Psychological and Personality Science* 1, no. 1 (Jan. 2010): 19–25.

248 *And as East Asia has modernized:* Sapolsky, *Behave*, 276–77.

248 *Indeed, those same Puritan forefathers:* David Lancy, *The Anthropology of Childhood: Cherubs, Chattel, Changelings* (Cambridge, UK: Cambridge University Press: 2008), 249.

248 *As late as 1900, 10 percent:* Golden, *Babies Made Us Modern*, 207.

249 *Ten percent of babies born in Afghanistan still don't reach their first birthdays:* Sapolsky, *Behave*, 272.

249 *Two out of three mothers:* Emily Smith-Greenaway and Jenny Trinitapoli, "Maternal cumulative prevalence measures of child mortality show heavy burden in sub-Saharan Africa," *PNAS* 117, no. 8 (Feb. 25, 2020): 4027–33.

249 *Today just three in five hundred American newborns die:* "Why American infant mortality rates are so high," ScienceDaily, Oct. 13, 2016, https://www.sciencedaily.com/releases /2016/10/161013103132.htm.

249   *This is likely a big part of why:* Giudice and Belsky, "Parent–Child Relationships," 75.

249   *Even our use of motherese, which feels:* Lancy, *Raising Children*, 38.

250   *When Jennifer Hahn-Holbrook of the University of California, Merced:* Jennifer Hahn-Holbrook, Taylor Cornwell-Hinrichs, and Itzel Anaya, "Economic and Health Predictors of National Postpartum Depression Prevalence: A Systematic Review, Meta-analysis, and Meta-Regression of 291 Studies from 56 Countries," *Frontiers in Psychiatry* 8, no. 248 (Feb. 1, 2017).

250   *As the anthropologist David F. Lancy points out:* David F. Lancy, "Accounting for Variability in Mother-Child Play," *American Anthropologist* 109, no. 2 (June 2007).

252   *American six- and twelve-month-olds:* Jimin Sung et al., "Exploring temperamental differences in infants from the USA and the Netherlands," *European Journal of Developmental Psychology* 12, no. 1 (2015): 15–28.

252   *This news did not come entirely out of the blue:* Peter Adamson, "Child Well-being in Rich Countries: A Comparative Overview," *Innocenti Report Card* no. 11, UNICEF, Apr. 2013.

252   *To get to the heart of the matter, Gartstein ran a second study:* Maria A. Gartstein et al., "Is prenatal maternal distress context-dependent? Comparing United States and the Netherlands," *Journal of Affective Disorders* 260 (Jan. 1, 2020): 710–15.

253   *This most insidious of all social stressors:* Tyan Parker Dominguez et al., "Racial differences in birth outcomes: The role of general, pregnancy, and racism stress," *Health Psychology* 27, no. 2 (Mar. 2008): 194–203; Clayton J. Hilmert, "Lifetime racism and blood pressure changes during pregnancy: Implications for fetal growth," *Healthy Psychology* 33, no. 1 (2014): 43–51; Linda Villarosa, "Why America's Black Mothers and Babies Are in a Life-or-Death Crisis," *New York Times Magazine*, Apr. 11, 2018.

253   *The chromosomes in the placental cells:* Christopher W. Jones et al., "Differences in placental telomere length suggest a link between racial disparities in birth outcomes and cellular aging," *American Journal of Obstetrics and Gynecology* 26, no. 3 (Mar. 1, 2017): 294.e1–294.e8.

254   *Black moms are less likely to get treatment:* Nina Feldman and Aneri Pattani, "Black mothers get less treatment for postpartum depression than other moms," KHN, Dec. 6, 2019, https://khn.org/news/black-mothers-get-less-treatment-for-postpartum-depression-than-other-moms/; Chelsea O. McKinney et al., "Racial and Ethnic Differences in Breastfeeding," *Pediatrics* 138, no. 2 (Aug. 2016): e20152388.

254   *Drug users of both sexes exhibit dampened reward responses:* Helena J. V. Rutherford et al., "Disruption of maternal parenting circuitry by addictive process: rewiring of reward and stress systems," *Frontiers in Psychiatry* 2, no. 37 (July 6, 2011); Sohye Kim et al., "Mothers with substance addictions show reduced reward responses when viewing their own infant's face," *Human Brain Mapping* 38, no. 11 (Nov. 2017): 5421–39.

254   *Today, more opioid addicts are giving birth:* S. C. Haight et al., "Opioid Use Disorder Documented at Delivery Hospitalization—United States, 1999–2014," *Morbidity and Mortality Weekly Report* 67 (2018): 845–49.

254   *Nearly 2 percent of new mothers:* Alex F. Peahl et al., "Rates of New Persistent Opioid Use After Vaginal or Cesarean Birth Among US Women," *Obstetrics and Gynecology* 2, no. 7 (July 26, 2019): e197863.

254   *so-called "technoference" can change:* Brandon T. McDaniel and Jenny S. Radesky, "Technoference: Parent Distraction With Technology and Association With Child Behavior Problems," *Child Development* 89, no. 1 (Jan./Feb. 2018): 100–109; Tiffany G. Munzer et al., "Differences in Parent-Toddler Interactions with Electronic Versus Print Books," *Pediatrics* 143, no. 4 (Apr. 1, 2019); Jenny Radesky et al., "Maternal Mobile Device Use During a Structured Parent-Child Interaction Task," *Academic Pediatrics* 15, no. 2 (Mar. 1, 2015): 238–44.

254   *Around 2007, emergency rooms saw a marked rise in child traumatic brain injuries:* Joanne
      N. Wood et al., "Local Macroeconomic Trends and Hospital Admissions for Child
      Abuse, 2000–2009," *Pediatrics* 130, no. 2 (Aug. 2012): e358–e364; William Schnei-
      der, Jane Waldfogel, and Jeanne Brooks-Gunn, "The Great Recession and risk for child
      abuse and neglect," *Children and Youth Services Review* 72 (Jan. 2017): 71–81.

254   *A Yale study, though:* Beth Greenfield, "The Surprising Reason More Kids Are Getting
      Hurt at the Playground," Yahoo Parenting, Nov. 13, 2014, https://www.yahoo.com
      /news/the-surprising-reason-more-kids-are-getting-hurt-at-the-102543542767.html.

255   *In Italy, I might—according to the recently proposed:* Sabina Castelfranco, "Italy Launches
      'Land for Children' Plan to Fight Declining Birthrate," VOA News, Nov. 2, 2018,
      https://www.voanews.com/europe/italy-launches-land-children-plan-fight-declining
      -birthrate.

255   *In Hungary, the birth of my fourth:* Holly Elyatt, "Have four or more babies in Hungary
      and you'll pay no income tax for life, prime minister says," CNBC, Feb. 11 2019, https://
      www.cnbc.com/2019/02/11/have-four-or-more-babies-in-hungary-and-youll-pay
      -no-income-tax-for-life.html; "Poland to grant pensions to stay-at-home mums of four,"
      Reuters, Jan. 22, 2019, https://www.reuters.com/article/us-poland-benefit/poland
      -to-grant-pensions-to-stay-at-home-mums-of-four-idUSKCN1PG1RM.

255   *But ideas to make it more effective:* Dylan Matthews, "Mitt Romney and Michael Bennet
      just unveiled a basic income plan for kids," Vox, Dec. 16, 2019, https://www.vox.com
      /future-perfect/2019/12/16/21024222/mitt-romney-michael-bennet-basic-income
      -kids-child-allowance.

255   *Labor and delivery ward quality varies wildly:* Katy Backes Kozhimannil, Michael R. Law,
      and Beth A. Virnig, "Cesarean Delivery Rates Vary Tenfold Among US Hospitals; Reduc-
      ing Variation May Address Quality and Cost Issues," *Health Affairs* 32, no. 3 (Mar. 2013).

255   *Often the most scalpel-happy facilities cluster in the poorest neighborhoods:* Carine Milcent
      and Saad Zbiri, "Prenatal care and socioeconomic status: effect on cesarean delivery,"
      Health Economics Review 8, no. 7 (2018).

256   *Humans are no different . . . mothers in private rooms are less stressed:* Rachelle Jones,
      Liz Jones, and Anne-Marie Feary, "The Effects of Single-Family Rooms on Parenting
      Behavior and Maternal Psychological Factors," *Journal of Obstetric, Gynecological &
      Neonatal Nursing* 45, no. 3 (May–June 2016): 359–70; Bente Silnes Tandberg et al.,
      "Parent-Infant Closeness, Parents' Participation, and Nursing Support in Single-Family
      Room and Open Bay NICUs," *Journal of Perinatal & Neonatal Nursing* 32, no. 4 (Oct./
      Dec. 2018): e22–e32; Nancy Feeley, et al., "A comparative study of mothers of infants
      hospitalized in an open ward neonatal intensive care unit and a combined pod and
      single-family room design," *BMC Pediatrics* 20, no. 38 (2020).

256   *By contrast, nations like Israel:* Debra Kamin, "These upscale Israeli hotels are designed
      for new moms and babies," Times of Israel, Feb. 4, 2017, https://www.timesofisrael
      .com/these-upscale-israeli-hotels-are-designed-for-new-moms-and-babies/.

256   *Several observant Jewish communities:* "Women's Resort Opens Lakewood," COLlive,
      May 1, 2018, https://collive.com/womens-resort-opens-lakewood/.

257   *This is done in the name of science, by "baby-friendly" hospitals:* Carrie Arnold, "Do 'Baby-
      Friendly' Hospitals Work for All Moms?," *New York Times*, Apr. 18, 2020.

257   *In Finland, though, moms are kitted out with:* Sarah Gardner, "Finland's 'baby box' is a
      tradition full of nudges," Marketplace, Dec. 28, 2016, https://www.marketplace.org
      /2016/12/28/baby-box/.

258   *What women do with leave time after birth varies by culture:* Catherine Pearson, "What the
      French Get So Right About Taking Care of Newborns," *HuffPost*, Jan. 17, 2017, https://
      www.huffpost.com/entry/what-the-french-get-so-right-about-taking-care-of-new
      -moms_n_587d27b4e4b086022ca939c4.

258 *Paid leave boosts breastfeeding and encourages:* Maureen Sayres Van Niel et al., "The Impact of Paid Maternity Leave on the Mental and Physical Health of Mothers and Children: A Review of the Literature and Policy Implications," *Harvard Review of Psychiatry* 28, no. 2 (Mar./Apr. 2020): 113–26.

258 *Meanwhile, taking less than eight weeks is linked to increased depression:* ibid.

258 *Estonia offers up to eighty-five:* ibid.

258 *Norway ponies up ninety-one:* Christopher Ingraham, "The world's richest countries guarantee mothers more than a year of paid maternity leave. The U.S. guarantees them nothing," *Washington Post,* Feb. 5, 2018.

258 *Finland takes the:* ibid.

259 *Yes, most of us are entitled to schedule a:* Nina Martin, "Redesigning Maternal Care: OB-GYNs Are Urged to See New Mothers Sooner And More Often," NPR/ProPublica, Apr. 23, 2018, https://www.npr.org/2018/04/23/605006555/redesigning-maternal-care-ob-gyns-are-urged-to-see-new-mothers-sooner-and-more-often.

259 *In the Netherlands, though, each recovering mother:* Charlotte Hutting, "What Is a Kraam-verzorgster and Where Can I Find One?," Amsterdam Mamas, https://amsterdam-mamas.nl/articles/what-kraamverzorgster-and-where-can-i-find-one; Gaby Hinsliff, "Here's What It's Like to Live In A Country That Actually Cares About Mothers," *Huff-Post,* July 17, 2019, https://www.huffpost.com/entry/maternity-leave-postpartum-america-best-countries_n_5d1dc5f4e4b0f312567f5277.

259 *In Australia, new moms avail themselves:* For example, see https://www.tresillian.org.au/about-us/what-we-do/residential-stay/.

260 *Next door in New Zealand—which was, perhaps not coincidentally:* Kimberly Paterson, "Plunket nurse," Kiwi Families, https://www.kiwifamilies.co.nz/articles/plunket-nurse/.

261 *With rates of maternal depression and insomnia through the roof:* Margie H. Davenport et al., "Moms Are Not OK: COVID-19 and Maternal Mental Health," *Frontiers in Global Women's Health* 1, no. 1 (June 19, 2020); Tim Henderson, "Mothers Are 3 Times More Likely Than Fathers to Have Lost Jobs in Pandemic," Pew Stateline, Sept. 28, 2020, https://www.pewtrusts.org/en/research-and-analysis/blogs/stateline/2020/09/28/mothers-are-3-times-more-likely-than-fathers-to-have-lost-jobs-in-pandemic; Kim Elsesser, "Moms Cut Work Hours Four Times More Than Dads During Pandemic," Forbes.com, July 19, 2020, https://www.forbes.com/sites/kimelsesser/2020/07/17/moms-cut-work-hours-four-times-more-than-dads-during-pandemic/?sh=376fe98a49ca.; Ghadir Zreih et al., "Maternal perceptions of sleep problems among children and mothers during the coronavirus disease 2019 (COVID-19) pandemic in Israel," *Journal of Sleep Research* (Sept. 29, 2020): e13201.

264 *while I have a natural penchant:* Maressa Brown, "How Clear C-Section Drapes Let Moms Meet Their Babies in a New Way," *Parents,* Sept. 16, 2019.

Index

About the Author

Abigail Tucker's work has been featured in the Best American Science and Nature Writing series. She is the *New York Times* bestselling author of *The Lion in the Living Room: How House Cats Tamed Us and Took Over the World,* named a Best Science Book of 2016 by *Library Journal* and *Forbes* and a Barnes and Noble Discover Great New Writers selection, now translated into thirteen languages. A correspondent for *Smithsonian* magazine, she lives in New Haven, Connecticut, with her husband and four (equally amazing!) children.

By Diana Gabaldon

SEVEN STONES TO STAND OR FALL

SEVEN STONES TO STAND OR FALL

A Collection of Outlander Fiction

THE CUSTOM OF THE ARMY
THE SPACE BETWEEN | A PLAGUE OF ZOMBIES
A LEAF ON THE WIND OF ALL HALLOWS
VIRGINS | A FUGITIVE GREEN | BESIEGED

DIANA GABALDON

DELACORTE PRESS

NEW YORK

Copyright © 2017 by Diana Gabaldon
"The Custom of the Army" copyright © 2010 by Diana Gabaldon
"The Space Between" copyright © 2013 by Diana Gabaldon
"A Plague of Zombies" copyright © 2011 by Diana Gabaldon
"A Leaf on the Wind of All Hallows" copyright © 2010 by Diana Gabaldon
"Virgins" copyright © 2013 by Diana Gabaldon

All rights reserved.

Published in the United States by Delacorte Press, an imprint of Random House, a division of Penguin Random House LLC, New York.

DELACORTE PRESS and the HOUSE colophon are registered trademarks of Penguin Random House LLC.

"The Custom of the Army" was originally published in *Warriors,* edited by George R. R. Martin and Gardner Dozois, published by Tor Books, a division of Macmillan, in 2010. "The Space Between" was originally published in *The Mad Scientist's Guide to World Domination: Original Short Fiction for the Modern Evil Genius,* edited by John Joseph Adams, published by Tor Books, a division of Macmillan, in 2013. "A Plague of Zombies" was originally published as "Lord John and the Plague of Zombies" in *Down These Strange Streets: All-New Stories of Urban Fantasy,* edited by George R. R. Martin and Gardner Dozois, published by Ace Books, a division of Penguin, in 2011. "A Leaf on the Wind of All Hallows" was originally published in *Songs of Love and Death: All Original Tales of Star-Crossed Love,* edited by George R. R. Martin, published by Gallery Books, a division of Simon & Schuster, in 2010. "Virgins" was originally published in *Dangerous Women,* edited by George R. R. Martin and Gardner Dozois, published by Tor Books, a division of Macmillan, in 2013.

Hardback ISBN 978-0-399-59342-0
Ebook ISBN 978-0-399-59344-4

Printed in the United States of America on acid-free paper

randomhousebooks.com

2 4 6 8 9 7 5 3 1

First Edition

Book design by Virginia Norey

This book is dedicated with the greatest respect and gratitude to Karen Henry, Rita Meistrell, Vicki Pack, Sandy Parker, and Mandy Tidwell (collectively known as "the Cadre of Eyeball-Numbing Nitpickery") for their invaluable help in spotting errors, inconsistencies, and assorted rubbish.

(Any errors remaining in the text are purely the responsibility of the author, who not only blithely ignores inconsistencies on occasion, but has been known to deliberately perpetrate others.)

CONTENTS

INTRODUCTION

A Chronology of the *Outlander* Series

I f you picked this book up under the misapprehension that it's the ninth novel in the main *Outlander* series, it's not. I apologize.

So, if it's not the ninth novel, what is it? Well, it's a collection of seven . . . er . . . things, of varying length and content, but all having to do with the *Outlander* universe. As for the title . . . basically, it's the result of my editor not liking my original title choice, *Salmagundi.** Not that I couldn't see her point . . . Anyway, there was a polite request via my agent for something more in line with the "resonant, poetic" nature of the main titles.

Without going too much into the mental process that led to this (words like "sausage-making" and "rock-polishing" come to mind), I wanted a title that at least suggested that there were a number of elements in this book (hence the *Seven*), and *Seven Stones* just came naturally, and that was nice ("stone" is always a weighty word) and suitably alliterative but not a complete poetic thought (or rhythm). So, a bit more thinkering (no, that's not a typo), and I came up with *to Stand or Fall*, which sounded suitably portentous.

It took a bit of *ex post facto* thought to figure out what the heck that *meant*, but things usually do mean something if you think long enough. In this instance, the "stand or fall" has to do with people's response to grief and adversity: to wit, if you aren't killed outright by whatever happened, you have a choice in how the rest of your life is lived—you keep standing, though battered and worn by time and elements, still a buttress and a signpost—or you fall and return quietly to the earth from which you sprang, your elements giving succor to those who come after you.

---

* *Salmagundi: 1) A collection of disparate elements, or 2) a dish composed of meats, fruits, vegetables, and/or any other items the cook has on hand, often provided as an* ad hoc *accompaniment to an insufficient meal.*

SO. THIS IS (as the front cover suggests) a collection of seven novellas (fiction shorter than a novel but longer than a short story), though all of them are indeed part of the *Outlander* universe and do intersect with the main novels.

Five of the novellas included in this book were originally written for various anthologies over the last few years; two are brand-new and have never been published before: "A Fugitive Green" and "Besieged."

Owing to differences among publishers in different countries, some of the previously published novellas may subsequently have been published in print form as a four-story collection (in the UK and Germany) or as separate ebooks (in the United States). *Seven Stones* provides a complete print collection for those readers who like tactile books and includes the two new stories.

Since the novellas fit into the main series at different points (and involve a number of different characters), below is an overall chronology of the *Outlander* series, to explain Who, What, and When.

THE *OUTLANDER* SERIES includes three kinds of stories:

The Big, Enormous Books of the main series, which have no discernible genre (or all of them);

The Shorter, Less Indescribable Novels, which are more or less historical mysteries (though dealing also with battles, eels, and assorted sexual practices);

And

The Bulges,—these being short(er) pieces that fit somewhere inside the story lines of the novels, much in the nature of squirming prey swallowed by a large snake. These deal frequently—but not exclusively—with secondary characters, are prequels or sequels, and/or fill some lacuna left in the original story lines.

The Big Books of the main series deal with the lives and times of Claire and Jamie Fraser. The shorter novels focus on the adventures of Lord John Grey but intersect with the larger books (*The Scottish Prisoner*, for example, features both Lord John and Jamie Fraser in a shared story). The novellas all feature people from the main series, including Jamie and/or Claire on occasion. The description below explains which characters appear in which stories.

Most of the shorter Lord John novels and novellas (so far) fit within a large lacuna left in the middle of *Voyager,* in the years between 1756 and 1761. Some of the Bulges also fall in this period; others don't.

So, for the reader's convenience, the detailed listing here shows the sequence of the various elements in terms of the story line. *However, it should be noted that the shorter novels and novellas are all designed in such a way that they may be read alone,* without reference either to one another or to the Big, Enormous Books—should you be in the mood for a light literary snack instead of the nine-course meal with wine pairings and dessert trolley.

(For your added convenience, the description of each story includes the dates covered in it, and,—if it has been published before, the original anthology title and year of publication are also given. This information will be mostly useful to collectors and hardcore bibliophiles, but we aim to please as many people as possible.)

"Virgins" (novella)—Set in 1740 in France. In which Jamie Fraser (aged nineteen) and his friend Ian Murray (aged twenty) become young mercenaries. [Originally published in the anthology *Dangerous Women,* eds. George R. R. Martin and Gardner Dozois, 2012.]

Outlander (novel)—If you've never read any of the series, I'd suggest starting here. If you're unsure about it, open the book anywhere and read three pages; if you can put it down again, I'll give you a dollar. (1946/1743)

Dragonfly in Amber (novel)—It doesn't start where you think it's going to. And it doesn't end how you think it's going to, either. Just keep reading; it'll be fine. (1968/1744–46)

"A Fugitive Green" (novella)—Set in 1744–45 in Paris, London, and Amsterdam, this is the story of Lord John's elder brother, Hal (Harold, Earl Melton and Duke of Pardloe), and his (eventual) wife, Minnie—at the time of this story a seventeen-year-old dealer in rare books with a sideline in forgery, blackmail, and burglary. Jamie Fraser also appears in this one.

Voyager (novel)—This won an award from *EW* magazine for "Best Opening Line." (To save you having to find a copy just to read the opening, it was: *He was dead. However, his nose throbbed painfully, which he thought odd in the circumstances.*) If you're reading the series in order rather than piecemeal, you do want to read this book before tackling the novellas. (1968/1746–67)

Lord John and the Hand of Devils, "Lord John and the Hellfire Club" (short story)—Just to add an extra layer of confusion, *The Hand of Devils* is a collection that includes three novellas. The first one, "Lord John and the Hellfire Club," is set in London in 1756 and deals with a red-haired man who approaches Lord John Grey with an urgent plea for help, just before dying in front of him. [Originally published in the anthology *Past Poisons,* ed. Maxim Jakubowski, 1998.]

Lord John and the Private Matter (novel)—Set in London in 1757, this is a historical mystery steeped in blood and even less-savory substances, in which Lord John meets (in short order) a valet, a traitor, an apothecary with a sure cure for syphilis, a bumptious German, and an unscrupulous merchant prince.

Lord John and the Hand of Devils, "Lord John and the Succubus" (novella)—The second novella in the *Hand of Devils* collection finds Lord John in Germany in 1757, having unsettling dreams about Jamie Fraser, unsettling encounters with Saxon princesses, night hags, and a really disturbing encounter with a big blond Hanoverian graf. [Originally published in the anthology *Legends II,* ed. Robert Silverberg, 2003.]

Lord John and the Brotherhood of the Blade (novel)—The second full-length novel focused on Lord John (though Jamie Fraser also appears) is set in 1758, deals with a twenty-year-old family scandal, and sees Lord John engaged at close range with exploding cannon and even more dangerously explosive emotions.

Lord John and the Hand of Devils, "Lord John and the Haunted Soldier" (novella)—The third novella in this collection is set in 1758, in London and the Woolwich Arsenal. In which Lord John faces a court of inquiry into the explosion of a cannon and learns that there are more-dangerous things in the world than gunpowder.

"The Custom of the Army" (novella)—Set in 1759. In which his lordship attends an electric-eel party in London and consequently ends up at the Battle of Quebec. He's just the sort of person things like that happen to. [Originally published in *Warriors,* eds. George R. R. Martin and Gardner Dozois, 2010.]

The Scottish Prisoner (novel)—This one's set in 1760, in the Lake District, London, and Ireland. A sort of hybrid novel, it's divided

evenly between Jamie Fraser and Lord John Grey, who are recount-
ing their different perspectives in a tale of politics, corruption, murder, opium dreams, horses, and illegitimate sons.

"A Plague of Zombies" (novella)—Set in 1761 in Jamaica, when
Lord John is sent in command of a battalion to put down a slave
rebellion and discovers a hitherto unsuspected affinity for snakes,
cockroaches, and zombies. [Originally published in *Down These
Strange Streets*, eds. George R. R. Martin and Gardner Dozois,
2011.]

Drums of Autumn (novel)—The fourth novel of the main series,
this one begins in 1767, in the New World, where Jamie and Claire
find a foothold in the mountains of North Carolina, and their daughter, Brianna, finds a whole lot of things she didn't expect, when a
sinister newspaper clipping sends her in search of her parents.
(1969–1970/1767–70)

The Fiery Cross (novel)—The historical background to this, the
fifth novel of the main series, is the War of the Regulation in North
Carolina (1767–1771), which was more or less a dress rehearsal for
the oncoming Revolution. In which Jamie Fraser becomes a reluctant Rebel, his wife, Claire, becomes a conjure-woman, and their
grandson, Jeremiah, gets drunk on cherry bounce. Something Much
Worse happens to Brianna's husband, Roger, but I'm not telling you
what. This won several awards for "Best Last Line," but I'm not telling you that, either. (1770–1772)

A Breath of Snow and Ashes (novel)—Sixth novel of the main series, this book won the 2006 Corine International Prize for Fiction
and a Quill Award (this book beat novels by both George R. R.
Martin *and* Stephen King, which I thought was pretty entertaining;
I mean, how often does *that* happen?). All the books have an internal
"shape" that I see while I'm writing them. This one looks like the
Hokusai print titled "The Great Wave Off Kanagawa." Think
tsunami—two of them. (1773–1776/1980)

An Echo in the Bone (novel)—Set in America, London, Canada,
and Scotland, this is the seventh novel of the main series. The book's
cover image reflects the internal shape of the novel: a caltrop. That's
an ancient military weapon that looks like a child's jack with sharp
points; the Romans used them to deter elephants, and the highway
patrol still uses them to stop fleeing perps in cars. This book has four

major story lines: Jamie and Claire; Roger and Brianna (and family); Lord John and William; and Young Ian, all intersecting in the nexus of the American Revolution—and all the stories have sharp points. (1776–1778/1980)

Written in My Own Heart's Blood (novel)—The eighth book of the main series, *Blood* begins where *An Echo in the Bone* leaves off, in the summer of 1778 (and the autumn of 1980). The American Revolution is in full roar, and a lot of fairly horrifying things are happening in Scotland in the 1980s, too.

"A Leaf on the Wind of All Hallows" (short story [no, really, it is])—Set (mostly) in 1941–43, this is the story of What Really Happened to Roger MacKenzie's parents. [Originally published in the anthology *Songs of Love and Death,* eds. George R. R. Martin and Gardner Dozois, 2010.]

"The Space Between" (novella)—Set in 1778, mostly in Paris, this novella deals with Michael Murray (Young Ian's elder brother), Joan MacKimmie (Marsali's younger sister), the Comte St. Germain (who is Not Dead After All), Mother Hildegarde, and a few other persons of interest. The space between *what?* It depends who you're talking to. [Originally published in the anthology *The Mad Scientist's Guide to World Domination,* ed. John Joseph Adams, 2013]

"Besieged" (novella)—Set in 1762 in Jamaica and Havana. Lord John, about to leave his post as temporary military governor of Jamaica, learns that his mother is in Havana, Cuba. Which would be fine, save that the British Navy is on its way to lay siege to the city. Attended by his valet, Tom Byrd, an ex-zombie named Rodrigo, and Rodrigo's homicidally inclined wife, Azeel, Lord John sets out to rescue the erstwhile Dowager Duchess of Pardloe before the warships arrive.

NOW, REMEMBER . . .

You can read the short novels and novellas by themselves, or in any order you like. I would recommend reading the Big, Enormous Books of the main series in order, though. Hope you enjoy them all!

SEVEN STONES TO STAND OR FALL

THE CUSTOM OF THE ARMY

INTRODUCTION

ONE OF THE PLEASURES OF WRITING HISTORICAL fiction is that the best parts aren't made up. This particular story came about as the result of my having read Wendy Moore's excellent biography of Dr. John Hunter, *The Knife Man*—and my having read at the same time a brief facsimile book printed by the National Park Service, detailing regulations of the British Army during the American Revolution.

I wasn't *looking* for anything in particular in either of these books; just reading for background, general information on the period, and the always alluring chance of stumbling across something fascinating, like electric eel parties in London (these, along with Dr. Hunter himself—who appears briefly in this story—are a matter of historical record).

As for British Army regulations, a little of that stuff goes a long way; as a novelist, you want to resist the temptation to tell people things just because you happen to know them. Still, that book too had its little nuggets, such as the information that the word "bomb" was common in the eighteenth century, and that (in addition to merely meaning "an explosive device") it referred also to a wrapped and tarred parcel of shrapnel shot from a cannon (though we must be careful not to use the word "shrapnel," as it's derived from Lt. Henry Shrapnel of the Royal Artillery, who took the original "bomb" concept and developed the "shrapnel shell," a debris-filled bomb filled also with gunpowder and designed to explode in mid-air after being fired from a cannon; unfortunately, he did this in 1784, which was inconvenient, as "shrapnel" is a pretty good word to have when writing about warfare).

Among the other bits of interesting trivia, though, I was struck by a brief description of the procedure for courts-martial: *The custom of the army is that a court-martial be presided over by a senior officer and such a number of other officers as he shall think fit to serve as council, these being generally four in number, but can be more but not generally less than three. . . . The person accused shall have the right to*

call witnesses in his support, and the council shall question these, as well as any other persons whom they may wish, and shall thus determine the circumstances, and if conviction ensue, the sentence to be imposed.

And that was it. No elaborate procedures for the introduction of evidence, no standards for conviction, no sentencing guidelines, no requirements for who could or should serve as "council" to a court-martial—just "the custom of the army." The phrase—rather obviously—stuck in my head.

This story is for Karen Henry, Aedile Curule,
and Chief Bumblebee-Herder

ALL THINGS CONSIDERED, it was probably the fault of the electric eel. John Grey could—and for a time, did—blame the Honorable Caroline Woodford, as well. And the surgeon. And certainly that blasted poet. Still . . . no, it was the eel's fault.

The party had been at Lucinda Joffrey's house. Sir Richard was absent; a diplomat of his stature could not have countenanced something so frivolous. Electric-eel parties were a mania in London just now, but owing to the scarcity of the creatures, a private party was a rare occasion. Most such parties were held at public theaters, with the fortunate few selected for encounter with the eel summoned onstage, there to be shocked and sent reeling like ninepins for the entertainment of the audience.

"The record is forty-two at once!" Caroline had told him, her eyes wide and shining as she looked up from the creature in its tank.

"Really?" It was one of the most peculiar things he'd seen, though not very striking. Nearly three feet long, it had a heavy, squarish body with a blunt head, which looked to have been inexpertly molded out of sculptor's clay, and tiny eyes like dull glass beads. It had little in common with the lashing, lithesome eels of the fish market—and certainly did not seem capable of felling forty-two people at once.

The thing had no grace at all, save for a small thin ruffle of a fin that ran the length of its lower body, undulating as a gauze curtain does in the wind. Lord John expressed this observation to the Honorable Caroline and was accused in consequence of being poetic.

"Poetic?" said an amused voice behind him. "Is there no end to our gallant major's talents?"

Lord John turned, with an inward grimace and an outward smile, and bowed to Edwin Nicholls.

"I should not think of trespassing upon your province, Mr. Nicholls," he said politely. Nicholls wrote execrable verse, mostly upon the subject of love, and was much admired by young women of a certain turn of mind. The Honorable Caroline wasn't one of them; she'd written a very clever

parody of his style, though Grey thought Nicholls had not heard about it. He hoped not.

"Oh, don't you?" Nicholls raised one honey-colored brow at him and glanced briefly but meaningfully at Miss Woodford. His tone was jocular, but his look was not, and Grey wondered just how much Mr. Nicholls had had to drink. Nicholls was flushed of cheek and glittering of eye, but that might be only the heat of the room, which was considerable, and the excitement of the party.

"Do you think of composing an ode to our friend?" Grey asked, ignoring Nicholls's allusion and gesturing toward the large tank that contained the eel.

Nicholls laughed, too loudly—yes, quite a bit the worse for drink—and waved a dismissive hand.

"No, no, Major. How could I think of expending my energies upon such a gross and insignificant creature, when there are angels of delight such as this to inspire me?" He leered—Grey did not wish to impugn the fellow, but he undeniably leered—at Miss Woodford, who smiled, with compressed lips, and tapped him rebukingly with her fan.

Where was Caroline's uncle? Grey wondered. Simon Woodford shared his niece's interest in natural history and would certainly have escorted her. . . . Oh, there. Simon Woodford was deep in discussion with Dr. Hunter, the famous surgeon—what had possessed Lucinda to invite *him*? Then he caught sight of Lucinda, viewing Dr. Hunter over her fan with narrowed eyes, and realized that she *hadn't* invited him.

John Hunter was a famous surgeon—and an infamous anatomist. Rumor had it that he would stop at nothing to bag a particularly desirable body—whether human or not. He did move in society, but not in the Joffreys' circles.

Lucinda Joffrey had most expressive eyes. Her one claim to beauty, they were almond-shaped, clear gray in color, and capable of sending remarkably minatory messages across a crowded room.

Come here! they said. Grey smiled and lifted his glass in salute to her but made no move to obey. The eyes narrowed further, gleaming dangerously, then cut abruptly toward the surgeon, who was edging toward the tank, his face alight with curiosity and acquisitiveness.

The eyes whipped back to Grey.

Get rid of him! they said.

Grey glanced at Miss Woodford. Mr. Nicholls had seized her hand in his and appeared to be declaiming something; she looked as though she wanted the hand back. Grey looked back at Lucinda and shrugged, with a small gesture toward Mr. Nicholls's ochre-velvet back, expressing regret that social responsibility prevented his carrying out her order.

"Not only the face of an angel," Nicholls was saying, squeezing Caro-

line's fingers so hard that she squeaked, "but the skin, as well." He stroked her hand, the leer intensifying. "What do angels smell like in the morning, I wonder?"

Grey measured him up thoughtfully. One more remark of that sort, and he might be obliged to invite Mr. Nicholls to step outside. Nicholls was tall and heavily built, outweighed Grey by a couple of stone, and had a reputation for bellicosity. *Best try to break his nose first,* Grey thought, shifting his weight, *then run him headfirst into a hedge. He won't come back in if I make a mess of him.*

"What are you looking at?" Nicholls inquired unpleasantly, catching Grey's gaze upon him.

Grey was saved from reply by a loud clapping of hands—the eel's proprietor calling the party to order. Miss Woodford took advantage of the distraction to snatch her hand away, cheeks flaming with mortification. Grey moved at once to her side and put a hand beneath her elbow, fixing Nicholls with an icy stare.

"Come with me, Miss Woodford," he said. "Let us find a good place from which to watch the proceedings."

"Watch?" said a voice beside him. "Why, surely you don't mean to *watch,* do you, sir? Are you not curious to try the phenomenon yourself?"

It was Hunter himself, bushy hair tied carelessly back, though decently dressed in a damson-red suit, and grinning up at Grey; the surgeon was broad-shouldered and muscular but quite short—barely five foot two, to Grey's five-six. Evidently he had noted Grey's wordless exchange with Lucinda.

"Oh, I think—" Grey began, but Hunter had his arm and was tugging him toward the crowd gathering round the tank. Caroline, with an alarmed glance at the glowering Nicholls, hastily followed him.

"I shall be most interested to hear your account of the sensation," Hunter was saying chattily. "Some people report a remarkable euphoria, a momentary disorientation . . . shortness of breath or dizziness—sometimes pain in the chest. You have not a weak heart, I hope, Major? Or you, Miss Woodford?"

"Me?" Caroline looked surprised.

Hunter bowed to her.

"I should be particularly interested to see your own response, ma'am," he said respectfully. "So few women have the courage to undertake such an adventure."

"She doesn't want to," Grey said hurriedly.

"Well, perhaps I *do,*" she said, and gave him a little frown, before glancing at the tank and the long gray form inside it. She gave a brief shiver— but Grey recognized it, from long acquaintance with the lady, as a shiver of anticipation rather than revulsion.

Dr. Hunter recognized it, too. He grinned more broadly and bowed again, extending his arm to Miss Woodford.

"Allow me to secure you a place, ma'am."

Grey and Nicholls both moved purposefully to prevent him, collided, and were left scowling at each other as Dr. Hunter escorted Caroline to the tank and introduced her to the eel's owner, a small dark-looking creature named Horace Suddfield.

Grey nudged Nicholls aside and plunged into the crowd, elbowing his way ruthlessly to the front. Hunter spotted him and beamed.

"Have you any metal remaining in your chest, Major?"

"Have I—what?"

"Metal," Hunter repeated. "Arthur Longstreet described to me the operation in which he removed thirty-seven pieces of metal from your chest—most impressive. If any bits remain, though, I must advise you against trying the eel. Metal conducts electricity, you see, and the chance of burns—"

Nicholls had made his way through the throng, as well, and gave an unpleasant laugh, hearing this.

"A good excuse, Major," he said, a noticeable jeer in his voice. He was very drunk indeed, Grey thought. Still—

"No, I haven't," he said abruptly.

"Excellent," Suddfield said politely. "A soldier, I understand you are, sir? A bold gentleman, I perceive—who better to take first place?"

And before Grey could protest, he found himself next to the tank, Caroline Woodford's hand clutching his, her other held by Nicholls, who was glaring malevolently.

"Are we all arranged, ladies and gentlemen?" Suddfield cried. "How many, Dobbs?"

"Forty-five!" came a call from his assistant in the next room, through which the line of participants snaked, joined hand-to-hand and twitching with excitement, the rest of the party standing well back, agog.

"All touching, all touching?" Suddfield cried. "Take a firm grip of your friends, please, a very firm grip!" He turned to Grey, his small face alight. "Go ahead, sir! Grip it tightly, please—just there, just there before the tail!"

Disregarding his better judgment and the consequences to his lace cuff, Grey set his jaw and plunged his hand into the water.

In the split second when he grasped the slimy thing, he expected something like the snap one got from touching a Leyden jar and making it spark. Then he was flung violently backward, every muscle in his body contorted, and he found himself on the floor, thrashing like a landed fish, gasping in a vain attempt to recall how to breathe.

The surgeon, Mr. Hunter, squatted next to him, observing him with bright-eyed interest.

"How do you feel?" he inquired. "Dizzy at all?"

Grey shook his head, mouth opening and closing like a goldfish's, and with some effort thumped his chest. Thus invited, Mr. Hunter leaned down at once, unbuttoned Grey's waistcoat, and pressed an ear to his shirtfront. Whatever he heard—or didn't—seemed to alarm him, for he jerked up, clenched both fists together, and brought them down on Grey's chest with a thud that reverberated to his backbone.

This blow had the salutary effect of forcing breath out of his lungs; they filled again by reflex, and suddenly he remembered how to breathe. His heart also seemed to have been recalled to a sense of its duty, and began beating again. He sat up, fending off another blow from Mr. Hunter, and sat blinking at the carnage round him.

The floor was filled with bodies. Some still writhing, some lying still, limbs outflung in abandonment; some already recovered and being helped to their feet by friends. Excited exclamations filled the air, and Suddfield stood by his eel, beaming with pride and accepting congratulations. The eel itself seemed annoyed; it was swimming round in circles, angrily switching its heavy body.

Edwin Nicholls was on hands and knees, Grey saw, rising slowly to his feet. He reached down to grasp Caroline Woodford's arms and help her to rise. This she did, but so awkwardly that she lost her balance and fell face-first into Mr. Nicholls. He in turn lost his own balance and sat down hard, the Honorable Caroline atop him. Whether from shock, excitement, drink, or simple boorishness, he seized the moment—and Caroline—and planted a hearty kiss upon her astonished lips.

Matters thereafter were somewhat confused. He had a vague impression that he *had* broken Nicholls's nose—and there was a set of burst and swollen knuckles on his right hand to give weight to the supposition. There was a lot of noise, though, and he had the disconcerting feeling of not being altogether firmly confined within his own body. Parts of him seemed to be constantly drifting off, escaping the outlines of his flesh.

What *did* remain inside was distinctly jangled. His hearing—still somewhat impaired from the cannon explosion a few months before—had given up entirely under the strain of electric shock. That is, he could hear, but what he heard made no sense. Random words reached him through a fog of buzzing and ringing, but he could not connect them sensibly to the moving mouths around him. He wasn't at all sure that his own voice was saying what he meant it to, for that matter.

He was surrounded by voices, faces—a sea of feverish sound and movement. People touched him, pulled him, pushed him. He flung out an arm,

trying as much to discover where it was as to strike anyone, but felt the impact of flesh. More noise. Here and there a face he recognized: Lucinda, shocked and furious; Caroline, distraught, her red hair disheveled and coming down, all its powder lost.

The net result of everything was that he was not positive whether he had called Nicholls out or the reverse. Surely Nicholls must have challenged him? He had a vivid recollection of Nicholls, gore-soaked handkerchief held to his nose and a homicidal light in his narrowed eyes. But then he'd found himself outside, in his shirtsleeves, standing in the little park that fronted the Joffreys' house, with a pistol in his hand. He wouldn't have chosen to fight with a strange pistol, would he?

Maybe Nicholls had insulted him, and he had called Nicholls out without quite realizing it?

It had rained earlier, was chilly now; wind was whipping his shirt round his body. His sense of smell was remarkably acute; it seemed to be the only thing working properly. He smelled smoke from the chimneys, the damp green of the plants, and his own sweat, oddly metallic. And something faintly foul—something redolent of mud and slime. By reflex, he rubbed the hand that had touched the eel against his breeches.

Someone was saying something to him. With difficulty, he fixed his attention on Mr. Hunter, standing by his side, still with that look of penetrating interest. *Well, of course. They'd need a surgeon,* he thought dimly. *Have to have a surgeon at a duel.*

"Yes," he said, seeing Hunter's eyebrows raised in inquiry of some sort. Then, seized by a belated fear that he had just promised his body to the surgeon were he killed, seized Hunter's coat with his free hand.

"You . . . don't . . . touch me," he said. "No . . . knives. Ghoul," he added for good measure, finally locating the word. Hunter nodded, seeming unoffended.

The sky was overcast, the only light shed by the distant torches at the house's entrance. Nicholls was a whitish blur, coming closer.

Someone grabbed Grey, turned him forcibly about, and he found himself back-to-back with Nicholls, the bigger man's heat startling, so near.

Shit, he thought suddenly. *Is he any kind of a shot?*

Someone spoke and he began to walk—he thought he was walking—until an outthrust arm stopped him, and he turned in answer to someone pointing urgently behind him.

Oh, hell, he thought wearily, seeing Nicholls's arm come down. *I don't care.*

He blinked at the muzzle flash—the report was lost in the shocked gasp from the crowd—and stood for a moment, wondering whether he'd been hit. Nothing seemed amiss, though, and someone nearby was urging him to fire.

Frigging poet, he thought. *I'll delope and have done. I want to go home.* He raised his arm, aiming straight up into the air, but his arm lost contact with his brain for an instant, and his wrist sagged. He jerked, correcting it, and his hand tensed on the trigger. He had barely time to jerk the barrel aside, firing wildly.

To his surprise, Nicholls staggered a bit, then sank down onto the grass. He sat propped on one hand, the other clutched dramatically to his shoulder, head thrown back.

It had begun to rain, quite hard. Grey blinked water off his lashes and shook his head. The air tasted sharp, like cut metal, and for an instant he had the impression that it smelled . . . purple.

"That can't be right," he said aloud, and found that his ability to speak seemed to have come back. He turned to speak to Hunter, but the surgeon had, of course, darted across to Nicholls, was peering down the neck of the poet's shirt. There was blood on it, Grey saw, but Nicholls was refusing to lie down, gesturing vigorously with his free hand. Blood was running down his face from his nose; perhaps that was it.

"Come away, sir," said a quiet voice at his side. "It'll be bad for Lady Joffrey else."

"What?" He looked, surprised, to find Richard Tarleton, who had been his ensign in Germany, now in the uniform of a Lancers lieutenant. "Oh. Yes, it will." Dueling was illegal in London; for the police to arrest Lucinda's guests in the park before her house would be a scandal—not something that would please her husband, Sir Richard, at all.

The crowd had already melted away, as though the rain had rendered them soluble. The torches by the door had been extinguished. Nicholls was being helped off by Hunter and someone else, lurching away through the increasing rain. Grey shivered. God knew where his coat or cloak was.

"Let's go, then," he said.

GREY OPENED HIS eyes.

"Did you say something, Tom?"

Tom Byrd, his valet, had produced a cough like a chimney sweep's, at a distance of approximately one foot from Grey's ear. Seeing that he had obtained his employer's attention, he presented the chamber pot at port arms.

"His Grace is downstairs, me lord. With her ladyship."

Grey blinked at the window behind Tom, where the open drapes showed a dim square of rainy light.

"Her ladyship? What, the duchess?" What could have happened? It couldn't be past nine o'clock. His sister-in-law never paid calls before af-

ternoon, and he had never known her to go anywhere with his brother during the day.

"No, me lord. The little 'un."

"The little—oh. My goddaughter?" He sat up, feeling well but strange, and took the utensil from Tom.

"Yes, me lord. His Grace said as he wants to speak to you about 'the events of last night.'" Tom had crossed to the window and was looking censoriously at the remnants of Grey's shirt and breeches, these stained with grass, mud, blood, and powder stains, and flung carelessly over the back of the chair. He turned a reproachful eye on Grey, who closed his own, trying to recall exactly what the events of last night had been.

He felt somewhat odd. Not drunk, he hadn't been drunk; he had no headache, no uneasiness of digestion. . . .

"Last night," he repeated, uncertain. Last night had been confused, but he did remember it. The eel party. Lucinda Joffrey, Caroline . . . Why on earth ought Hal to be concerned with . . . what, the duel? Why should his brother care about such a silly affair—and even if he did, why appear at Grey's door at the crack of dawn with his six-month-old daughter?

It was more the time of day than the child's presence that was unusual; his brother often did take his daughter out, with the feeble excuse that the child needed air. His wife accused him of wanting to show the baby off—she was beautiful—but Grey thought the cause somewhat more straight-forward. His ferocious, autocratic, dictatorial brother—Colonel of his own regiment, terror of both his own troops and his enemies—had fallen in love with his daughter. The regiment would leave for its new posting within a month's time. Hal simply couldn't bear to have her out of his sight.

Thus he found the Duke of Pardloe seated in the morning room, Lady Dorothea Jacqueline Benedicta Grey cradled in his arm and gnawing on a rusk her father held for her. Her wet silk bonnet, her tiny rabbit-fur bunting, and two letters, one open, one still sealed, lay upon the table at the duke's elbow.

Hal glanced up at him.

"I've ordered your breakfast. Say hallo to Uncle John, Dottie." He turned the baby gently round. She didn't remove her attention from the rusk but made a small chirping noise.

"Hallo, sweetheart." John leaned over and kissed the top of her head, covered with a soft blond down and slightly damp. "Having a nice outing with Daddy in the pouring rain?"

"We brought you something." Hal picked up the opened letter and, raising an eyebrow at his brother, handed it to him.

Grey raised an eyebrow back and began to read.

"What?!" He looked up from the sheet, mouth open.

"Yes, that's what I said," Hal agreed cordially, "when it was delivered to my door, just before dawn." He reached for the sealed letter, carefully balancing the baby. "Here, this one's yours. It came just after dawn."

Grey dropped the first letter as though it were on fire and seized the second, ripping it open.

Oh, John, it read without preamble, *forgive me, I couldn't stop him, I really couldn't, I'm SO sorry. I told him, but he wouldn't listen. I'd run away, but I don't know where to go. Please, please do something!* It wasn't signed but didn't need to be. He'd recognized the Honorable Caroline Woodford's writing, scribbled and frantic as it was. The paper was blotched and puckered—with tearstains?

He shook his head violently, as though to clear it, then picked up the first letter again. It was just as he'd read it the first time—a formal demand from Alfred, Lord Enderby, to His Grace the Duke of Pardloe, for satisfaction regarding the injury to the honor of his sister, the Honorable Caroline Woodford, by the agency of His Grace's brother, Lord John Grey.

Grey glanced from one document to the other, several times, then looked at his brother.

"What the devil?"

"I gather you had an eventful evening," Hal said, grunting slightly as he bent to retrieve the rusk Dottie had dropped on the carpet. "No, darling, you don't want that anymore."

Dottie disagreed violently with this assertion and was distracted only by Uncle John picking her up and blowing in her ear.

"Eventful," he repeated. "Yes, it was, rather. But I didn't do anything to Caroline Woodford save hold her hand whilst being shocked by an electric eel, I swear it. Gleeglgleeglgleegl-pppppssssshhhhh," he added to Dottie, who shrieked and giggled in response. He glanced up to find Hal staring at him.

"Lucinda Joffrey's party," he amplified. "Surely you and Minnie were invited?"

Hal grunted. "Oh. Yes, we were, but I had a prior engagement. Minnie didn't mention the eel. What's this I hear about you fighting a duel over the girl, though?"

"What? It wasn't—" He stopped, trying to think. "Well, perhaps it was, come to think. Nicholls—you know, that swine who wrote the ode to Minnie's feet?—he kissed Miss Woodford, and she didn't want him to, so I punched him. Who told you about the duel?"

"Richard Tarleton. He came into White's cardroom late last night and said he'd just seen you home."

"Well, then, you likely know as much about it as I do. Oh, you want

Daddy back now, do you?" He handed Dottie to his brother and brushed at a damp patch of saliva on the shoulder of his coat.

"I suppose that's what Enderby's getting at." Hal nodded at the earl's letter. "That you made the poor girl publicly conspicuous and compromised her virtue by fighting a scandalous duel over her. I suppose he's got a point."

Dottie was now gumming her father's knuckle, making little growling noises. Hal dug in his pocket and came out with a silver teething ring, which he offered her in lieu of his finger, meanwhile giving Grey a sidelong look.

"You don't want to marry Caroline Woodford, do you? That's what Enderby's demand amounts to."

"God, no." Caroline was a good friend—bright, pretty, and given to mad escapades—but marriage? Him?

Hal nodded.

"Lovely girl, but you'd end in Newgate or Bedlam within a month."

"Or dead," Grey said, gingerly picking at the bandage Tom had insisted on wrapping round his knuckles. "How's Nicholls this morning, do you know?"

"Ah." Hal rocked back a little, drawing a deep breath. "Well . . . dead, actually. I had rather a nasty letter from his father, accusing you of murder. That one came over breakfast; didn't think to bring it. Did you mean to kill him?"

Grey sat down quite suddenly, all the blood having left his head.

"No," he whispered. His lips felt stiff and his hands had gone numb. "Oh, Jesus. No."

Hal swiftly pulled his snuffbox from his pocket, one-handed, dumped out the vial of smelling salts he kept in it, and handed it to his brother. Grey was grateful; he hadn't been going to faint, but the assault of ammoniac fumes gave him excuse for watering eyes and congested breathing.

"Jesus," he repeated, and sneezed explosively several times in a row. "I didn't aim to kill—I swear it, Hal. I deloped. Or tried to," he added honestly.

Lord Enderby's letter now made more sense, as did Hal's presence. What had been a silly affair that should have disappeared with the morning dew had become—or would, directly the gossip had time to spread—not merely a scandal but quite possibly something worse. It was not unthinkable that he *might* be arrested for murder. Quite without warning, the figured carpet yawned at his feet, an abyss into which his life might vanish.

Hal nodded and gave him his own handkerchief.

"I know," he said quietly. "Things . . . happen sometimes. That you don't intend—that you'd give your life to have back."

Grey wiped his face, glancing at his brother under cover of the gesture.

Hal looked suddenly older than his years, his face drawn by more than worry over Grey.

"Nathaniel Twelvetrees, you mean?" Normally he wouldn't have mentioned that matter, but both men's guards were down.

Hal gave him a sharp look, then glanced away.

"No, not Twelvetrees. I hadn't any choice about that. And I did mean to kill him. I meant . . . what led to that duel." He grimaced. "Marry in haste, repent at leisure." He looked at the note on the table and shook his head. His hand passed gently over Dottie's head. "I won't have you repeat my mistakes, John," he said quietly.

Grey nodded, wordless. Hal's first wife had been seduced by Nathaniel Twelvetrees. Hal's mistakes notwithstanding, Grey had never intended marriage with anyone and didn't now.

Hal frowned, tapping the folded letter on the table in thought. He darted a glance at John and sighed, then set the letter down, reached into his coat, and withdrew two further documents, one clearly official, from its seal.

"Your new commission," he said, handing it over. "For Crefeld," he said, raising an eyebrow at his brother's look of blank incomprehension. "You were brevetted lieutenant-colonel. You didn't remember?"

"I—well . . . not exactly." He had a vague feeling that someone— probably Hal—had told him about it, soon after Crefeld, but he'd been badly wounded then and in no frame of mind to think about the army, let alone to care about battlefield promotion. Later—

"Wasn't there some confusion over it?" Grey took the commission and opened it, frowning. "I thought they'd changed their minds."

"Oh, you do remember, then," Hal said, eyebrow still cocked. "General Wiedman gave it to you after the battle. The confirmation was held up, though, because of the inquiry into the cannon explosion, and then the . . . ah . . . kerfuffle over Adams."

"Oh." Grey was still shaken by the news of Nicholls's death, but mention of Adams started his brain functioning again. "Adams. Oh. You mean Twelvetrees held up the commission?" Colonel Reginald Twelvetrees, of the Royal Artillery: brother to Nathaniel and cousin to Bernard Adams, the traitor awaiting trial in the Tower as a result of Grey's efforts the preceding autumn.

"Yes. Bastard," Hal added dispassionately. "I'll have him for breakfast, one of these days."

"Not on my account, I hope," Grey said dryly.

"Oh, no," Hal assured him, jiggling his daughter gently to prevent her fussing. "It will be a purely personal pleasure."

Grey smiled at that, despite his disquiet, and put down the commission. "Right," he said, with a glance at the fourth document, which still lay

folded on the table. It was an official-looking letter and had been opened; the seal was broken. "A proposal of marriage, a denunciation for murder, and a new commission—what the devil's that one? A bill from my tailor?"

"Ah, that. I didn't mean to show it to you," Hal said, leaning carefully to hand it over without dropping Dottie. "But under the circumstances . . ."

Hal waited, noncommittal, as Grey opened the letter and read it. It was a request—or an order, depending how you looked at it—for the attendance of Major Lord John Grey at the court-martial of one Captain Charles Carruthers, to serve as witness of character for the same. In . . .

"In Canada?" John's exclamation startled Dottie, who crumpled up her face and threatened to cry.

"Hush, sweetheart." Hal jiggled faster, hastily patting her back. "It's all right; only Uncle John being an ass."

Grey ignored this, waving the letter at his brother.

"What the devil is Charlie Carruthers being court-martialed for? And why on earth am I being summoned as a character witness?"

"Failure to suppress a mutiny," Hal said. "As to why you—he asked for you, apparently. An officer under charges is allowed to call his own witnesses, for whatever purpose. Didn't you know that?"

Grey supposed that he had, in an academic sort of way. But he had never attended a court-martial himself; it wasn't a common proceeding, and he had no real idea of the shape of the proceedings. He glanced sideways at Hal.

"You say you didn't mean to show it to me?"

Hal shrugged and blew softly over the top of his daughter's head, making the short blond hairs furrow and rise like wheat in the wind.

"No point. I meant to write back and say that as your commanding officer I required you here; why should you be dragged off to the wilds of Canada? But given your talent for awkward situations . . . What did it feel like?" he inquired curiously.

"What did—oh, the eel." Grey was accustomed to his brother's lightning shifts of conversation and made the adjustment easily. "Well, it was rather a shock."

He laughed—if tremulously—at Hal's glower, and Dottie squirmed round in her father's arms, reaching out her own plump little arms appealingly to her uncle.

"Flirt," he told her, taking her from Hal. "No, really, it was remarkable. You know how it feels when you break a bone? That sort of jolt that goes right through you before you feel the pain, and you go blind for a moment and feel as if someone's driven a nail through your belly? It was like that, only much stronger, and it went on for longer. Stopped my breath," he admitted. "Quite literally. And my heart, too, I think. Dr. Hunter—

you know, the anatomist?—was there and pounded on my chest to get it started again."

Hal was listening with close attention and asked several questions, which Grey answered automatically, his mind occupied with this latest surprising communiqué.

Charlie Carruthers. They'd been young officers together, though from different regiments. Fought beside each other in Scotland, gone round London together for a bit on their next leave. They'd had—well, you couldn't call it an affair. Three or four brief encounters—sweating, breathless quarters of an hour in dark corners that could be conveniently forgotten in daylight or shrugged off as the result of drunkenness, not spoken of by either party.

That had been in the Bad Time, as he thought of it: those years after Hector's death, when he'd sought oblivion wherever he could find it— and found it often—before slowly recovering himself.

Likely he wouldn't have recalled Carruthers at all, save for the one thing.

Carruthers had been born with an interesting deformity—he had a double hand. While Carruthers's right hand was normal in appearance and worked quite as usual, there was another, dwarf hand that sprang from his wrist and nestled neatly against its larger partner. Dr. Hunter would probably pay hundreds for that hand, Grey thought, with a mild lurch of the stomach.

The dwarf hand had only two short fingers and a stubby thumb—but Carruthers could open and close it, though not without also opening and closing the larger one. The shock when Carruthers had closed both of them simultaneously on Grey's prick had been nearly as extraordinary as had the electric eel's.

"Nicholls hasn't been buried yet, has he?" he asked abruptly, the thought of the eel party and Dr. Hunter causing him to interrupt some remark of Hal's.

Hal looked surprised.

"Surely not. Why?" He narrowed his eyes at Grey. "You don't mean to attend the funeral, do you?"

"No, no," Grey said hastily. "I was only thinking of Dr. Hunter. He, um, has a certain reputation, and Nicholls did go off with him. After the duel."

"A reputation as what, for God's sake?" Hal demanded impatiently.

"As a body snatcher," Grey blurted.

There was a sudden silence, awareness dawning in Hal's face. He'd gone pale.

"You don't think—no! How could he?"

"A . . . um . . . hundredweight or so of stones substituted just prior to

the coffin's being nailed shut is the usual method—or so I've heard," Grey said, as well as he could with Dottie's fist poked up his nose.

Hal swallowed. Grey could see the hairs rise on his wrist.

"I'll ask Harry," Hal said, after a short silence. "The funeral can't have been arranged yet, and if . . ."

Both brothers shuddered reflexively, imagining all too exactly the scene as an agitated family member insisted upon raising the coffin lid, to find . . .

"Maybe better not," Grey said, swallowing. Dottie had left off trying to remove his nose and was patting her tiny hand over his lips as he talked. The feel of it on his skin . . .

He peeled her gently off and gave her back to Hal.

"I don't know what use Charles Carruthers thinks I might be to him— but, all right, I'll go." He glanced at Lord Enderby's note, Caroline's crumpled missive. "After all, I suppose there are worse things than being scalped by red Indians."

Hal nodded, sober.

"I've arranged your sailing. You leave tomorrow." He stood and lifted Dottie. "Here, sweetheart. Kiss your Uncle John goodbye."

A MONTH LATER, Grey found himself, Tom Byrd at his side, climb- ing off the *Harwood* and into one of the small boats that would land them and the battalion of Louisbourg grenadiers with whom they had been traveling on a large island near the mouth of the St. Lawrence River.

He had never seen anything like it. The river itself was larger than any he had ever seen, nearly half a mile across, running wide and deep, a dark blue-black under the sun. Great cliffs and undulating hills rose on either side of the river, so thickly forested that the underlying stone was nearly invisible. It was hot, and the sky arched brilliant overhead, much brighter and much wider than any sky he had seen before. A loud hum echoed from the lush growth—insects, he supposed, birds, and the rush of the water, though it felt as if the wilderness were singing to itself, in a voice heard only in his blood. Beside him, Tom was fairly vibrating with excite- ment, his eyes out on stalks, not to miss anything.

"Cor, is that a red Indian?" he whispered, leaning close to Grey in the boat.

"I don't suppose he can be anything else," Grey replied, as the gentle- man loitering by the landing was naked save for a breechclout, a striped blanket slung over one shoulder, and a coating of what—from the shim- mer of his limbs—appeared to be grease of some kind.

"I thought they'd be redder," Tom said, echoing Grey's own thought. The Indian's skin was considerably darker than Grey's own, to be sure, but

a rather pleasant soft brown in color, something like dried oak leaves. The Indian appeared to find them nearly as interesting as they had found him; he was eyeing Grey in particular with intent consideration.

"It's your hair, me lord," Tom hissed in Grey's ear. "I told you you ought to have worn a wig."

"Nonsense, Tom." At the same time, Grey experienced an odd frisson up the back of the neck, constricting his scalp. Vain of his hair, which was blond and thick, he didn't commonly wear a wig, choosing instead to bind and powder his own for formal occasions. The present occasion wasn't formal in the least. With the advent of freshwater aboard, Tom had insisted upon washing Grey's hair that morning, and it was still spread loose upon his shoulders, though it had long since dried.

The boat crunched on the shingle, and the Indian flung aside his blanket and came to help the men run it up the shore. Grey found himself next to the man, close enough to smell him. He smelled quite unlike anyone Grey had ever encountered: gamy, certainly—he wondered, with a small thrill, whether the grease the man wore might be bear fat—but with the tang of herbs and a sweat like fresh-sheared copper.

Straightening up from the gunwale, the Indian caught Grey's eye and smiled.

"You be careful, Englishman," he said, in a voice with a noticeable French accent, and, reaching out, ran his fingers quite casually through Grey's loose hair. "Your scalp would look good on a Huron's belt."

This made the soldiers from the boat all laugh, and the Indian, still smiling, turned to them.

"They are not so particular, the Abenaki who work for the French. A scalp is a scalp—and the French pay well for one, no matter what color." He nodded genially to the grenadiers, who had stopped laughing. "You come with me."

THERE WAS A small camp on the island already, a detachment of infantry under a Captain Woodford—whose name gave Grey a slight wariness but who turned out to be no relation, thank God, to Lord Enderby's family.

"We're fairly safe on this side of the island," he told Grey, offering him a flask of brandy outside his own tent after supper. "But the Indians raid the other side regularly—I lost four men last week, three killed and one carried off."

"You have your own scouts, though?" Grey asked, slapping at the mosquitoes that had begun to swarm in the dusk. He had not seen the Indian who had brought them to the camp again, but there were several more in

camp. Most clustered together around their own fire, but one or two squatted, bright-eyed and watchful, among the Louisbourg grenadiers who had crossed with Grey on the *Harwood*.

"Yes, and trustworthy for the most part," Woodford said, answering Grey's unasked question. He laughed, though not with any humor. "At least we hope so."

Woodford gave him supper, and they had a hand of cards, Grey exchanging news of home for gossip of the current campaign.

General Wolfe had spent no little time at Montmorency, below the town of Quebec, but had nothing but disappointment from his attempts there, and so had abandoned that post, regathering the main body of his troops some miles upstream from the Citadel of Quebec. The so-far impregnable fortress, perched on sheer cliffs above the river, commanded both the river and the plains to the west with her cannon, obliging English warships to steal past under cover of night—and not always successfully.

"Wolfe'll be champing at the bit, now his grenadiers are come," Woodford predicted. "He puts great store by those fellows; fought with 'em at Louisbourg. Here, Colonel, you're being eaten alive—try a bit of this on your hands and face." He dug about in his campaign chest and came up with a tin of strong-smelling grease, which he pushed across the table.

"Bear grease and mint," he explained. "The Indians use it—that, or cover themselves with mud."

Grey helped himself liberally; the scent wasn't quite the same as what he had smelled earlier on the scout, but it was very similar, and he felt an odd sense of disturbance in its application. Though it did discourage the biting insects.

He had made no secret of the reason for his presence and now asked openly about Carruthers.

"Where is he held, do you know?"

Woodford frowned and poured more brandy.

"He's not. He's paroled; has a billet in the town at Gareon, where Wolfe's headquarters are."

"Ah?" Grey was mildly surprised—but, then, Carruthers was not charged with mutiny but rather with failure to suppress one—a rare charge. "Do you know the particulars of the case?"

Woodford opened his mouth, as though to speak, but then drew a deep breath, shook his head, and drank brandy. From which Grey deduced that probably everyone knew the particulars but that there was something fishy about the affair. Well, time enough. He'd hear about the matter directly from Carruthers.

Conversation became general, and after a time Grey said good night. The grenadiers had been busy; a new little city of canvas tents had sprung

up at the edge of the existing camp, and the appetizing smells of fresh meat roasting and tea brewing were rising on the air.

Tom had doubtless managed to raise his own tent, somewhere in the mass. Grey was in no hurry to find it, though; he was enjoying the novel sensations of firm footing and solitude, after weeks of crowded shipboard life. He cut outside the orderly rows of new tents, walking just beyond the glow of the firelight, feeling pleasantly invisible, though still close enough for safety—or at least he hoped so. The forest stood only a few yards away, the outlines of trees and bushes still visible, the dark not quite complete.

A drifting spark of green drew his eye, and he felt delight well up in him. There was another . . . another . . . ten, a dozen, and the air was suddenly full of fireflies, soft green sparks that winked on and off, glowing like tiny distant candles among the dark foliage. He'd seen fireflies once or twice before, in Germany, but never in such abundance. They were simple magic, pure as moonlight.

He could not have said how long he watched them, wandering slowly along the edge of the encampment, but at last he sighed and turned toward the center, full-fed, pleasantly tired, and with no immediate responsibility to do anything. He had no troops under his command, no reports to write . . . nothing, really, to do until he reached Gareon and Charlie Carruthers.

With a sigh of peace, he closed the flap of his tent and shucked his outer clothing.

He was roused abruptly from the edge of sleep by screams and shouts, and sat bolt upright. Tom, who had been asleep on his bed sack at Grey's feet, sprang up like a frog onto hands and knees, scrabbling madly for pistol and shot in the chest.

Not waiting, Grey seized the dagger he had hung on the tent peg before retiring and, flinging back the flap, peered out. Men were rushing to and fro, colliding with tents, shouting orders, yelling for help. There was a glow in the sky, a reddening of the low-hanging clouds.

"Fireships!" someone shouted. Grey shoved his feet into his shoes and joined the throng of men now rushing toward the water.

Out in the center of the broad dark river stood the bulk of the *Harwood*, at anchor. And coming slowly down upon her were one, two, and then three blazing vessels. A raft, stacked with flammable waste, doused with oil and set afire. A small boat, its mast and sail flaming bright against the night. Something else—an Indian canoe, with a heap of burning grass and leaves? Too far to see, but it was coming closer.

He glanced at the ship and saw movement on deck—too far to make out individual men, but things were happening. The ship couldn't raise anchor and sail away, not in time—but she was lowering her boats, sailors

setting out to try to deflect the fireships, keep them away from the *Har-wood*.

Absorbed in the sight, he had not noticed the shrieks and shouts still coming from the other side of the camp. But now, as the men on the shore fell silent, watching the fireships, they began to stir, realizing belatedly that something else was afoot.

"Indians," the man beside Grey said suddenly, as a particularly high, ululating screech split the air. "Indians!"

This cry became general, and everyone began to rush in the other direction.

"Stop! Halt!" Grey flung out an arm, catching a man across the throat and knocking him flat. He raised his voice in the vain hope of stopping the rush. "You! You and you—seize your neighbor, come with me!" The man he had knocked down bounced up again, white-eyed in the starlight.

"It may be a trap!" Grey shouted. "Stay here! Stand to your arms!"

"Stand! Stand!" A short gentleman in his nightshirt took up the cry in a cast-iron bellow, adding to its effect by seizing a dead branch from the ground and laying about himself, turning back those trying to get past him to the encampment.

Another spark grew upstream, and another beyond it: more fireships. The boats were in the water now, mere dots in the darkness. If they could fend off the fireships, the *Harwood* might be saved from immediate destruction; Grey's fear was that whatever was going on in the rear of the encampment was a ruse designed to pull men away from the shore, leaving the ship protected only by her marines. The French could then send down a barge loaded with explosives, or a boarding craft, hoping to elude detection while everyone was dazzled or occupied by the blazing fireships and the raid.

The first of the fireships had drifted harmlessly onto the far shore and was burning itself out on the sand, brilliant and beautiful against the night. The short gentleman with the remarkable voice—clearly he was a sergeant, Grey thought—had succeeded in rallying a small group of soldiers, whom he now presented to Grey with a brisk salute.

"Will they go and fetch their muskets, all orderly, sir?"

"They will," Grey said. "And hurry. Go with them, Sergeant—it is Sergeant?"

"Sergeant Aloysius Cutter, sir," the short gentleman replied with a nod, "and pleased to know an officer what has a brain in his head."

"Thank you, Sergeant. And fetch back as many more men as fall conveniently to hand, if you please. With arms. A rifleman or two, if you can find them."

Matters thus momentarily attended to, he turned his attention once more to the river, where two of the *Harwood*'s small boats were herding

one of the fireships away from the transport, circling it and pushing water with their oars; he caught the splash of their efforts and the shouts of the sailors.

"Me lord?"

The voice at his elbow nearly made him swallow his tongue. He turned with an attempt at calmness, ready to reproach Tom for venturing out into the chaos, but before he could summon words, his young valet stooped at his feet, holding something.

"I've brought your breeches, me lord," Tom said, voice trembling. "Thought you might need 'em, if there was fighting."

"Very thoughtful of you, Tom," he assured his valet, fighting an urge to laugh. He stepped into the breeches and pulled them up, tucking in his shirt. "What's been happening in the camp, do you know?"

He could hear Tom swallow hard.

"Indians, me lord," Tom said. "They came screaming through the tents, set one or two afire. They killed one man I saw, and . . . and scalped him." His voice was thick, as though he might be about to vomit. "It was nasty."

"I daresay." The night was warm, but Grey felt the hairs rise on arms and neck. The chilling screams had stopped, and while he could still hear considerable hubbub in the camp, it was of a different tone now: no random shouting, just the calls of officers, sergeants, and corporals ordering the men, beginning the process of assembly, of counting noses and reckoning damage.

Tom, bless him, had brought Grey's pistol, shot bag, and powder, as well as his coat and stockings. Aware of the dark forest and the long, narrow trail between the shore and the camp, Grey didn't send Tom back but merely told him to keep out of the way as Sergeant Cutter—who, with good military instinct, had also taken time to put his breeches on—came up with his armed recruits.

"All present, sir," Cutter said, saluting. "'Oom 'ave I the honor of h'addressing, sir?"

"I am Lieutenant-Colonel Grey. Set your men to watch the ship, please, Sergeant, with particular attention to dark craft coming downstream, and then come back to report what you know of matters in camp."

Cutter saluted and promptly vanished with a shout of "Come on, you shower o' shit! Look lively, look lively!"

Tom gave a brief, strangled scream, and Grey whirled, drawing his dagger by reflex, to find a dark shape directly behind him.

"Don't kill me, Englishman," said the Indian who had led them to the camp earlier. He sounded mildly amused. "*Le capitaine* sent me to find you."

"Why?" Grey asked shortly. His heart was still pounding from the shock. He disliked being taken at a disadvantage, and disliked even more

the thought that the man could easily have killed him before Grey knew he was there.

"The Abenaki set your tent on fire; he supposed they might have dragged you and your servant into the forest."

Tom uttered an extremely coarse expletive and made as though to dive directly into the trees, but Grey stopped him with a hand on his arm.

"Stay, Tom. It doesn't matter."

"The bloody hell you say," Tom replied heatedly, agitation depriving him of his normal manners. "I daresay I can find you more smallclothes, not as that will be easy, but what about your cousin's painting of her and the little 'un she sent for Captain Stubbs? What about your good hat with the gold lace?!?"

Grey had a brief moment of alarm—his young cousin Olivia had sent a miniature of herself and her newborn son, charging Grey to deliver this to her husband, Captain Malcolm Stubbs, presently with Wolfe's troops. He clapped a hand to his side, though, and felt with relief the oval shape of the miniature in its wrappings, safe in his pocket.

"That's all right, Tom; I've got it. As to the hat . . . we'll worry about that later, I think. Here—what is your name, sir?" he inquired of the Indian, unwilling to address him simply as "you."

"Manoke," said the Indian, still sounding amused.

"Quite. Will you take my servant back to the camp?" He saw the small, determined figure of Sergeant Cutter appear at the mouth of the trail and, firmly overriding Tom's protests, shooed him off in care of the Indian.

IN THE EVENT, all five fireships either drifted or were steered away from the *Harwood*. Something that might—or might not—have been a boarding craft did appear upstream but was frightened off by Grey's impromptu troops on the shore, firing volleys—though the range was woefully short; there was no possibility of hitting anything.

Still, the *Harwood* was secure, and the camp had settled into a state of uneasy watchfulness. Grey had seen Woodford briefly upon his return, near dawn, and learned that the raid had resulted in the deaths of two men and the capture of three more, dragged off into the forest. Three of the Indian raiders had been killed, another wounded—Woodford intended to interview this man before he died but doubted that any useful information would result.

"They never talk," he'd said, rubbing at his smoke-reddened eyes. His face was pouchy and gray with fatigue. "They just close their eyes and start singing their damned death songs. Not a blind bit of difference what you do to 'em—they just keep on singing."

Grey had heard it, or thought he had, as he crawled wearily into his borrowed shelter toward daybreak. A faint, high-pitched chant that rose and fell like the rush of the wind in the trees overhead. It kept up for a bit, then stopped abruptly, only to resume again, faint and interrupted, as he teetered on the edge of sleep.

What was the man saying? he wondered. Did it matter that none of the men hearing him knew what he said? Perhaps the scout—Manoke, that was his name—was there; perhaps he would know.

Tom had found Grey a small tent at the end of a row. Probably he had ejected some subaltern, but Grey wasn't inclined to object. It was barely big enough for the canvas bed sack that lay on the ground and a box that served as table, on which stood an empty candlestick, but it was shelter. It had begun to rain lightly as he walked up the trail to camp, and the rain was now pattering busily on the canvas overhead, raising a sweet, musty scent. If the death song continued, it was no longer audible over the sound of the rain.

Grey turned over once, the grass stuffing of the bed sack rustling softly beneath him, and fell at once into sleep.

HE WOKE ABRUPTLY, face-to-face with an Indian. His reflexive flurry of movement was met with a low chuckle and a slight withdrawal, rather than a knife across the throat, though, and he broke through the fog of sleep in time to avoid doing serious damage to the scout Manoke.

"What?" he muttered, and rubbed the heel of his hand across his eyes. "What is it?" *And why the devil are you lying on my bed?*

In answer to this, the Indian put a hand behind his head, drew him close, and kissed him. The man's tongue ran lightly across his lower lip, darted like a lizard's into his mouth, and then was gone.

So was the Indian.

He rolled over onto his back, blinking. A dream. It was still raining, harder now. He breathed in deeply; he could smell bear grease, of course, on his own skin, and mint—was there any hint of metal? The light was stronger—it must be day; he heard the drummer passing through the aisles of tents to rouse the men, the rattle of his sticks blending with the rattle of the rain, the shouts of corporals and sergeants—but still faint and gray. He could not have been asleep for more than half an hour, he thought.

"Christ," he muttered, and, turning himself stiffly over, pulled his coat over his head and sought sleep once again.

THE *HARWOOD* TACKED slowly upriver, with a sharp eye out for French marauders. There were a few alarms, including another raid by hostile Indians while camped on shore. This one ended more happily, with four marauders killed and only one cook wounded, not seriously. They were obliged to loiter for a time, waiting for a cloudy night, in order to steal past the fortress of Quebec, menacing on its cliffs. They were spotted, in fact, and one or two cannon fired in their direction, but to no effect. And at last came into port at Gareon, the site of General Wolfe's headquarters.

The town itself had been nearly engulfed by the growing military encampment that surrounded it, acres of tents spreading upward from the settlement on the riverbank, the whole presided over by a small French Catholic mission, whose tiny cross was just visible at the top of the hill that lay behind the town. The French inhabitants, with the political indifference of merchants everywhere, had given a Gallic shrug and set about happily overcharging the occupying forces.

The general himself was elsewhere, Grey was informed, fighting inland, but would doubtless return within the month. A lieutenant-colonel without brief or regimental affiliation was simply a nuisance; he was provided with suitable quarters and politely shooed away. With no immediate duties to fulfill, he gave a shrug of his own and set out to discover the whereabouts of Captain Carruthers.

It wasn't difficult to find him. The *patron* of the first tavern Grey visited directed him at once to the habitat of *le capitaine,* a room in the house of a widow named Lambert, near the mission church. Grey wondered whether he would have received the information as readily from any other tavern-keeper in the village. Charlie had liked to drink when Grey knew him, and evidently he still did, judging from the genial attitude of the *patron* when Carruthers's name was mentioned. Not that Grey could blame him, under the circumstances.

The widow—young, chestnut-haired, and quite attractive—viewed the English officer at her door with a deep suspicion, but when he followed his request for Captain Carruthers by mentioning that he was an old friend of the captain's, her face relaxed.

"*Bon,*" she said, swinging the door open abruptly. "He needs friends."

He ascended two flights of narrow stairs to Carruthers's attic, feeling the air about him grow warmer. It was pleasant at this time of day but must grow stifling by mid-afternoon. He knocked and felt a small shock of pleased recognition at hearing Carruthers's voice bid him enter.

Carruthers was seated at a rickety table in shirt and breeches, writing, an inkwell made from a gourd at one elbow, a pot of beer at the other. He looked at Grey blankly for an instant, then joy washed across his features, and he rose, nearly upsetting both.

"John!"

Before Grey could offer his hand, he found himself embraced—and returned the embrace wholeheartedly, a wash of memory flooding through him as he smelled Carruthers's hair, felt the scrape of his unshaven cheek against Grey's own. Even in the midst of this sensation, though, he felt the slightness of Carruthers's body, the bones that pressed through his clothes.

"I never thought you'd come," Carruthers was repeating, for perhaps the fourth time. He let go and stepped away smiling as he dashed the back of his hand across his eyes, which were unabashedly wet.

"Well, you have an electric eel to thank for my presence," Grey told him, smiling himself.

"A what?" Carruthers stared at him blankly.

"Long story—tell you later. For the moment, though—what the devil have you been doing, Charlie?"

The happiness faded somewhat from Carruthers's lean face but didn't disappear altogether.

"Ah. Well. That's a long story, too. Let me send Martine for more beer." He waved Grey toward the room's only stool and went out before Grey could protest. He sat, gingerly, lest the stool collapse, but it held his weight. Besides the stool and table, the attic was very plainly furnished; a narrow cot, a chamber pot, and an ancient washstand with an earthenware basin and ewer completed the ensemble. It was very clean, but there was a faint smell of something in the air—something sweet and sickly, which he traced at once to a corked bottle standing at the back of the washstand.

Not that he had needed the smell of laudanum; one look at Carruthers's gaunt face told him enough. He glanced at the papers Carruthers had been working on. They appeared to be notes in preparation for the court-martial; the one on top was an account of an expedition undertaken by troops under Carruthers's command, on the orders of a Major Gerald Siverly.

Our orders instructed us to march to a village called Beaulieu, there to ransack and fire the houses, driving off such animals as we encountered. This we did. Some men of the village offered us resistance, armed with scythes and other implements. Two of these were shot, the others fled. We returned with two wagons filled with flour, cheeses, and small household goods, three cows, and two good mules.

Grey got no further before the door opened. Carruthers came in and sat on the bed, nodding toward the papers.

"I thought I'd best write everything down. Just in case I don't live long enough for the court-martial." He spoke matter-of-factly and, seeing the look on Grey's face, smiled faintly. "Don't be troubled, John. I've always

known I'd not make old bones. This"—he turned his right hand upward, letting the drooping cuff of his shirt fall back—"isn't all of it." He tapped his chest gently with his left hand.

"More than one doctor's told me I have some gross defect of the heart. Don't know, quite, if I have two of those, too"—he grinned, the sudden, charming smile Grey remembered so well—"or only half of one, or what. Used to be I just went faint now and then, but it's getting worse. Sometimes I feel it stop beating and just flutter in my chest, and everything begins to go all black and breathless. So far, it's always started beating again—but one of these days it isn't going to."

Grey's eyes were fixed on Charlie's hand, the small dwarf hand curled against its larger fellow, looking as though Charlie held a strange flower cupped in his palm. As Grey watched, both hands opened slowly, the fingers moving in strangely beautiful synchrony.

"All right," he said quietly. "Tell me."

Failure to suppress a mutiny was a rare charge—difficult to prove and thus unlikely to be brought, unless other factors were involved. Which, in the present instance, they undoubtedly were.

"Know Siverly, do you?" Carruthers asked, taking the papers onto his knee.

"Not at all. I gather he's a bastard." Grey gestured at the papers. "What kind of bastard, though?"

"A corrupt one." Carruthers tapped the pages square, carefully evening the edges, eyes fixed on them. "That—what you read—it wasn't Siverly. It's General Wolfe's directive. I'm not sure whether the point is to deprive the fortress of provisions, in hopes of starving them out eventually, or to put pressure on Montcalm to send out troops to defend the countryside, where Wolfe could get at them—possibly both. But he means deliberately to terrorize the settlements on both sides of the river. No, we did this under the general's orders." His face twisted a little, and he looked up suddenly at Grey. "You remember the Highlands, John?"

"You know that I do." No one involved in Cumberland's cleansing of the Highlands would ever forget. He had seen many Scottish villages like Beaulieu.

Carruthers took a deep breath.

"Yes. Well. The trouble was that Siverly took to appropriating the plunder we took from the countryside—under the pretext of selling it in order to make an equitable distribution among the troops."

"What?" This was contrary to the normal custom of the army, whereby any soldier was entitled to what plunder he seized. "Who does he think he is, an admiral?" The navy did divide shares of prize money among the crew, according to formula—but the navy was the navy; crews acted much

more as single entities than did army companies, and there were Admiralty courts set up to deal with the sale of captured prize ships.

Carruthers laughed at the question.

"His brother's a commodore. Perhaps that's where he got the notion. At any rate," he added, sobering, "he never did distribute the funds. Worse—he began withholding the soldiers' pay. Paying later and later, stopping pay for petty offenses, claiming that the pay chest hadn't been delivered—when several men had seen it unloaded from the coach with their own eyes.

"Bad enough—but the soldiers were still being fed and clothed adequately. But then he went too far."

Siverly began to steal from the commissary, diverting quantities of supplies and selling them privately.

"I had my suspicions," Carruthers explained, "but no proof. I'd begun to watch him, though—and he knew I was watching him, so he trod carefully for a bit. But he couldn't resist the rifles."

A shipment of a dozen new rifles, vastly superior to the ordinary Brown Bess musket, and very rare in the army.

"I think it must have been a clerical oversight that sent them to us in the first place. We hadn't any riflemen, and there was no real need for them. That's probably what made Siverly think he could get away with it."

But he hadn't. Two private soldiers had unloaded the box and, curious at the weight, had opened it. Excited word had spread—and excitement had turned to disgruntled surprise when, instead of new rifles, muskets showing considerable wear were later distributed. The talk—already angry—had escalated.

"Egged on by a hogshead of rum we confiscated from a tavern in Levi," Carruthers said with a sigh. "They drank all night—it was January; the nights are damned long in January here—and made up their minds to go and find the rifles. Which they did—under the floor in Siverly's quarters."

"And where was Siverly?"

"In his quarters. He was rather badly used, I'm afraid." A muscle by Carruthers's mouth twitched. "Escaped through a window, though, and made his way through the snow to the next garrison. It was twenty miles. Lost a couple of toes to frostbite but survived."

"Too bad."

"Yes, it was." The muscle twitched again.

"What happened to the mutineers?"

Carruthers blew out his cheeks, shaking his head.

"Deserted, most of them. Two were caught and hanged pretty promptly; three more rounded up later; they're in prison here."

"And you—"

"And I." Carruthers nodded. "I was Siverly's company adjutant. I didn't know about the mutiny—one of the ensigns ran to fetch me when the men started to move toward Siverly's quarters—but I did arrive before they'd finished."

"Not a great deal you could do under those circumstances, was there?"

"I didn't try," Carruthers said bluntly.

"I see," Grey said.

"Do you?" Carruthers gave him a crooked smile.

"Certainly. I take it Siverly is still in the army and still holds a command? Yes, of course. He might have been furious enough to prefer the original charge against you, but you know as well as I do that, under normal circumstances, the matter would likely have been dropped as soon as the general facts were known. You insisted on a court-martial, didn't you? So that you can make what you know public." Given Carruthers's state of health, the knowledge that he risked a long imprisonment if convicted apparently didn't trouble him.

The smile straightened and became genuine.

"I knew I chose the right man," Carruthers said.

"I am exceedingly flattered," Grey said dryly. "Why me, though?"

Carruthers had laid aside his papers and now rocked back a little on the cot, hands linked around one knee.

"Why you, John?" The smile had vanished, and Carruthers's gray eyes were level on his. "You know what we do. Our business is chaos, death, destruction. But you know why we do it, too."

"Oh? Perhaps you'd have the goodness to tell me, then. I've always wondered."

Humor lighted Charlie's eyes, but he spoke seriously.

"Someone has to keep order, John. Soldiers fight for all kinds of reasons, most of them ignoble. You and your brother, though—" He broke off, shaking his head. Grey saw that his hair was streaked with gray, though he knew Carruthers was no older than himself.

"The world is chaos and death and destruction. But people like you—you don't stand for that. If there is any order in the world, any peace—it's because of you, John, and those very few like you."

Grey felt he should say something but was at a loss as to what that might be. Carruthers rose and came to Grey, putting a hand—the left—on his shoulder, the other gently against his face.

"What is it the Bible says?" Carruthers said quietly. "Blessed are they who hunger and thirst for justice, for they shall be satisfied? I hunger, John," he whispered. "And you thirst. You won't fail me." The fingers of Charlie's secret moved on his skin, a plea, a caress.

*The custom of the army is that a court-martial be presided over by a
senior officer and such a number of other officers as he shall think fit to
serve as council, these being generally four in number, but can be more
but not generally less than three. The person accused shall have the
right to call witnesses in his support, and the council shall question
these, as well as any other persons whom they may wish, and shall thus
determine the circumstances and, if conviction ensue, the sentence to
be imposed.*

THAT RATHER VAGUE statement was evidently all that existed in
terms of written definition and directive regarding the operations of
courts-martial—or was all that Hal had turned up for him in the brief period prior to his departure. There were no formal laws governing such
courts, nor did the law of the land apply to them. In short, the army
was—as always, Grey thought—a law unto itself.

That being so, he might have considerable leeway in accomplishing
what Charlie Carruthers wanted—or not, depending upon the personalities and professional alliances of the officers who composed the court. It
would behoove him to discover these men as soon as possible.

In the meantime, he had another small duty to discharge.

"Tom," he called, rummaging in his trunk, "have you discovered Captain Stubbs's billet?"

"Yes, me lord. And if you'll give over ruining your shirts there, I'll tell
you." With a censorious look at his master, Tom nudged him deftly aside.
"What you a-looking for in there, anyway?"

"The miniature of my cousin and her child." Grey stood back, permit-
ting Tom to bend over the open chest, tenderly patting the abused shirts
back into their tidy folds. The chest itself was rather scorched, but the
soldiers had succeeded in rescuing it—and Grey's wardrobe, to Tom's
relief.

"Here, me lord." Tom withdrew the packet and handed it gently to
Grey. "Give me best to Captain Stubbs. Reckon he'll be glad to get that.
The little 'un's got quite the look of him, don't he?"

It took some time, even with Tom's direction, to discover Malcolm
Stubbs's billet. The address—insofar as it could be called one—lay in the
poorer section of the town, somewhere down a muddy lane that ended
abruptly at the river. Grey was surprised at this; Stubbs was a most sociable
sort, and a conscientious officer. Why was he not billeted at an inn, or a
good private house, near his troops?

By the time Grey found the lane, he had an uneasy feeling; this grew
markedly as he poked his way through the ramshackle sheds and the knots
of filthy, polyglot children that broke from their play, brightening at the

novel sight. They followed him, hissing unintelligible speculations to one another but staring blankly at him, mouths open, when he asked after Captain Stubbs, pointing at his own uniform by way of illustration, with a questioning wave at their surroundings.

He had made his way all the way down the lane, and his boots were caked with mud, dung, and a thick plastering of the leaves that drifted lazily from the giant trees, before he discovered someone willing to answer him. This was an ancient Indian sitting peacefully on a rock at the river's edge, wrapped in a striped British trade blanket, fishing. The man spoke a mixture of three or four languages, only two of which Grey understood, but this basis of understanding was adequate.

"*Un, deux, trois,* in back," the ancient told him, pointing a thumb up the lane, then jerking this appendage sideways. Something in an aboriginal tongue followed, in which Grey thought he detected a reference to a woman—doubtless the owner of the house where Stubbs was billeted. A concluding reference to *"le bon capitaine"* seemed to reinforce this impression, and, thanking the gentleman in both French and English, Grey retraced his steps to the third house up the lane, still trailing a line of curious urchins like the ragged tail of a kite.

No one answered his knock, but he went round the house—followed by the children—and discovered a small hut behind it, smoke coming from its gray stone chimney.

The day was beautiful, with a sky the color of sapphires, and the air was suffused with the ripeness of late summer. The door of the hut was ajar, to admit the fresh air, but he did not push it open. Instead, he drew his dagger from his belt and knocked with the hilt—to admiring gasps from his audience at the appearance of the knife. He repressed the urge to turn round and bow to them.

He heard no footsteps from within, but the door opened suddenly, revealing a young Indian woman, whose face blazed with joy at beholding him.

He blinked, startled, and in that blink of an eye, the joy disappeared and the young woman clutched at the doorjamb for support, her other hand fisted into her chest.

"*Batinse!?*" she gasped, clearly terrified. "*Qu'est-ce qui s'passe?*"

"*Rien,*" he replied, equally startled. "*Ne vous inquietez pas, madame. Est-ce que Capitaine Stubbs habite ici?*" *Don't perturb yourself, madame. Does Captain Stubbs live here?*

Her eyes, already huge, rolled back in her head, and he seized her arm, fearing lest she faint at his feet. The largest of the urchins following him rushed forward and pushed the door open, and he put an arm round the woman's waist and half-dragged, half-carried her into the house.

Taking this as invitation, the rest of the children crowded in behind

him, murmuring in what appeared to be sympathy, as he lugged the young woman to the bed and deposited her thereon. A small girl, wearing little more than a pair of drawers snugged round her insubstantial waist with a piece of string, pressed in beside him and said something to the young woman. Not receiving an answer, the girl behaved as though she had, turning and racing out of the door.

Grey hesitated, not sure what to do. The woman was breathing, though pale, and her eyelids fluttered.

"Voulez-vous un petit eau?" he inquired, turning about in search of water. He spotted a bucket of water near the hearth, but his attention was distracted by an object propped beside it: a cradleboard, with a swaddled infant bound to it, blinking large, curious eyes in his direction.

He knew already, of course, but knelt down before the infant and waggled a tentative forefinger at it. The baby's eyes were big and dark, like its mother's, and the skin a paler shade of her own. The hair, though, was not straight, thick, and black. It was the color of cinnamon and exploded from the child's skull in a nimbus of the same curls that Malcolm Stubbs kept rigorously clipped to his scalp and hidden beneath his wig.

"Wha' happen with *le capitaine?*" a peremptory voice demanded behind him. He turned on his heels and, finding a rather large woman looming over him, rose to his feet and bowed.

"Nothing whatever, madame," he assured her. *Not yet, it hasn't.* "I was merely seeking Captain Stubbs to give him a message."

"Oh." The woman—French, but plainly the younger woman's mother or aunt—left off glowering at him and seemed to deflate somewhat, settling back into a less threatening shape. "Well, then. *D'un urgence,* this message?" She eyed him; clearly, other British officers were not in the habit of visiting Stubbs at home. Most likely Stubbs had an official billet elsewhere, where he conducted his regimental business. No wonder they thought he'd come to say that Stubbs was dead or injured. *Not yet,* he added grimly to himself.

"No," he said, feeling the weight of the miniature in his pocket. "Important, but not urgent." He left then. None of the children followed him.

NORMALLY, IT WAS not difficult to discover the whereabouts of a particular soldier, but Malcolm Stubbs seemed to have disappeared into thin air. Over the course of the next week, Grey combed headquarters, the military encampment, and the village, but no trace of his disgraceful cousin-by-marriage could be found. Still odder, no one appeared to have missed the captain. The men of Stubbs's immediate company merely

shrugged in confusion, and his superior officer had evidently gone off upriver to inspect the state of various postings. Frustrated, Grey retired to the riverbank to think.

Two logical possibilities presented themselves—no, three. One, Stubbs had heard about Grey's arrival, supposed that Grey would discover exactly what he had discovered, and had in consequence panicked and deserted. Two, he'd fallen afoul of someone in a tavern or back alley, been killed, and was presently decomposing quietly under a layer of leaves in the woods. Or, three—he'd been sent somewhere to do something, quietly.

Grey doubted the first exceedingly; Stubbs wasn't prone to panic, and if he had heard of Grey's arrival, Malcolm's first act would have been to come and find him, thus preventing his poking about in the village and finding what he'd found. He dismissed that possibility accordingly.

He dismissed the second still more promptly. Had Stubbs been killed, either deliberately or by accident, the alarm would have been raised. The army did generally know where its soldiers were, and if they weren't where they were meant to be, steps were taken. The same held true for desertion.

Right, then. If Stubbs was gone and no one was looking for him, it naturally followed that the army had sent him to wherever he'd gone. Since no one seemed to know where that was, his mission was presumably secret. And given Wolfe's current position and present obsession, that almost certainly meant that Malcolm Stubbs had gone downriver, searching for some way to attack Quebec. Grey sighed, satisfied with his deductions. Which in turn meant that—barring his being caught by the French, scalped or abducted by hostile Indians, or eaten by a bear—Stubbs would be back eventually. There was nothing to do but wait.

He leaned against a tree, watching a couple of fishing canoes make their way slowly downstream, hugging the bank. The sky was overcast and the air light on his skin, a pleasant change from the day's earlier heat. Cloudy skies were good for fishing; his father's gamekeeper had told him that. He wondered why—were the fish dazzled by sun, and thus sought murky hiding places in the depths, but rose toward the surface in dimmer light?

He thought suddenly of the electric eel, which Suddfield had told him lived in the silt-choked waters of the Amazon. The thing did have remarkably small eyes, and its proprietor had opined that it was able to use its electrical abilities in some way to discern, as well as to electrocute, its prey.

He couldn't have said what made him raise his head at that precise moment, but he looked up to find one of the canoes hovering in the shallow water a few feet from him. The Indian paddling the canoe gave him a brilliant smile.

"Englishman!" he called. "You want to fish with me?"

A small jolt of electricity ran through him and he straightened up. Ma-

noke's eyes were fixed on his, and he felt in memory the touch of lips and tongue and the scent of fresh-sheared copper. His heart was racing—go off in company with an Indian he barely knew? It might easily be a trap. He could end up scalped or worse. But electric eels were not the only ones to discern things by means of a sixth sense, he thought.

"Yes!" he called. "Meet you at the landing!"

TWO WEEKS LATER, he stepped out of Manoke's canoe onto the landing, thin, sunburned, cheerful, and still in possession of his hair. Tom Byrd would be beside himself, he reflected; he'd left word as to what he was doing but naturally had been able to give no estimate of his return. Doubtless poor Tom would be thinking he'd been captured and dragged off into slavery or scalped, his hair sold to the French.

In fact, they had drifted slowly downriver, pausing to fish wherever the mood took them, camping on sandbars and small islands, grilling their catch and eating their supper in smoke-scented peace, beneath the leaves of oak and alder. They had seen other craft now and then—not only canoes but many French packet boats and brigs, as well as two English warships, tacking slowly up the river, sails bellying, the distant shouts of the sailors as foreign to him just then as the tongues of the Iroquois.

And in the late summer dusk of the first day, Manoke had wiped his fingers after eating, stood up, casually untied his breechclout, and let it fall. Then waited, grinning, while Grey fought his way out of shirt and breeches.

They'd swum in the river to refresh themselves before eating; the Indian was clean, his skin no longer greasy. And yet he seemed to taste of wild game, the rich, uneasy tang of venison. Grey had wondered whether it was the man's race that was responsible or only his diet?

"What do I taste like?" he'd asked, out of curiosity.

Manoke, absorbed in his business, had said something that might have been "cock" but might equally have been some expression of mild disgust, so Grey thought better of pursuing this line of inquiry. Besides, if he *did* taste of beef and biscuit or Yorkshire pudding, would the Indian recognize that? For that matter, did he really want to know, if he did? He did not, he decided, and they enjoyed the rest of the evening without benefit of conversation.

He scratched the small of his back where his breeches rubbed, uncomfortable with mosquito bites and the peel of fading sunburn. He'd tried the native style of dress, seeing its convenience, but had scorched his bum by lying too long in the sun one afternoon and thereafter resorted to

breeches, not wishing to hear any further jocular remarks regarding the whiteness of his arse.

Thinking such pleasant but disjointed thoughts, he'd made his way halfway through the town before noticing that there were many more soldiers in evidence than there had been when he'd left. Drums were pattering up and down the sloping, muddy streets, calling men from their billets, the rhythm of the military day making itself felt. His own steps fell naturally into the beat of the drums; he straightened and felt the army reach out suddenly, seizing him, shaking him out of his sunburned bliss.

He glanced involuntarily up the hill and saw the flags fluttering above the large inn that served as field headquarters. Wolfe had returned.

GREY FOUND HIS own quarters, reassured Tom as to his well-being, submitted to having his hair forcibly untangled, combed, perfumed, and tightly bound up in a formal queue, and, with his clean uniform chafing his sunburned skin, went to present himself to the general, as courtesy demanded. He knew James Wolfe by sight—Wolfe was about his own age, had fought at Culloden, been a junior officer under Cumberland during the Highland campaign—but did not know him personally. He'd heard a great deal about him, though.

"Grey, is it? Pardloe's brother, are you?" Wolfe lifted his long nose in Grey's direction, as though sniffing at him, in the manner of one dog inspecting another's backside. Grey trusted he would not be required to reciprocate and instead bowed politely.

"My brother's compliments, sir."

Actually, what his brother had had to say was far from complimentary.

"Melodramatic ass" was what Hal had said, hastily briefing him before his departure. "Showy, bad judgment, terrible strategist. Has the devil's own luck, though, I'll give him that. *Don't* follow him into anything stupid."

Wolfe nodded amiably enough.

"And you've come as a witness for who is it—Captain Carruthers?"

"Yes, sir. Has a date been set for the court-martial?"

"Dunno. Has it?" Wolfe asked his adjutant, a tall, spindly creature with a beady eye.

"No, sir. Now that his lordship is here, though, we can proceed. I'll tell Brigadier Lethbridge-Stewart; he's to chair the proceeding."

Wolfe waved a hand.

"No, wait a bit. The brigadier will have other things on his mind. 'Til after . . ."

The adjutant nodded and made a note.

"Yes, sir."

Wolfe was eyeing Grey, in the manner of a small boy bursting to share some secret.

"D'you understand Highlanders, Colonel?"

Grey blinked, surprised.

"Insofar as such a thing is possible, sir," he replied politely, and Wolfe brayed with laughter.

"Good man." The general turned his head to one side and appraised Grey. "I've got a hundred or so of the creatures; been thinking what use they might be. I think I've found one—a small adventure."

The adjutant smiled despite himself, then quickly erased the smile.

"Indeed, sir?" Grey said cautiously.

"Somewhat dangerous," Wolfe went on carelessly. "But, then, it's the Highlanders—no great mischief should they fall. Would you care to join us?"

"Don't follow him into anything stupid." Right, Hal, he thought. Any suggestions on how to decline an offer like that from one's titular commander?

"I should be pleased, sir," he said, feeling a brief ripple of unease down his spine. "When?"

"In two weeks—at the dark of the moon." Wolfe was all but wagging his tail in enthusiasm.

"Am I permitted to know the nature of the . . . er . . . expedition?"

Wolfe exchanged a look of anticipation with his adjutant, then turned eyes shiny with excitement on Grey.

"We're going to take Quebec, Colonel."

SO WOLFE THOUGHT he had found his *point d'appui*. Or, rather, his trusted scout, Malcolm Stubbs, had found it for him. Grey returned briefly to his quarters, put the miniature of Olivia and little Cromwell in his pocket, and went to find Stubbs.

He didn't bother thinking what to say to Malcolm. It was as well, he thought, that he hadn't found Stubbs immediately after his discovery of the Indian mistress and her child; he might simply have knocked Stubbs down, without the bother of explanation. But time had elapsed, and his blood was cooler now. He was detached.

Or so he thought, until he entered a prosperous tavern—Malcolm had elevated tastes in wine—and found his cousin-by-marriage at a table, relaxed and jovial among his friends. Stubbs was aptly named, being

approximately five foot four in both dimensions, a fair-haired fellow with an inclination to become red in the face when deeply entertained or deep in drink.

At the moment, he appeared to be experiencing both conditions, laughing at something one of his companions had said, waving his empty glass in the barmaid's direction. He turned back, spotted Grey coming across the floor, and lit up like a beacon. He'd been spending a good deal of time out of doors, Grey saw; he was nearly as sunburned as Grey himself.

"Grey!" he cried. "Why, here's a sight for sore eyes! What the devil brings you to the wilderness?" Then he noticed Grey's expression, and his joviality faded slightly, a puzzled frown growing between his thick brows.

It hadn't time to grow far. Grey lunged across the table, scattering glasses, and seized Stubbs by the shirtfront.

"You come with me, you bloody swine," he whispered, face shoved up against the younger man's, "or I'll kill you right here, I swear it."

He let go then and stood, blood hammering in his temples. Stubbs rubbed at his chest, affronted, startled—and afraid. Grey could see it in the wide blue eyes. Slowly, Stubbs got up, motioning to his companions to stay.

"No bother, chaps," he said, making a good attempt at casualness. "My cousin—family emergency, what?"

Grey saw two of the men exchange knowing glances, then look at Grey, wary. They knew, all right.

Stiffly, he gestured for Stubbs to precede him, and they passed out of the door in a pretense of dignity. Once outside, though, he grabbed Stubbs by the arm and dragged him round the corner into a small alleyway. He pushed Stubbs hard, so that he lost his balance and fell against the wall; Grey kicked his legs out from under him, then knelt on his thigh, digging his knee viciously into the thick muscle. Stubbs uttered a strangled noise, not quite a scream.

Grey dug in his pocket, hand trembling with fury, and brought out the miniature, which he showed briefly to Stubbs before grinding it into the man's cheek. Stubbs yelped, grabbed at it, and Grey let him have it, rising unsteadily off the man.

"How dare you?" he said, low-voiced and vicious. "How dare you dishonor your wife, your son?"

Malcolm was breathing hard, one hand clutching his abused thigh, but was regaining his composure.

'It's nothing," he said. "Nothing to do with Olivia at all." He swallowed, wiped a hand across his mouth, and took a cautious glance at the miniature in his hand. "That the sprat, is it? Good . . . good-looking lad. Looks like me, don't he?"

Grey kicked him brutally in the stomach.

"Yes, and so does your *other* son," he hissed. "How could you do such a thing?"

Malcolm's mouth opened, but nothing came out. He struggled for breath like a landed fish. Grey watched without pity. He'd have the man split and grilled over charcoal before he was done. He bent and took the miniature from Stubbs's unresisting hand, tucking it back in his pocket.

After a long moment, Stubbs achieved a whining gasp, and the color of his face, which had gone puce, subsided back toward its normal brick color. Saliva had collected at the corners of his mouth; he licked his lips, spat, then sat up, breathing heavily, and looked at Grey.

"Going to hit me again?"

"Not just yet."

"Good." He stretched out a hand, and Grey took it, grunting as he helped Stubbs to his feet. Malcolm leaned against the wall, still panting, and eyed him.

"So, who made you God, Grey? Who are you to sit in judgment of me, eh?"

Grey nearly hit him again but desisted.

"Who am *I*?" he echoed. "Olivia's fucking cousin, that's who! The nearest male relative she's got on this continent! And you, need I remind you—and evidently I do—are her fucking husband. Judgment? What the devil d'you mean by that, you filthy lecher?"

Malcolm coughed and spat again.

"Yes. Well. As I said, it's nothing to do with Olivia—and so it's nothing to do with you." He spoke with apparent calmness, but Grey could see the pulse hammering in his throat, the nervous shiftiness of his eyes. "It's nothing out of the ordinary—it's the bloody custom, for God's sake. Everybody—"

He kneed Stubbs in the balls.

"Try again," he advised Stubbs, who had fallen down and was curled into a fetal position, moaning. "Take your time; I'm not busy."

Aware of eyes upon him, Grey turned to see several soldiers gathered at the mouth of the alley, hesitating. He was still wearing his dress uniform, though—somewhat the worse for wear but clearly displaying his rank—and when he gave them an evil look, they hastily dispersed.

"I should kill you here and now, you know," he said to Stubbs after a few moments. The rage that had propelled him was draining away, though, as he watched the man retch and heave at his feet, and he spoke wearily. "Better for Olivia to have a dead husband, and whatever property you leave, than a live scoundrel, who will betray her with her friends—likely with her own maid."

Stubbs muttered something indistinguishable, and Grey bent, grasping him by the hair, and pulled his head up.

"What was that?"

"Wasn't . . . like that." Groaning and clutching himself, Malcolm maneuvered gingerly into a sitting position, knees drawn up. He gasped for a bit, head on his knees, before being able to go on.

"You don't know, do you?" He spoke low-voiced, not raising his head. "You haven't seen the things I've seen. Not . . . done what I've had to do."

"What do you mean?"

"The . . . the killing. Not . . . battle. Not an honorable thing. Farmers. Women . . ." Grey saw Stubbs's heavy throat move, swallowing. "I—we— for months now. Looting the countryside, burning farms, villages." He sighed, broad shoulders slumping. "The men, they don't mind. Half of them are brutes to begin with." He breathed. "Think . . . nothing of shooting a man on his doorstep and taking his wife next to his body." He swallowed. "'Tisn't only Montcalm who pays for scalps," he said in a low voice. Grey couldn't avoid hearing the rawness in his voice, a pain that wasn't physical.

"Every soldier's seen such things, Malcolm," he said after a short silence, almost gently. "You're an officer. It's your job to keep them in check." *And you know damned well it isn't always possible,* he thought.

"I know," Malcolm said, and began to cry. "I couldn't."

Grey waited while he sobbed, feeling increasingly foolish and uncomfortable. At last, the broad shoulders heaved and subsided. After a moment, Malcolm said, in a voice that quivered only a little, "Everybody finds a way, don't they? And there're not that many ways. Drink, cards, or women." He raised his head and shifted a bit, grimacing as he eased into a more comfortable position. "But you don't go in much for women, do you?" he added, looking up.

Grey felt the bottom of his stomach drop but realized in time that Malcolm had spoken matter-of-factly, with no tone of accusation.

"No," he said, and drew a deep breath. "Drink, mostly."

Malcolm nodded, wiping his nose on his sleeve.

"Drink doesn't help me," he said. "I fall asleep, but I don't forget. I just dream about . . . things. And whores—I—well, I didn't want to get poxed and maybe . . . well, Olivia," he muttered, looking down. "No good at cards," he said, clearing his throat. "But sleeping in a woman's arms—I can sleep then."

Grey leaned against the wall, feeling nearly as battered as Malcolm Stubbs. Pale green aspen leaves drifted through the air, whirling round them, settling in the mud.

"All right," he said eventually. "What do you mean to do?"

"Dunno," Stubbs said, in a tone of flat resignation. "Think of something, I suppose."

Grey reached down and offered a hand; Stubbs got carefully to his feet and, nodding to Grey, shuffled toward the alley's mouth, bent over and holding himself as though his insides might fall out. Halfway there, though, he stopped and looked back over his shoulder. There was an anxious look on his face, half embarrassed.

"Can I . . . The miniature? They are still mine, Olivia and the—my son."

Grey heaved a sigh that went to the marrow of his bones; he felt a thousand years old.

"Yes, they are," he said, and, digging the miniature out of his pocket, tucked it carefully into Stubbs's coat. "Remember it, will you?"

TWO DAYS LATER, a convoy of troop ships arrived, under the command of Admiral Holmes. The town was flooded afresh with men hungry for unsalted meat, fresh baked bread, liquor, and women. And a messenger arrived at Grey's quarters, bearing a parcel for him from his brother, with Admiral Holmes's compliments.

It was small but packaged with care, wrapped in oilcloth and tied about with twine, the knot sealed with his brother's crest. That was unlike Hal, whose usual communiqués consisted of hastily dashed-off notes, generally employing slightly fewer than the minimum number of words necessary to convey his message. They were seldom signed, let alone sealed.

Tom Byrd appeared to think the package slightly ominous, too; he had set it by itself, apart from the other mail, and weighted it down with a large bottle of brandy, apparently to prevent it escaping. That, or he suspected Grey might require the brandy to sustain him in the arduous effort of reading a letter consisting of more than one page.

"Very thoughtful of you, Tom," he murmured, smiling to himself and reaching for his penknife.

In fact, the letter within occupied less than a page, bore neither salutation nor signature, and was completely Hal-like.

Minnie wishes to know whether you are starving, though I don't know what she proposes to do about it, should the answer be yes. The boys wish to know whether you have taken any scalps—they are confident that no red Indian would succeed in taking yours; I share this opinion. You had better bring three tommyhawks when you come home.

Here is your paperweight; the jeweler was most impressed by the quality of the stone. The other thing is a copy of Adams's confession. They hanged him yesterday.

The other contents of the parcel consisted of a small washleather pouch and an official-looking document on several sheets of good parchment, this folded and sealed—this time with the insignia of George II. Grey left it lying on the table, fetched one of the pewter cups from his campaign chest, and filled it to the brim with brandy, wondering anew at his valet's perspicacity.

Thus fortified, he sat down and took up the little pouch, from which he decanted into his hand a small, heavy gold paperweight, made in the shape of a half-moon set among ocean waves. It was set with a faceted—and very large—sapphire, which glowed like the evening star in its setting. Where had James Fraser acquired such a thing?

He turned it in his hand, admiring the workmanship, but then set it aside. He sipped his brandy for a bit, watching the official document as though it might explode. He was reasonably sure it would.

He weighed the document in his hand and felt the breeze from his window lift the pages a little, like the flap of a sail just before it fills and bellies with a snap.

Waiting wouldn't help. And Hal plainly knew what it said, anyway; he'd tell Grey eventually, whether he wanted to know or not. Sighing, he put by his brandy and broke the seal.

> *I, Bernard Donald Adams, do make this confession of my own free will . . .*

Was it? he wondered. He did not know Adams's handwriting, could not tell whether the document had been written or dictated—no, wait. He flipped over the sheets and examined the signature. Same hand. All right, he had written it himself.

He squinted at the writing. It seemed firm. Probably not extracted under torture, then. Perhaps it was the truth.

"Idiot," he said under his breath. "Read the goddamned thing and have done with it!"

He drank the rest of his brandy at a gulp, flattened the pages upon the stone of the parapet, and read, at last, the story of his father's death.

THE DUKE HAD suspected the existence of a Jacobite ring for some time and had identified three men whom he thought involved in it. Still, he made no move to expose them until the warrant was issued for his own arrest, upon the charge of treason. Hearing of this, he had sent at once to Adams, summoning him to the duke's country home at Earlingden.

Adams did not know how much the duke knew of his own involvement but did not dare to stay away, lest the duke, under arrest, denounce him. So he armed himself with a pistol and rode by night to Earlingden, arriving just before dawn.

He had come to the conservatory's outside doors and been admitted by the duke. Whereupon "some conversation" had ensued.

> *I had learned that day of the issuance of a warrant for arrest upon the charge of treason, to be served upon the body of the Duke of Pardloe. I was uneasy at this, for the duke had questioned both myself and some colleagues previously, in a manner that suggested to me that he suspected the existence of a secret movement to restore the Stuart throne.*
>
> *I argued against the duke's arrest, as I did not know the extent of his knowledge or suspicion, and feared that, if placed in exigent danger himself, he might be able to point a finger at myself or my principal colleagues, these being Victor Arbuthnot, Lord Creemore, and Sir Edwin Bellman. Sir Edwin was urgent upon the point, though, saying that it would do no harm; any accusations made by Pardloe could be dismissed as simple attempts to save himself, with no grounding in fact—while the fact of his arrest would naturally cause a widespread assumption of guilt and would distract any attentions that might at present be directed toward us.*
>
> *The duke, hearing of the warrant, sent to my lodgings that evening and summoned me to call upon him at his country home immediately. I dared not spurn this summons, not knowing what evidence he might possess, and therefore rode by night to his estate, arriving soon before dawn.*

Adams had met the duke there, in the conservatory. Whatever the form of this conversation, its result had been drastic.

> *I had brought with me a pistol, which I had loaded outside the house. I meant this only for protection, as I did not know what the duke's demeanor might be.*

Dangerous, evidently. Gerard Grey, Duke of Pardloe, had also come armed to the meeting. According to Adams, the duke had withdrawn his pistol from the recesses of his jacket—whether to attack or merely threaten was not clear—whereupon Adams had drawn his own pistol in panic. Both men fired; Adams thought the duke's pistol had misfired, since the duke could not have missed at the distance.

Adams's shot did not misfire, nor did it miss its target, and seeing the blood upon the duke's bosom, Adams had panicked and run. Looking back, he had seen the duke, mortally stricken but still upright, seize the branch of the peach tree beside him for support, whereupon the duke had used the last of his strength to hurl his own useless weapon at Adams before collapsing.

John Grey sat still, slowly rubbing the parchment sheets between his fingers. He wasn't seeing the neat strokes in which Adams had set down his bloodless account. He saw the blood. A dark red, beautiful as a jewel where the sun through the glass of the roof struck it suddenly. His father's hair, tousled as it might be after hunting. And the peach, fallen to those same tiles, its perfection spoiled and ruined.

He set the papers down on the table; the wind stirred them, and, by reflex, he reached for his new paperweight to hold them down.

What was it Carruthers had called him? Someone who keeps order. *"You and your brother,"* he'd said. *"You don't stand for that. If there is any order in the world, any peace—it's because of you, John, and those very few like you."*

Perhaps. He wondered if Carruthers knew the cost of peace and order—but then recalled Charlie's haggard face, its youthful beauty gone, nothing left in it now save the bones and the dogged determination that kept him breathing.

Yes, he knew.

NEARLY TWO WEEKS later, just after full dark, they boarded the ships. The convoy included Admiral Holmes's flagship, the *Lowestoff;* three men of war: the *Squirrel, Sea Horse,* and *Hunter;* a number of armed sloops; others loaded with ordnance, powder, and ammunition; and a number of transports for the troops—1,800 men in all. The *Sutherland* had been left below, anchored just out of firing range of the fortress, to keep an eye on the enemy's motions; the river there was littered with floating batteries and prowling small French craft.

Grey traveled with Wolfe and the Highlanders aboard *Sea Horse* and spent the journey on deck, too keyed up to bear being below.

His brother's warning kept recurring in the back of his mind—"Don't *follow him into anything stupid*"—but it was much too late to think of that, and, to block it out, he challenged one of the other officers to a whistling contest. Each party was to whistle the entirety of "The Roast Beef of Old England," the loser the man who laughed first. He lost, but did not think of his brother again.

Just after midnight, the big ships quietly furled their sails, dropped anchor, and lay like slumbering gulls on the dark river. Anse au Foulon, the landing spot that Malcolm Stubbs and his scouts had recommended to General Wolfe, lay seven miles downriver, at the foot of sheer and crumbling slate cliffs that led upward to the Plains of Abraham.

"Is it named for the biblical Abraham, do you think?" Grey had asked curiously, hearing the name, but had been informed that, in fact, the cliff top comprised a farmstead belonging to an ex-pilot named Abraham Martin.

On the whole, he thought this prosaic origin just as well. There was likely to be drama enough enacted on that ground, without thought of ancient prophets, conversations with God, nor any calculation of how many just men might be contained within the fortress of Quebec.

With a minimum of fuss, the Highlanders and their officers, Wolfe and his chosen troops—Grey among them—debarked into the small *bateaux* that would carry them silently down to the landing point.

The sounds of oars were mostly drowned by the river's rushing, and there was little conversation in the boats. Wolfe sat in the prow of the lead boat, facing his troops, looking now and then over his shoulder at the shore. Quite without warning, he began to speak. He didn't raise his voice, but the night was so still that those in the boat had little trouble in hearing him. To Grey's astonishment, he was reciting "Elegy Written in a Country Churchyard."

Melodramatic ass, Grey thought—and yet could not deny that the recitation was oddly moving. Wolfe made no show of it. It was as though he was simply talking to himself, and a shiver went over Grey as Wolfe intoned:

> *The boast of heraldry, the pomp of power,*
> *And all that beauty, all that wealth e'er gave,*
> *Awaits alike the inevitable hour.*

"*The paths of glory lead but to the grave,*" Wolfe ended, so low-voiced that only the three or four men closest heard him. Grey was near enough to hear him clear his throat, with a small "hem" noise, and saw his shoulders lift.

"Gentlemen," Wolfe said, lifting his voice, as well, "I should rather have written those lines than have taken Quebec."

There was a faint stir and a breath of laughter among the men.

So would I, Grey thought. *The poet who wrote them is likely sitting by his cozy fire in Cambridge, eating buttered crumpets, not preparing to fall from a great height or get his arse shot off.*

He didn't know whether this was simply more of Wolfe's characteristic drama. Possibly—possibly not, he thought. He'd met Colonel Walsing by the latrines that morning, and Walsing had mentioned that Wolfe had given him a pendant the night before, with instructions to deliver it to Miss Landringham, to whom Wolfe was engaged.

But, then, it was nothing out of the ordinary for men to put their personal valuables into the care of a friend before a hot battle. Were you killed or badly injured, your body might be looted before your comrades managed to retrieve you, and not everyone had a trustworthy servant with whom to leave such items. Grey himself had often carried snuffboxes, pocket watches, or rings into battle for friends—he'd had a reputation for luck, prior to Crefeld. No one had asked him to carry anything tonight.

He shifted his weight by instinct, feeling the current change, and Simon Fraser, next to him, swayed in the opposite direction, bumping him.

"Pardon," Fraser murmured. Wolfe had made them all recite poetry in French round the dinner table the night before, and it was agreed that Fraser had the most authentic accent, he having fought with the French in Holland some years prior. Should they be hailed by a sentry, it would be his job to reply. Doubtless, Grey thought, Fraser was now thinking frantically in French, trying to saturate his mind with the language, lest any stray bit of English escape in panic.

"De rien," Grey murmured back, and Fraser chuckled, deep in his throat.

It was cloudy, the sky streaked with the shredded remnants of retreating rain clouds. That was good; the surface of the river was broken, patched with faint light, fractured by stones and drifting tree branches. Even so, a decent sentry could scarcely fail to spot a train of boats.

Cold numbed his face, but his palms were sweating. He touched the dagger at his belt again; he was aware that he touched it every few minutes, as if needing to verify its presence, but couldn't help it and didn't worry about it. He was straining his eyes, looking for anything—the glow of a careless fire, the shifting of a rock that was not a rock . . . Nothing.

How far? he wondered. Two miles, three? He'd not yet seen the cliffs himself, was not sure how far below Gareon they lay.

The rush of water and the easy movement of the boat began to make him sleepy, tension notwithstanding, and he shook his head, yawning exaggeratedly to throw it off.

"Quel est ce bateau?" *What boat is that?* The shout from the shore seemed anticlimactic when it came, barely more remarkable than a night bird's call. But the next instant Simon Fraser's hand crushed his, grinding the bones together, as Fraser gulped air and shouted, *"Celui de la Reine!!"*

Grey clenched his teeth, not to let any blasphemous response escape. If

the sentry demanded a password, he'd likely be crippled for life, he thought. An instant later, though, the sentry shouted, *"Passez!"* and Fraser's death grip relaxed. Simon was breathing like a bellows but nudged him and whispered, *"Pardon,"* again.

"De fucking *rien,"* he muttered, rubbing his hand and tenderly flexing the fingers.

They were getting close. Men were shifting to and fro in anticipation, even more than Grey—checking their weapons, straightening coats, coughing, spitting over the side, readying themselves. Still, it was a nerve-wracking quarter-hour more before they began to swing toward shore—and another sentry called from the dark.

Grey's heart squeezed like a fist, and he nearly gasped with the twinge of pain from his old wounds.

"Qui etes-vous? Que sont ces bateaux?" a French voice demanded suspiciously. *Who are you? What boats are those?*

This time, he was ready and seized Fraser's hand himself. Simon held on and, leaning out toward the shore, called hoarsely, *"Des bateaux de provisions! Tasiez-vous—les anglais sont proches!"* *Provision boats! Be quiet—the British are nearby!* Grey felt an insane urge to laugh but didn't. In fact, the *Sutherland was* nearby, lurking out of cannon shot downstream, and doubtless the frogs knew it. In any case, the guard called, more quietly, *"Passez!"* and the train of boats slid smoothly past and round the final bend.

The bottom of the boat grated on sand, and half the men were over at once, tugging it farther up. Wolfe half-leapt, half-fell over the side in eagerness, all trace of somberness gone. They'd come aground on a small sandbar just offshore, and the other boats were beaching now, a swarm of black figures gathering like ants.

Twenty-four of the Highlanders were meant to try the ascent first, finding—and, insofar as possible, clearing, for the cliff was defended not only by its steepness but by abatis, nests of sharpened logs—a trail for the rest. Simon's bulky form faded into the dark, his French accent changing at once into the sibilant Gaelic as he hissed the men into position. Grey rather missed his presence.

He was not sure whether Wolfe had chosen the Highlanders for their skill at climbing or because he preferred to risk them rather than his other troops. The latter, he thought. Wolfe regarded the Highlanders with distrust and a certain contempt, as did most English officers. Those officers, at least, who'd never fought with them—or against them.

From his spot at the foot of the cliff, Grey couldn't see them, but he could hear them: the scuffle of feet, now and then a wild scrabble and a clatter of falling small stones, loud grunts of effort, and what he recog-

nized as Gaelic invocations of God, his mother, and assorted saints. One man near him pulled a string of beads from the neck of his shirt, kissed the tiny cross attached to it, and tucked it back; then, seizing a small sapling that grew out of the rock face, he leapt upward, kilt swinging, broadsword swaying from his belt in brief silhouette, before the darkness took him. Grey touched his dagger's hilt again, his own talisman against evil.

It was a long wait in the darkness; to some extent he envied the Highlanders, who, whatever else they might be encountering—and the scrabbling noises and half-strangled whoops as a foot slipped and a comrade grabbed a hand or arm suggested that the climb was just as impossible as it seemed—were not dealing with boredom.

A sudden rumble and crashing came from above, and the shore party scattered in panic as several sharpened logs plunged out of the dark, dislodged from an abatis. One of them had struck point-down no more than six feet from Grey and stood quivering in the sand. With no discussion, the shore party retreated to the sandbar.

The scrabblings and gruntings grew fainter and abruptly ceased. Wolfe, who had been sitting on a boulder, stood up, straining his eyes upward.

"They've made it," he whispered, and his fists curled in an excitement that Grey shared. "God, they've made it!"

Well enough, and the men at the foot of the cliff held their breaths; there was a guard post at the top of the cliff. Silence, bar the everlasting noise of tree and river. And then a shot.

Just one. The men below shifted, touching their weapons, ready, not knowing for what.

Were there sounds above? Grey could not tell and, out of sheer nervousness, turned aside to urinate against the side of the cliff. He was fastening his flies when he heard Simon Fraser's voice above.

"Got 'em, by God!" he said. "Come on, lads—the night's not long enough!"

The next few hours passed in a blur of the most arduous endeavor Grey had seen since he'd crossed the Scottish Highlands with his brother's regiment, bringing cannon to General Cope. No, actually, he thought, as he stood in darkness, one leg wedged between a tree and the rock face, thirty feet of invisible space below him and rope burning through his palms with an unseen deadweight of two hundred pounds or so on the end, this was worse.

The Highlanders had surprised the guard, shot their fleeing captain in the heel, and made all of them prisoner. That was the easy part. The next thing was for the rest of the landing party to ascend to the cliff top, now that the trail—if there was such a thing—had been cleared. There they would make preparations to raise not only the rest of the troops now coming down the river aboard the transports but also seventeen battering can-

non, twelve howitzers, three mortars, and all of the necessary encumbrances in terms of shell, powder, planks, and limbers necessary to make this artillery effective. At least, Grey reflected, by the time they were done, the vertical trail up the cliffside would likely have been trampled into a simple cow path.

As the sky lightened, Grey looked up for a moment from his spot at the top of the cliff, where he was now overseeing the last of the artillery as it was heaved over the edge, and saw the *bateaux* coming down again like a flock of swallows, they having crossed the river to collect an additional 1,200 troops that Wolfe had directed to march to Levi on the opposite shore, there to lie hidden in the woods until the Highlanders' expedient should have been proved.

A head, cursing freely, surged up over the edge of the cliff. Its attendant body lunged into view, tripped, and sprawled at Grey's feet.

"Sergeant Cutter!" Grey said, grinning as he bent to yank the little sergeant to his feet. "Come to join the party, have you?"

"Jesus fuck," replied the sergeant, belligerently brushing dirt from his coat. "We'd best win, that's all I can say." And, without waiting for reply, he turned round to bellow down the cliff, "Come ON, you bloody rascals! 'Ave you all eaten lead for breakfast, then? Shit it out and step lively! CLIMB, God damn your eyes!"

The net result of this monstrous effort being that, as dawn spread its golden glow across the Plains of Abraham, the French sentries on the walls of the Citadel of Quebec gaped in disbelief at the sight of more than four thousand British troops drawn up in battle array before them.

Through his telescope, Grey could see the sentries. The distance was too great to make out their facial expressions, but their attitudes of alarm and consternation were easy to read, and he grinned, seeing one French officer clutch his head briefly, then wave his arms like one dispelling a flock of chickens, sending his subordinates rushing off in all directions.

Wolfe was standing on a small hillock, long nose lifted as though to sniff the morning air. Grey thought he probably considered his pose noble and commanding; he reminded Grey of a dachshund scenting a badger; the air of alert eagerness was the same.

Wolfe wasn't the only one. Despite the ardors of the night, skinned hands, battered shins, twisted knees and ankles, and a lack of food and sleep, a gleeful excitement ran through the troops like wine. Grey thought they were all giddy with fatigue.

The sound of drums came faintly to him on the wind: the French, beating hastily to quarters. Within minutes, he saw horsemen streaking away from the fortress, and he smiled grimly. They were going to rally whatever troops Montcalm had within summoning distance, and Grey felt a tightening of the belly at the sight.

The matter hadn't really been in doubt; it was September, and winter was coming on. The town and fortress had been unable to provision themselves for a long siege, owing to Wolfe's scorched-earth policies. The French were there, the English before them—and the simple fact, apparent to both sides, was that the French would starve long before the English did. Montcalm would fight; he had no choice.

Many of the men had brought canteens of water, some a little food. They were allowed to relax sufficiently to eat, to ease their muscles— though none of them ever took their attention from the French gathering before the fortress. Employing his telescope further, Grey could see that, while the mass of milling men was growing, they were by no means all trained troops; Montcalm had called his militias from the countryside— farmers, fishermen, and *coureurs du bois,* by the look of them—and his Indians. Grey eyed the painted faces and oiled topknots warily, but his acquaintance with Manoke had deprived the Indians of much of their terrifying aspect—and they would not be nearly so effective on open ground, against cannon, as they were sneaking through the forest.

It took surprisingly little time for Montcalm to ready his troops, impromptu as they might be. The sun was no more than halfway up the sky when the French lines began their advance.

"HOLD your fucking fire, you villains! Fire before you're ordered, and I'll give your fuckin' heads to the artillery to use for cannonballs!" He heard the unmistakable voice of Sergeant Aloysius Cutter, some distance back but clearly audible. The same order was being echoed, if less picturesquely, through the British lines, and if every officer on the field had one eye firmly on the French, the other was fixed on General Wolfe, standing on his hillock, aflame with anticipation.

Grey felt his blood twitch and moved restlessly from foot to foot, trying to ease a cramp in one leg. The advancing French line stopped, knelt, and fired a volley. Another from the line standing behind them. Too far, much too far to have any effect. A deep rumble came from the British troops— something visceral and hungry.

Grey's hand had been on his dagger for so long that the wire-wrapped hilt had left its imprint on his fingers. His other hand was clenched upon a saber. He had no command here, but the urge to raise his sword, gather the eyes of his men, hold them, focus them, was overwhelming. He shook his shoulders to loosen them and glanced at Wolfe.

Another volley, close enough this time that several British soldiers in the front lines fell, knocked down by musket fire.

"Hold, hold!" The order rattled down the lines like gunfire. The brimstone smell of slow match was thick, pungent above the scent of powder smoke; the artillerymen held their fire, as well.

French cannon fired, and balls bounced murderously across the field, but they seemed puny, ineffectual, despite the damage they did. How many French? he wondered. Perhaps twice as many, but it didn't matter. It wouldn't matter.

Sweat ran down his face, and he rubbed a sleeve across to clear his eyes. "Hold!"

Closer, closer. Many of the Indians were on horseback; he could see them in a knot on the left, milling. Those would bear watching. . . .

"Hold!"

Wolfe's arm rose slowly, sword in hand, and the army breathed deep. His beloved grenadiers were next to him, solid in their companies, wrapped in sulfurous smoke from the match tubes at their belts.

"Come on, you buggers," the man next to Grey was muttering. "Come on, come on!"

Smoke was drifting over the field, low white clouds. Forty paces. Effective range.

"Don't fire, don't fire, don't fire . . ." someone was chanting to himself, struggling against panic.

Through the British lines, sun glinted on the rising swords, the officers echoing Wolfe's order.

"Hold . . . hold . . ."

The swords fell as one.

"FIRE!" and the ground shook.

A shout rose in Grey's throat, part of the roar of the army, and he was charging with the men near him, swinging his saber with all his might, finding flesh.

The volley had been devastating; bodies littered the ground. He leapt over a fallen Frenchman, brought his saber down upon another, caught halfway in the act of loading, took him in the cleft between neck and shoulder, yanked his saber free of the falling man, and went on.

The British artillery was firing as fast as the guns could be served. Each boom shook his flesh. He gritted his teeth, squirmed aside from the point of a half-seen bayonet, and found himself panting, eyes watering from the smoke, standing alone.

Chest heaving, he turned round in a circle, disoriented. There was so much smoke around him that he could not for a moment tell where he was. It didn't matter.

An enormous blur of something passed him, shrieking, and he dodged by instinct and fell to the ground as the horse's feet churned by. Grey heard as an echo the Indian's grunt, the rush of the tomahawk blow that had missed his head.

"Shit," he muttered, and scrambled to his feet.

The grenadiers were hard at work nearby; he heard their officers' shouts, the bang and pop of their explosions as they worked their way stolidly through the French like the small mobile batteries they were.

A grenade struck the ground a few feet away, and he felt a sharp pain in his thigh; a metal fragment had sliced through his breeches, drawing blood.

"Christ," he said, belatedly aware that being in the vicinity of a company of grenadiers was not a good idea. He shook his head to clear it and made his way away from them.

He heard a familiar sound that made him recoil for an instant from the force of memory—wild Highland screams, filled with rage and berserk glee. The Highlanders were hard at work with their broadswords—he saw two of them appear from the smoke, bare legs churning beneath their kilts, pursuing a pack of fleeing Frenchman, and felt laughter bubble up through his heaving chest.

He didn't see the man in the smoke. His foot struck something heavy and he fell, sprawling across the body. The man screamed, and Grey scrambled hastily off him.

"Sorry. Are you—Christ, Malcolm!"

He was on his knees, bending low to avoid the smoke. Stubbs was gasping, grasping desperately at Grey's coat.

"Jesus." Malcolm's right leg was gone below the knee, flesh shredded and the white bone splintered, butcher-stained with spurting blood. Or . . . no. It wasn't gone. It—the foot, at least—was lying a little way beyond, still clad in shoe and tattered stocking.

Grey turned his head and threw up.

Bile stinging the back of his nose, he choked and spat, turned back, and grappled with his belt, wrenching it free.

"Don't—" Stubbs gasped, putting out a hand as Grey began wrapping the belt round his thigh. His face was whiter than the bone of his leg. "Don't. Better—better if I die."

"The devil you will," Grey replied briefly.

His hands were shaking, slippery with blood. It took three tries to get the end of the belt through the buckle, but it went at last, and he jerked it tight, eliciting a yell from Stubbs.

"Here," said an unfamiliar voice by his ear. "Let's get him off. I'll—shit!" He looked up, startled, to see a tall British officer lunge upward, blocking the musket butt that would have brained Grey. Without thinking, he drew his dagger and stabbed the Frenchman in the leg. The man screamed, his leg buckling, and the strange officer pushed him over, kicked him in the face, and stamped on his throat, crushing it.

"I'll help," the man said calmly, bending to take hold of Malcolm's arm, pulling him up. "Take the other side; we'll get him to the back."

They got Malcolm up, his arms round their shoulders, and dragged him, paying no heed to the Frenchman thrashing and gurgling on the ground behind them.

Malcolm lived, long enough to make it to the rear of the lines, where the army surgeons were already at work. By the time Grey and the other officer had turned him over to the surgeons, the battle was over.

Grey turned to see the French scattered and demoralized, fleeing toward the fortress. British troops were flooding across the trampled field, cheering, overrunning the abandoned French cannon.

The entire battle had lasted less than a quarter of an hour.

He found himself sitting on the ground, his mind quite blank, with no notion how long he had been there, though he supposed it couldn't have been much time at all.

He noticed an officer standing near him and thought vaguely that the man seemed familiar. Who . . . Oh, yes. Wolfe's adjutant. He'd never learned the man's name.

He stood up slowly, stiff as a nine-day pudding.

The adjutant was simply standing there. His eyes were turned in the direction of the fortress and the fleeing French, but Grey could tell that he wasn't really seeing either. Grey glanced over his shoulder, toward the hillock where Wolfe had stood earlier, but the general was nowhere in sight.

"General Wolfe?" he said.

"The general . . ." the adjutant said, and swallowed thickly. "He was struck."

Of course he was, silly ass, Grey thought uncharitably. *Standing up there like a bloody target, what could he expect?* But then he saw the tears standing in the adjutant's eyes and understood.

"Dead, then?" he asked, stupidly, and the adjutant—why had he never thought to ask the man's name?—nodded, rubbing a smoke-stained sleeve across a smoke-stained countenance.

"He . . . In the wrist first. Then in the body. He fell and crawled—then he fell again. I turned him over . . . told him the battle was won, the French were scattered."

"He understood?"

The adjutant nodded and took a deep breath that rattled in his throat. "He said—" He stopped and coughed, then went on more firmly. "He said that in knowing he had conquered, he was content to die."

"Did he?" Grey said blankly. He'd seen men die, often, and imagined it much more likely that if James Wolfe had managed anything beyond an inarticulate groan, his final word had likely been either "shit," or "oh, God," depending upon the general's religious leanings, of which Grey had no notion.

"Yes, good," he said meaninglessly, and turned toward the fortress. Ant trails of men were streaming toward it, and in the midst of one such stream he saw Montcalm's colors, fluttering in the wind. Below the colors, small in the distance, a man in general's uniform rode his horse, hatless, hunched and swaying in the saddle, his officers bunched close on either side, anxious lest he fall.

The British lines were reorganizing, though it was clear no further fighting would be required. Not today. Nearby, he saw the tall officer who had saved his life and helped him to drag Malcolm Stubbs to safety, limping back toward his troops.

"The major over there," he said, nudging the adjutant and nodding. "Do you know his name?"

The adjutant blinked, then firmed his shoulders.

"Yes, of course. That's Major Siverly."

"Oh. Well, it would be, wouldn't it?"

ADMIRAL HOLMES, third in command after Wolfe, accepted the surrender of Quebec five days later, Wolfe and his second, Brigadier Monckton, having perished in battle. Montcalm was dead, too; had died the morning following the battle. There was no way out for the French save surrender; winter was coming on, and the fortress and its city would starve long before its besiegers.

Two weeks after the battle, John Grey returned to Gareon and found that smallpox had swept through the village like an autumn wind. The mother of Malcolm Stubbs's son was dead; her mother offered to sell him the child. He asked her politely to wait.

Charlie Carruthers had perished, too, the smallpox not waiting for the weakness of his body to overcome him. Grey had the body burned, not wishing Carruthers's hand to be stolen, for both the Indians and the local *habitants* regarded such things superstitiously. He took a canoe by himself and, on a deserted island in the St. Lawrence, scattered his friend's ashes to the wind.

He returned from this expedition to discover a letter, forwarded by Hal, from Dr. John Hunter, surgeon and anatomist. He checked the level of brandy in the decanter and opened it with a sigh.

My dear Lord John,

I have heard some recent conversation regarding the unfortunate death of Mr. Nicholls, including comments indicating a public perception that you were responsible for his death. In case you shared

*this perception, I thought it might ease your mind to know that in
fact you were not.*

Grey sank slowly onto a stool, eyes glued to the sheet.

> *It is true that your ball did strike Mr. Nicholls, but this accident
> contributed little or nothing to his demise. I saw you fire upward
> into the air—I said as much to those present at the time, though most
> of them did not appear to take much notice. The ball apparently
> went up at a slight angle and then fell upon Mr. Nicholls from above.
> At this point, its power was quite spent, and, the missile itself being
> negligible in size and weight, it barely penetrated the skin above
> his collarbone, where it lodged against the bone, doing no further
> damage.*
>
>   *The true cause of his collapse and death was an aortic aneurysm,
> a weakness in the wall of one of the great vessels emergent from
> the heart; such weaknesses are often congenital. The stress of the
> electric shock and the emotion of the duello that followed apparently
> caused this aneurysm to rupture. Such an occurrence is untreatable
> and invariably fatal, I am afraid. There is nothing that could
> have saved him.*
>
>   *Your servant,*
>   *John Hunter, Surgeon*

Grey was conscious of a most extraordinary array of sensations. Relief—
yes, there was a sense of profound relief, as of waking from a nightmare.
There was also a sense of injustice, colored by the beginnings of indignation; by God, he had nearly been married! He might, of course, also have
been maimed or killed as a result of the imbroglio, but that seemed relatively inconsequent; he was a soldier, after all—such things happened.

His hand trembled slightly as he set the note down. Beneath relief,
gratitude, and indignation was a growing sense of horror.

I thought it might ease your mind . . . He could see Hunter's face saying
this; sympathetic, intelligent, and cheerful. It was a straightforward remark but one fully cognizant of its own irony.

Yes, he was pleased to know he had not caused Edwin Nicholls's death.
But the means of that knowledge . . . Gooseflesh rose on his arms and he
shuddered involuntarily, imagining—

"Oh, God," he said. He'd been once to Hunter's house—to a poetry
reading, held under the auspices of Mrs. Hunter, whose salons were famous. Dr. Hunter did not attend these but sometimes would come down
from his part of the house to greet guests. On this occasion, he had done
so and, falling into conversation with Grey and a couple of other scientifi-

cally minded gentlemen, had invited them up to see some of the more interesting items of his famous collection: the rooster with a transplanted human tooth growing in its comb, the child with two heads, the fetus with a foot protruding from its stomach.

Hunter had made no mention of the walls of jars, these filled with eyeballs, fingers, sections of livers . . . or of the two or three complete human skeletons that hung from the ceiling, fully articulated and fixed by a bolt through the tops of their skulls. It had not occurred to Grey at the time to wonder where—or how—Hunter had acquired these.

Nicholls had had an eyetooth missing, the front tooth beside the empty space badly chipped. If he ever visited Hunter's house again, might he come face-to-face with a skull with a missing tooth?

He seized the brandy decanter, uncorked it, and drank directly from it, swallowing slowly and repeatedly, until the vision disappeared.

His small table was littered with papers. Among them, under his sapphire paperweight, was the tidy packet that the widow Lambert had handed him, her face blotched with weeping. He put a hand on it, feeling Charlie's doubled touch, gentle on his face, soft around his heart.

"You won't fail me."

"No," he said softly. "No, Charlie, I won't."

WITH MANOKE'S HELP as translator, Grey bought the child, after prolonged negotiation, for two golden guineas, a brightly colored blanket, a pound of sugar, and a small keg of rum. The grandmother's face was sunken, not with grief, he thought, but with dissatisfaction and weariness. With her daughter dead of the smallpox, her life would be harder. The English, she conveyed to Grey through Manoke, were cheap bastards; the French were much more generous. He resisted the impulse to give her another guinea.

It was full autumn now, and the leaves had all fallen. The bare branches of the trees spread black ironwork flat against a pale-blue sky as he made his way upward through the town, to the French mission. There were several small buildings surrounding the tiny church, with children playing outside; some of them paused to look at him, but most of them ignored him—British soldiers were nothing new.

Father LeCarré took the bundle gently from him, turning back the blanket to look at the child's face. The boy was awake; he pawed at the air, and the priest put out a finger for him to grasp.

"Ah," he said, seeing the clear signs of mixed blood, and Grey knew the priest thought the child was his. He started to explain, but, after all, what did it matter?

"We will baptize him as a Catholic, of course," Father LeCarré said, looking up at Grey. The priest was a young man, rather plump, dark, and clean-shaven, but with a gentle face. "You do not mind that?"

"No." Grey drew out a purse. "For his maintenance. I will send an additional five pounds each year, if you will advise me once a year of his continued welfare. Here—the address to which to write." A sudden inspiration struck him—not that he did not trust the good father, he assured himself, only . . . "Send me a lock of his hair," he said. "Every year."

He was turning to go when the priest called him back, smiling.

"Has the infant a name, sir?"

"A—" He stopped dead. The boy's mother had surely called him something, but Malcolm Stubbs hadn't thought to tell Grey what it was before being shipped back to England. What should he call the child? Malcolm, for the father who had abandoned him? Hardly.

Charles, maybe, in memory of Carruthers . . .

"*. . . one of these days, it isn't going to.*"

"His name is John," he said abruptly, and cleared his throat. "John Cinnamon."

"*Mais oui,*" the priest said, nodding. "*Bon voyage, Monsieur—et voyez avec le Bon Dieu.*"

"Thank you," he said politely, and went away, not looking back, down to the riverbank where Manoke waited to bid him farewell.

AUTHOR'S NOTES

THE BATTLE OF QUEBEC IS JUSTLY FAMOUS AS ONE of the great military triumphs of the eighteenth-century British Army. If you go today to the battlefield at the Plains of Abraham (in spite of this poetic name, it really was just named for the farmer who owned the land, one Abraham Martin; I suppose "The Plains of Martin" just didn't have the same ring to it), you'll see a plaque at the foot of the cliff there, commemorating the heroic achievement of the Highland troops who climbed this sheer cliff from the river below, clearing the way for the entire army—*and* their cannon, mortars, howitzers, and accompanying impedimenta—to make a harrowing overnight ascent and confront General Montcalm with a jaw-dropping spectacle by the dawn's early light.

If you go up onto the field itself, you'll find another plaque, this one put up by the French, explaining (in French) what a dirty, unsportsmanlike trick this was for those sneaky British to have played on the noble troops defending the Citadel. Ah, perspective.

General James Wolfe, along with Montcalm, was of course a real historical character, as was Brigadier Simon Fraser (whom you will have met—or will meet later—in *An Echo in the Bone*). My own rule of thumb when dealing with historical persons in the context of fiction is to try not to portray them as having done anything worse than what I *know* they did, according to the historical record.

In General Wolfe's case, Hal's opinion of his character and abilities is one commonly held and recorded by a number of contemporary military commentators. And there is documentary proof of his attitude toward the Highlanders, whom he used for this endeavor, in the form of the letter quoted in the story: ". . . no great mischief if they fall." (Allow me to recommend a wonderful novel by Alistair MacLeod, titled *No Great Mischief*. It isn't about Wolfe; it's a novelized history of a family of Scots who settle in Nova Scotia, beginning

in the eighteenth century and carrying on through the decades, but it is from Wolfe's letter that the book takes its title, and he's mentioned.)

Wolfe's policy with regard to the *habitant* villages surrounding the Citadel (looting, burning, general terrorizing of the populace) is a matter of record. It wasn't (and isn't) an unusual thing for an invading army to do.

General Wolfe's dying words are also a matter of historical record, but like Lord John, I take leave to doubt that that's really what he said. He *is* reported by several sources to have recited Gray's "Elegy Written in a Country Churchyard" in the boat on the way to battle—and I think that's a sufficiently odd thing to have done, that the reports are probably true.

As for Simon Fraser, he's widely reported to have been the British officer who fooled the French lookouts by calling out to them in French as the boats went by in the darkness—and he undoubtedly spoke excellent French, having campaigned in France years before. As for the details of exactly what he said—accounts vary, and that's not really an important detail, so I rolled my own.

Now, speaking of French . . . Brigadier Fraser spoke excellent French. I don't. I can read that language, but I can't speak or write it, possess absolutely no grammar, and have a really low tolerance for diacritical marks. So for the purposes of this story I did as I always do in such cases; I solicit the opinions of several native speakers of French for those bits of dialogue that occur in that language. What you see in this story is due to the assistance of these kind and helpful speakers. I fully expect—because it happens every time I include French in a story—to receive indignant email from assorted French speakers denouncing the French dialogue. If the French was provided to me by a Parisian, someone from Montreal will tell me *that's* not right; if the original source was *Quebecois,* outraged screams emanate from the mother country. And if it came from a textbook or (*quelle horreur*) an academic source . . . well, *bonne chance* with that. There's also the consideration that it's very difficult to spot typographical errors in a language you can't speak. But we do our best. My apologies for anything egregious.

Now, you may notice that John Hunter is referred to in various places either as "Mr. Hunter," or as "Dr. Hunter." By long-standing tradition, English surgeons are (and were) addressed as "Mr." rather than "Dr."—presumably a nod to their origins as barbers with a sanguinary sideline. However, John Hunter, with his brother William, was also a formally trained physician, as well as an eminent scientist and anatomist—hence entitled also to the honorific "Dr."

THE SPACE BETWEEN

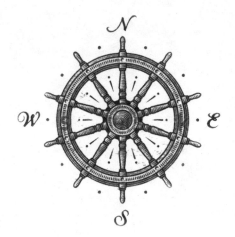

INTRODUCTION

The Comte St. Germain

THERE WAS AN HISTORICAL PERSON (QUITE POSSI-bly more than one) who went by this name. There are also numerous reports (mostly unverified) of a person by this name who appears in various parts of Europe over parts of two centuries. These observations have led some to speculate that the Comte (or *a* Comte of that name) was a practitioner of the occult, a mystic, or even a time traveler.

Let's put it this way: The Comte St. Germain in this story is not intended to portray the documented historical person of that name.

H E STILL DIDN'T KNOW why the frog hadn't killed him. Paul Rakoczy, Comte St. Germain, picked up the vial, pulled the cork, and sniffed cautiously, for the third time, but then recorked it, still dissatisfied. Maybe. Maybe not. The scent of the dark-gray powder in the vial held the ghost of something familiar—but it had been thirty years.

He sat for a moment, frowning at the array of jars, bottles, flasks, and pelicans on his workbench. It was late afternoon, and the early spring sun of Paris was like honey, warm and sticky on his face, but glowing in the rounded globes of glass, throwing pools of red and brown and green on the wood from the liquids contained therein. The only discordant note in this peaceful symphony of light was the body of a large rat, lying on its back in the middle of the workbench, a pocket watch open beside it.

The comte put two fingers delicately on the rat's chest and waited patiently. It didn't take so long this time; he was used to the coldness as his mind felt its way into the body. Nothing. No hint of light in his mind's eye, no warm red of a pulsing heart. He glanced at the watch: half an hour.

He took his fingers away, shaking his head.

"Mélisande, you evil bitch," he murmured, not without affection. "You didn't think I'd try anything *you* sent me on myself, did you?"

Still . . . he himself had stayed dead a great while longer than half an hour when the frog had given him the dragon's blood. It had been early evening when he went into Louis's Star Chamber thirty years before, heart beating with excitement at the coming confrontation—a duel of wizards, with a king's favor as the stakes—and one he'd thought he'd win. He remembered the purity of the sky, the beauty of the stars just visible, Venus bright on the horizon, and the joy of it in his blood. Everything always had a greater intensity when you knew life could cease within the next few minutes.

And an hour later he thought his life *had* ceased, the cup falling from his numbed hand, the coldness rushing through his limbs with amazing speed, freezing the words "I've lost," an icy core of disbelief in the center of his mind. He hadn't been looking at the frog; the last thing he had seen through darkening eyes was the woman—La Dame Blanche—her face over the cup she'd given him appalled and white as bone. But what he recalled, and recalled again now, with the same sense of astonishment and avidity, was the great flare of blue, intense as the color of the evening sky beyond Venus, that had burst from her head and shoulders as he died.

He didn't recall any feeling of regret or fear, just astonishment. This was nothing, however, to the astonishment he'd felt when he regained his senses, naked on a stone slab in a revolting subterranean chamber next to a drowned corpse. Luckily, there had been no one alive in that disgusting grotto, and he had made his way—reeling and half blind, clothed in the drowned man's wet and stinking shirt—out into a dawn more beautiful than any twilight could ever be. So—ten to twelve hours from the moment of apparent death to revival.

He glanced at the rat, then put out a finger and lifted one of the small, neat paws. Nearly twelve hours. Limp; the rigor had already passed. It was warm up here at the top of the house. Then he turned to the counter that ran along the far wall of the laboratory, where a line of rats lay, possibly insensible, probably dead. He walked slowly along the line, prodding each body. Limp, limp, stiff. Stiff. Stiff. All dead, without doubt. Each had had a smaller dose than the last, but all had died—though he couldn't yet be positive about the latest. Wait a bit more, then, to be sure.

He needed to know. Because the Court of Miracles was talking. And they said the frog was back.

The English Channel

THEY DID SAY that red hair was a sign of the devil. Joan eyed her escort's fiery locks consideringly. The wind on deck was fierce enough to make her eyes water, and it jerked bits of Michael Murray's hair out of its binding so they did dance round his head like flames, a bit. You might expect his face to be ugly as sin if he was one of the devil's, though, and it wasn't.

Lucky for him, he looked like his mother in the face, she thought critically. His younger brother, Ian, wasn't so fortunate, and that without the heathen tattoos. Michael's was a fairly pleasant face, for all it was blotched with windburn and the lingering marks of sorrow, and no wonder, him having just lost his father, and his wife dead in France no more than a month before that.

But she wasn't braving this gale in order to watch Michael Murray, even if he might burst into tears or turn into Auld Horny on the spot. She touched her crucifix for reassurance, just in case. It had been blessed by the priest, and her mother'd carried it all the way to St. Ninian's Spring and dipped it in the water there, to ask the saint's protection. And it was her mother she wanted to see, as long as ever she could.

She pulled her kerchief off and waved it, keeping a tight grip lest the wind make off with it. Her mother was growing smaller on the quay, waving madly, too, Joey behind her with his arm round her waist to keep her from falling into the water.

Joan snorted a bit at sight of her new stepfather but then thought better and touched the crucifix again, muttering a quick Act of Contrition in penance. After all, it was she herself who'd made that marriage happen, and a good thing, too. If not, she'd still be stuck to home at Balriggan, not on her way at last to be a Bride of Christ in France.

A nudge at her elbow made her glance aside, to see Michael offering her a handkerchief. Well, so. If her eyes were streaming—aye, *and* her nose—it was no wonder, the wind so fierce as it was. She took the scrap of cloth with a curt nod of thanks, scrubbed briefly at her cheeks, and waved her kerchief harder.

None of his family had come to see Michael off, not even his twin sister, Janet. But they were taken up with all there was to do in the wake of Old Ian Murray's death, and no wonder. No need to see Michael to the ship, either—Michael Murray was a wine merchant in Paris, and a wonderfully well-traveled gentleman. She took some comfort from the knowledge that he knew what to do and where to go and had said he would see her safely delivered to the Convent of Angels, because the thought of making her way through Paris alone and the streets full of people all speaking French— though she knew French quite well, of course. She'd been studying it all the winter, and Michael's mother helping her—though perhaps she had better not tell the reverend mother about the sorts of French novels Jenny Murray had in her bookshelf, because . . .

"Voulez-vous descendre, mademoiselle?"

"Eh?" She glanced at him, to see him gesturing toward the hatchway that led downstairs. She turned back, blinking—but the quay had vanished, and her mother with it.

"No," she said. "Not yet. I'll just . . ." She wanted to see the land so long as she could. It would be her last sight of Scotland, ever, and the thought made her wame curl into a small, tight ball. She waved a vague hand toward the hatchway. "You go, though. I'm all right by myself."

He didn't go but came to stand beside her, gripping the rail. She turned away from him a little, so he wouldn't see her weep, but on the whole she wasn't sorry he'd stayed.

Neither of them spoke, and the land sank slowly, as though the sea swallowed it, and there was nothing round them now but the open sea, glassy gray and rippling under a scud of clouds. The prospect made her dizzy, and she closed her eyes, swallowing.

Dear Lord Jesus, don't let me be sick!

A small shuffling noise beside her made her open her eyes, to find Michael Murray regarding her with some concern.

"Are ye all right, Miss Joan?" He smiled a little. "Or should I call ye Sister?"

"No," she said, taking a grip on her nerve and her stomach and drawing herself up. "I'm no a nun yet, am I?"

He looked her up and down, in the frank way Hieland men did, and smiled more broadly.

"Have ye ever *seen* a nun?" he asked.

"I have not," she said, as starchily as she could. "I havena seen God or the Blessed Virgin, either, but I believe in them, too."

Much to her annoyance, he burst out laughing. Seeing the annoyance, though, he stopped at once, though she could see the urge still trembling there behind his assumed gravity.

"I do beg your pardon, Miss MacKimmie," he said. "I wasna questioning the existence of nuns. I've seen quite a number of the creatures with my own eyes." His lips were twitching, and she glared at him.

"Creatures, is it?"

"A figure of speech, nay more, I swear it! Forgive me, Sister—I ken not what I do!" He held up a hand, cowering in mock terror. The urge to laugh made her that much more cross, but she contented herself with a simple "mmphm" of disapproval.

Curiosity got the better of her, though, and after a few moments spent inspecting the foaming wake of the ship, she asked, not looking at him, "When ye saw the nuns, then—what were they doing?"

He'd got control of himself by now and answered her seriously.

"Well, I see the Sisters of Notre Dame, who work among the poor all the time in the streets. They always go out by twos, ken, and both nuns will be carrying great huge baskets, filled with food, I suppose—maybe medicines? They're covered, though—the baskets—so I canna say for sure what's in them. Perhaps they're smuggling brandy and lace down to the docks—" He dodged aside from her upraised hand, laughing.

"Oh, ye'll be a rare nun, Sister Joan! *Terror daemonum, solatium miserorum . . .*"

She pressed her lips tight together, not to laugh. Terror of demons— the cheek of him!

"Not Sister Joan," she said. "They'll give me a new name, likely, at the convent."

"Oh, aye?" He wiped hair out of his eyes, interested. "D'ye get to choose the name yourself?"

"I don't know," she admitted.

"Well, though—what name would ye pick, if ye had the choosing?"

"Er . . . well . . ." She hadn't told anyone, but, after all, what harm could it do? She wouldn't see Michael Murray again once they reached Paris. "Sister Gregory," she blurted.

Rather to her relief, he didn't laugh.

"Oh, that's a good name," he said. "After St. Gregory the Great, is it?"

"Well . . . aye. Ye don't think it's presumptuous?" she asked, a little anxious.

"Oh, no!" he said, surprised. "I mean, how many nuns are named Mary? If it's not presumptuous to be named after the mother o' God, how can it be highfalutin to call yourself after a mere pope?" He smiled at that, so merrily that she smiled back.

"How many nuns *are* named Mary?" she asked, out of curiosity. "It's common, is it?"

"Oh, aye, ye said ye'd not seen a nun." He'd stopped making fun of her, though, and answered seriously. "About half the nuns I've met seem to be called Sister Mary Something—ye ken, Sister Mary Polycarp, Sister Mary Joseph . . . like that."

"And ye meet a great many nuns in the course o' your business, do ye?" Michael Murray was a wine merchant, the junior partner of Fraser *et Cie*—and, judging from the cut of his clothes, did well enough at it.

His mouth twitched, but he answered seriously.

"Well, I do, really. Not every day, I mean, but the sisters come round to my office quite often—or I go to them. Fraser *et Cie* supplies wine to most o' the monasteries and convents in Paris, and some will send a pair of nuns to place an order or to take away something special—otherwise, we deliver it, of course. And even the orders who dinna take wine themselves—and most of the Parisian houses do, they bein' French, aye?—need sacramental wine for their chapels. And the begging orders come round like clockwork to ask alms."

"Really." She was fascinated: sufficiently so as to put aside her reluctance to look ignorant. "I didna ken . . . I mean . . . so the different orders do quite different things, is that what ye're saying? What other kinds are there?"

He shot her a brief glance but then turned back, narrowing his eyes against the wind as he thought.

"Well . . . there's the sort of nun that prays all the time—contemplative, I think they're called. I see them in the cathedral all hours of the day and night. There's more than one order of that sort, though; one kind wears gray habits and prays in the chapel of St. Joseph, and another wears black;

ye see them mostly in the chapel of Our Lady of the Sea." He glanced at her, curious. "Will it be that sort of nun that you'll be?"

She shook her head, glad that the wind-chafing hid her blushes.

"No," she said, with some regret. "That's maybe the holiest sort of nun, but I've spent a good bit o' my life being contemplative on the moors, and I didna like it much. I think I havena got the right sort of soul to do it verra well, even in a chapel."

"Aye," he said, and wiped back flying strands of hair from his face. "I ken the moors. The wind gets into your head after a bit." He hesitated for a moment. "When my uncle Jamie—your da, I mean—ye ken he hid in a cave after Culloden?"

"For seven years," she said, a little impatient. "Aye, everyone kens that story. Why?"

He shrugged.

"Only thinking. I was no but a wee bairn at the time, but I went now and then wi' my mam, to take him food there. He'd be glad to see us, but he wouldna talk much. And it scared me to see his eyes."

Joan felt a small shiver pass down her back, nothing to do with the stiff breeze. She saw—suddenly *saw*, in her head—a thin, dirty man, the bones starting in his face, crouched in the dank, frozen shadows of the cave.

"Da?" she scoffed, to hide the shiver that crawled up her arms. "How could anyone be scairt of him? He's a dear, kind man."

Michael's wide mouth twitched at the corners.

"I suppose it would depend whether ye'd ever seen him in a fight. But—"

"Have you?" she interrupted, curious. "Seen him in a fight?"

"I have, aye. BUT—" he said, not willing to be distracted, "I didna mean *he* scared me. It was that I thought he was haunted. By the voices in the wind."

That dried up the spit in her mouth, and she worked her tongue a little, hoping it didn't show. She needn't have worried; he wasn't looking at her.

"My own da said it was because Jamie spent so much time alone, that the voices got into his head and he couldna stop hearing them. When he'd feel safe enough to come to the house, it would take hours sometimes before he could start to hear *us* again—Mam wouldna let us talk to him until he'd had something to eat and was warmed through." He smiled, a little ruefully. "She said he wasna human 'til then—and, looking back, I dinna think she meant that as a figure of speech."

"Well," she said, but stopped, not knowing how to go on. She wished fervently that she'd known this earlier. Her da and his sister were coming on to France later, but she might not see him. She could maybe have talked to Da, asked him just what the voices in his head were like—what they said. Whether they were anything like the ones she heard.

NEARLY TWILIGHT, and the rats were still dead. The comte heard the bells of Notre Dame calling *sept* and glanced at his pocket watch. The bells were two minutes before their time, and he frowned. He didn't like sloppiness. He stood up and stretched himself, groaning as his spine cracked like the ragged volley of a firing squad. No doubt about it, he *was* aging, and the thought sent a chill through him.

If. If he could find the way forward, then perhaps . . . but you never knew, that was the devil of it. For a little while, he'd thought—hoped— that traveling back in time stopped the process of aging. That initially seemed logical, like rewinding a clock. But, then again, it *wasn't* logical, because he'd always gone back farther than his own lifetime. Only once he'd tried to go back just a few years, to his early twenties. *That* was a mistake, and he still shivered at the memory.

He went to the tall gabled window that looked out over the Seine.

That particular view of the river had changed barely at all in the last two hundred years; he'd seen it at several different times. He hadn't always owned this house, but it had stood in this street since 1620, and he always managed to get in briefly, if only to reestablish his own sense of reality after a passage.

Only the trees changed in his view of the river, and sometimes a strange- looking boat would be there. But the rest was always the same and no doubt always would be: the old fishermen, catching their supper off the landing in stubborn silence, each guarding his space with outthrust el- bows, the younger ones, barefoot and slump-shouldered with exhaustion, laying out their nets to dry, naked little boys diving off the quay. It gave him a soothing sense of eternity, watching the river. Perhaps it didn't mat- ter so much if he must one day die?

"The devil it doesn't," he murmured to himself, and glanced up at the sky. Venus shone bright. He should go.

Pausing conscientiously to place his fingers on each rat's body and en- sure that no spark of life remained, he passed down the line, then swept them all into a burlap bag. If he was going to the Court of Miracles, at least he wouldn't arrive empty-handed.

JOAN WAS STILL reluctant to go below, but the light was fading, the wind getting up regardless, and a particularly spiteful gust that blew her petticoats right up round her waist and grabbed her arse with a chilly hand made her yelp in a very undignified way. She smoothed her skirts hastily and made for the hatchway, followed by Michael Murray.

Seeing him cough and chafe his hands at the bottom of the ladder made her sorry; here she'd kept him freezing on deck, too polite to go below and leave her to her own devices, and her too selfish to see he was cold, the poor man. She made a hasty knot in her handkerchief, to remind her to say an extra decade of the rosary for penance, when she got to it.

He saw her to a bench and said a few words to the woman sitting next to her, in French. Obviously he was introducing her, she understood that much—but when the woman nodded and said something in reply, she could only sit there openmouthed. She didn't understand a word. Not a word!

Michael evidently grasped the situation, for he said something to the woman's husband, which drew her attention away from Joan, and engaged them in a conversation that let Joan sink quietly back against the wooden wall of the ship, sweating with embarrassment.

Well, she'd get into the way of it, she reassured herself. Bound to. She settled herself with determination to listen, picking out the odd word here and there in the conversation. It was easier to understand Michael; he spoke slower and didn't swallow the back half of each word.

She was trying to puzzle out the probable spelling of a word that *sounded* like "pwufgweemiarniere" but surely couldn't be, when her eye caught a slight movement from the bench opposite, and the gurgling vowels caught in her throat.

A man sat there, maybe close to her own age, which was twenty-five. He was good-looking, if a bit thin in the face, decently dressed—and he was going to die.

There was a gray shroud over him, the same as if he were wrapped in mist, so his face showed through it. She'd seen that same thing—the grayness lying on someone's face like fog—seen it twice before and knew it at once for death's shadow. Once it had been on an elderly man, and that might have been only what anybody could see, because Angus MacWheen *was* ill, but then again, and only a few weeks after, she'd seen it on the second of Vhairi Fraser's little boys, and him a rosy-faced wee bairn with dear chubby legs.

She hadn't wanted to believe it. Either that she saw it or what it meant. But four days later, the wean was crushed in the lane by an ox that was maddened by a hornet's sting. She'd vomited when they told her, and couldn't eat for days after, for sheer grief and terror. Because could she have stopped it if she'd said? And what—dear Lord, *what*—if it happened again?

Now it had, and her wame twisted. She leapt to her feet and blundered toward the companionway, cutting short some slowly worded speech from the Frenchman. ·

Not again, not again! she thought in agony. *Why show me such things? What can I do?*

She pawed frantically at the ladder, climbing as fast as she could, gasping for air, needing to be away from the dying man. How long might it be, dear Lord, until she reached the convent, and safety?

THE MOON WAS rising over the Île de la Cité, glowing through the haze of cloud. He glanced at it, estimating the time; no point in arriving at Madame Fabienne's house before the girls had taken their hair out of curling papers and rolled on their red stockings. There were other places to go first, though: the obscure drinking places where the professionals of the court fortified themselves for the night ahead. One of those was where he had first heard the rumors—he'd see how far they had spread and would judge the safety of asking openly about Maître Raymond.

That was one advantage to hiding in the past, rather than going to Hungary or Sweden—life at this court tended to be short, and there were not so many who knew either his face or his history, though there would still be stories. Paris held on to its *histoires.* He found the iron gate—rustier than it had been; it left red stains on his palm—and pushed it open with a creak that would alert whatever now lived at the end of the alley.

He had to *see* the frog. Not meet him, perhaps—he made a brief sign against evil—but see him. Above all else, he needed to know: had the man—if he was a man—aged?

"Certainly he's a man," he muttered to himself, impatient. "What else could he be, for heaven's sake?"

He could be something like you, was the answering thought, and a shiver ran up his spine. *Fear?* He wondered. *Anticipation of an intriguing philosophical mystery? Or possibly . . . hope?*

"WHAT A WASTE of a wonderful arse," Monsieur Brechin remarked in French, watching Joan's ascent from the far side of the cabin. "And, *mon Dieu,* those legs! Imagine those wrapped around your back, eh? Would you have her keep the striped stockings on? I would."

It hadn't occurred to Michael to imagine that, but he was now having a hard time dismissing the image. He coughed into his handkerchief to hide the reddening of his face.

Madame Brechin gave her husband a sharp elbow in the ribs. He grunted but seemed undisturbed by what was evidently a normal form of marital communication.

"Beast," she said, with no apparent heat. "Speaking so of a Bride of Christ. You will be lucky if God himself doesn't strike you dead with a lightning bolt."

"Well, she isn't his bride yet," Monsieur protested. "And who created that arse in the first place? Surely God would be flattered to hear a little sincere appreciation of his handiwork. From one who is, after all, a connoisseur in such matters." He leered affectionately at Madame, who snorted.

A faint snigger from the young man across the cabin indicated that Monsieur was not alone in his appreciation, and Madame turned a reproving glare on the young man. Michael wiped his nose carefully, trying not to catch Monsieur's eye. His insides were quivering, and not entirely from either amusement or the shock of inadvertent lust. He felt very queer.

Monsieur sighed as Joan's striped stockings disappeared through the hatchway.

"Christ will not warm her bed," he said, shaking his head.

"Christ will not fart in her bed, either," said Madame, taking out her knitting.

"Pardonnez-moi . . ." Michael said in a strangled voice, and, clapping his handkerchief to his mouth, made hastily for the ladder, as though seasickness might be catching.

It wasn't *mal de mer* that was surging up from his belly, though. He caught sight of Joan, dim in the evening light at the rail, and turned quickly, going to the other side, where he gripped the rail as though it were a life raft and let the overwhelming waves of grief wash through him. It was the only way he'd been able to manage these last few weeks. Hold on as long as he could, keeping a cheerful face, until some small unexpected thing, some bit of emotional debris, struck him through the heart like a hunter's arrow, and then hurry to find a place to hide, curling up in mindless pain until he could get a grip on himself.

This time, it was Madame's remark that had come out of the blue, and he grimaced painfully, laughing in spite of the tears that poured down his face, remembering Lillie. She'd eaten eels in garlic sauce for dinner—those always made her fart with a silent deadliness like poison swamp gas. As the ghastly miasma had risen up round him, he'd sat bolt upright in bed, only to find her staring at him, a look of indignant horror on her face.

"How *dare* you?" she'd said, in a voice of offended majesty. "*Really,* Michel."

"You *know* it wasn't me!"

Her mouth had dropped open, outrage added to horror and distaste.

"Oh!" she gasped, gathering her pug-dog to her bosom. "You not only fart like a rotting whale, you attempt to blame it on my poor puppy! *Cochon!*" Whereupon she had begun to shake the bedsheets delicately, using

her free hand to waft the noxious odors in his direction, addressing censorious remarks to Plonplon, who gave Michael a sanctimonious look before turning to lick his mistress's face with great enthusiasm.

"Oh, Jesus," he whispered, and, sinking down, pressed his face against the rail. "Oh, God, lass, I love you!"

He shook silently, head buried in his arms, aware of sailors passing now and then behind him, but none of them took notice of him in the dark. At last the agony eased a little, and he drew breath.

All right, then. He'd be all right now, for a time. And he thanked God, belatedly, that he had Joan—or Sister Gregory, if she liked—to look after for a bit. He didn't know how he'd manage to walk through the streets of Paris to his house, alone. Go in, greet the servants—would Jared be there?—face the sorrow of the household, accept their sympathy for his father's death, order a meal, sit down . . . and all the time wanting just to throw himself on the floor of their empty bedroom and howl like a lost soul. He'd have to face it, sooner or later—but not just yet. And right now he'd take the grace of any respite that was offered.

He blew his nose with resolution, tucked away his mangled handkerchief, and went downstairs to fetch the basket his mother had sent. He couldn't swallow a thing himself, but feeding Joan would maybe keep his mind off things for that one minute more.

"That's how ye do it," his brother Ian had told him, as they leant together on the rail of their mother's sheep pen, the winter's wind cold on their faces, waiting for their da to find his way through dying. "Ye find a way to live for that one more minute. And then another. And another." Ian had lost a wife, too, and knew.

He'd wiped his face—he could weep before Ian, while he couldn't with his elder brother or the girls, certainly not in front of his mother—and asked, "And it gets better after a time, is that what ye're telling me?"

His brother had looked at him straight on, the quiet in his eyes showing through the outlandish Mohawk tattoos.

"No," he'd said softly. "But after a time, ye find ye're in a different place than ye were. A different person than ye were. And then ye look about and see what's there with ye. Ye'll maybe find a use for yourself. *That* helps."

"Aye, fine," he said, under his breath, and squared his shoulders. "We'll see, then."

TO RAKOCZY'S SURPRISE, there was a familiar face behind the rough bar. If Maximilian the Great was surprised to see him, the Spanish dwarf gave no indication of it. The other drinkers—a pair of jugglers, each

missing an arm (but the opposing arm), a toothless hag who smacked and muttered over her mug of arrack, and something that looked like a ten-year-old girl but almost certainly wasn't—turned to stare at him but, seeing nothing remarkable in his shabby clothing and burlap bag, turned back to the business of getting sufficiently drunk as to do what needed to be done tonight.

He nodded to Max and pulled up one of the splintering kegs to sit on.

"What's your pleasure, *señor*?"

Rakoczy narrowed his eyes; Max had never served anything but arrack. But times had changed; there was a stone bottle of something that might be beer and a dark glass bottle with a chalk scrawl on it, standing next to the keg of rough brandy.

"Arrack, please, Max," he said—better the devil you know—and was surprised to see the dwarf's eyes narrow in return.

"You knew my honored father, I see, *señor*," the dwarf said, putting the cup on the board. "It's some time since you've been in Paris?"

"*Pardonnez,*" Rakoczy said, accepting it and tossing it back. If you could afford more than one cup, you didn't let it linger on the tongue. "Your honored . . . late father? Max?"

"Maximiliano el Maximo," the dwarf corrected him firmly.

"To be sure." Rakoczy gestured for another drink. "And whom have I the honor to address?"

The Spaniard—though perhaps his accent wasn't as strong as Max's had been—drew himself up proudly. "Maxim Le Grand, *a su servicio!*"

Rakoczy saluted him gravely and threw back the second cup, motioning for a third and, with a gesture, inviting Maxim to join him.

"It has been some time since I was last here," he said. No lie there. "I wonder if another old acquaintance might be still alive—*Maître* Raymond, otherwise called the frog?"

There was a tiny quiver in the air, a barely perceptible flicker of attention, gone almost as soon as he'd sensed it—somewhere behind him?

"A frog," Maxim said, meditatively pouring himself a drink. "I don't know any frogs myself, but should I hear of one, who shall I say is asking for him?"

Should he give his name? No, not yet.

"It doesn't matter," he said. "But word can be left with Madame Fabienne. You know the place? In the Rue Antoine?"

The dwarf's sketchy brows rose, and his mouth turned up at one corner.

"I know it."

Doubtless he did, Rakoczy thought. "El Maximo" hadn't referred to Max's stature, and probably "Le Grand" didn't, either. God had a sense of justice, as well as a sense of humor.

"Bon." He wiped his lips on his sleeve and put down a coin that would have bought the whole keg. *"Merci."*

He stood up, the hot taste of the brandy bubbling at the back of his throat, and belched. Two more places to visit, maybe, before he went to Fabienne's. He couldn't visit more than that and stay upright; he *was* getting old.

"Good night." He bowed to the company and gingerly pushed open the cracked wooden door; it was hanging by one leather hinge, and that looked ready to give way at any moment.

"Ribbit," someone said very softly, just before the door closed behind him.

MADELEINE'S FACE LIGHTED when she saw him, and his heart warmed. She wasn't very bright, poor creature, but she was pretty and amiable and had been a whore long enough to be grateful for small kindnesses.

"Monsieur Rakoczy!" She flung her arms about his neck, nuzzling affectionately.

"Madeleine, my dear." He cupped her chin and kissed her gently on the lips, drawing her close so that her belly pressed against his. He held her long enough, kissing her eyelids, her forehead, her ears—so that she made high squeaks of pleasure—that he could feel his way inside her, hold the weight of her womb in his mind, evaluate her ripening.

It felt warm, the color in the heart of a dark crimson rose, the kind called *sang de dragon*. A week before, it had felt solid, compact as a folded fist; now it had begun to soften, to hollow slightly as she readied. Three more days? he wondered. Four?

He let her go, and when she pouted prettily at him, he laughed and raised her hand to his lips, feeling the same small thrill he had felt when he first found her, as the faint blue glow rose between her fingers in response to his touch. She couldn't see it—he'd raised their linked hands to her face before and she had merely looked puzzled—but it was there.

"Go and fetch some wine, *ma belle*," he said, squeezing her hand gently. "I need to talk to Madame."

Madame Fabienne was not a dwarf, but she was small, brown, and mottled as a toadstool—and as watchful as a toad, round yellow eyes seldom blinking, never closed.

"Monsieur le Comte," she said graciously, nodding him to a damask chair in her *salon*. The air was scented with candle wax and flesh—flesh of a far better quality than that on offer in the court. Even so, Madame had come from that court and kept her connections there alive; she made no

bones about that. She didn't blink at his clothes, but her nostrils flared at him, as though she picked up the scent of the dives and alleys he had come from.

"Good evening, Madame," he said, smiling at her, and lifted the burlap bag. "I brought a small present for Leopold. If he's awake?"

"Awake and cranky," she said, eyeing the bag with interest. "He's just shed his skin—you don't want to make any sudden moves."

Leopold was a remarkably handsome—and remarkably large—python; an albino, quite rare. Opinion of his origins was divided; half of Madame Fabienne's clientele held that she had been given the snake by a noble client—some said the late King himself—whom she had cured of impotence. Others said the snake had once *been* a noble client, who had refused to pay her for services rendered. Rakoczy had his own opinions on that one, but he liked Leopold, who was ordinarily tame as a cat and would sometimes come when called—as long as you had something he regarded as food in your hand when you called.

"Leopold! Monsieur le Comte has brought you a treat!" Fabienne reached across to an enormous wicker cage and flicked the door open, withdrawing her hand with sufficient speed as to indicate just what she meant by "cranky."

Almost at once, a huge yellow head poked out into the light. Snakes had transparent eyelids, but Rakoczy could swear the python blinked irritably, swaying up a coil of its monstrous body for a moment before plunging out of the cage and swarming across the floor with amazing rapidity for such a big creature, tongue flicking in and out like a seamstress's needle.

He made straight for Rakoczy, jaws yawning as he came, and Rakoczy snatched up the bag just before Leopold tried to engulf it—or Rakoczy—whole. He jerked aside, hastily seized a rat, and threw it. Leopold flung a coil of his body on top of the rat with a thud that rattled Madame's spoon in her teabowl, and before the company could blink, he had whipped the rat into a half-hitch knot of coil.

"Hungry as well as ill-tempered, I see," Rakoczy remarked, trying for nonchalance. In fact, the hairs were prickling over his neck and arms. Normally, Leopold took his time about feeding, and the violence of the python's appetite at such close quarters had shaken him.

Fabienne was laughing, almost silently, her tiny sloping shoulders quivering beneath the green Chinese silk tunic she wore.

"I thought for an instant he'd have you," she remarked at last, wiping her eyes. "If he had, I shouldn't have had to feed him for a month!"

Rakoczy bared his teeth in an expression that might have been taken for a smile.

"We cannot let Leopold go hungry," he said. "I wish to make a special

arrangement for Madeleine—it should keep the worm up to his yellow arse in rats for some time."

Fabienne put down her handkerchief and regarded him with interest.

"Leopold has two cocks, but I can't say I've ever noticed an arse. Twenty *écus* a day. Plus two extra if she needs clothes."

He waved an easy hand, dismissing this.

"I had in mind something longer." He explained what he had in mind and had the satisfaction of seeing Fabienne's face go quite blank with stupefaction. It didn't stay that way more than a few moments; by the time he had finished, she was already laying out her initial demands.

When they finally came to agreement, they had drunk half a bottle of decent wine, and Leopold had swallowed the rat. It made a small bulge in the muscular tube of the snake's body but hadn't slowed him appreciably; the coils slithered restlessly over the painted canvas floorcloth, glowing like gold, and Rakoczy saw the patterns of his skin like trapped clouds beneath the scales.

"He *is* beautiful, no?" Fabienne saw his admiration and basked a little in it. "Did I ever tell you where I got him?"

"Yes, more than once. And more than one story, too." She looked startled, and he compressed his lips. He'd been patronizing her establishment for no more than a few weeks, this time. He'd known her fifteen years before—though only a couple of months, that time. He hadn't given his name then, and a madam saw so many men that there was little chance of her recalling him. On the other hand, he also thought it unlikely that she troubled to recall to whom she'd told which story, and this seemed to be the case, for she lifted one shoulder in a surprisingly graceful shrug and laughed.

"Yes, but this one is true."

"Oh, well, then." He smiled and, reaching into the bag, tossed Leopold another rat. The snake moved more slowly this time and didn't bother to constrict its motionless prey, merely unhinging its jaw and engulfing it in a single-minded way.

"He is an old friend, Leopold," she said, gazing affectionately at the snake. "I brought him with me from the West Indies, many years ago. He is a *Mystère*, you know."

"I didn't, no." Rakoczy drank more wine; he had sat long enough that he was beginning to feel almost sober again. "And what is that?" He was interested—not so much in the snake but in Fabienne's mention of the West Indies. He'd forgotten that she claimed to have come from there, many years ago, long before he'd known her the first time.

The *afile* powder had been waiting in his laboratory when he'd come back; no telling how many years it had sat there—the servants couldn't recall. Mélisande's brief note—*Try this. It may be what the frog used*—had

not been dated, but there was a brief scrawl at the top of the sheet, saying, *Rose Hall, Jamaica*. If Fabienne retained any connections in the West Indies, perhaps . . .

"Some call them *loa*"—her wrinkled lips pursed as she kissed the word—"but those are the Africans. A *Mystère* is a spirit, one who is an intermediary between the Bondye and us. Bondye is *le bon Dieu*, of course," she explained to him. "The African slaves speak very bad French. Give him another rat; he's still hungry, and it scares the girls if I let him hunt in the house."

Another two rats and the snake was beginning to look like a fat string of pearls. He was showing an inclination to lie still, digesting. The tongue still flickered, tasting the air, but lazily now.

Rakoczy picked up the bag again, weighing the risks—but, after all, if news came from the Court of Miracles, his name would soon be known in any case.

"I wonder, Madame, as you know everyone in Paris"—he gave her a small bow, which she graciously returned—"are you acquainted with a certain man known as *Maître* Raymond? Some call him the frog," he added.

She blinked, then looked amused.

"You're looking for the frog?"

"Yes. Is that funny?" He reached into the sack, fishing for a rat.

"Somewhat. I should perhaps not tell you, but since you are so accommodating"—she glanced complacently at the purse he had put beside her teabowl, a generous deposit on account—"*Maître Grenouille* is looking for *you*."

He stopped dead, hand clutching a furry body.

"What? You've seen him?"

She shook her head and, sniffing distastefully at her cold tea, rang the bell for her maid.

"No, but I've heard the same from two people."

"Asking for me by name?" Rakoczy's heart beat faster.

"Monsieur le Comte St. Germain. That *is* you?" She asked with no more than mild interest; false names were common in her business.

He nodded, mouth suddenly too dry to speak, and pulled the rat from the sack. It squirmed suddenly in his hand, and a piercing pain in his thumb made him hurl the rodent away.

"*Sacrebleu!* It bit me!"

The rat, dazed by impact, staggered drunkenly across the floor toward Leopold, whose tongue began to flicker faster. Fabienne, though, uttered a sound of disgust and threw a silver-backed hairbrush at the rat. Startled by the clatter, the rat leapt convulsively into the air, landed on and raced

directly over the snake's astonished head, disappearing through the door into the foyer, where—by the resultant scream—it evidently encountered the maid before making its ultimate escape into the street.

"*Jésus Marie*," Madame Fabienne said, piously crossing herself. "A miraculous resurrection. Two weeks before Easter, too."

IT WAS A SMOOTH passage; the shore of France came into sight just after dawn the next day. Joan saw it, a low smudge of dark green on the horizon, and felt a little thrill at the sight, in spite of her tiredness.

She hadn't slept, though she'd reluctantly gone below after nightfall, there to wrap herself in her cloak and shawl, trying not to look at the young man with the shadow on his face. She'd lain all night, listening to the snores and groans of her fellow passengers, praying doggedly and wondering in despair whether prayer was all she could do.

She often wondered whether it was because of her name. She'd been proud of her name when she was small; it was a heroic name, a saint's name, but also a warrior's name. Her mother'd told her that, often and often. She didn't think her mother had considered that the name might also be haunted.

Surely it didn't happen to everyone named Joan, though, did it? She wished she knew another Joan to ask. Because if it *did* happen to them all, the others would be keeping it quiet, just as she did.

You didn't go round telling people that you heard voices that weren't there. Still less that you saw things that weren't there, either. You just *didn't*.

She'd heard of a seer, of course; everyone in the Highlands had. And nearly everyone she knew at least claimed to have seen the odd fetch or had a premonition that Angus MacWheen was dead when he didn't come home that time last winter. The fact that Angus MacWheen was a filthy auld drunkard and so yellow and crazed that it was heads or tails whether he'd die on any particular day, let alone when it got cold enough that the loch froze, didn't come into it.

But she'd never *met* a seer—there was the rub. How did you get into the way of it? Did you just tell folk, "*Here's a thing . . . I'm a seer,*" and they'd nod and say, "*Oh, aye, of course; what's like to happen to me next Tuesday?*" More important, though, how the devil—

"Ow!" She'd bitten her tongue fiercely as penance for the inadvertent blasphemy, and clapped a hand to her mouth.

"What is it?" said a concerned voice behind her. "Are ye hurt, Miss MacKimmie? Er . . . Sister Gregory, I mean?"

"Mm! No. No, I jutht . . . bit my tongue." She turned to Michael Murray, gingerly touching the injured tongue to the roof of her mouth.

"Well, that happens when ye talk to yourself." He took the cork from a bottle he was carrying and held the bottle out to her. "Here, wash your mouth wi' that; it'll help."

She took a large mouthful and swirled it round; it burned the bitten place, but not badly, and she swallowed, as slowly as possible, to make it last.

"Jesus, Mary, and Bride," she breathed. "Is that *wine?*" The taste in her mouth bore some faint kinship with the liquid she knew as wine—just as apples bore some resemblance to horse turds.

"Aye, it *is* pretty good," he said modestly. "German. Umm . . . have a wee nip more?"

She didn't argue and sipped happily, barely listening to his talk, telling about the wine, what it was called, how they made it in Germany, where he got it . . . on and on. Finally she came to herself enough to remember her manners, though, and reluctantly handed back the bottle, now half empty.

"I thank ye, sir," she said primly. "'Twas kind of ye. Ye needna waste your time in bearing me company, though; I shall be well enough alone."

"Aye, well . . . it's no really for your sake," he said, and took a reasonable swallow himself. "It's for mine."

She blinked against the wind. He was flushed, but not from drink or wind, she thought.

She managed a faint interrogative "Ah . . . ?"

"Well, what I want to ask," he blurted, and looked away, cheekbones burning red. "Will ye pray for me? Sister? And my—my wife. The repose of—of—"

"Oh!" she said, mortified that she'd been so taken up with her own worries as not to have seen his distress. *Think you're a seer, dear Lord, ye dinna see what's under your neb; you're no but a fool, and a selfish fool at that.* She put her hand over his where it lay on the rail and squeezed tight, trying to channel some sense of God's goodness into his flesh. "To be sure I will!" she said. "I'll remember ye at every Mass, I swear it!" She wondered briefly whether it was proper to swear to something like that, but after all . . . "And your poor wife's soul, of course I will! What . . . er . . . what was her name? So as I'll know what to say when I pray for her," she explained hurriedly, seeing his eyes narrow with pain.

"Lilliane," he said, so softly that she barely heard him over the wind. "I called her Lillie."

"Lilliane," she repeated carefully, trying to form the syllables like he did. It was a soft, lovely name, she thought, slipping like water over the rocks at the top of a burn. *You'll never see a burn again,* she thought with

a pang, but dismissed this, turning her face toward the growing shore of France. "I'll remember."

He nodded in mute thanks, and they stood for some little while, until she realized that her hand was still resting on his and drew it back with a jerk. He looked startled, and she blurted—because it was the thing on the top of her mind—"What was she like? Your wife?"

The most extraordinary mix of emotions flooded over his face. She couldn't have said what was uppermost—grief, laughter, or sheer bewilderment—and she realized suddenly just how little of his true mind she'd seen before.

"She was . . ." He shrugged and swallowed. "She was my wife," he said, very softly. "She was my life."

She should know something comforting to say to him, but she didn't.

She's with God? That was the truth, she hoped, and yet clearly to this young man, the only thing that mattered was that his wife was not with *him*.

"What happened to her?" she asked instead, baldly, only because it seemed necessary to say something.

He took a deep breath and appeared to sway a little; he'd finished the rest of the wine, she saw, and she took the empty bottle from his hand, tossing it overboard.

"The influenza. They said it was quick. Didn't feel quick to me—and yet, it was, I suppose it was. It took two days, and God kens well that I recall every second of those days—yet it seems that I lost her between one heartbeat and the next. And I—I keep lookin' for her there, in that space between."

He swallowed. "She—she was . . ." The words "with child" came so quietly that she barely heard them.

"Oh," Joan said softly, very moved. "Oh, *a chuisle*." "Heart's blood," it meant, and what *she* meant was that his wife had been that to him—dear Lord, she hoped he hadn't thought she meant—no, he hadn't, and the tight-wound spring in her backbone relaxed a little, seeing the look of gratitude on his face. He did know what she'd meant and seemed glad that she'd understood.

Blinking, she looked away—and caught sight of the young man with the shadow on him, leaning against the railing a little way down. The breath caught in her throat at sight of him.

The shadow was darker in the morning light. The sun was beginning to warm the deck, frail white clouds swam in the blue of clear French skies, and yet the mist now swirled and thickened, obscuring the young man's face, wrapping round his shoulders like a shawl.

Dear Lord, tell me what to do! Her body jerked, wanting to go to the young man, speak to him. But to say what? *"You're in danger, be careful"*?

He'd think she was mad. And if the danger was a thing he couldn't help, like with wee Ronnie and the ox, what difference might her speaking make?

She was dimly aware of Michael staring at her, curious. He said something to her, but she wasn't listening, listening hard instead inside her head. Where were the damned voices when you bloody *needed* one?

But the voices were stubbornly silent, and she turned to Michael, the muscles of her arm jumping, she'd held so tight to the ship's rigging.

"I'm sorry," she said. "I wasna listening properly. I just—thought of something."

"If it's a thing I can help ye with, Sister, ye've only to ask," he said, smiling faintly. "Oh! And speak of that, I meant to say—I said to your mam, if she liked to write to you in care of Fraser *et Cie,* I'd see to it that ye got the letters." He shrugged, one-shouldered. "I dinna ken what the rules are at the convent, aye? About getting letters from outside."

Joan didn't know that, either, and had worried about it. She was so relieved to hear this that a huge smile split her face.

"Oh, it's that kind of ye!" she said. "And if I could—maybe write back . . . ?"

His smile grew wider, the marks of grief easing in his pleasure at doing her a service.

"Anytime," he assured her. "I'll see to it. Perhaps I could—"

A ragged shriek cut through the air, and Joan glanced up, startled, thinking it one of the seabirds that had come out from shore to wheel round the ship, but it wasn't. The young man was standing on the rail, one hand on the rigging, and before she could so much as draw breath, he let go and was gone.

MICHAEL WAS WORRIED for Joan; she sat slumped in the coach, not bothering to look out of the window, until a faint waft of the cool breeze touched her face. The smell was so astonishing that it drew her out of the shell of shocked misery in which she had traveled from the docks.

"Mother o' God!" she said, clapping a hand to her nose. "What *is* that?"

Michael dug in his pocket and pulled out the grubby rag of his handkerchief, looking dubiously at it.

"It's the public cemeteries. I'm sorry, I didna think—"

"Moran taing." She seized the damp cloth from him and held it over her face, not caring. "Do the French not *bury* folk in their cemeteries?" Because, judging from the smell, a thousand corpses had been thrown out on wet ground and left to rot, and the sight of darting, squabbling flocks of black corbies in the distance did nothing to correct this impression.

"They do." Michael felt exhausted—it had been a terrible morning—but struggled to pull himself together. "It's all marshland over there, though; even coffins buried deep—and most of them aren't—work their way through the ground in a few months. When there's a flood—and there's a flood whenever it rains—what's left of the coffins falls apart, and . . ." He swallowed, just as pleased that he'd not eaten any breakfast.

"There's talk of maybe moving the bones at least, putting them in an ossuary, they call it. There are mine workings, old ones, outside the city—over there"—he pointed with his chin—"and perhaps . . . but they havena done anything about it yet," he added in a rush, pinching his nose fast to get a breath in through his mouth. It didn't matter whether you breathed through your nose or your mouth, though; the air was thick enough to taste.

She looked as ill as he felt, or maybe worse, her face the color of spoilt

custard. She'd vomited when the crew had finally pulled the suicide aboard, pouring gray water and slimed with the seaweed that had wrapped round his legs and drowned him. There were still traces of sick down her front, and her dark hair was lank and damp, straggling out from under her cap. She hadn't slept at all, of course—neither had he.

He couldn't take her to the convent in this condition. The nuns maybe wouldn't mind, but she would. He stretched up and rapped on the ceiling of the carriage.

"Monsieur?"

"Au château, vite!"

He'd take her to his house first. It wasn't much out of the way, and the convent wasn't expecting her at any particular day or hour. She could wash, have something to eat, and put herself to rights. And if it saved him from walking into his house alone, well, they did say a kind deed carried its own reward.

BY THE TIME they'd reached the Rue Trémoulins, Joan had forgotten—partly—her various reasons for distress, in the sheer excitement of being in Paris. She had never seen so many people in one place at the same time—and that was only the folk coming out of Mass at a parish church! Round the corner, a pavement of fitted stones stretched wider than the whole River Ness, and those stones covered from one side to the other in barrows and wagons and stalls, rioting with fruit and vegetables and flowers and fish and meat . . . She'd given Michael back his filthy handkerchief and was panting like a dog, turning her face to and fro, trying to draw all the wonderful smells into herself at once.

"Ye look a bit better," Michael said, smiling at her. He was still pale himself, but he, too, seemed happier. "Are ye hungry yet?"

"I'm famished!" She cast a starved look at the edge of the market. "Could we stop, maybe, and buy an apple? I've a bit of money. . . ." She fumbled for the coins in her stocking top, but he stopped her.

"Nay, there'll be food a-plenty at the house. They were expecting me this week, so everything will be ready."

She stared longingly at the market for a brief moment, then turned obligingly in the direction he pointed, craning out the carriage window to see his house as they approached.

"That's the biggest house I've ever seen!" she exclaimed.

"Och, no," he said, laughing. "Lallybroch's bigger than that."

"Well . . . this one's *taller*," she replied. And it was—a good four stories, and a huge roof of lead slates and green-coppered seams, with what must be more than a score of glass windows set in, and . . .

She was still trying to count the windows when Michael helped her down from the carriage and offered her his arm to walk up to the door. She was goggling at the big yew trees set in brass pots and wondering how much trouble it must be to keep those polished, when she felt his arm go suddenly rigid as wood.

She glanced at Michael, startled, then looked where he was looking—toward the door of his house. The door had swung open, and three people were coming down the marble steps, smiling and waving, calling out.

"Who's that?" Joan whispered, leaning close to Michael. The one short fellow in the striped apron must be a butler; she'd read about butlers. But the other man was a gentleman, limber as a willow tree and wearing a coat and waistcoat striped in lemon and pink—with a hat decorated with . . . well, she supposed it must be a feather, but she'd pay money to see the bird it came off. By comparison, she had hardly noticed the woman, who was dressed in black. But now she saw that Michael had eyes only for the woman.

"Lé—" he began, and choked it back. "Lé—Léonie. Léonie is her name. My wife's sister."

Joan looked sharp then, because from the look of Michael Murray, he'd just seen his wife's ghost. But Léonie seemed flesh and blood, slender and pretty, though her own face bore the same marks of sorrow as did Michael's, and her face was pale under a small, neat black tricorne with a tiny curled blue feather.

"Michel," she said. "Oh, Michel!" And with tears brimming from eyes shaped like almonds, she threw herself into his arms.

Feeling extremely superfluous, Joan stood back a little and glanced at the gentleman in the lemon-striped waistcoat—the butler had tactfully withdrawn into the house.

"Charles Pépin, mademoiselle," he said, sweeping off his hat. Taking her hand, he bowed low over it, and now she saw the band of black mourning he wore around his bright sleeve. *"A votre service."*

"Oh," she said, a little flustered. "Um. Joan MacKimmie. *Je suis* . . . er . . . um . . ."

"Tell him not to do it," said a sudden small, calm voice inside her head, and she jerked her own hand away as though he'd bitten her.

"Pleased to meet you," she gasped. "Excuse me." And, turning, threw up into one of the bronze yew pots.

JOAN HAD BEEN afraid it would be awkward, coming to Michael's bereaved and empty house, but had steeled herself to offer comfort and support, as became a distant kinswoman and a daughter of God. She might

have been miffed, therefore, to find herself entirely supplanted in the department of comfort and support—quite relegated to the negligible position of guest, in fact, served politely and asked periodically if she wished more wine, a slice of ham, some gherkins . . . but otherwise ignored, while Michael's servants, sister-in-law, and . . . she wasn't quite sure of the position of M. Pépin, though he seemed to have something personal to do with Léonie—perhaps someone had said he was her cousin?—all swirled round Michael like perfumed bathwater, warm and buoyant, touching him, kissing him—well, all right, she'd heard of men kissing one another in France, but she couldn't help staring when M. Pépin gave Michael a big wet one on both cheeks—and generally making a fuss over him.

She was more than relieved, though, not to have to make conversation in French, beyond a simple *merci* or *s'il vous plaît* from time to time. It gave her a chance to settle her nerves—and her stomach, and she would say the wine was a wonder for that—and to keep a close eye on Monsieur Charles Pépin.

"Tell him not to do it." And just *what d'ye mean by that?* she demanded of the voice. She didn't get an answer, which didn't surprise her. The voices weren't much for details.

She couldn't tell whether the voices were male or female; they didn't seem either one, and she wondered whether they might maybe be angels—angels didn't have a sex, and doubtless that saved them a lot of trouble. Joan of Arc's voices had had the decency to introduce themselves, but not hers, oh, no. On the other hand, if they *were* angels and told her their names, she wouldn't recognize them anyway, so perhaps that's why they didn't bother.

Well, so. Did this particular voice mean that Charles Pépin was a villain? She squinted closely at him. He didn't look it. He had a strong, good-looking face, and Michael seemed to like him—after all, Michael must be a fair judge of character, she thought, and him in the wine business.

What was it Monsieur Charles Pépin oughtn't to do, though? Did he have some wicked crime in mind? Or might he be bent on doing away with himself, like that poor wee gomerel on the boat? There was still a trace of slime on her hand, from the seaweed.

She rubbed her hand inconspicuously against the skirt of her dress, frustrated. She hoped the voices would stop once she was in the convent. That was her nightly prayer. But if they didn't, at least she might be able to tell someone there about them without fear of being packed off to a madhouse or stoned in the street. She'd have a confessor, she knew that much. Maybe he could help her discover what God had meant, landing her with a gift like this, and no explanation what she was to do with it.

In the meantime, Monsieur Pépin would bear watching; she should

maybe say something to Michael before she left. *Aye, what?* she thought, helpless.

Still, she was glad to see that Michael grew less pale as they all carried on, vying to feed him tidbits, refill his glass, tell him bits of gossip. She was also pleased to find that she mostly understood what they were saying, as she relaxed. Jared—that would be Jared Fraser, Michael's elderly cousin, who'd founded the wine company, and whose house this was—was still in Germany, they said, but was expected at any moment. He had sent a letter for Michael, too; where was it? No matter, it would turn up . . . and Madame Nesle de La Tourelle had had a fit, a veritable *fit*, at court last Wednesday, when she came face-to-face with Mademoiselle de Perpignan wearing a confection in the particular shade of pea green that was de La Tourelle's alone, and God alone knew why, because she always looked like a cheese in it, and had slapped her own maid so hard for pointing this out that the poor girl flew across the rushes and cracked her head on one of the mirrored walls—and cracked the mirror, too, very bad luck that, but no one could agree whether the bad luck was de La Tourelle's, the maid's, or de Perpignan's.

Birds, Joan thought dreamily, sipping her wine. *They sound just like cheerful wee birds in a tree, all chattering away together.*

"The bad luck belongs to the seamstress who made the dress for de Perpignan," Michael said, a faint smile touching his mouth. "Once de La Tourelle finds out who it is." His eye lighted on Joan then, sitting there with a fork—an actual fork, and silver, too!—in her hand, her mouth half open in the effort of concentration required to follow the conversation.

"Sister Joan—Sister Gregory, I mean—I'm that sorry, I was forgetting. If ye've had enough to eat, will ye have a bit of a wash, maybe, before I deliver ye to the convent?"

He was already rising, reaching for a bell, and before she knew where she was, a maidservant had whisked her off upstairs, deftly undressed her, and, wrinkling her nose at the smell of the discarded garments, wrapped Joan in a robe of the most amazing green silk, light as air, and ushered her into a small stone room with a copper bath in it, then disappeared, saying something in which Joan caught the word *"eau."*

She sat on the wooden stool provided, clutching the robe about her nakedness, head spinning with more than wine. She closed her eyes and took deep breaths, trying to put herself in the way of praying. God was everywhere, she assured herself, embarrassing as it was to contemplate him being with her in a bathroom in Paris. She shut her eyes harder and firmly began the rosary, starting with the Joyful Mysteries.

She'd got through the Visitation before she began to feel steady again. This wasn't quite how she'd expected her first day in Paris to be. Still,

she'd have something to write home to Mam about, that was for sure. If they let her write letters in the convent.

The maid came in with two enormous cans of steaming water and upended these into the bath with a tremendous splash. Another came in on her heels, similarly equipped, and between them they had Joan up, stripped, and stepping into the tub before she'd so much as said the first word of the Lord's Prayer for the third decade.

They said French things to her, which she didn't understand, and held out peculiar-looking instruments to her in invitation. She recognized the small pot of soap and pointed at it, and one of them at once poured water on her head and began to wash her hair!

She had for months been bidding farewell to her hair whenever she combed it, quite resigned to its loss, for whether she must sacrifice it immediately, as a postulant, or later, as a novice, plainly it must go. The shock of knowing fingers rubbing her scalp, the sheer sensual delight of warm water coursing through her hair, the soft wet weight of it lying in ropes down over her breasts—was this God's way of asking if she'd truly thought it through? Did she know what she was giving up?

Well, she did, then. And she *had* thought about it. On the other hand . . . she couldn't make them stop, really; it wouldn't be mannerly. The warmth of the water was making the wine she'd drunk course faster through her blood, and she felt as though she were being kneaded like toffee, stretched and pulled, all glossy and falling into languid loops. She closed her eyes and gave up trying to remember how many Hail Marys she had yet to go in the third decade.

It wasn't until the maids had hauled her, pink and steaming, out of the bath and wrapped her in a most remarkable huge fuzzy kind of towel that she emerged abruptly from her sensual trance. The cold air coalesced in her stomach, reminding her that all this luxury was indeed a lure of the devil—for lost in gluttony and sinful bathing, she'd forgot entirely about the young man on the ship, the poor despairing sinner who had thrown himself into the sea.

The maids had gone for the moment. She dropped at once to her knees on the stone floor and threw off the coddling towels, exposing her bare skin to the full chill of the air in penance.

"Mea culpa, mea culpa, mea maxima culpa," she breathed, knocking a fist against her bosom in a paroxysm of sorrow and regret. The sight of the drowned young man was in her mind, soft brown hair fanned across his cheek, eyes half closed, seeing nothing—and what terrible thing was it that he'd seen, or thought of, before he jumped, that he'd screamed so?

She thought briefly of Michael, the look on his face when he spoke of his poor wife—perhaps the young brown-haired man had lost someone dear and couldn't face his life alone?

She should have spoken to him. That was the undeniable, terrible truth. It didn't matter that she didn't know what to say. She should have trusted God to give her words, as he had when she'd spoken to Michael.

"Forgive me, Father!" she said urgently, out loud. "Please—forgive me, give me strength!"

She'd betrayed that poor young man. And herself. *And* God, who'd given her the terrible gift of sight for a reason. And the voices . . .

"Why did ye not tell me?" she cried. "Have ye nothing to say for yourselves?" Here she'd thought the voices those of angels, and they weren't—just drifting bits of bog mist, getting into her head, pointless, useless . . . useless as she was, oh, Lord Jesus . . .

She didn't know how long she knelt there, naked, half drunk, and in tears. She heard the muffled squeaks of dismay from the French maids, who poked their heads in and just as quickly withdrew them, but paid no attention. She didn't know if it was right even to pray for the poor young man—for suicide was a mortal sin, and surely he'd gone straight to hell. But she couldn't give him up; she couldn't. She felt somehow that he'd been her charge, that she'd carelessly let him fall, and surely God would not hold the young man entirely responsible when it was she who should have been watching out for him.

And so she prayed, with all the energy of body and mind and spirit, asking mercy. Mercy for the young man, for wee Ronnie and wretched auld Angus—mercy for poor Michael, and for the soul of Lillie, his dear wife, and their babe unborn. And mercy for herself, this unworthy vessel of God's service.

"I'll do better!" she promised, sniffing and wiping her nose on the fluffy towel. "Truly, I will. I'll be braver. I will."

MICHAEL TOOK THE candlestick from the footman, said good night, and shut the door. He hoped Sister almost-Gregory was comfortable; he'd told the staff to put her in the main guest room. He was fairly sure she'd sleep well. He smiled wryly to himself; unaccustomed to wine, and obviously nervous in company, she'd sipped her way through most of a decanter of Jerez sherry before he noticed, and was sitting in the corner with unfocused eyes and a small inward smile that reminded him of a painting he had seen at Versailles, a thing the steward had called *La Gioconda*.

He couldn't very well deliver her to the convent in such a condition and had gently escorted her upstairs and given her into the hands of the chambermaids, both of whom regarded her with some wariness, as though a tipsy nun were a particularly dangerous commodity.

He'd drunk a fair amount himself in the course of the afternoon and more at dinner. He and Charles had sat up late, talking and drinking rum punch. Not talking of anything in particular; he had just wanted not to be alone. Charles had invited him to go to the gaming rooms—Charles was an inveterate gambler—but was kind enough to accept his refusal and simply bear him company.

The candle flame blurred briefly at thought of Charles's kindness. He blinked and shook his head, which proved a mistake; the contents shifted abruptly, and his stomach rose in protest at the sudden movement. He barely made it to the chamber pot in time and, once evacuated, lay numbly on the floor, cheek pressed to the cold boards.

It wasn't that he couldn't get up and go to bed. It was that he couldn't face the thought of the cold white sheets, the pillows round and smooth, as though Lillie's head had never dented them, the bed never known the heat of her body.

Tears ran sideways over the bridge of his nose and dripped on the floor. There was a snuffling noise, and Plonplon came squirming out from under the bed and licked his face, whining anxiously. After a little while, he sat up and, leaning against the side of the bed with the dog in one arm, reached for the decanter of port that the butler had left—by instruction—on the table beside it.

THE SMELL WAS appalling. Rakoczy had wrapped a woolen comforter about his lower face, but the odor seeped in, putrid and cloying, clinging to the back of the throat, so that even breathing through the mouth didn't preserve you from the stench. He breathed as shallowly as he could, though, picking his way carefully past the edge of the cemetery by the narrow beam of a dark lantern. The mine lay well beyond it, but the stench carried amazingly when the wind blew from the east.

The chalk mine had been abandoned for years; it was rumored to be haunted. It was. Rakoczy knew what haunted it. Never religious—he was a philosopher and a natural scientist, a rationalist—he still crossed himself by reflex at the head of the ladder that led down the shaft into those spectral depths.

At least the rumors of ghosts and earth demons and the walking dead would keep anyone from coming to investigate strange light glowing from the subterranean tunnels of the workings, if it was noticed at all. Though just in case . . . he opened the burlap bag, still redolent of rats, and fished out a bundle of pitchblende torches and the oiled-silk packet that held several lengths of cloth saturated with *salpêtre,* salts of potash, blue vitriol,

verdigris, butter of antimony, and a few other interesting compounds from his laboratory.

He found the blue vitriol by smell and wrapped the cloth tightly around the head of one torch, then—whistling under his breath—made three more torches, each impregnated with different salts. He loved this part. It was so simple, and so astonishingly beautiful.

He paused for a minute to listen, but it was well past dark and the only sounds were those of the night itself—frogs chirping and bellowing in the distant marshes by the cemetery, wind stirring the leaves of spring. A few hovels sat a half mile away, only one with firelight glowing dully from a smoke hole in the roof. *Almost a pity there's no one but me to see this.* He took the little clay firepot from its wrappings and touched a coal to the cloth-wrapped torch. A tiny green flame flickered like a serpent's tongue, then burst into life in a brilliant globe of ghostly color.

He grinned at the sight, but there was no time to lose; the torches wouldn't last forever, and there was work to be done. He tied the bag to his belt and, with the green fire crackling softly in one hand, climbed down into darkness.

He paused at the bottom, breathing deep. The air was clear, the dust long settled. No one had been down here recently. The dull white walls glowed soft, eerie under the green light, and the passage yawned before him, black as a murderer's soul. Even knowing the place as well as he did, and with light in his hand, it gave him a qualm to walk into it.

Is that what death is like? he wondered. A black void that you walked into with no more than a feeble glimmer of faith in your hand? His lips compressed. Well, he'd done *that* before, if less permanently. But he disliked the way that the notion of death seemed always to be lurking in the back of his mind these days.

The main tunnel was large, big enough for two men to walk side by side, and the roof was high enough above him that the roughly excavated chalk lay in shadow, barely touched by his torch. The side tunnels were smaller, though. He counted the ones on the left and, despite himself, hurried his step a little as he passed the fourth. That was where *it* lay, down the side tunnel, a turn to the left, another to the left—was it "widdershins" the English called it, turning against the direction of the sun? He thought that was what Mélisande had called it when she'd brought him here. . . .

The sixth. His torch had begun to gutter already, and he pulled another from the bag and lit it from the remains of the first, which he dropped on the floor at the entrance to the side tunnel, leaving it to flare and smolder behind him, the smoke catching at his throat. He knew his way, but even so, it was as well to leave landmarks, here in the realm of everlasting night. The mine had deep rooms, one far back that showed strange paintings on

the wall, of animals that didn't exist but had an astonishing vividness, as though they would leap from the wall and stampede down the passages. Sometimes—rarely—he went all the way down into the bowels of the earth, just to look at them.

The fresh torch burned with the warm light of natural fire, and the white walls took on a rosy glow. So did the painting at the end of the corridor, this one different: a crude but effective rendering of the Annunciation. He didn't know who had made the paintings that appeared unexpectedly here and there in the mines—most were of religious subjects, a few most emphatically *not*—but they were useful. There was an iron ring in the wall by the Annunciation, and he set his torch into it.

Turn back at the Annunciation, then three paces . . . He stamped his foot, listening for the faint echo, and found it. He'd brought a trowel in his bag, and it was the work of a few moments to uncover the sheet of tin that covered his cache.

The cache itself was three feet deep and three feet square—he found satisfaction in the knowledge of its perfect cubicity whenever he saw it; any alchemist was by profession a numerologist, as well. It was half full, the contents wrapped in burlap or canvas, not things he wanted to carry openly through the streets. It took some prodding and unwrapping to find the pieces he wanted. Madame Fabienne had driven a hard bargain but a fair one: two hundred *écus* a month times four months for the guaranteed exclusive use of Madeleine's services.

Four months would surely be enough, he thought, feeling a rounded shape through its wrappings. In fact, he thought one night would be enough, but his man's pride was restrained by a scientist's prudence. And even if . . . there was always some chance of early miscarriage; he wanted to be sure of the child before he undertook any more personal experiments with the space between times. If he knew that something of himself—someone with his peculiar abilities—might be left, just in case *this* time . . .

He could feel *it* there, somewhere in the smothered dark behind him. He knew he couldn't hear it now; it was silent, save on the days of solstice and equinox or when you actually walked into it . . . but he felt the sound of it in his bones, and it made his hands tremble on the wrappings.

The gleam of silver, of gold. He chose two gold snuffboxes, a filigreed necklace, and—with some hesitation—a small silver salver. Why did the void not affect metal? he wondered for the thousandth time. In fact, carrying gold or silver eased the passage—or at least he thought so. Mélisande had told him it did. But jewels were always destroyed by the passage, though they gave the most control and protection.

That made some sense; everyone knew that gemstones had a specific vibration that corresponded to the heavenly spheres, and the spheres

themselves of course affected the earth: *As above, so below.* He still had no idea exactly *how* the vibrations should affect the space, the portal . . . *it.* But thinking about it gave him a need to touch them, to reassure himself, and he moved wrapped bundles out of the way, digging down to the left-hand corner of the wood-lined cache, where pressing on a particular nail-head caused one of the boards to loosen and turn sideways, rotating smoothly on spindles. He reached into the dark space thus revealed and found the small washleather bag, feeling his sense of unease dissipate at once when he touched it.

He opened it and poured the contents into his palm, glittering and sparking in the dark hollow of his hand. Reds and blues and greens, the brilliant white of diamonds, the lavender and violet of amethyst, and the golden glow of topaz and citrine. Enough?

Enough to travel back, certainly. Enough to steer himself with some accuracy, to choose how far he went. But enough to go forward?

He weighed the glittering handful for a moment, then poured them carefully back. Not yet. But he had time to find more; he wasn't going anywhere for at least four months. Not until he was sure that Madeleine was well and truly with child.

"JOAN." MICHAEL PUT his hand on her arm, keeping her from leaping out of the carriage. "Ye're *sure,* now? I mean, if ye didna feel quite ready, ye're welcome to stay at my house until—"

"I'm ready." She didn't look at him, and her face was pale as a slab of lard. "Let me go, please."

He reluctantly let go of her arm but insisted upon getting down with her and ringing the bell at the gate, stating their business to the portress. All the time, though, he could feel her shaking, quivering like a blancmange. Was it fear, though, or just understandable nerves? He'd feel a bit cattywampus himself, he thought with sympathy, were he making such a shift, beginning a new life so different from what had gone before.

The portress went away to fetch the mistress of postulants, leaving them in the little enclosure by the gatehouse. From here, he could see across a sunny courtyard with a cloister walk on the far side and what looked like extensive kitchen gardens to the right. To the left was the looming bulk of the hospital run by the order and, beyond that, the other buildings that belonged to the convent. It was a beautiful place, he thought—and hoped the sight of it would settle her fears.

She made an inarticulate noise, and he glanced at her, alarmed to see what looked like tears slicking her cheeks.

"Joan," he said more quietly, and handed her his fresh handkerchief.

"Dinna be afraid. If ye need me, send for me, anytime; I'll come. And I meant it about the letters."

He would have said more, but just then the portress reappeared with Sister Eustacia, the postulant mistress, who greeted Joan with a kind motherliness that seemed to comfort her, for the girl sniffed and straightened herself and, reaching into her pocket, pulled out a little folded square, obviously kept with care through her travels.

"J'ai une lettre," she said in halting French. *"Pour Madame le . . . pour . . .* Reverend Mother?" she said in a small voice. "Mother Hildegarde?"

"Oui?" Sister Eustacia took the note with the same care with which it was proffered.

"It's from . . . her," Joan said to Michael, having plainly run out of French. She still wouldn't look at him. "Da's . . . er . . . wife. You know. Claire."

"Jesus Christ!" Michael blurted, making both the portress and the postulant mistress stare reprovingly at him.

"She said she was a friend of Mother Hildegarde. And if she was still alive . . ." She stole a look at Sister Eustacia, who appeared to have followed this.

"Oh, Mother Hildegarde is certainly alive," she assured Joan, in English. "And I'm sure she will be most interested to speak with you." She tucked the note into her own capacious pocket and held out a hand. "Now, my dear child, if you are quite ready . . ."

"Je suis prêt," Joan said, shaky but dignified. And so Joan MacKimmie of Balriggan passed through the gates of the Convent of Angels, still clutching Michael Murray's clean handkerchief and smelling strongly of his dead wife's scented soap.

MICHAEL HAD DISMISSED his carriage and wandered restlessly about the city after leaving Joan at the convent, not wanting to go home. He hoped they would be good to her, hoped that she'd made the right decision.

Of course, he comforted himself, she wouldn't actually be a nun for some time. He didn't know quite how long it took, from entering as a postulant to becoming a novice to taking the final vows of poverty, chastity, and obedience, but at least a few years. There would be time for her to be sure. And at least she was in a place of safety; the look of terror and distress on her face as she'd shot through the gates of the convent still haunted him. He strolled toward the river, where the evening light glowed

on the water like a bronze mirror. The deckhands were tired and the day's shouting had died away. In this light, the reflections of the boats gliding homeward seemed more substantial than the boats themselves.

He'd been surprised at the letter and wondered whether that had anything to do with Joan's distress. He'd had no notion that his uncle's wife had anything to do with le Couvent des Anges—though now he cast his mind back, he did recall Jared mentioning that Uncle Jamie had worked in Paris in the wine business for a short time, back before the Rising. He supposed Claire might have met Mother Hildegarde then . . . but it was all before he was born.

He felt an odd warmth at the thought of Claire; he couldn't really think of her as his auntie, though she was. He'd not spent much time with her alone at Lallybroch—but he couldn't forget the moment when she'd met him, alone at the door. Greeted him briefly and embraced him on impulse. And he'd felt an instant sense of relief, as though she'd taken a heavy burden from his heart. Or maybe lanced a boil on his spirit, as she might one on his bum.

That thought made him smile. He didn't know what she was—the talk near Lallybroch painted her as everything from a witch to an angel, with most of the opinion hovering cautiously around "faerie," for the Auld Ones were dangerous, and you didn't talk too much about them—but he liked her. So did Da and Young Ian, and that counted for a lot. And Uncle Jamie, of course—though everyone said, very matter-of-fact, that Uncle Jamie was bewitched. He smiled wryly at that. Aye, if being mad in love with your wife was bewitchment.

If anyone outside the family kent what she'd told them—he cut that thought short. It wasn't something he'd forget, but it wasn't something he wanted to think about just yet, either. The gutters of Paris running with blood . . . He glanced down involuntarily, but the gutters were full of the usual assortment of animal and human sewage, dead rats, and bits of rubbish too far gone to be salvaged for food even by the street beggars.

He walked, making his way slowly through the crowded streets, past La Chapelle and the Tuileries. If he walked enough, sometimes he could fall asleep without too much wine.

He sighed, elbowing his way through a group of buskers outside a tavern, turning back toward the Rue Trémoulins. Some days, his head was like a bramble patch: thorns catching at him no matter which way he turned, and no path leading out of the tangle.

Paris wasn't a large city, but it was a complicated one; there was always somewhere else to walk. He crossed the Place de la Corcorde, thinking of what Claire had told them, seeing there in his mind the tall shadow of a terrible machine.

JOAN HAD HAD her dinner with Mother Hildegarde, a lady so ancient and holy that Joan had feared to breathe too heavily, lest Mother Hildegarde fragment like a stale croissant and go straight off to heaven in front of her. Mother Hildegarde had been delighted with the letter Joan had delivered, though; it brought a faint flush to her face.

"From my . . . er . . ." Martha, Mary, and Lazarus, what was the French word for "stepmother"? "Ahh . . . the wife of my . . ." Fittens, she didn't know the word for "stepfather," either! "The wife of my father," she ended weakly.

"You are the daughter of my good friend Claire!" Mother had exclaimed. "And how is she?"

"Bonny, er . . . *bon,* I mean, last I saw her," said Joan, and then tried to explain, but there was a lot of French being spoken very fast, and she gave up and accepted the glass of wine that Mother Hildegarde offered her. She was going to be a sot long before she took her vows, she thought, trying to hide her flushed face by bending down to pat Mother's wee dog, a fluffy, friendly creature the color of burnt sugar, named Bouton.

Whether it was the wine or Mother's kindness, her wobbly spirit steadied. Mother had welcomed her to the community and kissed her forehead at the end of the meal, before sending her off in the charge of Sister Eustacia to see the convent.

Now she lay on her narrow cot in the dormitory, listening to the breathing of a dozen other postulants. It sounded like a byre full of cows and had much the same warm, humid scent—bar the manure. Her eyes filled with tears, the vision of the homely stone byre at Balriggan sudden and vivid in her mind. She swallowed them back, though, pinching her lips together. A few of the girls sobbed quietly, missing home and family, but she wouldn't be one of them. She was older than most—a few were nay more than fourteen—and she'd promised God to be brave.

It hadn't been bad during the afternoon. Sister Eustacia had been very kind, taking her and a couple of other new postulants round the walled estate, showing them the big gardens where the convent grew medicinal herbs and fruit and vegetables for the table, the chapel where devotions were held six times a day, plus Mass in the mornings, the stables and kitchens, where they would take turns working—and the great Hôpital des Anges, the order's main work. They had seen the *hôpital* only from the outside, though; they would see the inside tomorrow, when Sister Marie-Amadeus would explain their duties.

It was strange, of course—she still understood only half what people said to her and was sure from the looks on their faces that they understood

much less of what she tried to say to *them*—but wonderful. She loved the spiritual discipline, the hours of devotion, with the sense of peace and unity that came upon the sisters as they chanted and prayed together. Loved the simple beauty of the chapel, amazing in its clean elegance, the solid lines of granite and the grace of carved wood, a faint smell of incense in the air, like the breath of angels.

The postulants prayed with the others but did not yet sing. They would be trained in music—such excitement! Mother Hildegarde had been a famous musician in her youth, it was rumored, and considered it one of the most important forms of devotion.

The thought of the new things she'd seen, and the new things to come, distracted her mind—a little—from thoughts of her mother's voice, the wind off the moors, the . . . She shoved these hastily away and reached for her new rosary, this a substantial thing with smooth wooden beads, lovely and comforting in the fingers.

Above all, there was peace. She hadn't heard a word from the voices, hadn't seen anything peculiar or alarming. She wasn't foolish enough to think she'd escaped her dangerous gift, but at least there might be help at hand if—when—it came back.

And at least she already knew enough Latin to say her rosary properly; Da had taught her. *"Ave, Maria,"* she whispered, *"gratia plena, Dominus tecum,"* and closed her eyes, the sobs of the homesick fading in her ears as the beads moved slow and silent through her fingers.

Next day

MICHAEL MURRAY STOOD in the aisle of the aging shed, feeling puny and unreal. He'd waked with a terrible headache, the result of having drunk a great deal of mixed spirits on an empty stomach, and while the headache had receded to a dull throb at the back of his skull, it had left him feeling trampled and left for dead. His cousin Jared, owner of Fraser *et Cie,* looked at him with the cold eye of long experience, shook his head and sighed deeply, but said nothing, merely taking the list from his nerveless fingers and beginning the count on his own.

He wished Jared had rebuked him. Everyone still tiptoed round him, careful of him. And like a wet dressing on a wound, their care kept the wound of Lillie's loss open and weeping. The sight of Léonie didn't help, either—so much like Lillie to look at, so different in character. She said they must help and comfort each other and, to that end, came to visit every other day, or so it seemed. He really wished she would . . . just go away, though the thought shamed him.

"How's the wee nun, then?" Jared's voice, dry and matter-of-fact as always, drew him out of his bruised and soggy thoughts. "Give her a good send-off to the convent?"

"Aye. Well—aye. More or less." Michael mustered up a feeble smile. He didn't really want to think about Sister Gregory this morning, either.

"What did ye give her?" Jared handed the checklist to Humberto, the Italian shed-master, and looked Michael over appraisingly. "I hope it wasna the new Rioja that did that to ye."

"Ah . . . no." Michael struggled to focus his attention. The heady atmosphere of the shed, thick with the fruity exhalations of the resting casks, was making him dizzy. "It was Moselle. Mostly. And a bit of rum punch."

"Oh, I see." Jared's ancient mouth quirked up on one side. "Did I never tell ye not to mix wine wi' rum?"

"Not above two hundred times, no." Jared was moving, and Michael followed him perforce down the narrow aisle, the casks in their serried ranks rising high above on either side.

"Rum's a demon. But whisky's a virtuous dram," Jared said, pausing by a rack of small blackened casks. "So long as it's a good make, it'll never turn on ye. Speakin' of which"—he tapped the end of one cask, which gave off the resonant deep *thunk* of a full barrel—"what's this? It came up from the docks this morning."

"Oh, aye." Michael stifled a belch and smiled painfully. "That, cousin, is the Ian Alastair Robert MacLeod Murray memorial *uisge baugh*. My da and Uncle Jamie made it during the winter. They thought ye might like a wee cask for your personal use."

Jared's brows rose and he gave Michael a swift sideways glance. Then he turned back to examine the cask, bending close to sniff at the seam between the lid and staves.

"I've tasted it," Michael assured him. "I dinna think it will poison ye. But ye should maybe let it age a few years."

Jared made a rude noise in his throat, and his hand curved gently over the swell of the staves. He stood thus for a moment as though in benediction, then turned suddenly and took Michael into his arms. His own breathing was hoarse, congested with sorrow. He was years older than Da and Uncle Jamie but had known the two of them all their lives.

"I'm sorry for your faither, lad," he said after a moment, and let go, patting Michael on the shoulder. He looked at the cask and sniffed deeply. "I can tell it will be fine." He paused, breathing slowly, then nodded once, as though making up his mind to something.

"I've a thing in mind, *a charaid*. I'd been thinking, since ye went to Scotland—and now that we've a kinswoman in the church, so to speak . . . Come back to the office with me, and I'll tell ye."

IT WAS CHILLY in the street, but the goldsmith's back room was cozy as a womb, with a porcelain stove throbbing with heat and woven wool hangings on the walls. Rakoczy hastily unwound the comforter about his neck. It didn't do to sweat indoors; the sweat chilled the instant one went out again, and next thing you knew, it would be *la grippe* at the best, pleurisy or pneumonia at the worst.

Rosenwald himself was comfortable in shirt and waistcoat, without even a wig, only a plum-colored turban to keep his polled scalp warm. The goldsmith's stubby fingers traced the curves of the octofoil salver, turned it over—and stopped dead. Rakoczy felt the tingle of warning at the base of his spine and deliberately relaxed himself, affecting a nonchalant self-confidence.

"Where did you get this, monsieur, if I may ask?" Rosenwald looked up at him, but there was no accusation in the goldsmith's aged face—only a wary excitement.

"It was an inheritance," Rakoczy said, glowing with earnest innocence. "An elderly aunt left it—and a few other pieces—to me. Is it worth anything more than the value of the silver?"

The goldsmith opened his mouth, then shut it, glancing at Rakoczy. Was he honest? Rakoczy wondered with interest. *He's already told me it's something special. Will he tell me why, in hopes of getting other pieces? Or lie, to get this one cheap?* Rosenwald had a good reputation, but he was a Jew.

"Paul de Lamerie," Rosenwald said reverently, his index finger tracing the hallmark. "This was made by Paul de Lamerie."

A shock ran up Rakoczy's backbone. *Merde!* He'd brought the wrong one!

"Really?" he said, striving for simple curiosity. "Does that mean something?"

It means I'm a fool, he thought, and wondered whether to snatch the thing back and leave instantly. The goldsmith had carried it away, though, to look at it more closely under the lamp.

"De Lamerie was one of the very best goldsmiths ever to work in London—perhaps in the world," Rosenwald said, half to himself.

"Indeed," Rakoczy said politely. He was sweating freely. *Nom d'une pipe!* Wait, though—Rosenwald had said "was." De Lamerie was dead, then, thank God. Perhaps the Duke of Sandringham, from whom he'd stolen the salver, was dead, too? He began to breathe more easily.

He never sold anything identifiable within a hundred years of his acquisition of it; that was his principle. He'd taken the other salver from a rich merchant in a game of cards in the Low Countries in 1630; he'd stolen this one in 1745—much too close for comfort. Still . . .

His thoughts were interrupted by the chime of the silver bell over the door, and he turned to see a young man come in, removing his hat to reveal a startling head of dark-red hair. He was dressed *à la mode* and addressed the goldsmith in perfect Parisian French, but he didn't look French. A long-nosed face with faintly slanted eyes. There was a slight sense of familiarity about that face, yet Rakoczy was sure he'd never seen this man before.

"Please, sir, go on with your business," the young man said with a courteous bow. "I meant no interruption."

"No, no," Rakoczy said, stepping forward. He motioned the young man toward the counter. "Please, go ahead. Monsieur Rosenwald and I are merely discussing the value of this object. It will take some thought." He snaked out an arm and seized the salver, feeling a little better with it clasped to his bosom. He wasn't sure; if he decided it was too risky to sell, he could slink out quietly while Rosenwald was busy with the redheaded young man.

The Jew looked surprised but, after a moment's hesitation, nodded and turned to the young man, who introduced himself as one Michael Murray, partner in Fraser *et Cie,* the wine merchants.

"I believe you are acquainted with my cousin Jared Fraser?"

Rosenwald's round face lighted at once. "Oh, to be sure, sir! A man of the most exquisite taste and discrimination. I made him a wine cistern with a motif of sunflowers, not a year past!"

"I know." The young man smiled, a smile that creased his cheeks and narrowed his eyes, and that small bell of recognition rang again. But the name held no familiarity to Rakoczy—only the face, and that only vaguely.

"My uncle has another commission for you, if it's agreeable?"

"I never say no to honest work, monsieur." From the pleasure apparent on the goldsmith's rubicund face, honest work that paid very well was even more welcome.

"Well, then—if I may?" The young man pulled a folded paper from his pocket but half-turned toward Rakoczy, eyebrow cocked in inquiry. Rakoczy motioned him to go on and turned himself to examine a music box that stood on the counter—an enormous thing the size of a cow's head, crowned with a nearly naked nymph festooned with the airiest of gold draperies and dancing on mushrooms and flowers, in company with a large frog.

"A chalice," Murray was saying, the paper laid flat on the counter. From the corner of his eye, Rakoczy could see that it held a list of names. "It's a presentation to the chapel at le Couvent des Anges, to be given in memory of my late father. A young cousin of mine has just entered the convent there as a postulant," he explained. "So Monsieur Fraser thought that the best place."

"An excellent choice." Rosenwald picked up the list. "And you wish all of these names inscribed?"

"Yes, if you can."

"Monsieur!" Rosenwald waved a hand, professionally insulted. "These are your father's children?"

"Yes, these at the bottom." Murray bent over the counter, his finger tracing the lines, speaking the outlandish names carefully. "At the top, these are my parents' names: Ian Alastair Robert MacLeod Murray, and Janet Flora Arabella Fraser Murray. Now, also, I—we, I mean—we want these two names, as well: James Alexander Malcolm MacKenzie Fraser, and Claire Elizabeth Beauchamp Fraser. Those are my uncle and aunt; my uncle was very close to my father," he explained. "Almost a brother."

He went on saying something else, but Rakoczy wasn't listening. He grasped the edge of the counter, vision flickering so that the nymph seemed to leer at him.

Claire Fraser. That had been the woman's name, and her husband, James, a Highland lord from Scotland. That was who the young man resembled, though he was not so imposing as . . . But La Dame Blanche! It was her, it had to be.

And in the next instant, the goldsmith confirmed this, straightening up from the list with an abrupt air of wariness, as though one of the names might spring off the paper and bite him.

"That name—your aunt, she'd be? Did she and your uncle live in Paris at one time?"

"Yes," Murray said, looking mildly surprised. "Maybe thirty years ago—only for a short time, though. Did you know her?"

"Ah. Not to say I was personally acquainted," Rosenwald said, with a crooked smile. "But she was . . . known. People called her La Dame Blanche."

Murray blinked, clearly surprised to hear this.

"Really?" He looked rather appalled.

"Yes, but it was all a long time ago," Rosenwald said hastily, clearly thinking he'd said too much. He waved a hand toward his back room. "If you'll give me a moment, monsieur, I have a chalice actually here, if you would care to see it—and a paten, too; we might make some accommodation of price, if you take both. They were made for a patron who died suddenly, before the chalice was finished, so there is almost no decoration—plenty of room for the names to be applied, and perhaps we might put the, um, aunt and uncle on the paten?"

Murray nodded, interested, and, at Rosenwald's gesture, went round the counter and followed the old man into his back room. Rakoczy put the octofoil salver under his arm and left, as quietly as possible, head buzzing with questions.

JARED EYED MICHAEL over the dinner table, shook his head, and bent to his plate.

"I'm not drunk!" Michael blurted, then bent his own head, face flaming. He could feel Jared's eyes boring into the top of his head.

"Not now, ye're not." Jared's voice wasn't accusing. In fact, it was quiet, almost kindly. "But ye have been. Ye've not touched your dinner, and ye're the color of rotten wax."

"I—" The words caught in his throat, just as the food had. Eels in garlic sauce. The smell wafted up from the dish, and he stood up suddenly, lest he either vomit or burst into tears.

"I've nay appetite, cousin," he managed to say, before turning away. "Excuse me."

He would have left, but he hesitated that moment too long, not wanting to go up to the room where Lillie no longer was but not wanting to look petulant by rushing out into the street. Jared rose and came round to him with a decided step.

"I'm nay verra hungry myself, *a charaid*," Jared said, taking him by the arm. "Come sit wi' me for a bit and take a dram. It'll settle your wame."

He didn't much want to, but there was nothing else he could think of doing, and within a few moments he found himself in front of a fragrant applewood fire, with a glass of his father's whisky in hand, the warmth of both easing the tightness of chest and throat. It wouldn't cure his grief, he knew, but it made it possible to breathe.

"Good stuff," Jared said, sniffing cautiously but approvingly. "Even raw as it is. It'll be wonderful aged a few years."

"Aye. Uncle Jamie kens what he's about; he said he'd made whisky a good many times in America."

Jared chuckled.

"Your uncle Jamie usually kens what he's about," he said. "Not that knowing it keeps him out o' trouble." He shifted, making himself more comfortable in his worn leather chair. "Had it not been for the Rising, he'd likely have stayed here wi' me. Aye, well . . ." The old man sighed with regret and lifted his glass, examining the spirit. It was still nearly as pale as water—it hadn't been casked above a few months—but had the slightly viscous look of a fine strong spirit, as if it might climb out of the glass if you took your eye off it.

"And if he had, I suppose I'd not be here myself," Michael said dryly.

Jared glanced at him, surprised.

"Och! I didna mean to say ye were but a poor substitute for Jamie, lad." He smiled crookedly, and his hooded eyes grew moist. "Not at all. Ye've been the best thing ever to come to me. You and dear wee Lillie,

and . . ." He cleared his throat. "I . . . well, I canna say anything that will help, I ken that. But . . . it won't always be like this."

"Won't it?" Michael said bleakly. "Aye, I'll take your word for it." A silence fell between them, broken only by the hissing and snap of the fire. The mention of Lillie was like an awl digging into his breastbone, and he took a deeper sip of the whisky to quell the ache. Maybe Jared was right to mention the drink to him. It helped, but not enough. And the help didn't last. He was tired of waking to grief and headache both.

Shying away from thoughts of Lillie, his mind fastened on Uncle Jamie instead. He'd lost his wife, too, and from what Michael had seen of the aftermath, it had torn his soul in two. Then she'd come back to him, and he was a man transformed. But in between . . . he'd managed. He'd found a way to be.

Thinking of Auntie Claire gave him a slight feeling of comfort: as long as he didn't think too much about what she'd told the family, who—or what—she was, and where she'd been while she was gone those twenty years. The brothers and sisters had talked among themselves about it afterward; Young Jamie and Kitty didn't believe a word of it, Maggie and Janet weren't sure—but Young Ian believed it, and that counted for a lot with Michael. And she'd looked at him—right at him—when she said what was going to happen in Paris. He felt the same small thrill of horror now, remembering. *The Terror. That's what it will be called, and that's what it will be. People will be arrested for no cause and beheaded in the Place de la Concorde. The streets will run with blood, and no one—no one—will be safe.*

He looked at his cousin; Jared was an old man, though still hale enough. Michael knew there was no way he could persuade Jared to leave Paris and his wine business. But it would be some time yet—if Auntie Claire was right. No need to think about it now. But she'd seemed so sure, like a seer, talking from a vantage point after everything had happened, from a safer time.

And yet she'd come back from that safe time, to be with Uncle Jamie again.

For a moment, he entertained the wild fantasy that Lillie wasn't dead but only swept away into a distant time. He couldn't see or touch her, but the knowledge that she was doing things, was alive . . . maybe it was knowing that, thinking that, that had kept Uncle Jamie whole. He swallowed, hard.

"Jared," he said, clearing his own throat. "What did ye think of Auntie Claire? When she lived here?"

Jared looked surprised but lowered his glass to his knee, pursing his lips in thought.

"She was a bonny lass, I'll tell ye that," he said. "Verra bonny. A tongue

like the rough side of a rasp, if she took against something, though—and decided opinions." He nodded, twice, as though recalling a few, and grinned suddenly. "Verra decided indeed!"

"Aye? The goldsmith—Rosenwald, ye ken?—mentioned her when I went to commission the chalice and he saw her name on the list. He called her La Dame Blanche." This last was not phrased as a question, but he gave it a slight rising inflection, and Jared nodded, his smile widening into a grin.

"Oh, aye, I mind that! 'Twas Jamie's notion. She'd find herself now and then in dangerous places without him—ken how some folk are just the sort as things happen to—so he put it about that she was La Dame Blanche. Ken what a White Lady is, do ye?"

Michael crossed himself, and Jared followed suit, nodding.

"Aye, just so. Make any wicked sod with villainy in mind think twice. A White Lady can strike ye blind or shrivel a man's balls, and likely a few more things than that, should she take the notion. And I'd be the last to say that Claire Fraser couldn't, if she'd a mind to." Jared raised the glass absently to his lips, took a bigger sip of the raw spirit than he'd meant to, and coughed, spraying droplets of memorial whisky halfway across the room.

Rather to his own shock, Michael laughed.

Jared wiped his mouth, still coughing, but then sat up straight and lifted his glass, which still held a few drops.

"To your da. *Slàinte mhath!*"

"*Slàinte!*" Michael echoed, and drained what remained in his own glass. He set it down with finality and rose. He'd drink nay more tonight.

"*Oidhche mhath, mo bràthair-athair no mathar.*"

"Good night, lad," said Jared. The fire was burning low but still cast a warm ruddy glow on the old man's face. "Fare ye well."

Next night

MICHAEL DROPPED HIS key several times before finally managing to turn it in the old-fashioned lock. It wasn't drink; he'd not had a drop since the wine at supper. Instead, he'd walked the length of the city and back, accompanied only by his thoughts; his whole body quivered and he felt mindless with exhaustion, but he was sure he would sleep. Jean-Baptiste had left the door unbarred, according to his orders, but one of the footmen was sprawled on a settle in the entryway, snoring. He smiled a little, though it was an effort to raise the corners of his mouth.

"Bolt the door and go to bed, Alphonse," he whispered, bending and

shaking the man gently by the shoulder. The footman stirred and snorted, but Michael didn't wait to see whether he woke entirely. There was a tiny oil lamp burning on the landing of the stairs, a little round glass globe in the gaudy colors of Murano. It had been there since the first day he came from Scotland to stay with Jared, years before, and the sight of it soothed him and drew his aching body up the wide, dark stair.

The house creaked and talked to itself at night; all old houses did. To-night, though, it was silent, the big copper-seamed roof gone cold and its massive timbers settled into somnolence.

He flung off his clothes and crawled naked into bed, head spinning. Tired as he was, his flesh quivered and twitched, his legs jerking like a spit-ted frog's, before he finally relaxed enough to fall headfirst into the seeth-ing cauldron of dreams that awaited him.

She was there, of course. Laughing at him, playing with her ridiculous pug. Running a hand filled with desire across his face, down his neck, eas-ing her body close, and closer. Then they were somehow in bed, with the wind blowing cool through gauzy curtains, too cool, he felt cold, but then her warmth came close, pressed against him. He felt a terrible desire but at the same time feared her. She felt utterly familiar, utterly strange—and the mixture thrilled him.

He reached for her and realized that he couldn't raise his arms, couldn't move. And yet she was against him, writhing in a slow squirm of need, greedy and tantalizing. In the way of dreams, he was at the same time in front of her, behind her, touching, and seeing from a distance. Candle glow on naked breasts, the shadowed weight of solid buttocks, falling drapes of parting white, one round, firm leg protruding, a pointed toe rooting gently between his legs. Urgency.

She was curled behind him then, kissing the back of his neck, and he reached back, groping, but his hands were heavy, drifting; they slid help-less over her. Hers on him were firm, more than firm—she had him by the cock, was working him. Working him hard, fast and hard. He bucked and heaved, suddenly released from the dream swamp of immobility. She loosed her grip, tried to pull away, but he folded his hand round hers and rubbed their folded hands hard up and down with joyous ferocity, spilling himself convulsively, hot wet spurts against his belly, running thick over their clenched knuckles.

She made a sound of horrified disgust, and his eyes flew open. Staring into them were a pair of huge, bugging eyes, over a gargoyle's mouth full of tiny, sharp teeth. He shrieked.

Plonplon leaped off the bed and ran to and fro, barking hysterically. There was a body behind him. Michael flung himself off the bed, tangled in a winding sheet of damp, sticky bedclothes, then fell and rolled in panic.

"Jesus, Jesus, Jesus!"

On his knees, he gaped, rubbed his hands hard over his face, shook his head. Could *not* make sense of it, couldn't.

"Lillie," he gasped. "Lillie!"

But the woman in his bed, tears running down her face, wasn't Lillie; he realized it with a wrench that made him groan, doubling up in the desolation of fresh loss.

"Oh, Jesus!"

"Michel, Michel, please, please forgive me!"

"You . . . what . . . for God's *sake* . . . !"

Léonie was weeping frantically, reaching out toward him.

"I couldn't help it. I'm so lonely, I wanted you so much!"

Plonplon had ceased barking and now came up behind Michael, nosing his bare backside with a blast of hot, moist breath.

"Va-t'en!"

The pug backed up and started barking again, eyes bulging with offense.

Unable to find any words suitable to the situation, he grabbed the dog and muffled it with a handful of sheet. He got unsteadily to his feet, still holding the squirming pug.

"I—" he began. "You—I mean . . . oh, Jesus Christ!" He leaned over and put the dog carefully on the bed. Plonplon instantly wriggled free of the sheet and rushed to Léonie, licking her solicitously. Michael had thought of giving her the dog after Lillie's death, but for some reason this had seemed a betrayal of the pug's former mistress and brought Michael near to weeping.

"I can't," he said simply. "I just can't. You go to sleep now, lass. We'll talk about it later, aye?"

He went out, walking carefully, as though very drunk, and closed the door gently behind him. He got halfway down the main stair before realizing he was naked. He stood there, his mind blank, watching the colors of the Murano lamp fade as the daylight grew outside, until Paul saw him and ran up to wrap him in a cloak and lead him off to a bed in one of the guest rooms.

RAKOCZY'S FAVORITE gaming club was the Golden Cockerel, and the wall in the main salon was covered by a tapestry featuring one of these creatures, worked in gold thread, wings spread, and throat swollen as it crowed in triumph at the winning hand of cards laid out before it. It was a cheerful place, catering to a mix of wealthy merchants and

lesser nobility, and the air was spicy with the scents of candle wax, powder, perfume, and money.

He'd thought of going to the offices of Fraser *et Cie,* making some excuse to speak to Michael Murray, and maneuvering his way into an inquiry about the whereabouts of the young man's aunt. Upon consideration, though, he thought such a move might make Murray wary—and possibly lead to word getting back to the woman, if she was somewhere in Paris. That was the last thing he wanted to happen.

Better, perhaps, to instigate his inquiries from a more discreet distance. He'd learned that Murray occasionally came to the Cockerel, though he himself had never seen him there. But if he was known . . .

It took several evenings of play, wine, and conversation before he found Charles Pépin. Pépin was a popinjay, a reckless gambler, and a man who liked to talk. And to drink. He was also a good friend of the young wine merchant's.

"Oh, the nun!" he said, when Rakoczy had—after the second bottle—mentioned having heard that Murray had a young relative who had recently entered the convent. Pépin laughed, his handsome face flushed.

"A less likely nun I've never seen—an arse that would make the archbishop of Paris forget his vows, and he's eighty-six if he's a day. Doesn't speak any sort of French, poor thing—the girl, not the archbishop. Not that I for one would be wanting to carry on a lot of conversation if I had her to myself, you understand. . . . She's Scotch; terrible accent . . ."

"Scotch, you say." Rakoczy held a card consideringly, then put it down. "She is Murray's cousin—would she perhaps be the daughter of his uncle James?"

Pépin looked blank for a moment.

"I don't really—oh, yes, I do know!" He laughed heartily, and laid down his own losing hand. "Dear me. Yes, she did say her father's name was Jay-mee, the way the Scotches do; that must be James."

Rakoczy felt a ripple of anticipation go up his spine. *Yes!* This sense of triumph was instantly succeeded by a breathless realization. The girl was the daughter of La Dame Blanche.

"I see," he said casually. "And which convent did you say the girl has gone to?"

To his surprise, Pépin gave him a suddenly sharp look.

"Why do you want to know?"

Rakoczy shrugged, thinking fast.

"A wager," he said, with a grin. "If she is as luscious as you say . . . I'll bet you five hundred *louis* that I can get her into bed before she takes her first vows."

Pépin scoffed.

"Oh, never! She's tasty, but she doesn't know it. And she's virtuous, I'd swear it. And if you think you can seduce her inside the convent . . . !"

Rakoczy lounged back in his chair and motioned for another bottle.

"In that case . . . what do you have to lose?"

Next day

SHE COULD SMELL the *hôpital* long before the small group of new postulants reached the door. They walked two by two, practicing custody of the eyes, but she couldn't help a quick glance upward at the building, a three-story chateau, originally a noble house that had—rumor said—been given to Mother Hildegarde by her father, as part of her dowry when she joined the church. It had become a convent house and then gradually had been given over more and more to the care of the sick, the nuns moving to the new chateau built in the park.

It was a lovely old house—on the outside. The odor of sickness, of urine and shit and vomit, hung about it like a cloying veil, though, and she hoped she wouldn't vomit, too. The little postulant next to her, Sister Miséricorde de Dieu (known to all simply as Mercy), was as white as her veil, eyes fixed on the ground but obviously not seeing it: she stepped smack on a slug and gave a small cry of horror as it squished under her sandal.

Joan looked hastily away; she would never master custody of the eyes, she was sure. Nor yet custody of thought.

It wasn't the notion of sick people that troubled her. She'd seen sick people before, and they wouldn't be expecting her to do more than wash and feed them; she could manage that easily. It was fear of seeing those who were about to die—for surely there would be a great many of those in a hospital. And what might the voices tell her about *them*?

As it was, the voices had nothing to say. Not a word, and after a little she began to lose her nervousness. She *could* do this and in fact, to her surprise, quite enjoyed the sense of competence, the gratification of being able to ease someone's pain, give them at least a little attention—and if her French made them laugh (and it did), that at least took their minds off pain and fear for a moment.

There were those who lay under the veil of death. Only a few, though, and it seemed somehow much less shocking here than when she had seen it on Vhairi's lad or the young man on the ship. Maybe it was resignation, perhaps the influence of the angels for whom the *hôpital* was named . . . Joan didn't know, but she found that she wasn't afraid to speak to or touch the ones she knew were going to die. For that matter, she observed that the other sisters, even the orderlies, behaved gently toward these peo-

ple, and it occurred to her that no particular sight was needed to know that the man with the wasting sickness, whose bones poked through his skin, was not long for this world.

Touch him, said a soft voice inside her head. *Comfort him.*

All right, she said, taking a deep breath. She had no idea how to comfort anyone, but she bathed him, as gently as she could, and coaxed him to take a few spoonsful of porridge. Then she settled him in his bed, straightening his nightshirt and the thin blanket over him.

"Thank you, Sister," he said, and, taking her hand, kissed it. "Thank you for your sweet touch."

She went back to the postulants' dormitory that evening feeling thoughtful, but with a strange sense of being on the verge of discovering something important.

That night

RAKOCZY LAY WITH his head on Madeleine's bosom, eyes closed, breathing the scent of her body, feeling the whole of her between his palms, a slowly pulsing entity of light. She was a gentle gold, traced with veins of incandescent blue, her heart deep as lapis beneath his ear, a living stone. And, deep inside, her red womb, open, soft. Refuge and succor. Promise.

Mélisande had shown him the rudiments of sexual magic, and he'd read about it with great interest in some of the older alchemical texts. He'd never tried it with a whore, though—and, in fact, hadn't been trying to do it this time. And yet it had happened. Was happening. He could see the miracle unfolding slowly before him, under his hands.

How odd, he thought dreamily, watching the tiny traces of green energy spread upward through her womb, slowly but inexorably. He'd thought it happened instantly, that a man's seed found its root in the woman and there you were. But that wasn't what was happening at all. There were *two* types of seed, he now saw. She had one; he felt it plainly, a brilliant speck of light, glowing like a fierce, tiny sun. His own—the tiny green animalcula—were being drawn toward it, bent on immolation.

"Happy, *chéri?*" she whispered, stroking his hair. "Did you have a good time?"

"Most happy, sweetheart." He wished she wouldn't talk, but an unexpected sense of tenderness toward her made him sit up and smile at her. She also began to sit up, reaching for the clean rag and douching syringe, and he put a hand on her shoulder, urging her to lie back down.

"Don't douche this time, *ma belle,*" he said. "A favor to me."

"But—" She was confused; usually he was insistent upon cleanliness.

"Do you *want* me to get with child?" For he had stopped her using the wine-soaked sponge beforehand, too.

"Yes, of course," he said, surprised. "Did Madame Fabienne not tell you?"

Her mouth dropped open.

"She did *not*. What—why, for God's sake?" In agitation, she squirmed free of his restraining hand and swung her legs out of bed, reaching for her wrapper. "You aren't—what do you mean to do with it?"

"Do with it?" he said, blinking. "What do you mean, do with it?"

She had the wrapper on, pulled crookedly round her shoulders, and had backed up against the wall, hands plastered against her stomach, regarding him with open fear.

"You're a *magicien;* everyone knows that. You take newborn children and use their blood in your spells!"

"What?" he said, rather stupidly. He reached for his breeches but changed his mind. He got up and went to her instead, putting his hands on her shoulders.

"No," he said, bending down to look her in the eye. "No, I do no such thing. Never." He used all the force of sincerity he could summon, pushing it into her, and felt her waver a little, still fearful but less certain. He smiled at her.

"Who told you I was a *magicien,* for heaven's sake? I am a *philosophe, chérie*—an inquirer into the mysteries of nature, no more. And I can swear to you, by my hope of heaven"—this being more or less nonexistent, but why quibble?—"that I have never, not once, used anything more than the water of a man-child in any of my investigations."

"What, little boys' piss?" she said, diverted. He let his hands relax but kept them on her shoulders.

"Certainly. It's the purest water one can find. Collecting it is something of a chore, mind you"—she smiled at that; good—"but the process does not the slightest harm to the infant, who will eject the water whether anyone has a use for it or not."

"Oh." She was beginning to relax a little, but her hands were still pressed protectively over her belly, as though she felt the imminent child already. *Not yet,* he thought, pulling her against him and feeling his way gently into her body. *But soon!* He wondered if he should remain with her until it happened; the idea of feeling it as it happened inside her—to be an intimate witness to the creation of life itself! But there was no telling how long it might take. From the progress of his animalcula, it could be a day, even two.

Magic, indeed.

Why do men never think of that? he wondered. Most men—himself

included—regarded the engendering of babies as necessity, in the case of inheritance, or nuisance, but *this* . . . But then, most men would never know what he now knew or see what he had seen.

Madeleine had begun to relax against him, her hands at last leaving her belly. He kissed her, with a real feeling of affection.

"It will be beautiful," he whispered to her. "And once you are well and truly with child, I will buy your contract from Fabienne and take you away. I will buy you a house."

"A *house?*" Her eyes went round. They were green, a deep, clear emerald, and he smiled at her again, stepping back.

"Of course. Now, go and sleep, my dear. I shall come again tomorrow."

She flung her arms around him, and he had some difficulty in extracting himself, laughing, from her embraces. Normally he left a whore's bed with no feeling save physical relief. But what he had done had made a connection with Madeleine that he had not experienced with any woman save Mélisande.

Mélisande. A sudden thought ran through him like the spark from a Leyden jar. *Mélisande.*

He looked hard at Madeleine, now crawling happily naked and white-rumped into bed, her wrapper thrown aside. That bottom . . . the eyes, the soft blond hair, the gold-white of fresh cream.

"Chérie," he said, as casually as he might, pulling on his breeches, "how old are you?"

"Eighteen," she said, without hesitation. "Why, monsieur?"

"Ah. A wonderful age to become a mother." He pulled the shirt over his head and kissed his hand to her, relieved. He had known Mélisande Robicheaux in 1744. He had not, in fact, just committed incest with his own daughter.

It was only as he passed Madame Fabienne's parlor on his way out that it occurred to him that Madeleine *might* possibly still be his granddaughter. That thought stopped him short, but he had no time to dwell on it, for Fabienne appeared in the doorway and motioned to him.

"A message, monsieur," she said, and something in her voice touched his nape with a cold finger.

"Yes?"

"Maître Grenouille begs the favor of your company at midnight tomorrow. In the square before Notre Dame de Paris."

THEY DIDN'T HAVE to practice custody of the eyes in the market. In fact, Sister George—the stout nun who oversaw these expeditions,

warned them in no uncertain terms to keep a sharp eye out for short weight and uncivil prices, to say nothing of pickpockets.

"Pickpockets, Sister?" Mercy had said, her blond eyebrows all but vanishing into her veil. "But we are nuns—more or less," she added hastily. "We have nothing to steal!"

Sister George's big red face got somewhat redder, but she kept her patience.

"Normally that would be true," she agreed. "But we—or I, rather—have the money with which to buy our food, and once we've bought it, you will be carrying it. A pickpocket steals to eat, *n'est-ce pas?* They don't care whether you have money or food, and most of them are so depraved that they would willingly steal from God himself, let alone a couple of chick-headed postulants."

For Joan's part, she wanted to see *everything*, pickpockets included. To her delight, the market was the one she'd passed with Michael on her first day in Paris. True, the sight of it brought back the horrors and doubts of that first day, too—but, for the moment, she pushed those aside and followed Sister George into the fascinating maelstrom of color, smells, and shouting.

Filing away a particularly entertaining expression that she planned to make Sister Philomène explain to her—Sister Philomène was a little older than Joan, but painfully shy and with such delicate skin that she blushed like an apple at the least excuse—she followed Sister George and Sister Mathilde through the fishmonger's section, where Sister George bargained shrewdly for a great quantity of sand dabs, scallops, tiny gray translucent shrimp, and an enormous sea salmon, the pale spring light shifting through its scales in colors that faded so subtly from pink to blue to silver and back that some of them had no name at all—so beautiful even in its death that it made Joan catch her breath with joy at the wonder of creation.

"Oh, *bouillabaisse* tonight!" said Mercy, under her breath. *"Délicieuse!"*

"What is *bouillabaisse?*" Joan whispered back.

"Fish stew—you'll like it, I promise!" Joan had no doubt of it; brought up in the Highlands during the poverty-stricken years following the Rising, she'd been staggered by the novelty, deliciousness, and sheer abundance of the convent's food. Even on Fridays, when the community fasted during the day, supper was simple but mouthwatering, toasted sharp cheese on nutty brown bread with sliced apples.

Luckily, the salmon was so huge that Sister George arranged for the fish seller to deliver it to the convent, along with the other briny purchases; thus they had room in their baskets for fresh vegetables and fruit and so passed from Neptune's realm to that of Demeter. Joan hoped it wasn't sacrilegious to think of Greek gods, but she couldn't forget the book of

myths that Da had read to Marsali and her when they were young, with wonderful hand-colored illustrations.

After all, she told herself, you needed to know about the Greeks if you studied medicine. She had some trepidation at the thought of working in the hospital, but God called people to do things, and if it was his will, then—

The thought stopped short as she caught sight of a neat dark tricorne with a curled blue feather bobbing slowly through the tide of people. Was it—it was! Léonie, the sister of Michael Murray's dead wife. Moved by curiosity, Joan glanced at Sister George, who was engrossed in a huge display of fungus—dear God, people *ate* such things?—and slipped around a barrow billowing with green sallet herbs.

She meant to speak to Léonie, ask her to tell Michael that she needed to talk to him. Perhaps he could contrive a way to visit the convent . . . But before Joan could get close enough, Léonie looked furtively over her shoulder, as though fearing discovery, then ducked behind a curtain that hung across the back of a small caravan.

Joan had seen gypsies before, though not often. A dark-skinned man loitered nearby, talking with a group of others; their eyes passed over her habit without pausing, and she sighed with relief. Being a nun was as good as having a cloak of invisibility in most circumstances, she thought.

She looked round for her companions and saw that Sister Mathilde had been called into consultation regarding a big warty lump of something that looked like the excrement of a seriously diseased hog. Good, she could wait for a minute longer.

In fact, it took very little more than that before Léonie slipped out from behind the curtain, tucking something into the small basket on her arm. For the first time, it struck Joan as unusual that someone like Léonie should be shopping without a servant to push back crowds and carry purchases—or even be in a public market. Michael had told her about his own household during the voyage—how Madame Hortense, the cook, went to the markets at dawn to be sure of getting the freshest things. What would a lady like Léonie be buying, alone?

Joan slithered as best she could through the rows of stalls and wagons, following the bobbing blue feather. A sudden stop allowed her to come up behind Léonie, who had paused by a flower stall, fingering a bunch of white jonquils.

It occurred suddenly to Joan that she had no idea what Léonie's last name was, but she couldn't worry about politeness now.

"Ah . . . madame?" she said tentatively. "Mademoiselle, I mean?" Léonie swung round, eyes huge and face pale. Finding herself faced with a nun, she blinked, confused.

"Er . . . it's me," Joan said, diffident, resisting the impulse to pull off

her veil. "Joan MacKimmie?" It felt odd to say it, as though "Joan MacKimmie" were truly someone else. It took a moment for the name to register, but then Léonie's shoulders relaxed a little.

"Oh." She put a hand to her bosom and mustered a small smile. "Michael's cousin. Of course. I didn't . . . er . . . How nice to see you!" A small frown wrinkled the skin between her brows. "Are you . . . alone?"

"No," Joan said hurriedly. "And I mustn't stop. I only saw you, and I wanted to ask—" It seemed even stupider than it had a moment ago, but no help for it. "Would you tell Monsieur Murray that I must talk to him? I know something—something important—that I have to tell him."

"Soeur Gregory?" Sister George's stentorian tones boomed through the higher-pitched racket of the market, making Joan jump. She could see the top of Sister Mathilde's head, with its great white sails, turning to and fro in vain search.

"I have to go," she said to the astonished Léonie. "Please. Please tell him!" Her heart was pounding, and not only from the sudden meeting. She'd been looking at Léonie's basket, where she caught the glint of a brown glass bottle half hidden beneath a thick bunch of what even Joan recognized as black hellebores. Lovely cup-shaped flowers of an eerie greenish-white—and deadly poison.

She dodged back across the market to arrive breathless and apologizing at Sister Mathilde's side, wondering if . . . She hadn't spent much time at all with Da's wife—but she *had* heard her talking with Da as she wrote down receipts in a book, and she'd mentioned black hellebore as something women used to make themselves miscarry. If Léonie were pregnant . . . Holy Mother of God, could she be with child by *Michael*? The thought struck her like a blow in the stomach.

No. No, she couldn't believe it. He was still in love with his wife, anyone could see that, and even if not, she'd swear he wasn't the sort to . . . But what did she ken about men, after all?

Well, she'd ask him when she saw him, she decided, her mouth clamping tight. And 'til then . . . Her hand went to the rosary at her waist and she said a quick, silent prayer for Léonie. Just in case.

As she was bargaining doggedly in her execrable French for six aubergines (wondering meanwhile what on earth they were for, medicine or food?), she became aware of someone standing at her elbow. A handsome man of middle age, taller than she was, in a well-cut dove-gray coat. He smiled at her and, touching one of the peculiar vegetables, said in slow, simple French, "You don't want the big ones. They're tough. Get small ones, like that." A long finger tapped an aubergine half the size of the ones the vegetable seller had been urging on her, and the vegetable seller burst into a tirade of abuse that made Joan step back, blinking.

Not so much because of the expressions being hurled at her—she didn't

understand one word in ten—but because a voice in plain English had just said clearly, *"Tell him not to do it."*

She felt hot and cold at the same time.

"I . . . er . . . *je suis* . . . um . . . *merci beaucoup, monsieur!*" she blurted, and, turning, ran, scrambling back between piles of paper narcissus bulbs and fragrant spikes of hyacinth, her shoes skidding on the slime of trodden leaves.

"Soeur Gregory!" Sister Mathilde loomed up so suddenly in front of her that she nearly ran into the massive nun. "What are you doing? Where is Sister Miséricorde?"

"I . . . oh." Joan swallowed, gathering her wits. "She's—over there." She spoke with relief, spotting Mercy's small head in the forefront of a crowd by the meat-pie wagon. "I'll get her!" she blurted, and walked hastily off before Sister Mathilde could say more.

"Tell him not to do it." That's what the voice had said about Charles Pépin. What was going on? she thought wildly. Was M. Pépin engaged in something awful with the man in the dove-gray coat?

As though thought of the man had reminded the voice, it came again.

"Tell him not to do it," the voice repeated in her head, with what seemed like particular urgency. *"Tell him he must not!"*

"Hail Mary, full of grace, the Lord is with thee, blessed art thou among women . . ." Joan clutched at her rosary and gabbled the words, feeling the blood leave her face. There he was, the man in the dove-gray coat, looking curiously at her over a stall of Dutch tulips and sprays of yellow forsythia.

She couldn't feel the pavement under her feet but was moving toward him. *I have to,* she thought. *It doesn't matter if he thinks I'm mad. . . .*

"Don't do it," she blurted, coming face-to-face with the astonished gentleman. "You mustn't do it!"

And then she turned and ran, rosary in hand, apron and veil flapping like wings.

HE COULDN'T HELP thinking of the cathedral as an entity. An immense version of one of its own gargoyles, crouched over the city. In protection or threat?

Notre Dame de Paris rose black above him, solid, obliterating the light of the stars, the beauty of the night. Very appropriate. He'd always thought that the church blocked one's sight of God. Nonetheless, the sight of the monstrous stone creature made him shiver as he passed under its shadow, despite the warm cloak.

Perhaps it was the cathedral's stones themselves that gave him the sense

of menace? He stopped, paused for a heartbeat, and then strode up to the church's wall and pressed his palm flat against the cold limestone. There was no immediate sense of anything, just the cold roughness of the rock. Impulsively, he shut his eyes and tried to feel his way into the rock. At first, nothing. But he waited, pressing with his mind, a repeated question. *Are you there?*

He would have been terrified to receive an answer but was obscurely disappointed not to. Even so, when he finally opened his eyes and took his hands away, he saw a trace of blue light, the barest trace, glowing briefly between his knuckles. That frightened him, and he hurried away, hiding his hands beneath the shelter of the cloak.

Surely not, he assured himself. He'd done that before, made the light happen when he held the jewels he used for travel and said the words over them—his own version of consecration, he supposed. He didn't know if the words were necessary, but Mélisande had used them; he was afraid not to. And yet. He had felt *something* here. The sense of something heavy, inert. Nothing resembling thought, let alone speech, thank God. By reflex, he crossed himself, then shook his head, rattled and irritated.

But something. Something immense and very old. Did God have the voice of a stone? He was further unsettled by the thought. The stones there in the chalk mine, the noise they made—was it after all God that he'd glimpsed, there in that space between?

A movement in the shadows banished all such thoughts in an instant. The frog! Rakoczy's heart clenched like a fist.

"Monsieur le Comte," said an amused, gravelly voice. "I see the years have been kind to you."

Raymond stepped into the starlight, smiling. The sight of him was disconcerting; Rakoczy had imagined this meeting for so long that the reality seemed oddly anticlimactic. Short, broad-shouldered, with long, loose hair that swept back from a massive forehead. A broad, almost lipless mouth. Raymond the frog.

"Why are you here?" Rakoczy blurted.

Maître Raymond's brows were black—surely they had been white thirty years ago? One of them lifted in puzzlement.

"I was told that you were looking for me, monsieur." He spread his hands, the gesture graceful. "I came!"

"Thank you," Rakoczy said dryly, beginning to regain some composure. "I meant—why are you in Paris?"

"Everyone has to be somewhere, don't they? They can't be in the same place." This should have sounded like badinage but didn't. It sounded serious, like a statement of scientific principle, and Rakoczy found it unsettling.

"Did you come looking for me?" he asked boldly. He moved a little,

trying to get a better view of the man. He was nearly sure that the frog appeared *younger* than he had when last seen. Surely his flowing hair was darker, his step more elastic? A spurt of excitement bubbled in his chest.

"For you?" The frog seemed amused for a moment, but then the look faded. "No. I'm searching for a lost daughter."

Rakoczy was surprised and disconcerted.

"Yours?"

"More or less." Raymond seemed uninterested in explaining further. He moved a little to one side, eyes narrowing as he sought to make out Rakoczy's face in the darkness. "You can hear stones, then, can you?"

"I—what?"

Raymond nodded at the façade of the cathedral. "They do speak. They move, too, but very slowly."

An icy chill shot up Rakoczy's spine at the thought of the grinning gargoyles perched high above him and the implication that one might at any moment choose to spread its silent wings and hurtle down upon him, teeth still bared in carnivorous hilarity. Despite himself, he looked up, over his shoulder.

"Not that fast." The note of amusement was back in the frog's voice. "You would never see them. It takes them millennia to move the slightest fraction of an inch—unless of course they are propelled or melted. But you don't want to see them do that, of course. Much too dangerous."

This kind of talk struck him as frivolous, and Rakoczy was bothered by it but for some reason not irritated. Troubled, with a sense that there was something under it, something that he simultaneously wanted to know—and wanted very much to avoid knowing. The sensation was novel, and unpleasant.

He cast caution to the wind and demanded boldly, "Why did you not kill me?"

Raymond grinned at him; Rakoczy could see the flash of teeth and felt yet another shock: he was sure—almost sure—that the frog had *had* no teeth when last seen.

"If I had wanted you dead, son, you wouldn't be here talking to me," he said. "I wanted you to be out of the way, that's all; you obliged me by taking the hint."

"And just why did you want me 'out of the way'?" Had he not needed to find out, Rakcozy would have taken offense at the man's tone.

The frog lifted one shoulder.

"You were something of a threat to the lady."

Sheer astonishment brought Rakoczy to his full height.

"The lady? You mean the woman—La Dame Blanche?"

"They did call her that." The frog seemed to find the notion amusing.

It was on the tip of Rakoczy's tongue to tell Raymond that La Dame

Blanche still lived, but he hadn't lived as long as he had by blurting out everything he knew—and he didn't want Raymond thinking that he himself might be still a threat to her.

"What is the ultimate goal of an alchemist?" the frog said very seriously.

"To transform matter," Rakoczy replied automatically.

The frog's face split in a broad amphibian grin.

"Exactly!" he said. And vanished.

He *had* vanished. No puffs of smoke, no illusionist's tricks, no smell of sulfur—the frog was simply gone. The square stretched empty under the starlit sky; the only thing that moved was a cat that darted mewing out of the shadows and brushed past Rakoczy's leg.

WORN OUT WITH constant walking, Michael slept like the dead these days, without dreams or motion, and woke when the sun came up. His valet, Robert, heard him stir and came in at once, one of the *femmes de chambre* on his heels with a bowl of coffee and some pastry.

He ate slowly, suffering himself to be brushed, shaved, and tenderly tidied into fresh linen. Robert kept up a soothing murmur of the sort of conversation that doesn't require response and smiled encouragingly when presenting the mirror. Rather to Michael's surprise, the image in the mirror looked quite normal. Hair neatly clubbed—he wore his own, without powder—suit modest in cut but of the highest quality. Robert hadn't asked him what he required but had dressed him for an ordinary day of business. He supposed that was all right. What, after all, did clothes matter? It wasn't as though there was a costume *de rigueur* for calling upon the sister of one's deceased wife, who had come uninvited into one's bed in the middle of the night.

He had spent the last two days trying to think of some way never to see or speak to Léonie again, but, really, there was no help for it. He'd have to see her.

But what was he to say to her, he wondered, as he made his way through the streets toward the house where Léonie lived with an aged aunt, Eugenie Galantine. He wished he could talk the situation over with Sister Joan, but that wouldn't be appropriate, even were she available.

He'd hoped that walking would give him time to come up at least with a *point d'appui*, if not an entire statement of principle, but instead he found himself obsessively counting the flagstones of the market as he crossed it, counting the bongs of the public horologe as it struck the hour of three, and—for lack of anything else—counting his own footsteps as he approached her door. *Six hundred and thirty-seven, six hundred and thirty-eight . . .*

As he turned into the street, though, he abruptly stopped counting. He stopped walking, too, for an instant—then began to run. Something was wrong at the house of Madame Galantine.

He pushed his way through the crowd of neighbors and vendors clustered near the steps and seized the butler, whom he knew, by a sleeve.

"What?" he barked. "What's happened?" The butler, a tall, cadaverous man named Hubert, was plainly agitated but settled a bit on seeing Michael.

"I don't know, sir," he said, though a sideways slide of his eyes made it clear that he did. "Mademoiselle Léonie . . . she's ill. The doctor . . ."

He could smell the blood. Not waiting for more, he pushed Hubert aside and sprinted up the stairs, calling for Madame Eugenie, Léonie's aunt.

Madame Eugenie popped out of a bedroom, her cap and wrapper neat in spite of the uproar.

"Monsieur Michel!" she said, blocking him from entering the room. "It's all right, but you must not go in."

"Yes, I must." His heart was thundering in his ears, and his hands felt cold.

"You may *not*," she said firmly. "She's ill. It isn't proper."

"Proper? A young woman tries to make away with herself and you tell me it isn't *proper*?"

A maid appeared in the doorway, a basket piled with bloodstained linen in her arms, but the look of shock on Madame Eugenie's broad face was more striking.

"Make away with herself?" The old lady's mouth hung open for a moment, then snapped shut like a turtle's. "Why would you think such a thing?" She was regarding him with considerable suspicion. "And what are you doing here, for that matter? Who told you she was ill?"

A glimpse of a man in a dark robe, who must be the doctor, decided Michael that little was to be gained by engaging further with Madame Eugenie. He took her gently but firmly by the elbows, picked her up—she uttered a small shriek of surprise—and set her aside.

He went in and shut the bedroom door behind him.

"Who are you?" The doctor looked up, surprised. He was wiping out a freshly used bleeding-bowl, and his case lay open on the boudoir's settee. Léonie's bedroom must lie beyond; the door was open, and Michael caught a glimpse of the foot of a bed but could not see the bed's inhabitant.

"It doesn't matter. How is she?"

The doctor eyed him narrowly, but after a moment nodded.

"She will live. As for the child . . ." He made an equivocal motion of the hand. "I've done my best. She took a great deal of the—"

"The *child?*" The floor shifted under his feet, and the dream of the night before flooded him, that queer sense of something half wrong, half familiar. It was the feeling of a small, hard swelling pressed against his bum; that's what it was. Lillie had not been far gone with child when she died, but he remembered all too well the feeling of a woman's body in early pregnancy.

"It's yours? I beg your pardon, I shouldn't ask." The doctor put away his bowl and fleam and shook out his black velvet turban.

"I want—I need to talk to her. Now."

The doctor opened his mouth in automatic protest but then glanced thoughtfully over his shoulder.

"Well . . . you must be careful not to—" But Michael was already inside the bedroom, standing by the bed.

She was pale. They had always been pale, Lillie and Léonie, with the soft glow of cream and marble. This was the paleness of a frog's belly, of a rotting fish, blanched on the shore.

Her eyes were ringed with black, sunk in her head. They rested on his face, flat, expressionless, as still as the ringless hands that lay limp on the coverlet.

"Who?" he said quietly. "Charles?"

"Yes." Her voice was as dull as her eyes, and he wondered whether the doctor had drugged her.

"Was it his idea—to try to foist the child off on me? Or yours?"

She did look away then, and her throat moved.

"His." The eyes came back to him. "I didn't want to, Michel. Not— not that I find you disgusting, not that . . ."

"Merci," he muttered, but she went on, disregarding him.

"You were Lillie's husband. I didn't envy her you," she said frankly, "but I envied what you had together. It couldn't be like that between you and me, and I didn't like betraying her. But"—her lips, already pale, compressed to invisibility—"I didn't have much choice."

He was obliged to admit that she hadn't. Charles couldn't marry her; he had a wife. Bearing an illegitimate child was not a fatal scandal in high court circles, but the Galantines were of the emerging bourgeoisie, where respectability counted for almost as much as money. Finding herself pregnant, she would have had two alternatives: find a complaisant husband quickly, or . . . He tried not to see that one of her hands rested lightly across the slight swell of her stomach.

The child . . . He wondered what he would have done had she come to him and told him the truth, asked him to marry her for the sake of the child. But she hadn't. And she wasn't asking now.

It would be best—or at least easiest—were she to lose the child. And she might yet.

"I couldn't wait, you see," she said, as though continuing a conversation. "I would have tried to find someone else, but I thought she knew. She'd tell you as soon as she could manage to see you. So I had to, you see, before you found out."

"She? Who? Tell me what?"

"The nun," Léonie said, and sighed deeply, as though losing interest. "She saw me in the market and rushed up to me. She said she had to talk to you—that she had something important to tell you. I saw her look into my basket, though, and her face . . . thought she must realize . . ."

Her eyelids were fluttering, whether from drugs or fatigue, he couldn't tell. She smiled faintly, but not at him; she seemed to be looking at something a long way off.

"So funny," she murmured. "Charles said it would solve everything— that the comte would pay him such a lot for her, it would solve everything. But how can you solve a baby?"

Michael jerked as though her words had stabbed him.

"What? Pay for whom?"

"The nun."

He grabbed her by the shoulders.

"Sister Joan? What do you mean, pay for her? What did Charles tell you?"

She made a whiny sound of protest. Michael wanted to shake her hard enough to break her neck but forced himself to withdraw his hand. She settled into the pillow like a bladder losing air, flattening under the bedclothes. Her eyes were closed, but he bent down, speaking directly into her ear.

"The comte, Léonie. What is his name? Tell me his name."

A faint frown rippled the flesh of her brow, then passed.

"St. Germain," she murmured, scarcely loud enough to be heard. "The Comte St. Germain."

HE WENT INSTANTLY to Rosenwald and, by dint of badgering and the promise of extra payment, got him to finish the engraving on the chalice at once. Michael waited impatiently while it was done and, scarcely pausing for the cup and paten to be wrapped in brown paper, flung money to the goldsmith and made for les Couvent des Anges, almost running.

With great difficulty, he restrained himself while making the presentation of the chalice, and with great humility, he inquired whether he might ask the great favor of seeing Sister Gregory, that he might convey a message to her from her family in the Highlands. Sister Eustacia looked surprised and somewhat disapproving—postulants were not normally per-

mitted visits—but after all . . . in view of Monsieur Murray's and Monsieur Fraser's great generosity to the convent . . . perhaps just a few moments, in the visitor's parlor, and in the presence of Sister herself . . .

HE TURNED AND blinked once, his mouth opening a little. He looked shocked. Did she look so different in her robe and veil?

"It's me," Joan said, and tried to smile reassuringly. "I mean . . . still me."

His eyes fixed on her face, and he let out a deep breath and smiled, as if she'd been lost and he'd found her again.

"Aye, so it is," he said softly. "I was afraid it was Sister Gregory. I mean, the . . . er . . . ." He made a sketchy, awkward gesture indicating her gray robes and white postulant's veil.

"It's only clothes," she said, and put a hand to her chest, defensive.

"Well, no," he said, looking her over carefully, "I dinna think it is, quite. It's more like a soldier's uniform, no? Ye're doing your job when ye wear it, and everybody as sees it kens what ye are and knows what ye do."

Kens what I am. I suppose I should be pleased it doesn't *show,* she thought, a little wildly.

"Well . . . aye, I suppose." She fingered the rosary at her belt. She coughed. "In a way, at least."

Ye've got to tell him. It wasn't one of the voices, just the voice of her own conscience, but that was demanding enough. She could feel her heart beating, so hard that she thought the bumping must show through the front of her habit.

He smiled encouragingly at her.

"Léonie told me ye wanted to see me."

"Michael . . . can I tell ye something?" she blurted.

He seemed surprised. "Well, of course ye can," he said. "Whyever not?"

"Whyever not," she said, half under her breath. She glanced over his shoulder, but Sister Eustacia was on the far side of the room, talking to a very young, frightened-looking French girl and her parents.

"Well, it's like this, see," she said, in a determined voice. "I hear voices." She stole a look at him, but he didn't appear shocked. Not yet.

"In my head, I mean."

"Aye?" He sounded cautious. "Um . . . what do they say, then?"

She realized she was holding her breath, and let a little of it out.

"Ah . . . different things. But they now and then tell me something's going to happen. More often, they tell me I should say thus-and-so to someone."

"Thus-and-so," he repeated attentively, watching her face. "What . . . *sort* of thus-and-so?"

"I wasna expecting the Spanish Inquisition," she said, a little testily. "Does it matter?"

His mouth twitched.

"Well, I dinna ken, now, do I?" he pointed out. "It might give a clue as to who's talkin' to ye, might it not? Or do ye already know that?"

"No, I don't," she admitted, and felt a sudden lessening of tension. "I—I was worrit—a bit—that it might be demons. But it doesna really . . . well, they dinna tell me *wicked* sorts of things. Just . . . more like when something's going to happen to a person. And sometimes it's no a good thing—but sometimes it is. There was wee Annie MacLaren, her wi' a big belly by the third month, and by six lookin' as though she'd burst, and she was frightened she was goin' to die come her time, like her ain mother did, wi' a babe too big to be born—I mean, *really* frightened, not just like all women are. And I met her by St. Ninian's Spring one day, and one of the voices said to me, '*Tell her it will be as God wills and she will be delivered safely of a son.*'"

"And ye did tell her that?"

"Yes. I didna say how I knew, but I must have sounded like I *did* know, because her poor face got bright all of a sudden, and she grabbed on to my hands and said, 'Oh! From your lips to God's ear!'"

"And was she safely delivered of a son?"

"Aye—and a daughter, too." Joan smiled, remembering the glow on Annie's face.

Michael glanced aside at Sister Eustacia, who was bidding farewell to the new postulant's family. The girl was white-faced and tears ran down her cheeks, but she clung to Sister Eustacia's sleeve as though it were a lifeline.

"I see," he said slowly, and looked back at Joan. "Is that why—is it the voices told ye to be a nun, then?"

She blinked, surprised by his apparent acceptance of what she'd told him but more so by the question.

"Well . . . no. They never did. Ye'd think they would have, wouldn't ye?"

He smiled a little.

"Maybe so." He coughed, then looked up, a little shyly. "It's no my business, but what *did* make ye want to be a nun?"

She hesitated, but why not? She'd already told him the hardest bit.

"Because of the voices. I thought maybe—maybe I wouldna hear them in here. Or . . . if I still did, maybe somebody—a priest, maybe?—could tell me what they were and what I should do about them."

Sister Eustacia was comforting the new girl, half-sunk on one knee to bring her big, homely, sweet face close to the girl's. Michael glanced at them, then back at Joan, one eyebrow raised.

"I'm guessing ye havena told anyone yet," he said. "Did ye reckon ye'd practice on me first?"

Her own mouth twitched.

"Maybe." His eyes were dark but had a sort of warmth to them, as if they drew it from the heat of his hair. She looked down; her hands were pleating the edge of her blouse, which had come untucked. "It's no just that, though."

He made the sort of noise in his throat that meant, *"Aye, then, go on."* Why didn't French people do like that? she wondered. So much easier. But she pushed the thought aside; she'd made up her mind to tell him, and now was the time to do it.

"I told ye because—that man," she blurted. "The Comte." He squinted at her. "The Comte St. Germain?"

"Well, I dinna ken his name, now, do I?" she snapped. "But when I saw him, one of the voices pops up and says to me, *'Tell him not to do it. Tell him he must not.'* "

"It did?"

"Aye, and it was verra firm about it. I mean—they are, usually. It's no just an opinion, take it or leave it. But this one truly meant it." She spread her hands, helpless to explain the feeling of dread and urgency. She swallowed.

"And then . . . your friend. Monsieur Pépin. The first time I saw him, one o' the voices said *'Tell him not to do it.'* "

Michael's thick red eyebrows drew together.

"D'ye think it's the same thing they're not supposed to do?" He sounded startled. "Well, I don't know, now, do I?" she said, a little exasperated. "The voices didn't say. But I saw that the man on the ship was going to die, and I didna say anything, because I couldn't think what to say. And then he *did* die, and maybe he wouldn't have if I'd spoken . . . so I—well, I thought I'd best say something to *someone.*"

He thought about that for a moment, then nodded uncertainly.

"Aye. All right. I'll—well, I dinna ken what to do about it, either, to be honest. But I'll talk to them both and I'll have that in my mind, so maybe I'll think of something. D'ye want me to tell them, *'Don't do it'?*"

She grimaced and looked at Sister Eustacia. There wasn't much time.

"I already told the comte. Just . . . maybe. If ye think it might help. Now—" Her hand darted under her apron and she passed him the slip of paper, fast. "We're only allowed to write to our families twice a year," she said, lowering her voice. "But I wanted Mam to know I was all right.

Could ye see she gets that, please? And . . . and maybe tell her a bit, your-self, that I'm weel and—and happy. Tell her I'm happy," she repeated, more firmly.

Sister Eustacia was now standing by the door, emanating an intent to come and tell them it was time for Michael to leave.

"I will," he said. He couldn't touch her, he knew that, so bowed in-stead and bowed deeply to Sister Eustacia, who came toward them, look-ing benevolent.

"I'll come to Mass at the chapel on Sundays, how's that?" he said rap-idly. "If I've a letter from your mam, or ye have to speak to me, gie me a wee roll of the eyes or something—I'll figure something out."

TWENTY-FOUR HOURS LATER, Sister Gregory, postulant in the Convent of Angels, regarded the bum of a large cow. The cow in ques-tion was named Mirabeau and was of uncertain temper, as evidenced by the nervously lashing tail.

"She's kicked three of us this week," said Sister Anne-Joseph, eyeing the cow resentfully. "*And* spilt the milk twice. Sister Jeanne-Marie was most upset."

"Well, we canna have that, now, can we?" Joan murmured in English. "*N'inquiétez-vous pas,*" she added in French, hoping that was at least somewhat grammatical. "Let me do it."

"Better you than me," Sister Anne-Joseph said, crossing herself, and vanished before Sister Joan might think better of the offer.

A week spent working in the cowshed was intended as punishment for her flighty behavior in the marketplace, but Joan was grateful for it. There was nothing better for steadying the nerves than cows.

Granted, the convent's cows were not quite like her mother's sweet-tempered, shaggy red Hieland coos, but if you came right down to it, a cow was a cow, and even a French-speaking wee besom like the present Mirabeau was no match for Joan MacKimmie, who'd driven kine to and from the shielings for years and fed her mother's kine in the byre beside the house with sweet hay and the leavings from supper.

With that in mind, she circled Mirabeau thoughtfully, eyeing the steadily champing jaws and the long slick of blackish-green drool that hung down from slack pink lips. She nodded once, slipped out of the cow-shed, and made her way down the allée behind it, picking what she could find. Mirabeau, presented with a bouquet of fresh grasses, tiny daisies, and—delicacy of all delicacies—fresh sorrel, bulged her eyes half out of her head, opened her massive jaw, and inhaled the sweet stuff. The ominous

tail ceased its lashing and the massive creature stood as if turned to stone, aside from the ecstatically grinding jaws.

Joan sighed in satisfaction, sat down, and, resting her head on Mirabeau's monstrous flank, got down to business. Her mind, released, took up the next worry of the day.

Had Michael spoken to his friend Pépin? And if so, had he told him what she'd said, or just asked whether he kent the Comte St. Germain? Because if *"tell him not to do it"* referred to the same thing, then plainly the two men must be acquent with each other.

She had got thus far in her own ruminations when Mirabeau's tail began to switch again. She hurriedly stripped the last of the milk from Mirabeau's teats and snatched the bucket out of the way, standing up in a hurry. Then she saw what had disturbed the cow.

The man in the dove-gray coat was standing in the door to the shed, watching her. She hadn't noticed before, in the market, but he had a handsome dark face, though rather hard about the eyes, and with a chin that brooked no opposition. He smiled pleasantly at her, though, and bowed.

"Mademoiselle. I must ask you, please, to come with me."

MICHAEL WAS IN the warehouse, stripped to his shirtsleeves and sweating in the hot, wine-heady atmosphere, when Jared appeared, looking disturbed.

"What is it, cousin?" Michael wiped his face on a towel, leaving black streaks; the crew was clearing the racks on the southeast wall, and there were years of filth and cobwebs behind the most ancient casks.

"Ye haven't got that wee nun in your bed, have ye, Michael?" Jared lifted a beetling gray brow at him.

"Have I what?"

"I've just had a message from the Mother Superior of le Couvent des Anges, saying that one Sister Gregory appears to have been abducted from their cowshed, and wanting to know whether you might possibly have anything to do with the matter."

Michael stared at his cousin for a moment, unable to take this in.

"Abducted?" he said stupidly. "Who would be kidnapping a nun? What for?"

"Well, now, there ye have me." Jared was carrying Michael's coat over his arm and at this point handed it to him. "But maybe best ye go to the convent and find out."

"FORGIVE ME, MOTHER," Michael said carefully. Mother Hildegarde looked as though a breath would make her roll across the floor, wizened as a winter apple. "Did ye think . . . is it possible that Sister J— Sister Gregory might have . . . left of her own accord?"

The old nun gave him a look that revised his opinion of her state of health instantly.

"We did," she said dryly. "It happens. However"—she raised a sticklike finger—"one: there were signs of a considerable struggle in the cowshed. A full bucket of milk not merely spilt but apparently *thrown* at something, the manger overturned, the door left open, and two of the cows escaped into the herb garden." Another finger. "Two: had Sister Gregory experienced doubt regarding her vocation, she was quite free to leave the convent after speaking with me, and she knew that."

One more finger, and the old nun's black eyes bored into his. "And three: had she felt it necessary to leave suddenly and without informing us, where would she go? To you, Monsieur Murray. She knows no one else in Paris, does she?"

"I—well, no, not really." He was flustered, almost stammering, confusion and a burgeoning alarm for Joan making it difficult to think.

"But you have not seen her since you brought us the chalice and paten—and I thank you and your cousin with the deepest sentiments of gratitude, monsieur—which would be yesterday afternoon?"

"No." He shook his head, trying to clear it. "No, Mother."

Mother Hildegarde nodded, her lips nearly invisible, pressed together amid the lines of her face.

"Did she say anything to you on that occasion? Anything that might assist us in discovering her?"

"I—well . . ." Jesus, should he tell her what Joan had said about the voices she heard? It couldn't have anything to do with this, surely, and it wasna his secret to share. On the other hand, Joan *had* said she meant to tell Mother Hildegarde about them . . .

"You'd better tell me, my son." The reverend mother's voice was somewhere between resignation and command. "I see she told you *something.*"

"Well, she did, then, Mother," he said, rubbing a hand over his face in distraction. "But I canna see how it has anything to do—she hears voices," he blurted, seeing Mother Hildegarde's eyes narrow dangerously.

The eyes went round.

"She what?"

"Voices," he said helplessly. "They come and say things to her. She thinks maybe they're angels, but she doesn't know. And she can see when folk are going to die. Sometimes," he added dubiously. "I don't know whether she can always say."

"Par le sang sacré de Jésus Christ," the old nun said, sitting up straight

as an oak sapling. "Why did she not—well, never mind about that. Does anyone else know this?"

He shook his head. "She was afraid to tell anyone. That's why—well, one reason why—she came to the convent. She thought you might believe her."

"I might," Mother Hildegarde said dryly. She shook her head rapidly, making her veil flap. "*Nom de Dieu!* Why did her mother not tell me this?"

"Her mother?" Michael said stupidly.

"Yes! She brought me a letter from her mother, very kind, asking after my health and recommending Joan to me—but surely her mother would have known!"

"I don't think she—wait." He remembered Joan fishing out the carefully folded note from her pocket. "The letter she brought—it was from Claire Fraser. That's the one you mean?"

"Of course!"

He took a deep breath, a dozen disconnected pieces falling suddenly into a pattern. He cleared his throat and raised a tentative finger.

"One, Mother: Claire Fraser is the wife of Joan's stepfather. But she's not Joan's mother."

The sharp black eyes blinked once.

"And two: my cousin Jared tells me that Claire Fraser was known as a—a White Lady, when she lived in Paris many years ago."

Mother Hildegarde clicked her tongue angrily.

"She was no such thing. Stuff! But it is true that there was a common rumor to that effect," she admitted grudgingly. She drummed her fingers on the desk; they were knobbed with age but surprisingly nimble, and he remembered that Mother Hildegarde was a musician.

"Mother . . ."

"Yes?"

"I don't know if it has anything to do—do you know of a man called the Comte St. Germain?"

The old nun was already the color of parchment; at this, she went white as bone and her fingers gripped the edge of the desk.

"I do," she said. "Tell me—and quickly—what he has to do with Sister Gregory."

JOAN GAVE THE very solid door one last kick, for form's sake, then turned and collapsed with her back against it, panting. The room was huge, extending across the entire top floor of the house, though pillars

and joists here and there showed where walls had been knocked down. It smelled peculiar and looked even more peculiar.

"Blessed Michael, protect me," she whispered to herself, reverting to the Gaelic in her agitation. There was a very fancy bed in one corner, piled with feather pillows and bolsters, with writhing corner posts and heavy swags and curtains of cloth embroidered in what looked like gold and silver thread. Did the comte—he'd told her his name, or at least his title, when she asked—haul young women up here for wicked ends on a regular basis? For surely he hadn't set up this establishment solely in anticipation of her arrival—the area near the bed was equipped with all kinds of solid, shiny furniture with marble tops and alarming gilt feet that looked like they'd come off some kind of beast or bird with great curving claws.

He'd told her in the most matter-of-fact way that he was a sorcerer, too, and not to touch anything. She crossed herself and averted her gaze from the table with the nastiest-looking feet; maybe he'd charmed the furniture, and it came to life and walked round after dark. The thought made her move hastily off to the farther end of the room, rosary clutched tight in one hand.

This side of the room was scarcely less alarming, but at least it didn't look as though any of the big colored glass balls and jars and tubes could move on their own. It *was* where the worst smells were coming from, though: something that smelled like burnt hair and treacle, and something else very sharp that curled the hairs in your nose, like it did when someone dug out a jakes for the saltpeter. But there *was* a window near the long table where all this sinister stuff was laid out, and she went to this at once.

The big river—the Seine, Michael had called it—was right there, and the sight of boats and people made her feel a bit steadier. She put a hand on the table to lean closer but set it on something sticky and jerked it back. She swallowed and leaned in more gingerly. The window was barred on the inside. Glancing round, she saw that all the others were, too.

What in the name of the Blessed Virgin did that man expect would try to get in? Gooseflesh raced right up the curve of her spine and spread down her arms, her imagination instantly conjuring a vision of flying demons hovering over the street in the night, beating leathery wings against the window. *Or—dear Lord in heaven!—was it to keep the furniture* in?

There was a fairly normal-looking stool; she sank down on this and, closing her eyes, prayed with great fervor. After a bit, she remembered to breathe, and after a further bit, began to be able to think again, shuddering only occasionally.

He hadn't threatened her. Nor had he hurt her, really, just put a hand over her mouth and his other arm round her body and pulled her along, then boosted her into his coach with a shockingly familiar hand under her

bottom, though it hadn't been done with any sense that he was wanting to interfere with her.

In the coach, he'd introduced himself, apologized briefly for the inconvenience—*inconvenience? The cheek of him*—and then had grasped both her hands in his, staring intently into her face as he clasped them tighter and tighter. He'd raised her hands to his face, so close she'd thought he meant to smell them or kiss them, but then had let go, his brow deeply furrowed.

He'd ignored all her questions and her insistence upon being returned to the convent. In fact, he almost seemed to forget she was there, leaving her huddled in the corner of the seat while he thought intently about something, lips pursing in and out. And then he had lugged her up here, told her briefly that she wouldn't be hurt, added the bit about being a sorcerer in a very offhand sort of way, and locked her in!

She was terrified, and indignant, too. But now that she'd calmed down a wee bit, she thought that she wasn't really afraid of *him,* and that seemed odd. Surely she should be?

But she'd believed him when he said he meant her no harm. He hadn't threatened her or tried to frighten her. But if that was true . . . what did he want of her?

He likely wants to know what ye meant by rushing up to him in the market and telling him not to do it, her common sense—lamentably absent to this point—remarked.

"Oh," she said aloud. That made some sense. Naturally, he'd be curious about that.

She got up again and explored the room, thinking. She couldn't tell him any more than she had, though; that was the thing. Would he believe her, about the voices? Even if so, he'd try to find out more, and there wasn't any more to find out. What then?

Don't wait about to see, advised her common sense.

Having already come to this conclusion, she didn't bother replying. She'd found a heavy marble mortar and pestle; that might do. Wrapping the mortar in her apron, she went to the window that overlooked the street. She'd break the glass, then shriek 'til she got someone's attention. Even so high up, she thought, someone would hear. Pity it was a quiet street. But—

She stiffened like a bird dog. A coach was stopped outside one of the houses opposite, and Michael Murray was getting out of it! He was just putting on his hat—no mistaking that flaming red hair.

"Michael!" she shouted at the top of her lungs. But he didn't look up; the sound wouldn't pierce glass. She swung the cloth-wrapped mortar at the window, but it bounced off the bars with a ringing *clang!* She took a deep breath and a better aim; this time, she hit one of the panes and

cracked it. Encouraged, she tried again, with all the strength of muscular arms and shoulders, and was rewarded with a small crash, a shower of glass, and a rush of mud-scented air from the river.

"*Michael!*" But he had disappeared. A servant's face showed briefly in the open door of the house opposite, then vanished as the door closed. Through a red haze of frustration, she noticed the swag of black crepe hanging from the knob. Who was dead?

CHARLES'S WIFE, EULALIE, was in the small parlor, surrounded by a huddle of women. All of them turned to see who had come, many of them lifting their handkerchiefs automatically in preparation for a fresh outbreak of tears. All of them blinked at Michael, then turned to Eulalie, as though for an explanation.

Eulalie's eyes were red but dry. She looked as though she had been dried in an oven, all the moisture and color sucked out of her, her face paper-white and drawn tight over her bones. She, too, looked at Michael, but without much interest. He thought she was too much shocked for anything to matter much. He knew how she felt.

"Monsieur Murray," she said tonelessly, as he bowed over her hand. "How kind of you to call."

"I . . . offer my condolences, madame, mine and my cousin's. I hadn't . . . heard. Of your grievous loss." He was almost stuttering, trying to grasp the reality of the situation. What the devil had happened to Charles?

Eulalie's mouth twisted.

"Grievous loss," she repeated. "Yes. Thank you." Then her dull self-absorption cracked a little and she looked at him more sharply. "You hadn't heard. You mean—you didn't know? You came to *see* Charles?"

"Er . . . yes, madame," he said awkwardly. A couple of the women gasped, but Eulalie was already on her feet.

"Well, you might as well see him, then," she said, and walked out of the room, leaving him with no choice but to follow her.

"They've cleaned him up," she remarked, opening the door to the large parlor across the hall. She might have been talking about a messy domestic incident in the kitchen.

Michael thought it must in fact have been very messy. Charles lay on the large dining table, this adorned with a cloth and wreaths of greenery and flowers. A woman clad in gray was sitting by the table, weaving more wreaths from a basket of leaves and grasses; she glanced up, her eyes going from Eulalie to Michael and back.

"Leave," said Eulalie with a flip of the hand, and the woman got up at

once and went out. Michael saw that she'd been making a wreath of laurel leaves and had the sudden absurd thought that she meant to crown Charles with it, in the manner of a Greek hero.

"He cut his throat," Eulalie said. "The coward." She spoke with an eerie calmness, and Michael wondered what might happen when the shock that surrounded her began to dissipate.

He made a respectful sort of noise in his throat and, touching her arm gently, went past her to look down at his friend.

"Tell him not to do it."

The dead man didn't look peaceful. There were lines of stress in his countenance that hadn't yet smoothed out, and he appeared to be frowning. The undertaker's people had cleaned the body and dressed him in a slightly worn suit of dark blue; Michael thought that it was probably the only thing he'd owned that was in any way appropriate in which to appear dead, and suddenly missed his friend's frivolity with a surge that brought unexpected tears to his eyes.

"Tell him not to do it." He hadn't come in time. *If I'd come right away, when she told me—would it have stopped him?*

He could smell the blood, a rusty, sickly smell that seeped through the freshness of the flowers and leaves. The undertaker had tied a white neck-cloth for Charles—he'd used an old-fashioned knot, nothing that Charles himself would have worn for a moment. The black stitches showed above it, though, the wound harsh against the dead man's livid skin.

His own shock was beginning to fray, and stabs of guilt and anger poked through it like needles.

"Coward?" he said softly. He didn't mean it as a question, but it seemed more courteous to say it that way. Eulalie snorted, and, looking up, Michael met the full charge of her eyes. No, not shocked any longer.

"You'd know, wouldn't you," she said, and it wasn't at all a question, the way she said it. "You knew about your slut of a sister-in-law, didn't you? And Babette?" Her lips curled away from the name. "His *other* mistress?"

"I—no. I mean . . . Léonie told me yesterday. That was why I came to talk to Charles." Well, he would certainly have mentioned Léonie. And he wasn't going anywhere near the mention of Babette, whom he'd known about for quite some time. But, Jesus, what did the woman think he could have done about it?

"Coward," she said, looking down at Charles's body with contempt. "He made a mess of everything—*everything!*—and then couldn't deal with it, so he runs off and leaves me alone, with children, penniless!"

"Tell him not to do it."

Michael looked to see if this was an exaggeration, but it wasn't. She was burning now, but with fear as much as anger, her frozen calm quite vanished.

"The . . . house . . . ?" he began, with a rather vague wave around the expensive, stylish room. He knew it was her family house; she'd brought it to the marriage.

She snorted.

"He lost it in a card game last week," she said bitterly. "If I'm lucky, the new owner will let me bury him before we have to leave."

"Ah." The mention of card games jolted him back to an awareness of his reason for coming here. "I wonder, madame, do you know an acquaintance of Charles's—the Comte St. Germain?" It was crude, but he hadn't time to think of a graceful way to come to it.

Eugenia blinked, nonplussed.

"The comte? Why do you want to know about *him*?" Her expression sharpened into eagerness. "Do you think he owes Charles money?"

"I don't know, but I'll certainly find out for you," Michael promised her. "If you can tell me where to find Monsieur le Comte."

She didn't laugh, but her mouth quirked in what might in another mood have been humor.

"He lives across the street." She pointed toward the window. "In that big pile of—where are you going?"

But Michael was already through the door and into the hallway, boot heels clattering on the parquet in his haste.

THERE WERE FOOTSTEPS coming up the stairs; Joan started away from the window but then craned back, desperately willing the door across the street to open and let Michael out. What was he *doing* there?

That door didn't open, but a key rattled in the lock of the door to the room. In desperation, she tore the rosary from her belt and pushed it through the hole in the window, then dashed across the room and threw herself into one of the repulsive chairs.

It was the comte. He glanced round, worried for an instant, and then his face relaxed when he saw her. He came toward her, holding out his hand.

"I'm sorry to have kept you waiting, mademoiselle," he said, very courtly. "Come, please. I have something to show you."

"I don't want to see it." She stiffened a little and tucked her feet under her, to make it harder for him to pick her up. If she could just delay him until Michael came out! But he might well not see her rosary or, even if he did, know it was hers. Why should he? All nuns' rosaries looked the same!

She strained her ears, hoping to hear the sounds of departure on the other side of the street—she'd scream her lungs out. In fact . . .

The comte sighed a little but bent and took her by the elbows, lifting

her straight up, her knees still absurdly bent. He was really very strong. She put her feet down, and there she was, her hand tucked into the crook of his elbow, being led across the room toward the door, docile as a cow on its way to be milked! She made her mind up in an instant, yanked free, and ran to the smashed window.

"HELP!" she bellowed through the broken pane. "Help me, help me! *Au secours,* I mean! *AU SECOU*—" The comte's hand clapped across her mouth, and he said something in French that she was sure must be bad language. He scooped her up, so fast that the wind was knocked out of her, and had her through the door before she could make another sound.

MICHAEL DIDN'T PAUSE for hat or cloak but burst into the street, so fast that his driver started out of a doze and the horses jerked and neighed in protest. He didn't pause for that, either, but shot across the cobbles and pounded on the door, a big bronze-coated affair that boomed under his fists.

It couldn't have been very long but seemed an eternity. He fumed, pounded again, and, pausing for breath, caught sight of the rosary on the pavement. He ran to catch it up, scratched his hand, and saw that it lay in a scatter of glass fragments. At once he looked up, searching, and saw the broken window just as the big door opened.

He sprang at the butler like a wildcat, seizing him by the arms.

"Where is she? Where, damn you?"

"She? But there is no 'she,' monsieur. . . . Monsieur le Comte lives quite alone. You—"

"Where is Monsieur le Comte?" Michael's sense of urgency was so great, he felt that he might strike the man. The man apparently felt he might, too, because he turned pale and, wrenching himself loose, fled into the depths of the house. With no more than an instant's hesitation, Michael pursued him.

The butler, his feet fueled by fear, flew down the hall, Michael in grim pursuit. The man burst through the door to the kitchen; Michael was dimly aware of the shocked faces of cooks and maids, and then they were out into the kitchen garden. The butler slowed for an instant going down the steps, and Michael launched himself at the man, knocking him flat.

They rolled together on the graveled path, then Michael got on top of the smaller man, seized him by the shirtfront, and, shaking him, shouted, "WHERE IS HE?"

Thoroughly undone, the butler covered his face with one arm and pointed blindly toward a gate in the wall.

Michael leapt off the supine body and ran. He could hear the rumble of coach wheels, the rattle of hooves—he flung open the gate in time to see the back of a coach rattling down the allée and a gaping servant paused in the act of sliding to the doors of a carriage house. He ran, but it was clear that he'd never catch the coach on foot.

"JOAN!" he bellowed after the vanishing equipage. "I'm coming!"

He didn't waste time in questioning the servant but ran back, pushing his way through the maids and footmen gathered round the cowering butler, and burst out of the house, startling his own coachman afresh.

"That way!" he shouted, pointing toward the distant conjunction of the street and the allée, where the comte's coach was just emerging. "Follow that coach! *Vite!*"

"*VITE!*" THE COMTE urged his coachman on, then sank back, letting fall the hatch in the roof. The light was fading; his errand had taken longer than he'd expected, and he wanted to be out of the city before night fell. The city streets were dangerous at night.

His captive was staring at him, her eyes enormous in the dim light. She'd lost her postulant's veil, and her dark hair was loose on her shoulders. She looked charming but very scared. He reached into the bag on the floor and pulled out a flask of brandy.

"Have a little of this, *chérie*." He removed the cork and handed it to her. She took it but looked uncertain what to do with it, nose wrinkling at the hot smell.

"Really," he assured her. "It will make you feel better."

"That's what they all say," she said in her slow, awkward French.

"All of whom?" he asked, startled.

"The Auld Ones. I don't know what you call them in French, exactly. The folk that live in the hills—*souterrain*?" she added doubtfully. "Underground?"

"Underground? And they give you brandy?" He smiled at her, but his heart gave a sudden thump of excitement. Perhaps she *was*. He'd doubted his instincts when his touch failed to kindle her, but clearly she was *something*.

"They give you food and drink," she said, putting the flask down between the squab and the wall. "But if you take any, you lose time."

The spurt of excitement came again, stronger.

"Lose time?" he repeated, encouraging. "How do you mean?"

She struggled to find words, smooth brow furrowed with the effort.

"They . . . you . . . one who is enchanted by them—he, it? No, he—

goes into the hill, and there's music and feasting and dancing. But in the morning, when he goes . . . back, it's two hundred years later than it was when he went to feast with the . . . the Folk. Everybody he knew has turned to dust."

"How interesting!" he said. It was. He also wondered, with a fresh spasm of excitement, whether the old paintings, the ones far back in the bowels of the chalk mine, might have been made by these Folk, whoever they were.

She observed him narrowly, apparently for an indication that he was a faerie. He smiled at her, though his heart was now thumping audibly in his ears. *Two hundred years!* For that was what Mélisande—*Damn her,* he thought briefly, with a pang at the reminder of Madeleine—had told him was the usual period when one traveled through stone. It could be changed by use of gemstones or blood, she said, but that was the usual. And it had been, the first time he went back.

"Don't worry," he said to the girl, hoping to reassure her. "I only want you to look at something. Then I'll take you back to the convent—assuming that you still want to go there?" He lifted an eyebrow, half-teasing. It really wasn't his intent to frighten her, though he already had, and he feared that more fright was unavoidable. He wondered just what she might do when she realized that he was in fact planning to take her underground.

MICHAEL KNELT ON the seat, his head out the window of the coach, urging it on by force of will and muscle. It was nearly full dark, and the comte's coach was visible only as a distantly moving blot. They were out of the city, though; there were no other large vehicles on the road, nor likely to be—and there were very few turnings where such a large equipage might leave the main road.

The wind blew in his face, tugging strands of hair loose so they beat about his face. It blew the faint scent of decay, too—they'd pass the cemetery in a few minutes.

He wished passionately that he'd thought to bring a pistol, a smallsword—anything! But there was nothing in the coach with him, and he had nothing on his person save his clothes and what was in his pockets: this consisting, after a hasty inventory, of a handful of coins, a used handkerchief—the one Joan had given back to him, in fact, and he crumpled it tightly in one hand—a tinderbox, a mangled paper spill, a stub of sealing wax, and a small stone he'd picked up in the street, pinkish with a yellow stripe. Perhaps he could improvise a sling with the handkerchief, he

thought wildly, and paste the comte in the forehead with the stone, à la David and Goliath. And then cut off the comte's head with the penknife he discovered in his breast pocket, he supposed.

Joan's rosary was also in that pocket; he took it out and wound it round his left hand, holding the beads for comfort—he was too distracted to pray, beyond the words he repeated silently over and over, hardly noticing what he said.

Let me find her in time!

"TELL ME," THE COMTE asked curiously, "why did you speak to me in the market that day?"

"I wish I hadn't," Joan replied briefly. She didn't trust him an inch— still less since he'd offered her the brandy. It hadn't struck her before that that he really *might* be one of the Auld Ones. They could walk about, looking just like people. Her own mother had been convinced for years— and even some of the Murrays thought so—that Da's wife, Claire, was one. She herself wasn't sure; Claire had been kind to her, but no one said the Folk *couldn't* be kind if they wanted to.

Da's wife. A sudden thought paralyzed her: the memory of her first meeting with Mother Hildegarde, when she'd given the Reverend Mother Claire's letter. She'd said, *"ma mère,"* unable to think of a word that might mean "stepmother." It hadn't seemed to matter; why should anyone care?

"Claire Fraser," she said aloud, watching the comte carefully. "Do you know her?"

His eyes widened, showing white in the gloaming. Oh, aye, he kent her, all right!

"I do," he said, leaning forward. "Your mother, is she not?"

"No!" Joan said, with great force, and repeated it in French, several times for emphasis. "No, she's not!"

But she observed, with a sinking heart, that her force had been misplaced. He didn't believe her; she could tell by the eagerness in his face. He thought she was lying to put him off.

"I told you what I did in the market because the voices told me to!" she blurted, desperate for anything that might distract him from the horrifying notion that she was one of the Folk. Though if *he* was one, her common sense pointed out, he ought to be able to recognize her. Oh, Jesus, Lamb of God—that's what he'd been trying to do, holding her hands so tight and staring into her face.

"Voices?" he said, looking rather blank. "What voices?"

"The ones in my head," she said, heaving an internal sigh of exaspera-

tion. "They tell me things now and then. About other people, I mean. You know," she went on, encouraging him, "I'm a—a"—St. Jerome on a bannock, what was the *word*?!?—"someone who sees the future," she ended weakly. "Er . . . some of it. Sometimes. Not always."

The comte was rubbing a finger over his upper lip; she didn't know if he was expressing doubt or trying not to laugh, but either way it made her angry.

"So one of them told me to tell ye that, and I did!" she said, lapsing into Scots. "I dinna ken what it is ye're no supposed to do, but I'd advise ye not to do it!"

It occurred to her belatedly that perhaps killing her was the thing he wasn't supposed to do, and she was about to put this notion to him, but by the time she had disentangled enough grammar to have a go at it, the coach was slowing, bumping from side to side as it turned off the main road. A sickly smell seeped into the air, and she sat up straight, her heart in her throat.

"Mary, Joseph, and Bride," she said, her voice no more than a squeak. "Where *are* we?"

MICHAEL LEAPT FROM the coach almost before it had stopped moving. He daren't let them get too far ahead of him; his driver had nearly missed the turning, as it was, and the comte's coach had come to a halt minutes before his own reached it.

"Talk to the other driver," he shouted at his own, half visible on the box. "Find out why the comte has come here! Find out what he's doing!"

Nothing good. He was sure of that. Though he couldn't imagine why anyone would kidnap a nun and drag her out of Paris in the dark, only to stop at the edge of a public cemetery. Unless . . . half-heard rumors of depraved men who murdered and dismembered their victims, even those who *ate* . . . His wame rose and he nearly vomited, but it wasn't possible to vomit and run at the same time, and he could see a pale splotch on the darkness that he thought—he hoped, he feared—must be Joan.

Suddenly the night burst into flower. A huge puff of green fire bloomed in the darkness, and by its eerie glow he saw her clearly, her hair flying in the wind.

He opened his mouth to shout, to call out to her, but he had no breath, and before he could recover it she vanished into the ground, the comte following her, torch in hand.

He reached the shaft moments later, and he saw below the faintest green glow, just vanishing down a tunnel. Without an instant's hesitation, he flung himself down the ladder.

"DO YOU HEAR anything?" the comte kept asking her as they stumbled along the white-walled tunnels, he grasping her so hard by the arm that he'd surely leave bruises on her skin.

"No," she gasped. "What . . . am I listening for?"

He merely shook his head in a displeased way, but more as though he was listening for something himself than because he was angry with her for not hearing it.

She had some hopes that he'd meant what he said and would take her back. He did mean to go back himself; he'd lit several torches and left them burning along their way. So he wasn't about to disappear into the hill altogether, taking her with him to the lighted ballroom where people danced all night with the Fine Folk, unaware that their own world slipped past beyond the stones of the hill.

The comte stopped abruptly, hand squeezing harder round her arm.

"Be still," he said very quietly, though she wasn't making any noise. "Listen."

She listened as hard as possible—and thought she did hear something. What she thought she heard, though, was footsteps, far in the distance. Behind them. Her heart seized up for a moment.

"What—what do *you* hear?" she thought of asking. He glanced down at her, but not as though he really saw her.

"Them," he said. "The stones. They make a buzzing sound, most of the time. If it's close to a fire feast or a sun feast, though, they begin to sing."

"Do they?" she said faintly. He was hearing *something,* and evidently it wasn't the footsteps she'd heard. The footsteps had stopped now, as though whoever followed was waiting, maybe stealing along, one step at a time, careful to make no sound.

"Yes," he said, and his face was intent. He looked at her sharply again, and this time he saw her.

"You don't hear them," he said with certainty, and she shook her head. He pressed his lips tight together but after a moment lifted his chin, gesturing toward another tunnel, where there seemed to be something painted on the chalk.

He paused there to light another torch—this one burned a brilliant yellow and stank of sulfur—and she saw by its light the wavering shape of the Virgin and Child. Her heart lifted at the sight, for surely faeries would have no such thing in their lair.

"Come," he said, and now took her by the hand. His own was cold.

MICHAEL CAUGHT A glimpse of them as they moved into a side tunnel. The comte had lit another torch, a red one this time—how did he do that?—and it was easy to follow its glow.

How far down in the bowels of the earth were they? He had long since lost track of the turnings, though he might be able to get back by following the torches—assuming they hadn't all burned out.

He still had no plan in mind, other than to follow them until they stopped. Then he'd make himself known and . . . well, take Joan away, by whatever means proved necessary.

Swallowing hard, rosary still wrapped around his left hand and penknife in his right, he stepped into the shadows.

<hr />

THE CHAMBER WAS round and quite large. Big enough that the torchlight didn't reach all the edges, but it lit the pentagram inscribed into the floor in the center.

The noise was making Rakoczy's bones ache, and as often as he had heard it, it never failed to make his heart race and his hands sweat. He let go of the nun's hand for a moment to wipe his palm on the skirts of his coat, not wanting to disgust her. She looked scared but not terrified, and if she heard it, surely she—

Her eyes had widened suddenly.

"Who's *that?*" she said.

He whirled, to see Raymond standing tranquilly in the center of the pentagram.

"Bon soir, mademoiselle," the frog said, bowing politely.

"Ah . . . *bon soir,*" the girl replied faintly.

"What the devil are you doing here?" Rakoczy interposed his body between Raymond and the nun.

"Very likely the same thing you are," the frog replied. "Might you introduce your *petite amie,* sir?"

Shock, anger, and sheer confusion robbed Rakoczy of speech for a moment. What was the infernal creature *doing* here? Wait—the girl! The lost daughter he'd mentioned: the nun was the daughter! He'd discovered her whereabouts and somehow followed them to this place. Rakoczy took hold of the girl's arm again, firmly.

"She is a Scotch," he said. "And, as you see, a nun. No concern of yours."

The frog looked amused, cool and unruffled. Rakoczy was sweating, the noise beating against his skin in waves. He could feel the little bag of stones in his pocket, a hard lump against his heart. They seemed to be warm, warmer even than his skin.

"I doubt that she is, really," said Raymond. "Why is she a concern of yours, though?"

"That's also none of your business." He was trying to think. He couldn't lay out the stones, not with the damned frog standing there. Could he just leave with the girl? But if the frog meant him harm . . . and if the girl truly wasn't . . .

Raymond ignored the incivility and bowed again to the girl.

"I am Master Raymond, my dear," he said. "And you?"

"Joan Mac—" she said. "Er . . . Sister Gregory, I mean." She tried to pull away from Rakoczy's grip. "Um. If I'm not the concern of either of you gentlemen—"

"She's my concern, gentlemen." The voice was high with nerves, but firm. Rakoczy looked round, shocked to see the young wine merchant walk into the chamber, disheveled and dirty but eyes fixed on the girl. At Rakoczy's side, the nun gasped.

"Sister." The merchant bowed. He was white-faced but not sweating. He looked as though the chill of the cavern had seeped into his bones, but he put out a hand, from which the beads of a wooden rosary swung. "You dropped your rosary."

JOAN THOUGHT SHE might faint from sheer relief. Her knees wobbled from terror and exhaustion, but she summoned enough strength to wrench free of the comte and run, stumbling, into Michael's arms. He grabbed her and hauled her away from the comte, half-dragging her.

The comte made an angry sound and took a step in Joan's direction, but Michael said, "Stop right there, ye wicked bugger!" just as the little froggy-faced man said sharply, "Stop!"

The comte swung toward first one and then the other. He looked . . . crazed. Joan swallowed and nudged Michael, urging him toward the chamber's door, only then noticing the penknife in his hand.

"What were ye going to do wi' *that*?" she whispered. "Shave him?"

"Let the air out of him," Michael muttered. He lowered his hand but didn't put the knife away and kept his eyes on the two men.

"Your daughter," the comte said hoarsely to the man who called himself Master Raymond. "You were looking for a lost daughter. I've found her for you."

Raymond's brows shot up, and he glanced at Joan.

"Mine?" he said, astonished. "She isn't one of mine. Can't you tell?"

The comte drew a breath so deep it cracked in his throat.

"Tell? But—"

The frog looked impatient.

"Can you not see auras? The electrical fluid that surrounds people," he elucidated, waving a hand around his own head.

The comte rubbed a hand hard over his face. "I can't—"

"For goodness sake, come in here!" Raymond stepped to the edge of the star, reached across, and seized the comte's hand.

RAKOCZY STIFFENED AT the touch. Blue light exploded from their linked hands, and he gasped, feeling a surge of energy such as he had never before experienced. Raymond pulled hard, and Rakoczy stepped across the line into the pentagram.

Silence. The buzzing had stopped. He nearly wept with the relief of it.

"I—you—" he stammered, looking at the linked hands.

"You didn't know?" Raymond looked surprised.

"That you were a—" He waved at the pentagram. "I thought you might be."

"Not that," Raymond said, almost gently. "That you were one of mine."

"Yours?" Rakoczy looked down again; the blue light was pulsing gently now, surrounding their fingers.

"Everyone has an aura of some kind," Raymond said. "But only my . . . people . . . have *this*."

In the blessed silence, it was possible to think again. And the first thing that came to mind was the Star Chamber, the king looking on as they had faced each other over a poisoned cup. And now he knew why the frog hadn't killed him.

HIS MIND BUBBLED with questions. La Dame Blanche, blue light, Mélisande, and Madeleine . . . Thought of Madeleine and what grew in her womb nearly stopped him, but the urge to find out, to *know* at last, was too strong.

"Can you—can we—go forward?"

Raymond hesitated a moment, then nodded.

"Yes. But it's not safe. Not safe at all."

"Will you show me?"

"I mean it." The frog's grip tightened on his. "It's not a safe thing to know, let alone to do."

Rakoczy laughed, feeling all at once exhilarated, full of joy. Why should he fear knowledge? Perhaps the passage would kill him—but he had a

pocket full of gems, and, besides, what was the point of waiting to die slowly?

"Tell me!" he said, squeezing the other's hand. "For the sake of our shared blood!"

JOAN STOOD STOCK-STILL, amazed. Michael's arm was still around her, but she scarcely noticed.

"He *is*!" she whispered. "He truly is! They both are!"

"Are what?" Michael gaped at her.

"Auld Folk! Faeries!"

He looked wildly back at the scene before them. The two men stood face-to-face, hands locked together, their mouths moving in animated conversation—in total silence. It was like watching mimes but even less interesting.

"I dinna care *what* they are. Loons, criminals, demons, angels . . . Come on!" He dropped his arm and seized her hand, but she was planted solid as an oak sapling, her eyes growing wide and wider.

She gripped his hand hard enough to grind the bones and shrieked at the top of her lungs, *"Don't do it!!"*

He whirled round just in time to see them vanish.

THEY STUMBLED TOGETHER down the long, pale passages, bathed in the flickering light of dying torches, red, yellow, blue, green, a ghastly purple that made Joan's face look drowned.

"Des feux d'artifice," Michael said. His voice sounded queer, echoing in the empty tunnels. "A conjurer's trick."

"What?" Joan looked drugged, her eyes black with shock.

"The fires. The . . . colors. Have ye never heard of fireworks?"

"No."

"Oh." It seemed too much a struggle to explain, and they went on in silence, hurrying as much as they could, to reach the shaft before the light died entirely.

At the bottom, he paused to let her go first, thinking too late that he should have gone first—she'd think he meant to look up her dress. . . . He turned hastily away, face burning.

"D'ye think he was? That *they* were?" She was hanging on to the ladder, a few feet above him. Beyond her, he could see the stars, serene in a velvet sky.

"Were what?" He looked at her face, so as not to risk her modesty. She was looking better now but very serious.

"Were they Auld Folk? Faeries?"

"I suppose they must ha' been." His mind was moving very slowly; he didn't want to have to try to think. He motioned to her to climb and followed her up, his eyes tightly shut. If they were Auld Ones, then likely so was Auntie Claire. He truly didn't want to think about *that*.

He drew the fresh air gratefully into his lungs. The wind was toward the city now, coming off the fields, full of the resinous cool scent of pine trees and the breath of grass and cattle. He felt Joan breathe it in, sigh deeply, and then she turned to him, put her arms around him, and rested her forehead on his chest. He put his arms round her and they stood for some time, in peace.

Finally, she stirred and straightened up.

"Ye'd best take me back, then," she said. "The sisters will be half out o' their minds."

He was conscious of a sharp sense of disappointment but turned obediently toward the coach, standing in the distance. Then he turned back.

"Ye're sure?" he said. "Did your voices tell ye to go back?"

She made a sound that wasn't quite a rueful laugh.

"I dinna need a voice to tell me that." She brushed a hand through her hair, smoothing it off her face. "In the Highlands, if a man's widowed, he takes another wife as soon as he can get one; he's got to have someone to mend his shirt and rear his bairns. But Sister Philomène says it's different in Paris; that a man might mourn for a year."

"He might," he said, after a short silence. Would a year be enough, he wondered, to heal the great hole where Lillie had been? He knew he would never forget—never stop looking for her—but he didn't forget what Ian had told him, either.

"But after a time, ye find ye're in a different place than ye were. A different person than ye were. And then ye look about and see what's there with ye. Ye'll maybe find a use for yourself."

Joan's face was pale and serious in the moonlight, her mouth gentle.

"It's a year before a postulant makes up her mind. Whether to stay and become a novice—or . . . or leave. It takes time. To know."

"Aye," he said softly. "Aye, it does."

He turned to go, but she stopped him, a hand on his arm.

"Michael," she said. "Kiss me, aye? I think I should maybe know *that*, before I decide."

A PLAGUE OF ZOMBIES

INTRODUCTION

THE THING ABOUT LORD JOHN'S SITUATION AND career—unmarried, no fixed establishment, discreet political connections, fairly high-ranking officer—is that he can easily take part in far-flung adventures rather than being bound to a pedestrian daily life. To be honest, once I started doing "bulges" (that is, shorter pieces of fiction) involving him, I just looked at which year it was and then consulted one of my historical timeline references to see what kinds of interesting events happened in that year. That's how he happened to find himself in Quebec for the battle there.

In terms of this story, though, the impetus came from two different sources, both "trails" leading back from the main book of the series—*Voyager*, in this case. To wit: I knew that Lord John was the governor of Jamaica in 1766, when Claire met him aboard the *Porpoise;* it wasn't by any means impossible for a man with connections and no experience to be appointed to such a post—but it was more likely for a man who *had* had experience in the territory to which he was appointed. "Plague" is set in 1761, and is the story of how Lord John gained that experience. I knew also that Geillis Duncan wasn't dead and where she was. And, after all, with a story set in Jamaica, how could I possibly resist zombies?

THERE WAS A SNAKE on the drawing-room table. A small snake, but still. Lord John Grey wondered whether to say anything about it.

The governor, appearing quite oblivious of the coiled reptile's presence, picked up a cut-crystal decanter that stood not six inches from the snake. Perhaps it was a pet, or perhaps the residents of Jamaica were accustomed to keeping a tame snake in residence, to kill rats. Judging from the number of rats Grey had seen since leaving the ship, this was sensible— though this particular snake didn't appear large enough to take on even your average mouse.

The wine was decent, but served at body heat, and it seemed to pass directly through Grey's gullet and into his blood. He'd had nothing to eat since before dawn and felt the muscles of his lower back begin to tingle and relax. He put the glass down; he wanted a clear head.

"I cannot tell you, sir, how happy I am to receive you," said the governor, putting down his own glass, empty. "The position is acute."

"So you said in your letter to Lord North. The situation has not changed appreciably since then?" It had been nearly three months since that letter was written; a lot could change in three months.

He thought Governor Warren shuddered, despite the temperature in the room.

"It has become worse," the governor said, picking up the decanter. "Much worse."

Grey felt his shoulders tense, but spoke calmly.

"In what way? Have there been more—" He hesitated, searching for the right word. "More demonstrations?" It was a mild word to describe the burning of cane fields, the looting of plantations, and the wholesale liberation of slaves.

Warren gave a hollow laugh. His handsome face was beading with sweat. There was a crumpled handkerchief on the arm of his chair, and he picked it up to mop at his skin. He hadn't shaved this morning—or, quite possibly, yesterday; Grey could hear the faint rasp of his dark whiskers on the cloth.

"Yes. More destruction. They burnt a sugar press last month, though still in the remoter parts of the island. Now, though . . ." He paused, licking dry lips as he poured more wine. He made a cursory motion toward Grey's glass, but Grey shook his head.

"They've begun to move toward Kingston," Warren said. "It's deliberate; you can see it. One plantation after another, in a line coming straight down the mountain." He sighed. "I shouldn't say straight. Nothing in this bloody place is straight, starting with the landscape."

That was true enough; Grey had admired the vivid green peaks that soared up from the centre of the island, a rough backdrop for the amazingly blue lagoon and the white-sand shore.

"People are terrified," Warren went on, seeming to get a grip on himself, though his face was once again slimy with sweat, and his hand shook on the decanter. It occurred to Grey, with a slight shock, that the *governor* was terrified. "I have merchants—and their wives—in my office every day, begging, demanding protection from the blacks."

"Well, you may assure them that protection will be provided them," Grey said, sounding as reassuring as possible. He had half a battalion with him—three hundred infantry troops and a company of artillery, equipped with small cannon. Enough to defend Kingston, if necessary. But his brief from Lord North was not merely to reassure the merchants and defend the shipping of Kingston and Spanish Town—nor even to provide protection to the larger sugar plantations. He was charged with putting down the slave rebellion entirely. Rounding up the ringleaders and stopping the violence altogether.

The snake on the table moved suddenly, uncoiling itself in a languid manner. It startled Grey, who had begun to think it was a decorative sculpture. It was exquisite: only seven or eight inches long and a beautiful pale yellow marked with brown, a faint iridescence in its scales like the glow of good Rhenish wine.

"It's gone further now, though," Warren was going on. "It's not just burning and property destruction. Now it's come to murder."

That brought Grey back with a jerk.

"Who has been murdered?" he demanded.

"A planter named Abernathy. Murdered in his own house, last week. His throat cut."

"Was the house burnt?"

"No, it wasn't. The maroons ransacked it but were driven off by Abernathy's own slaves before they could set fire to the place. His wife survived by submerging herself in a spring behind the house, concealed by a patch of reeds."

"I see." He could imagine the scene all too well. "Where is the plantation?"

"About ten miles out of Kingston. Rose Hall, it's called. Why?" A bloodshot eye swivelled in Grey's direction, and he realised that the glass of wine the governor had invited him to share had not been his first of the day. Nor, likely, his fifth.

Was the man a natural sot? he wondered. Or was it only the pressure of the current situation that had caused him to take to the bottle in such a blatant manner? He surveyed the governor covertly; the man was perhaps in his late thirties and, while plainly drunk at the moment, showed none of the signs of habitual indulgence. He was well built and attractive; no bloat, no soft belly straining at his silk waistcoat, no broken veins in cheeks or nose . . .

"Have you a map of the district?" Surely it hadn't escaped Warren that if indeed the maroons were burning their way straight toward Kingston, it should be possible to predict where their next target lay and to await them with several companies of armed infantry?

Warren drained the glass and sat panting gently for a moment, eyes fixed on the tablecloth, then pulled himself together.

"Map," he repeated. "Yes, of course. Dawes—my secretary—he'll . . . he'll find you one."

Motion caught Grey's eye. Rather to his surprise, the tiny snake, after casting to and fro, tongue tasting the air, had started across the table in what appeared a purposeful, if undulant, manner, headed straight for him. By reflex, he put up a hand to catch the little thing, lest it plunge to the floor.

The governor saw it, uttered a loud shriek, and flung himself back from the table. Grey looked at him in astonishment, the tiny snake curling over his fingers.

"It's not venomous," he said, as mildly as he could. At least, he didn't *think* so. His friend Oliver Gwynne was a natural philosopher and mad for snakes; Gwynne had shown him all the prizes of his collection during the course of one hair-raising afternoon, and Grey seemed to recall Gwynne telling him that there were no venomous reptiles at all on the island of Jamaica. Besides, the nasty ones all had triangular heads, while the harmless kinds were blunt-headed, like this fellow.

Warren was indisposed to listen to a lecture on the physiognomy of snakes. Shaking with terror, he backed against the wall.

"Where?" he gasped. "Where did it come from?"

"It's been sitting on the table since I came in. I . . . um . . . thought it was . . ." Well, plainly it wasn't a pet, let alone an intended part of the table décor. He coughed and got up, meaning to put the snake outside through the French doors that led onto the terrace.

Warren mistook his intent, though, and, seeing Grey come closer, snake writhing through his fingers, he burst through the French doors, crossed the terrace in a mad leap, and pelted down the flagstoned walk, coattails flying as though the devil himself were in pursuit.

Grey was still staring after him in disbelief when a discreet cough from the inner door made him turn.

"Gideon Dawes, sir." The governor's secretary was a short, tubby man with a round pink face that probably was rather jolly by nature. At the moment, it bore a look of profound wariness. "You are Lieutenant-Colonel Grey?"

Grey thought it unlikely that there were a plethora of men wearing the uniform and insignia of a lieutenant-colonel on the premises of King's House at that very moment but nonetheless bowed, murmuring, "Your servant, Mr. Dawes. I'm afraid Mr. Warren has been taken . . . er . . ." He nodded toward the open French doors. "Perhaps someone should go after him?"

Mr. Dawes closed his eyes with a look of pain, then sighed and opened them again, shaking his head.

"He'll be all right," he said, though his tone lacked any real conviction. "I've just been discussing commissary and billeting requirements with your Major Fettes; he wishes you to know that all the arrangements are quite in hand."

"Oh. Thank you, Mr. Dawes." In spite of the unnerving nature of the governor's departure, Grey felt a sense of pleasure. He'd been a major himself for years; it was astonishing how pleasant it was to know that someone else was now burdened with the physical management of troops. All *he* had to do was give orders.

That being so, he gave one, though it was phrased as a courteous request, and Mr. Dawes promptly led him through the corridors of the rambling house to a small clerk's hole near the governor's office, where maps were made available to him.

He could see at once that Warren had been right regarding both the devious nature of the terrain and the trail of attacks. One of the maps was marked with the names of plantations, and small notes indicated where maroon raids had taken place. It was far from being a straight line, but, nonetheless, a distinct sense of direction was obvious.

The room was warm, and he could feel sweat trickling down his back.

Still, a cold finger touched the base of his neck lightly when he saw the name *Twelvetrees* on the map.

"Who owns this plantation?" he asked, keeping his voice level as he pointed at the paper.

"What?" Dawes had fallen into a sort of dreamy trance, looking out the window into the green of the jungle, but blinked and pushed his spectacles up, bending to peer at the map. "Oh, Twelvetrees. It's owned by Philip Twelvetrees—a young man; inherited the place from a cousin only recently. Killed in a duel, they say—the cousin, I mean," he amplified helpfully.

"Ah. Too bad." Grey's chest tightened unpleasantly. He could have done without *that* complication. If . . . "The cousin—was he named Edward Twelvetrees, by chance?"

Dawes looked mildly surprised.

"I do believe that was the name. I didn't know him, though; no one here did. He was an absentee owner; ran the place through an overseer."

"I see." He wanted to ask whether Philip Twelvetrees had come from London to take possession of his inheritance, but didn't. He didn't want to draw any attention by singling out the Twelvetrees family. Time enough for that.

He asked a few more questions regarding the timing of the raids, which Mr. Dawes answered promptly, but when it came to an explanation of the inciting causes of the rebellion, the secretary proved suddenly unhelpful—which Grey thought interesting.

"Really, sir, I know almost nothing of such matters," Mr. Dawes protested, when pressed on the subject. "You would be best advised to speak with Captain Cresswell. He's the superintendent in charge of the maroons."

Grey was surprised at this.

"Escaped slaves? They have a superintendent?"

"Oh. No, sir." Dawes seemed relieved to have a more straightforward question with which to deal. "The maroons are not escaped slaves. Or rather," he corrected himself, "they are *technically* escaped slaves, but it is a pointless distinction. These maroons are the descendants of slaves who escaped during the last century and took to the mountain uplands. They have settlements up there. But as there is no way of identifying any current owner . . ." And as the government lacked any means of finding them and dragging them back, the Crown had wisely settled for installing a white superintendent, as was usual for dealing with native populations. The superintendent's business was to be in contact with the maroons and deal with any matter that might arise pertaining to them.

Which raised a question, Grey thought: why had this Captain Cresswell not been brought to meet him at once? He had sent word of his arrival as

soon as the ship docked at daylight, not wishing to take Derwent Warren unawares.

"Where is Captain Cresswell presently?" he asked, still polite. Mr. Dawes looked unhappy.

"I, um, am afraid I don't know, sir," he said, casting down his gaze behind his spectacles.

There was a momentary silence, in which Grey could hear the calling of some bird from the jungle nearby.

"Where is he *normally?*" Grey asked, with slightly less politesse.

Dawes blinked.

"I don't know, sir. I believe he has a house near the base of Guthrie's Defile—there is a small village there. But he would of course go up into the maroon settlements from time to time, to meet with the . . ." He waved a small, fat hand, unable to find a suitable word. "The headmen. He did buy a new hat in Spanish Town earlier this month," Dawes added, in the tones of someone offering a helpful observation.

"A *hat?*"

"Yes. Oh—but of course you would not know. It is customary among the maroons, when some agreement of importance is made, that the persons making the agreement shall exchange hats. So you see—"

"Yes, I do," Grey said, trying not to let annoyance show in his voice. "Will you be so kind, Mr. Dawes, as to send to Guthrie's Defile, then—and to any other place in which you think Captain Cresswell might be discovered? Plainly I must speak with him, and as soon as possible."

Dawes nodded vigorously, but before he could speak, the rich sound of a small gong came from somewhere in the house below. As though it had been signaled, Grey's stomach emitted a loud gurgle.

"Dinner in half an hour," Mr. Dawes said, looking happier than Grey had yet seen him. He almost scurried out the door, Grey in his wake.

"Mr. Dawes," he said, catching up at the head of the stair. "Governor Warren. Do you think—"

"Oh, he will be present at dinner," Dawes assured him. "I'm sure he is quite recovered now; these small fits of excitement never last very long."

"What causes them?" A savoury smell, rich with currants, onion, and spice, wafted up the stair, making Grey hasten his step.

"Oh . . ." Dawes, hastening along as well, glanced sideways at him. "It is nothing. Only that His Excellency has a, um, somewhat morbid fancy concerning reptiles. Did he see a snake in the drawing room or hear something concerning one?"

"He did, yes—though a remarkably small and harmless one." Vaguely, Grey wondered what had happened to the little yellow snake. He thought he must have dropped it in the excitement of the governor's abrupt exit and hoped it hadn't been injured.

Mr. Dawes looked troubled and murmured something that sounded like, "Oh, dear, oh, dear . . ." but then he merely shook his head and sighed.

GREY MADE HIS way to his room, meaning to freshen himself before dinner; the day was warm, and he smelled strongly of ship's reek— this composed in equal parts of sweat, seasickness, and sewage, well marinated in salt water—and horse, having ridden up from the harbour to Spanish Town. With any luck, his valet would have clean linen aired for him by now.

King's House, as all royal governors' residences were known, was a rambling old wreck of a mansion, perched on a high spot of ground on the edge of Spanish Town. Plans were afoot for an immense new Palladian building, to be erected in the town's centre, but it would be another year at least before construction could commence. In the meantime, efforts had been made to uphold His Majesty's dignity by means of beeswax polish, silver, and immaculate linen, but the dingy printed wallpaper peeled from the corners of the rooms, and the dark-stained wood beneath exhaled a mouldy breath that made Grey want to hold his own whenever he walked inside.

One good feature of the house, though, was that it was surrounded on all four sides by a broad terrace and was overhung by large, spreading trees that cast lacy shadows on the flagstones. A number of the rooms opened directly onto this terrace—Grey's did—and it was therefore possible to step outside and draw a clean breath, scented by the distant sea or the equally distant upland jungles. There was no sign of his valet, but there *was* a clean shirt on the bed. He shucked his coat, changed his shirt, and then threw the French doors open wide.

He stood for a moment in the centre of the room, mid-afternoon sun spilling through the open doors, and enjoyed the sense of a solid surface under his feet after seven weeks at sea and seven hours on horseback. Enjoyed even more the transitory sense of being alone. Command had its prices, and one of those was a nearly complete loss of solitude. He therefore seized it when he found it, knowing it wouldn't last for more than a few moments, but valuing it all the more for that.

Sure enough, it didn't last more than two minutes this time. He called out, "Come," at a rap on the door frame and, turning, was struck by a visceral sense of attraction such as he had not experienced in months.

The man was young, perhaps twenty, and slender in his blue and gold livery, but with a breadth of shoulder that spoke of strength and a head and neck that would have graced a Greek sculpture. Perhaps because of

the heat, he wore no wig, and his tight-curled hair was clipped so close that the finest modelling of his skull was apparent.

"Your servant, sah," he said to Grey, bowing respectfully. "The governor's compliments, and dinner will be served in ten minutes. May I see you to the dining room?"

"You may," Grey said, reaching hastily for his coat. He didn't doubt that he could find the dining room unassisted, but the chance to watch this young man walk . . .

"You *may*," Tom Byrd corrected, entering with his hands full of grooming implements, "once I've put his lordship's hair to rights." He fixed Grey with a minatory eye. "You're not a-going in to dinner like that, me lord, and don't you think it. You sit down there." He pointed sternly to a stool, and Lieutenant-Colonel Grey, commander of His Majesty's forces in Jamaica, meekly obeyed the dictates of his twenty-one-year-old valet. He didn't *always* allow Tom free rein but in the current circumstance was just as pleased to have an excuse to sit still in the company of the young black servant.

Tom laid out all his implements neatly on the dressing table, from a pair of silver hairbrushes to a box of powder and a pair of curling tongs, with the care and attention of a surgeon arraying his knives and saws. Selecting a hairbrush, he leaned closer, peering at Grey's head, then gasped. "Me lord! There's a big huge spider—walking right up your temple!"

Grey smacked his temple by reflex, and the spider in question—a clearly visible brown thing nearly a half inch long—shot off into the air, striking the looking glass with an audible tap before dropping to the surface of the dressing table and racing for its life.

Tom and the black servant uttered identical cries of horror and lunged for the creature, colliding in front of the dressing table and falling over in a thrashing heap. Grey, strangling an almost irresistible urge to laugh, stepped over them and dispatched the fleeing spider neatly with the back of his other hairbrush.

He pulled Tom to his feet and dusted him off, allowing the black servant to scramble up by himself. He brushed off all apologies, as well, but asked whether the spider had been a deadly one.

"Oh, yes, sah," the servant assured him fervently. "Should one of those bite you, sah, you would suffer excruciating pain at once. The flesh around the wound would putrefy, you would commence to be fevered within an hour, and, in all likelihood, you would not live past dawn."

"Oh, I see," Grey said mildly, his flesh creeping briskly. "Well, then. Perhaps you would not mind looking about the room while Tom is at his work? In case such spiders go about in company?"

Grey sat and let Tom brush and plait his hair, watching the young man

as he assiduously searched under the bed and dressing table, pulled out Grey's trunk, and pulled up the trailing curtains and shook them.

"What is your name?" he asked the young man, noting that Tom's fingers were trembling badly and hoping to distract him from thoughts of the hostile wildlife with which Jamaica undoubtedly teemed. Tom was fearless in the streets of London and perfectly willing to face down ferocious dogs or foaming horses. Spiders, though, were quite another matter.

"Rodrigo, sah," said the young man, pausing in his curtain-shaking to bow. "Your servant, sah."

He seemed quite at ease in company and conversed with them about the town, the weather—he confidently predicted rain in the evening, at about ten o'clock, leading Grey to think that he had likely been employed as a servant in good families for some time. Was the man a slave, he wondered, or a free black?

His admiration for Rodrigo was, he assured himself, the same that he might have for a marvellous piece of sculpture, an elegant painting. And one of his friends did in fact possess a collection of Greek amphorae decorated with scenes that gave him quite the same sort of feeling. He shifted slightly in his seat, crossing his legs. He would be going in to dinner soon. He resolved to think of large, hairy spiders and was making some progress with this subject when something huge and black dropped down the chimney and rushed out of the disused hearth.

All three men shouted and leapt to their feet, stamping madly. This time it was Rodrigo who felled the intruder, crushing it under one sturdy shoe.

"What the devil was that?" Grey asked, bending over to peer at the thing, which was a good three inches long, gleamingly black, and roughly ovoid, with ghastly long, twitching antennae.

"Only a cockroach, sah," Rodrigo assured him, wiping a hand across a sweating ebony brow. "They will not harm you, but they *are* most disagreeable. If they come into your bed, they feed upon your eyebrows."

Tom uttered a small, strangled cry. The cockroach, far from being destroyed, had merely been inconvenienced by Rodrigo's shoe. It now extended thorny legs, heaved itself up, and was proceeding about its business, though at a somewhat slower pace. Grey, the hairs prickling on his arms, seized the ash shovel from among the fireplace implements and, scooping up the insect on its blade, jerked open the door and flung the nasty creature as far as he could—which, given his state of mind, was some considerable distance.

Tom was pale as custard when Grey came back in, but he picked up his employer's coat with trembling hands. He dropped it, though, and with a mumbled apology bent to pick it up again, only to utter a strangled shriek,

drop it once more, and run backwards, slamming so hard against the wall that Grey heard a crack of laths and plaster.

"What the devil?" He bent, reaching gingerly for the fallen coat.

"Don't touch it, me lord!" Tom cried, but Grey had seen what the trouble was: a tiny yellow snake slithered out of the crimson-velvet folds, head moving to and fro in slow curiosity.

"Well, hallo, there." He reached out a hand, and the little snake tasted his skin with a flickering tongue, then wove its way up into the palm of his hand. He stood up, cradling it carefully.

Tom and Rodrigo were standing like men turned to stone, staring at him.

"It's quite harmless," he assured them. "At least I think so. It must have fallen into my pocket earlier."

Rodrigo was regaining a bit of his nerve. He came forward and looked at the snake but declined an offer to touch it, putting both hands firmly behind his back.

"That snake likes you, sah," he said, glancing curiously from the snake to Grey's face, as though trying to distinguish a reason for such odd particularity.

"Possibly." The snake had made its way upwards and was now wrapped round two of Grey's fingers, squeezing with remarkable strength. "On the other hand, I believe he may be attempting to kill and eat me. Do you know what his natural food might be?"

Rodrigo laughed at that, displaying very beautiful white teeth, and Grey had such a vision of those teeth, those soft mulberry lips, applied to—he coughed, hard, and looked away.

"He would eat anything that did not try to eat him first, sah," Rodrigo assured him. "It was probably the sound of the cockroach that made him come out. He would hunt those."

"What a very admirable sort of snake. Could we find him something to eat, do you think? To encourage him to stay, I mean."

Tom's face suggested strongly that if the snake was staying, he was not. On the other hand . . . he glanced toward the door, whence the cockroach had made its exit, and shuddered. With great reluctance, he reached into his pocket and extracted a rather squashed bread roll containing ham and pickle.

The snake was placed on the floor with this object before it. It inspected the roll gingerly, ignored the bread and pickle, but twined itself carefully about a chunk of ham, squeezing it fiercely into limp submission. Then, opening its jaw to an amazing extent, the snake engulfed its prey, to general cheers. Even Tom clapped his hands, and, if not ecstatic at Grey's suggestion that the snake might be accommodated in the dark space beneath the bed for the sake of preserving Grey's eyebrows, he uttered no

objections to this plan, either. The snake being ceremoniously installed and left to digest its meal, Grey was about to ask Rodrigo further questions regarding the natural fauna of the island but was forestalled by the faint sound of a distant gong.

"Dinner!" he exclaimed, reaching for his now snakeless coat.

"Me lord! Your hair's not even powdered!"

Grey refused to wear a wig, to Tom's ongoing dismay, but was obliged in the present instance to submit to powder. This toiletry accomplished in haste, he shrugged into his coat and fled, before Tom could suggest any further refinements to his appearance.

THE GOVERNOR APPEARED, as Mr. Dawes had predicted, calm and dignified at the dinner table. All trace of sweat, hysteria, and drunkenness had vanished, and beyond a brief word of apology for his abrupt disappearance, no reference was made to his earlier departure.

Major Fettes and Grey's adjutant, Captain Cherry, also appeared at table. A quick glance at them assured Grey that all was well with the troops. Fettes and Cherry couldn't be more diverse physically—the latter resembling a ferret and the former a block of wood—but both were extremely competent and well liked by the men.

There was little conversation to begin with; the three soldiers had been eating ship's biscuit and salt beef for weeks. They settled down to the feast before them with the single-minded attention of ants presented with a loaf of bread; the magnitude of the challenge had no effect upon their earnest willingness. As the courses gradually slowed, though, Grey began to instigate conversation—his prerogative, as senior guest and commanding officer.

"Mr. Dawes explained to me the position of superintendent," he said, keeping his attitude superficially pleasant. "How long has Captain Cresswell held this position, sir?"

"For approximately six months, Colonel," the governor replied, wiping crumbs from his lips with a linen napkin. The governor was quite composed, but Grey had Dawes in the corner of his eye and thought the secretary stiffened a little. That was interesting; he must get Dawes alone again and go into this matter of superintendents more thoroughly.

"And was there a superintendent before Captain Cresswell?"

"Yes . . . in fact, there were two of them, were there not, Mr. Dawes?"

"Yes, sir. Captain Ludgate and Captain Perriman." Dawes was assiduously not meeting Grey's eye.

"I should like very much to speak with those gentlemen," Grey said pleasantly.

Dawes jerked as though someone had run a hatpin into his buttock. The governor finished chewing a grape, swallowed, and said, "I'm so sorry, Colonel. Both Ludgate and Perriman have left their offices."

"Why?" John Fettes asked bluntly. The governor hadn't been expecting that, and blinked.

"I expect Major Fettes wishes to know whether they were replaced in their offices because of some peculation or corruption," Bob Cherry put in chummily. "And if that be the case, were they allowed to leave the island rather than face prosecution? And if so—"

"Why?" Fettes put in neatly.

Grey repressed a smile. Should peace break out on a wide scale and an army career fail them, Fettes and Cherry could easily make a living as a music-hall knockabout cross-talk act. As interrogators, they could reduce almost any suspect to incoherence, confusion, and confession in nothing flat.

Governor Warren, though, appeared to be made of tougher stuff than the usual regimental miscreant. Either that or he had nothing to hide, Grey considered, listening to him explain with tired patience that Ludgate had retired because of ill health and that Perriman had inherited money and gone back to England.

No. He watched the governor's hand twitch and hover indecisively over the fruit bowl. *He's got something to hide. And so does Dawes. Is it the same thing, though? And has it got anything to do with the present trouble?*

The governor could easily be hiding some peculation or corruption of his own—and likely was, Grey thought dispassionately, taking in the lavish display of silver on the sideboard. Such corruption was—within limits—considered more or less a perquisite of office. But if that were the case, it was not Grey's concern—unless it was in some way connected to the maroons and their rebellion.

Entertaining as it was to watch Fettes and Cherry at their work, he cut them off with a brief nod and turned the conversation firmly back to the rebellion.

"What communications have you had from the rebels, sir?" he asked the governor. "For I believe that, in these cases, rebellion arises usually from some distinct source of grievance. What is it?"

Warren looked at him, jaw agape. He closed his mouth, slowly, and hesitated for a moment before replying. Grey surmised he was considering how much Grey might discover from other avenues of inquiry.

Everything I bloody can, Grey thought, assuming an expression of neutral interest.

"Why, as to that, sir . . . the incident that began the . . . um . . . the difficulties . . . was the arrest of two young maroons, accused of stealing from

a warehouse in Kingston." The two had been whipped in the town square and committed to prison, after which—

"Following a trial?" Grey interrupted.

The governor's gaze rested on him, red-rimmed but cool. "No, Colonel. They had no right to a trial."

"You had them whipped and imprisoned on the word of . . . whom? The affronted merchant?"

Warren drew himself up a little and lifted his chin. Grey saw that he had been shaved, but a patch of black whisker had been overlooked; it showed in the hollow of his cheek like a blemish, a hairy mole.

"*I* did not, no, sir," he said coldly. "The sentence was imposed by the magistrate in Kingston."

"Who is?"

Dawes had closed his eyes with a small grimace.

"Judge Samuel Peters."

Grey nodded thanks.

"Captain Cherry will visit Mr. Judge Peters tomorrow," he said pleasantly. "And the prisoners, as well. I take it they are still in custody?"

"No, they aren't," Mr. Dawes put in, suddenly emerging from his impersonation of a dormouse. "They escaped, within a week of their capture."

The governor shot a brief, irritated glance at his secretary but nodded reluctantly. With further prodding, it was admitted that the maroons had sent a protest at the treatment of the prisoners, via Captain Cresswell. The prisoners having escaped before the protest was received, though, it had not seemed necessary to do anything about it.

Grey wondered briefly whose patronage had got Warren his position but dismissed the thought in favour of further explorations. The first violence had come without warning, he was told, with the burning of cane fields on a remote plantation. Word of it had reached Spanish Town several days later, by which time another plantation had suffered similar depredation.

"Captain Cresswell rode at once to investigate the matter, of course," Warren said, lips tight.

"And?"

"He didn't return. The maroons have not demanded ransom for him, nor have they sent word that he is dead. He may be with them; he may not. We simply don't know."

Grey could not help looking at Dawes, who appeared unhappy but gave the ghost of a shrug. It wasn't his place to tell more than the governor wanted told, was it?

"Let me understand you, sir," Grey said, not bothering to hide the

edge in his voice. "You have had *no* communication with the rebels since their initial protest? And you have taken no action to achieve any?"

Warren seemed to swell slightly but replied in an even tone.

"In fact, Colonel, I have. I sent for you." He smiled, very slightly, and reached for the decanter.

THE EVENING AIR hung damp and viscid, trembling with distant thunder. Unable to bear the stifling confines of his uniform any longer, Grey flung it off, not waiting for Tom's ministrations, and stood naked in the middle of the room, eyes closed, enjoying the touch of air from the terrace on his bare skin.

There was something remarkable about the air. Warm as it was, it had a silken touch that spoke of the sea and clear blue water, even indoors. He couldn't see the water from his room; even had it been visible from Spanish Town, his room faced a hillside covered with jungle. He could feel it, though, and had a sudden longing to wade out through surf and immerse himself in the clean coolness of the ocean. The sun had nearly set now, and the cries of parrots and other birds were growing intermittent.

He peered underneath the bed but didn't see the snake. Perhaps it was far back in the shadows; perhaps it had gone off in search of more ham. He straightened, stretched luxuriously, then shook himself and stood blinking, feeling stupid from too much wine and food and lack of sleep—he had slept barely three hours out of the preceding four-and-twenty, what with the arrival, disembarkation, and the journey to King's House.

His mind appeared to have taken French leave for the moment; no matter; it would be back shortly. Meanwhile, though, its abdication had left his body in charge—not at all a responsible course of action.

He felt exhausted, but restless, and scratched idly at his chest. The wounds there were solidly healed, slightly raised pink weals under his fingers, crisscrossing through the blond hair. One had passed within an inch of his left nipple; he'd been lucky not to lose it.

An immense pile of gauze cloth lay upon his bed. This must be the mosquito netting described to him by Mr. Dawes at dinner—a draped contraption meant to enclose the entire bed, thus protecting its occupant from the depredations of bloodthirsty insects.

He'd spent some time with Fettes and Cherry after dinner, laying plans for the morrow. Cherry would call upon Judge Peters and obtain details of the maroons who had been captured. Fettes would send men into Kingston in a search for the location of the retired Mr. Ludgate, erstwhile superintendent; if Ludgate could be found, Grey would like to know this

gentleman's opinion of his successor. As for that successor—if Dawes did not manage to unearth Captain Cresswell by the end of tomorrow . . . Grey yawned involuntarily, then shook his head, blinking. Enough.

The troops would all be billeted by now, some granted their first liberty in months. He spared a glance at the small sheaf of maps and reports he had extracted from Mr. Dawes earlier, but those could wait till morning, and better light. He'd think more clearly after a good night's sleep.

He leaned against the frame of the open door, after a quick glance down the terrace showed him that the rooms nearby seemed unoccupied. Clouds were beginning to drift in from the sea, and he remembered what Rodrigo had said about the rain at night. He thought perhaps he could feel a slight coolness in the air, whether from rain or oncoming night, and the hair on his body prickled and rose.

From here he could see nothing but the deep green of a jungle-clad hill, glowing like a sombre emerald in the twilight. From the other side of the house, though, as he left dinner, he'd seen the sprawl of Spanish Town below, a puzzle of narrow, aromatic streets. The taverns and the brothels would be doing a remarkable business tonight, he imagined.

The thought brought with it a rare feeling of something that wasn't quite resentment. Any one of the soldiers he'd brought, from the lowliest private soldier to Fettes himself, could walk into any brothel in Spanish Town—and there were a good many, Cherry had told him—and relieve the stresses caused by a long voyage without the slightest comment or even the slightest attention. Not him.

His hand had dropped lower as he watched the light fade, idly kneading his flesh. There were accommodations for men such as himself in London, but it had been many years since he'd had recourse to such a place.

He had lost one lover to death, another to betrayal. The third . . . His lips tightened. Could you call a man who would never touch you—would recoil from the very thought of touching you—your lover? No. But at the same time, what would you call a man whose *mind* touched yours, whose prickly friendship was a gift, whose character, whose very existence, helped to define your own?

Not for the first time—and surely not for the last—he wished briefly that Jamie Fraser were dead. It was an automatic wish, though, at once dismissed from mind. The colour of the jungle had died to ash, and insects were beginning to whine past his ears.

He went in and began to worry the folds of the gauze on his bed, until Tom came in to take it away from him, hang the mosquito netting, and ready him for the night.

HE COULDN'T SLEEP. Whether it was the heavy meal, the unac-customed place, or simply the worry of his new and so-far-unknown com-mand, his mind refused to settle, and so did his body. He didn't waste time in useless thrashing, though; he'd brought several books. Reading a bit of *The History of Tom Jones, A Foundling* would distract his mind and let sleep steal in upon him.

The French doors were covered with sheer muslin curtains, but the moon was nearly full, and there was enough light by which to find his tinderbox, striker, and candlestick. The candle was good beeswax, and the flame rose pure and bright—and instantly attracted a small cloud of in-quisitive gnats, mosquitoes, and tiny moths. He picked it up, intending to take it to bed with him, but then thought better.

Was it preferable to be gnawed by mosquitoes or to be incinerated? Grey debated the point for all of three seconds, then set the lit candlestick back on the desk. The gauze netting would go up in a flash if the candle fell over in bed.

Still, he needn't face death by bloodletting or be covered in itching bumps simply because his valet didn't like the smell of bear grease. He wouldn't get it on his clothes, in any case.

He flung off his nightshirt and knelt to rummage in his trunk, with a guilty look over his shoulder. Tom, though, was safely tucked up some-where amid the attics or outbuildings of King's House and almost cer-tainly sound asleep. Tom suffered badly with seasickness, and the voyage had been hard on him.

The heat of the Indies hadn't done the battered tin of bear grease any good, either; the rancid fat nearly overpowered the scent of the pepper-mint and other herbs mixed into it. Still, he reasoned, if it repelled him, how much more a mosquito, and he rubbed it into as much of his flesh as he could reach. Despite the stink, he found it not unpleasant. There was enough of the original smell left as to remind him of his usage of the stuff in Canada. Enough to remind him of Manoke, who had given it to him. Anointed him with it, in a cool blue evening on a deserted sandy isle in the St. Lawrence River.

Finished, he put down the tin and touched his rising prick. He didn't suppose he'd ever see Manoke again. But he did remember. Vividly.

A little later, he lay gasping on the bed under his netting, heart thump-ing slowly in counterpoint to the echoes of his flesh. He opened his eyes, feeling pleasantly relaxed, his head finally clear. The room was close; the servants had shut the windows, of course, to keep out the dangerous night air, and sweat misted his body. He felt too slack to get up and open the French doors onto the terrace, though; in a moment would do.

He closed his eyes again—then opened them abruptly and leapt out of bed, reaching for the dagger he'd laid on the table. The servant called

Rodrigo stood pressed against the door, the whites of his eyes showing in his black face.

"What do you want?" Grey put the dagger down but kept his hand on it, his heart still racing.

"I have a message for you, sah," the young man said. He swallowed audibly.

"Yes? Come into the light, where I can see you." Grey reached for his banyan and slid into it, still keeping an eye on the man.

Rodrigo peeled himself off the door with evident reluctance, but he'd come to say something, and say it he would. He advanced into the dim circle of candlelight, hands at his sides, nervously clutching air.

"Do you know, sah, what an Obeah man is?"

"No."

That disconcerted Rodrigo visibly. He blinked and twisted his lips, obviously at a loss as how to describe this entity. Finally, he shrugged his shoulders helplessly and gave up.

"He says to you, beware."

"Does he?" Grey said dryly. "Of anything specific?"

That seemed to help; Rodrigo nodded vigorously.

"You don't be close to the governor. Stay right away, as far as you can. He's going to—I mean . . . something bad might happen. Soon. He—" The servant broke off, apparently realising that he could be dismissed—if not worse—for talking about the governor in this loose fashion. Grey was more than curious, though, and sat down, motioning to Rodrigo to take the stool, which he did with obvious reluctance.

Whatever an Obeah man was, Grey thought, he clearly had considerable power, to force Rodrigo to do something he so plainly didn't want to do. The young man's face shone with sweat, and his hands clenched mindlessly on the fabric of his coat.

"Tell me what the Obeah man said," Grey said, leaning forward, intent. "I promise you, I will tell no one."

Rodrigo gulped but nodded. He bent his head, looking at the table as though he might find the right words written in the grain of the wood.

"Zombie," he muttered, almost inaudibly. "The zombie come for him. For the governor."

Grey had no notion what a zombie might be, but the word was spoken in such a tone as to make a chill flicker over his skin, sudden as distant lightning.

"Zombie," he said carefully. Mindful of the governor's reaction earlier, he asked, "Is a zombie perhaps a snake of some kind?"

Rodrigo gasped but then seemed to relax a little.

"No, sah," he said seriously. "Zombie are dead people." He stood up then, bowed abruptly, and left, his message delivered.

NOT SURPRISINGLY, Grey did not fall asleep immediately in the wake of this visit.

Having encountered German night-hags, Indian ghosts, and having spent a year or two in the Scottish Highlands, he had more acquaintance than most with picaresque superstition. While he wasn't inclined to give instant credence to local custom and belief, neither was he inclined to discount such belief out of hand. Belief made people do things that they otherwise wouldn't—and whether the belief had substance or not, the consequent actions certainly did.

Obeah men and zombies notwithstanding, plainly there was some threat to Governor Warren—and Grey rather thought the governor knew what it was.

How exigent was the threat, though? He pinched out the candle flame and sat in darkness for a moment, letting his eyes adjust, then rose and went soft-footed to the French doors, through which Rodrigo had vanished.

The guest bedchambers of King's House were merely a string of boxes, all facing the long terrace and each opening directly onto it through a pair of French doors. Grey paused for a moment, hand on the muslin drape; if anyone was watching his room, they would see the curtain being drawn aside.

Instead, he turned and went to the inner door of the room. This opened onto a narrow service corridor, completely dark at the moment—and completely empty, if his senses could be trusted. He closed the door quietly, glancing over his shoulder at the French doors. It was interesting, he thought, that Rodrigo had come to the front door, so to speak, when he could have approached Grey unseen.

But Rodrigo had said the Obeah man sent him. Plainly he wanted it to be seen that he had obeyed his order. Which in turn meant that someone was likely watching to see that he had.

The logical conclusion would be that the same someone—or someones—was watching to see what Grey might do next.

His body had reached its own conclusions already and was reaching for breeches and shirt before he had quite decided that if something were about to happen to Warren, it was clearly his duty to stop it, zombies or not. He stepped out of the French doors onto the terrace, moving quite openly.

There was an infantryman posted at either end of the terrace, as he'd expected; Robert Cherry was nothing if not meticulous. On the other hand, the bloody sentries had plainly not seen Rodrigo entering his room,

and he wasn't at all pleased about that. Recriminations could wait, though; the nearer sentry saw *him* and challenged him with a sharp "Who goes there?"

"It's me," Grey said briefly, and, without ceremony, dispatched the sentry with orders to alert the other soldiers posted around the house, then send two men into the house, where they should wait in the hall until summoned.

Grey then went back into his room, through the inner door, and down the dark service corridor. He found a dozing black servant behind a door at the end of it, minding the fire under the row of huge coppers that supplied hot water to the household.

The man blinked and stared when shaken awake but eventually nodded in response to Grey's demand to be taken to the governor's bedchamber. He led Grey into the main part of the house and up a darkened stair lit only by the moonlight streaming through the tall casements. Everything was quiet on the upper floor save for slow, regular snoring coming from what the slave said was the governor's room.

The man was swaying with weariness; Grey dismissed him, with orders to let in and send up the soldiers who should now be at the door. The man yawned hugely, and Grey watched him stumble down the stairs into the murk of the hall below, hoping he would not fall and break his neck. The house was very quiet. He was beginning to feel somewhat foolish. And yet . . .

The house seemed to breathe around him, almost as though it were a sentient thing and aware of him. He found the fancy unsettling.

Ought he to wake Warren? he wondered. Warn him? Question him? No, he decided. There was no point in disturbing the man's rest. Questions could wait for the morning.

The sound of feet coming up the stair dispelled his sense of uneasiness, and he gave his orders quietly. The sentries were to keep guard on this door until relieved in the morning; at any sound of disturbance within, they were to enter at once. Otherwise . . .

"Stay alert. If you see or hear *anything*, I wish to know about it."

He paused, but Warren continued to snore, so he shrugged and made his way downstairs, out into the silken night, and back to his own room.

He smelled it first. For an instant he thought he had left the tin of bear-grease ointment uncovered—and then the reek of sweet decay took him by the throat, followed instantly by a pair of hands that came out of the dark and fastened on said throat.

He fought back in blind panic, striking and kicking wildly, but the grip on his windpipe didn't loosen, and bright lights began to flicker at the corners of what would have been his vision if he'd had any. With a tremen-

dous effort of will, he made himself go limp. The sudden weight surprised his assailant and jerked Grey free of the throttling grasp as he fell. He hit the floor and rolled.

Bloody hell, where was the man? If it was a man. For even as his mind reasserted its claim to reason, his more-visceral faculties were recalling Rodrigo's parting statement: *'Zombie are dead people.'* And whatever was here in the dark with him seemed to have been dead for several days, judging from its smell.

He could hear the rustling of something moving quietly toward him. Was it breathing? He couldn't tell for the rasp of his own breath, harsh in his throat, and the blood-thick hammering of his heart in his ears.

He was lying at the foot of a wall, his legs half under the dressing table's bench. There was light in the room, now that his eyes were accustomed; the French doors were pale rectangles in the dark, and he could make out the shape of the thing that was hunting him. It was man-shaped but oddly hunched and swung its head and shoulders from side to side, almost as though it meant to smell him out. Which wouldn't take it more than two more seconds, at most.

He sat up abruptly, seized the small padded bench, and threw it as hard as he could at the thing's legs. It made a startled *oof!* noise that was undeniably human, then it staggered, waving its arms for balance. The noise reassured Grey, and he rolled up onto one knee and launched himself at the creature, bellowing incoherent abuse.

He butted it around chest height, felt it fall backwards, then lunged for the pool of shadow where he thought the table was. It was there and, feeling frantically over the surface, he found his dagger, still where he'd left it. He snatched it up and turned just in time to face the thing, which closed on him at once, reeking and making a disagreeable gobbling noise. He slashed at it and felt his knife skitter down the creature's forearm, bouncing off bone. It screamed, releasing a blast of foul breath directly into his face, then turned and rushed for the French doors, bursting through them in a shower of glass and flying muslin.

Grey charged after it, onto the terrace, shouting for the sentries. But the sentries, as he recalled belatedly, were in the main house, keeping watch over the governor, lest that worthy's rest be disturbed by . . . whatever sort of thing this was. Zombie?

Whatever it was, it was gone.

He sat down abruptly on the stones of the terrace, shaking with reaction. No one had come out in response to the noise. Surely no one could have slept through that; perhaps no one else was housed on this side of the mansion.

He felt ill and breathless and rested his head for a moment on his knees, before jerking it up to look round, lest something else be stealing up on

him. But the night was still and balmy. The only noise was an agitated rustling of leaves in a nearby tree, which for a shocked second he thought might be the creature, climbing from branch to branch in search of refuge. Then he heard soft chitterings and hissing squeaks. *Bats,* said the calmly rational part of his mind—what was left of it.

He gulped and breathed, trying to get clean air into his lungs to replace the disgusting stench of the creature. He'd been a soldier most of his life; he'd seen the dead on battlefields, and smelled them, too. Had buried fallen comrades in trenches and burned the bodies of his enemies. He knew what graves and rotting flesh smelled like. And the thing that had had its hands round his throat had almost certainly come from a recent grave.

He was shivering violently, despite the warmth of the night. He rubbed a hand over his left arm, which ached from the struggle; he had been badly wounded three years before, at Crefeld, and had nearly lost the arm. It worked but was still a good deal weaker than he'd like. Glancing at it, though, he was startled. Dark smears befouled the pale sleeve of his banyan, and, turning over his right hand, he found it wet and sticky.

"Jesus," he murmured, and brought it gingerly to his nose. No mistaking *that* smell, even overlaid as it was by grave reek and the incongruous scent of night-blooming jasmine from the vines that grew in tubs by the terrace. Rain was beginning to fall, pungent and sweet—but even that could not obliterate the smell.

Blood. Fresh blood. Not his, either.

He rubbed the rest of the blood from his hand with the hem of his banyan, and the cold horror of the last few minutes faded into a glowing coal of anger, hot in the pit of his stomach.

He'd been a soldier most of his life; he'd killed. He'd seen the dead on battlefields. And one thing he knew for a fact. Dead men don't bleed.

FETTES AND CHERRY had to know, of course. So did Tom, as the wreckage of his room couldn't be explained as the result of a nightmare. The four of them gathered in Grey's room, conferring by candlelight as Tom went about tidying the damage, white to the lips.

"You've never heard of zombie—or zombies? I have no idea whether the term is plural or not." Heads were shaken all round. A large square bottle of excellent Scotch whisky had survived the rigours of the voyage in the bottom of his trunk, and he poured generous tots of this, including Tom in the distribution.

"Tom—will you ask among the servants tomorrow? Carefully, of course. Drink that; it will do you good."

"Oh, I'll be careful, me lord," Tom assured him fervently. He took an obedient gulp of the whisky before Grey could warn him. His eyes bulged and he made a noise like a bull that has sat on a bumblebee, but managed somehow to swallow the mouthful, after which he stood still, opening and closing his mouth in a stunned sort of way.

Bob Cherry's mouth twitched, but Fettes maintained his usual stolid imperturbability.

"Why the attack upon you, sir, do you suppose?"

"If the servant who warned me about the Obeah man was correct, I can only suppose that it was a consequence of my posting sentries to keep guard upon the governor. But you're right." He nodded at Fettes's implication. "That means that whoever was responsible for this"—he waved a hand to indicate the disorder of his chamber, which still smelled of its recent intruder, despite the rain-scented wind that came through the shattered doors and the burnt-honey smell of the whisky—"either was watching the house closely, or—"

"Or lives here," Fettes said, and took a meditative sip. "Dawes, perhaps?"

Grey's eyebrows rose. That small, tubby, genial man? And yet he'd known a number of small, wicked men.

"Well," he said slowly, "it was not he who attacked me; I can tell you that much. Whoever it was was taller than I am and of a very lean build—not corpulent at all."

Tom made a hesitant noise, indicating that he had had a thought, and Grey nodded at him, giving permission to speak.

"You're quite sure, me lord, as the man who went for you . . . er . . . *wasn't* dead? Because by the smell of him, he's been buried for a week, at least."

A reflexive shudder went through all of them, but Grey shook his head.

"I am positive," he said, as firmly as he could. "It was a live man—though certainly a peculiar one," he added, frowning.

"Ought we to search the house, sir?" Cherry suggested.

Grey shook his head reluctantly.

"He—or it—went away into the garden. He left discernible footmarks." He did not add that there had been sufficient time for the servants—if they were involved—to hide any traces of the creature by now. If there was involvement, he thought, the servant Rodrigo was his best avenue of inquiry—and it would not serve his purposes to alarm the house and focus attention on the young man ahead of time.

"Tom," he said, turning to his valet. "Does Rodrigo appear to be approachable?"

"Oh, yes, me lord. He was friendly to me over supper," Tom assured him, brush in hand. "D'ye want me to talk to him?"

"Yes, if you will. Beyond that . . ." He rubbed a hand over his face, feeling the sprouting beard-stubble on his jaw. "I think we will proceed with the plans for tomorrow. But, Captain Cherry, will you also find time to question Mr. Dawes? You may tell him what transpired here tonight; I should find his response to that most interesting."

"Yes, sir." Cherry finished his whisky, coughed, and sat blinking for a moment, then cleared his throat. "The, um, the governor, sir . . . ?"

"I'll speak to him myself," Grey said. "And then I propose to ride up into the hills, to pay a visit to a couple of plantations, with an eye to defensive postings. For we must be seen to be taking prompt and decisive action. If there's offensive action to be taken against the maroons, it will wait until we see what we're up against." Fettes and Cherry nodded; lifelong soldiers, they had no urgent desire to rush into combat.

The meeting dismissed, Grey sat down with a fresh glass of whisky, sipping it as Tom finished his work in silence.

"You're sure as you want to sleep in this room tonight, me lord?" he said, putting the dressing-table bench neatly back in its spot. "I could find you another place, I'm sure."

Grey smiled at him with affection.

"I'm sure you could, Tom. But so could our recent friend, I expect. No, Captain Cherry will post a double guard on the terrace, as well as inside the house. It will be perfectly safe." And even if it wasn't, the thought of hiding, skulking away from whatever the thing was that had visited him . . . No. He wouldn't allow them—whoever they were—to think they had shaken his nerve.

Tom sighed and shook his head but reached into his shirt and drew out a small cross, woven of wheat stalks and somewhat battered, suspended on a bit of leather string.

"All right, me lord. But you'll wear this, at least."

"What is it?"

"A charm, me lord. Ilsa gave it to me, in Germany. She said it would protect me against evil—and so it has."

"Oh, no, Tom—surely you must keep—"

Mouth set in an expression of obstinacy that Grey knew well, Tom leaned forward and put the leather string over Grey's head. The mouth relaxed.

"There, me lord. Now *I* can sleep, at least."

GREY'S PLAN TO speak to the governor at breakfast was foiled, as that gentleman sent word that he was indisposed. Grey, Cherry, and Fettes all exchanged looks across the breakfast table, but Grey said merely,

"Fettes? *And* you, Captain Cherry, please." They nodded, a look of subdued satisfaction passing between them. He hid a smile; they loved questioning people.

The secretary, Dawes, was present at breakfast but said little, giving all his attention to the eggs and toast on his plate. Grey inspected him carefully, but he showed no sign, either of nocturnal excursions or of clandestine knowledge. Grey gave Cherry an eye. Both Fettes and Cherry brightened perceptibly.

For the moment, though, his own path lay clear. He needed to make a public appearance, as soon as possible, and to take such action as would make it apparent to the public that the situation was under control—and would make it apparent to the maroons that attention was being paid and that their destructive activities would no longer be allowed to pass unchallenged.

He summoned one of his other captains after breakfast and arranged for an escort. Twelve men should make enough of a show, he decided.

"And where will you be going, sir?" Captain Lossey asked, squinting as he made mental calculations regarding horses, pack mules, and supplies.

Grey took a deep breath and grasped the nettle.

"A plantation called Twelvetrees," he said. "Twenty miles or so into the uplands above Kingston."

PHILIP TWELVETREES was young, perhaps in his mid-twenties, and good-looking in a sturdy sort of way. He didn't stir Grey personally, but nonetheless Grey felt a tightness through his body as he shook hands with the man, studying his face carefully for any sign that Twelvetrees recognised his name or attributed any importance to his presence beyond the present political situation.

Not a flicker of unease or suspicion crossed Twelvetrees's face, and Grey relaxed a little, accepting the offer of a cooling drink. This turned out to be a mixture of fruit juices and wine, tart but refreshing.

"It's called *sangria*," Twelvetrees remarked, holding up his glass so the soft light fell glowing through it. "Blood, it means. In Spanish."

Grey did not speak much Spanish but did know that. However, blood seemed as good a *point d'appui* as any, concerning his business.

"So you think we might be next?" Twelvetrees paled noticeably beneath his tan. He hastily swallowed a gulp of sangria and straightened his shoulders, though. "No, no. I'm sure we'll be all right. Our slaves are loyal, I'd swear to that."

"How many have you? And do you trust them with arms?"

"One hundred and sixteen," Twelvetrees replied automatically. Plainly he was contemplating the expense and danger of arming some fifty men—for at least half his slaves must be women or children—and setting them essentially at liberty upon his property. Not to mention the vision of an unknown number of maroons, also armed, coming suddenly out of the night with torches. He drank a little more sangria. "Perhaps . . . what did you have in mind?" he asked abruptly, setting down his glass.

Grey had just finished laying out his suggested plans, which called for the posting of two companies of infantry at the plantation, when a flutter of muslin at the door made him lift his eyes.

"Oh, Nan!" Philip put a hand over the papers Grey had spread out on the table and shot Grey a quick warning look. "Here's Colonel Grey come to call. Colonel, my sister, Nancy."

"Miss Twelvetrees." Grey had risen at once and now took two or three steps toward her, bowing over her hand. Behind him, he heard the rustle as Twelvetrees hastily shuffled maps and diagrams together.

Nancy Twelvetrees shared her brother's genial sturdiness. Not pretty in the least, she had intelligent dark eyes—and these sharpened noticeably at her brother's introduction.

"Colonel Grey," she said, waving him gracefully back to his seat as she took her own. "Would you be connected with the Greys of Ilford, in Sussex? Or perhaps your family are from the London branch . . . ?"

"My brother has an estate in Sussex, yes," he said hastily. Forbearing to add that it was his half-brother Paul, who was not in fact a Grey, having been born of his mother's first marriage. Forbearing also to mention that his elder full brother was the Duke of Pardloe, and the man who had shot one Nathaniel Twelvetrees twenty years before. Which would logically expose the fact that Grey himself . . .

Philip Twelvetrees rather obviously did not want his sister alarmed by any mention of the present situation. Grey gave him the faintest of nods in acknowledgement, and Twelvetrees relaxed visibly, settling down to exchange polite social conversation.

"And what it is that brings you to Jamaica, Colonel Grey?" Miss Twelvetrees asked eventually. Knowing this was coming, Grey had devised an answer of careful vagueness, having to do with the Crown's concern for shipping. Halfway through this taradiddle, though, Miss Twelvetrees gave him a very direct look and demanded, "Are you here because of the governor?"

"Nan!" said her brother, shocked.

"Are you?" she repeated, ignoring her brother. Her eyes were very bright, and her cheeks flushed.

Grey smiled at her.

"What makes you think that that might be the case, may I ask, ma'am?"

"Because if you haven't come to remove Derwent Warren from his office, then *someone* should!"

"Nancy!" Philip was nearly as flushed as his sister. He leaned forward, grasping her wrist. "Nancy, please!"

She made as though to pull away, but then, seeing his pleading face, contented herself with a simple "Hmph!" and sat back in her chair, mouth set in a thin line.

Grey would dearly have liked to know what lay behind Miss Twelvetrees's animosity toward the governor, but he couldn't well inquire directly. Instead, he guided the conversation smoothly away, inquiring of Philip regarding the operations of the plantation and of Miss Twelvetrees regarding the natural history of Jamaica, for which she seemed to have some feeling, judging by the rather good watercolours of plants and animals that hung about the room, all neatly signed *N. T.*

Gradually, the sense of tension in the room relaxed, and Grey became aware that Miss Twelvetrees was focusing her attentions upon him. Not quite flirting—she was not built for flirtation—but definitely going out of her way to make him aware of her as a woman. He didn't quite know what she had in mind—he was presentable enough but didn't think she was truly attracted to him. Still, he made no move to stop her; if Philip should leave them alone together, he might be able to find out why she had said that about Governor Warren.

A quarter hour later, a mulatto man in a well-made suit put his head in at the door to the drawing room and asked if he might speak with Philip. He cast a curious eye toward Grey, but Twelvetrees made no move to introduce them, instead excusing himself and taking the visitor—who, Grey conceived, must be an overseer of some kind—to the far end of the large, airy room, where they conferred in low voices.

He at once seized the opportunity to fix his attention on Miss Nancy, in hopes of turning the conversation to his own ends.

"I collect you are acquainted with the governor, Miss Twelvetrees?" he asked, to which she gave a short laugh.

"Better than I might wish, sir."

"Really?" he said, in as inviting a tone as possible.

"Really," she said, and smiled unpleasantly. "But let us not waste time in discussing a . . . a person of such low character." The smile altered, and she leaned towards him, touching his hand, which surprised him. "Tell me, Colonel, does your wife accompany you? Or does she remain in London, from fear of fevers and slave uprisings?"

"Alas, I am unmarried, ma'am," he said, thinking that she likely knew a good deal more than her brother wished her to.

"Really," she said again, in an altogether different tone.

Her touch lingered on his hand, a fraction of a moment too long. Not long enough to be blatant, but long enough for a normal man to perceive it—and Grey's reflexes in such matters were much better developed than a normal man's, from necessity.

He barely thought consciously but smiled at her, then glanced at her brother, then back, with the tiniest of regretful shrugs. He forbore to add the lingering smile that would have said, *"Later."*

She sucked her lower lip in for a moment, then released it, wet and reddened, and gave him a look under lowered lids that said, *"Later,"* and a good deal more. He coughed, and out of the sheer need to say *something* completely free of suggestion asked abruptly, "Do you by chance know what an Obeah man is, Miss Twelvetrees?"

Her eyes sprang wide, and she lifted her hand from his arm. He managed to move out of her easy reach without actually appearing to shove his chair backwards and thought she didn't notice; she was still looking at him with great attention, but the nature of that attention had changed. The sharp vertical lines between her brows deepened into a harsh eleven.

"Where did you encounter that term, Colonel, may I ask?" Her voice was quite normal, her tone light—but she also glanced at her brother's turned back, and she spoke quietly.

"One of the governor's servants mentioned it. I see you are familiar with the term—I collect it is to do with Africans?"

"Yes." Now she was biting her upper lip, but the intent was not sexual. "The Koromantyn slaves—you know what those are?"

"No."

"Negroes from the Gold Coast," she said, and putting her hand once more on his sleeve, pulled him up and drew him a little away, toward the far end of the room. "Most planters want them, because they're big and strong and usually very well formed." Was it—no, he decided, it was *not* his imagination; the tip of her tongue had darted out and touched her lip in the fraction of an instant before she'd said "well formed." He thought Philip Twelvetrees had best find his sister a husband, and quickly.

"Do you have Koromantyn slaves here?"

"A few. The thing is, Koromantyns tend to be intractable. Very aggressive and hard to control."

"Not a desirable trait in a slave, I collect," he said, making an effort to keep any edge out of his tone.

"Well, it can be," she said, surprising him. She smiled briefly. "If your slaves are loyal—and ours are, I'd swear it—then you don't mind them being a bit bloody-minded toward . . . anyone who might want to come and cause trouble."

He was sufficiently shocked at her language that it took him a moment to absorb her meaning. The tongue tip flickered out again, and had she had dimples, she would certainly have employed them.

"I see," he said carefully. "But you were about to tell me what an Obeah man is. Some figure of authority, I take it, among the Koromantyns?"

The flirtatiousness vanished abruptly, and she frowned again.

"Yes. *Obi* is what they call their . . . religion, I suppose one must call it. Though from what little I know of it, no minister or priest would allow it that name."

Loud screams came from the garden below, and Grey glanced out, to see a flock of small, brightly coloured parrots swooping in and out of a big, lacy tree with yellowish fruit. Like clockwork, two small black children, naked as eggs, shot out of the shrubbery and aimed slingshots at the birds. Rocks spattered harmless among the branches, but the birds rose in a feathery vortex of agitation and flapped off, shrieking their complaints.

Miss Twelvetrees ignored the interruption, resuming her explanation directly the noise subsided.

"An Obeah man talks to the spirits. He, or she—there are Obeah women, too—is the person that one goes to, to . . . arrange things."

"What sorts of things?"

A faint hint of her former flirtatiousness reappeared.

"Oh . . . to make someone fall in love with you. To get with child. To get *without* child"—and here she looked to see whether she had shocked him again, but he merely nodded—"or to curse someone. To cause them ill luck or ill health. Or death."

This was promising.

"And how is this done, may I ask? Causing illness or death?"

Here, however, she shook her head.

"I don't know. It's really not safe to ask," she added, lowering her voice still further, and now her eyes were serious. "Tell me—the servant who spoke to you, what did he say?"

Aware of just how quickly gossip spreads in rural places, Grey wasn't about to reveal that threats had been made against Governor Warren. Instead, he asked, "Have you ever heard of zombies?"

She went quite white.

"No," she said abruptly.

It was a risk, but he took her hand to keep her from turning away.

"I cannot tell you why I need to know," he said, very low-voiced, "but please believe me, Miss Twelvetrees—Nancy." Callously, he pressed her hand. "It's extremely important. Any help that you can give me would be . . . well, I should appreciate it extremely."

Her hand was warm; the fingers moved a little in his, and not in an effort to pull away. Her colour was coming back.

"I truly don't know much," she said, equally low-voiced. "Only that zombies are dead people who have been raised by magic to do the bidding of the person who made them."

"The person who made them—this would be an Obeah man?"

"Oh! No," she said, surprised. "The Koromantyns don't make zombies. In fact, they think it quite an unclean practice."

"I'm entirely of one mind with them," he assured her. "Who *does* make zombies?"

"Nancy!" Philip had concluded his conversation with the overseer and was coming toward them, a hospitable smile on his broad, perspiring face. "I say, can we not have something to eat? I'm sure the colonel must be famished, and I'm most extraordinarily clemmed myself."

"Yes, of course," Miss Twelvetrees said, with a quick warning glance at Grey. "I'll tell Cook." Grey tightened his grip momentarily on her fingers, and she smiled at him.

"As I was saying, Colonel, you must call on Mrs. Abernathy at Rose Hall. She would be the person best equipped to inform you."

"Inform you?" Twelvetrees, curse him, chose this moment to become inquisitive. "About what?"

"Customs and beliefs among the Ashanti, my dear," his sister said blandly. "Colonel Grey has a particular interest in such things."

Twelvetrees snorted briefly.

"Ashanti, my left foot! Ibo, Fulani, Koromantyn—baptise 'em all proper Christians and let's hear no more about what heathen beliefs they may have brought with 'em. From the little *I* know, you don't want to hear about that sort of thing, Colonel. Though if you *do*, of course," he added hastily, recalling that it was not his place to tell the lieutenant-colonel who would be protecting Twelvetrees's life and property his business, "then my sister's quite right—Mrs. Abernathy would be best placed to advise you. Almost all her slaves are Ashanti. She . . . er . . . she's said to . . . um . . . take an interest."

To Grey's own interest, Twelvetrees's face went a deep red, and he hastily changed the subject, asking Grey fussy questions about the exact disposition of his troops. Grey evaded direct answers, beyond assuring Twelvetrees that two companies of infantry would be dispatched to his plantation as soon as word could be sent to Spanish Town.

He wished to leave at once, for various reasons, but was obliged to remain for tea, an uncomfortable meal of heavy, stodgy food, eaten under the heated gaze of Miss Twelvetrees. For the most part, he thought he had handled her with tact and delicacy, but toward the end of the meal she began to give him little pursed-mouth jabs. Nothing one could—or should—overtly notice, but he saw Philip blink at her once or twice in frowning bewilderment.

"Of course, I could not pose as an authority regarding any aspect of life on Jamaica," she said, fixing Grey with an unreadable look. "We have lived here barely six months."

"Indeed," he said politely, a wodge of undigested Savoy cake settling heavily in his stomach. "You seem very much at home—and a very lovely home it is, Miss Twelvetrees. I perceive your most harmonious touch throughout."

This belated attempt at flattery was met with the scorn it deserved; the eleven was back, hardening her brow.

"My brother inherited the plantation from his cousin, Edward Twelvetrees. Edward lived in London himself." She levelled a look like the barrel of a musket at him. "Did you know him, Colonel?"

And just what would the bloody woman do if he told her the truth? he wondered. Clearly, she thought she knew something, but . . . No, he thought, watching her closely. She couldn't know the truth but had heard some rumour. So this poking at him was an attempt—and a clumsy one—to get him to say more.

"I know several Twelvetrees casually," he said very amiably. "But if I met your cousin, I do not think I had the pleasure of speaking with him at any great length." *You bloody murderer!* and *Fucking sodomite!* not really constituting conversation, if you asked Grey.

Miss Twelvetrees blinked at him, surprised, and he realised what he should have seen much earlier. She was drunk. He had found the sangria light, refreshing—but had drunk only one glass himself. He had not noticed her refill her own, and yet the pitcher stood nearly empty.

"My dear," said Philip, very kindly. "It is warm, is it not? You look a trifle pale and indisposed." In fact, she was flushed, her hair beginning to come down behind her rather large ears—but she did indeed look indisposed. Philip rang the bell, rising to his feet, and nodded to the black maid who came in.

"I am not indisposed," Nancy Twelvetrees said, with some dignity. "I'm—I simply—that is—" But the black maid, evidently used to this office, was already hauling Miss Twelvetrees toward the door, though with sufficient skill as to make it look as though she merely assisted her mistress.

Grey rose, perforce, and took Miss Nancy's hand, bowing over it.

"Your servant, Miss Twelvetrees," he said. "I hope—"

"We know," she said, staring at him from large, suddenly tear-filled eyes. "Do you hear me? *We know.*" Then she was gone, the sound of her unsteady steps a ragged drumbeat on the parquet floor.

There was a brief, awkward silence between the two men. Grey cleared his throat just as Philip Twelvetrees coughed.

"Didn't really like cousin Edward," he said.

"Oh," said Grey.

They walked together to the yard, where Grey's horse browsed under a tree, its sides streaked with parrot droppings.

"Don't mind Nancy, will you?" Twelvetrees said quietly, not looking at him. "She had . . . a disappointment, in London. I thought she might get over it more easily here, but—well, I made a mistake, and it's not easy to unmake." He sighed, and Grey had a strong urge to pat him sympathetically on the back.

In lieu of that, he made an indeterminate noise in his throat, nodded, and mounted.

"The troops will be here the day after tomorrow, sir," he said. "You have my word upon it."

GREY HAD INTENDED to return to Spanish Town, but instead paused on the road, pulled out the chart Dawes had given him, and calculated the distance to Rose Hall. It would mean camping on the mountain overnight, but they were prepared for that—and beyond the desirability of hearing firsthand the details of a maroon attack, he was now more than curious to speak with Mrs. Abernathy regarding zombies.

He called his aide, wrote out instructions for the dispatch of troops to Twelvetrees, then sent two men back to Spanish Town with the message and two more on ahead to discover a good campsite. They reached this as the sun was beginning to sink, glowing like a flaming pearl in a soft pink sky.

"What is that?" he asked, glancing up abruptly from the cup of gunpowder tea Corporal Sansom had handed him. Sansom looked startled, too, and stared up the slope where the sound had come from.

"Don't know, sir," he said. "It sounds like a horn of some kind."

It did. Not a trumpet or anything of a standard military nature. Definitely a sound of human origin, though. The men stood quiet, waiting. A moment or two, and the sound came again.

"That's a different one," Sansom said, sounding alarmed. "It came from over there"—pointing up the slope—"didn't it?"

"Yes, it did," Grey said absently. "Hush!"

The first horn sounded again, a plaintive bleat almost lost in the noises of the birds settling for the night, and then fell silent.

Grey's skin tingled, his senses alert. They were not alone in the jungle. Someone—some*ones*—were out there in the oncoming night, signalling to each other. Quietly, he gave orders for the building of a hasty fortification, and the camp fell at once into the work of organising defence. The

men with him were mostly veterans and, while wary, not at all panicked. Within a very short time, a redoubt of stone and brush had been thrown up, sentries were posted in pairs around camp, and every man's weapon was loaded and primed, ready for an attack.

Nothing came, though, and while the men lay on their arms all night, there was no further sign of human presence. Such presence was there, though; Grey could feel it. Them. Watching.

He ate his supper and sat with his back against an outcrop of rock, dagger in his belt and loaded musket to hand. Waiting.

But nothing happened, and the sun rose. They broke camp in an orderly fashion, and if horns sounded in the jungle, the sound was lost in the shriek and chatter of the birds.

HE HAD NEVER been in the presence of anyone who repelled him so acutely. He wondered why that was; there was nothing overtly ill-favoured or ugly about her. If anything, she was a handsome Scotchwoman of middle age, fair-haired and buxom. And yet the widow Abernathy chilled him, despite the warmth of the air on the terrace where she had chosen to receive him at Rose Hall.

She was not dressed in mourning, he saw, nor did she make any obvious acknowledgement of the recent death of her husband. She wore white muslin, embroidered in blue about the hems and cuffs.

"I understand that I must congratulate you upon your survival, madam," he said, taking the seat she gestured him to. It was a somewhat callous thing to say, but she looked hard as nails; he didn't think it would upset her, and he was right.

"Thank you," she said, leaning back in her own wicker chair and looking him frankly up and down in a way that he found unsettling. "It was bloody cold in that spring, I'll tell ye that for nothing. Like to died myself, frozen right through."

He inclined his head courteously.

"I trust you suffered no lingering ill effects from the experience? Beyond, of course, the lamentable death of your husband," he hurried to add.

She laughed coarsely.

"Glad to get shot o' the wicked sod."

At a loss how to reply to this, Grey coughed and changed the subject.

"I am told, madam, that you have an interest in some of the rituals practised by slaves."

Her somewhat bleared green glance sharpened at that.

"Who told you that?"

"Miss Nancy Twelvetrees." There was no reason to keep the identity of his informant secret, after all.

"Oh, wee Nancy, was it?" She seemed amused by that, and shot him a sideways look. "I expect she liked *you*, no?"

He couldn't see what Miss Twelvetrees's opinion of him might have to do with the matter, and politely said so.

Mrs. Abernathy merely smirked, waving a hand. "Aye, well. What is it ye want to know, then?"

"I want to know how zombies are made."

Shock wiped the smirk off her face, and she blinked at him stupidly for a moment before picking up her glass and draining it.

"Zombies," she said, and looked at him with a certain wary interest. "Why?"

He told her. From careless amusement, her attitude changed, interest piqued. She made him repeat the story of his encounter with the thing in his room, asking pointed questions regarding its smell particularly.

"Decayed flesh," she said. "Ye'd ken what that smells like, would ye?"

It must have been her accent that brought back the battlefield at Culloden and the stench of burning corpses. He shuddered, unable to stop himself.

"Yes," he said abruptly. "Why?"

She pursed her lips in thought.

"There are different ways to go about it, aye? One way is to give the *afile* powder to the person, wait until they drop, and then bury them atop a recent corpse. Ye just spread the earth lightly over them," she explained, catching his look. "And make sure to put leaves and sticks over the face afore sprinkling the earth, so as the person can still breathe. When the poison dissipates enough for them to move again and sense things, they see they're buried, they smell the reek, and so they ken they must be dead." She spoke matter-of-factly, as though she had been telling him her private recipe for apple pandowdy or treacle cake. Weirdly enough, that steadied him, and he was able to speak calmly past his revulsion.

"Poison. That would be the *afile* powder? What sort of poison is it, do you know?"

Seeing the spark in her eye, he thanked the impulse that had led him to add "do you know?" to that question—for if not for pride, he thought, she might not have told him. As it was, she shrugged and answered offhand.

"Oh . . . herbs. Ground bones—bits o' other things. But the main thing, the one thing ye *must* have, is the liver of a *fugu* fish."

He shook his head, not recognising the name. "Describe it, if you

please." She did; from her description, he thought it must be one of the odd puffer fish that blew themselves up like bladders if disturbed. He made a silent resolve never to eat one. In the course of the conversation, though, something was becoming apparent to him.

"But what you are telling me—your pardon, madam—is that in fact a zombie is *not* a dead person at all? That they are merely drugged?"

Her lips curved; they were still plump and red, he saw, younger than her face would suggest. "What good would a dead person be to anyone?"

"But plainly the widespread belief is that zombies *are* dead."

"Aye, of course. The zombies think they're dead, and so does everyone else. It's not true, but it's effective. Scares folk rigid. As for 'merely drugged,' though . . ." She shook her head. "They don't come back from it, ye ken. The poison damages their brains and their nervous systems. They can follow simple instructions, but they've no real capacity for thought anymore—and they mostly move stiff and slow."

"Do they?" he murmured. The creature—well, the man, he was now sure of that—who had attacked him had not been stiff and slow, by any means. Ergo . . .

"I'm told, madam, that most of your slaves are Ashanti. Would any of them know more about this process?"

"No," she said abruptly, sitting up a little. "I learnt what I ken from a *houngan*—that would be a sort of . . . practitioner, I suppose ye'd say. He wasna one of my slaves, though."

"A practitioner of *what*, exactly?"

Her tongue passed slowly over the tips of her sharp teeth, yellowed but still sound.

"Of magic," she said, and laughed softly, as though to herself. "Aye, magic. African magic. Slave magic."

"You believe in magic?" He asked it as much from curiosity as anything else.

"Don't you?" Her brows rose, but he shook his head.

"I do not. And from what you have just told me yourself, the process of creating—if that's the word—a zombie is *not* in fact magic but merely the administration of poison over a period of time, added to the power of suggestion." Another thought struck him. "Can a person recover from such poisoning? You say it does not kill them."

She shook her head.

"The poison doesn't, no. But they always die. They starve, for one thing. They lose all notion of will and canna do anything save what the *houngan* tells them to do. Gradually they waste away to nothing, and—" Her fingers snapped silently.

"Even were they to survive," she went on practically, "the people would kill them. Once a person's been made a zombie, there's nay way back."

Throughout the conversation, Grey had become aware that Mrs. Abernathy spoke from what seemed a much closer acquaintance with the notion than one might acquire from an idle interest in natural philosophy. He wanted to get away from her but obliged himself to sit still and ask one more question.

"Do you know of any particular significance attributed to snakes, madam? In African magic, I mean."

She blinked, somewhat taken aback by that.

"Snakes," she repeated slowly. "Aye. Well . . . snakes ha' wisdom, they say. And some o' the *loas* are snakes."

"*Loas?*"

She rubbed absently at her forehead, and he saw, with a small prickle of revulsion, the faint stippling of a rash. He'd seen that before: the sign of advanced syphilitic infection.

"I suppose ye'd call them spirits," she said, and eyed him appraisingly. "D'ye see snakes in your dreams, Colonel?"

"Do I—no. I don't." He didn't, but the suggestion was unspeakably disturbing. She smiled.

"A *loa* rides a person, aye? Speaks through them. And I see a great huge snake lyin' on your shoulders, Colonel." She heaved herself abruptly to her feet.

"I'd be careful what ye eat, Colonel Grey."

THEY RETURNED TO Spanish Town two days later. The ride back gave Grey time for thought, from which he drew certain conclusions. Among these conclusions was the conviction that maroons had not, in fact, attacked Rose Hall. He had spoken to Mrs. Abernathy's overseer, who seemed reluctant and shifty, very vague on the details of the presumed attack. And later . . .

After his conversations with the overseer and several slaves, he had gone back to the house to take formal leave of Mrs. Abernathy. No one had answered his knock, and he had walked round the house in search of a servant. What he had found instead was a path leading downward from the house, with a glimpse of water at the bottom.

Out of curiosity, he had followed this path and found the infamous spring in which Mrs. Abernathy had presumably sought refuge from the murdering intruders. Mrs. Abernathy was in the spring, naked, swimming with slow composure from one side to the other, white-streaked fair hair streaming out behind her.

The water was crystalline; he could see the fleshy pumping of her buttocks, moving like a bellows that propelled her movements—and glimpsed

the purplish hollow of her sex, exposed by the flexion. There were no banks of concealing reeds or other vegetation; no one could have failed to see the woman if she'd been in the spring—and, plainly, the temperature of the water was no dissuasion to her.

So she'd lied about the maroons. He had a cold certainty that Mrs. Abernathy had murdered her husband, or arranged it—but there was little he was equipped to do with that conclusion. Arrest her? There were no witnesses—or none who could legally testify against her, even if they wanted to. And he rather thought that none of her slaves would want to; those he had spoken with had displayed extreme reticence with regard to their mistress. Whether that was the result of loyalty or fear, the effect would be the same.

What the conclusion *did* mean to him was that the maroons were in fact likely not guilty of murder, and that was important. So far, all reports of mischief involved only property damage—and that, only to fields and equipment. No houses had been burned, and while several plantation owners had claimed that their slaves had been taken, there was no proof of this; the slaves in question might simply have taken advantage of the chaos of an attack to run.

This spoke to him of a certain amount of care on the part of whoever led the maroons. Who did? he wondered. What sort of man? The impression he was gaining was not that of a rebellion—there had been no declaration, and he would have expected that—but of the boiling over of a long-simmering frustration. He *had* to speak with Captain Cresswell. And he hoped that bloody secretary had managed to find the superintendent by the time he reached King's House.

IN THE EVENT, he reached King's House long after dark and was informed by the governor's butler—appearing like a black ghost in his nightshirt—that the household were asleep.

"All right," he said wearily. "Call my valet, if you will. And tell the governor's servant in the morning that I will require to speak to His Excellency after breakfast, no matter what his state of health may be."

Tom was sufficiently pleased to see Grey in one piece as to make no protest at being awakened and had him washed, nightshirted, and tucked up beneath his mosquito netting before the church bells of Spanish Town tolled midnight. The doors of his room had been repaired, but Grey made Tom leave the window open and fell asleep with a silken wind caressing his cheeks and no thought of what the morning might bring.

He was roused from an unusually vivid erotic dream by an agitated

banging. He pulled his head out from under the pillow, the feel of rasping red hairs still rough on his lips, and shook his head violently, trying to re-orient himself in space and time. Bang, bang, bang, bang, *bang!* Bloody hell . . . ? Oh. Door.

"What? Come in, for God's sake! What the devil—oh. Wait a moment, then." He struggled out of the tangle of bedclothes and discarded nightshirt—good Christ, had he really been doing what he'd been dreaming about doing?—and flung his banyan over his rapidly detumescing flesh.

"What?" he demanded, finally getting the door open. To his surprise, Tom stood there, saucer-eyed and trembling, next to Major Fettes.

"Are you all right, me lord?" Tom burst out, cutting off Major Fettes's first words.

"Do I appear to be spurting blood or missing any necessary append-ages?" Grey demanded, rather irritably. "What's happened, Fettes?"

Now that he'd got his eyes properly open, he saw that Fettes looked almost as disturbed as Tom. The major—veteran of a dozen major cam-paigns, decorated for valour, and known for his coolness—swallowed vis-ibly and braced his shoulders.

"It's the governor, sir. I think you'd best come and see."

⁓

"WHERE ARE THE MEN who were assigned to guard him?" Grey asked calmly, stepping out of the governor's bedroom and closing the door gently behind him. The doorknob slid out of his fingers, slick under his hand. He knew the slickness was his own sweat, and not blood, but his stomach gave a lurch and he rubbed his fingers convulsively against the leg of his breeches.

"They're gone, sir." Fettes had got his voice, if not quite his face, back under control. "I've sent men to search the grounds."

"Good. Would you please call the servants together? I'll need to ques-tion them."

Fettes took a deep breath.

"They're gone, too."

"What? All of them?"

"Yes, sir."

He took a deep breath himself—and let it out again, fast. Even outside the room, the stench was gagging. He could feel the smell, thick on his skin, and rubbed his fingers on his breeches once again, hard. He swal-lowed and, holding his breath, jerked his head to Fettes—and to Cherry, who had joined them, shaking his head mutely in answer to Grey's raised

brow. No sign of the vanished sentries, then. God damn it; a search would have to be made for their bodies. The thought made him cold, despite the growing warmth of the morning.

He went down the stairs, his officers only too glad to follow. By the time he reached the foot, he had decided where to begin, at least. He stopped and turned to Fettes and Cherry.

"Right. The island is under military law as of this moment. Notify the officers, but tell them there is to be no public announcement yet. And *don't* tell them why." Given the flight of the servants, it was more than likely that news of the governor's death would reach the inhabitants of Spanish Town within hours—if it hadn't already. But if there was the slightest chance that the populace might remain in ignorance of the fact that Governor Warren had been killed and partially devoured in his own residence, while under the guard of His Majesty's army, Grey was taking it.

"What about the secretary?" he asked abruptly, suddenly remembering. "Dawes. Is he gone, too? Or dead?"

Fettes and Cherry exchanged a guilty look.

"Don't know, sir," Cherry said gruffly. "I'll go and look."

"Do that, if you please."

He nodded in return to their salutes and went outside, shuddering in relief at the touch of the sun on his face, the warmth of it through the thin linen of his shirt. He walked slowly around the terrace toward his room, where Tom had doubtless already managed to assemble and clean his uniform.

Now what? Dawes, if the man was still alive—and he hoped to God he was . . . A surge of saliva choked him, and he spat several times on the terrace, unable to swallow for the memory of that throat-clenching smell.

"Tom," he said urgently, coming into the room. "Did you have an opportunity to speak to the other servants? To Rodrigo?"

"Yes, me lord." Tom waved him onto the stool and knelt to put his stockings on. "They all knew about zombies—said they were dead people, just like Rodrigo said. A *houngan*—that's a . . . well, I don't quite know, but folk are right scared of 'em. Anyway, one of those who takes against somebody—or what's paid to do so, I reckon—will take the somebody and kill them, then raise 'em up again to be his servant, and that's a zombie. They were all dead scared of the notion, me lord," he said earnestly, looking up.

"I don't blame them in the slightest. Did any of them know about my visitor?"

Tom shook his head.

"They said not, but I think they did, me lord. They weren't a-going to

say. I got Rodrigo off by himself and he admitted he knew about it, but he said he didn't think it was a zombie what came after you, because I told him how you fought it and what a mess it made of your room." He narrowed his eyes at the dressing table, with its cracked mirror.

"Really? What did he think it was?"

"He wouldn't quite say, but I pestered him a bit, and he finally let on as it might have been a *houngan,* just pretending to be a zombie."

Grey digested that possibility for a moment. Had the creature who attacked him meant to kill him? If so, why? But if not, the attack might only have been meant to pave the way for what had now happened, by making it seem that there were zombies lurking about King's House in some profusion. That made a certain amount of sense, save for the fact . . .

"But I'm told that zombies are slow and stiff in their movements. Could one of them have done what . . . was done to the governor?" He swallowed.

"I dunno, me lord. Never met one." Tom grinned briefly at him, rising from fastening his knee buckles. It was a nervous grin, but Grey smiled back, heartened by it.

"I suppose I will have to go and look at the body again," he said, rising. "Will you come with me, Tom?" His valet was very observant, especially in matters pertaining to the body, and had been of help to him before in interpreting postmortem phenomena.

Tom paled noticeably but gulped and nodded and, squaring his shoulders, followed Lord John out onto the terrace.

On their way to the governor's room, they met Major Fettes, gloomily eating a slice of pineapple scavenged from the kitchen.

"Come with me, Major," Grey ordered. "You can tell me what discoveries you and Cherry have made in my absence."

"I can tell you one such, sir," Fettes said, putting down the pineapple and wiping his hands on his waistcoat. "Judge Peters has gone to Eleuthera."

"What the devil for?" That was a nuisance; he'd been hoping to discover more about the original incident that had incited the rebellion, and as he was obviously not going to learn anything from Warren . . . He waved a hand at Fettes; it hardly mattered why Peters had gone.

"Right. Well, then—" Breathing through his mouth as much as possible, Grey pushed open the door. Tom, behind him, made an involuntary sound but then stepped carefully up and squatted beside the body.

Grey squatted beside him. He could hear thickened breathing behind him.

"Major," he said, without turning round. "If Captain Cherry has found Mr. Dawes, would you be so kind as to fetch him in here?"

THEY WERE HARD at it when Dawes came in, accompanied by both Fettes and Cherry, and Grey ignored all of them.

"The bite marks *are* human?" he asked, carefully turning one of Warren's lower legs toward the light from the window. Tom nodded, wiping the back of his hand across his mouth.

"Sure of it, me lord. I been bitten by dogs—nothing like this. Besides—" He inserted his forearm into his mouth and bit down fiercely, then displayed the results to Grey. "See, me lord? The teeth go in a circle, like."

"No doubt of it." Grey straightened and turned to Dawes, who was sagging at the knees to such an extent that Captain Cherry was obliged to hold him up. "Do sit down, please, Mr. Dawes, and give me your opinion of matters here."

Dawes's round face was blotched, his lips pale. He shook his head and tried to back away but was prevented by Cherry's grip on his arm.

"I know nothing, sir," he gasped. "Nothing at all. Please, may I go? I, I . . . really, sir, I grow faint!"

"That's all right," Grey said pleasantly. "You can lie down on the bed if you can't stand up."

Dawes glanced at the bed, went white, and sat down heavily on the floor. Saw what was on the floor beside him and scrambled hurriedly to his feet, where he stood swaying and gulping.

Grey nodded at a stool, and Cherry propelled the little secretary, not ungently, onto it.

"What's he told you, Fettes?" Grey asked, turning back toward the bed. "Tom, we're going to wrap Mr. Warren up in the counterpane, then lay him on the floor and roll him up in the carpet. To prevent leakage."

"Right, me lord." Tom and Captain Cherry set gingerly about this process, while Grey walked over and stood looking down at Dawes.

"Pled ignorance, for the most part," Fettes said, joining Grey and giving Dawes a speculative look. "He did tell us that Derwent Warren had seduced a woman called Nancy Twelvetrees, in London. Threw her over, though, and married the heiress to the Atherton fortune."

"Who had better sense than to accompany her husband to the West Indies, I take it? Yes. Did he know that Miss Twelvetrees and her brother had inherited a plantation on Jamaica and were proposing to emigrate here?"

"No, sir." Dawes's voice was little more than a croak. He cleared his throat and spoke more firmly. "He was entirely surprised to meet the Twelvetrees at his first assembly."

"I daresay. Was the surprise mutual?"

"It was. Miss Twelvetrees went white, then red, then removed her shoe and set about the governor with the heel of it."

"I wish I'd seen that," Grey said, with real regret. "Right. Well, as you can see, the governor is no longer in need of your discretion. I, on the other hand, am in need of your loquacity. You can start by telling me why he was afraid of snakes."

"Oh." Dawes gnawed his lower lip. "I cannot be sure, you under-stand—"

"Speak up, you lump," growled Fettes, leaning menacingly over Dawes, who recoiled.

"I—I—" he stammered. "Truly, I don't know the details. But it—it had to do with a young woman. A young black woman. He—the gover-nor, that is—women were something of a weakness for him . . ."

"And?" Grey prodded.

The young woman, it appeared, was a slave in the household. And not disposed to accept the governor's attentions. The governor was not ac-customed to take "no" for an answer—and didn't. The young woman had vanished the next day, run away, and had not been recaptured as yet. But the day after, a black man in a turban and loincloth had come to King's House and had requested audience.

"He wasn't admitted, of course. But he wouldn't go away, either." Dawes shrugged. "Just squatted at the foot of the front steps and waited."

When Warren had at length emerged, the man had risen, stepped for-ward, and in formal tones informed the governor that he was herewith cursed.

"Cursed?" said Grey, interested. "How?"

"Well, now, there my knowledge reaches its limits, sir," Dawes replied. He had recovered some of his self-confidence by now and straightened up a little. "For, having pronounced the fact, he then proceeded to speak in an unfamiliar tongue—I think some of it may have been Spanish, though it wasn't all like that. I must suppose that he was, er, administering the curse, so to speak?"

"I'm sure I don't know." By now Tom and Captain Cherry had com-pleted their disagreeable task, and the governor reposed in an innocuous cocoon of carpeting. "I'm sorry, gentlemen, but there are no servants to assist us. We're going to take him down to the garden shed. Come, Mr. Dawes; you can be assistant pallbearer. And tell us on the way where the snakes come into it."

Panting and groaning, with the occasional near slip, they manhandled the unwieldy bundle down the stairs. Mr. Dawes, making ineffectual grabs at the carpeting, was prodded by Captain Cherry into further discourse.

"Well, I *thought* that I caught the word 'snake' in the man's tirade," he said. "And then . . . the snakes began to come."

Small snakes, large snakes. A snake was found in the governor's bath. Another appeared under the dining table, to the horror of a merchant's lady who was dining with the governor and who had hysterics all over the dining room before fainting heavily across the table. Mr. Dawes appeared to find something amusing in this, and Grey, perspiring heavily, gave him a glare that returned him more soberly to his account.

"Every day, it seemed, and in different places. We had the house searched, repeatedly. But no one could—or would, perhaps—detect the source of the reptiles. And while no one was bitten, still the nervous strain of not knowing whether you would turn back your coverlet to discover something writhing amongst your bedding . . ."

"Quite. Ugh!" They paused and set down their burden. Grey wiped his forehead on his sleeve. "And how did you make the connection, Mr. Dawes, between this plague of snakes and Mr. Warren's mistreatment of the slave girl?"

Dawes looked surprised and pushed his spectacles back up his sweating nose.

"Oh, did I not say? The man—I was told later that he was an Obeah man, whatever that may be—spoke her name, in the midst of his denunciation. Azeel, it was."

"I see. All right, ready? One, two, three—up!"

Dawes had given up any pretence of helping but scampered down the garden path ahead of them to open the shed door. He had quite lost any lingering reticence and seemed anxious to provide any information he could.

"He did not tell me directly, but I believe he had begun to dream of snakes and of the girl."

"How do—you know?" Grey grunted. "That's my foot, Major!"

"I heard him . . . er . . . speaking to himself. He had begun to drink rather heavily, you see. Quite understandable under the circumstances, don't you think?"

Grey wished he could drink heavily but had no breath left with which to say so.

There was a sudden cry of startlement from Tom, who had gone in to clear space in the shed, and all three officers dropped the carpet with a thump, reaching for nonexistent weapons.

"Me lord, me lord! Look who I found, a-hiding in the shed!" Tom was leaping up the path toward Grey, face abeam with happiness, the youth Rodrigo coming warily behind him. Grey's heart leapt at the sight, and he felt a most unaccustomed smile touch his face.

"Your servant, sah." Rodrigo, very timid, made a deep bow.

"I'm very pleased to see you, Rodrigo. Tell me—did you see anything of what passed here last night?"

The young man shuddered and turned his face away.

"No, sah," he said, so low-voiced Grey could barely hear him. "It was zombies. They . . . eat people. I heard them, but I know better than to look. I ran down into the garden and hid myself."

"You heard them?" Grey said sharply. "What did you hear, exactly?"

Rodrigo swallowed, and if it had been possible for a green tinge to show on skin such as his, he would undoubtedly have turned the shade of a sea turtle.

"Feet, sah," he said. "Bare feet. But they don't walk, *step-step*, like a person. They only shuffle, *sh-sh, sh-sh.*" He made small pushing motions with his hands in illustration, and Grey felt a slight lifting of the hairs on the back of his neck.

"Could you tell how many . . . men . . . there were?"

Rodrigo shook his head. "More than two, from the sound."

Tom pushed a little forward, round face intent. "Was there anybody else with 'em, d'you think? Somebody with a regular step, I mean?"

Rodrigo looked startled and then horrified.

"You mean a *houngan*? I don't know." He shrugged. "Maybe. I didn't hear shoes. But . . ."

"Oh. Because—" Tom stopped abruptly, glanced at Grey, and coughed. "Oh."

Despite more questions, this was all that Rodrigo could contribute, and so the carpet was picked up again—this time, with the servant helping—and bestowed in its temporary resting place. Fettes and Cherry chipped away a bit more at Dawes, but the secretary was unable to offer any further information regarding the governor's activities, let alone speculate as to what malign force had brought about his demise.

"Have you heard of zombies before, Mr. Dawes?" Grey inquired, mopping his face with the remains of his handkerchief.

"Er . . . yes," the secretary replied cautiously. "But surely you don't believe what the servant . . . Oh, surely not!" He cast an appalled glance at the shed.

"Are zombies in fact reputed to devour human flesh?"

Dawes resumed his sickly pallor.

"Well, yes. But . . . oh, dear!"

"Sums it up nicely," muttered Cherry, under his breath. "I take it you don't mean to make a public announcement of the governor's demise, then, sir?"

"You are correct, Captain. I don't want public panic over a plague of zombies at large in Spanish Town, whether that is actually the case or not. Mr. Dawes, I believe we need trouble you no more for the moment; you

are excused." He watched the secretary stumble off, before beckoning his officers closer. Tom moved a little away, discreet as always, and took Rodrigo with him.

"Have you discovered anything else that might have bearing on the present circumstance?"

They glanced at each other, and Fettes, wheezing gently, nodded to Cherry. Cherry strongly resembled that eponymous fruit, but, being younger and more slender than Fettes, had more breath than his superior.

"Yes, sir. I went looking for Ludgate, the old superintendent. Didn't find him—he's buggered off to Canada, they said—but I got a right earful concerning the present superintendent."

Grey groped for a moment for the name.

"Cresswell?"

"That's him."

"Peculation or corruption" appeared to sum up the subject of Captain Cresswell's tenure as superintendent very well, according to Cherry's informants in Spanish Town and Kingston. Amongst other abuses, he had arranged trade between the maroons on the uplands and the merchants below, in the form of bird skins, snakeskins, and other exotica, timber from the upland forests, and so on—but had, by report, accepted payment on behalf of the maroons but failed to deliver it.

"Had he any part in the arrest of the two young maroons accused of theft?"

Cherry's teeth flashed in a grin.

"Odd you should ask, sir. Yes, they said—well, some of them did—that the two young men had come down to complain about Cresswell's behaviour, but the governor wouldn't see them. They were heard to declare they would take back their goods by force—so when a substantial chunk of the contents of one warehouse went missing, it was assumed that was what they'd done. They—the maroons—insisted they hadn't touched the stuff, but Cresswell seized the opportunity and had them arrested for theft."

Grey closed his eyes, enjoying the momentary coolness of a breeze from the sea.

"The governor wouldn't see the young men, you said. Is there any suggestion of an improper connection between the governor and Captain Cresswell?"

"Oh, yes," said Fettes, rolling his eyes. "No proof yet—but we haven't been looking long, either."

"I see. And we still do not know the whereabouts of Captain Cresswell?"

Cherry and Fettes shook their heads in unison.

"The general conclusion is that Accompong scragged him," Cherry said.

"Who?"

"Oh. Sorry, sir," Cherry apologised. "That's the name of the maroon's headman, so they say. *Captain* Accompong, he calls himself, if you please." Cherry's lips twisted a little.

Grey sighed. "All right. No reports of any further depredations by the maroons, by whatever name?"

"Not unless you count murdering the governor," said Fettes.

"Actually," Grey said slowly, "I don't think that the maroons are responsible for this particular death." He was somewhat surprised to hear himself say so, in truth—and yet he found that he *did* think it.

Fettes blinked, this being as close to an expression of astonishment as he ever got, and Cherry looked openly sceptical. Grey did not choose to go into the matter of Mrs. Abernathy nor yet to explain his conclusions about the maroons' disinclination for violence. Strange, he thought. He had heard Captain Accompong's name only moments before, but with that name his thoughts began to coalesce around a shadowy figure. Suddenly there was a mind out there, someone with whom he might engage.

In battle, the personality and temperament of the commanding officer was nearly as important as the number of troops he commanded. So. He needed to know more about Captain Accompong, but that could wait for the moment.

He nodded to Tom, who approached respectfully, Rodrigo behind him.

"Tell them what you discovered, Tom."

Tom cleared his throat and folded his hands at his waist.

"Well, we . . . er . . . disrobed the governor"—Fettes flinched, and Tom cleared his throat again before going on—"and had a close look. And the long and the short of it, sir, and sir," he added, with a nod to Cherry, "is that Governor Warren was stabbed in the back."

Both officers looked blank.

"But . . . the place is covered with blood and filth and nastiness," Cherry protested. "It smells like that place where they put the bloaters they drag out of the Thames!"

"Footprints," Fettes said, giving Tom a faintly accusing look. "There were footprints. Big, bloody, *bare* footprints."

"I do not deny that something objectionable was present in that room," Grey said dryly. "But whoever—or whatever—gnawed the governor probably did not kill him. He was almost certainly dead when the . . . er . . . subsequent damage occurred."

Rodrigo's eyes were huge. Fettes was heard to observe under his breath

that he would be damned, but both Fettes and Cherry were good men and did not argue with Grey's conclusions, any more than they had taken issue with his order to hide Warren's body: they could plainly perceive the desirability of suppressing rumour of a plague of zombies.

"The point, gentlemen, is that after several months of incident, there has been nothing for the last month. Perhaps Mr. Warren's death is meant to be incitement, but if it was not the work of the maroons, then the question is . . . what are the maroons waiting for?"

Tom lifted his head, eyes wide.

"Why, me lord, I'd say—they're waiting for *you*. What else?"

WHAT ELSE, INDEED. Why had he not seen that at once? Of course Tom was right. The maroons' protest had gone unanswered, their complaint unremedied. So they had set out to attract attention in the most noticeable—if not the best—way open to them. Time had passed, nothing was done in response, and then they had heard that soldiers were coming. Lieutenant-Colonel Grey had now appeared. Naturally they were waiting to see what he would do.

What had he done so far? Sent troops to guard the plantations that were the most likely targets of a fresh attack. That was not likely to encourage the maroons to abandon their present plan of action, though it might cause them to direct their efforts elsewhere.

He walked to and fro in the wilderness of the King's House garden, thinking, but there were few alternatives.

He summoned Fettes and informed him that he, Fettes, was, until further notice, acting governor of the island of Jamaica.

Fettes looked more like a block of wood than usual.

"Yes, sir," he said. "If I might ask, sir . . . where are you going?"

"I'm going to talk to Captain Accompong."

"ALONE, SIR?" Fettes was appalled. "Surely you cannot mean to go up there *alone*!"

"I won't be," Grey assured him. "I'm taking my valet and the servant boy. I'll need someone who can translate for me, if necessary."

Seeing the mulish cast settling upon Fettes's brow, he sighed.

"To go there in force, Major, is to invite battle, and that is not what I want."

"No, sir," Fettes said dubiously, "but surely a proper escort . . . !"

"No, Major." Grey was courteous but firm. "I wish to make it clear that

I am coming to *speak* with Captain Accompong, and nothing more. I go alone."

"Yes, sir." Fettes was beginning to look like a block of wood that someone had set about with a hammer and chisel.

"As you wish, sir."

Grey nodded and turned to go into the house, but then paused and turned back.

"Oh, there is one thing that you might do for me, Major."

Fettes brightened slightly.

"Yes, sir?"

"Find me a particularly excellent hat, would you? With gold lace, if possible."

THEY RODE FOR nearly two days before they heard the first of the horns. A high, melancholy sound in the twilight, it seemed far away, and only a sort of metallic note made Grey sure that it was not in fact the cry of some large exotic bird.

"Maroons," Rodrigo said under his breath, and crouched a little, as though trying to avoid notice, even in the saddle. "That's how they talk to one another. Every group has a horn; they all sound different."

Another long, mournful falling note. Was it the same horn? Grey wondered. Or a second, answering the first?

"Talk to one another, you say. Can you tell what they're saying?"

Rodrigo had straightened up a bit in his saddle, putting a hand automatically behind him to steady the leather box that held the most ostentatious hat available in Spanish Town.

"Yes, sah. They're telling one another we're here."

Tom muttered something under his own breath, which sounded like, "Could have told you that meself for free," but declined to repeat or expand upon his sentiment when invited to do so.

They camped for the night under the shelter of a tree, so tired that they merely sat in silence as they ate, watching the nightly rainstorm come in over the sea, then crawled into the canvas tent Grey had brought. The young men fell asleep instantly to the pattering of rain above them.

Grey lay awake for a little, fighting tiredness, his mind reaching upwards. He had worn uniform, though not full dress, so that his identity would be apparent. And his gambit so far had been accepted; they had not been challenged, let alone attacked. Apparently Captain Accompong would receive him.

Then what? He wasn't sure. He did hope that he might recover his men—the two sentries who had disappeared on the night of Governor

Warren's murder. Their bodies had not been discovered, nor had any of their uniform or equipment turned up—and Captain Cherry had had the whole of Spanish Town *and* Kingston turned over in the search. If they had been taken alive, though, that reinforced his impression of Accompong—and gave him some hope that this rebellion might be resolved in some manner not involving a prolonged military campaign fought through jungles and rocks and ending in chains and executions. But if . . . Sleep overcame him, and he lapsed into incongruous dreams of bright birds, whose feathers brushed his cheeks as they flew silently past.

Grey woke in the morning to the feel of sun on his face. He blinked for a moment, confused, and then sat up. He was alone. Truly alone.

He scrambled to his feet, heart thumping, reaching for his dagger. It was there in his belt, but that was the only thing still where it should be. His horse—all the horses—were gone. So was his tent. So was the pack mule and its panniers. And so were Tom and Rodrigo.

He saw this at once—the blankets in which they'd lain the night before were still there, tumbled into the bushes—but he called for them anyway, again and again, until his throat was raw with shouting.

From somewhere high above him, he heard one of the horns, a long-drawn-out hoot that sounded mocking to his ears.

He understood the present message instantly. *You took two of ours; we have taken two of yours.*

"And you don't think I'll come and get them?" he shouted upwards into the dizzying sea of swaying green. "Tell Captain Accompong I'm coming! I'll have my young men back, and back *safe*—or I'll have his head!"

Blood rose in his face, and he thought he might burst but had better sense than to punch something, which was his very strong urge. He was alone; he couldn't afford to damage himself. He had to arrive among the maroons with everything that still remained to him, if he meant to rescue Tom and resolve the rebellion—and he did mean to rescue Tom, no matter what. It didn't matter that this might be a trap; he was going.

He calmed himself with an effort of will, stamping round in a circle in his stockinged feet until he had worked off most of his anger. That's when he saw them, sitting neatly side by side under a thorny bush.

They'd left him his boots. They did expect him to come.

HE WALKED FOR three days. He didn't bother trying to follow a trail; he wasn't a particularly skilled tracker, and finding any trace among the rocks and dense growth was a vain hope in any case. He simply climbed and listened for the horns.

The maroons hadn't left him any supplies, but that didn't matter. There were numerous small streams and pools, and while he was hungry, he didn't starve. Here and there he found trees of the sort he had seen at Twelvetrees, festooned with small yellowish fruits. If the parrots ate them, he reasoned, the fruits must be at least minimally comestible. They were mouth-puckeringly sour, but they didn't poison him.

The horns had increased in frequency since dawn. There were now three or four of them, signaling back and forth. Clearly, he was getting close. To what, he didn't know, but close.

He paused, looking up. The ground had begun to level out here; there were open spots in the jungle, and in one of these small clearings he saw what were plainly crops: mounds of curling vines that might be yams, beanpoles, the big yellow flowers of squash or gourds. At the far edge of the field, a tiny curl of smoke rose against the green. Close.

He took off the crude hat he had woven from palm leaves against the strong sun and wiped his face on the tail of his shirt. That was as much preparation as it was possible to make. The gaudy gold-laced hat he'd brought was presumably still in its box—wherever that was. He put his palm-leaf hat back on and limped toward the curl of smoke.

As he walked, he became aware of people fading slowly into view. Dark-skinned people, dressed in ragged clothing, coming out of the jungle to watch him with big, curious eyes. He'd found the maroons.

A SMALL GROUP of men took him further upwards. It was just before sunset, and the sunlight slanted gold and lavender through the trees when they led him into a large clearing, where there was a compound consisting of a number of huts. One of the men accompanying Grey shouted, and from the largest hut emerged a man who announced himself with no particular ceremony as Captain Accompong.

Captain Accompong was a surprise. He was very short, very fat, and hunchbacked, his body so distorted that he did not so much walk as proceed by a sort of sideways lurching. He was attired in the remnants of a splendid coat, now buttonless and with its gold lace half missing, the cuffs filthy with wear.

He peered from under the drooping brim of a ragged felt hat, eyes bright in its shadow. His face was round and much creased, lacking a good many teeth—but giving the impression of great shrewdness and perhaps good humor. Grey hoped so.

"Who are you?" Accompong asked, peering up at Grey like a toad under a rock.

Everyone in the clearing very plainly knew his identity; they shifted

from foot to foot and nudged one another, grinning. He paid no attention to them, though, and bowed very correctly to Accompong.

"I am the man responsible for the two young men who were taken on the mountain. I have come to get them back—along with my soldiers."

A certain amount of scornful hooting ensued, and Accompong let it go on for a few moments before lifting his hand. He sat down, carefully, sighing as he settled.

"You say so? Why you think I have anything to do with these young men?"

"I do not say that you do. But I know a great leader when I see one— and I know that you can help me to find my young men. If you will."

"Phu!" Accompong's face creased into a gap-toothed smile. "You think you flatter me and I help?"

Grey could feel some of the smaller children stealing up behind him; he heard muffled giggles but didn't turn round.

"I ask for your help. But I do not offer you only my good opinion in return."

A small hand reached under his coat and rudely tweaked his buttock. There was an explosion of laughter and mad scampering behind him. He didn't move.

Accompong chewed slowly at something in the back of his capacious mouth, one eye narrowed.

"Yes? What do you offer, then? Gold?" One corner of his thick lips turned up.

"Do you have any need of gold?" Grey asked. The children were whispering and giggling again behind him, but he also heard shushing noises from some of the women—they were getting interested. Maybe.

Accompong thought for a moment, then shook his head.

"No. What else you offer?"

"What do you want?" Grey parried.

"Captain Cresswell's head!" said a woman's voice, very clearly. There was a shuffle and smack, a man's voice rebuking in Spanish, a heated crackle of women's voices in return. Accompong let it go on for a minute or two, then raised one hand. Silence fell abruptly.

It lengthened. Grey could feel the pulse beating in his temples, slow and labouring. Ought he to speak? He came as a suppliant already; to speak now would be to lose face, as the Chinese put it. He waited.

"The governor is dead?" Accompong asked at last.

"Yes. How do you know of it?"

"You mean did I kill him?" The bulbous yellowed eyes creased.

"No," Grey said patiently. "I mean do you know how he died?"

"The zombies kill him." The answer came readily—and seriously. There was no hint of humor in those eyes now.

"Do you know who made the zombies?"

A most extraordinary shudder ran through Accompong, from his ragged hat to the horny soles of his bare feet.

"You do know," Grey said softly, raising a hand to prevent the automatic denial. "But it wasn't you, was it? Tell me."

The captain shifted uneasily from one buttock to the other but didn't reply. His eyes darted toward one of the huts, and after a moment he raised his voice, calling something in the maroons' patois, wherein Grey thought he caught the word "Azeel." He was puzzled momentarily, finding the word familiar but not knowing why. Then the young woman emerged from the hut, ducking under the low doorway, and he remembered.

Azeel. The young slave woman whom the governor had taken and misused, whose flight from King's House had presaged the plague of serpents.

Seeing her as she came forward, he couldn't help but see what had inspired the governor's lust, though it was not a beauty that spoke to him. She was small but not inconsequential. Perfectly proportioned, she stood like a queen, and her eyes burned as she turned her face to Grey. There was anger in her face—but also something like a terrible despair.

"Captain Accompong says that I will tell you what I know—what happened."

Grey bowed to her.

"I should be most grateful to hear it, madam."

She looked hard at him, obviously suspecting mockery, but he'd meant it, and she saw that. She gave a brief, nearly imperceptible nod.

"Well, then. You know that beast"—she spat neatly on the ground—"forced me? And I left his house?"

"Yes. Whereupon you sought out an Obeah man, who invoked a curse of snakes upon Governor Warren, am I correct?"

She glared at him and gave a short nod. "The snake is wisdom, and that man had none. None!"

"I think you're quite right about that. But the zombies?"

There was a general intake of breath among the crowd. Fear, distaste—and something else. The girl's lips pressed together, and tears glimmered in her large dark eyes.

"Rodrigo," she said, and choked on the name. "He—and I—" Her jaw clamped hard; she couldn't speak without weeping and would not weep in front of him. He cast down his gaze to the ground, to give her what privacy he could. He could hear her breathing through her nose, a soft, snuffling noise. Finally, she heaved a deep breath.

"He was not satisfied. He went to a *houngan*. The Obeah man warned him, but—" Her entire face contorted with the effort to hold in her feelings. "The *houngan*. He had zombies. Rodrigo paid him to kill the beast."

Grey felt as though he had been punched in the chest. Rodrigo. Rodrigo, hiding in the garden shed at the sound of shuffling bare feet in the night—or Rodrigo, warning his fellow servants to leave, then unbolting the doors, following a silent horde of ruined men in clotted rags up the stairs . . . or running up before them, in apparent alarm, summoning the sentries, drawing them outside, where they could be taken.

"And where is Rodrigo now?" Grey asked sharply. There was a deep silence in the clearing. None of the people even glanced at one another; every eye was fixed on the ground. He took a step toward Accompong. "Captain?"

Accompong stirred. He raised his misshapen face to Grey and a hand toward one of the huts.

"We do not like zombies, Colonel," he said. "They are unclean. And to kill a man using them . . . this is a great wrong. You understand this?"

"I do, yes."

"This man, Rodrigo . . ." Accompong hesitated, searching out words. "He is not one of us. He comes from Hispaniola. They . . . do such things there."

"Such things as make zombies? But presumably it happens here, as well." Grey spoke automatically; his mind was working furiously in light of these revelations. The thing that had attacked him in his room—it would be no great trick for a man to smear himself with grave dirt and wear rotted clothing . . .

"Not among us," Accompong said very firmly. "Before I say more, my colonel—do you believe what you have heard so far? Do you believe that we—that *I*—had nothing to do with the death of your governor?"

Grey considered that one for a moment. There was no evidence, only the story of the slave girl. Still . . . he did have evidence. The evidence of his own observations and conclusions regarding the nature of the man who sat before him.

"Yes," he said abruptly. "So?"

"Will your king believe it?"

Well, not as baldly stated, no, Grey thought. The matter would need a little tactful handling . . .

Accompong snorted faintly, seeing the thoughts cross Grey's face.

"This man, Rodrigo. He has done us great harm by taking his private revenge in a way that . . . that . . ." Accompong groped for the word.

"That incriminates you," Grey finished for him. "Yes, I see that. What have you done with him?"

"I cannot give this man to you," Accompong said at last. His thick lips pressed together briefly, but he met Grey's eye. "He is dead."

The shock hit Grey like a musket ball, a thump that knocked him off-balance and the sickening knowledge of irrevocable damage done.

"How?" he said, short and sharp. "What happened to him?"

The clearing was still silent. Accompong stared at the ground in front of him. After a long moment, a sigh, a whisper, drifted from the crowd.

"Zombie."

"Where?" he barked. "Where is he? Bring him to me. Now!"

The crowd shrank away from the hut, and a sort of moan ran through them. Women snatched up their children, pushed back so hastily that they stepped on the feet of their companions. The door opened.

"Anda!" said a voice from inside. *Walk,* it meant, in Spanish. Grey's numbed mind had barely registered this when the darkness inside the hut changed and a form appeared at the door.

It was Rodrigo. But then again—it wasn't. The glowing skin had gone pale and muddy, almost waxen. The firm, soft mouth hung loose, and the eyes—oh, God, the eyes! They were sunken, glassy, and showed no comprehension, no movement, not the least sense of awareness. They were a dead man's eyes. And yet . . . he walked.

This was the worst of all. Gone was every trace of Rodrigo's springy grace, his elegance. This creature moved stiffly, shambling, feet dragging, almost lurching from foot to foot. Its clothing hung upon its bones like a scarecrow's rags, smeared with clay and stained with dreadful liquids. The odour of putrefaction reached Grey's nostrils, and he gagged.

"Alto," said the voice softly, and Rodrigo stopped abruptly, arms hanging like a marionette's. Grey looked up then at the hut. A tall, dark man stood in the doorway, burning eyes fixed on Grey.

The sun was all but down; the clearing lay in deep shadow, and Grey felt a convulsive shiver go through him. He lifted his chin and, ignoring the horrid thing standing stiff before him, addressed the tall man.

"Who are you, sir?"

"Call me Ishmael," said the man, in an odd, lilting accent. He stepped out of the hut, and Grey was conscious of a general shrinking, everyone pulling away from the man, as though he suffered from some deadly contagion. Grey wanted to step back, too, but didn't.

"You did . . . this?" Grey asked, flicking a hand at the remnant of Rodrigo.

"I was paid to do it, yes." Ishmael's eyes flicked toward Accompong, then back to Grey.

"And Governor Warren—you were paid to kill him, as well, were you? By this man?" A brief nod at Rodrigo; he could not bear to look directly at him.

"The zombies think they're dead, and so does everyone else."

A frown drew Ishmael's brows together, and with the change of expression, Grey noticed that the man's faced was scarred, with apparent deliberation, long channels cut in cheeks and forehead. He shook his head.

"No. This"—he nodded at Rodrigo—"paid me to bring my zombies. He says to me that he wishes to terrify a man. And zombies will do that," he added, with a wolfish smile. "But when I brought them into the room and the *buckra* turned to flee, this one"—the flick of a hand toward Rodrigo—"sprang upon him and stabbed him. The man fell dead, and Rodrigo then *ordered* me"—his tone of voice made it clear what he thought of anyone ordering him to do anything—"to make my zombies feed upon him. And I did," he ended abruptly.

Grey swung round to Captain Accompong, who had sat silently through this testimony.

"And then you paid this . . . this—"

"Houngan," Ishmael put in helpfully.

"—to do *that*?!" He pointed at Rodrigo, and his voice shook with outraged horror.

"Justice," said Accompong, with simple dignity. "Don't you think so?"

Grey found himself temporarily bereft of speech. While he groped for something possible to say, the headman turned to a lieutenant and said, "Bring the other one."

"The *other*—" Grey began, but before he could speak further, there was another stir among the crowd, and from one of the huts a maroon emerged, leading another man by a rope around his neck. The man was wild-eyed and filthy, his hands bound behind him, but his clothes had originally been very fine. Grey shook his head, trying to dispel the remnants of horror that clung to his mind.

"Captain Cresswell, I presume?" he said.

"Save me!" the man panted, and collapsed on his knees at Grey's feet. "I beg you, sir—whoever you are—save me!"

Grey rubbed a hand wearily over his face and looked down at the erstwhile superintendent, then at Accompong.

"Does he need saving?" he asked. "I don't want to—I know what he's done—but it *is* my duty."

Accompong pursed his lips, thinking.

"You know what he is, you say. If I give him to you, what would you do with him?"

At least there was an answer to that one.

"Charge him with his crimes and send him to England for trial. If he is convicted, he would be imprisoned—or possibly hanged. What would happen to him here?" he asked curiously.

Accompong turned his head, looking thoughtfully at the *houngan,* who grinned unpleasantly.

"No!" gasped Cresswell. "No, please! Don't let him take me! I can't—I can't—oh, GOD!" He glanced, appalled, at the stiff figure of Rodrigo, then fell face-first onto the ground at Grey's feet, weeping convulsively.

Numbed with shock, Grey thought for an instant that it *would* probably resolve the rebellion . . . but no. Cresswell couldn't countenance the possibility of being handed over to Ishmael, and neither could Grey.

"Right," said Grey, and swallowed before turning to Accompong. "He *is* an Englishman, and, as I said, it's my duty to see that he's subject to English laws. I must therefore ask that you give him into my custody and take my word that I will see he receives justice. Our sort of justice," he added, giving the evil look back to the *houngan.*

"And if I don't?" Accompong asked, blinking genially at him.

"Well, I suppose I'll have to fight you for him," Grey said. "But I'm bloody tired and I really don't want to." Accompong laughed at this, and Grey followed swiftly up with "I will, of course, appoint a new superintendent—and, given the importance of the office, I will bring the new superintendent here so that you may meet him and approve of him."

"If I don't approve?"

"There are a bloody lot of Englishmen on Jamaica," Grey said, impatient. "You're bound to like *one* of them."

Accompong laughed out loud, his little round belly jiggling under his coat.

"I like you, Colonel," he said. "You want to be superintendent?"

Grey suppressed the natural answer to this and instead said, "Alas, I have a duty to the army which prevents my accepting the offer, amazingly generous though it is." He coughed. "You have my word that I will find you a suitable candidate, though."

The tall lieutenant who stood behind Captain Accompong lifted his voice and said something sceptical in a patois that Grey didn't understand—but from the man's attitude, his glance at Cresswell, and the murmur of agreement that greeted his remark, Grey had no trouble in deducing what had been said.

What is the word of an Englishman worth?

Grey gave Cresswell, grovelling and snivelling at his feet, a look of profound disfavour. It would serve the man right if—then he caught the faint reek of corruption wafting from Rodrigo's still form, and shuddered. No, nobody deserved *that.*

Putting aside the question of Cresswell's fate for the moment, Grey turned to the question that had been in the forefront of his mind since he'd come in sight of that first curl of smoke.

"My men," he said. "I want to see my men. Bring them out to me, please. At once." He didn't raise his voice, but he knew how to make a command sound like one.

Accompong tilted his head a little to one side, as though considering, but then waved a hand casually. There was a stirring in the crowd, an expectation. A turning of heads, then bodies, and Grey looked toward the

rocks where their focus lay. An explosion of shouts, catcalls, and laughter, and the two soldiers and Tom Byrd came out of the defile. They were roped together by the necks, their ankles hobbled and hands tied in front of them, and they shuffled awkwardly, bumping into one another, turning their heads to and fro like chickens, in a vain effort to avoid the spitting and the small clods of earth thrown at them.

Grey's outrage at this treatment was overwhelmed by his relief at seeing Tom and his young soldiers, all plainly scared but uninjured. He stepped forward at once so they could see him, and his heart was wrung by the pathetic relief that lighted their faces.

"Now, then," he said, smiling. "You didn't think I would leave you, surely?"

"*I* didn't, me lord," Tom said stoutly, already yanking at the rope about his neck. "I told 'em you'd be right along, the minute you got your boots on!" He glared at the little boys, naked but for shirts, who were dancing round him and the soldiers, shouting, *"Buckra! Buckra!"* and making not-quite-pretend jabs at the men's genitals with sticks. "Can you make 'em leave off that filthy row, me lord? They been at it ever since we got here."

Grey looked at Accompong and politely raised his brows. The headman barked a few words of something not quite Spanish, and the boys reluctantly fell back, though they continued to make faces and rude arm-pumping gestures.

Captain Accompong put out a hand to his lieutenant, who hauled the fat little headman to his feet. He dusted fastidiously at the skirts of his coat, then walked slowly around the small group of prisoners, stopping at Cresswell. He contemplated the man, who had now curled himself into a ball, then looked up at Grey.

"Do you know what a *loa* is, my colonel?" he asked quietly.

"I do, yes," Grey replied warily. "Why?"

"There is a spring, quite close. It comes from deep in the earth, where the *loas* live, and sometimes they will come forth and speak. If you will have back your men—I ask you to go there and speak with whatever *loa* may find you. Thus we will have truth, and I can decide."

Grey stood for a moment, looking back and forth from the fat old man to Cresswell, whose back heaved with silent sobs, to the young girl Azeel, who had turned her head to hide the hot tears coursing down her cheeks. He didn't look at Tom. There didn't seem much choice.

"All right," he said, turning back to Accompong. "Let me go now, then."

Accompong shook his head.

"In the morning," he said. "You do not want to go there at night."

"Yes, I do," Grey said. "Now."

"QUITE CLOSE" was a relative term, apparently. Grey thought it must be near midnight by the time they arrived at the spring—Grey, the *houngan* Ishmael, and four maroons bearing torches and armed with the long cane knives called machetes.

Accompong hadn't told him it was a *hot* spring. There was a rocky overhang and what looked like a cavern beneath it, from which steam drifted out like dragon's breath. His attendants—or guards, as one chose to look at it—halted as one, a safe distance away. He glanced at them for instruction, but they were silent.

He'd been wondering what the *houngan*'s role in this peculiar undertaking was. The man was carrying a battered canteen; now he uncorked this and handed it to Grey. It smelled hot, though the tin of the heavy canteen was cool in his hands. Raw rum, he thought, from the sweetly searing smell of it—and doubtless a few other things.

". . . Herbs. Ground bones—bits o' other things. But the main thing, the one thing ye must have, is the liver of a fugu fish . . . They don't come back from it, ye ken. The poison damages their brains . . ."

"Now we drink," Ishmael said. "And we enter the cave."

"Both of us?"

"Yes. I will summon the *loa*. I am a priest of Damballa." The man spoke seriously, with none of the hostility or smirking he had displayed earlier. Grey noticed, though, that their escort kept a safe distance from the *houngan*, and a wary eye upon him.

"I see," said Grey, though he didn't. "This . . . Damballa. He, or she—"

"Damballa is the great serpent," Ishmael said, and smiled, teeth flashing briefly in the torchlight. "I am told that snakes speak to you." He nodded at the canteen. "Drink."

Repressing the urge to say, "You first," Grey raised the canteen to his lips and drank slowly. It was *very* raw rum, with a strange taste, sweetly acrid, rather like the taste of fruit ripened to the edge of rot. He tried to keep any thought of Mrs. Abernathy's casual description of *afile* powder out of mind—she hadn't, after all, mentioned how the stuff might taste. And surely Ishmael wouldn't simply poison him . . . ? He hoped not.

He sipped the liquid until a slight shift of the *houngan*'s posture told him it was enough, then he handed the canteen to Ishmael, who drank from it without hesitation. Grey supposed he should find this comforting, but his head was beginning to swim in an unpleasant manner, his heartbeat throbbing audibly in his ears, and something odd was happening to his vision: it went intermittently dark, then returned with a brief flash of light, and when he looked at one of the torches, it had a halo of coloured rings around it.

He barely heard the *clunk* of the canteen, dropped on the ground, and watched, blinking, as the *houngan*'s white-clad back wavered before him. A dark blur of face as Ishmael turned to him.

"Come." The man disappeared into the veil of water.

"Right," he muttered. "Well, then . . ." He removed his boots, un-buckled the knee bands of his breeches, and peeled off his stockings. Then Grey shucked his coat and stepped cautiously into the steaming water.

It was hot enough to make him gasp, but within a few moments he had got used to the temperature and made his way across a shallow, steaming pool toward the mouth of the cavern, shifting gravel hard under his bare feet. He heard whispering from his guards, but no one offered any alternative suggestions.

Water poured from the overhang but not in the manner of a true waterfall—slender streams, like jagged teeth. The guards had pegged the torches into the ground at the edge of the spring; the flames danced like rainbows in the drizzle of the falling water as he passed beneath the over-hang.

The hot, wet air pressed his lungs and made it hard to breathe. After a short while he couldn't feel any difference between his skin and the moist air through which he walked; it was as though he had melted into the darkness of the cavern.

And it *was* dark. Completely. A faint glow came from behind him, but he could see nothing at all before him and was obliged to feel his way, one hand on the rough rock wall. The sound of falling water grew fainter, re-placed by the heavy thump of his own heartbeat, struggling against the pressure on his chest. Once he stopped and pressed his fingers against his eyelids, taking comfort in the coloured patterns that appeared there; he wasn't blind, then. When he opened his eyes again, though, the darkness was still complete.

He thought the walls were narrowing—he could touch them on both sides by stretching out his arms—and had a nightmare moment when he seemed to *feel* them drawing in upon him. He forced himself to breathe, a deep, explosive gasp, and forced the illusion back.

"Stop there." The voice was a whisper. He stopped.

There was silence for what felt like a long time.

"Come forward," said the whisper, sounding suddenly quite near him. "There is dry land, just before you."

He shuffled forward, felt the floor of the cave rise beneath him, and stepped out carefully onto bare rock. Walked slowly forward until again the voice bade him stop.

Silence. He thought he could make out breathing but wasn't sure; the sound of the water was still faintly audible in the distance. *All right,* he thought. *Come along, then.*

It hadn't been precisely an invitation, but what came into his mind was Mrs. Abernathy's intent green eyes, staring at him as she said, *"I see a great, huge snake lyin' on your shoulders, Colonel."*

With a convulsive shudder, he realised that he felt a weight on his shoulders. Not a dead weight but something live. It moved, just barely.

"Jesus," he whispered, and thought he heard the ghost of a laugh from somewhere in the cave. He stiffened and fought back against the mental image, for surely this was nothing more than imagination, fuelled by rum. Sure enough, the illusion of green eyes vanished—but the weight rested on him still, though he couldn't tell whether it lay upon his shoulders or his mind.

"So," said the low voice, sounding surprised. "The *loa* has come already. The snakes *do* like you, *buckra.*"

"And if they do?" he asked. He spoke in a normal tone of voice; his words echoed from the walls around him.

The voice chuckled briefly, and he felt rather than heard movement nearby, the rustle of limbs and a soft thump as something struck the floor near his right foot. His head felt immense, throbbing with rum, and waves of heat pulsed through him, though the depths of the cave were cool.

"See if this snake likes you, *buckra,*" the voice invited. "Pick it up."

He couldn't see a thing but slowly moved his foot, feeling his way over the silty floor. His toes touched something and he stopped. Whatever he had touched moved abruptly, recoiling from him. Then he felt the tiny flicker of a snake's tongue on his toe, tasting him.

Oddly, the sensation steadied him. Surely this wasn't his friend, the tiny yellow constrictor—but it was a serpent much like that one in general size, so far as he could tell. Nothing to fear from that.

"Pick it up," the voice invited again. "The krait will tell us if you speak the truth."

"Will he, indeed?" Grey said dryly. "How?"

The voice laughed, and he thought he heard two or three more chuckling behind it—but perhaps it was only echoes.

"If you die . . . you lied."

Grey gave a small, contemptuous snort. There were no venomous snakes on Jamaica. He cupped his hand and bent at the knee, but hesitated. He had an instinctive aversion to being bitten by a snake, venomous or not. And how did he know how the man—or men—sitting in the shadows would take it if the thing *did* bite him?

"I trust this snake," said the voice softly. "Krait comes with me from Africa. Long time now."

Grey's knees straightened abruptly. Africa! Now he placed the name, and cold sweat broke out on his face. *Krait.* A fucking *African* krait. Gwynne had had one. Small, no bigger than the circumference of a man's

little finger. *"Bloody deadly,"* Gwynne had crooned, stroking the thing's back with the tip of a goose quill—an attention to which the snake, a slender, nondescript brown thing, had seemed oblivious.

This one was squirming languorously over the top of Grey's foot; he had to restrain a strong urge to kick it away and stamp on it. What the devil was it about him that attracted *snakes*, of all ungodly things? He supposed it could be worse; it might be cockroaches. Instantly he felt a hideous crawling sensation upon his forearms and rubbed them hard reflexively, seeing—yes, he bloody *saw* them, here in the dark—thorny jointed legs and wriggling, inquisitive antennae brushing his skin.

He might have cried out. Someone laughed.

If he thought at all, he wouldn't be able to do it. He stooped and snatched the thing and, rising, hurled it into the darkness. There was a yelp and a scrabbling, then a brief, shocked scream.

He stood panting and trembling from reaction, checking and rechecking his hand—but felt no pain, could find no puncture wounds. The scream had been succeeded by a low stream of unintelligible curses, punctuated by the deep gasps of a man in terror. The voice of the *houngan*—if that's who it was—came urgently, followed by another voice, doubtful, fearful. Behind him, before him? He had no sense of direction anymore.

Something brushed past him, the heaviness of a body, and he fell against the wall of the cave, scraping his arm. He welcomed the pain; it was something to cling to, something real.

More urgency in the depths of the cave, sudden silence. And then a swishing *thunk!* as something struck hard into flesh, and the sheared-copper smell of fresh blood came strong over the scent of hot rock and rushing water. No further sound.

He was sitting on the muddy floor of the cave; he could feel the cool dirt under him. He pressed his hands flat against it, getting his bearings. After a moment, he heaved himself to his feet and stood, swaying and dizzy.

"I don't lie," he said, into the dark. "And I *will* have my men."

Dripping with sweat and water, he turned back, toward the rainbows.

THE SUN HAD barely risen when he came back into the mountain compound. The smoke of cooking fires hung among the huts, and the smell of food made his stomach clench painfully, but all that could wait. He strode as well as he might—his feet were so badly blistered that he hadn't been able to get his boots back on and had walked back barefoot, over rocks and thorns—to the largest hut, where Captain Accompong sat placidly waiting for him.

Tom and the soldiers were there, too, no longer roped together but still bound, kneeling by the fire. And Cresswell, a little way apart, appearing wretched but at least upright.

Accompong looked at one of his lieutenants, who stepped forward with a big cane knife and cut the prisoners' bonds with a series of casual but fortunately accurate swipes.

"Your men, my colonel," he said magnanimously, flipping one fat hand in their direction. "I give them back to you."

"I am deeply obliged to you, sir." Grey bowed. "There is one missing, though. Where is Rodrigo?"

There was a sudden silence. Even the shouting children hushed instantly, melting back behind their mothers. Grey could hear the trickling of water down the distant rock face and the pulse beating in his ears.

"The zombie?" Accompong said at last. He spoke mildly, but Grey sensed some unease in his voice. "He is not yours."

"Yes," Grey said firmly. "He is. He came to the mountain under my protection—and he will leave the same way. It is my duty."

The squatty headman's expression was hard to interpret. None of the crowd moved or murmured, though Grey caught glimpses from the corner of his eyes of the faint turning of heads, as folk asked silent questions of one another.

"It is my duty," Grey repeated. "I cannot go without him." He carefully omitted any suggestion that it might not be his choice whether to go or not. Still, why would Accompong return the white men to him if he planned to kill or imprison Grey?

The headman pursed fleshy lips, then turned his head and said something questioning. Movement in the hut where Ishmael had emerged the night before. There was a considerable pause, but, once more, the *houngan* came out.

His face was pale, and one of his feet was wrapped in a bloodstained wad of fabric, bound tightly. Amputation, Grey thought with interest, recalling the metallic *thunk* that had seemed to echo through his own flesh in the cave. It was the only sure way to keep a snake's venom from spreading through the body.

"Ah," said Grey, voice light. "So the krait liked me better, did he?"

He thought Accompong laughed under his breath, but he didn't really pay attention. The *houngan*'s eyes flashed hate at him, and Grey regretted his wit, fearing that it might cost Rodrigo more than had already been taken from him.

Despite his shock and horror, though, he clung to what Mrs. Abernathy had told him. The young man was *not* truly dead. He swallowed. Could Rodrigo perhaps be restored? The Scotchwoman had said not—but perhaps she was wrong. Clearly Rodrigo had not been a zombie for more

than a few days. And she did say that the drug dissipated over time. Perhaps . . .

Accompong spoke sharply, and the *houngan* lowered his head.

"Anda," he said sullenly. There was stumbling movement in the hut, and he stepped aside, half-pushing Rodrigo out into the light, where he came to a stop, staring vacantly at the ground, mouth open.

"You want this?" Accompong waved a hand at Rodrigo. "What for? He's no good to you surely? Unless you want to take him to bed—he won't say no to you!"

Everyone thought that very funny; the clearing rocked with laughter. Grey waited it out. From the corner of his eye, he saw the girl Azeel watching him with something like a fearful hope in her eyes.

"He is under my protection," he repeated. "Yes, I want him."

Accompong nodded and took a deep breath, sniffing appreciatively at the mingled scents of cassava porridge, fried plantain, and frying pig meat.

"Sit down, Colonel," he said, "and eat with me."

Grey sank slowly down beside him, weariness throbbing through his legs. Looking around, he saw Cresswell dragged roughly off but left sitting on the ground against a hut, unmolested. Tom and the two soldiers, looking dazed, were being fed at one of the cook fires. Then he saw Rodrigo, still standing like a scarecrow, and struggled to his feet.

He took the young man's tattered sleeve and said, "Come with me." Rather to his surprise, Rodrigo did, turning like an automaton. He led the young man through the staring crowd to the girl Azeel, and said, "Stop." He lifted Rodrigo's hand and offered it to the girl, who, after a moment's hesitation, took firmly hold of it.

"Look after him, please," Grey said to her. Only as he turned away did it register upon him that the arm he had held was wrapped with a bandage. Ah. Dead men don't bleed.

Returning to Accompong's fire, he found a wooden platter of steaming food awaiting him. He sank down gratefully upon the ground again and closed his eyes—then opened them, startled, as he felt something descend upon his head and found himself peering out from under the drooping felt brim of the headman's ragged hat.

"Oh," he said. "Thank you." He hesitated, looking round, either for the leather hatbox or for his ragged palm-frond hat, but didn't see either one.

"Never mind," said Accompong, and, leaning forward, slid his hands carefully over Grey's shoulders, palms up, as though lifting something heavy. "I will take your snake, instead. You have carried him long enough, I think."

AUTHOR'S NOTES

MY SOURCE FOR THE THEORETICAL BASIS OF MAK-
ing zombies was *The Serpent and the Rainbow: A Harvard Scientist's Astonishing Journey into the Secret Societies of Haitian Voodoo,
Zombis, and Magic,* by Wade Davis, which I'd read many years ago.
Information on the maroons of Jamaica, the temperament, beliefs,
and behaviour of Africans from different regions, and on historical
slave rebellions came chiefly from *Black Rebellion: Five Slave Revolts,*
by Thomas Wentworth Higginson. This manuscript (originally a series of articles published in *Atlantic Monthly, Harper's* magazine,
and *Century*) also supplied a number of valuable details regarding
terrain and personalities.

Captain Accompong was a real maroon leader—I took his physical description from this source—and the custom of trading hats
upon conclusion of a bargain also came from *Black Rebellion.* General background, atmosphere, and the importance of snakes came
from Zora Neale Hurston's *Tell My Horse* and a number of less important books dealing with voodoo. (By the way, I now have most of
my reference collection—some 2,500 books—listed on LibraryThing
and cross-indexed by topic, in case you're interested in pursuing anything like, say, Scotland, magic, or the American Revolution.)

A LEAF ON THE WIND
OF ALL HALLOWS

INTRODUCTION

ONE OF THE INTERESTING THINGS YOU CAN DO with a "bulge" (i.e., one of the novellas or short stories in the Outlander universe) is to follow mysteries, hints, and loose ends from the main books of the series. One such trail follows the story of Roger MacKenzie's parents.

In *Outlander*, we learn that Roger was orphaned during World War II, and then adopted by his great-uncle, the Reverend Reginald Wakefield, who tells his friends, Claire and Frank Randall, that Roger's mother was killed in the Blitz, and that his father was a Spitfire pilot "shot down over the Channel."

In *Drums of Autumn*, Roger tells his wife, Brianna, the moving story of his mother's death in the collapse of a Tube station during the bombing of London.

But in *An Echo in the Bone*, there is a poignant conversation in the moonlight between Claire and Roger, during which we encounter *this* little zinger:

Her hands wrapped his, small and hard and smelling of medicine.

"I don't know what happened to your father," she said. "But it wasn't what they told you [. . .]

"Of course things happen," she said, as though able to read his thoughts. "Accounts get garbled, too, over time and distance. Whoever told your mother might have been mistaken; she might have said something that the reverend misconstrued. All those things are possible. But during the War, I had letters from Frank—he wrote as often as he could, up until they recruited him into MI6. After that, I often wouldn't hear anything for months. But just before that, he wrote to me, and mentioned—just as casual chat, you know—that he'd run into something strange in the reports he was handling. A Spitfire had gone down, crashed—not shot down; they thought it must have been an engine failure—in Northumbria, and while it hadn't burned, for a wonder, there was no sign of the pilot. None. And he did mention the

name of the pilot, because he thought Jeremiah rather an appropriately doomed sort of name."

"Jerry," Roger said, his lips feeling numb. "My mother always called him Jerry."

"Yes," she said softly. "And there are circles of standing stones scattered all over Northumbria."

So what *really* happened to Jerry MacKenzie and his wife, Marjorie (known to her husband as Dolly)? Read on.

To the RAF flyers:
"Never have so many owed so much to so few."

IT WAS TWO WEEKS yet to Hallowe'en, but the gremlins were already at work.

Jerry MacKenzie turned *Dolly II* onto the runway full throttle, shoulder hunched, blood thumping, already halfway up Green leader's arse—pulled back on the stick and got a choking shudder instead of the giddy lift of takeoff. Alarmed, he eased back, but before he could try again, there was a bang that made him jerk by reflex, smacking his head against the Perspex. It hadn't been a bullet, though; the off tyre had blown, and a sickening tilt looped them off the runway, bumping and jolting into the grass.

There was a strong smell of petrol, and Jerry popped the Spitfire's hood and hopped out in panic, envisioning imminent incineration, just as the last plane of Green flight roared past him and took wing, its engine fading to a buzz within seconds.

A mechanic was pelting down from the hangar to see what the trouble was, but Jerry'd already opened Dolly's belly and the trouble was plain: the fuel line was punctured. Well, thank Christ he hadn't got into the air with it, that was one thing, but he grabbed the line to see how bad the puncture was, and it came apart in his hands and soaked his sleeve nearly to the shoulder with high-test petrol. Good job the mechanic hadn't come loping up with a lit cigarette in his mouth.

He rolled out from under the plane, sneezing, and Gregory the mechanic stepped over him.

"Not flying her today, mate," Greg said, squatting down to look up into the engine, and shaking his head at what he saw.

"Aye, tell me something I don't know." He held his soaked sleeve gingerly away from his body. "How long to fix her?"

Greg shrugged, eyes squinted against the cold wind as he surveyed Dolly's guts.

"Half an hour for the tyre. You'll maybe have her back up tomorrow, if the fuel line's the only engine trouble. Anything else we should be looking at?"

"Aye, the left wing-gun trigger sticks sometimes. Gie' us a bit o' grease, maybe?"

"I'll see what the canteen's got in the way of leftover dripping. You best hit the showers, Mac. You're turning blue."

He was shivering, right enough, the rapidly evaporating petrol wicking his body heat away like candlesmoke. Still, he lingered for a moment, watching as the mechanic poked and prodded, whistling through his teeth.

"Go on, then," Greg said in feigned exasperation, backing out of the engine and seeing Jerry still there. "I'll take good care of her."

"Aye, I know. I just—aye, thanks." Adrenaline from the aborted flight was still surging through him, thwarted reflexes making him twitch. He walked away, suppressing the urge to look back over his shoulder at his wounded plane.

JERRY CAME OUT of the pilots' WC half an hour later, eyes stinging with soap and petrol, backbone knotted. Half his mind was on Dolly, the other half with his mates. Blue and Green were up this morning, Red and Yellow resting. Green flight would be out over Flamborough Head by now, hunting.

He swallowed, still restless, dry-mouthed by proxy, and went to fetch a cup of tea from the canteen. That was a mistake; he heard the gremlins laughing as soon as he walked in and saw Sailor Malan.

Malan was Group Captain and a decent bloke overall. South African, a great tactician—and the most ferocious, most persistent air fighter Jerry'd seen yet. Rat terriers weren't in it. Which was why he felt a beetle skitter briefly down his spine when Malan's deep-set eyes fixed on him.

"Lieutenant!" Malan rose from his seat, smiling. "The very man I had in mind!"

The Devil he had, Jerry thought, arranging his face into a look of respectful expectancy. Malan couldn't have heard about Dolly's spot of bother yet, and without that, Jerry would have scrambled with A squadron on their way to hunt 109s over Flamborough Head. Malan hadn't been looking for Jerry; he just thought he'd do, for whatever job was up. And the fact that the Group Captain had called him by his rank, rather than his name, meant it probably wasn't a job anyone would volunteer for.

He didn't have time to worry about what that might be, though; Malan was introducing the other man, a tallish chap in army uniform with dark hair and a pleasant, if sharp, look about him. Eyes like a good sheepdog's, he thought, nodding in reply to Captain Randall's greeting. Kindly, maybe, but he won't miss much.

"Randall's come over from Ops at Ealing," Sailor was saying over his shoulder. He hadn't waited for them to exchange polite chat, but was already leading them out across the tarmac, heading for the Flight Command offices. Jerry grimaced and followed, casting a longing glance downfield at Dolly, who was being towed ignominiously into the hangar. The rag doll painted on her nose was blurred, the black curls partially dissolved by weather and spilled petrol. Well, he'd touch it up later, when he'd heard the details of whatever horrible job the stranger had brought.

His gaze rested resentfully on Randall's neck, and the man turned suddenly, glancing back over his shoulder as though he'd felt the stress of Jerry's regard. Jerry felt a qualm in the pit of his stomach, as half-recognised observations—the lack of insignia on the uniform, that air of confidence peculiar to men who kept secrets—gelled with the look in the stranger's eye.

Ops at Ealing, my Aunt Fanny, he thought. He wasn't even surprised, as Sailor waved Randall through the door, to hear the Group Captain lean close and murmur in his ear, "Careful—he's a funny bugger."

Jerry nodded, stomach tightening. Malan didn't mean Captain Randall was either humorous or a Freemason. "Funny bugger" in this context meant only one thing. MI6.

CAPTAIN RANDALL WAS from the secret arm of British Intelligence. He made no bones about it, once Malan had deposited them in a vacant office and left them to it.

"We're wanting a pilot—a good pilot," he added with a faint smile, "to fly solo reconnaissance. A new project. Very special."

"Solo? Where?" Jerry asked warily. Spitfires normally flew in four-plane flights, or in larger configurations, all the way up to an entire squadron, sixteen planes. In formation, they could cover one another to some extent against the heavier Heinkels and Messerschmitts. But they seldom flew alone by choice.

"I'll tell you that a bit later. First—are you fit, do you think?"

Jerry reared back a bit at that, stung. What did this bloody boffin think he— Then he caught a glance at his reflection in the windowpane. Eyes red as a mad boar's, his wet hair sticking up in spikes, a fresh red bruise spreading on his forehead, and his blouson stuck to him in damp patches where he hadn't bothered to dry off before dressing.

"Extremely fit," he snapped. "Sir."

Randall lifted a hand half an inch, dismissing the need for *sirs*.

"I meant your knee," he said mildly.

"Oh," Jerry said, disconcerted. "That. Aye, it's fine."

He'd taken two bullets through his right knee a year before, when he'd dived after a 109 and neglected to see another one that popped out of nowhere behind him and peppered his arse.

On fire, but terrified of bailing out into a sky filled with smoke, bullets, and random explosions, he'd ridden his burning plane down, both of them screaming as they fell out of the sky, *Dolly I*'s metal skin so hot it had seared his left forearm through his jacket, his right foot squelching in the blood that filled his boot as he stamped the pedal. Made it, though, and had been on the sick-and-hurt list for two months. He still limped very noticeably, but he didn't regret his smashed patella; he'd had his second month's sick leave at home—and wee Roger had come along nine months later.

He smiled broadly at the thought of his lad, and Randall smiled back in involuntary response.

"Good," he said. "You're all right to fly a long mission, then?"

Jerry shrugged. "How long can it be in a Spitfire? Unless you've thought up a way to refuel in the air." He'd meant that as a joke, and was further disconcerted to see Randall's lips purse a little, as though thinking whether to tell him they *had*.

"It is a Spitfire ye mean me to fly?" he asked, suddenly uncertain. Christ, what if it was one of the experimental birds they heard about now and again? His skin prickled with a combination of fear and excitement. But Randall nodded.

"Oh, yes, certainly. Nothing else is maneuverable enough, and there may be a good bit of ducking and dodging. What we've done is to take a Spitfire II, remove one pair of wing guns, and refit it with a pair of cameras."

"One pair?"

Again, that slight pursing of lips before Randall replied.

"You might need the second pair of guns."

"Oh. Aye. Well, then . . ."

The immediate notion, as Randall explained it, was for Jerry to go to Northumberland, where he'd spend two weeks being trained in the use of the wing cameras, taking pictures of selected bits of landscape at different altitudes. And where he'd work with a support team who were meant to be trained in keeping the cameras functioning in bad weather. They'd teach him how to get the film out without ruining it, just in case he had to. After which . . .

"I can't tell you yet exactly where you'll be going," Randall said. His manner through the conversation had been intent, but friendly, joking now and then. Now all trace of joviality had vanished; he was dead serious. "Eastern Europe is all I can say just now."

Jerry felt his inside hollow out a little and took a deep breath to fill the

empty space. He could say no. But he'd signed up to be an RAF flier, and that's what he was.

"Aye, right. Will I—maybe see my wife once, before I go, then?"

Randall's face softened a little at that, and Jerry saw the Captain's thumb touch his own gold wedding ring in reflex.

"I think that can be arranged."

MARJORIE MACKENZIE—Dolly, to her husband—opened the blackout curtains. No more than an inch . . . well, two inches. It wouldn't matter; the inside of the little flat was dark as the inside of a coal scuttle. London outside was equally dark; she knew the curtains were open only because she felt the cold glass of the window through the narrow crack. She leaned close, breathing on the glass, and felt the moisture of her breath condense, cool near her face. Couldn't see the mist, but felt the squeak of her fingertip on the glass as she quickly drew a small heart there, the letter *J* inside.

It faded at once, of course, but that didn't matter; the charm would be there when the light came in, invisible but there, standing between her husband and the sky.

When the light came, it would fall just so, across his pillow. She'd see his sleeping face in the light: the jackstraw hair, the fading bruise on his temple, the deep-set eyes, closed in innocence. He looked so young, asleep. Almost as young as he really was. Only twenty-two; too young to have such lines in his face. She touched the corner of her mouth but couldn't feel the crease the mirror showed her—her mouth was swollen, tender, and the ball of her thumb ran across her lower lip, lightly, to and fro.

What else, what else? What more could she do for him? He'd left her with something of himself. Perhaps there would be another baby—something he gave her, but something she gave him, as well. Another baby. Another child to raise alone?

"Even so," she whispered, her mouth tightening, face raw from hours of stubbled kissing; neither of them had been able to wait for him to shave. "Even so."

At least he'd got to see Roger. Hold his little boy—and have said little boy sick up milk all down the back of his shirt. Jerry'd yelped in surprise, but hadn't let her take Roger back; he'd held his son and petted him until the wee mannie fell asleep, only then laying him down in his basket and stripping off the stained shirt before coming to her.

It was cold in the room, and she hugged herself. She was wearing nothing but Jerry's string vest—he thought she looked erotic in it, "lewd," he

said, approving, his Highland accent making the word sound really dirty—and the thought made her smile. The thin cotton clung to her breasts, true enough, and her nipples poked out something scandalous, if only from the chill.

She wanted to go crawl in next to him, longing for his warmth, longing to keep touching him for as long as they had. He'd need to go at eight, to catch the train back; it would barely be light then. Some puritanical impulse of denial kept her hovering there, though, cold and wakeful in the dark. She felt as though if she denied herself, her desire, offered that denial as sacrifice, it would strengthen the magic, help to keep him safe and bring him back. God knew what a minister would say to that bit of superstition, and her tingling mouth twisted in self-mockery. And doubt.

Still, she sat in the dark, waiting for the cold blue light of the dawn that would take him.

Baby Roger put an end to her dithering, though; babies did. He rustled in his basket, making the little waking-up grunts that presaged an outraged roar at the discovery of a wet nappy and an empty stomach, and she hurried across the tiny room to his basket, breasts swinging heavy, already letting down her milk. She wanted to keep him from waking Jerry, but stubbed her toe on the spindly chair, and sent it over with a bang.

There was an explosion of bedclothes as Jerry sprang up with a loud "FUCK!" that drowned her own muffled "Damn!" and Roger topped them both with a shriek like an air-raid siren. Like clockwork, old Mrs. Munns in the next flat thumped indignantly on the thin wall.

Jerry's naked shape crossed the room in a bound. He pounded furiously on the partition with his fist, making the wallboard quiver and boom like a drum. He paused, fist still raised, waiting. Roger had stopped screeching, impressed by the racket.

Dead silence from the other side of the wall, and Marjorie pressed her mouth against Roger's round little head to muffle her giggling. He smelled of baby scent and fresh pee, and she cuddled him like a large hot-water bottle, his immediate warmth and need making her notions of watching over her men in the lonely cold seem silly.

Jerry gave a satisfied grunt and came across to her.

"Ha," he said, and kissed her.

"What d'ye think you are?" she whispered, leaning into him. "A gorilla?"

"Yeah," he whispered back, taking her hand and pressing it against him. "Want to see my banana?"

"Dzień dobry."

JERRY HALTED IN the act of lowering himself into a chair, and stared at a smiling Frank Randall.

"Oh, aye," he said. "Like that, is it? *Niech sie pan odpierdoli.*" It meant, "Fuck off, sir," in formal Polish, and Randall, taken by surprise, broke out laughing.

"Like that," he agreed. He had a wodge of papers with him, official forms, all sorts, the bumf, as the pilots called it—Jerry recognised the one you signed that named who your pension went to, and the one about what to do with your body if there was one and anyone had time to bother. He'd done all that when he signed up, but they made you do it again, if you went on special service. He ignored the forms, though, eyes fixed instead on the maps Randall had brought.

"And here's me thinkin' you and Malan picked me for my bonny face," he drawled, exaggerating his accent. He sat and leaned back, affecting casualness. "It is Poland, then?" So it hadn't been coincidence, after all—or only the coincidence of Dolly's mishap sending him into the building early. In a way, that was comforting; it wasn't the bloody Hand of Fate tapping him on the shoulder by puncturing the fuel line. The Hand of Fate had been in it a good bit earlier, putting him in Green flight with Andrej Kolodziewicz.

Andrej was a real guid yin, a good friend. He'd copped it a month before, spiralling up away from a Messerschmitt. Maybe he'd been blinded by the sun, maybe just looking over the wrong shoulder. Left wing shot to hell, and he'd spiralled right back down and into the ground. Jerry hadn't seen the crash, but he'd heard about it. And got drunk on vodka with Andrej's brother after.

"Poland," Randall agreed. "Malan says you can carry on a conversation in Polish. That true?"

"I can order a drink, start a fight, or ask directions. Any of that of use?"

"The last one might be," Randall said, very dry. "But we'll hope it doesn't come to that."

The MI6 agent had pushed aside the forms and unrolled the maps. Despite himself, Jerry leaned forward, drawn as by a magnet. They were official maps, but with markings made by hand—circles, *X*'s.

"It's like this," Randall said, flattening the maps with both hands. "The Nazis have had labour camps in Poland for the last two years, but it's not common knowledge among the public, either home or abroad. It would be very helpful to the war effort if it *were* common knowledge. Not just the camps' existence, but the kind of thing that goes on there." A shadow crossed the dark, lean face—anger, Jerry thought, intrigued. Apparently, Mr. MI6 knew what kind of thing went on there, and he wondered how.

"If we want it widely known and widely talked about—and we do—

we need documentary evidence," Randall said matter-of-factly. "Photographs."

There'd be four of them, he said, four Spitfire pilots. A flight—but they wouldn't fly together. Each one of them would have a specific target, geographically separate, but all to be hit on the same day.

"The camps are guarded, but not with anti-aircraft ordnance. There are towers, though; machine-guns." And Jerry didn't need telling that a machine-gun was just as effective in someone's hands as it was from an enemy plane. To take the sort of pictures Randall wanted would mean coming in low—low enough to risk being shot from the towers. His only advantage would be the benefit of surprise; the guards might spot him, but they wouldn't be expecting him to come diving out of the sky for a low pass just above the camp.

"Don't try for more than one pass, unless the cameras malfunction. Better to have fewer pictures than none at all."

"Yes, sir." He'd reverted to "sir," as Group Captain Malan was present at the meeting, silent but listening intently. Got to keep up appearances.

"Here's the list of the targets you'll practise on in Northumberland. Get as close as you think reasonable, without risking—" Randall's face did change at that, breaking into a wry smile. "Get as close as you can manage with a chance of coming back, all right? The cameras may be worth even more than you are."

That got a faint chuckle from Malan. Pilots—especially trained pilots—were valuable. The RAF had plenty of planes now, but nowhere near enough pilots to fly them.

He'd be taught to use the wing cameras and to unload the film safely. If he was shot down but was still alive and the plane didn't burn, he was to get the film out and try to get it back over the border.

"Hence the Polish." Randall ran a hand through his hair, and gave Jerry a crooked smile. "If you have to walk out, you may need to ask directions." They had two Polish-speaking pilots, he said—one Pole and a Hungarian who'd volunteered, and an Englishman with a few words of the language, like Jerry.

"And it is a volunteer mission, let me reiterate."

"Aye, I know," Jerry said irritably. "Said I'd go, didn't I? Sir."

"You did." Randall looked at him for a moment, dark eyes unreadable, then lowered his gaze to the maps again. "Thanks," he said softly.

THE CANOPY SNICKED shut over his head. It was a dank, damp Northumberland day, and his breath condensed on the inside of the Perspex hood within seconds. He leaned forward to wipe it away, emitting a

sharp yelp as several strands of his hair were ripped out. He'd forgotten to duck. Again. He shoved the canopy release with a muttered oath and the light brown strands that had caught in the seam where the Perspex closed flew away, caught up by the wind. He closed the canopy again, crouching, and waiting for the signal for takeoff.

The signalman wig-wagged him, and he turned up the throttle, feeling the plane begin to move.

He touched his pocket automatically, whispering, "Love you, Dolly," under his breath. Everyone had his little rituals, those last few moments before takeoff. For Jerry MacKenzie, it was his wife's face and his lucky stone that usually settled the worms in his belly. She'd found it in a rocky hill on the Isle of Lewis, where they'd spent their brief honeymoon—a rough sapphire, she said, very rare.

"Like you," he'd said, and kissed her.

No need for worms just the now, but it wasn't a ritual if you only did it sometimes, was it? And even if it wasn't going to be combat today, he'd need to be paying attention.

He went up in slow circles, getting the feel of the new plane, sniffing to get her scent. He wished they'd let him fly *Dolly II,* her seat stained with his sweat, the familiar dent in the console where he'd slammed his fist in exultation at a kill—but they'd already modified this one with the wing cameras and the latest thing in night sights. It didn't do to get attached to the planes, anyway; they were almost as fragile as the men flying them, though the parts could be reused.

No matter; he'd sneaked out to the hangar the evening before and done a quick rag doll on the nose to make it his. He'd know *Dolly III* well enough by the time they went into Poland.

He dived, pulled up sharp, and did Dutch rolls for a bit, wig-wagging through the cloud layer, then complete rolls and Immelmanns, all the while reciting Malan's Rules to focus his mind and keep from getting airsick.

The Rules were posted in every RAF barracks now: the Ten Commandments, the fliers called them—and not as a joke.

TEN OF MY RULES FOR AIR FIGHTING, the poster said in bold black type. Jerry knew them by heart.

"Wait until you see the whites of his eyes," he chanted under his breath. "Fire short bursts of one to two seconds only when your sights are definitely 'ON.'" He glanced at his sights, suffering a moment's disorientation. The camera wizard had relocated them. Shite.

"Whilst shooting, think of nothing else, brace the whole of your body: have both hands on the stick: concentrate on your ring sight." Well, away to fuck, then. The buttons that operated the camera weren't on

the stick; they were on a box connected to a wire that ran out the window; the box itself was strapped to his knee. He'd be bloody looking out the window anyway, not using sights—unless things went wrong and he had to use the guns. In which case . . .

"Always keep a sharp lookout. 'Keep your finger out.'" Aye, right, that one was still good.

"Height gives you the initiative." Not in this case. He'd be flying low, under the radar, and not be looking for a fight. Always the chance one might find him, though. If any German craft found him flying solo in Poland, his best chance was likely to head straight for the sun and fall in. That thought made him smile.

"Always turn and face the attack." He snorted and flexed his bad knee, which ached with the cold. Aye, if you saw it coming in time.

"Make your decisions promptly. It is better to act quickly even though your tactics are not the best." He'd learnt that one fast. His body often was moving before his brain had even notified his consciousness that he'd seen something. Nothing to see just now, nor did he expect to, but he kept looking by reflex.

"Never fly straight and level for more than thirty seconds in the combat area." Definitely out. Straight and level was just what he was going to have to do. And slowly.

"When diving to attack, always leave a proportion of your formation above to act as a top guard." Irrelevant; he wouldn't have a formation—and that was a thought that gave him the cold grue. He'd be completely alone, no help coming if he got into bother.

"INITIATIVE, AGGRESSION, AIR DISCIPLINE, and TEAM WORK are words that MEAN something in Air Fighting." Yeah, they did. What meant something in reconnaissance? Stealth, Speed, and Bloody Good Luck, more like. He took a deep breath, and dived, shouting the last of the Ten Commandments so it echoed in his Perspex shell.

"Go in quickly—Punch hard—GET OUT!"

Rubbernecking, they called it, but Jerry usually ended a day's flying feeling as though he'd been cast in concrete from the shoulder blades up. He bent his head forward now, ferociously massaging the base of his skull to ease the growing ache. He'd been practising since dawn, and it was nearly teatime. *Ball bearings, set, for the use of pilots, one,* he thought. Ought to add that to the standard equipment list. He shook his head like a wet dog, hunched his shoulders, groaning, then resumed the sector-by-sector scan of the sky around him that every pilot did religiously, three hundred and sixty degrees, every moment in the air. All the live ones, anyway.

Dolly'd given him a white silk scarf as a parting present. He didn't

know how she'd managed the money for it and she wouldn't let him ask, just settled it round his neck inside his flight jacket. Somebody'd told her the Spitfire pilots all wore them, to save the constant collar chafing, and she meant him to have one. It felt nice, he'd admit that. Made him think of her touch when she'd put it on him. He pushed the thought hastily aside; the last thing he could afford to do was start thinking about his wife, if he ever hoped to get back to her. And he did mean to get back to her.

Where was that bugger? Had he given up?

No, he'd not; a dark spot popped out from behind a bank of cloud just over his left shoulder and dived for his tail. Jerry turned, a hard, high spiral, up and into the same clouds, the other after him like stink on shite. They played at dodgem for a few moments, in and out of the drifting clouds—he had the advantage in altitude, could play the coming-out-of-the-sun trick, if there was any sun, but it was autumn in Northumberland and there hadn't been any sun in days . . .

Gone. He heard the buzzing of the other plane, faintly, for a moment—or thought he had. Hard to tell above the dull roar of his own engine. Gone, though; he wasn't where Jerry'd expected him to be.

"Oh, like that, is it?" He kept on looking, ten degrees of sky every second; it was the only way to be sure you didn't miss any— A glimpse of something dark, and his heart jerked along with his hand. Up and away. It was gone then, the black speck, but he went on climbing, slowly now, looking. Wouldn't do to get too low, and he wanted to keep the altitude . . .

The cloud was thin here, drifting waves of mist, but getting thicker. He saw a solid-looking bank of cloud moving slowly in from the west, but still a good distance away. It was cold, too; his face was chilled. He might be picking up ice if he went too hi— There.

The other plane, closer and higher than he'd expected. The other pilot spotted him at the same moment and came roaring down on him, too close to avoid. He didn't try.

"Aye, wait for it, ye wee bugger," he murmured, hand tight on the stick. One second, two, almost on him—and he buried the stick in his balls, jerked it hard left, turned neatly over, and went off in a long, looping series of barrel rolls that put him right away out of range.

His radio crackled and he heard Paul Rakoczy chortling through his hairy nose.

"*Kurwa twoja mac!* Where you learn that, you Scotch fucker?"

"At my mammy's tit, *dupek*," he replied, grinning. "Buy me a drink, and I'll teach it to ye."

A burst of static obscured the end of an obscene Polish remark, and Rakoczy flew off with a wig-wag of farewell. Ah, well. Enough skylarking, then; back to the fucking cameras.

Jerry rolled his head, worked his shoulders and stretched as well as could be managed in the confines of a II's cockpit—it had minor improvements over the Spitfire I, but roominess wasn't one of them—had a glance at the wings for ice—no, that was all right—and turned farther inland.

It was too soon to worry over it, but his right hand found the trigger that operated the cameras. His fingers twiddled anxiously over the buttons, checking, rechecking. He was getting used to them, but they didn't work like the gun triggers; he didn't have them wired in to his reflexes yet. Didn't like the feeling, either. Tiny things, like typewriter keys, not the snug feel of the gun triggers.

He'd had the left-handed ones only since yesterday; before that, he'd been flying a plane with the buttons on the right. Much discussion with Flight and the MI6 button-boffin, whether it was better to stay with the right, as he'd had practice already, or change for the sake of his cack-handedness. When they'd finally got round to asking him which he wanted, it had been too late in the day to fix it straight off. So he'd been given a couple of hours' extra flying time today, to mess about with the new fix-up.

Right, there it was. The bumpy grey line that cut through the yellowing fields of Northumberland like a perforation, same as you might tear the countryside along it, separating north from south as neat as tearing a piece of paper. Bet the emperor Hadrian wished it was that easy, he thought, grinning, as he swooped down along the line of the ancient wall.

The cameras made a loud *clunk-clunk* noise when they fired. *Clunk-clunk, clunk-clunk!* Okay, sashay out, bank over, come down . . . *clunk-clunk, clunk-clunk* . . . He didn't like the noise, not the same satisfaction as the vicious short *Brrpt!* of his wing guns. Made him feel wrong, like something gone with the engine . . . Aye, there it was coming up, his goal for the moment.

Mile-castle 37.

A stone rectangle, attached to Hadrian's Wall like a snail on a leaf. The old Roman legions had made these small, neat forts to house the garrisons that guarded the wall. Nothing left now but the outline of the foundation, but it made a good target.

He circled once, calculating, then dived and roared over it at an altitude of maybe fifty feet, cameras clunking like an army of stampeding robots. Pulled up sharp and hared off, circling high and fast, pulling out to run for the imagined border, circling up again . . . and all the time his heart thumped and the sweat ran down his sides, imagining what it would be like when the real day came.

Mid-afternoon, it would be, like this. The winter light just going, but still enough to see clearly. He'd circle, find an angle that would let him

cross the whole camp and please God, one that would let him come out of the sun. And then he'd go in.

One pass, Randall had said. Don't risk more than one, unless the cameras malfunction.

The bloody things did malfunction, roughly every third pass. The buttons were slippery under his fingers. Sometimes they worked on the next try; sometimes they didn't.

If they didn't work on the first pass over the camp, or didn't work often enough, he'd have to try again.

"*Niech to szlag,*" he muttered, *Fuck the Devil,* and pressed the buttons again, one-two, one-two. "Gentle but firm, like you'd do it to a lady's privates," the boffin had told him, illustrating a brisk twiddle. He'd never thought of doing that . . . Would Dolly like it? he wondered. And where exactly did you do it? Aye, well, women did come with a button, maybe that was it—but then, two fingers? . . . *Clunk-clunk. Clunk-clunk. Crunch.*

He reverted to English profanity, and smashed both buttons with his fist. One camera answered with a startled *clunk!* but the other was silent.

He poked the button again and again, to no effect. "Bloody fucking arse-buggering . . ." He thought vaguely that he'd have to stop swearing once this was over and he was home again—bad example for the lad.

"FUCK!" he bellowed, and ripping the strap free of his leg, he picked up the box and hammered it on the edge of the seat, then slammed it back onto his thigh—visibly dented, he saw with grim satisfaction—and pressed the balky button.

Clunk, the camera answered meekly.

"Aye, well, then, just you remember that!" he said, and, puffing in righteous indignation, gave the buttons a good jabbing.

He'd not been paying attention during this small temper-tantrum, but had been circling upward—standard default for a Spitfire flier. He started back down for a fresh pass at the mile-castle, but within a minute or two, began to hear a knocking sound from the engine.

"No!" he said, and gave it more throttle. The knocking got louder; he could feel it vibrating through the fuselage. Then there was a loud *clang!* from the engine compartment right by his knee, and with horror he saw tiny droplets of oil spatter on the Perspex in front of his face. The engine stopped.

"Bloody, bloody . . ." He was too busy to find another word. His lovely agile fighter had suddenly become a very clumsy glider. He was going down and the only question was whether he'd find a relatively flat spot to crash in.

His hand groped automatically for the landing gear but then drew back—no time, belly landing, where was the bottom? Jesus, he'd been

distracted, hadn't seen that solid bank of cloud move in; it must have come faster than he . . . Thoughts flitted through his mind, too fast for words. He glanced at the altimeter, but what it told him was of limited use, because he didn't know what the ground under him was like: crags, flat meadow, water? He hoped and prayed for a road, a grassy flat spot, anything short of—God, he was at five hundred feet and still in cloud!

"Christ!"

The ground appeared in a sudden burst of yellow and brown. He jerked the nose up, saw the rocks of a crag dead ahead, swerved, stalled, nose-dived, pulled back, pulled back, not enough, *Oh, God*—

HIS FIRST CONSCIOUS thought was that he should have radioed base when the engine went.

"Stupid fucker," he mumbled. "Make your decisions promptly. It is better to act quickly even though your tactics are not the best. Clot-heid."

He seemed to be lying on his side. That didn't seem right. He felt cautiously with one hand—grass and mud. What, had he been thrown clear of the plane?

He had. His head hurt badly, his knee much worse. He had to sit down on the matted wet grass for a bit, unable to think through the waves of pain that squeezed his head with each heartbeat.

It was nearly dark, and rising mist surrounded him. He breathed deep, sniffing the dank, cold air. It smelt of rot and old mangelwurzels—but what it didn't smell of was petrol and burning fuselage.

Right. Maybe she hadn't caught fire when she crashed, then. If not, and if her radio was still working . . .

He staggered to his feet, nearly losing his balance from a sudden attack of vertigo, and turned in a slow circle, peering into the mist. There was nothing *but* mist to his left and behind him, but to his right, he made out two or three large, bulky shapes, standing upright.

Making his way slowly across the lumpy ground, he found that they were stones. Remnants of one of those prehistoric sites that littered the ground in northern Britain. Only three of the big stones were still standing, but he could see a few more, fallen or pushed over, lying like bodies in the darkening fog. He paused to vomit, holding on to one of the stones. Christ, his head was like to split! And he had a terrible buzzing in his ears . . . He pawed vaguely at his ear, thinking somehow he'd left his headset on, but felt nothing but a cold, wet ear.

He closed his eyes again, breathing hard, and leaned against the stone for support. The static in his ears was getting worse, accompanied by a

sort of whine. Had he burst an eardrum? He forced himself to open his eyes, and was rewarded with the sight of a large, dark irregular shape, well beyond the remains of the stone circle. Dolly!

The plane was barely visible, fading into the swirling dark, but that's what it had to be. Mostly intact, it looked like, though very much nose-down with her tail in the air—she must have ploughed into the earth. He staggered on the rock-strewn ground, feeling the vertigo set in again, with a vengeance. He waved his arms, trying to keep his balance, but his head spun, and Christ, the bloody *noise* in his head . . . He couldn't think, oh, Jesus, he felt as if his bones were dissolv—

IT WAS FULL DARK when he came to himself, but the clouds had broken and a three-quarter moon shone in the deep black of a country sky. He moved, and groaned. Every bone in his body hurt—but none was broken. That was something, he told himself. His clothes were sodden with damp, he was starving, and his knee was so stiff he couldn't straighten his right leg all the way, but that was all right; he thought he could make shift to hobble as far as a road.

Oh, wait. Radio. Yes, he'd forgotten. If Dolly's radio were intact, he could . . .

He stared blankly at the open ground before him. He'd have sworn it was—but he must have got turned round in the dark and fog—no.

He turned quite round, three times, before he stopped, afraid of becoming dizzy again. The plane was gone.

It *was* gone. He was sure it had lain about fifty feet beyond that one stone, the tallest one; he'd taken note of it as a marker, to keep his bearings. He walked out to the spot where he was sure Dolly had come down, walked slowly round the stones in a wide circle, glancing to one side and then the other in growing confusion.

Not only was the plane gone, it didn't seem ever to have been there. There was no trace, no furrow in the thick meadow grass, let alone the kind of gouge in the earth that such a crash would have made. Had he been imagining its presence? Wishful thinking?

He shook his head to clear it—but in fact, it *was* clear. The buzzing and whining in his ears had stopped, and while he still had bruises and a mild headache, he was feeling much better. He walked slowly back around the stones, still looking, a growing sense of deep cold curling through his wame. It wasn't fucking there.

HE WOKE IN the morning without the slightest notion where he was. He was curled up on grass; that much came dimly to him—he could smell it. Grass that cattle had been grazing, because there was a large cow pat just by him, and fresh enough to smell that, too. He stretched out a leg, cautious. Then an arm. Rolled onto his back, and felt a hair better for having something solid under him, though the sky overhead was a dizzy void.

It was a soft, pale blue void, too. Not a trace of cloud.

How long . . . ? A jolt of alarm brought him up onto his knees, but a bright yellow stab of pain behind his eyes sat him down again, moaning and cursing breathlessly.

Once more. He waited 'til his breath was coming steady, then risked cracking one eye open.

Well, it was certainly still Northumbria, the northern part, where England's billowing fields crash onto the inhospitable rocks of Scotland. He recognised the rolling hills, covered with sere grass and punctuated by towering rocks that shot straight up into sudden toothy crags. He swallowed, and rubbed both hands hard over his head and face, assuring himself he was still real. He didn't feel real. Even after he'd taken a careful count of fingers, toes, and private bits—counting the last twice, just in case—he still felt that something important had been misplaced, torn off somehow, and left behind.

His ears still rang, rather like they did after an especially active trip. Why, though? What had he heard?

He found that he could move a little more easily now, and managed to look all round the sky, sector by sector. Nothing up there. No memory of anything up there. And yet the inside of his head buzzed and jangled, and the flesh on his body rippled with agitation. He chafed his arms, hard, to make it go.

Horripilation. That's the proper word for gooseflesh; Dolly'd told him that. She kept a little notebook and wrote down words she came across in her reading; she was a great one for the reading. She'd already got wee Roger sitting in her lap to be read to after tea, round-eyed as Bonzo at the coloured pictures in his rag book.

The thought of his family got him up onto his feet, swaying, but all right now, better, yes, definitely better, though he still felt as though his skin didn't quite fit. The plane, where was that?

He looked round him. No plane was visible. Anywhere. Then it came back to him, with a lurch of the stomach. Real, it was real. He'd been sure in the night that he was dreaming or hallucinating, had lain down to recover himself, and must have fallen asleep. But he was awake now, no mistake; there was a bug of some kind down his back, and he slapped viciously to try to squash it.

His heart was pounding unpleasantly and his palms were sweating. He

wiped them on his trousers and scanned the landscape. It wasn't flat, but neither did it offer much concealment. No trees, no bosky dells. There was a small lake off in the distance—he caught the shine of water—but if he'd ditched in water, surely to God he'd be wet?

Maybe he'd been unconscious long enough to dry out, he thought. Maybe he'd imagined that he'd seen the plane near the stones. Surely he couldn't have walked this far from the lake and forgotten it? He'd started walking toward the lake, out of sheer inability to think of anything more useful to do. Clearly time had passed; the sky had cleared like magic. Well, they'd have little trouble finding him, at least; they knew he was near the wall. A truck should be along soon; he couldn't be more than two hours from the airfield.

"And a good thing, too," he muttered. He'd picked an especially god-forsaken spot to crash—there wasn't a farmhouse or a paddock anywhere in sight, not so much as a sniff of chimney smoke.

His head was becoming clearer now. He'd circle the lake—just in case—then head for the road. Might meet the support crew coming in.

"And tell them I've lost the bloody plane?" he asked himself aloud. "Aye, right. Come on, ye wee idjit, think! Now, where did ye see it last?"

HE WALKED FOR a long time. Slowly, because of the knee, but that began to feel easier after a while. His mind was not feeling easier. There was something wrong with the countryside. Granted, Northumbria was a ragged sort of place, but not *this* ragged. He'd found a road—but it wasn't the B road he'd seen from the air. It was a dirt track, pocked with stones and showing signs of being much travelled by hooved animals with a heavily fibrous diet.

Wished he hadn't thought of diet. His wame was flapping against his backbone. Thinking about breakfast was better than thinking about other things, though, and for a time, he amused himself by envisioning the powdered eggs and soggy toast he'd have got in the mess, then going on to the lavish breakfasts of his youth in the Highlands: huge bowls of steaming parritch, slices of black pudding fried in lard, bannocks with marmalade, gallons of hot, strong tea . . .

An hour later, he found Hadrian's Wall. Hard to miss, even grown over with grass and all-sorts like it was. It marched stolidly along, just like the Roman legions who'd built it, stubbornly workmanlike, a grey seam stitching its way up hill and down dale, dividing the peaceful fields to the south from those marauding buggers up north. He grinned at the thought and sat down on the wall—it was less than a yard high, just here—to massage his knee.

He hadn't found the plane, or anything else, and was beginning to doubt his own sense of reality. He'd seen a fox, any number of rabbits, and a pheasant that'd nearly given him heart failure by bursting out from right under his feet. No people at all, though, and that was giving him a queer feeling in his water.

Aye, there was a war on, right enough, and many of the menfolk were gone, but the farmhouses hadn't been sacrificed to the war effort, had they? The women were running the farms, feeding the nation, all that— he'd heard the PM on the radio praising them for it only last week. So where the bloody hell was everybody?

The sun was getting low in the sky when at last he saw a house. It was flush against the wall, and struck him as somehow familiar, though he knew he'd never seen it before. Stone-built and squat, but quite large, with a ratty-looking thatch. There was smoke coming from the chimney, though, and he limped toward it as fast as he could go.

There was a person outside—a woman in a ratty long dress and an apron, feeding chickens. He shouted, and she looked up, her mouth falling open at the sight of him.

"Hey," he said, breathless from hurry. "I've had a crash. I need help. Are ye on the phone, maybe?"

She didn't answer. She dropped the basket of chicken feed and ran right away, round the corner of the house. He sighed in exasperation. Well, maybe she'd gone to fetch her husband. He didn't see any sign of a vehicle, not so much as a tractor, but maybe the man was—

The man was tall, stringy, bearded, and snaggletoothed. He was also dressed in a dirty shirt and baggy short pants that showed his hairy legs and bare feet—and accompanied by two other men in similar comic attire. Jerry instantly interpreted the looks on their faces, and didn't stay to laugh.

"Hey, nay problem, mate," he said, backing up, hands out. "I'm off, right?"

They kept coming, slowly, spreading out to surround him. He hadn't liked the looks of them to start with, and was liking them less by the second. Hungry, they looked, with a speculative glitter in their eyes.

One of them said something to him, a question of some kind, but the Northumbrian accent was too thick for him to catch more than a word. "Who" was the word, and he hastily pulled his dog tags from the neck of his blouson, waving the red and green disks at them. One of the men smiled, but not in a nice way.

"Look," he said, still backing up. "I didna mean to—"

The man in the lead reached out a horny hand and took hold of Jerry's forearm. He jerked back, but the man, instead of letting go, punched him in the belly.

He could feel his mouth opening and shutting like a fish's, but no air

came in. He flailed wildly, but they all were on him then. They were calling out to each other, and he didn't understand a word, but the intent was plain as the nose he managed to butt with his head.

It was the only blow he landed. Within two minutes, he'd been efficiently beaten into pudding, had his pockets rifled, been stripped of his jacket and dog tags, been frog-marched down the road and heaved bodily down a steep, rocky slope.

He rolled, bouncing from one outcrop to the next, until he managed to fling out an arm and grab on to a scrubby thornbush. He came to a scraping halt and lay with his face in a clump of heather, panting and thinking incongruously of taking Dolly to the pictures, just before he'd joined up. They'd seen *The Wizard of Oz,* and he was beginning to feel creepily like the lass in that film—maybe it was the resemblance of the Northumbrians to scarecrows and lions.

"At least the fucking lion spoke English," he muttered, sitting up. "Jesus, now what?"

It occurred to him that it might be a good time to stop cursing and start praying.

London, two years later

SHE'D BEEN HOME from her work no more than five minutes. Just time to meet Roger's mad charge across the floor, shrieking "MUMMY!," she pretending to be staggered by his impact—not so much a pretence; he was getting big. Just time to call out to her own mum, hear the muffled reply from the kitchen, sniff hopefully for the comforting smell of tea, and catch a tantalising whiff of tinned sardines that made her mouth water—a rare treat.

Just time to sit down for what seemed the first time in days, and take off her high-heeled shoes, relief washing over her feet like seawater when the tide comes in. She noticed with dismay the hole in the heel of her stocking, though. Her last pair, too. She was just undoing her garter, thinking that she'd have to start using leg-tan like Maisie, drawing a careful seam up the back of each leg with an eyebrow pencil, when there came a knock at the door.

"Mrs. MacKenzie?" The man who stood at the door of her mother's flat was tall, a dark silhouette in the dimness of the hall, but she knew at once he was a soldier.

"Yes?" She couldn't help the leap of her heart, the clench of her stomach. She tried frantically to damp it down, deny it, the hope that had sprung up like a struck match. A mistake. There'd been a mistake. He hadn't been killed, he'd been lost somehow, maybe captured, and now

they'd found hi— Then she saw the small box in the soldier's hand and her legs gave way under her.

Her vision sparkled at the edges, and the stranger's face swam above her, blurred with concern. She could hear, though—hear her mum rush through from the kitchen, slippers slapping in her haste, voice raised in agitation. Heard the man's name, Captain Randall, Frank Randall. Hear Roger's small, husky voice warm in her ear, saying "Mummy? Mummy?" in confusion.

Then she was on the swaybacked davenport, holding a cup of hot water that smelt of tea—they could change the tea leaves only once a week, and this was Friday, she thought irrelevantly. He should have come on Sunday, her mum was saying, they could have given him a decent cuppa. But perhaps he didn't work on Sundays?

Her mum had put Captain Randall in the best chair, near the electric fire, and had switched on two bars as a sign of hospitality. Her mother was chatting with the Captain, holding Roger in her lap. Her son was more interested in the little box sitting on the tiny piecrust table; he kept reaching for it, but his grandmother wouldn't let him have it. Marjorie recognised the intent look on his face. He wouldn't throw a fit—he hardly ever did—but he wouldn't give up, either.

He didn't look a lot like his father, save when he wanted something badly. She pulled herself up a bit, shaking her head to clear the dizziness, and Roger looked up at her, distracted by her movement. For an instant, she saw Jerry look out of his eyes, and the world swam afresh. She closed her own, though, and gulped her tea, scalding as it was.

Mum and Captain Randall had been talking politely, giving her time to recover herself. Did he have children of his own? Mum asked.

"No," he said, with what might have been a wistful look at wee Roger. "Not yet. I haven't seen my wife in two years."

"Better late than never," said a sharp voice, and she was surprised to discover that it was hers. She put down the cup, pulled up the loose stocking that had puddled round her ankle, and fixed Captain Randall with a look. "What have you brought me?" she said, trying for a tone of calm dignity. Didn't work; she sounded brittle as broken glass, even to her own ears.

Captain Randall eyed her cautiously, but took up the little box and held it out to her.

"It's Lieutenant MacKenzie's," he said. "An MID oakleaf cluster. Awarded posthumously for—"

With an effort, she pushed herself away, back into the cushions, shaking her head.

"I don't want it."

"Really, Marjorie!" Her mother was shocked.

"And I don't like that word. Pos—posth—don't say it."

She couldn't overcome the notion that Jerry was somehow inside the box—a notion that seemed dreadful at one moment, comforting the next. Captain Randall set it down, very slowly, as though it might blow up.

"I won't say it," he said gently. "May I say, though . . . I knew him. Your husband. Very briefly, but I did know him. I came myself, because I wanted to say to you how very brave he was."

"Brave." The word was like a pebble in her mouth. She wished she could spit it at him.

"Of course he was," her mother said firmly. "Hear that, Roger? Your dad was a good man, and he was a brave one. You won't forget that."

Roger was paying no attention, struggling to get down. His gran set him reluctantly on the floor and he lurched over to Captain Randall, taking a firm grip on the Captain's freshly creased trousers with both hands—hands greasy, she saw, with sardine oil and toast crumbs. The Captain's lips twitched, but he didn't try to detach Roger, just patted his head.

"Who's a good boy, then?" he asked.

"Fith," Roger said firmly. "Fith!"

Marjorie felt an incongruous impulse to laugh at the Captain's puzzled expression, though it didn't touch the stone in her heart.

"It's his new word," she said. " 'Fish.' He can't say 'sardine.' "

"Thar . . . DEEM!" Roger said, glaring at her. "Fittttthhhhh!"

The Captain laughed out loud, and pulling out a handkerchief, carefully wiped the spittle off Roger's face, casually going on to wipe the grubby little paws as well.

"Of course it's a fish," he assured Roger. "You're a clever lad. And a big help to your mummy, I'm sure. Here, I've brought you something for your tea." He groped in the pocket of his coat and pulled out a small pot of jam. Strawberry jam. Marjorie's salivary glands contracted painfully. With the sugar rationing, she hadn't tasted jam in . . .

"He's a great help," her mother put in stoutly, determined to keep the conversation on a proper plane despite her daughter's peculiar behaviour. She avoided Marjorie's eyes. "A lovely boy. His name's Roger."

"Yes, I know." He glanced at Marjorie, who'd made a brief movement. "Your husband told me. He was—"

"Brave. You told me." Suddenly something snapped. It was her half-hooked garter, but the pop of it made her sit up straight, fists clenched in the thin fabric of her skirt. "Brave," she repeated. "They're all brave, aren't they? Every single one. Even you—or are you?"

She heard her mother's gasp, but went on anyway, reckless.

"You all have to be brave and noble and—and—perfect, don't you? Because if you were weak, if there were any cracks, if anyone looked like being not quite the thing, you know—well, it might all fall apart, mightn't

it? So none of you will, will you? Or if somebody did, the rest of you would cover it up. You won't ever not do something, no matter what it is, because you can't not do it; all the other chaps would think the worse of you, wouldn't they, and we can't have that, oh, no, we can't have that!"

Captain Randall was looking at her intently, his eyes dark with concern. Probably thought she was a nutter—probably she was, but what did it matter?

"Marjie, Marjie, love," her mother was murmuring, horribly embarrassed. "You oughtn't to say such things to—"

"You made him do it, didn't you?" She was on her feet now, looming over the Captain, making him look up at her. "He told me. He told me about you. You came and asked him to do—whatever it was that got him killed. Oh, don't trouble yourself, he didn't tell me your bloody precious secrets—not him, he wouldn't do that. He was a flier." She was panting with rage and had to stop to draw breath. Roger, she saw dimly, had shrunk into himself and was clinging to the Captain's leg; Randall put an arm about the boy automatically, as though to shelter him from his mother's wrath. With an effort she made herself stop shouting, and, to her horror, felt tears begin to course down her face.

"And now you come and bring me—and bring me . . ."

"Marjie." Her mother came up close beside her, her body warm and soft and comforting in her worn old pinny. She thrust a tea towel into Marjorie's hands, then moved between her daughter and the enemy, solid as a battleship.

"It's kind of you to've brought us this, Captain," Marjorie heard her saying, and felt her move away, bending to pick up the little box. Marjorie sat down blindly, pressing the tea towel to her face, hiding.

"Here, Roger, look. See how it opens? See how pretty? It's called— what did you say it was again, Captain? Oh, oakleaf cluster. Yes, that's right. Can you say 'medal,' Roger? *Meh-dul.* This is your dad's medal."

Roger didn't say anything. Probably scared stiff, poor little chap. She had to pull herself together. But she'd gone too far. She couldn't stop.

"He cried when he left me." She muttered the secret into the folds of the tea towel. "He didn't want to go." Her shoulders heaved with a convulsive, unexpected sob, and she pressed the towel hard against her eyes, whispering to herself, "You said you'd come back, Jerry, you said you'd come *back.*"

She stayed hidden behind her flour-sacking fortress, while renewed offers of tea were made and, to her vague surprise, accepted. She'd thought Captain Randall would seize the chance of her retreat to make his own. But he stayed, chatting calmly with her mother, talking slowly to Roger while her mother fetched the tea, ignoring her embarrassing performance entirely, keeping up a quiet, companionable presence in the shabby room.

The rattle and bustle of the tea tray's arrival gave her the opportunity to drop her cloth façade, and she meekly accepted a slice of toast spread with a thin scrape of margarine and a delectable spoonful of the strawberry jam.

"There, now," her mother said, looking on with approval. "You'll not have eaten anything since breakfast, I daresay. Enough to give anyone the wambles."

Marjorie shot her mother a look, but in fact it was true; she hadn't had any luncheon because Maisie was off with "female trouble"—a condition that afflicted her roughly every other week—and she'd had to mind the shop all day.

Conversation flowed comfortably around her, a soothing stream past an immoveable rock. Even Roger relaxed with the introduction of jam. He'd never tasted any before, and sniffed it curiously, took a cautious lick—and then took an enormous bite that left a red smear on his nose, his moss-green eyes round with wonder and delight. The little box, now open, sat on the piecrust table, but no one spoke of it or looked in that direction.

After a decent interval, Captain Randall got up to go, giving Roger a shiny sixpence in parting. Feeling it the least she could do, Marjorie got up to see him out. Her stockings spiralled down her legs, and she kicked them off with contempt, walking bare-legged to the door. She heard her mother sigh behind her.

"Thank you," she said, opening the door for him. "I . . . appreciate—"

To her surprise, he stopped her, putting a hand on her arm.

"I've no particular right to say this to you—but I will," he said, low-voiced. "You're right; they're not all brave. Most of them—of us—we're just . . . there, and we do our best. Most of the time," he added, and the corner of his mouth lifted slightly, though she couldn't tell whether it was in humor or bitterness.

"But your husband—" He closed his eyes for a moment and said, "The bravest are surely those who have the clearest vision of what is before them, glory and danger alike, and yet notwithstanding, go out to meet it. He did that, every day, for a long time."

"You sent him, though," she said, her voice as low as his. "You did."

His smile was bleak.

"I've done such things every day . . . for a long time."

The door closed quietly behind him, and she stood there swaying, eyes closed, feeling the draft come under it, chilling her bare feet. It was well into the autumn now, and the dark was smudging the windows, though it was just past teatime.

I've done what I do every day for a long time, too, she thought. *But they don't call it brave when you don't have a choice.*

Her mother was moving through the flat, muttering to herself as she closed the curtains. Or not so much to herself.

"He liked her. Anyone could see that. So kind, coming himself to bring the medal and all. And how does she act? Like a cat that's had its tail stepped on, all claws and caterwauling, that's how. How does she ever expect a man to—"

"I don't want a man," Marjorie said loudly. Her mother turned round, squat, solid, implacable.

"You need a man, Marjorie. And little Rog needs a father."

"He has a father," she said through her teeth. "Captain Randall has a wife. And I don't need anyone."

Anyone but Jerry.

Northumbria

HE LICKED HIS lips at the smell. Hot pastry, steaming, juicy meat. There was a row of fat little pasties ranged along the sill, covered with a clean cloth in case of birds, but showing plump and rounded through it, the odd spot of gravy soaking through the napkin.

His mouth watered so fiercely that his salivary glands ached and he had to massage the underside of his jaw to ease the pain.

It was the first house he'd seen in two days. Once he'd got out of the ravine, he'd circled well away from the mile-castle and eventually struck a small cluster of cottages, where the people were no more understandable, but did give him some food. That had lasted him a little while; beyond that, he'd been surviving on what he could glean from hedges and the odd vegetable patch. He'd found another hamlet, but the folk there had driven him away.

Once he'd got enough of a grip on himself to think clearly, it became obvious that he needed to go back to the standing stones. Whatever had happened to him had happened there, and if he really *was* somewhere in the past—and hard as he'd tried to find some alternative explanation, none was forthcoming—then his only chance of getting back where he belonged seemed to lie there, too.

He'd come well away from the drover's track, though, seeking food, and as the few people he met didn't understand him any more than he understood them, he'd had some difficulty in finding his way back to the wall. He thought he was quite close, now, though—the ragged country was beginning to seem familiar, though perhaps that was only delusion.

Everything else had faded into unimportance, though, when he smelt food.

He circled the house at a cautious distance, checking for dogs. No dog.

Aye, fine, then. He chose an approach from the side, out of view of any of the few windows. Darted swiftly from bush to ploughshare to midden to house, and plastered himself against the grey stone wall, breathing hard—and breathing in that delicious, savoury aroma. Shite, he was drooling. He wiped his sleeve hastily across his mouth, slithered round the corner, and reached out a hand.

As it happened, the farmstead did boast a dog, which had been attending its absent master in the barn. Both these worthies returning unexpectedly at this point, the dog at once spotted what it assumed to be jiggery-pokery taking place, and gave tongue in an altogether proper manner. Alerted in turn to felonious activity on his premises, the householder instantly joined the affray, armed with a wooden spade, with which he batted Jerry over the head.

As he staggered back against the wall of the house, he had just wit enough left to notice that the farmwife—now sticking out of her window and shrieking like the Glasgow Express—had knocked one of the pasties to the ground, where it was being devoured by the dog, who wore an expression of piety and rewarded virtue that Jerry found really offensive.

Then the farmer hit him again, and he stopped being offended.

IT WAS A well-built byre, the stones fitted carefully and mortared. He wore himself out with shouting and kicking at the door until his gammy leg gave way and he collapsed onto the earthen floor.

"Now bloody what?" he muttered. He was damp with sweat from his effort, but it was cold in the byre, with that penetrating damp cold peculiar to the British Isles, that seeps into your bones and makes the joints ache. His knee would give him fits in the morning. The air was saturated with the scent of manure and chilled urine. "Why would the bloody Jerries want the damn place?" he said, and, sitting up, huddled into his shirt. It was going to be a frigging long night.

He got up onto his hands and knees and felt carefully round inside the byre, but there was nothing even faintly edible—only a scurf of mouldy hay. Not even the rats would have that; the inside of the place was empty as a drum and silent as a church.

What had happened to the cows? he wondered. Dead of a plague, eaten, sold? Or maybe just not yet back from the summer pastures—though it was late in the year for that, surely.

He sat down again, back against the door, as the wood was marginally less cold than the stone walls. He'd thought about being captured in battle, made prisoner by the Germans—they all had, now and then, though chaps mostly didn't talk about it. He thought about POW camps, and

those camps in Poland, the ones he'd been meant to photograph. Were they as bleak as this? Stupid thing to think of, really.

But he'd got to pass the time 'til morning one way or another, and there were lots of things he'd rather not think about just now. Like what would happen once morning came. He didn't think breakfast in bed was going to be part of it.

The wind was rising. Whining past the corners of the cow byre with a keening noise that set his teeth on edge. He still had his silk scarf; it had slipped down inside his shirt when the bandits in the mile-castle had attacked him. He fished it out now and wrapped it round his neck, for comfort, if not warmth.

He'd brought Dolly breakfast in bed now and then. She woke up slow and sleepy, and he loved the way she scooped her tangled curly black hair off her face, peering out slit-eyed, like a small, sweet mole blinking in the light. He'd sit her up and put the tray on the table beside her, and then he'd shuck his own clothes and crawl in bed, too, cuddling close to her soft, warm skin. Sometimes sliding down in the bed, and her pretending not to notice, sipping tea or putting marmite on her toast while he burrowed under the covers and found his way up through the cottony layers of sheets and nightie. He loved the smell of her, always, but especially when he'd made love to her the night before, and she bore the strong, musky scent of him between her legs.

He shifted a little, roused by the memory, but the subsequent thought— that he might never see her again—quelled him at once.

Still thinking of Dolly, though, he put his hand automatically to his pocket, and was alarmed to find no lump there. He slapped at his thigh, but failed to find the small, hard bulge of the sapphire. Could he have put it in the other pocket by mistake? He delved urgently, shoving both hands deep into his pockets. No stone—but there was something in his right-hand pocket. Something powdery, almost greasy . . . what the devil?

He brought his fingers out, peering as closely at them as he could, but it was too dark to see more than a vague outline of his hand, let alone anything on it. He rubbed his fingers gingerly together; it felt something like the thick soot that builds up inside a chimney.

"Jesus," he whispered, and put his fingers to his nose. There was a distinct smell of combustion. Not petrolish at all, but a scent of burning so intense he could taste it on the back of his tongue. Like something out of a volcano. What in the name of God Almighty could burn a rock and leave the man who carried it alive?

The sort of thing he'd met among the standing stones, that was what.

He'd been doing all right with the not feeling too afraid until now, but . . . he swallowed hard, and sat down again, quietly.

"Now I lay me down to sleep," he whispered to the knees of his trousers. "I pray the Lord my soul to keep . . ."

He did in fact sleep eventually, in spite of the cold, from simple exhaustion. He was dreaming about wee Roger, who for some reason was a grown man now, but still holding his tiny blue bear, minuscule in a broad-palmed grasp. His son was speaking to him in Gaelic, saying something urgent that he couldn't understand, and he was growing frustrated, telling Roger over and over for Christ's sake to speak English, couldn't he?

Then he heard another voice through the fog of sleep and realised that someone was in fact talking somewhere close by.

He jerked awake, struggling to grasp what was being said and failing utterly. It took him several seconds to realise that whoever was speaking—there seemed to be two voices, hissing and muttering in argument—really was speaking in Gaelic.

He had only a smattering of it himself; his mother had had it, but—he was moving before he could complete the thought, panicked at the notion that potential assistance might get away.

"Hoy!" he bellowed, scrambling—or trying to scramble—to his feet. His much-abused knee wasn't having any, though, and gave way the instant he put weight on it, catapulting him face-first toward the door.

He twisted as he fell and hit it with his shoulder. The booming thud put paid to the argument; the voices fell silent at once.

"Help! Help me!" he shouted, pounding on the door. "Help!"

"Will ye for God's sake hush your noise?" said a low, annoyed voice on the other side of the door. "Ye want to have them all down on us? Here, then, bring the light closer."

This last seemed to be addressed to the voice's companion, for a faint glow shone through the gap at the bottom of the door. There was a scraping noise as the bolt was drawn, and a faint grunt of effort, then a *thunk!* as the bolt was set down against the wall. The door swung open, and Jerry blinked in a sudden shaft of light as the slide of a lantern grated open.

He turned his head aside and closed his eyes for an instant, deliberate, as he would if flying at night and momentarily blinded by a flare or by the glow of his own exhaust. When he opened them again, the two men were in the cow byre with him, looking him over with open curiosity.

Biggish buggers, both of them, taller and broader than he was. One fair, one black-haired as Lucifer. They didn't look much alike, and yet he had the feeling that they might be related—some fleeting glimpse of bone, a similarity of expression, maybe.

"What's your name, mate?" said the dark chap, softly. Jerry felt the nip of wariness at his nape, even as he felt a thrill in the pit of his stomach. It was regular speech, perfectly understandable. A Scots accent, but—

"MacKenzie, J. W.," he said, straightening up to attention. "Lieutenant, Royal Air Force. Service number—"

An indescribable expression flitted across the dark bloke's face. An urge to laugh, of all bloody things, and a flare of excitement in his eyes—really striking eyes, a vivid green that flashed suddenly in the light. None of that mattered to Jerry; what was important was that the man plainly knew. He *knew*.

"Who are you?" he asked, urgent. "Where d'ye come from?"

The two exchanged an unfathomable glance, and the other answered. "Inverness."

"Ye know what I mean!" He took a deep breath. *"When?"*

The two strangers were much of an age, but the fair one had plainly had a harder life; his face was deeply weathered and lined.

"A lang way from you," he said quietly, and, despite his own agitation, Jerry heard the note of desolation in his voice. "From now. Lost."

Lost. Oh, God. But still—

"Jesus. And where are we now? Wh-when?"

"Northumbria," the dark man answered briefly, "and I don't bloody know for sure. Look, there's no time. If anyone hears us—"

"Aye, right. Let's go, then."

The air outside was wonderful after the smells of the cow byre, cold and full of dying heather and turned earth. He thought he could even smell the moon, a faint green sickle above the horizon; he tasted cheese at the thought, and his mouth watered. He wiped a trickle of saliva away and hurried after his rescuers, hobbling as fast as he could.

The farmhouse was black, a squatty black blot on the landscape. The dark bloke grabbed him by the arm as he was about to go past it, quickly licked a finger, and held it up to test the wind.

"The dogs," he explained in a whisper. "This way."

They circled the farmhouse at a cautious distance, and found themselves stumbling through a ploughed field. Clods burst under Jerry's boots as he hurried to keep up, lurching on his bad knee with every step.

"Where we going?" he panted, when he thought it safe to speak.

"We're taking ye back to the stones near the lake," the dark man said tersely. "That has to be where ye came through." The fair one just snorted, as though this wasn't his notion—but he didn't argue.

Hope flared up in Jerry like a bonfire. They knew what the stones were, how it worked. They'd show him how to get back!

"How—how did ye find me?" He could hardly breathe, such a pace they kept up, but he had to know. The lantern was shut and he couldn't see their faces, but the dark man made a muffled sound that might have been a laugh.

"I met an auld wifie wearing your dog tags. Very proud of them, she was."

"Ye've got them?" Jerry gasped.

"Nay, she wouldna give them up." It was the fair man, sounding definitely amused. "Told us where she'd got them, though, and we followed your trail backward. Hey!" He caught Jerry's elbow, just as his foot twisted out from under him. The sound of a barking dog broke the night—some way away, but distinct. The fair man's hand clenched tight on his arm. "Come on, then—hurry!"

Jerry had a bad stitch in his side, and his knee was all but useless by the time the little group of stones came in sight, a pale huddle in the light of the waning moon. Still, he was surprised at how near the stones were to the farmhouse; he must have circled round more than he thought in his wanderings.

"Right," said the dark man, coming to an abrupt halt. "This is where we leave you."

"Ye do?" Jerry panted. "But—but you—"

"When ye came . . . through. Did ye have anything on you? A gemstone, any jewellery?"

"Aye," Jerry said, bewildered. "I had a raw sapphire in my pocket. But it's gone. It's like it—"

"Like it burnt up," the blond man finished for him, grim-voiced. "Aye. Well, so?" This last was clearly addressed to the dark man, who hesitated. Jerry couldn't see his face, but his whole body spoke of indecision. He wasn't one to dither, though—he stuck a hand into the leather pouch at his waist, pulled something out, and pressed it into Jerry's hand. It was faintly warm from the man's body, and hard in his palm. A small stone of some kind. Faceted, like the stone in a ring.

"Take this; it's a good one. When ye go through," the dark man was speaking urgently to him, "think about your wife, about Marjorie. Think hard; see her in your mind's eye, and walk straight through. Whatever the hell ye do, though, don't think about your son. Just your wife."

"What?" Jerry was gob-smacked. "How the bloody hell do you know my wife's name? And where've ye heard about my son?"

"It doesn't matter," the man said, and Jerry saw the motion as he turned his head to look back over his shoulder.

"Damn," said the fair one, softly. "They're coming. There's a light."

There was: a single light, bobbing evenly over the ground, as it would if someone carried it. But look as he might, Jerry could see no one behind it, and a violent shiver ran over him.

"*Tannasg,*" said the other man under his breath. Jerry knew that word well enough—spirit, it meant. And usually an ill-disposed one. A haunt.

"Aye, maybe." The dark man's voice was calm. "And maybe not. It's near Samhain, after all. Either way, ye need to go, man, and now. Remember, think of your wife."

Jerry swallowed, his hand closing tight around the stone.

"Aye. Aye . . . right. Thanks, then," he added awkwardly, and heard the breath of a rueful laugh from the dark man.

"Nay bother, mate," he said. And with that, they were both off, making their way across the stubbled meadow, two lumbering shapes in the moonlight.

Heart thumping in his ears, Jerry turned toward the stones. They looked just like they'd looked before. Just stones. But the echo of what he'd heard in there . . . He swallowed. It wasn't like there was much choice.

"Dolly," he whispered, trying to summon up a vision of his wife. "Dolly. Dolly, help me!"

He took a hesitant step toward the stones. Another. One more. Then nearly bit his tongue off as a hand clamped down on his shoulder. He whirled, fist up, but the dark man's other hand seized his wrist.

"I love you," the dark man said, his voice fierce. Then he was gone again, with the *shoof-shoof* sounds of boots in dry grass, leaving Jerry with his mouth agape.

He caught the other man's voice from the darkness, irritated, half-amused. He spoke differently from the dark man, a much thicker accent, but Jerry understood him without difficulty.

"Why did ye tell him a daft thing like that?"

And the dark one's reply, soft-spoken, in a tone that terrified him more than anything had so far.

"Because he isn't going to make it back. It's the only chance I'll ever have. Come on."

THE DAY WAS DAWNING when he came to himself again, and the world was quiet. No birds sang, and the air was cold with the chill of November and winter coming on. When he could stand up, he went to look, shaky as a newborn lamb.

The plane wasn't there, but there was still a deep gouge in the earth where it had been. Not raw earth, though; furred over with grass and meadow plants—not just furred, he saw, limping over to have a closer look. Matted. Dead stalks from earlier years' growth.

If he'd been where he thought he'd been, if he'd truly gone . . . back . . . then he'd come forward again, but not to the same place he'd left. How long? A year, two? He sat down on the grass, too drained to stand up any

longer. He felt as though he'd walked every second of the time between then and now.

He'd done what the green-eyed stranger had said. Concentrated fiercely on Dolly. But he hadn't been able to keep from thinking of wee Roger, not altogether. How could he? The picture he had most vividly of Dolly was her holding the lad, close against her breast; that's what he'd seen. And yet he'd made it. He thought he'd made it. Maybe.

What might have happened? he wondered. There hadn't been time to ask. There'd been no time to hesitate, either; more lights had come bobbing across the dark, with uncouth Northumbrian shouts behind them, hunting him, and he'd hurled himself into the midst of the standing stones and things went pear-shaped again, even worse. He hoped the strangers who'd rescued him had got away.

Lost, the fair man had said, and even now, the word went through him like a bit of jagged metal. He swallowed.

He thought he wasn't where he had been, but was he still lost, himself? Where was he now? Or rather, when?

He stayed for a bit, gathering his strength. In a few minutes, though, he heard a familiar sound—the low growl of engines, and the swish of tyres on asphalt. He swallowed hard, and, standing up, turned away from the stones, toward the road.

HE WAS LUCKY—for once, he thought wryly. There was a line of troop transports passing, and he swung aboard one without difficulty. The soldiers looked startled at his appearance—he was rumpled and stained, bruised and torn about and with a two-week beard—but they instantly assumed he'd been off on a tear and was now trying to sneak back to his base without being detected. They laughed and nudged him knowingly, but were sympathetic, and when he confessed he was skint, they had a quick whip-round for enough cash to buy a train ticket from Salisbury, where the transport was headed.

He did his best to smile and go along with the ragging, but soon enough they tired of him and turned to their own conversations, and he was allowed to sit, swaying on the bench, feeling the thrum of the engine through his legs, surrounded by the comfortable presence of comrades.

"Hey, mate," he said casually to the young soldier beside him. "What year is it?"

The boy—he couldn't be more than seventeen, and Jerry felt the weight of the five years between them as though they were fifty—looked at him wide-eyed, then whooped with laughter.

"What've you been having to drink, Dad? Bring any away with you?"

That led to more ragging, and he didn't try asking again.
Did it matter?

***

HE REMEMBERED almost nothing of the journey from Salisbury to London. People looked at him oddly, but no one tried to stop him. It didn't matter; nothing mattered but getting to Dolly. Everything else could wait.

London was a shock. There was bomb damage everywhere. Streets were scattered with shattered glass from shop windows, glinting in the pale sun, other streets blocked off by barriers. Here and there a stark black notice: Do Not Enter—UNEXPLODED BOMB.

He made his way from Saint Pancras on foot, needing to see, his heart rising into his throat fit to choke him as he did see what had been done. After a while, he stopped seeing the details, perceiving bomb craters and debris only as blocks to his progress, things stopping him from reaching home.

And then he did reach home.

The rubble had been pushed off the street into a heap, but not taken away. Great blackened lumps of shattered stone and concrete lay like a cairn where Montrose Terrace had once stood.

All the blood in his heart stopped dead, congealed by the sight. He groped, pawing mindlessly for the wrought-iron railing to keep himself from falling, but it wasn't there.

Of course not, his mind said, quite calmly. It's gone for the war, hasn't it? Melted down, made into planes. Bombs.

His knee gave way without warning, and he fell, landing hard on both knees, not feeling the impact, the crunch of pain from his badly mended kneecap quite drowned out by the blunt, small voice inside his head.

Too late. Ye went too far.

"Mr. MacKenzie, Mr. MacKenzie!" He blinked at the blurred thing above him, not understanding what it was. Something tugged at him, though, and he breathed, the rush of air in his chest ragged and strange.

"Sit up, Mr. MacKenzie, do." The anxious voice was still there, and hands—yes, it was hands—tugging at his arm. He shook his head, screwed his eyes shut hard, then opened them again, and the round thing became the houndlike face of old Mr. Wardlaw, who kept the corner shop.

"Ah, there you are." The old man's voice was relieved, and the wrinkles in his baggy old face relaxed their anxious lines. "Had a bad turn, did you?"

"I—" Speech was beyond him, but he flapped his hand at the wreckage.

He didn't think he was crying, but his face was wet. The wrinkles in Ward-law's face creased deeper in concern, then the old grocer realised what he meant, and his face lit up.

"Oh, dear!" he said. "Oh, no! No, no, no—they're all right, sir, your family's all right! Did you hear me?" he asked anxiously. "Can you breathe? Had I best fetch you some salts, do you think?"

It took Jerry several tries to make it to his feet, hampered both by his knee and by Mr. Wardlaw's fumbling attempts to help him, but by the time he'd got all the way up, he'd regained the power of speech.

"Where?" he gasped. "Where are they?"

"Why—your missus took the little boy and went to stay with her mother, sometime after you left. I don't recall quite where she said . . ." Mr. Wardlaw turned, gesturing vaguely in the direction of the river. "Camberwell, was it?"

"Bethnal Green." Jerry's mind had come back, though it felt still as though it was a pebble rolling round the rim of some bottomless abyss, its balance uncertain. He tried to dust himself off, but his hands were shaking. "She lives in Bethnal Green. You're sure—you're sure, man?"

"Yes, yes." The grocer was altogether relieved, smiling and nodding so hard that his jowls trembled. "She left—must be more than a year ago, soon after she—soon after she . . ." The old man's smile faded abruptly and his mouth slowly opened, a flabby dark hole of horror.

"But you're dead, Mr. MacKenzie," he whispered, backing away, hands held up before him. "Oh, God. You're dead."

"THE FUCK I AM, the fuck I am, the *fuck* I am!" He caught sight of a woman's startled face and stopped abruptly, gulping air like a landed fish. He'd been weaving down the shattered street, fists pumping, limping and staggering, muttering his private motto under his breath like the Hail Marys of a rosary. Maybe not as far under his breath as he'd thought.

He stopped, leaning against the marble front of the Bank of England, panting. He was streaming with sweat and the right leg of his trousers was heavily streaked with dried blood from the fall. His knee was throbbing in time with his heart, his face, his hands, his thoughts. *They're alive. So am I.*

The woman he'd startled was down the street, talking to a policeman; she turned, pointing at him. He straightened up at once, squaring his shoulders. Braced his knee and gritted his teeth, forcing it to bear his weight as he strode down the street, officerlike. The very last thing he wanted just now was to be taken up as drunk.

He marched past the policeman, nodding politely, touching his forehead in lieu of cap. The policeman looked taken aback, made to speak but couldn't quite decide what to say, and a moment later, Jerry was round the corner and away.

It was getting dark. There weren't many cabs in this area at the best of times—none at all, now, and he hadn't any money, anyway. The Tube. If the lines were open, it was the fastest way to Bethnal Green. And surely he could cadge the fare from someone. Somehow. He went back to limping, grimly determined. He had to reach Bethnal Green by dark.

IT WAS SO much changed. Like the rest of London. Houses damaged, halfway repaired, abandoned, others no more than a blackened depression or a heap of rubble. The air was thick with cold dust, stone dust, and the smells of paraffin and cooking grease, the brutal, acrid smell of cordite.

Half the streets had no signs, and he wasn't so familiar with Bethnal Green to begin with. He'd visited Dolly's mother just twice, once when they went to tell her they'd run off and got married—she hadn't been best pleased, Mrs. Wakefield, but she'd put a good face on it, even if the face had a lemon-sucking look to it.

The second time had been when he signed up with the RAF; he'd gone alone to tell her, to ask her to look after Dolly while he was gone. Dolly's mother had gone white. She knew as well as he did what the life expectancy was for fliers. But she'd told him she was proud of him, and held his hand tight for a long moment before she let him leave, saying only, "Come back, Jeremiah. She needs you."

He soldiered on, skirting craters in the street, asking his way. It was nearly full dark, now; he couldn't be on the streets much longer. His anxiety began to ease a little as he started to see things he knew, though. Close, he was getting close.

And then the sirens began, and people began to pour out of the houses.

He was being buffeted by the crowd, borne down the street as much by their barely controlled panic as by their physical impact. There was shouting, people calling for separated family members, wardens bellowing directions, waving their torches, their flat, white helmets pale as mushrooms in the gloom. Above it, through it, the air-raid siren pierced him like a sharpened wire, thrust him down the street on its spike, ramming him into others likewise skewered by fright.

The tide of it swept round the next corner, and he saw the red circle with its blue line over the entrance to the Tube station, lit up by a warden's flashlight. He was sucked in, propelled through sudden bright lights, hurtling down the stair, the next, onto a platform, deep into the earth,

into safety. And all the time the whoop and moan of the sirens still filling the air, barely muffled by the dirt above.

There were wardens moving among the crowd, pushing people back against the walls, into the tunnels, away from the edge of the track. He brushed up against a woman with two toddlers, picked one—a little girl with round eyes and a blue teddy bear—out of her arms and turned his shoulder into the crowd, making a way for them. He found a small space in a tunnel-mouth, pushed the woman into it, and gave her back the little girl. Her mouth moved in thanks, but he couldn't hear her above the noise of the crowd, the sirens, the creaking, the—

A sudden monstrous thud from above shook the station, and the whole crowd was struck silent, every eye on the high arched ceiling above them.

The tiles were white, and as they looked, a dark crack appeared suddenly between two rows of them. A gasp rose from the crowd, louder than the sirens. The crack seemed to stop, to hesitate—and then it zigzagged suddenly, parting the tiles, in different directions.

He looked down from the growing crack, to see who was below it—the people still on the stair. The crowd at the bottom was too thick to move, everyone stopped still by horror. And then he saw her, partway up the stair.

Dolly. *She's cut her hair,* he thought. It was short and curly, black as soot—black as the hair of the little boy she held in her arms, close against her, sheltering him. Her face was set, jaw clenched. And then she turned a bit, and saw him.

Her face went blank for an instant and then flared like a lit match, with a radiant joy that struck him in the heart and flamed through his being.

There was a much louder *thud!* from above, and a scream of terror rose from the crowd, louder, much louder than the sirens. Despite the shrieking, he could hear the fine rattle, like rain, as dirt began to pour from the crack above. He shoved with all his might, but couldn't get past, couldn't reach them. Dolly looked up, and he saw her jaw set hard again, her eyes ablaze with determination. She shoved the man in front of her, who stumbled and fell down a step, squashing into the people in front of him. She swung Roger down into the little space she'd made, and with a twist of her shoulders and the heave of her whole body, hurled the little boy up, over the rail—toward Jerry.

He saw what she was doing and was already leaning, pushing forward, straining to reach . . . The boy struck him high in the chest like a lump of concrete, little head smashing painfully into Jerry's face, knocking his head back. He had one arm round the child, falling back on the people behind him, struggling to find his footing, get a firmer hold—and then something gave way in the crowd around him, he staggered into an open space, and then his knee gave way and he plunged over the lip of the track.

He didn't hear the crack of his head against the rail or the screams of the people above; it was all lost in a roar like the end of the world as the roof over the stair fell in.

THE LITTLE BOY was still as death, but he wasn't dead; Jerry could feel his heartbeat, thumping fast against his own chest. It was all he could feel. Poor little bugger must have had his wind knocked out.

People had stopped screaming, but there was still shouting, calling out. There was a strange silence underneath all the racket. His blood had stopped pounding through his head, his own heart no longer hammering. Perhaps that was it.

The silence underneath felt alive, somehow. Peaceful, but like sunlight on water, moving, glittering. He could still hear the noises above the silence, feet running, anxious voices, bangs and creakings—but he was sinking gently into the silence; the noises grew distant, though he could still hear voices.

"Is that one—?"

"Nay, he's gone—look at his head, poor chap, caved in something horrid. The boy's well enough, I think, just bumps and scratches. Here, lad, come up . . . no, no, let go, now. It's all right, just let go. Let me pick you up, yes, that's good, it's all right now, hush, hush, there's a good boy . . ."

"What a look on that bloke's face. I never saw anything like—"

"Here, take the little chap. I'll see if the bloke's got any identification."

"Come on, big man, yeah, that's it, that's it, come with me. Hush, now, it's all right, it's all right . . . is that your daddy, then?"

"No tags, no service book. Funny, that. He's RAF, though, isn't he? AWOL, d'ye think?"

He could hear Dolly laughing at that, felt her hand stroke his hair. He smiled and turned his head to see her smiling back, the radiant joy spreading round her like rings in shining water . . .

"Rafe! The rest of it's going! Run! *Run!*"

AUTHOR'S NOTES

BEFORE Y'ALL GET TANGLED UP IN YOUR UNDER-
wear about it being All Hallows' Eve when Jeremiah leaves, and
"nearly Samhain" (aka All Hallows' Eve) when he returns—bear in
mind that Great Britain changed from the Julian to the Gregorian
calendar in 1752, this resulting in a "loss" of twelve days. And for
those of you who'd like to know more about the two men who rescue him, more of their story can be found in *An Echo in the Bone*.

"Never have so many owed so much to so few." This was Winston
Churchill's acknowledgement to the RAF pilots who protected Britain during World War II—and he was about right.

Adolph Gysbert Malan—known as Sailor (probably because Adolph
was not a popular name at the time)—was a South African flying ace
who became the leader of the famous No. 74 Squadron RAF. He
was known for sending German bomber pilots home with dead
crews, to demoralize the Luftwaffe, and I would have mentioned this
gruesomely fascinating detail in the story, had there been any good
way of getting it in, but there wasn't. His Ten Commandments for
Air Fighting are as given in the text.

While the mission that Captain Frank Randall recruits Jerry MacKenzie for is fictional, the situation wasn't. The Nazis did have labor
camps in Poland long before anyone in the rest of Europe became
aware of them, and the eventual revelation did much to rally anti-
Nazi feeling.

I'd like particularly to acknowledge the assistance of Maria Szybek in the delicate matter of Polish vulgarities (any errors in grammar, spelling, or accent marks are entirely mine), and of Douglas Watkins in the technical descriptions of small-plane maneuvers (also the valuable suggestion of the malfunction that brought Jerry's Spitfire down).

VIRGINS

INTRODUCTION

WHILE MANY OF THE STORIES IN THIS BOOK SHOW you an alternate view of events seen in the main novels or explore the stories of heretofore minor characters, "Virgins" is a straightforward prequel. Set about three years prior to the events recounted in *Outlander*, this story explains what happened to Jamie Fraser after his escape from Fort William. He's suddenly an outlaw, wounded and with a price on his head, his family and home left in shambles, and his only choice is to seek refuge outside of Scotland with his best friend and blood brother, Ian Murray.

As young mercenaries in France, neither Ian nor Jamie has yet killed a man or bedded a lass—but they're trying.

I AN MURRAY KNEW FROM the moment he saw his best friend's face that something terrible had happened. The fact that he was seeing Jamie Fraser's face at all was evidence enough of that, never mind the look of the man.

Jamie was standing by the armorer's wagon, his arms full of the bits and pieces Armand had just given him, white as milk and swaying back and forth like a reed on Loch Awe. Ian reached him in three paces and took him by the arm before he could fall over.

"Ian." Jamie looked so relieved at seeing him that Ian thought he might break into tears. "God, Ian."

Ian seized Jamie in an embrace and felt him stiffen and draw in his breath at the same instant that Ian felt the bandages beneath Jamie's shirt.

"Jesus!" he began, startled, but then coughed and said, "Jesus, man, it's good to see ye." He patted Jamie's back gently and let go. "Ye'll need a bit to eat, aye? Come on, then."

Plainly they couldn't talk now, but he gave Jamie a quick private nod, took half the equipment from him, and then led him to the fire, to be introduced to the others.

Jamie'd picked a good time of day to turn up, Ian thought. Everyone was tired but happy to sit down, looking forward to their supper and the daily ration of whatever was going in the way of drink. Ready for the possibilities a new fish offered for entertainment, but without the energy to include the more physical sorts of entertainment.

"That's Big Georges over there," Ian said, dropping Jamie's gear and gesturing toward the far side of the fire. "Next to him, the wee fellow wi' the warts is Juanito; doesna speak much French and nay English at all."

"Do any of them speak English?" Jamie likewise dropped his gear and

sat heavily on his bedroll, tucking his kilt absently down between his knees. His eyes flicked round the circle, and he nodded, half-smiling in a shy sort of way.

"I do." The captain leaned past the man next to him, extending a hand to Jamie. "I'm *le capitaine*—Richard D'Eglise. You'll call me Captain. You look big enough to be useful—your friend says your name is Fraser?"

"Jamie Fraser, aye."

Ian was pleased to see that Jamie knew to meet the captain's eye square and had summoned the strength to return the handshake with due force.

"Know what to do with a sword?"

"I do. And a bow, forbye." Jamie glanced at the unstrung bow by his feet and the short-handled ax beside it. "Havena had much to do wi' an ax before, save chopping wood."

"That's good," one of the other men put in, in French. "That's what you'll use it for." Several of the others laughed, indicating that they at least understood English, whether they chose to speak it or not.

"Did I join a troop of soldiers, then, or charcoal-burners?" Jamie asked, raising one brow. He said that in French—very good French, with a faint Parisian accent—and a number of eyes widened. Ian bent his head to hide a smile, in spite of his anxiety. The wean might be about to fall face-first into the fire, but nobody—save maybe Ian—was going to know it, if it killed him.

Ian *did* know it, though, and kept a covert eye on Jamie, pushing bread into his hand so the others wouldn't see it shake, sitting close enough to catch him if he should in fact pass out. The light was fading into gray now, and the clouds hung low and soft, pink-bellied. Going to rain, likely, by the morning. He saw Jamie close his eyes, just for an instant, saw his throat move as he swallowed, and felt the trembling of Jamie's thigh, near his own.

What the devil's happened? he thought in anguish. *Why are ye here?*

IT WASN'T UNTIL everyone had settled for the night that Ian got an answer.

"I'll lay out your gear," he whispered to Jamie, rising. "You stay by the fire that wee bit longer—rest a bit, aye?" The firelight cast a ruddy glow on Jamie's face, but Ian thought his friend was likely still white as a sheet; he hadn't eaten much.

Coming back, he saw the dark spots on the back of Jamie's shirt, blotches where fresh blood had seeped through the bandages. The sight filled him with fury, as well as fear. He'd seen such things; the wean had been flogged. Badly, and recently. *Who? How?*

"Come on, then," he said roughly, and, bending, slipped an arm under Jamie's and got him to his feet and away from the fire and the other men. He was alarmed to feel the clamminess of Jamie's hand and hear his shallow breath.

"What?" he demanded, the moment they were out of earshot. "What happened?"

Jamie sat down abruptly.

"I thought one joined a band of mercenaries because they didna ask ye questions."

Ian gave him the snort this statement deserved and was relieved to hear a breath of laughter in return.

"Eejit," he said. "D'ye need a dram? I've got a bottle in my sack."

"Wouldna come amiss," Jamie murmured. They were camped at the edge of a wee village, and D'Eglise had arranged for the use of a byre or two, but it wasn't cold out, and most of the men had chosen to sleep by the fire or in the field. Ian had put their gear down a little distance away and, with the possibility of rain in mind, under the shelter of a plane tree that stood at the side of a field.

Ian uncorked the bottle of whisky—it wasn't good, but it *was* whisky— and held it under his friend's nose. When Jamie reached for it, though, Ian pulled it away.

"Not a sip do ye get until ye tell me," he said. "And ye tell me *now, a charaid*."

Jamie sat hunched, a pale blur on the ground, silent. When the words came at last, they were spoken so softly that Ian thought for an instant he hadn't really heard them.

"My father's dead."

He tried to believe he *hadn't* heard, but his heart had; it froze in his chest.

"Oh, Jesus," he whispered. "Oh, God, Jamie." He was on his knees then, holding Jamie's head fierce against his shoulder, trying not to touch his hurt back. His thoughts were in confusion, but one thing was clear to him—Brian Fraser's death hadn't been a natural one. If it had, Jamie would be at Lallybroch. Not here, and not in this state.

"Who?" he said hoarsely, relaxing his grip a little. "Who killed him?"

More silence, then Jamie gulped air with a sound like fabric being ripped.

"I did," he said, and began to cry, shaking with silent, tearing sobs.

IT TOOK SOME TIME to winkle the details out of Jamie—and no wonder, Ian thought. He wouldn't want to talk about such things, either,

or to remember them. The English dragoons who'd come to Lallybroch to loot and plunder, who'd taken Jamie away with them when he'd fought them. And what they'd done to him then, at Fort William.

"A hundred lashes?" he said in disbelief and horror. "For protecting your *home*?"

"Only sixty the first time." Jamie wiped his nose on his sleeve. "For escaping."

"The *first* ti— Jesus, God, man! What . . . how . . ."

"Would ye let go my arm, Ian? I've got enough bruises; I dinna need any more." Jamie gave a small, shaky laugh, and Ian hastily let go but wasn't about to let himself be distracted.

"Why?" he said, low and angry. Jamie wiped his nose again, sniffing, but his voice was steadier.

"It was my fault," he said. "It—what I said before. About my . . ." He had to stop and swallow, but went on, hurrying to get the words out before they could bite him in a tender place. "I spoke chough to the commander. At the garrison, ken. He—well, it's nay matter. It was what I said to him made him flog me again, and Da—he—he'd come. To Fort William, to try to get me released, but he couldn't, and he—he was there, when they . . . did it."

Ian could tell from the thicker sound of his voice that Jamie was weeping again but trying not to, and he put a hand on the wean's knee and gripped it, not too hard, just so as Jamie would ken he was there, listening.

Jamie took a deep, deep breath and got the rest out.

"It was . . . hard. I didna call out, or let them see I was scairt, but I couldna keep my feet. Halfway through it, I fell into the post, just—just hangin' from the ropes, ken, wi' the blood . . . runnin' down my legs. They thought for a bit that I'd died—and Da must ha' thought so, too. They told me he put his hand to his head just then and made a wee noise, and then . . . he fell down. An apoplexy, they said."

"Mary, Mother o' God, have mercy on us," Ian said. "He—died right there?"

"I dinna ken was he dead when they picked him up or if he lived a bit after that." Jamie's voice was desolate. "I didna ken a thing about it; no one told me until days later, when Uncle Dougal got me away." He coughed and wiped the sleeve across his face again. "Ian . . . would ye let go my knee?"

"No," Ian said softly, though he did indeed take his hand away. Only so he could gather Jamie gently into his arms, though. "No. I willna let go, Jamie. Bide. Just . . . bide."

JAMIE WOKE DRY-MOUTHED, thickheaded, and with his eyes half swollen shut by midgie bites. It was also raining, a fine, wet mist coming down through the leaves above him. For all that, he felt better than he had in the last two weeks, though he didn't at once recall why that was—or where he was.

"Here." A piece of half-charred bread rubbed with garlic was shoved under his nose. He sat up and grabbed it.

Ian. The sight of his friend gave him an anchor, and the food in his belly another. He chewed slower now, looking about. Men were rising, stumbling off for a piss, making low rumbling noises, rubbing their heads and yawning.

"Where are we?" he asked. Ian gave him a look.

"How the devil did ye find us if ye dinna ken where ye are?"

"Murtagh brought me," he muttered. The bread turned to glue in his mouth as memory came back; he couldn't swallow and spat out the half-chewed bit. Now he remembered it all, and wished he didn't. "He found the band but then left; said it would look better if I came in on my own."

His godfather had said, in fact, *"The Murray lad will take care of ye now. Stay wi' him, mind—dinna come back to Scotland. Dinna come back, d'ye hear me?"* He'd heard. Didn't mean he meant to listen.

"Oh, aye. I wondered how ye'd managed to walk this far." Ian cast a worried look at the far side of the camp, where a pair of sturdy horses was being brought to the traces of a canvas-covered wagon. "*Can* ye walk, d'ye think?"

"Of course. I'm fine." Jamie spoke crossly, and Ian gave him the look again, even more slit-eyed than the last.

"Aye, right," he said, in tones of rank disbelief. "Well. We're near Bèguey, maybe twenty miles from Bordeaux; that's where we're going. We're takin' the wagon yon to a Jewish moneylender there."

"Is it full of money, then?" Jamie glanced at the heavy wagon, interested.

"No," Ian said. "There's a wee chest, verra heavy, so it's maybe gold, and there are a few bags that clink and might be silver, but most of its rugs."

"Rugs?" He looked at Ian in amazement. "What sort of rugs?"

Ian shrugged. "Couldna say. Juanito says they're Turkey rugs and verra valuable, but I dinna ken that he knows. He's Jewish, too," Ian added, as an afterthought. "Jews are—" He made an equivocal gesture, palm flattened. "But they dinna really hunt them in France, or exile them anymore, and the captain says they dinna even arrest them, so long as they keep quiet."

"And go on lending money to men in the government," Jamie said

cynically. Ian looked at him, surprised, and Jamie gave him the *I went to the Université in Paris and ken more than you do* smart-arse look, fairly sure that Ian wouldn't thump him, seeing he was hurt.

Ian looked tempted but had learned enough merely to give Jamie back the *I'm older than you and ye ken well ye havena sense enough to come in out of the rain, so dinna be trying it on* look instead. Jamie laughed, feeling better.

"Aye, right," he said, bending forward. "Is my shirt verra bloody?"

Ian nodded, buckling his sword belt. Jamie sighed and picked up the leather jerkin the armorer had given him. It would rub, but he wasn't wanting to attract attention.

HE MANAGED. The troop kept up a decent pace, but it wasn't anything to trouble a Highlander accustomed to hill-walking and running down the odd deer. True, he grew a bit light-headed now and then, and sometimes his heart raced and waves of heat ran over him—but he didn't stagger any more than a few of the men who'd drunk too much for breakfast.

He barely noticed the countryside but was conscious of Ian striding along beside him, and Jamie took pains now and then to glance at his friend and nod, in order to relieve Ian's worried expression. The two of them were close to the wagon, mostly because he didn't want to draw attention by lagging at the back of the troop but also because he and Ian were taller than the rest by a head or more, with a stride that eclipsed the others, and he felt a small bit of pride in that. It didn't occur to him that possibly the others didn't *want* to be near the wagon.

The first inkling of trouble was a shout from the driver. Jamie had been trudging along, eyes half closed, concentrating on putting one foot ahead of the other, but a bellow of alarm and a sudden loud *bang!* jerked him to attention. A horseman charged out of the trees near the road, slewed to a halt, and fired his second pistol at the driver.

"What—" Jamie reached for the sword at his belt, half fuddled but starting forward; the horses were neighing and flinging themselves against the traces, the driver cursing and on his feet, hauling on the reins. Several of the mercenaries ran toward the horseman, who drew his own sword and rode through them, slashing recklessly from side to side. Ian seized Jamie's arm, though, and jerked him round. "Not there! The back!"

He followed Ian at the run, and, sure enough, there was the captain on his horse at the back of the troop, in the middle of a mêlée, a dozen strangers laying about with clubs and blades, all shouting.

"*Caisteal DHOON!*" Ian bellowed, and swung his sword over his head

and flat down on the head of an attacker. It hit the man a glancing blow, but he staggered and fell to his knees, where Big Georges seized him by the hair and kneed him viciously in the face.

"Caisteal DHOON!" Jamie shouted as loud as he could, and Ian turned his head for an instant, a big grin flashing.

It was a bit like a cattle raid but lasting longer. Not a matter of hit hard and get away; he'd never been a defender before and found it heavy going. Still, the attackers were outnumbered and began to give way, some glancing over their shoulders, plainly thinking of running back into the wood.

They began to do just that, and Jamie stood panting, dripping sweat, his sword a hundredweight in his hand. He straightened, though, and caught the flash of movement from the corner of his eye.

"Dhoon!" he shouted, and broke into a lumbering, gasping run. Another group of men had appeared near the wagon and were pulling the driver's body quietly down from his seat, while one of their number grabbed at the lunging horses' bridles, pulling their heads down. Two more had got the canvas loose and were dragging out a long rolled cylinder—one of the rugs, he supposed.

He reached them in time to grab another man trying to mount the wagon, yanking him clumsily back onto the road. The man twisted, falling, and came to his feet like a cat, knife in hand. The blade flashed, bounced off the leather of Jamie's jerkin, and cut upward, an inch from his face. Jamie squirmed back, off-balance, narrowly keeping his feet, and two more of the bastards charged him.

"On your right, man!" Ian's voice came suddenly at his shoulder, and without a moment's hesitation Jamie turned to take care of the man to his left, hearing Ian's grunt of effort as he laid about with a broadsword.

Then something changed; he couldn't tell what, but the fight was over. The attackers melted away, leaving one or two of their number lying in the road.

The driver wasn't dead; Jamie saw him roll half over, an arm across his face. Then he himself was sitting in the dust, black spots dancing before his eyes. Ian bent over him, panting, hands braced on his knees. Sweat dripped from his chin, making dark spots in the dust that mingled with the buzzing spots that darkened Jamie's vision.

"All . . . right?" Ian asked.

He opened his mouth to say yes, but the roaring in his ears drowned it out, and the spots merged suddenly into a solid sheet of black.

HE WOKE TO find a priest kneeling over him, intoning the Lord's Prayer in Latin. Not stopping, the priest took up a little bottle and poured

oil into the palm of one hand, then dipped his thumb into the puddle and made a swift sign of the cross on Jamie's forehead.

"I'm no dead, aye?" Jamie said, then repeated this information in French. The priest leaned closer, squinting nearsightedly.

"Dying?" he asked.

"Not that, either."

The priest made a small, disgusted sound but went ahead and made crosses on the palms of Jamie's hands, his eyelids, and his lips. *"Ego te absolvo,"* he said, making a final quick sign of the cross over Jamie's supine form. "Just in case you've killed anyone." Then he rose swiftly to his feet and disappeared behind the wagon in a flurry of dark robes.

"All right, are ye?" Ian reached down a hand and hauled him into a sitting position.

"Aye, more or less. Who was that?" He nodded in the direction of the recent priest.

"Père Renault. This is a verra well-equipped outfit," Ian said, boosting him to his feet. "We've got our own priest, to shrive us before battle and give us Extreme Unction after."

"I noticed. A bit over-eager, is he no?"

"He's blind as a bat," Ian said, glancing over his shoulder to be sure the priest wasn't close enough to hear. "Likely thinks better safe than sorry, aye?"

"D'ye have a surgeon, too?" Jamie asked, glancing at the two attackers who had fallen. The bodies had been pulled to the side of the road; one was clearly dead, but the other was beginning to stir and moan.

"Ah," Ian said thoughtfully. "That would be the priest, as well."

"So if I'm wounded in battle, I'd best try to die of it, is that what ye're sayin'?"

"I am. Come on, let's find some water."

THEY FOUND A rock-lined irrigation ditch running between two fields, a little way off the road. Ian pulled Jamie into the shade of a tree and, rummaging in his rucksack, produced a spare shirt, which he shoved into his friend's hands.

"Put it on," he said, low-voiced. "Ye can wash yours out; they'll think the blood on it's from the fightin'." Jamie looked surprised, but grateful, and with a nod skimmed out of the leather jerkin and peeled the sweaty, stained shirt gingerly off his back. Ian grimaced; the bandages were filthy and coming loose, save where they stuck to Jamie's skin, crusted black with old blood and dried pus.

"Shall I pull them off?" he muttered in Jamie's ear. "I'll do it fast."

Jamie arched his back in refusal, shaking his head.

"Nay, it'll bleed more if ye do." There wasn't time to argue; several more of the men were coming. Jamie ducked hurriedly into the clean shirt and knelt to splash water on his face.

"Hey, Scotsman!" Alexandre called to Jamie. "What's that you two were shouting at each other?" He put his hands to his mouth and hooted, "GOOOOOON!" in a deep, echoing voice that made the others laugh.

"Have ye never heard a war cry before?" Jamie asked, shaking his head at such ignorance. "Ye shout it in battle, to call your kin and your clan to your side."

"Does it mean anything?" Petit Philippe asked, interested.

"Aye, more or less," Ian said. "Castle Dhuni's the dwelling place of the chieftain of the Frasers of Lovat. *Caisteal Dhuin* is what ye call it in the *Gàidhlig*—that's our own tongue."

"And that's our clan," Jamie clarified. "Clan Fraser, but there's more than one branch, and each one will have its own war cry and its own motto." He pulled his shirt out of the cold water and wrung it out; the bloodstains were still visible but faint brown marks now, Ian saw with approval. Then he saw Jamie's mouth opening to say more.

Don't say it! he thought, but, as usual, Jamie wasn't reading his mind, and Ian closed his eyes in resignation, knowing what was coming.

"Our clan motto's in French, though," Jamie said, with a small air of pride. *"Je suis prêt."*

It meant "I am ready" and was, as Ian had foreseen, greeted with gales of laughter and a number of crude speculations as to just what the young Scots might be ready for. The men were in good humor from the fight, and it went on for a bit. Ian shrugged and smiled, but he could see Jamie's ears turning red.

"Where's the rest of your queue, Georges?" Petit Philippe demanded, seeing Big Georges shaking off after a piss. "Someone trim it for you?"

"Your wife bit it off," Georges replied, in a tranquil tone indicating that this was common badinage. "Mouth like a sucking pig, that one. And a *cramouille* like a—"

This resulted in a further scatter of abuse, but it was clear from the sidelong glances that it was mostly performance for the benefit of the two Scots. Ian ignored it. Jamie had gone squiggle-eyed; Ian wasn't sure his friend had ever heard the word *"cramouille"* before, but he likely figured what it meant.

Before Jamie could get them in more trouble, though, the conversation by the stream was stopped dead by a strangled scream beyond the scrim of trees that hid them from the roadside.

"The prisoner," Alexandre murmured after a moment.

Ian knelt by Jamie, water dripping from his cupped hands. He knew

what was happening; it curdled his wame. He let the water fall and wiped his hands on his thighs.

"The captain," he said softly to Jamie. "He'll . . . need to know who they were. Where they came from."

"Aye." Jamie's lips pressed tight at the sound of muted voices, the sudden meaty smack of flesh and a loud grunt. "I know." He splashed water fiercely onto his face.

The jokes had stopped. There was little conversation now, though Alexandre and Josef-from-Alsace began a random argument, speaking loudly, trying to drown out the noises from the road. Most of the men finished their washing and drinking in silence and sat hunched in the shade, shoulders pulled in.

"Père Renault!" The captain's voice rose, calling for the priest. Père Renault had been performing his own ablutions a discreet distance from the men but stood at this summons, wiping his face on the hem of his robe. He crossed himself and headed for the road, but on the way he paused by Ian and motioned toward his drinking cup.

"May I borrow this from you, my son? Only for a moment."

"Aye, of course, Father," Ian said, baffled. The priest nodded, bent to scoop up a cup of water, and went on his way. Jamie looked after him, then at Ian, brows raised.

"They say he's a Jew," Juanito said nearby, very quietly. "They want to baptize him first." He knelt by the water, fists curled tight against his thighs.

Hot as the air was, Ian felt a spear of ice run right through his chest. He stood up fast and made as though to follow the priest, but Big Georges snaked out a hand and caught him by the shoulder.

"Leave it," he said. He spoke quietly, too, but his fingers dug hard into Ian's flesh.

He didn't pull away but stayed standing, holding Georges's eyes. He felt Jamie make a brief, convulsive movement, but said, "No!" under his breath, and Jamie stopped.

They could hear French cursing from the road, mingled with Père Renault's voice. *"In nomine Patris, et Filii . . ."* Then struggling, spluttering, and shouting, the prisoner, the captain, and Mathieu, and even the priest, all using such language as made Jamie blink. Ian might have laughed if not for the sense of dread that froze every man by the water.

"No!" shouted the prisoner, his voice rising above the others, anger lost in terror. "No, please! I told you all I—" There was a small sound, a hollow noise like a melon being kicked in, and the voice stopped.

"Thrifty, our captain," Big Georges said, under his breath. "Why waste a bullet?" He took his hand off Ian's shoulder, shook his head, and knelt down to wash his hands.

THERE WAS A ghastly silence under the trees. From the road, they could hear low voices—the captain and Mathieu speaking to each other, and over that, Père Renault repeating, *"In nomine Patris, et Filii . . ."* but in a very different tone. Ian saw the hairs on Jamie's arms rise, and Jamie rubbed the palms of his hands against his kilt, maybe feeling a slick from the chrism oil still there.

Jamie plainly couldn't stand to listen and turned to Big Georges at random.

"Queue?" he said with a raised brow. "That what ye call it in these parts, is it?"

Big Georges managed a crooked smile.

"And what do you call it? In your tongue?"

"Bot," Ian said, shrugging. There were other words, but he wasn't about to try one like *clipeachd* on them.

"Mostly just cock," Jamie said, shrugging, too.

"Or penis, if ye want to be all English about it," Ian chimed in.

Several of the men were listening now, willing to join in any sort of conversation to get away from the echo of the last scream, still hanging in the air like fog.

"Ha," Jamie said. "Penis isna even an English word, ye wee ignoramus. It's Latin. And even in Latin, it doesna mean a man's closest companion—it means 'tail.' "

Ian gave him a long, slow look.

"Tail, is it? So ye canna even tell the difference between your cock and your arse, and ye're preachin' to me about *Latin*?"

The men roared. Jamie's face flamed up instantly, and Ian laughed and gave him a good nudge with his shoulder. Jamie snorted but elbowed Ian back and laughed, too, reluctantly.

"Aye, all right, then." He looked abashed; he didn't usually throw his education in Ian's face. Ian didn't hold it against him; he'd floundered for a bit, too, his first days with the company, and that was the sort of thing you did, trying to get your feet under you by making a point of what you were good at. But if Jamie tried rubbing Mathieu's or Big Georges's face in his Latin and Greek, he'd be proving himself with his fists, and fast, too. Right this minute, he didn't look as though he could fight a rabbit and win.

The renewed murmur of conversation, subdued as it was, dried up at once with the appearance of Mathieu through the trees. Mathieu was a big man, though broad rather than tall, with a face like a mad boar and a character to match. Nobody called him "Pig-face" *to* his face.

"You, cheese rind—go bury that turd," he said to Jamie, adding with a

narrowing of red-rimmed eyes, "far back in the wood. And go before I put a boot in your arse. Move!"

Jamie got up—slowly—eyes fixed on Mathieu with a look Ian didn't care for. He came up quick beside Jamie and gripped him by the arm.

"I'll help," he said. "Come on."

"WHY DO THEY want this one buried?" Jamie muttered to Ian. "Giving him a *Christian* burial?" He drove one of the trenching spades Armand had lent them into the soft leaf mold, with a violence that would have told Ian just how churned up his friend was if he hadn't known already.

"Ye kent it's no a verra civilized life, *a charaid*," Ian said. He didn't feel any better about it himself, after all, and spoke sharp. "Not like the *Université*."

The blood flamed up Jamie's neck like tinder taking fire, and Ian held out a palm, in hopes of quelling him. He didn't want a fight, and Jamie couldn't stand one.

"We're burying him because D'Eglise thinks his friends might come back to look for him, and it's better they don't see what was done to him, aye? Ye can see by looking that the other fellow was just killed fightin'. Business is one thing; revenge is another."

Jamie's jaw worked for a bit, but gradually the hot flush faded and his clench on the shovel loosened.

"Aye," he muttered, and resumed digging. The sweat was running down his neck in minutes, and he was breathing hard. Ian nudged him out of the way with an elbow and finished the digging. Silent, they took the dead man by the oxters and ankles and dragged him into the shallow pit.

"D'ye think D'Eglise found out anything?" Jamie asked, as they scattered matted chunks of old leaves over the raw earth.

"I hope so," Ian replied, eyes on his work. "I wouldna like to think they did that for nothing."

He straightened up and they stood awkwardly for a moment, not quite looking at each other. It seemed wrong to leave a grave, even that of a stranger and a Jew, without a word of prayer. But it seemed worse to say a Christian prayer over the man—more insult than blessing, in the circumstances.

At last Jamie grimaced and, bending, dug about under the leaves, coming out with two small stones. He gave one to Ian, and one after the other, they squatted and placed the stones together atop the grave. It wasn't much of a cairn, but it was something.

IT WASN'T THE captain's way to make explanations or to give more than brief, explicit orders to his men. He had come back into camp at evening, his face dark and his lips pressed tight. But three other men had heard the interrogation of the Jewish stranger, and by the usual metaphysical processes that happen around campfires, everyone in the troop knew by the next morning what he had said.

"Ephraim bar-Sefer," Ian said to Jamie, who had come back late to the fire, after going off quietly to wash his shirt out again. "That was his name." Ian was a bit worrit about the wean. His wounds weren't healing as they should, and the way he'd passed out . . . He'd a fever now; Ian could feel the heat coming off his skin, but he shivered now and then, though the night wasn't bitter.

"Is it better to know that?" Jamie asked bleakly.

"We can pray for him by name," Ian pointed out. "That's better, is it not?"

Jamie wrinkled up his brow, but after a moment he nodded.

"Aye, it is. What else did he say, then?"

Ian rolled his eyes. Ephraim bar-Sefer had confessed that the band of attackers were professional thieves, mostly Jews, who—

"Jews?" Jamie interrupted. "Jewish *bandits*?" For some reason, the thought struck Jamie as funny, but Ian didn't laugh.

"Why not?" he asked briefly, and went on without waiting for an answer. The men gained advance knowledge of valuable shipments and made a practice of lying in wait, to ambush and rob. "It's mostly other Jews they rob, so there's nay much danger of being pursued by the French army or a local judge."

"Oh. And the advance knowledge—that's easier come by, too, I suppose, if the folk they rob are Jews. Jews live close by one another in groups," Jamie explained, seeing the look of surprise on Ian's face. "They all read and write, though, and they write letters all the time; there's a good bit of information passed to and fro between the groups. Wouldna be that hard to learn who the moneylenders and merchants are and intercept their correspondence, would it?"

"Maybe not," Ian said, giving Jamie a look of respect. "Bar-Sefer said they got notice from someone—he didna ken who it was himself—who kent a great deal about valuables comin' and goin'. The person who knew wasna one of their group, though; it was someone outside, who got a percentage o' the proceeds."

That, however, was the total of the information bar-Sefer had divulged. He wouldn't give up the names of any of his associates—D'Eglise didn't

care so much about that—and had died stubbornly insisting that he knew nothing of future robberies planned.

"D'ye think it might ha' been one of ours?" Jamie asked, low-voiced.

"One of—oh, our Jews, ye mean?" Ian frowned at the thought. There were three Spanish Jews in D'Eglise's band—Juanito, Big Georges, and Raoul—but all three were good men and fairly popular with their fellows. "I doubt it. All three o' them fought like fiends. When I noticed," he added fairly.

"What I want to know is how the thieves got away wi' that rug," Jamie said reflectively. "Must have weighed what, ten stone?"

"At least that," Ian assured him, flexing his shoulders at the memory. "I helped load the wretched things. I suppose they must have had a wagon somewhere nearby, for their booty. Why?"

"Well, but . . . *rugs*? Who steals rugs? Even valuable ones. And if they kent ahead of time that we were comin', presumably they kent what we carried."

"Ye're forgettin' the gold and silver," Ian reminded him. "It was in the front of the wagon, under the rugs. They had to pull the rugs out to get at it."

"Mmphm." Jamie looked vaguely dissatisfied—and it was true that the bandits had gone to the trouble to carry the rug away with them. But there was nothing to be gained by more discussion, and when Ian said he was for bed, Jamie came along without argument.

They settled down in a nest of long yellow grass, wrapped in their plaids, but Ian didn't sleep at once. He was bruised and tired, but the excitements of the day were still with him, and he lay looking up at the stars for some time, remembering some things and trying hard to forget others—like the look of Ephraim bar-Sefer's head. Maybe Jamie was right and it was better not to have kent his right name.

He forced his mind into other paths, succeeding to the extent that he was surprised when Jamie shifted, cursing under his breath as the movement hurt him.

"Have ye ever done it?" Ian asked suddenly.

There was a small rustle as Jamie hitched himself into a more comfortable position.

"Have I ever done what?" he asked. His voice sounded that wee bit hoarse but none so bad. "Killed anyone? No."

"Nay, lain wi' a lass."

"Oh, that."

"Aye, *that*. Gowk." Ian rolled toward Jamie and aimed a feint toward his middle.

Despite the darkness, Jamie caught his wrist before the blow landed. "Have you?"

"Oh, ye haven't, then." Ian detached the grip without difficulty. "I thought ye'd be up to your ears in whores and poetesses in Paris."

"Poetesses?" Jamie was beginning to sound amused. "What makes ye think women write poetry? Or that a woman who writes poetry would be wanton?"

"Well, o' course they are. Everybody kens that. The words get into their heads and drive them mad, and they go looking for the first man who—"

"Ye've bedded a poetess?" Jamie's fist struck him lightly in the middle of the chest. "Does your mam ken that?"

"Dinna be telling my mam anything about poetesses," Ian said firmly. "No, but Big Georges did, and he told everyone about her. A woman he met in Marseilles. He has a book of her poetry and read some out."

"Any good?"

"How would I ken? There was a good bit o' swooning and swellin' and bursting goin' on, but it seemed to do wi' flowers, mostly. There was a good wee bit about a bumblebee, though, doin' the business wi' a sunflower. Pokin' it, I mean. With its snout."

There was a momentary silence as Jamie absorbed the mental picture.

"Maybe it sounds better in French," he said.

"I'LL HELP YE," Ian said suddenly, in a tone that was serious to the bone.

"Help me . . . ?"

"Help ye kill this Captain Randall."

Jamie lay silent for a moment, feeling his chest go tight.

"Jesus, Ian," he said, very softly. He lay for several minutes, eyes fixed on the shadowy tree roots near his face.

"No," he said at last. "Ye can't. I need ye to do something else for me, Ian. I need ye to go home."

"Home? What—"

"I need ye to go home and take care of Lallybroch—and my sister. I—I canna go. Not yet." He bit his lower lip hard.

"Ye've got tenants and friends enough there," Ian protested. "Ye need me here, man. I'm no leavin' ye alone, aye? When ye go back, we'll go together." And he turned over in his plaid with an air of finality.

Jamie lay with his eyes tight closed, ignoring the singing and conversation near the fire, the beauty of the night sky over him, and the nagging pain in his back. He should perhaps be praying for the soul of the dead Jew, but he had no time for that just now. He was trying to find his father.

Brian Fraser's soul must still exist, and he was positive that his father

was in heaven. But surely there must be some way to reach him, to sense him. When first Jamie had left home, to foster with Dougal at Beannachd, he'd been lonely and homesick, but Da had told him he would be and not to trouble overmuch about it.

"Ye think of me, Jamie, and Jenny and Lallybroch. Ye'll not see us, but we'll be here nonetheless and thinking of you. Look up at night, and see the stars, and ken we see them, too."

He opened his eyes a slit, but the stars swam, their brightness blurred. He squeezed his eyes shut again and felt the warm glide of a single tear down his temple. He couldn't think about Jenny. Or Lallybroch. The homesickness at Dougal's had stopped. The strangeness when he went to Paris had eased. This wouldn't stop, but he'd have to go on living anyway.

Where are ye, Da? he thought in anguish. *Da, I'm sorry!*

HE PRAYED AS he walked next day, making his way doggedly from one Hail Mary to the next, using his fingers to count the rosary. For a time, it kept him from thinking and gave him a little peace. But eventually the slippery thoughts came stealing back, memories in small flashes, quick as sun on water. Some he fought off—Captain Randall's voice, playful as he took the cat in hand—the fearful prickle of the hairs on his body in the cold wind when he took his shirt off—the surgeon's *"I see he's made a mess of you, boy. . . ."*

But some memories he seized, no matter how painful they were. The feel of his da's hands, hard on his arms, holding him steady. The guards had been taking him somewhere—he didn't recall and it didn't matter—when suddenly his da was there before him, in the yard of the prison, and he'd stepped forward fast when he saw Jamie, a look of joy and eagerness on his face, this blasted into shock the next moment, when he saw what they'd done to him.

"Are ye bad hurt, Jamie?"

"No, Da, I'll be all right."

For a minute, he had been. So heartened by seeing his father, sure it would all come right—and then he'd remembered Jenny, taking yon *crochaire* into the house, sacrificing herself for—

He cut that one off short, too, saying "Hail Mary, full of grace, the Lord is with thee!" savagely out loud, to the startlement of Petit Philippe, who was scuttling along beside him on his short bandy legs.

"Blessed art thou amongst women . . ." Philippe chimed in obligingly. "Pray for us sinners, now and at the hour of our death, amen!"

"Hail Mary," said Père Renault's deep voice behind him, taking it up, and within seconds seven or eight of them were saying it, marching sol-

emnly to the rhythm, and then a few more . . . Jamie himself fell silent, unnoticed. But he found the wall of prayer a barricade between himself and the wicked sly thoughts and, closing his eyes briefly, felt his father walk beside him and Brian Fraser's last kiss soft as the wind on his cheek.

THEY REACHED BORDEAUX just before sunset, and D'Eglise took the wagon off with a small guard, leaving the other men free to explore the delights of the city—though such exploration was somewhat constrained by the fact that they hadn't yet been paid. They'd get their money after the goods were delivered next day.

Ian, who'd been in Bordeaux before, led the way to a large, noisy tavern with drinkable wine and large portions.

"The barmaids are pretty, too," he observed, as he stood watching one of these creatures wend her way deftly through a crowd of groping hands.

"Is it a brothel upstairs?" Jamie asked, out of curiosity, having heard a few stories.

"I dinna ken," Ian said, with what sounded like regret, though Jamie was almost sure he'd never been to a brothel, out of a mixture of penury and fear of catching the pox. "D'ye want to go and find out later?"

Jamie hesitated.

"I—well. No, I dinna think so." He turned his face toward Ian and spoke very quietly. "I promised Da I wouldna go wi' whores, when I went to Paris. And now . . . I couldna do it without . . . thinkin' of him, ken?"

Ian nodded, his face showing as much relief as disappointment.

"Time enough another day," he said philosophically, and signaled for another jug. The barmaid didn't see him, though, and Jamie snaked out a long arm and tugged at her apron. She whirled, scowling, but seeing Jamie's face, wearing its best blue-eyed smile, chose to smile back and take the order.

Several other men from D'Eglise's band were in the tavern, and this byplay didn't pass unnoticed.

Juanito, at a nearby table, glanced at Jamie, raised a derisive eyebrow, then said something to Raoul in the Jewish sort of Spanish they called Ladino; both men laughed.

"You know what causes warts, friend?" Jamie said pleasantly—in biblical Hebrew. "Demons inside a man, trying to emerge through the skin." He spoke slowly enough that Ian could follow this, and Ian in turn broke out laughing—as much at the looks on the two Jews' faces as at Jamie's remark.

Juanito's lumpy face darkened, but Raoul looked sharply at Ian, first at his face, then, deliberately, at his crotch. Ian shook his head, still grinning,

and Raoul shrugged but returned the smile, then took Juanito by the arm, tugging him off in the direction of the back room, where dicing was to be found.

"What did you say to him?" the barmaid asked, glancing after the departing pair, then looking back wide-eyed at Jamie. "And what tongue did you say it in?"

Jamie was glad to have the wide brown eyes to gaze into; it was causing his neck considerable strain to keep his head from tilting farther down in order to gaze into her décolletage. The charming hollow between her breasts drew the eye . . .

"Oh, nothing but a little *bonhomie,*" he said, grinning down at her. "I said it in Hebrew." He wanted to impress her, and he did, but not the way he'd meant to. Her half smile vanished, and she edged back a little.

"Oh," she said. "Your pardon, sir, I'm needed . . ." and with a vaguely apologetic flip of the hand, she vanished into the throng of customers, pitcher in hand.

"Eejit," Ian said. "What did ye tell her that for? Now she thinks ye're a Jew."

Jamie's mouth fell open in shock. "What, me? How, then?" he demanded, looking down at himself. He'd meant his Highland dress, but Ian looked critically at him and shook his head.

"Ye've got the lang neb and the red hair," he pointed out. "Half the Spanish Jews I've seen look like that, and some of them are a good size, too. For all yon lass kens, ye stole the plaid off somebody ye killed."

Jamie felt more nonplussed than affronted. Rather hurt, too.

"Well, what if I was a Jew?" he demanded. "Why should it matter? I wasna askin' for her hand in marriage, was I? I was only talkin' to her, for God's sake!"

Ian gave him that annoyingly tolerant look. He shouldn't mind, he knew; he'd lorded it over Ian often enough about things he kent and Ian didn't. He did mind, though; the borrowed shirt was too small and chafed him under the arms, and his wrists stuck out, bony and raw-looking. He didn't look like a Jew, but he looked like a gowk and he knew it. It made him cross-grained.

"Most o' the Frenchwomen—the Christian ones, I mean—dinna like to go wi' Jews. Not because they're Christ-killers, but because of their . . . um . . ." He glanced down, with a discreet gesture at Jamie's crotch. "They think it looks funny."

"It doesna look *that* different."

"It does."

"Well, aye, when it's . . . but when it's—I mean, if it's in a state that a lassie would be lookin' at it, it isna . . ." He saw Ian opening his mouth to ask just how he happened to know what an erect circumcised cock looked

like. "Forget it," he said brusquely, and pushed past his friend. "Let's be goin' down the street."

———

AT DAWN, the band gathered at the inn where D'Eglise and the wagon waited, ready to escort it through the streets to its destination—a warehouse on the banks of the Garonne. Jamie saw that the captain had changed into his finest clothes, plumed hat and all, and so had the four men—among the biggest in the band—who had guarded the wagon during the night. They were all armed to the teeth, and Jamie wondered whether this was only to make a good show or whether D'Eglise intended to have them stand behind him while he explained why the shipment was one rug short, to discourage complaint from the merchant receiving the shipment.

Jamie was enjoying the walk through the city, though keeping a sharp eye out, as he'd been instructed, against the possibility of ambush from alleys or thieves dropping from a roof or balcony onto the wagon. He thought the latter possibility remote but dutifully looked up now and then. Upon lowering his eyes from one of these inspections, he found that the captain had dropped back and was now pacing beside him on his big gray gelding.

"Juanito says you speak Hebrew," D'Eglise said, looking down at him as though he'd suddenly sprouted horns. "Is this true?"

"Aye," he said cautiously. "Though it's more I can read the Bible in Hebrew—a bit—there not bein' so many Jews in the Highlands to converse with." There had been a few in Paris, but he knew better than to talk about the *Université* and the study of philosophers like Maimonides. They'd scrag him before supper.

The captain grunted but didn't look displeased. He rode for a time in silence but kept his horse to a walk, pacing at Jamie's side. This made Jamie nervous, and after a few moments, impulse made him jerk his head to the rear and say, "Ian can, too. Read Hebrew, I mean."

D'Eglise looked down at him, startled, and glanced back. Ian was clearly visible, as he stood a head taller than the three men with whom he was conversing as he walked.

"Will wonders never cease?" the captain said, as though to himself. But he nudged his horse into a trot and left Jamie in the dust.

———

IT WASN'T UNTIL the next afternoon that this conversation returned to bite Jamie in the arse. They'd delivered the rugs and the gold

and silver to the warehouse on the river, D'Eglise had received his payment, and consequently the men were scattered down the length of an *allée* that boasted cheap eating and drinking establishments, many of these with a room above or behind where a man could spend his money in other ways.

Neither Jamie nor Ian said anything further regarding the subject of brothels, but Jamie found his mind returning to the pretty barmaid. He had his own shirt on now and had half a mind to find his way back and tell her he wasn't a Jew.

He had no idea what she might do with that information, though, and the tavern was clear on the other side of the city.

"Think we'll have another job soon?" he asked idly, as much to break Ian's silence as to escape from his own thoughts. There had been talk around the fire about the prospects; evidently there were no good wars at the moment, though it was rumored that the King of Prussia was beginning to gather men in Silesia.

"I hope so," Ian muttered. "Canna bear hangin' about." He drummed long fingers on the tabletop. "I need to be movin'."

"That why ye left Scotland, is it?" He was only making conversation and was surprised to see Ian dart him a wary glance.

"Didna want to farm, wasna much else to do. I make good money here. *And* I mostly send it home."

"Still, I dinna imagine your da was pleased." Ian was the only son; Auld John was probably still livid, though he hadn't said much in Jamie's hearing during the brief time he'd been home, before the redcoats—

"My sister's marrit. Her husband can manage, if . . ." Ian lapsed into a moody silence.

Before Jamie could decide whether to prod Ian or not, the captain appeared beside their table, surprising them both.

D'Eglise stood for a moment, considering them. Finally he sighed and said, "All right. The two of you, come with me."

Ian shoved the rest of his bread and cheese into his mouth and rose, chewing. Jamie was about to do likewise when the captain frowned at him.

"Is your shirt clean?"

He felt the blood rise in his cheeks. It was the closest anyone had come to mentioning his back, and it was too close. Most of the wounds had crusted over long since, but the worst ones were still infected; they broke open with the chafing of the bandages or if he bent too suddenly. He'd had to rinse his shirt almost every night—it was constantly damp, and that didn't help—and he knew fine that the whole band knew but nobody'd spoken of it.

"It is," he replied shortly, and drew himself up to his full height, staring down at D'Eglise, who merely said, "Good, then. Come on."

THE NEW POTENTIAL client was a physician named Dr. Hasdi, reputed to be a person of great influence among the Jews of Bordeaux. The last client had made the introduction, so apparently D'Eglise had managed to smooth over the matter of the missing rug.

Dr. Hasdi's house was discreetly tucked away in a decent but modest side street, behind a stuccoed wall and locked gates. Ian rang the bell, and a man dressed like a gardener promptly appeared to let them in, gesturing them up the walk to the front door. Evidently, they were expected.

"They don't flaunt their wealth, the Jews," D'Eglise murmured out of the side of his mouth to Jamie. "But they have it."

Well, these did, Jamie thought. A manservant greeted them in a plain tiled foyer but then opened the door into a room that made the senses swim. It was lined with books in dark-wood cases, carpeted thickly underfoot, and what little of the walls was not covered with books was adorned with small tapestries and framed tiles that he thought might be Moorish. But above all, the scent! He breathed it in to the bottom of his lungs, feeling slightly intoxicated, and, looking for the source of it, finally spotted the owner of this earthly paradise sitting behind a desk and staring—at him. Or maybe him and Ian both; the man's eyes flicked back and forth between them, round as sucked toffees.

He straightened up instinctively and bowed. "We greet thee, lord," Jamie said, in carefully rehearsed Hebrew. "Peace be on your house."

The man's mouth fell open. Noticeably so; he had a large, bushy dark beard, going white near the mouth. An indefinable expression—surely it wasn't amusement?—ran over what could be seen of his face.

A small sound that certainly *was* amusement drew Jamie's attention to one side. A small brass bowl sat on a round, tile-topped table, with smoke wandering lazily up from it through a bar of late-afternoon sun. Between the sun and the smoke, he could just make out the form of a woman standing in the shadows. She stepped forward, materializing out of the gloom, and his heart jumped.

She inclined her head gravely to the soldiers, addressing them impartially.

"I am Rebekah bat-Leah Hauberger. My grandfather bids me make you welcome to our home, gentlemen," she said, in perfect French, though the old gentleman hadn't spoken. Jamie drew in a great breath of relief; he wouldn't have to try to explain their business in Hebrew, after all. The breath was so deep, though, that it made him cough, the perfumed smoke tickling his chest.

As he tried to strangle the cough, he could feel his face going red and Ian glancing sidelong at him. The girl—yes, she was young, maybe his

own age—swiftly took up a cover and clapped it on the bowl, then rang a bell and told the servant something in what sounded like Spanish. *Ladino?* he thought.

"Do please sit, sirs," she said, waving gracefully toward a chair in front of the desk, then turning to fetch another standing by the wall.

"Allow me, mademoiselle!" Ian leapt forward to assist her. Jamie, still choking as quietly as possible, followed suit.

She had dark hair, very wavy, bound back from her brow with a rose-colored ribbon but falling loose down her back, nearly to her waist. He had actually raised a hand to stroke it before catching hold of himself. Then she turned around. Pale skin, big dark eyes, and an oddly knowing look in those eyes when she met his own—which she did, very directly, when he set the third chair down before her.

Annalise. He swallowed, hard, and cleared his throat. A wave of dizzy heat washed over him, and he wished suddenly that they'd open a window.

D'Eglise, too, was visibly relieved at having a more reliable interpreter than Jamie and launched into a gallant speech of introduction, much decorated with French flowers, bowing repeatedly to the girl and her grandfather in turn.

Jamie wasn't paying attention to the talk; he was still watching Rebekah. It was her passing resemblance to Annalise de Marillac, the girl he'd loved in Paris, that had drawn his attention—but now he came to look, she was quite different.

Quite different. Annalise had been tiny and fluffy as a kitten. This girl was small—he'd seen that she came no higher than his elbow; her soft hair had brushed his wrist when she sat down—but there was nothing either fluffy or helpless about her. She'd noticed him watching her and was now watching *him*, with a faint curve to her red mouth that made the blood rise in his cheeks. He coughed and looked down.

"What's amiss?" Ian muttered out of the side of his mouth. "Ye look like ye've got a cocklebur stuck betwixt your hurdies."

Jamie gave an irritable twitch, then stiffened as he felt one of the rawer wounds on his back break open. He could feel the fast-cooling spot, the slow seep of pus or blood, and sat very straight, trying not to breathe deep, in hopes that the bandages would absorb the liquid before it got onto his shirt.

This niggling concern had at least distracted his mind from Rebekah bat-Leah Hauberger, and to distract himself from the aggravation of his back, he returned to the three-way conversation between D'Eglise and the Jews.

The captain was sweating freely, whether from the hot tea or the strain of persuasion, but he talked easily, gesturing now and then toward his matched pair of tall, Hebrew-speaking Scots, now and then toward the

window and the outer world, where vast legions of similar warriors awaited, ready and eager to do Dr. Hasdi's bidding.

The doctor watched D'Eglise intently, occasionally addressing a soft rumble of incomprehensible words to his granddaughter. It did sound like the Ladino that Juanito spoke, more than anything else; certainly it sounded nothing like the Hebrew that Jamie had been taught in Paris.

Finally the old Jew glanced among the three mercenaries, pursed his lips thoughtfully, and nodded. He rose and went to a large blanket chest that stood under the window, where he knelt and carefully gathered up a long, heavy cylinder wrapped in oiled cloth. Jamie could see that it was remarkably heavy for its size, from the slow way the old man rose with it, and his first thought was that it must be a gold statue of some sort. His second thought was that Rebekah smelled like rose petals and vanilla pods. He breathed in, very gently, feeling the shirt stick to his back.

The thing, whatever it was, jingled and chimed softly as it moved. Some sort of Jewish clock? Dr. Hasdi carried the cylinder to the desk and set it down, then curled a finger to invite the soldiers to step near.

Unwrapped with a slow and solemn sense of ceremony, the object emerged from its layers of linen, canvas, and oiled cloth. It *was* gold, in part, and not unlike statuary but made of wood and shaped like a prism, with a sort of crown at one end. While Jamie was still wondering what the devil it might be, the doctor's arthritic fingers touched a small clasp and the box opened, revealing yet more layers of cloth, from which yet another delicate, spicy scent emerged. All three soldiers breathed deep, in unison, and Rebekah made that small sound of amusement again.

"The case is cedar wood," she said. "From Lebanon."

"Oh," D'Eglise said respectfully. "Of course!"

The bundle inside was dressed—there was no other word for it; it was wearing a sort of caped mantle and a belt, with a miniature buckle—in velvet and embroidered silk. From one end, two massive golden finials protruded like twin heads. They were pierced work and looked like towers, adorned in the windows and along their lower edges with a number of tiny bells.

"This is a *very* old Torah scroll," Rebekah said, keeping a respectful distance. "From Spain."

"A priceless object, to be sure," D'Eglise said, bending to peer closer.

Dr. Hasdi grunted and said something to Rebekah, who translated:

"Only to those whose Book it is. To anyone else, it has a very obvious and attractive price. If this were not so, I would not stand in need of your services." The doctor looked pointedly at Jamie and Ian. "A respectable man—a Jew—will carry the Torah. It may not be touched. But you will safeguard it—and my granddaughter."

"Quite so, your honor." D'Eglise flushed slightly but was too pleased

to look abashed. "I am deeply honored by your trust, sir, and I assure you . . ." But Rebekah had rung her bell again, and the manservant came in with wine.

The job offered was simple; Rebekah was to be married to the son of the chief rabbi of the Paris synagogue. The ancient Torah was part of her dowry, as was a sum of money that made D'Eglise's eyes glisten. The doctor wished to engage D'Eglise to deliver all three items—the girl, the scroll, and the money—safely to Paris; the doctor himself would travel there for the wedding but later in the month, as his business in Bordeaux detained him. The only things to be decided were the price for D'Eglise's services, the time in which they were to be accomplished, and the guarantees D'Eglise was prepared to offer.

The doctor's lips pursed over this last; his friend Ackerman, who had referred D'Eglise to him, had not been entirely pleased at having one of his valuable rugs stolen en route, and the doctor wished to be assured that none of *his* valuable property—Jamie saw Rebekah's soft mouth twitch as she translated this—would go missing between Bordeaux and Paris. The captain gave Ian and Jamie a stern look, then altered this to earnest sincerity as he assured the doctor that there would be no difficulty; his best men would take on the job, and he would offer whatever assurances the doctor required. Small drops of sweat stood out on his upper lip.

Between the warmth of the fire and the hot tea, Jamie was sweating, too, and could have used a glass of wine. But the old gentleman stood up abruptly and, with a courteous bow to D'Eglise, came out from behind his desk and took Jamie by the arm, pulling him up and tugging him gently toward a doorway.

He ducked, just in time to avoid braining himself on a low archway, and found himself in a small, plain room, with bunches of drying herbs hung from its beams. What—

But before he could formulate any sort of question, the old man had got hold of his shirt and was pulling it free of his plaid. He tried to step back, but there was no room, and willy-nilly, he found himself set down on a stool, the old man's horny fingers pulling loose the bandages. The doctor made a deep sound of disapproval, then shouted something in which the words *"agua caliente"* were clearly discernible, back through the archway.

He daren't stand up and flee—and risk D'Eglise's new arrangement. And so he sat, burning with embarrassment, while the physician probed, prodded, and—a bowl of hot water having appeared—scrubbed at his back with something painfully rough. None of this bothered Jamie nearly as much as the appearance of Rebekah in the doorway, her dark eyebrows raised.

"My grandfather says your back is a mess," she told him, translating a remark from the old man.

"Thank ye. I didna ken that," he muttered in English, but then repeated the remark more politely in French. His cheeks burned with mortification, but a small, cold echo sounded in his heart. *"I see he's made a mess of you, boy."*

The surgeon at Fort William had said it when the soldiers dragged Jamie to him after the flogging, legs too wobbly to stand by himself. The surgeon had been right, and so was Dr. Hasdi, but it didn't mean Jamie wanted to hear it again.

Rebekah, evidently interested to see what her grandfather meant, came round behind Jamie. He stiffened, and the doctor poked him sharply in the back of the neck, making him bend forward again. The two Jews were discussing the spectacle in tones of detachment; he felt the girl's small, soft fingers trace a line between his ribs and nearly shot off the stool, his skin erupting in goose flesh.

"Jamie?" Ian's voice came from the hallway, sounding worried. "Are ye all right?"

"Aye!" he managed, half strangled. "Don't—ye needn't come in."

"Your name is Jamie?" Rebekah was now in front of him, leaning down to look into his face. Her own was alive with interest and concern. "James?"

"Aye. James." He clenched his teeth as the doctor dug a little harder, clicking his tongue.

"Diego," she said, smiling at him. "That's what it would be in Spanish—or Ladino. And your friend?"

"He's called Ian. That's"—he groped for a moment and found the English equivalent—"John. That would be . . ."

"Juan. Diego and Juan." She touched him gently on the bare shoulder. "You're friends? Brothers? I can see you come from the same place—where is that?"

"Friends. From . . . Scotland. The—the—Highlands. A place called Lallybroch." He'd spoken unwarily, and a pang shot through him at the name, sharper than whatever the doctor was scraping his back with. He looked away; the girl's face was too close—he didn't want her to see.

She didn't move away. Instead, she crouched gracefully beside him and took his hand. Hers was very warm, and the hairs on his wrist rose in response, in spite of what the doctor was doing to his back.

"It will be done soon," she promised. "He's cleaning the infected parts; he says they will scab over cleanly now and stop draining." A gruff question from the doctor. "He asks, do you have fever at night? Bad dreams?"

Startled, he looked back at her, but her face showed only compassion. Her hand tightened on his in reassurance.

"I . . . yes. Sometimes."

A grunt from the doctor, more words, and Rebekah let go his hand with a little pat and went out, skirts a-rustle. He closed his eyes and tried to keep the scent of her in his mind—he couldn't keep it in his nose, as the doctor was now anointing him with something vile-smelling. He could smell himself, too, and his jaw prickled with embarrassment; he reeked of stale sweat, campfire smoke, and fresh blood.

He could hear D'Eglise and Ian talking in the parlor, low-voiced, discussing whether to come and rescue him. He would have called out to them, save that he couldn't bear the captain to see . . . He pressed his lips together tight. Aye, well, it was nearly done; he could tell from the doctor's slower movements, almost gentle now.

"Rebekah!" the doctor called, impatient, and the girl appeared an instant later, a small cloth bundle in one hand. The doctor let off a short burst of words, then pressed a thin cloth of some sort over Jamie's back; it stuck to the nasty ointment.

"Grandfather says the cloth will protect your shirt until the ointment is absorbed," she told him. "By the time it falls off—don't peel it off, let it come off by itself—the wounds will be scabbed, but the scabs should be soft and not crack."

The doctor took his hand off Jamie's shoulder, and Jamie shot to his feet, looking round for his shirt. Rebekah handed it to him. Her eyes were fastened on his naked chest, and he was—for the first time in his life—embarrassed by the fact that he possessed nipples. An extraordinary but not unpleasant tingle made the curly hairs on his body stand up.

"Thank you—ah, I mean . . . *gracias, señor.*" His face was flaming, but he bowed to the doctor with as much grace as he could muster. *"Muchas gracias."*

"De nada," the old man said gruffly, with a dismissive wave of one hand. He pointed at the small bundle in his granddaughter's hand. "Drink. No fever. No dream." And then, surprisingly, he smiled.

"Shalom," he said, and made a shooing gesture.

D'EGLISE, LOOKING PLEASED with the new job, left Ian and Jamie at a large tavern called *Le Poulet Gai,* where some of the other mercenaries were enjoying themselves—in various ways. The Cheerful Chicken most assuredly did boast a brothel on the upper floor, and slatternly women in various degrees of undress wandered freely through the lower rooms, picking up new customers with whom they vanished upstairs.

The two tall young Scots provoked a certain amount of interest from

the women, but when Ian solemnly turned his empty purse inside out in front of them—he having put his money inside his shirt for safety—they left the lads alone.

"Couldna look at one of those," Ian said, turning his back on the whores and devoting himself to his ale. "Not after seein' the wee Jewess up close. Did ye ever seen anything like?"

Jamie shook his head, deep in his own drink. It was sour and fresh and went down a treat, parched as he was from the ordeal in Dr. Hasdi's surgery. He could still smell the ghost of Rebekah's scent, vanilla and roses, a fugitive fragrance among the reeks of the tavern. He fumbled in his sporran, bringing out the little cloth bundle Rebekah had given him.

"She said—well, the doctor said—I was to drink this. How, d'ye think?" The bundle held a mixture of broken leaves, small sticks, and a coarse powder, and smelled strongly of something he'd never smelled before. Not bad; just odd.

Ian frowned at it. "Well . . . ye'd brew a tea of it, I suppose," he said. "How else?"

"I havena got anything to brew it in," Jamie said. "I was thinkin' . . . maybe put it in the ale?"

"Why not?"

IAN WASN'T PAYING much attention; he was watching Mathieu Pig-face, who was standing against a wall, summoning whores as they passed by, looking them up and down, and occasionally fingering the merchandise before sending each one on with a smack on the rear.

He wasn't really tempted—the women scairt him, to be honest—but he was curious. If he ever *should* . . . how did ye start? Just grab, like Mathieu was doing, or did ye need to ask about the price first, to be sure you could afford it? And was it proper to bargain, like ye did for a loaf of bread or a flitch of bacon, or would the woman kick ye in the privates and find someone less mean?

He shot a glance at Jamie, who, after a bit of choking, had got his herbed ale down all right and was looking a little glazed. He didn't think Jamie knew, either, but he didn't want to ask, just in case he did.

"I'm goin' to the privy," Jamie said abruptly, and stood up. He looked pale.

"Have ye got the shits?"

"Not yet." With this ominous remark, he was off, bumping into tables in his haste, and Ian followed, pausing long enough to thriftily drain the last of Jamie's ale as well as his own.

Mathieu had found one he liked; he leered at Ian and said something obnoxious as he ushered his choice toward the stairs. Ian smiled cordially and said something much worse in *Gàidhlig*.

By the time Ian got to the yard at the back of the tavern, Jamie had disappeared. Figuring Jamie would be back as soon as he rid himself of his trouble, Ian leaned tranquilly against the back wall of the building, enjoying the cool night air and watching the folk in the yard.

There were a couple of torches burning, stuck in the ground, and it looked a bit like a painting he'd seen of the Last Judgment, with angels on the one side blowing trumpets, and sinners on the other going down to hell in a tangle of naked limbs and bad behavior. It was mostly sinners out here, though now and then he thought he saw an angel floating past the corner of his eye. He licked his lips thoughtfully, wondering what was in the stuff Dr. Hasdi had given Jamie.

Jamie himself emerged from the privy at the far side of the yard, looking a little more settled in himself. Spotting Ian, he made his way through the small knots of drinkers sitting on the ground singing and the others wandering to and fro, smiling vaguely as they looked for something, not knowing what they were looking for.

Ian was seized by a sudden sense of revulsion, almost terror: a fear that he would never see Scotland again, would die here, among strangers.

"We should go home," he said abruptly, as soon as Jamie was in earshot. "As soon as we've finished this job."

"Home?" Jamie looked strangely at Ian, as though he were speaking some incomprehensible language.

"Ye've business there, and so have I. We—"

A skelloch and the thud and clatter of a falling table with its burden of dishes interrupted them. The back door of the tavern burst open and a woman ran out, yelling in a sort of French that Ian didn't understand but knew fine was bad words from the tone of it. Similar words in a loud male voice, and Mathieu charged out after her.

He caught her by the shoulder, spun her round, and cracked her across the face with the back of one meaty hand. Ian flinched at the sound, and Jamie's hand tightened on his wrist.

"What—" Jamie began, but then stopped dead.

"Putain de . . . merde . . . tu fais . . . chien," Mathieu panted, slapping her with each word. She shrieked some more, trying to get away, but he had her by the arm and now jerked her round and pushed her hard in the back, knocking her to her knees.

Jamie's hand loosened, and Ian grabbed his arm, tight.

"Don't," he said tersely, and yanked Jamie back into the shadow.

"I wasn't," Jamie said, but under his breath and not noticing much

what he was saying, because his eyes were fixed on what was happening, as much as Ian's were.

The light from the door spilled over the woman, glowing off her hanging breasts, bared in the ripped neck of her shift. Glowing off her wide round buttocks, too; Mathieu had shoved her skirts up to her waist and was behind her, jerking at his flies one-handed, the other hand twisted in her hair so her head pulled back, throat straining and her face white-eyed as a panicked horse.

"*Pute!*" he said, and gave her arse a loud smack, open-handed. "Nobody says no to me!" He'd got his cock out now, in his hand, and shoved it into the woman with a violence that made her hurdies wobble and knotted Ian from knees to neck.

"*Merde,*" Jamie said, still under his breath. Other men and a couple of women had come out into the yard and were gathered round with the others, enjoying the spectacle as Mathieu set to work in a businesslike manner. He let go of the woman's hair in order to grasp her by the hips, and her head hung down, hair hiding her face. She grunted with each thrust, panting bad words that made the onlookers laugh.

Ian was shocked—and shocked as much at his own arousal as at what Mathieu was doing. He'd not seen open coupling before, only the heaving and giggling of things happening under a blanket, now and then a wee flash of pale flesh. This . . . He ought to look away, he knew that fine. But he didn't.

Jamie took in a breath, but no telling whether he meant to say something. Mathieu threw back his big head and howled like a wolf, and the watchers all cheered. Then his face convulsed, gapped teeth showing in a grin like a skull's, and he made a noise like a pig gives out when you knock it clean on the head and collapsed on top of the whore.

The whore squirmed out from under his bulk, abusing him roundly. Ian understood what she was saying now and would have been shocked anew if he'd had any capacity for being shocked left. She hopped up, evidently not hurt, and kicked Mathieu in the ribs once, then twice, but having no shoes on, she didn't hurt him. She reached for the purse still tied at his waist, stuck her hand in, and grabbed a handful of coins, then kicked him once more for luck and stomped off into the house, holding up the neck of her shift. Mathieu lay sprawled on the ground, his breeks around his thighs, laughing and wheezing.

Ian heard Jamie swallow and realized he was still gripping Jamie's arm. Jamie didn't seem to have noticed. Ian let go. His face was burning all the way down to the middle of his chest, and he didn't think it was just torchlight on Jamie's face, either.

"Let's . . . go someplace else," he said.

"I WISH WE'D . . . done something," Jamie blurted. They hadn't spoken at all after leaving *Le Poulet Gai*. They'd walked clear to the other end of the street and down a side alley, eventually coming to rest in a small tavern, fairly quiet. Juanito and Raoul were there, dicing with some locals, but gave Ian and Jamie no more than a glance.

"I dinna see what we *could* have done," Ian said reasonably. "I mean, we could maybe have taken on Mathieu together and got off with only bein' maimed. But ye ken it would ha' started a kebbie-lebbie, wi' all the others there." He hesitated and gave Jamie a quick glance before returning his gaze to his cup. "And . . . she *was* a whore. I mean, she wasna a—"

"I ken what ye mean." Jamie cut him off. "Aye, ye're right. And she did go with the man, to start. God knows what he did to make her take against him, but there's likely plenty to choose from. I wish—ah, feck it. D'ye want something to eat?"

Ian shook his head. The barmaid brought them a jug of wine, glanced at them, and dismissed them as negligible. It was rough wine that took the skin off the insides of your mouth, but it had a decent taste to it, under the resin fumes, and wasn't too much watered. Jamie drank deep and faster than he generally did; he was uneasy in his skin, prickling and irritable, and wanted the feeling to go away.

There were a few women in the place, not many. Jamie had to think that whoring maybe wasn't a profitable business, wretched as most of the poor creatures looked, raddled and half toothless. Maybe it wore them down, having to . . . He turned away from the thought and, finding the jug empty, waved to the barmaid for another.

Juanito gave a joyful whoop and said something in Ladino. Looking in that direction, Jamie saw one of the whores who'd been lurking in the shadows come gliding purposefully in, bending down to give Juanito a congratulatory kiss as he scooped in his winnings. Jamie snorted a little, trying to blow the smell of her out of his neb—she'd passed by close enough that he'd got a good whiff of her, a stink of rancid sweat and dead fish. Alexandre had told him that was from unclean privates, and he believed it.

He went back to the wine. Ian was matching him, cup for cup, and likely for the same reason. His friend wasn't usually irritable or crankit, but if he was well put out, he'd often stay that way until the next dawn—a good sleep erased his bad temper, but 'til then you didn't want to rile him.

He shot a sidelong glance at Ian. He couldn't tell Ian about Jenny. He just . . . couldn't. But neither could he think about her, left alone at Lallybroch . . . maybe with ch—

"Oh, God," he said, under his breath. "No. Please. No."

"Dinna come back," Murtagh had said, and plainly meant it. Well, he *would* go back—but not yet a while. It wouldn't help his sister, him going back just now and bringing Randall and the redcoats straight to her like flies to a fresh-killed deer . . . He shoved that analogy hastily out of sight, horrified. The truth was, it made him sick with shame to think about Jenny, and he tried not to—and was the more ashamed because he mostly succeeded.

Ian's gaze was fixed on another of the harlots. She was old, in her thirties at least, but had most of her teeth and was cleaner than most. She was flirting with Juanito and Raoul, too, and Jamie wondered whether she'd mind if she found out they were Jews. Maybe a whore couldn't afford to be choosy.

His treacherous mind at once presented him with a picture of his sister, obliged to follow that walk of life to feed herself, made to take any man who . . . Blessed Mother, what would the folk, the tenants, the servants, do to her if they found out what had happened? The talk . . . He shut his eyes tight, hoping to block the vision.

"That one's none sae bad," Ian said meditatively, and Jamie opened his eyes. The better-looking whore had bent over Juanito, deliberately rubbing her breast against his warty ear. "If she doesna mislike a Jew, maybe she'd . . ."

The blood flamed up in Jamie's face.

"If ye've got any thought to my sister, ye're no going to—to—pollute yourself wi' a French whore!"

Ian's face went blank but then flooded with color in turn.

"Oh, aye? And if I said your sister wasna worth it?"

Jamie's fist caught him in the eye and he flew backward, overturning the bench and crashing into the next table. Jamie scarcely noticed, the agony in his hand shooting fire and brimstone from his crushed knuckles up his forearm. He rocked to and fro, injured hand clutched between his thighs, cursing freely in three languages.

Ian sat on the floor, bent over, holding his eye and breathing through his mouth in short gasps. After a minute, he straightened up. His eye was puffing already, leaking tears down his lean cheek. He got up, shaking his head slowly, and put the bench back in place. Then he sat down, picked up his cup and took a deep gulp, put it down and blew out his breath. He took the snot-rag Jamie was holding out to him and dabbed at his eye.

"Sorry," Jamie managed. The agony in his hand was beginning to subside, but the anguish in his heart wasn't.

"Aye," Ian said quietly, not meeting his eye. "I wish we'd done something, too. Ye want to share a bowl o' stew?"

TWO DAYS LATER, they set off for Paris. After some thought, D'Eglise had decided that Rebekah and her maid, Marie, would travel by coach, escorted by Jamie and Ian. D'Eglise and the rest of the troop would take the money, with some men sent ahead in small groups to wait, both to check the road and so that they could ride in shifts, not stopping anywhere along the way. The women obviously would have to stop, but if they had nothing valuable with them, they'd be in no danger.

It was only when they went to collect the women at Dr. Hasdi's residence that they learned the Torah scroll and its custodian, a sober-looking man of middle age introduced to them as Monsieur Peretz, would be traveling with Rebekah. "I trust my greatest treasures to you, gentlemen," the doctor told them, through his granddaughter, and gave them a formal little bow.

"May you find us worthy of trust, lord," Jamie managed in halting Hebrew, and Ian bowed with great solemnity, hand on his heart. Dr. Hasdi looked from one to the other, gave a small nod, and then stepped forward to kiss Rebekah on the forehead.

"Go with God, child," he whispered, in something close enough to Spanish that Jamie understood it.

ALL WENT WELL for the first day and the first night. The autumn weather held fine, with no more than a pleasant tang of chill in the air, and the horses were sound. Dr. Hasdi had provided Jamie with a purse to cover the expenses of the journey, and they all ate decently and slept at a very respectable inn—Ian being sent in first to inspect the premises and insure against any nasty surprises.

The next day dawned cloudy, but the wind came up and blew the clouds away before noon, leaving the sky clean and brilliant as a sapphire overhead. Jamie was riding in the van, Ian post, and the coach was making good time, in spite of a rutted, winding road.

As they reached the top of a small rise, though, Jamie brought his horse to a sudden stop, raising a hand to halt the coach, and Ian reined up alongside him. A small stream had run through the roadbed in the dip below, making a bog some ten feet across.

"What—" Jamie began, but was interrupted. The driver had pulled his team up for an instant, but at a peremptory shout from inside the coach, now snapped the reins over the horses' backs and the coach lunged forward, narrowly missing Jamie's horse, which shied violently, flinging its rider off into the bushes.

"Jamie! Are ye all right?" Torn between concern for his friend and for his duty, Ian held his horse, glancing to and fro.

"Stop them! Get them! *Ifrinn!*" Jamie scuttled crabwise out of the weeds, face scratched and bright red with fury. Ian didn't wait but kicked his horse and lit out in pursuit of the heavy coach, this now lurching from side to side as it ran down into the boggy bottom. Shrill feminine cries of protest from inside were drowned by the driver's exclamation of "*Ladrones!*"

That was one word he kent in Spanish—"thieves." One of the *ladrones* was already skittering up the side of the coach like an eight-legged cob, and the driver promptly dived off the box, hit the ground, and ran for it.

"Coward!" Ian bellowed, and gave out with a Hieland screech that set the coach horses dancing, flinging their heads to and fro, and giving the would-be kidnapper fits with the reins. He forced his own horse—which hadn't liked the screeching any better than the coach horses did—through the narrow gap between the brush and the coach and, as he came even with the driver, had his pistol out. He drew down on the fellow—a young chap with long yellow hair—and shouted at him to pull up.

The man glanced at him, crouched low, and slapped the reins on the horses' backs, shouting at them in a voice like iron. Ian fired and missed—but the delay had let Jamie catch them up; he saw Jamie's red head poke up as he climbed the back of the coach, and there were more screams from inside as Jamie pounded across the roof and launched himself at the yellow-haired driver.

Leaving that bit of trouble to Jamie to deal with, Ian kicked his horse forward, meaning to get ahead and seize the reins, but another of the thieves had beat him to it and was hauling down on one horse's head. Aye, well, it worked once. Ian inflated his lungs as far as they'd go and let rip.

The coach horses bolted in a spray of mud. Jamie and the yellow-haired driver fell off the box, and the whoreson in the road disappeared, possibly trampled into the mire. Ian hoped so. Blood in his eye, he reined up his own agitated mount, drew his broadsword, and charged across the road, shrieking like a *ban-sidhe* and slashing wildly. Two thieves stared up at him openmouthed, then broke and ran for it.

He chased them a wee bit into the brush, but the going was too thick for his horse, and he turned back to find Jamie rolling about in the road, earnestly hammering the yellow-haired laddie. Ian hesitated—help him, or see to the coach? A loud crash and horrible screams decided him at once, and he charged down the road.

The coach, driver-less, had run off the road, hit the bog, and fallen sideways into a ditch. From the clishmaclaver coming from inside, he thought the women were likely all right and, swinging off his horse, wrapped the reins hastily round a tree and went to take care of the coach horses before they killed themselves.

It took no little while to disentangle the mess single-handed—luckily

the horses had not managed to damage themselves significantly—and his efforts were not aided by the emergence from the coach of two agitated and very disheveled women carrying on in an incomprehensible mix of French and Ladino.

Just as well, he thought, giving them a vague wave of a hand he could ill spare at the moment. *It wouldna help to hear what they're saying.* Then he picked up the word "dead" and changed his mind. Monsieur Peretz was normally so silent that Ian had in fact forgotten his presence, in the confusion of the moment. He was even more silent now, Ian learned, having broken his neck when the coach overturned.

"Oh, Jesus," he said, running to look. But the man was undeniably dead, and the horses were still creating a ruckus, slipping and stamping in the mud of the ditch. He was too busy for a bit to worry about how Jamie was faring, but as he got the second horse detached from the coach and safely tethered to a tree, he did begin to wonder where the wean was.

He didn't think it safe to leave the women; the banditti might come back, and a right numpty he'd look if they did. There was no sign of their driver, who had evidently abandoned them out of fright. He told the ladies to sit down under a sycamore tree and gave them his canteen to drink from, and, after a bit, they stopped talking quite so fast.

"Where is Diego?" Rebekah said, quite intelligibly.

"Och, he'll be along presently," Ian said, hoping it was true. He was beginning to be worrit himself.

"Perhaps he's been killed, too," said the maidservant, who shot an ill-tempered glare at her mistress. "How would you feel then?"

"I'm sure he wouldn't—I mean, he's not. I'm sure," Rebekah repeated, not sounding all that sure.

She was right, though; no sooner had Ian decided to march the women back along the road to have a keek when Jamie came shambling around the bend and sank down in the dry grass, closing his eyes.

"Are you all right?" Rebekah asked, bending down anxiously to look at him from under the brim of her straw traveling hat. He didn't look very peart, Ian thought.

"Aye, fine." He touched the back of his head, wincing slightly. "Just a wee dunt on the heid. The fellow who fell down in the road," he explained to Ian, closing his eyes again. "He got up again and hit me from behind. Didna knock me clean out, but it distracted me for a wee bit, and when I got my wits back, they'd both gone—the fellow that hit me, and the one I was hittin'."

"Mmphm," said Ian, and, squatting in front of his friend, thumbed up one of Jamie's eyelids and peered intently into the bloodshot blue eye behind it. He had no idea what to look for, but he'd seen Père Renault do that, after which he usually applied leeches somewhere. As it was, both

that eye and the other one looked fine to him; just as well, as he hadn't any leeches. He handed Jamie the canteen and went to look the horses over.

"Two of them are sound enough," he reported, coming back. "The light bay's lame. Did the bandits take your horse? And what about the driver?"

Jamie looked surprised.

"I forgot I had a horse," he confessed. "I dinna ken about the driver—didna see him lyin' in the road, at least." He glanced vaguely round. "Where's Monsieur Pickle?"

"Dead. Stay there, aye?"

Ian sighed, got up, and loped back down the road, where there was no sign of the driver, though he walked to and fro calling for a while. Fortunately, he did come across Jamie's horse peaceably cropping grass by the verge. He rode it back and found the women on their feet, discussing something in low voices, now and then looking down the road or standing on their toes in a vain attempt to see through the trees.

Jamie was still sitting on the ground, eyes closed—but at least upright.

"Can ye ride, man?" Ian asked softly, squatting down by his friend. To his relief, Jamie opened his eyes at once.

"Oh, aye. Ye're thinkin' we should ride into Saint-Aulaye and send someone back to do something about the coach and Peretz?"

"What else is there to do?"

"Nothing I can think of. I dinna suppose we can take him with us." Jamie got to his feet, swaying a little but without needing to hold on to the tree. "Can the women ride, d'ye think?"

Marie could, it turned out—at least a little. Rebekah had never been on a horse. After more discussion than Ian would have believed possible on the subject, he got the late M. Peretz decently laid out on the coach's seat with a handkerchief over his face against flies, and the rest of them finally mounted: Jamie on his horse with the Torah scroll in its canvas wrappings bound behind his saddle—between the profanation of its being touched by a Gentile and the prospect of its being left in the coach for anyone happening by to find, the women had reluctantly allowed the former—the maid on one of the coach horses, with a pair of saddlebags fashioned from the covers of the coach's seats, these filled with as much of the women's luggage as they could cram in, and Ian with Rebekah on the saddle before him.

Rebekah looked like a wee dolly, but she was surprisingly solid, as he found when she put her foot in his hands and he tossed her up into the saddle. She didn't manage to swing her leg over and instead lay across the saddle like a dead deer, waving her arms and legs in agitation. Wrestling her into an upright position and getting himself set behind her left him red-faced and sweating, far more than dealing with the horses had.

Jamie gave him a raised eyebrow, as much jealousy as amusement in it, and he gave Jamie a squinted eye in return and put his arm round Rebekah's waist to settle her against him, hoping that he didn't stink too badly.

IT WAS DARK by the time they made it into Saint-Aulaye and found an inn that could provide them with two rooms. Ian talked to the landlord and arranged that someone should go in the morning to retrieve M. Peretz's body and bury it; the women weren't happy about the lack of proper preparation of the body, but as they insisted he must be buried before the next sundown, there wasn't much else to be done. Then he inspected the women's room, looked under the beds, rattled the shutters in a confident manner, and bade them good night. They looked that wee bit frazzled.

Going back to the other room, he heard a sweet chiming sound and found Jamie on his knees, pushing the bundle that contained the Torah scroll under the single bed.

"That'll do," he said, sitting back on his heels with a sigh. He looked nearly as done up as the women, Ian thought, but didn't say so.

"I'll go and have some supper sent up," he said. "I smelled a joint roasting. Some of that, and maybe—"

"Whatever they've got," Jamie said fervently. "Bring it all."

THEY ATE HEARTILY, and separately, in their rooms. Jamie was beginning to feel that the second helping of *tarte tatin* with clotted cream had been a mistake, when Rebekah came into the men's room, followed by her maid carrying a small tray with a jug on it, wisping aromatic steam. Jamie sat up straight, restraining a small cry as pain flashed through his head. Rebekah frowned at him, gull-winged brows lowering in concern.

"Your head hurts very much, Diego?"

"No, it's fine. No but a wee bang on the heid." He was sweating and his wame was wobbly, but he pressed his hands flat on the table and was sure he looked steady. She appeared not to agree and came close, bending down to gaze searchingly into his eyes.

"I don't think so," she said. "You look . . . clammy."

"Oh. Aye?" he said, rather feebly.

"If she means ye resemble a fresh-shucked clam, then, aye, ye do," Ian informed him. "Shocked, ken? All pale and wet and—"

"I ken what clammy means, aye?" He glowered at Ian, who gave him half a grin—damn, he must look awful; Ian was actually worried. He swal-

lowed, groping for something witty to say in reassurance, but his gorge rose suddenly and he was obliged to shut both mouth and eyes tightly, concentrating fiercely to make it go back down.

"Tea," Rebekah was saying firmly. She took the jug from her maid and poured a cup, then folded Jamie's hands about it and, holding his hands with her own, guided the cup to his mouth. "Drink. It will help."

He drank, and it did. At least he felt less queasy at once. He recognized the taste of the tea, though he thought this cup had a few other things in it, too.

"Again." Another cup was presented; he managed to drink this one alone, and by the time it was down, he felt a good bit better. His head still throbbed with his heartbeat, but the pain seemed to be standing a little apart from him, somehow.

"You shouldn't be left alone for a while," Rebekah informed him, and sat down, sweeping her skirts elegantly around her ankles. He opened his mouth to say that he wasn't alone, Ian was there—but caught Ian's eye in time and stopped.

"The bandits," she was saying to Ian, her pretty brow creased, "who do you think that they were?"

"Ah . . . well, depends. If they kent who ye were and wanted to abduct ye, that's one thing. But could be they were no but random thieves and saw the coach and thought they'd chance it for what they might get. Ye didna recognize any of them, did ye?"

Her eyes sprang wide. They weren't quite the color of Annalise's, Jamie thought hazily. A softer brown . . . like the breast feathers on a grouse.

"Know who I was?" she whispered. "Wanted to abduct me?" She swallowed. "You . . . think that's possible?" She gave a little shudder.

"Well, I dinna ken, of course. Here, *a nighean,* ye ought to have a wee nip of that tea, I'm thinkin'." Ian stretched out a long arm for the jug, but she moved it back, shaking her head.

"No, it's medicine—and Diego needs it. Don't you?" she said, leaning forward to peer earnestly into Jamie's eyes. She'd taken off the hat but had her hair tucked up—mostly—in a lacy white cap with pink ribbon. He nodded obediently.

"Marie—bring some brandy, please. The shock . . ." She swallowed again and wrapped her arms briefly around herself. Jamie noticed the way it pushed her breasts up, so they swelled just a little above her stays. There was a bit of tea left in his cup; he drank it automatically.

Marie came with the brandy and poured a glass for Rebekah—then one for Ian, at Rebekah's gesture, and when Jamie made a small polite noise in his throat, half-filled his cup, pouring in more tea on top of it. The taste was peculiar, but he didn't really mind. The pain had gone off to the far side of the room; he could see it sitting over there, a wee glowering sort

of purple thing with a bad-tempered expression on its face. He laughed at it, and Ian frowned at him.

"What are ye giggling at?"

Jamie couldn't think how to describe the pain beastie, so he just shook his head, which proved a mistake—the pain looked suddenly gleeful and shot back into his head with a noise like tearing cloth. The room spun and he clutched the table with both hands.

"Diego!" Chairs scraped and there was a good bit of clishmaclaver that he paid no attention to. Next thing he knew, he was lying on the bed, staring at the ceiling beams. One of them seemed to be twining slowly, like a vine growing.

". . . and he told the captain that there was someone among the Jews who kent about . . ." Ian's voice was soothing, earnest and slow so Rebekah would understand him—though Jamie thought she maybe understood more than she said. The twining beam was slowly sprouting small green leaves, and he had the faint thought that this was unusual, but a great sense of tranquillity had come over him and he didn't mind it a bit.

Rebekah was saying something now, her voice soft and worried, and with some effort he turned his head to look. She was leaning over the table toward Ian, and he had both big hands wrapped round hers, reassuring her that he and Jamie would let no harm come to her.

A different face came suddenly into his view: the maid, Marie, frowning down at him. She rudely pulled back his eyelid and peered into his eye, so close he could smell the garlic on her breath. He blinked hard, and she let go with a small "hmph!" then turned to say something to Rebekah, who replied in quick Ladino. The maid shook her head dubiously but left the room.

Her face didn't leave with her, though. He could still see it, frowning down at him from above. It had become attached to the leafy beam, and he now realized that there was a snake up there, a serpent with a woman's head, and an apple in its mouth—that couldn't be right, surely it should be a pig?—and it came slithering down the wall and right over his chest, pressing the apple close to his face. It smelled wonderful, and he wanted to bite it, but before he could, he felt the weight of the snake change, going soft and heavy, and he arched his back a little, feeling the distinct imprint of big round breasts squashing against him. The snake's tail—she was mostly a woman now, but her back end seemed still to be snake-ish—was delicately stroking the inside of his thigh.

He made a very high-pitched noise, and Ian came hurriedly to the bed.

"Are ye all right, man?"

"I—oh. Oh! Oh, Jesus, do that again."

"Do *what*—" Ian was beginning, when Rebekah appeared, putting a hand on Ian's arm.

"Don't worry," she said, looking intently at Jamie. "He's all right. The medicine—it gives men strange dreams."

"He doesna look like he's asleep," Ian said dubiously.

In fact, Jamie was squirming—or thought he was squirming—on the bed, trying to persuade the lower half of the snake woman to change, too. He *was* panting; he could hear himself.

"It's a waking dream," Rebekah said reassuringly. "Come, leave him. He'll fall quite asleep in a bit, you'll see."

Jamie didn't think he'd fallen asleep, but it was evidently some time later that he emerged from a remarkable tryst with the snake demon—he didn't know how he knew she was a demon, but clearly she was—who had not changed her lower half but had a very womanly mouth about her. The tryst also included a number of her friends, these being small female demons who licked his ears—and other things—with great enthusiasm.

He turned his head on the pillow to allow one of these better access and saw, with no sense of surprise, Ian kissing Rebekah. The brandy bottle had fallen over, empty, and Jamie seemed to see the wraith of its perfume rise swirling through the air like smoke, wrapping the two of them in a mist shot with rainbows.

He closed his eyes again, the better to attend to the snake lady, who now had a number of new and interesting acquaintances. When he opened his eyes some time later, Ian and Rebekah were gone.

At one point he heard Ian give a sort of strangled cry and wondered dimly what had happened, but it didn't seem important, and the thought drifted away. He slept.

HE WOKE FEELING limp as a frostbitten cabbage leaf, but the pain in his head was gone. He just lay there for a bit, enjoying the feeling. It was dark in the room, and it was some time before he realized from the smell of brandy that Ian was lying beside him.

Memory came back to him. It took a little while to disentangle the real memories from the memory of dreams, but he was quite sure he'd seen Ian embracing Rebekah—and her, him. What the devil had happened *then?*

Ian wasn't asleep; he could tell. His friend lay rigid as one of the tomb figures in the crypt at Saint Denis, and his breathing was rapid and shaky, as though he'd just run a mile uphill. Jamie cleared his throat, and Ian jerked as though stabbed with a brooch pin.

"Aye, so?" he whispered, and Ian's breathing stopped abruptly. He swallowed audibly.

"If ye breathe a word of this to your sister," he said in an impassioned

whisper, "I'll stab ye in your sleep, cut off your heid, and kick it to Arles and back."

Jamie didn't want to think about his sister, and he did want to hear about Rebekah, so he merely repeated, "Aye. So?"

Ian made a small grunting noise, indicative of thinking how best to begin, and turned over in his plaid, facing Jamie.

"Aye, well. Ye raved a bit about the naked she-devils ye were havin' it away with, and I didna think the lass should have to be hearing that manner o' thing, so I said we should go into the other room, and—"

"Was this before or after ye started kissing her?" Jamie asked.

Ian inhaled strongly through his nose. "After," he said tersely. "And she was kissin' me back, aye?"

"Aye, I noticed that. So then . . . ?" He could feel Ian squirming slowly, like a worm on a hook, but Jamie waited. It often took Ian a moment to find words, but it was usually worth waiting for. Certainly in this instance.

He was a little shocked—and frankly envious—and he did wonder what might happen when the lass's affianced discovered she wasn't a virgin, but he supposed the man might not find out; she seemed a clever lass. It might be wise to leave D'Eglise's troop, though, and head south, just in case . . .

"D'ye think it hurts a lot to be circumcised?" Ian asked suddenly.

"I do. How could it not?" His hand sought out his own member, protectively rubbing a thumb over the bit in question. True, it wasn't a very big bit, but . . .

"Well, they do it to wee bairns," Ian pointed out. "Canna be that bad, can it?"

"Mmphm," Jamie said, unconvinced, though fairness made him add, "Aye, well, and they did it to Christ, too."

"Aye?" Ian sounded startled. "Aye, I suppose so—I hadna thought o' that."

"Well, ye dinna think of Him bein' a Jew, do ye? But He was, to start."

There was a momentary, meditative silence before Ian spoke again.

"D'ye think Jesus ever did it? Wi' a lass, I mean, before he went to preachin'?"

"I think Père Renault's goin' to have ye for blasphemy, next thing."

Ian twitched, as though worried that the priest might be lurking in the shadows.

"Père Renault's nowhere near here, thank God."

"Aye, but ye'll need to confess yourself to him, won't ye?"

Ian shot upright, clutching his plaid around him.

"What?"

"Ye'll go to hell, else, if ye get killed," Jamie pointed out, feeling rather smug. There was moonlight through the window and he could see Ian's

face drawn in anxious thought, his deep-set eyes darting right and left from Scylla to Charybdis. Suddenly Ian turned his head toward Jamie, having spotted the possibility of an open channel between the threats of hell and Père Renault.

"I'd only go to hell if it was a mortal sin," he said. "If it's no but venial, I'd only have to spend a thousand years or so in purgatory. That wouldna be so bad."

"Of course it's a mortal sin," Jamie said, cross. "Anybody kens fornication's a mortal sin, ye numpty."

"Aye, but . . ." Ian made a "wait a bit" gesture with one hand, deep in thought. "To be a *mortal* sin, though, ye've got the three things. Requirements, like." He put up an index finger. "It's got to be seriously wrong." Middle finger. "Ye've got to *know* it's seriously wrong." Ring finger. "And ye've got to give full consent to it. That's the way of it, aye?" He put his hand down and looked at Jamie, brows raised.

"Aye, and which part of that did ye not do? The full consent? Did she rape ye?" He was chaffing, but Ian turned his face away in a manner that gave him a sudden doubt. "Ian?"

"Noo . . ." his friend said, but it sounded doubtful, too. "It wasna like that—exactly. I meant more the seriously wrong part. I dinna think it was . . ." His voice trailed off.

Jamie flung himself over, raised on one elbow.

"Ian," he said, steel in his voice. "What did ye *do* to the lass? If ye took her maidenheid, it's seriously wrong. Especially with her betrothed. Oh—" A thought occurred to him, and he leaned a little closer, lowering his voice. "Was she no a virgin? Maybe that's different." If the lass was an out-and-out wanton, perhaps . . . she probably *did* write poetry, come to think . . .

Ian had now folded his arms on his knees and was resting his forehead on them, his voice muffled in the folds of his plaid. ". . . dinna ken . . ." emerged in a strangled croak.

Jamie reached out and dug his fingers hard into Ian's calf. His friend unfolded with a startled cry that made someone in a distant chamber shift and grunt in their sleep.

"What d'ye mean ye dinna ken? How could ye not notice?" he gibed.

"Ah . . . well . . . she . . . erm . . . she did me wi' her hand," Ian blurted. "Before I could . . . well."

"Oh." Jamie rolled onto his back, somewhat deflated in spirit, if not in flesh. His cock seemed still to want to hear the details.

"Is that seriously wrong?" Ian asked, turning his face toward Jamie again. "Or—well, I canna say I really gave full *consent* to it, because that wasna what I had in mind doing at all, but . . ."

"I think ye're headed for the Bad Place," Jamie assured him. "Ye meant to do it, whether ye managed or not. And how did it happen, come to that? Did she just . . . take hold?"

Ian let out a long, long sigh and sank his head in his hands. He looked as though it hurt.

"Well, we kissed for a bit, and there was more brandy—lots more. She . . . er . . . she'd take a mouthful and kiss me and, er . . . put it into my mouth, and—"

"*Ifrinn!*"

"Will ye not say, 'Hell!' like that, please? I dinna want to think about it."

"Sorry. Go on. Did she let ye feel her breasts?"

"Just a bit. She wouldna take her stays off, but I could feel her nipples through her shift—did ye say something?"

"No," Jamie said with an effort. "What then?"

"Well, she put her hand under my kilt and then pulled it out again like she'd touched a snake."

"And had she?"

"She had, aye. She was shocked. Will ye no snort like that?" he said, annoyed. "Ye'll wake the whole house. It was because it wasna circumcised."

"Oh. Is that why she wouldna . . . er . . . the regular way?"

"She didna say so, but maybe. After a bit, though, she wanted to look at it, and that's when . . . well."

"Mmphm." Naked demons versus the chance of damnation or not, Jamie thought Ian had had well the best of it this evening. A thought occurred to him. "Why did ye ask if being circumcised hurts? Ye werena thinking of doing it, were ye? For her, I mean?"

"I wouldna say the thought hadna occurred to me," Ian admitted. "I mean . . . I thought I should maybe marry her, under the circumstances. But I suppose I couldna become a Jew, even if I got up the nerve to be circumcised—my mam would tear my heid off if I did."

"No, ye're right," Jamie agreed. "She would. *And* ye'd go to hell." The thought of the rare and delicate Rebekah churning butter in the yard of a Highland croft or waulking urine-soaked wool with her bare feet was slightly more ludicrous than the vision of Ian in a skullcap and whiskers—but not by much. "Besides, ye havena got any money, have ye?"

"A bit," Ian said thoughtfully. "Not enough to go and live in Timbuktu, though, and I'd have to go at least that far."

Jamie sighed and stretched, easing himself. A meditative silence fell—Ian no doubt contemplating perdition, Jamie reliving the better bits of his opium dreams but with Rebekah's face on the snake lady. Finally he broke the silence, turning to his friend.

"So . . . was it worth the chance of goin' to hell?"

Ian sighed long and deep once more, but it was the sigh of a man at peace with himself.

"Oh, aye."

⁂

JAMIE WOKE AT DAWN, feeling altogether well and in a much better frame of mind. Some kindly soul had brought a jug of sour ale and some bread and cheese. He refreshed himself with these as he dressed, pondering the day's work.

He'd have to collect a few men to go back and deal with the coach.

He thought the coach wasn't badly damaged; they might get it back up on the road again by noon. . . . How far might it be to Bonnes? That was the next town with an inn. If it was too far, or the coach too badly hurt, or he couldn't find a Jew to dispose decently of M. Peretz, they'd need to stay the night here again. He fingered his purse but thought he had enough for another night and the hire of men; the doctor had been generous.

He was beginning to wonder what was keeping Ian and the women. Though he kent women took more time to do anything than a man would, let alone getting dressed—well, they had stays and the like to fret with, after all . . . He sipped ale, contemplating a vision of Rebekah's stays and the very vivid images his mind had been conjuring ever since Ian's description of his encounter with the lass. He could all but see her nipples through the thin fabric of her shift, smooth and round as pebbles . . .

Ian burst through the door, wild-eyed, his hair standing on end.

"They're gone!"

Jamie choked on his ale.

"What? How?"

Ian understood what he meant and was already heading for the bed.

"No one took them. There's nay sign of a struggle, and their things are gone. The window's open, and the shutters aren't broken."

Jamie was on his knees alongside Ian, thrusting first his hands and then his head and shoulders under the bed. There was a canvas-wrapped bundle there, and he was flooded with a momentary relief—which disappeared the instant Ian dragged it into the light. It made a noise, but not the gentle chime of golden bells. It rattled, and when Jamie seized the corner of the canvas and unrolled it, the contents were shown to be naught but sticks and stones, these hastily wrapped in a woman's petticoat to give the bundle the appropriate bulk.

"*Cramouille!*" he said, this being the worst word he could think of on

short notice. And very appropriate, too, if what he thought had happened really had. He turned on Ian.

"She drugged me and seduced you, and her bloody maid stole in here and took the thing whilst ye had your fat heid buried in her . . . er . . ."

"Charms," Ian said succinctly, and flashed him a brief, evil grin. "Ye're only jealous. Where d'ye think they've gone?"

It was the truth, and Jamie abandoned any further recriminations, rising and strapping on his belt, hastily arranging dirk, sword, and ax in the process.

"Not to Paris, would be my guess. Come on, we'll ask the ostler."

The ostler confessed himself at a loss; he'd been the worse for drink in the hay shed, he said, and if someone had taken two horses from the shelter, he hadn't waked to see it.

"Aye, right," said Jamie, impatient, and, grabbing the man's shirtfront, lifted him off his feet and slammed him into the inn's stone wall. The man's head bounced once off the stones and he sagged in Jamie's grip, still conscious but dazed. Jamie drew his dirk left-handed and pressed the edge of it against the man's weathered throat.

"Try again," he suggested pleasantly. "I dinna care about the money they gave you—keep it. I want to know which way they went and when they left."

The man tried to swallow but abandoned the attempt when his Adam's apple hit the edge of the dirk.

"About three hours past moonrise," he croaked. "They went toward Bonnes. There's a crossroads no more than three miles from here," he added, now trying urgently to be helpful.

Jamie dropped him with a grunt. "Aye, fine," he said in disgust. "Ian—oh, ye've got them." For Ian had gone straight for their own horses while Jamie dealt with the ostler and was already leading one out, bridled, the saddle over his arm. "I'll settle the bill, then."

The women hadn't made off with his purse; that was something. Either Rebekah bat-Leah Hauberger had some vestige of conscience—which he doubted very much—or she just hadn't thought of it.

IT WAS STILL EARLY; the women had perhaps six hours' lead.

"Do we believe the ostler?" Ian asked, settling himself in the saddle.

Jamie dug in his purse, pulled out a copper penny, and flipped it, catching it on the back of his hand.

"Tails we do, heads we don't?" He took his hand away and peered at the coin. "Heads."

"Aye, but the road back is straight all the way through Yvrac," Ian

pointed out. "And it's nay more than three miles to the crossroads, he said. Whatever ye want to say about the lass, she's no a fool."

Jamie considered that one for a moment, then nodded. Rebekah couldn't have been sure how much lead she'd have—and unless she'd been lying about her ability to ride (which he wouldn't put past her, but such things weren't easy to fake and she was gey clumsy in the saddle), she'd want to reach a place where the trail could be lost before her pursuers could catch up with her. Besides, the ground was still damp with dew; there might be a chance . . .

"Aye, come on, then."

LUCK WAS WITH THEM. No one had passed the inn during the late-night watches, and while the roadbed was trampled with hoof marks, the recent prints of the women's horses showed clear, edges still crumbling in the damp earth. Once sure they'd got upon the track, the men galloped for the crossroads, hoping to reach it before other travelers obscured the marks.

No such luck. Farm wagons were already on the move, loaded with produce headed for Parcoul or La Roche-Chalais, and the crossroads was a maze of ruts and hoofprints. But Jamie had the bright thought of sending Ian down the road that lay toward Parcoul, while he took the one toward La Roche-Chalais, catching up the incoming wagons and questioning the drivers. Within an hour, Ian came pelting back with the news that the women had been seen, riding slowly and cursing volubly at each other, toward Parcoul.

"And *that*," he said, panting for breath, "is not all."

"Aye? Well, tell me while we ride."

Ian did. He'd been hurrying back to find Jamie when he'd met Josef-from-Alsace, just short of the crossroads, come in search of them.

"D'Eglise was held up near La Teste-de-Buch," Ian reported in a shout. "The same band of men that attacked us at Bèguey—Alexandre and Raoul both recognized some of them. Jewish bandits."

Jamie was shocked and slowed for a moment to let Ian catch him up. "Did they get the dowry money?"

"No, but they had a hard fight. Three men wounded badly enough to need a surgeon, and Paul Martan lost two fingers of his left hand. D'Eglise pulled them into La Teste-de-Buch and sent Josef to see if all was well wi' us."

Jamie's heart bounced into his throat. "Jesus. Did ye tell him what happened?"

"I did not," Ian said tersely. "I told him we'd had an accident wi' the

coach, and ye'd gone ahead with the women; I was comin' back to fetch something left behind."

"Aye, good." Jamie's heart dropped back into his chest. The last thing he wanted was to have to tell the captain that they'd lost the girl and the Torah scroll. And he'd be damned if he would.

THEY TRAVELED FAST, stopping only to ask questions now and then, and by the time they pounded into the village of Aubeterre-sur-Dronne, they were sure that their quarry lay no more than an hour ahead of them—if the women had passed on through the village.

"Oh, those two?" said a woman, pausing in the act of scrubbing her steps. She stood up slowly, stretching her back. "I saw them, yes. They rode right by me and went down the lane there." She pointed.

"I thank you, madame," Jamie said, in his best Parisian French. "What lies down that lane, please?"

She looked surprised that they didn't know and frowned a little at such ignorance.

"Why, the chateau of the Vicomte Beaumont, to be sure!"

"To be sure," Jamie repeated, smiling at her, and Ian saw a dimple appear in her cheek in reply. *"Merci beaucoup, madame!"*

"WHAT THE DEVIL . . . ?" Ian murmured. Jamie reined up beside him, pausing to look at the place. It was a small manor house, somewhat run down but pretty in its bones. And the last place anyone would think to look for a runaway Jewess, he'd say that for it.

"What shall we do now, d'ye think?" he asked, and Jamie shrugged and kicked his horse.

"Go knock on the door and ask, I suppose."

Ian followed his friend up to the door, feeling intensely conscious of his grubby clothes, sprouting beard, and general state of uncouthness. Such concerns vanished, though, when Jamie's forceful knock was answered.

"Good day, gentlemen!" said the yellow-haired bugger Ian had last seen locked in combat in the roadbed with Jamie the day before. The man smiled broadly at them, cheerful despite an obvious black eye and a freshly split lip. He was dressed in the height of fashion, in a plum velvet suit; his hair was curled and powdered, and his yellow beard was neatly trimmed. "I hoped we would see you again. Welcome to my home!" he said, stepping back and raising his hand in a gesture of invitation.

"I thank you, monsieur . . . ?" Jamie said slowly, giving Ian a sidelong

glance. Ian lifted one shoulder in the ghost of a shrug. Did they have a choice?

The yellow-haired bugger bowed. "Pierre Robert Heriveaux d'Anton, Vicomte Beaumont, by the grace of the Almighty, for one more day. And you, gentlemen?"

"James Alexander Malcolm MacKenzie Fraser," Jamie said, with a good attempt at matching the other's grand manner. Only Ian would have noticed the faint hesitation or the slight tremor in his voice when he added, "Laird of Broch Tuarach."

"Ian Alastair Robert MacLeod Murray," Ian said, with a curt nod, and straightened his shoulders. "His . . . er . . . the laird's . . . tacksman."

"Come in, please, gentlemen." The yellow-haired bugger's eyes shifted just a little, and Ian heard the crunch of gravel behind them, an instant before he felt the prick of a dagger in the small of his back. No, they didn't have a choice.

Inside, they were relieved of their weapons, then escorted down a wide hallway and into a commodious parlor. The wallpaper was faded, and the furniture was good but shabby. By contrast, the big Turkey carpet on the floor glowed like it was woven from jewels. A big roundish thing in the middle was green and gold and red, and concentric circles with wiggly edges surrounded it in waves of blue and red and cream, bordered in a soft, deep red, and the whole of it so ornamented with unusual shapes it would take you a day to look at them all. He'd been so taken with it the first time he saw it, he'd spent a quarter of an hour looking at them, before Big Georges caught him at it and shouted at him to roll the thing up, they hadn't all day.

"Where did ye get this?" Ian asked abruptly, interrupting something the vicomte was saying to the two rough-clad men who'd taken their weapons.

"What? Oh, the carpet! Yes, isn't it wonderful?" The vicomte beamed at him, quite unself-conscious, and gestured the two roughs away toward the wall. "It's part of my wife's dowry."

"Your wife," Jamie repeated carefully. He darted a sideways glance at Ian, who took the cue.

"That would be Mademoiselle Hauberger, would it?" he asked. The vicomte blushed—actually blushed—and Ian realized that the man was no older than he and Jamie were.

"Well. It—we—we have been betrothed for some time, and in Jewish custom, that is almost like being married."

"Betrothed," Jamie echoed again. "Since . . . when, exactly?"

The vicomte sucked in his lower lip, contemplating them. But whatever caution he might have had was overwhelmed in what were plainly very high spirits.

"Four years," he said. And, unable to contain himself, he beckoned them to a table near the window and proudly showed them a fancy document covered with colored scrolly sorts of things and written in some very odd language that was all slashes and tilted lines.

"This is our *ketubah*," he said, pronouncing the word very carefully. "Our marriage contract."

Jamie bent over to peer closely at it. "Aye, verra nice," he said politely. "I see it's no been signed yet. The marriage hasna taken place, then?"

Ian saw Jamie's eyes flick over the desk and could sense him passing the possibilities through his mind: grab the letter opener off the desk and take the vicomte hostage? Then find the sly wee bitch, roll her up in one of the smaller rugs, and carry her to Paris? That would doubtless be his own job, Ian thought.

A slight movement as one of the roughs shifted his weight, catching Ian's eye, and he thought, *Don't do it, eejit!* at Jamie, as hard as he could. For once, the message seemed to get through; Jamie's shoulders relaxed a little and he straightened up.

"Ye do ken the lass is meant to be marrying someone else?" he asked baldly. "I wouldna put it past her not to tell ye."

The vicomte's color became higher. "Certainly I know!" he snapped. "She was promised to me first, by her father!"

"How long have ye been a Jew?" Jamie asked carefully, edging round the table. "I dinna think ye were born to it. I mean—ye *are* a Jew now, aye? For I kent one or two, in Paris, and it's my understanding that they dinna marry people who aren't Jewish." His eyes flicked round the solid, handsome room. "It's my understanding that they mostly aren't aristocrats, either."

The vicomte was quite red in the face by now. With a sharp word, he sent the roughs out—though they were disposed to argue. While the brief discussion was going on, Ian edged closer to Jamie and whispered rapidly to him about the rug in *Gàidhlig*.

"Holy God," Jamie muttered in the same language. "I didna see him or either of those two at Bèguey, did you?"

Ian had no time to reply and merely shook his head, as the roughs reluctantly acquiesced to Vicomte Beaumont's imperious orders and shuffled out with narrowed eyes aimed at Ian and Jamie. One of them had Jamie's dirk in his hand and drew this slowly across his neck in a meaningful gesture as he left.

Aye, they might manage in a fight, Ian thought, returning the slit-eyed glare, *but not that wee velvet gomerel.* Captain D'Eglise wouldn't have taken on the vicomte, and neither would a band of professional highwaymen, Jewish or not.

"All right," the vicomte said abruptly, leaning his fists on the desk. "I'll tell you."

And he did. Rebekah's mother, the daughter of Dr. Hasdi, had fallen in love with a Christian man and run away with him. The doctor had declared his daughter dead, as was the usual way in such a situation, and done formal mourning for her. But she was his only child, and he had not been able to forget her. He had arranged to have information brought to him and knew about Rebekah's birth.

"Then her mother died. That's when I met her—about that time, I mean. Her father was a judge, and my father knew him. She was fourteen and I sixteen; I fell in love with her. And she with me," he added, giving the Scots a hard eye, as though daring them to disbelieve it. "We were betrothed, with her father's blessing. But then her father caught a flux and died in two days. And—"

"And her grandfather took her back," Jamie finished. "And she became a Jew?"

"By Jewish belief, she was born Jewish; it descends through the mother's line. And . . . her mother had told her, privately, about her lost heritage. She embraced it, once she went to live with her grandfather."

Ian stirred and cocked a cynical eyebrow. "Aye? Why did ye not convert then, if ye're willing to do it now?"

"I said I would!" The vicomte had one fist curled round his letter opener as though he would strangle it. "The miserable old wretch said he did not believe me. He thought I would not give up my—my—this life." He waved a hand dismissively around the room, encompassing, presumably, his title and property, both of which would be confiscated by the government the moment his conversion became known.

"He said it would be a sham conversion and that the moment I had her I would become a Christian again and force Rebekah to be Christian, too. Like her father," he added darkly.

Despite the situation, Ian was beginning to have some sympathy for the wee popinjay. It was a very romantic tale, and he was partial to those. Jamie, however, was still reserving judgment. He gestured at the rug beneath their feet.

"Her dowry, ye said?"

"Yes," said the vicomte, but sounded much less certain. "She says it belonged to her mother. She had some men bring it here last week, along with a chest and a few other things. Anyway," he said, resuming his self-confidence and glowering at them, "when the old beast arranged her marriage to that fellow in Paris, I made up my mind to—to—"

"To abduct her. By arrangement, aye? Mmphm," Jamie said, his noise indicating his opinion of the vicomte's skills as a highwayman. He raised

one red brow at Pierre's black eye but forbore to make any more remarks, thank God. It hadn't escaped Ian that they were prisoners, though it maybe had Jamie.

"May we speak with Mademoiselle Hauberger?" Ian asked politely. "Just to be sure she's come of her own free will, aye?"

"Rather plainly she did, since you followed her here." The vicomte hadn't liked Jamie's noise. "No, you may not. She's busy." He raised his hands and clapped them sharply, and the rough fellows came back in, along with a half dozen or so male servants as reinforcement, led by a tall, severe-looking butler armed with a stout walking-stick.

"Go with Ecrivisse, gentlemen. He'll see to your comfort."

"COMFORT" PROVED TO be the chateau's wine cellar, which was fragrant but cold. Also dark. The vicomte's hospitality did not extend so far as a candle.

"If he meant to kill us, he'd have done it already," Ian reasoned.

"Mmphm." Jamie sat on the stairs, the fold of his plaid pulled up around his shoulders against the chill. There was music coming from somewhere outside: the faint sound of a fiddle and the tap of a little hand drum. It started, then stopped, then started again.

Ian wandered restlessly to and fro; it wasn't a very large cellar. If he didn't mean to kill them, what did the vicomte intend to do with them?

"He's waiting for something to happen," Jamie said suddenly, answering the thought. "Something to do wi' the lass, I expect."

"Aye, reckon." Ian sat down on the stairs, nudging Jamie over. "*A Dhia*, that's cold!"

"Mm," said Jamie absently. "Maybe they mean to run. If so, I hope he leaves someone behind to let us out and doesna leave us here to starve."

"We wouldna starve," Ian pointed out logically. "We could live on wine for a good long time. Someone would come before it ran out." He paused a moment, trying to imagine what it would be like to stay drunk for several weeks.

"That's a thought." Jamie got up, a little stiff from the cold, and went off to rummage the racks. There was no light to speak of, save what seeped through the crack at the bottom of the door to the cellar, but Ian could hear Jamie pulling out bottles and sniffing the corks.

He came back in a bit with a bottle and, sitting down again, drew the cork with his teeth and spat it to one side. He took a sip, then another, and tilted back the bottle for a generous swig, then handed it to Ian.

"No bad," he said.

It wasn't, and there wasn't much conversation for the next little while. Eventually, though, Jamie set the empty bottle down, belched gently, and said, "It's her."

"What's her? Rebekah, ye mean. I daresay." Then after a moment, "What's her?"

"It's her," Jamie repeated. "Ken what the Jew said—Ephraim bar-Sefer? About how his gang knew where to strike, because they got information from some outside source? It's her. She told them."

Jamie spoke with such certainty that Ian was staggered for a moment, but then he marshaled his wits.

"That wee lass? Granted, she put one over on us—and I suppose she at least kent about Pierre's abduction, but . . ."

Jamie snorted. "Aye, Pierre. Does the mannie strike ye either as a criminal or a great schemer?"

"No, but—"

"Does she?"

"Well . . ."

"Exactly."

Jamie got up and wandered off into the racks again, this time returning with what smelled to Ian like one of the very good local red wines. It was like drinking his mam's strawberry preserves on toast with a cup of strong tea, he thought approvingly.

"Besides," Jamie went on, as though there'd been no interruption in his train of thought, "mind what ye told me the maid said, when I got my heid half-stove in? *'Perhaps he's been killed, too. How would you feel then?'* Nay, she'd planned the whole thing—to have Pierre and his lads stop the coach and make away with the women and the scroll and doubtless Monsieur Pickle, too. *But,*" he added, sticking up a finger in front of Ian's face to stop him interrupting, "then Josef-from-Alsace tells ye that thieves— and the *same* thieves as before, or some of them—attacked the band wi' the dowry money. Ye ken well that canna have been Pierre. It had to be her who told them."

Ian was forced to admit the logic of this. Pierre had enthusiasm but couldn't possibly be considered a professional highwayman.

"But a lass . . ." he said helplessly. "How could she—"

Jamie grunted. "D'Eglise said Dr. Hasdi's a man much respected among the Jews of Bordeaux. And plainly he's kent as far as Paris, or how else did he make the match for his granddaughter? But he doesna speak French. Want to bet me that she didna manage his correspondence?"

"No," Ian said, and took another swallow. "Mmphm."

Some minutes later he said, "That rug. And the other things Monsieur le Vicomte mentioned—her *dowry.*"

Jamie made an approving noise. "Aye. Her percentage of the take, more like. Ye can see our lad Pierre hasna got much money, and he'd lose all his property when he converted. She was feathering their nest, like— makin' sure they'd have enough to live on. Enough to live *well* on."

"Well, then," Ian said, after a moment's silence. "There ye are."

THE AFTERNOON dragged on. After the second bottle, they agreed to drink no more for the time being, in case a clear head should be necessary if or when the door at last opened, and aside from going off now and then to have a pee behind the farthest wine racks, they stayed huddled on the stairs.

Jamie was singing softly along to the fiddle's distant tune when the door finally *did* open. He stopped abruptly and lunged awkwardly to his feet, nearly falling, his knees stiff with cold.

"Monsieurs?" said the butler, peering down at them. "If you will be so kind as to follow me, please?"

To their surprise, the butler led them straight out of the house and down a small path, in the direction of the distant music. The air outside was fresh and wonderful after the must of the cellar, and Jamie filled his lungs with it, wondering what the devil . . . ?

Then they rounded a bend in the path and saw a garden court before them, lit by torches driven into the ground. Somewhat overgrown, but with a fountain tinkling away in the center—and just by the fountain a sort of canopy, its cloth glimmering pale in the dusk. There was a little knot of people standing near it, talking, and as the butler paused, holding them back with one hand, Vicomte Beaumont broke away from the group and came toward them, smiling.

"My apologies for the inconvenience, gentlemen," he said, a huge smile splitting his face. He looked drunk, but Jamie thought he wasn't—no smell of spirits. "Rebekah had to prepare herself. And we wanted to wait for nightfall."

"To do what?" Ian asked suspiciously, and the vicomte giggled. Jamie didn't mean to wrong the man, but it was a giggle. He gave Ian an eye and Ian gave it back. Aye, it was a giggle.

"To be married," Pierre said, and while his voice was still full of *joie de vivre,* he said the words with a sense of deep reverence that struck Jamie somewhere in the chest. Pierre turned and waved a hand toward the darkening sky, where the stars were beginning to prick and sparkle. "For luck, you know—that our descendants may be as numerous as the stars."

"Mmphm," Jamie said politely.

"But come with me, if you will." Pierre was already striding back to the knot of . . . well, Jamie supposed they must be wedding guests. The vicomte beckoned to the Scots to follow.

Marie the maid was there, along with a few other women; she gave Jamie and Ian a wary look. But it was the men with whom the vicomte was concerned. He spoke a few words to his guests, and three men with enormous beards came back with him, all dressed formally, if somewhat oddly, with little velvet skullcaps decorated with beads.

"May I present Monsieur Gershom Sanders and Monsieur Levi Champfleur. Our witnesses. And Reb Cohen, who will officiate."

The men shook hands, murmuring politeness. Jamie and Ian exchanged looks. Why were *they* here?

The vicomte caught the look and interpreted it correctly.

"I wish you to return to Dr. Hasdi," he said, the effervescence in his voice momentarily supplanted by a note of steel, "and tell him that everything—everything!—was done in accordance with proper custom and according to the Law. This marriage will not be undone. By anyone."

"Mmphm," said Ian, less politely.

And so it was that a few minutes later they found themselves standing among the male wedding guests—the women stood on the other side of the canopy—watching as Rebekah came down the path, jingling faintly. She wore a dress of deep red silk; Jamie could see the torchlight shift and shimmer through its folds as she moved. There were gold bracelets on both wrists, and she had a veil over her head and face, with a little headdress sort of thing made of gold chains that dipped across her forehead, strung with small medallions and bells—it was this that made the jingling sound. It reminded him of the Torah scroll, and he stiffened a bit at the thought.

Pierre stood with the rabbi under the canopy; as Rebekah approached, he stepped apart, and she came to him. She didn't touch him, though, but proceeded to walk round him. And round him, and round him. Seven times she circled him, and the hairs rose a little on the back of Jamie's neck; it had the faint sense of magic about it—or witchcraft. Something she did to bind the man.

She came face-to-face with Jamie as she made each turn and plainly could see him in the light of the torches, but her eyes were fixed straight ahead; she made no acknowledgment of anyone—not even Pierre.

But then the circling was done and she came to stand by the vicomte's side. The rabbi said a few words of welcome to the guests and then, turning to the bride and groom, poured out a cup of wine, and said what appeared to be a Hebrew blessing over it. Jamie made out the beginning—*"Blessed are you, Adonai our God"*—but then lost the thread.

Pierre reached into his pocket when Reb Cohen stopped speaking, removed a small object—clearly a ring—and, taking Rebekah's hand in his, put it on the forefinger of her right hand, smiling down into her face with a tenderness that, despite everything, rather caught at Jamie's heart. Then Pierre lifted her veil, and Jamie caught a glimpse of the answering tenderness on Rebekah's face in the instant before her husband kissed her.

The congregation sighed as one.

The rabbi picked up a sheet of parchment from a little table nearby. The thing he'd called a *ketubah,* Jamie saw—the wedding contract.

The rabbi read the thing out, first in a language Jamie didn't recognize, and then again in French. It wasn't so different from the few marriage contracts he'd seen, laying out the disposition of property and what was due to the bride and all—though he noted with disapproval that it provided for the possibility of divorce. His attention wandered a bit then; Rebekah's face glowed in the torchlight like pearl and ivory, and the roundness of her bosom showed clearly as she breathed. In spite of everything he thought he now knew about her, he experienced a brief wave of envy toward Pierre.

The contract read and carefully laid aside, the rabbi recited a string of blessings; Jamie kent it was blessings because he caught the words "Blessed are you, Adonai . . ." over and over, though the subject of the blessings seemed to be everything from the congregation to Jerusalem, so far as he could tell. The bride and groom had another sip of wine.

A pause then, and Jamie expected some official word from the rabbi, uniting husband and wife, but it didn't come. Instead, one of the witnesses took the wineglass, wrapped it in a linen napkin, and placed it on the ground in front of Pierre. To the Scots' astonishment, he promptly stamped on the thing—and the crowd burst into applause.

For a few moments, everything seemed quite like a country wedding, with everyone crowding round, wanting to congratulate the happy couple. But within moments, the happy couple was moving off toward the house, while the guests all streamed toward tables that had been set up at the far side of the garden, laden with food and drink.

"Come on," Jamie muttered, and caught Ian by the arm. They hastened after the newly wedded pair, Ian demanding to know what the devil Jamie thought he was doing. "I want to talk to her—alone. You stop him, keep him talking for as long as ye can."

"I—how?"

"How would I know? Ye'll think of something."

They had reached the house, and ducking in close upon Pierre's heels, Jamie saw that by good luck the man had stopped to say something to a servant. Rebekah was just vanishing down a long hallway; he saw her put her hand to a door.

"The best of luck to ye, man!" Jamie said, clapping Pierre so heartily on the shoulder that the groom staggered.

Before Pierre could recover, Ian, very obviously commending his soul to God, stepped up and seized him by the hand, which he wrung vigorously, meanwhile giving Jamie a private *"Hurry the bloody hell up!"* sort of look.

Grinning, Jamie ran down the short hallway to the door where he'd seen Rebekah disappear. The grin faded as his hand touched the doorknob, though, and the face he presented to her as he entered was as grim as he could make it.

Her eyes widened in shock and indignation at sight of him.

"What are you doing here? No one is supposed to come in here but my husband and me!"

"He's on his way," Jamie assured her. "The question is—will he get here?"

Her little fist curled up in a way that would have been comical if he didn't know as much about her as he did.

"Is that a threat?" she said, in a tone as incredulous as it was menacing. "Here? You dare threaten me *here?*"

"Aye, I do. I want that scroll."

"Well, you're not getting it," she snapped. He saw her glance flicker over the table, probably in search of either a bell to summon help or something to bash him on the head with, but the table held nothing but a platter of stuffed rolls and exotic sweeties. There *was* a bottle of wine, and he saw her eye light on that with calculation, but he stretched out a long arm and got hold of it before she could.

"I dinna want it for myself," he said. "I mean to take it back to your grandfather."

"Him?" Her face hardened. "No. It's worth more to him than *I* am," she added bitterly, "but at least that means I can use it for protection. As long as I have it, he won't try to hurt Pierre or drag me back, for fear I might damage it. I'm keeping it."

"I think he'd be a great deal better off without ye, and doubtless he kens that fine," Jamie informed her, and had to harden himself against the sudden look of hurt in her eyes. He supposed even spiders might have feelings, but that was neither here nor there.

"Where's Pierre?" she demanded. "If you've harmed a hair on his head, I'll—"

"I wouldna touch the poor gomerel, and neither would Ian—Juan, I mean. When I said the question was whether he got to ye or not, I meant whether he thinks better of his bargain."

"What?" He thought she paled a little, but it was hard to tell.

"You give me the scroll to take back to your grandfather—a wee letter

of apology to go with it wouldna come amiss, but I willna insist on that—or Ian and I take Pierre out back and have a frank word regarding his new wife."

"Tell him what you like!" she snapped. "He wouldn't believe any of your made-up tales!"

"Oh, aye? And if I tell him exactly what happened to Ephraim bar-Sefer? And why?"

"Who?" she said, but now she really had gone pale to the lips and put out a hand to the table to steady herself.

"Do ye ken yourself what happened to him? No? Well, I'll tell ye, lass." And he did so, with a terse brutality that made her sit down suddenly, tiny pearls of sweat appearing round the gold medallions that hung across her forehead.

"Pierre already kens at least a bit about your wee gang, I think—but maybe not what a ruthless, grasping wee besom ye really are."

"It wasn't me! I didn't kill Ephraim!"

"If not for you, he'd no be dead, and I reckon Pierre would see that. I can tell him where the body is," he added, more delicately. "I buried the man myself."

Her lips were pressed so hard together that nothing showed but a straight white line.

"Ye havena got long," he said, quietly now, but keeping his eyes on hers. "Ian canna hold him off much longer, and if he comes in—then I tell him everything, in front of you, and ye do what ye can then to persuade him I'm a liar."

She stood up abruptly, her chains and bracelets all a-jangle, and stamped to the door of the inner room. She flung it open, and Marie jerked back, shocked.

Rebekah said something to her in Ladino, sharp, and with a small gasp the maid scurried off.

"All *right*," Rebekah said through gritted teeth, turning back to him. "Take it and be damned, you *dog*."

"Indeed I will, ye bloody wee bitch," he replied with great politeness.

Her hand closed round a stuffed roll, but instead of throwing it at him, she merely squeezed it into paste and crumbs, slapping the remains back on the tray with a small exclamation of fury.

The sweet chiming of the Torah scroll presaged Marie's hasty arrival, the precious thing clasped in her arms. She glanced at her mistress and, at Rebekah's curt nod, delivered it with great reluctance into the arms of the Christian dog.

Jamie bowed, first to maid and then mistress, and backed toward the door.

"Shalom," he said, and closed the door an instant before the silver platter hit it with a ringing thud.

"DID IT HURT a lot?" Ian was asking Pierre with interest, when Jamie
came up to them.

"My God, you have no idea," Pierre replied fervently. "But it was worth
it." He divided a beaming smile between Ian and Jamie and bowed to
them, not even noticing the canvas-wrapped bundle in Jamie's arms. "You
must excuse me, gentlemen; my bride awaits me!"

"Did what hurt a lot?" Jamie inquired, leading the way hastily out
through a side door. No point in attracting attention, after all.

"Ye ken he was born a Christian but converted in order to marry the
wee besom," Ian said. "So he had to be circumcised." He crossed himself
at the thought, and Jamie laughed.

"What is it they call the stick-insect things where the female one bites
off the head of the male one after he's got the business started?" Jamie
asked, nudging the door open with his bum.

Ian's brow creased for an instant. "Praying mantis, I think. Why?"

"I think our wee friend Pierre may have a more interesting wedding
night than he expects. Come on."

Bordeaux

IT WASN'T THE worst thing he'd ever had to do, but he wasn't looking forward to it. Jamie paused outside the gate of Dr. Hasdi's house, the
Torah scroll in its wrappings in his arms. Ian was looking a bit worm-
eaten, and Jamie reckoned he kent why. Having to tell the doctor what
had happened to his granddaughter was one thing; telling him to his face
with the knowledge of what said granddaughter's nipples felt like fresh in
the mind . . . or the hand . . .

"Ye dinna have to come in, man," he said to Ian. "I can do it alone."

Ian's mouth twitched, but he shook his head and stepped up next to
Jamie.

"On your right, man," he said simply.

Jamie smiled. When he was five years old, Ian's da, Auld John, had
persuaded his own da to let Jamie handle a sword cack-handed, as he was
wont to do. "And you, lad," he'd said to Ian, very serious, "it's your duty
to stand on your laird's right hand and guard his weak side."

"Aye," Jamie said. "Right, then." And rang the bell.

AFTERWARD, THEY WANDERED slowly through the streets of Bordeaux, making their way toward nothing in particular, not speaking much.

Dr. Hasdi had received them courteously, though with a look of mingled horror and apprehension on his face when he saw the scroll. This look had faded to one of relief at hearing—the manservant had had enough French to interpret for them—that his granddaughter was safe, then to shock, and finally to a set expression that Jamie couldn't read. Was it anger, sadness, resignation?

When Jamie had finished the story, they sat uneasily, not sure what to do next. Dr. Hasdi sat at his desk, head bowed, his hands resting gently on the scroll. Finally, he raised his head and nodded to them both, one and then the other. His face was calm now, giving nothing away.

"Thank you," he said in heavily accented French. *"Shalom."*

"ARE YE HUNGRY?" Ian motioned toward a small *boulangerie,* whose trays bore filled rolls and big, fragrant round loaves. He was starving himself, though half an hour ago, his wame had been in knots.

"Aye, maybe." Jamie kept walking, though, and Ian shrugged and followed.

"What d'ye think the captain will do when we tell him?" Ian wasn't all that bothered. There was always work for a good-sized man who kent what to do with a sword. And he owned his own weapons. They'd have to buy Jamie a sword, though. Everything he was wearing, from pistols to ax, belonged to D'Eglise.

He was busy enough calculating the cost of a decent sword against what remained of their pay that he didn't notice Jamie not answering him. He did notice that his friend was walking faster, though, and, hurrying to catch up, he saw what they were heading for. The tavern where the pretty brown-haired barmaid had taken Jamie for a Jew.

Oh, like that, is it? he thought, and hid a grin. Aye, well, there was one sure way the lad could prove to the lass that he wasn't a Jew.

The place was moiling when they walked in, and not in a good way; Ian sensed it instantly. There were soldiers there, army soldiers, and other fighting men, mercenaries like themselves, and no love wasted between them. You could cut the air with a knife, and judging from a splotch of half-dried blood on the floor, somebody had already tried.

There were women but fewer than before, and the barmaids kept their eyes on their trays, not flirting tonight.

Jamie wasn't taking heed of the atmosphere; Ian could see him looking round for her, but the brown-haired lass wasn't on the floor. They might have asked after her—if they'd known her name.

"Upstairs, maybe?" Ian said, leaning in to half-shout into Jamie's ear over the noise. Jamie nodded and began forging through the crowd, Ian bobbing in his wake, hoping they found the lass quickly so he could eat while Jamie got on with it.

THE STAIRS WERE crowded—with men coming down. Something was amiss up there, and Jamie shoved someone into the wall with a thump, pushing past. Some nameless anxiety shot jolts down his spine, and he was half prepared before he pushed through a little knot of onlookers at the head of the stairs and saw them.

Mathieu and the brown-haired girl. There was a big open room here, with a hallway lined with tiny cubicles leading back from it; Mathieu had the girl by the arm and was boosting her toward the hallway with a hand on her bum, despite her protests.

"Let go of her!" Jamie said, not shouting but raising his voice well enough to be heard easily. Mathieu paid not the least attention, though everyone else, startled, turned to look at Jamie.

He heard Ian mutter, "Joseph, Mary, and Bride preserve us," behind him, but he paid no heed. He covered the distance to Mathieu in three strides and kicked him in the arse.

He ducked, by reflex, but Mathieu merely turned and gave him a hot eye, ignoring the whoops and guffaws from the spectators.

"Later, little boy," he said. "I'm busy now."

He scooped the young woman into one big arm and kissed her sloppily, rubbing his stubbled face hard over hers so she squealed and pushed at him to get away.

Jamie drew the pistol from his belt.

"I said, let her go." The noise dropped suddenly, but he barely noticed for the roaring of blood in his ears.

Mathieu turned his head, incredulous. Then he snorted with contempt, grinned unpleasantly, and shoved the girl into the wall so her head struck with a thump, pinning her there with his bulk.

The pistol was primed.

"Salop!" Jamie roared. *"*Don't touch her! Let her go!" He clenched his teeth and aimed with both hands, rage and fright making his hands tremble.

Mathieu didn't even look at him. The big man half-turned away, a casual hand on the girl's breast. She squealed as he twisted it, and Jamie

fired. Mathieu whirled, the pistol he'd had concealed in his own belt now in hand, and the air shattered in an explosion of sound and white smoke.

There were shouts of alarm, excitement—and another pistol went off, somewhere behind Jamie. *Ian?* he thought dimly, but, no, Ian was running toward Mathieu, leaping for the massive arm rising, the second pistol's barrel making circles as Mathieu struggled to fix it on Jamie. It discharged, and the ball hit one of the lanterns that stood on the tables, which exploded with a *whuff* and a bloom of flame.

Jamie had reversed his pistol and was hammering at Mathieu's head with the butt before he was conscious of having crossed the room. Mathieu's mad-boar eyes were almost invisible, slitted with the glee of fighting, and the sudden curtain of blood that fell over his face did nothing but enhance his grin, blood running down between his teeth. He shook Ian off with a shove that sent him crashing into the wall, then wrapped one big arm almost casually around Jamie's body and, with a snap of his head, butted him in the face.

Jamie had turned his head reflexively and thus avoided a broken nose, but the impact crushed the flesh of his jaw into his teeth, and his mouth filled with blood. His head was spinning with the force of the blow, but he got a hand under Mathieu's jaw and shoved upward with all his strength, trying to break the man's neck. His hand slipped off the sweat-greased flesh, though, and Mathieu let go his grip in order to try to knee Jamie in the stones. A knee like a cannonball struck Jamie a numbing blow in the thigh as he squirmed free, and he staggered, grabbing Mathieu's arm just as Ian came dodging in from the side, seizing the other. Without a moment's hesitation, Mathieu's huge forearms twisted; he seized the Scots by the scruffs of their necks and cracked their heads together.

Jamie couldn't see and could barely move but kept moving anyway, groping blindly. He was on the floor, could feel boards, wetness . . . His pawing hand struck flesh, and he lunged forward and bit Mathieu as hard as he could in the calf of the leg. Fresh blood filled his mouth, hotter than his own, and he gagged but kept his teeth locked in the hairy flesh, clinging stubbornly as the leg kicked in frenzy. His ears were ringing, he was vaguely aware of screaming and shouting, but it didn't matter.

Something had come upon him, and nothing mattered. Some small remnant of his consciousness registered surprise, and then that was gone, too. No pain, no thought. He was a red thing, and while he saw other things—faces, blood, bits of room—they didn't matter. Blood took him, and when some sense of himself came back, he was kneeling astride the man, hands locked around the big man's neck, fingers throbbing with a pounding pulse—his or his victim's, he couldn't tell.

Him. Him. He'd lost the man's name. His eyes were bulging, the ragged mouth slobbered and gaped, and there was a small, sweet *crack* as

something broke under Jamie's thumbs. He squeezed with all he had, squeezed and squeezed and felt the huge body beneath him go strangely limp.

He went on squeezing, couldn't stop, until a hand seized him by the arm and shook him, hard.

"Stop," a voice croaked, hot in his ear. "Jamie. Stop."

He blinked up at the white bony face, unable to put a name to it. Then drew breath—the first he could remember drawing for some time—and with it came a thick stink, blood and shit and reeking sweat, and he became suddenly aware of the horrible spongy feeling of the body he was sitting on. He scrambled awkwardly off, sprawling on the floor as his muscles spasmed and trembled.

Then he saw her.

She was lying crumpled against the wall, curled into herself, her brown hair spilling across the boards. He got to his knees, crawling to her.

He was making a small whimpering noise, trying to talk, having no words. Got to the wall and gathered her into his arms, limp, her head lolling, striking his shoulder, her hair soft against his face, smelling of smoke and her own sweet musk.

"*A nighean,*" he managed. "Christ, *a nighean*. Are ye . . ."

"Jesus," said a voice by his side, and he felt the vibration as Ian—thank God, the name had come back, of course it was Ian—collapsed next to him. His friend had a bloodstained dirk still clutched in his hand. "Oh, Jesus, Jamie."

He looked up, puzzled, desperate, and then looked down as the girl's body slipped from his grasp and fell back across his knees with impossible boneless grace, the small dark hole in her white breast stained with only a little blood. Not much at all.

HE'D MADE JAMIE come with him to the cathedral of Saint André and insisted he go to Confession. Jamie had balked—no great surprise.

"No. I can't."

"We'll go together." Ian had taken him firmly by the arm and very literally dragged him over the threshold. He was counting on the atmosphere of the place to keep Jamie there, once inside.

His friend stopped dead, the whites of his eyes showing as he glanced warily around.

The stone vault of the ceiling soared into shadow overhead, but pools of colored light from the stained-glass windows lay soft on the worn slates of the aisle.

"I shouldna be here," Jamie muttered under his breath.

"Where better, eejit? Come on," Ian muttered back, and pulled Jamie down the side aisle to the chapel of Saint Estèphe. Most of the side chapels were lavishly furnished, monuments to the importance of wealthy families. This one was a tiny, undecorated stone alcove, containing little more than an altar, a faded tapestry of a faceless saint, and a small stand where candles could be placed.

"Stay here." Ian planted Jamie dead in front of the altar and ducked out, going to buy a candle from the old woman who sold them near the main door. He'd changed his mind about trying to make Jamie go to Confession; he knew fine when ye could get a Fraser to do something and when ye couldn't.

He worried a bit that Jamie would leave and hurried back to the chapel, but Jamie was still there, standing in the middle of the tiny space, head down, staring at the floor.

"Here, then," Ian said, pulling him toward the altar. He plunked the candle—an expensive one, beeswax and large—on the stand and pulled the paper spill the old lady had given him out of his sleeve, offering it to Jamie. "Light it. We'll say a prayer for your da. And . . . and for her."

He could see tears trembling on Jamie's lashes, glittering in the red glow of the sanctuary lamp that hung above the altar, but Jamie blinked them back and firmed his jaw.

"All right," he said, low-voiced, but he hesitated.

Ian sighed, took the spill out of his hand, and, standing on tiptoe, lit it from the sanctuary lamp. "Do it," he whispered, handing the spill to Jamie, "or I'll gie ye a good one in the kidney, right here."

Jamie made a sound that might have been the breath of a laugh and lowered the lit spill to the candle's wick. The fire rose up, a pure high flame with blue at its heart, then settled as Jamie pulled the spill away and shook it out in a plume of smoke.

They stood for some time, hands clasped loosely in front of them, watching the candle burn. Ian prayed for his mam and da, his sister and her bairns . . . with some hesitation (was it proper to pray for a Jew?) for Rebekah bat-Leah, and, with a sidelong glance at Jamie to be sure he wasn't looking, for Jenny Fraser. Then for the soul of Brian Fraser . . . and finally, eyes tight shut, for the friend beside him.

The sounds of the church faded, the whispering stones and echoes of wood, the shuffle of feet and the rolling gabble of the pigeons on the roof. Ian stopped saying words but was still praying. And then that stopped, too, and there was only peace and the soft beating of his heart.

He heard Jamie sigh, from somewhere deep inside, and opened his eyes. Without speaking, they went out, leaving the candle to keep watch.

"Did ye not mean to go to Confession yourself?" Jamie asked, stopping near the church's main door. There was a priest in the confessional; two or

three people stood a discreet distance away from the carved wooden stall, out of earshot, waiting.

"It'll bide," Ian said, with a shrug. "If ye're goin' to hell, I might as well go, too. God knows, ye'll never manage alone."

Jamie smiled—a wee bit of a smile, but still—and pushed the door open into sunlight.

They strolled aimlessly for a bit, not talking, and found themselves eventually on the river's edge, watching the Garonne's dark waters flow past, carrying debris from a recent storm.

"It means 'peace,'" Jamie said at last. "What he said to me. The doctor. *'Shalom.'*"

Ian kent that fine. "Aye," he said. "But peace is no our business now, is it? We're soldiers." He jerked his chin toward the nearby pier, where a packet boat rode at anchor. "I hear the King of Prussia needs a few good men."

"So he does," said Jamie, and squared his shoulders. "Come on, then."

A FUGITIVE GREEN

1

SURVIVAL

Paris, April 1744

MINNIE RENNIE HAD SECRETS. Some were for sale and some were strictly her own. She touched the bosom of her dress and glanced toward the latticework door at the rear of the shop. Still closed, the blue curtains behind it drawn firmly shut.

Her father had secrets, too; Andrew Rennie (as he called himself in Paris) was outwardly a dealer in rare books but more privately a collector of letters whose writers had never meant them to be read by any but the addressee. He also kept a stock of more fluid information, this soaked out of his visitors with a combination of tea, wine, small amounts of money, and his own considerable charm.

Minnie had a good head for wine, needed no money, and was impervious to her father's magnetism. She did, however, have a decently filial respect for his powers of observation.

The murmur of voices from the back room didn't have the rhythm of leave-taking, no scraping of chairs . . . She nipped across the book-crammed shop to the shelves of tracts and sermons.

Taking down a red-calf volume with marbled endpapers, titled *Collected Sermons of the Reverend George V. Sykes,* she snatched the letter from the bosom of her dress, tucked it between the pages, and slid the book back into place. Just in time: there was movement in the back room, the putting down of cups, the slight raising of voices.

Heart thumping, she took one more glance at the Reverend Sykes and saw to her horror that she'd disturbed the dust on the shelf—there was a clear track pointing to the oxblood-leather spine. She darted back to the main counter, seized the feather duster kept under it, and had the entire section flicked over in a matter of moments.

She took several deep breaths; she mustn't look flushed or flustered. Her father was an observant man—a trait that had (he often said, when instructing her in the art) kept him alive on more than one occasion.

But it was all right; the voices had changed again—some new point had come up.

She strolled composedly along the shelves and paused to look through the stacks of unsorted volumes that sat on a large table against the west wall. A strong scent of tobacco rose from the books, along with the usual smell of leather, buckram, glue, paper, and ink. This batch had plainly belonged to a man who liked a pipe when he read. She was paying little attention to the new stock, though; her mind was still on the letter.

The carter who had delivered this latest assemblage of books—the library of a deceased professor of history from Exeter—had given her a nod and a wink, and she'd slipped out with a market basket, meeting him round the corner by a fruiterer's shop. A *livre tournois* to the carter, and five *sous* for a wooden basket of strawberries, and she'd been free to read the letter in the shelter of the alley before sauntering back to the shop, fruit in hand to explain her absence.

No salutation, no signature, as she'd requested—only the information:

Have found her, it read simply. *Mrs. Simpson, Chapel House, Parson's Green, Peterborough Road, London.*

Mrs. Simpson. A name, at last. A name and a place, mysterious though both were.

Mrs. Simpson.

It had taken months, months of careful planning, choosing the men among the couriers her father used who might be amenable to making a bit extra on the side and a bit more for keeping her inquiries quiet.

She didn't know what her father might do should he find out that she'd been looking for her mother. But he'd refused for the last seventeen years to say a word about the woman; it was reasonable to assume he wouldn't be pleased.

Mrs. Simpson. She said it silently, feeling the syllables in her mouth. Mrs. Simpson . . . Was her mother married again, then? Did she have other children?

Minnie swallowed. The thought that she might have half brothers or sisters was at once horrifying, intriguing . . . and startlingly painful. That someone else might have had her mother—hers!—for all those years . . .

"This will *not* do," she said aloud, though under her breath. She had no idea of *Mrs. Simpson*'s personal circumstances, and it was pointless to waste emotion on something that might not exist. She blinked hard to refocus her mind and suddenly saw it.

The thing sitting atop a pigskin-bound edition of Volume III of *History of the Papacy* (Antwerp) was as long as her thumb and, for a cockroach, remarkably immobile. Minnie had been staring at it unwittingly for nearly a minute, and it hadn't so much as twitched an antenna. Perhaps it was

dead? She picked a ratty quill out of the collection in the Chinese jar and gingerly poked the thing with the quill's pointy end.

The thing hissed like a teakettle and she let out a small yelp, dropping the quill and leaping backward. The roach, disturbed, turned round in a slow, huffy circle, then settled back on the gilt-embossed capital "P" and tucked its thorny legs back under itself, obviously preparing to resume its nap.

"Oh, I don't *think* so," she said to it, and turned to the shelves in search of something heavy enough to smash it but with a cover that wouldn't show the stain. She'd set her hand on a Vulgate Bible with a dark-brown pebble-grain cover when the secret door beside the shelves opened, revealing her father.

"Oh, you've met Frederick?" he said, stepping forward and taking the Bible out of her hand. "You needn't worry, my dear; he's quite tame."

"Tame? Who would trouble to domesticate a cockroach?"

"The inhabitants of Madagascar, or so I'm told. Though the trait is heritable; Frederick here is the descendant of a long and noble line of hissing cockroaches but has never set foot on the soil of his native land. He was born—or hatched, I suppose—in Bristol."

Frederick had suspended his nap long enough to nuzzle inquiringly at her father's thumb, extended as one might hold out one's knuckles to a strange dog. Evidently finding the scent acceptable, the roach strolled up the thumb and onto the back of her father's hand. Minnie twitched, unable to keep the gooseflesh from rippling up her arms.

Mr. Rennie edged carefully toward the big shelves on the east wall, hand cradled next to his chest. These shelves contained the salable but less-valuable books: a jumble of everything from *Culpeper's Herbal* to tattered copies of Shakespeare's plays and—by far the most popular—a large collection of the more lurid gallows confessions of an assortment of highwaymen, murderers, forgers, and husband-slayers. Amid the volumes and pamphlets was scattered a miscellany of small curiosities, ranging from a toy bronze cannon and a handful of sharp-edged stones said to be used at the dawn of time for scraping hides to a Chinese fan that showed erotic scenes when spread. Her father picked a wicker cricket cage from the detritus and decanted Frederick neatly into it.

"Not before time, either, old cock," he said to the roach, now standing on its hind legs and peering out through the wickerwork. "Here's your new master, just coming."

Minerva peered round her father and her heart jumped a little; she recognized that tall, broad-shouldered silhouette automatically ducking beneath the lintel in order to avoid being brained.

"Lord Broch Tuarach!" Her father stepped forward, beaming, and inclined his head to the customer.

"Mr. Fraser will do," he said, as always, extending a hand. "Your servant, sir."

He'd brought a scent of the streets inside with him: the sticky sap of the plane trees, dust, manure and offal, and Paris's pervasive smell of piss, lightly perfumed by the orange-sellers outside the theater down the street. He carried his own deep tang of sweat, wine, and oak casks, as well; he often came from his warehouse. She inhaled appreciatively, then let her breath out as he turned, smiling, from her father toward her.

"Mademoiselle Rennie," he said, in a deep Scotch accent that rolled the "R" delightfully. He seemed a bit surprised when she held out her hand, but he obligingly bent over it, breathing courteously on her knuckles. *If I were married, he'd kiss it,* she thought, her grip tightening unconsciously on his. He blinked, feeling it, but straightened up and bowed to her, as elegantly as any courtier.

Her father made a slight sound in his throat and tried to catch her eye, but she ignored him, picking up the feather duster and heading industriously for the shelves behind the counter—the ones containing a select assortment of erotica from a dozen different countries. She knew perfectly well what his glance would have said.

"Frederick?" she heard Mr. Fraser say, in a bemused tone of voice. "Does he answer to his name?"

"I—er—I must admit that I've never called him to heel," her father replied, a little startled. "But he's very tame; will come to your hand." Evidently her father had unlatched the cricket cage in order to demonstrate Frederick's talents, for she heard a slight shuffle of feet.

"Nay, dinna bother," Mr. Fraser—his Christian name was James; she'd seen it on a bill of sale for a calf-bound octavo of *Persian Letters* with gilt impressions—said, laughing. "The beastie's not my pet. A gentleman of my acquaintance wants something exotic to present to his mistress—she's a taste for animals, he says."

Her sensitive ear easily picked up the delicate hesitation before "gentleman of my acquaintance." So had her father, for he invited James Fraser to take coffee with him, and in the next instant the two of them had vanished behind the latticework door that concealed her father's private lair and she was blinking at Frederick's stubby antennae, waving inquisitively from the cricket cage her father had dropped onto the shelf in front of her.

"Put up a bit of food for Mr. Fraser to take along," her father called back to her from behind the screen. "For Frederick, I mean."

"What does he eat?" she called.

"Fruit!" came a faint reply, and then a door closed behind the screen.

She caught one more glimpse of Mr. Fraser when he left half an hour later, giving her a smile as he took the parcel containing Frederick and the insect's breakfast of strawberries. Then he ducked once more beneath the

lintel, the afternoon sun glinting off his bright hair, and was gone. She stood staring at the empty door.

Her father had emerged from the back room, as well, and was regarding her, not without sympathy.

"Mr. Fraser? He'll never marry you, my dear—he has a wife, and quite a striking woman she is, too. Besides, while he's the best of the Jacobite agents, he doesn't have the scope you'd want. He's only concerned with the Stuarts, and the Scottish Jacobites will never amount to anything. Come, I've something to discuss with you." Without waiting, he turned and headed for the Chinese screen.

A wife. Striking, eh? While the word "wife" was undeniably a blow to the liver, Minnie's next thought was that she didn't necessarily need to *marry* Jamie Fraser. And if it came to striking, she could deal a man a good, sharp buffet in the cods herself. She twirled a lock of ripe-wheat hair around one finger and tucked it behind her ear.

She followed her father, finding him at the little satinwood table. The coffee cups had been pushed aside, and he was pouring wine; he handed her a glass and nodded for her to sit.

"Don't you think of it, my girl." Her father was watching her over his own glass, not unkindly. "After you're married, you do what you like. But you need to keep your virginity until we've got you settled. The English are notorious bores about virginity, and I have my heart set on an Englishman for you."

She made a dismissive noise with her lips and took a delicate sip of the wine.

"What makes you think I haven't already . . . ?"

He lifted one eyebrow and tapped the side of his nose.

"*Ma chère,* I could smell a man on you a mile away. And even when I'm not here . . . I'm here." He lifted the other eyebrow and stared at her. She sniffed, drained her glass, and poured another.

Was he? She sat back and examined him, her own face carefully bland. True, he had informants everywhere; after listening to him do business all day behind the latticework, she dreamed of spiders all night, busy in their webs. Spinning, climbing, hunting along the sleek silk paths that ran hidden through the sticky stuff. And sometimes just hanging there, round as marbles in the air, motionless. Watching with their thousands of eyes.

But the spiders had their own concerns, and for the most part she wasn't one of them. She smiled suddenly at her father, dimpling, and was pleased to see a flicker of unease in his eyes. She lowered her lashes and buried the smile in her wine.

He coughed.

"So," he said, sitting up straight. "How would you like to visit London, my darling?"

London . . .

She tilted her head from side to side, considering.

"The food's terrible, but the beer's not bad. Still, it rains all the time."

"You could have a new dress."

That was interesting—not purely a book-buying excursion, then—but she feigned indifference.

"Only one?"

"That depends somewhat on your success. You might need . . . something special."

That made something twitch behind her ears.

"Why do you bother with this nonsense?" she demanded, putting her glass down with a thump. "You know you can't cozen me into things anymore. Just tell me what you have in mind, and we'll discuss it. Like rational beings."

That made him laugh but not unkindly.

"You do know that women aren't rational, don't you?"

"I do. Neither are men."

"Well, you have a point," he admitted, patting a dribble of wine off his chin with a napkin. "But they do have patterns. And women's patterns are . . ." He paused, squinting over the gold rims of his spectacles, in search of the word.

"More complex?" she suggested, but he shook his head.

"No, no—superficially they seem chaotic, but in fact women's patterns are brutally simple."

"If you mean the influence of the moon, I might point out that every lunatic I've met has been a man."

His eyebrows rose. They were beginning to thicken and gray, to grow unruly; she saw of a sudden that he was becoming elderly, and her heart gave a small lurch at the thought.

He didn't ask how many lunatics she'd met—in the book business, such people were a weekly occurrence—but shook his head.

"No, no, such things are mere physical calendar-keeping. I mean the patterns that cause women to do what they do. And those all come down to survival."

"The day I marry a man merely to survive . . ." She didn't bother finishing the sentence but flicked her fingers scornfully and rose to take the steaming kettle off its spirit lamp and refresh the teapot. Two glasses of wine were her strict limit—particularly when dealing with her father—and today of all days she wanted her wits about her.

"Well, you do have rather higher standards than most women." Her father took the cup of tea she brought him, smiling at her over it. "And—I flatter myself—more resources with which to support them. But the fact

remains that you are a woman. Which means that you can conceive. And that, my dear, is where a woman's pattern becomes brutal indeed."

"Really," she said, but not in a tone to invite him to expand upon his point. It was London she wanted to hear about. She'd need to be careful, though.

"What are we looking for, then?" she asked, pouring tea into her own cup so she could keep her eyes fixed on the amber stream. "In London, I mean."

"Not we," her father corrected. "Not this time. I have a bit of business to do in Sweden—speaking of Jacobites. You—"

"There are Swedish Jacobites?"

Her father sighed and rubbed his temples with the forefingers of each hand.

"My dear, you have no idea. They spring up like weeds—and like the grass of the field, in the evening they are cut down and wither. Just when you think they're finally dead, though, something happens, and suddenly—but that's of no matter to you. You're to deliver a package to a particular gentleman and to receive information from a list of contacts that I'll give you. You needn't question them, just take whatever they hand over. And naturally—"

"Tell them nothing," she finished. She dropped a sugar lump into her own tea. "Of course not, Father; what sort of nincompoop do you think I am?"

That made him laugh, deep lines of amusement creasing his eyes almost shut.

"Where did you get that word?"

"Everyone says nincompoop," she informed him. "You hear it in the street in London a dozen times a day."

"Oh, I doubt it," he said. "Know where it comes from, do you?"

"Samuel Johnson told me it was from *non compos mentis.*"

"Oh, that's where you got it." He'd stopped laughing but still looked amused. "Well, Mr. Johnson would know. You're still corresponding with him? He's an Englishman, I grant you, but not at all what I have in mind for you, my girl. Bats in the belfry and not a penny to his name. Married, too," he added as an afterthought. "Lives on his wife's money."

That surprised her, and not in a pleasant way. But he was entirely straightforward; his tone was the same as he used when instructing her closely in some important aspect of the work. They didn't fence or mess each other about when it came to the work, and she sat back a little, indicating by the inclination of her head that she was ready to listen.

"Mind you," her father said, raising one ink-stained finger, "many folk would tell you that women have nothing on their minds but clothes, or

parties, or what Lady Whatnot said about Sir Fart-Catcher at yesterday's salon. And that's a reasonable observation, but it's only an observation. When you see something like that, you ask what's behind it. Or, perhaps, under it," he admitted judiciously. "Push the wine over, sweetheart. I'm done with business for the day."

"I daresay you are," she said tartly, and plunked the decanter of Madeira in front of him. He'd been out all morning, nominally visiting booksellers and collectors of rarities but in reality talking—talking and listening. And he never drank alcohol when working.

He refilled his glass and made to top hers up, as well, but she shook her head and reached for the teapot. She'd been right about needing her wits.

"Chalk up another woman's pattern there," she said, sardonic. "They can't hold drink in the quantities that men do—but they're much less likely to become drunk."

"Clearly you've never been down Gropecunt Lane in London after dark, my dear," her father said imperturbably. "Not that I recommend it, mind. Women drink for the same reasons men do: in order to ignore circumstance or to obliterate themselves. Given the right circumstance, either sex will drown itself. Women care much more about staying alive than men do, though. But enough talk—cut me a fresh pen, my dear, and let me tell you who you'll see in London."

He reached into one of the pigeonholes along the wall and came out with a shabby notebook.

"Ever heard of the Duke of Pardloe?"

Précis: Harold Grey, Duke of Pardloe

Family background: Gerard Grey, Earl of Melton, was given the Title Duke of Pardloe (with considerable Estates) in Reward for his raising of a Regiment (46th Foot, which served with distinction during the Jacobite Rebellions of 1715 and 1719, seeing Combat at Preston and Sheriffmuir). However, the Duke's Allegiance to the Crown appeared to waver during the Reign of George II, and Gerard Grey was implicated in the Cornbury Plot. While he escaped Arrest at that time, a later Plot caused a Warrant for his Arrest on a Charge of Treason to be issued. Hearing of this, Pardloe shot himself in the Conservatory of his Country Estate before the Arrest could be made.

Pardloe's eldest son, Harold Grey, succeeded to the Title at the Age of twenty-one, upon his Father's Death. While the Title was not formally attainted, the younger Grey considered the Title stained

with Treason and refused to adopt it, preferring to be known by the older Family Title, Earl of Melton. Married to Esmé Dufresne (a younger Daughter of the Marquis de Robillard) shortly before his Father's Suicide.

The present Duke has publicly and violently rejected all Jacobite Associations (from necessity), but this does not mean such Associations have rejected him, nor that such Rejection reflects his true Inclination. There is considerable Interest in some Quarters as to the Duke's political Leanings and Affiliations, and any Letters, known Meetings with Persons of interest (List attached), or Private Conversations that might give Indications of Jacobite Leanings would be valuable.

Précis: Sir Robert Abdy, Baronet

Succeeded to the Title at the Age of Three, and while living a personally (and regrettably) virtuous Life, became heavily involved in Jacobite Politics and, Last Year, was so injudicious as to sign his Name to a Petition sent to Louis of France, urging French Invasion of Britain in support of a Stuart Restoration. Needless to say, this is not generally known in Britain, and it would not be a good Idea to mention it directly to Sir Robert. Neither should you approach him, though he is active in Society and you may encounter him. If so, we are particularly interested in his present Associations—names only, for the Present. Don't get too close.

Précis: Henry Scudamore, Duke of Beaufort

The fourth-richest Man in England, and likewise a Signatory to the French Petition. Very much seen in Society and makes little Secret of his political Inclinations.

His private Life is much less virtuous than Sir Robert's, I'm afraid. Having adopted his Wife's Surname by Act of Parliament, he sued last Year to divorce her on Grounds of Adultery (true: she was having an adulterous Relationship with William Talbot, Heir to Earl Talbot, and she wasn't discreet about it). The Lady—her Name is Frances—

promptly countersued on Grounds that the Duke was impotent. The Duke, who is no shrinking Violet, demonstrated before several Court-appointed Examiners that he was capable of having an Erection, won his Divorce, and is now presumably enjoying his Freedom.

Don't get too close. Associates, Names only for the Present.

Précis: Mr. Robert Willimot

Lord Mayor of London until 1741. Presently associated with . . .

2

COLD HONEY AND SARDINES

London, May 1744
Argus House, Residence of the Duke of Pardloe

THE ROOM SMELLED OF dead flowers. It was raining heavily, but Hal seized the window sash and shoved, regardless. The action *was* regardless; the wood had swelled with the damp and the window remained shut. He tried twice more, then stood breathing heavily.

The chiming of the little carriage clock on the mantelpiece brought him to an awareness that he'd been standing in front of the closed window with his mouth half open, watching rain run down the glass, for a quarter of an hour, unable to make up his mind whether to call a footman to open the damn thing or just put his fist through it.

He turned away and, chilled, made his way by instinct toward the fire. He'd felt as though he were moving through cold honey ever since he'd forced himself out of bed, and now he collapsed joint by joint into his father's chair.

His father's chair. Blast. He closed his eyes, trying to summon the will to stand up and move. The leather was cold and stiff under his fingers, under his legs, hard against his back. He could feel the fire, a few feet away in its hearth, but the heat didn't reach him.

"I've brought your coffee, my lord." Nasonby's voice cut through the cold honey, as did the smell of the coffee. Hal opened his eyes. The footman had already put the tray down on the little marquetry table and was setting out the spoons, unlidding the sugar bowl, placing the tongs just so, tenderly removing the napkin folded about the jug of warm milk—the cream was in its twin jug at the other side, keeping cold. He found the symmetry and Nasonby's quiet, deft movements soothing.

"Thank you," he managed to say, and made a small gesture indicating that Nasonby should see to the details. This Nasonby did, and the cup was placed in his limp, waiting hands. He took a mouthful—it was perfect, very hot but not so much as to burn his mouth, sweet and milky—and nodded. Nasonby vanished.

For a little while, he could just drink coffee. He didn't have to think. Halfway through the cup, he briefly considered getting up and sitting in another chair, but by then the leather had warmed and molded to his body. He could almost imagine his father's touch on his shoulder, the brief squeeze the duke had always used to express affection for his sons. *Damn you.* His throat closed suddenly, and he set the cup down.

How was John managing? he wondered. He'd be safe enough in Aberdeen, surely. Still, he ought to write to his brother. Cousin Kenneth and Cousin Eloise were incredible bores, so rigidly Presbyterian that they didn't even countenance card-playing and disapproved of any activity on the Sabbath other than reading the Bible.

On the one occasion he and Esmé had stayed with them, Eloise had politely asked Esmé to read to them after the stodgy Sunday dinner of roast mutton and bashed neeps. Ignoring the text for the day, bookmarked with a handmade lace strip, Em had blithely thumbed through the book and settled on the story of Jephthah, who had sworn that if the Lord would grant him victory in battle against the sons of Ammon, then Jephthah would sacrifice to the Lord the first thing to greet him when he returned home.

"Really," said Esmé, swallowing the "R" in a particularly fetching French way. She looked up, frowning. "What if it should have been his dog? What do you say, Mercy," she said, addressing Hal's twelve-year-old cousin, Mercy. "If your papa should come home one day and announce that he was going to kill Jasper there"—the spaniel looked up from his rug, hearing his name—"just because he'd told God he would, what would you do?"

Mercy's eyes went round with horror and her lip quivered as she looked at the dog.

"But—but—he wouldn't," she said. But then she glanced at her father, doubt in her eyes. "You *wouldn't*, would you, Papa?"

"But if you had promised God?" Esmé put in helpfully, looking up at

Kenneth with her large blue eyes. Hal was enjoying the look on Kenneth's face, but Eloise was going a bit red round the jowls, so he coughed—and with a distinct, exhilarating sense that he was driving a carriage over a cliff said, "But Jephthah didn't meet his dog, did he? What *did* happen? Do remind me—been some time since I've read the Old Testament." In fact, he'd never read it, but Esmé loved to read it and tell him the stories—with her own inimitable commentary.

Esmé had carefully not looked at him but turned the page with delicate fingers and cleared her throat.

"And Jephthah came to Mizpeh unto his house, and, behold, his daughter came out to meet him with timbrels and with dances: and she was his only child; beside her he had neither son nor daughter.

"And it came to pass, when he saw her, that he rent his clothes, and said, Alas, my daughter! thou hast brought me very low, and thou art one of them that trouble me: for I have opened my mouth unto the Lord, and I cannot go back.

"And she said unto him, My father, if thou hast opened thy mouth unto the Lord, do to me according to that which hath proceeded out of thy mouth; forasmuch as the Lord hath taken vengeance for thee of thine enemies, even of the children of Ammon.

"And she said unto her father, Let this thing be done for me: let me alone two months, that I may go up and down upon the mountains, and bewail my virginity, I and my fellows.

"And he said, Go. And he sent her away for two months: and she went with her companions, and bewailed her virginity upon the mountains.

"And it came to pass at the end of two months, that she returned unto her father, who did with her according to his vow which he had vowed: and she knew no man. And it was a custom in Israel that the daughters of Israel went yearly to lament the daughter of Jephthah the Gileadite four days in a year."

Then she'd laughed, closing the book.

"I don't think I would have bewailed my virginity for long, me. I would have come home without it"—*then* she'd met his eyes, with a spark that had ignited his vitals—"and see whether my dear papa still considered me a suitable sacrifice."

His eyes were closed; he was breathing hard and dimly aware that tears were leaking out between his lids.

"You bitch," he whispered. "Em, you *bitch*!"

He breathed until the memory passed and the echo of her voice faded

from his ear. When he opened his eyes, he found that his chin was resting in his hands, elbows on his knees, and that he was staring at the hearth rug. An expensive bit of carpet for such a use. Soft white wool, tufted, with the Grey family coat of arms in the center and an extravagantly worked "H" and "E" in black silk on either side. She'd had it made for him—a wedding present.

He'd given her a diamond pendant. And buried it with her and her child, a month ago.

He closed his eyes again. And breathed.

AFTER A TIME, he got up and wandered down the hall to the nook he'd taken over as his study. It was cramped as an eggshell, but he didn't need much space—and the close confines seemed to help him think better, shutting out some of the outside world.

He plucked a quill from the jar and bit it absently, tasting the bitter tang of dried ink. He should cut a new one but couldn't summon up the energy to find his penknife, and after all, what did it matter? John wouldn't mind a few blots.

Paper . . . There was a half quire of the parchment sheets he'd used to reply to the expressions of sympathy about Esmé. They'd come in by the bushelful—unlike the spatter of embarrassed notes that had followed his father's suicide three years before. He'd written the replies himself, in spite of his mother's offer to help. He'd been filled with something like the electric fluid natural philosophers talked about, something that numbed him to any natural need like food or sleep, that filled his brain and body with a manic need to move, to do something—though God knew there was nothing more he could have done after killing Nathaniel Twelvetrees. Not that he hadn't tried . . .

The paper felt gritty with dust; he didn't let anyone touch his desk. He held up a sheet and blew at it, shook it a bit, and set it down, then dipped his quill.

J— he wrote, and stopped dead. What was there to say? *I hope to God you're not dead? Have you seen anyone strange asking questions? How are you finding Aberdeen? Other than cold, wet, dreary, and gray . . .*

After twiddling the quill for a while, he gave up, wrote, *Luck. —H,* sanded the sheet, folded it, and, taking up the candle, dribbled smoke-stained wax onto the paper and stamped it firmly with his signet. A swan, flying, neck outstretched, across a full moon.

He was still sitting at his desk an hour later. There was progress: John's letter sat there, squared to the corner of the desk, sealed and with the

Armstrongs' direction in Aberdeen neatly written—with a freshly cut pen. The quire of parchment had been shaken free of dust, tapped into alignment, and put away in a drawer. And he'd found the source of the dead-flower smell: a bunch of rotting carnations left in a pottery mug on the windowsill. He'd managed to open *that* window and throw them out and then had summoned a footman to take the mug away to be washed. He was exhausted.

He became aware of noises in the distance: the sound of the front door opening, voices. That was all right; Sylvester would take care of whoever it was.

To his surprise, the butler seemed to have been overcome by the intruder; there were raised voices and a determined step coming rapidly toward his sanctum.

"What the devil are you doing, Melton?" The door was flung open and Harry Quarry's broad face glowered in at him.

"Writing letters," Hal said, with what dignity he could summon. "What does it look like?"

Harry strode into the room, lit a taper from the fire, and touched it to the candlestick on the desk. Hal hadn't noticed it growing dark, but it must be teatime, at least. His friend lifted the candlestick and examined him critically by its light.

"You don't want to know what you look like," said Harry, shaking his head. He put down the candle. "You didn't recall that you were meant to be meeting with Washburn this afternoon, I take it."

"Wash—oh, Jesus." He'd risen halfway out of his chair at the name and now sank back, feeling hollow at mention of his solicitor.

"I've spent the last hour with him, after meeting with Anstruther and Josper—you remember, the adjutant from the Fourteenth?" He spoke with a strong note of sarcasm.

"I do," Hal said shortly, and rubbed a hand hard over his face, trying to rouse his wits.

"I'm sorry, Harry," he said, and shook his head. He rose, pulling his banyan round him. "Call Nasonby, will you? Have him bring us tea in the library. I have to change and wash."

Washed, dressed, brushed, and feeling some semblance of ability, he came into the library a quarter hour later to find the tea trolley already in place; a wisp of aromatic steam rose from the teapot's spout to mingle with the spicy scents of ham and sardines and the unctuous sweetness of a currant sponge, oozing cream and butter.

"When's the last time you ate anything?" Harry demanded, watching Hal consume sardines on toast with the single-mindedness of a starving cat.

"Yesterday. Maybe. I forget." He reached for his cup and washed the

sardines down far enough to make cake feasible as the next step. "Tell me what Washburn said."

Harry disposed of his own cake, swallowed, and replied.

"Well, you can't actually be tried in open court. Whatever you think about your damned title—no, don't tell me, I've heard it." He held out the palm of his hand in prevention, picking up a gherkin with the other.

"Whether you choose to call yourself the Duke of Pardloe, the Earl of Melton, or plain Harold Grey, you're still a peer. You can't be tried by anything save a jury *of* your peers—to wit, the House of Lords. And I didn't really require Washburn to tell me that the odds of a hundred noblemen agreeing that you should be either imprisoned or hanged for challenging the man who seduced your wife to a duel, and killing him as a result, is roughly a thousand to one—but he did tell me so."

"Oh." Hal hadn't given the matter a moment's thought but if he had would likely have reached a similar conclusion. Still, he felt some relief at hearing that the Honorable Lawrence Washburn, KC, shared it.

"Mind you—are you going to eat that last slice of ham?"

"Yes." Hal took it and reached for the mustard pot. Harry took an egg sandwich instead.

"Mind you," he repeated, mouth half full of deviled egg and thin white bread, "that doesn't mean you aren't in trouble."

"You mean with Reginald Twelvetrees, I suppose." Hal kept his eyes on his plate, carefully cutting the ham into pieces. "That isn't news to me, Harry."

"I shouldn't have thought so, no," Harry agreed. "I meant with the king."

Hal set down his fork and stared at Harry.

"The king?"

"Or, to be more exact, the army." Harry delicately plucked an almond biscuit from the wreckage of the tea trolley. "Reginald Twelvetrees has sent a petition to the secretary at war, asking that you be brought to a court-martial for the unlawful killing of his brother and, further, that you be removed as colonel of the Forty-sixth and the regiment refused permanent re-commission, on grounds that your behavior is so deranged as to constitute a danger to the readiness and ability of said regiment. That being where His Majesty comes in."

"Balderdash," Hal said shortly. But his hand trembled slightly as he lifted the teapot, and the lid rattled. He saw Harry notice, and he set it down carefully.

What the king giveth, the king also taketh away. It had taken months of painstaking work to have his father's regiment provisionally re-commissioned and more—much more—to find decent officers willing to join it.

"The scribblers—" Harry began, but Hal made a quick, violent gesture, cutting him off.

"I know."

"No, you don't—"

"I do! Don't bloody talk about it."

Harry made a soft growling noise but subsided. He picked up the pot and filled both cups, pushing Hal's toward him.

"Sugar?"

"Please."

The regiment—in its resurrected form—had not yet seen service anywhere; it had barely half its complement of men, and most of those didn't know one end of a musket from the other. He had only a skeleton staff, and while most of his officers were good, solid men, only a handful, like Harry Quarry, had any personal allegiance to him. Any pressure, any hint of scandal—well, any *more* scandal—and the whole structure could collapse. The remnants to be greedily scooped up or trampled on by Reginald Twelvetrees, Hal's father's blackened memory left forever dishonored as a traitor, and his own name dragged further through the mud—painted by the scribblers of the press not only as a cuckold but a murderer and lunatic.

The handle of his porcelain teacup broke off suddenly and shot across the table, striking the pot with a *tink!* The cup itself had cracked right through, and tea ran down his arm, soaking his cuff.

He carefully put down the two pieces of the cup and shook tea off his hand. Harry said nothing but raised one bushy black brow at him.

Hal closed his eyes and breathed through his nose for several moments.

"All right," he said, and opened his eyes. "One—Twelvetrees's petition. It hasn't been granted yet?"

"It has not." Harry was beginning to relax a little, which gave Hal a bit more confidence in his own assumption of composure.

"Well, then. That's the first thing—stop that petition. Do you know the secretary personally?"

Harry shook his head. "You?"

"I've met him once, at Ascot. Friendly wager. I won, though."

"Ah. Too bad." Harry drummed his fingers on the cloth for a moment, then darted a glance at Hal. "Ask your mother?"

"Absolutely not. She's in France, anyway, and she's not coming back."

Harry knew *why* the Dowager Countess of Melton was in France—and why John was in Aberdeen—and nodded reluctantly. Benedicta Grey knew a great many people, but the suicide of her husband on the eve of his being arrested as a Jacobite traitor had barred her from the sort of circles where Hal might otherwise have found influence.

There was a long silence, unbroken by Nasonby's appearance with a

new teacup. He filled this, took up the shattered bits of the old one, and vanished as he'd come, soft-footed as a cat.

"What does this petition say, exactly?" Hal asked finally.

Harry grimaced but settled himself to answer.

"That you killed Nathaniel Twelvetrees because you had conceived the unfounded notion that he had been, er . . . dallying with your wife. In the grip of this delusion, you then assassinated him. And thus you are plainly mentally unfit to hold command over—"

"Unfounded?" Hal said blankly. *"Assassinated?"*

Harry reached out quickly and took the cup from his hand.

"You know as well as I do, Melton—it's not what's true; it's what you can make people believe." He set the full cup gingerly on its saucer. "The hound was damned discreet about it, and apparently so was Esmé. There wasn't a breath of gossip until the news that you'd shot him on his own croquet lawn."

"*He* chose the ground! *And* the weapons!"

"I know that," Harry said patiently. "I was there, remember?"

"What do you think I am?" Hal snapped. "An idiot?"

Harry ignored that.

"I'll say what I know, of course—that it was a legitimate challenge and that Nathaniel Twelvetrees accepted it. But his second—that chap Buxton—was killed last month in a carriage accident near Smithfield. And no one else was on that croquet lawn. That's doubtless what gave Reginald the notion of trying to nobble you this way—no independent witnesses."

"Oh . . . *hell*." The sardines were stirring in his guts.

Harry took a breath that strained the seams of his uniform and looked down at the table.

"I—forgive me. But . . . is there any proof?"

Hal managed a laugh, dry as sawdust.

"Of the affair? Do you think I'd have killed him if I hadn't been sure?"

"No, of course not. I only mean—well . . . bloody hell . . . did she just . . . *tell* you? Or perhaps you . . . er . . . saw . . ."

"No." Hal was feeling dizzy. He shook his head, closed his eyes, and tried a deep breath of his own. "No, I never caught them together. And she didn't—didn't *quite* tell me. There were—there were letters."

She'd left them where she knew he'd find them. But why? That was one of the things that killed him, over and over again. She'd never told him why. Was it simple guilt? Had she grown tired of the affair but lacked the courage to end it herself? Worse—had she *wanted* him to kill Nathaniel?

No. Her face when he'd come back that day, when he told her what he'd done . . .

His face was resting on the white cloth and there were black and white

spots swarming before his eyes. He could smell starch and spilled tea, sardines with their tang of the sea. Of Esmé's birth waters. And her blood. *Oh, God, don't let me vomit. . . .*

3

IRISH ROVERS

London, May 1744

MINNIE LAY IN BED, the remains of breakfast on a tray beside her, and contemplated the shape of her first day in London. She'd arrived late the night before and had barely taken notice of the rooms her father had engaged for her—she had a suite in a townhouse on Great Ryder Street, "convenient to everything," as he'd assured her, complete with a housemaid and meals provided from the kitchen in the basement.

She had been filled with an intoxicating sense of freedom from the moment she'd taken an affectionate leave of her father on the dock at Calais. She could still feel the pleasure of it, bubbling in the slow, pleasant fashion of a crock of fermenting cabbage under her stays, but her innate caution kept a lid on it.

She'd done small jobs on her own before, sometimes outside of Paris, but those had been simple things like calling on the relatives of a dead bibliophile and sympathetically relieving them of their burdensome inheritance—she'd noticed that almost no one felt that a library was much of a legacy—and even then she'd had an escort, usually a stout, middle-aged, long-married man still capable of hoisting boxes and deflecting nuisances but unlikely to make improper advances to a young woman of seventeen.

Monsieur Perpignan would not, of course, be a suitable escort for London. Aside from a tendency to seasickness, a fondness for his wife, and a disgust for British cooking, he didn't speak English and had no sense of direction. She'd been a bit surprised that her father would let her stay in London entirely on her own—but of course he hadn't. He had Made Arrangements: his specialty.

"I've arranged a chaperone for you," her father had said, handing over

a neat docket of notes, addresses, maps, and English money. "A Lady Buford, a widow of slender means but good connections. She'll arrange a social life for you, introduce you to the right sorts of people, take you to plays and salons, that sort of thing."

"What fun," she'd said politely, and he laughed.

"Oh, I expect you'll find some, my dear," he said. "That's why I've also arranged two . . . shall we call them bodyguards?"

"So much more tactful than minders, or wardens. Two?"

"Yes, indeed. They'll run errands for you, as well as accompany you when you visit clients." He reached into one of the pigeonholes of his desk and drew out a folded sheet of paper, which he handed her. "This is a précis of what I told you about the Duke of Pardloe—and a few others. I didn't mention him to Lady Buford, and you should be somewhat discreet about your interest in him. There's a great deal of scandal attached to that family, and you—"

"Don't touch pitch until you're ready to set light to it," she finished, with no more than a slight roll of the eyes.

"Travel safe, my dear." He'd kissed her forehead and embraced her briefly. "I'll miss you."

"I'll miss you, too, Papa," she murmured now, climbing out of bed. "But not *that* much."

She glanced at the secretaire, where she'd put all the lists and documents. Time enough for the chaste Duke of Pardloe and the randy Duke of Beaufort when she came within sight of them. Lady Buford had left a card, saying that she would meet Minnie at Rumm's Tea-Room in Piccadilly at four o'clock for tea. *Wear something pretty, modest, and not over-elaborate,* Lady Buford had added, with welcome practicality. The pink muslin, then, with the little jacket.

There were three appointments already scheduled for the early afternoon—routine book business—and the two bodyguards were meant to come and introduce themselves at eleven. She glanced at her little traveling clock, which showed half-eight. A quick wash, a simple dress, stout boots for walking, and London was hers—alone!—for two hours.

THEY'D LIVED IN London for a time, when she was much younger. And she'd come with her father twice for brief visits, when she was fourteen and fifteen. She had a general idea of the city's shape but had never needed to find her own way.

She was accustomed to exploring a new place, though, and within the first hour had discovered a decent-looking ordinary for quick meals out-

side her rooms, a baker's shop for cakes, and the nearest church. Her father had nothing to do with religion, and so far as she knew, she'd never even been christened—but it was as well to look the part you played, and pious, modest young women went to church on Sunday. Besides, she liked the music.

The day was bright, the air tangy with spring sap, and the streets were full of an exuberant bustle, quite different from Paris or Prague. There was really no place like London. Particularly as no other city contained her mother. But that small matter would need to wait for a bit; much as she longed to rush off to Parson's Green at once and see this Mrs. Simpson, it was too important. She needed to reconnoiter, to calculate her approach. To be hasty or importunate might ruin everything.

She headed toward Piccadilly, which housed a good many booksellers. On the way, though, were Regent and then Oxford Streets, charmingly studded with expensive shops. She must ask Lady Buford about dressmakers.

She had a little French watch pinned to her fichu—it didn't do to be late for appointments—and when it told her in a tiny silver voice that it was now half-ten, she sighed and turned back toward Great Ryder Street. As she crossed the corner of Upper St. James's Park, though, she began to have an odd feeling at the back of her neck.

She reached the corner, made as though to step into the street, then suddenly darted sideways, across a lane, and into the park itself. She whipped behind a large tree and stood in the shadow, frozen, watching. Sure enough, a young man came hurtling into the lane, looking sharp from side to side. He was roughly dressed, brown hair tied back with string—perhaps an apprentice or a laborer.

He halted for an instant, then walked fast down the lane, out of sight. She was just about to slide out of her shelter and run for the street when she heard him whistle loudly. An answering whistle came *from* the street, and she pressed herself against the tree, heart hammering.

Bloody, bloody hell, she thought. *If I'm raped and murdered, I'll never hear the end of it!*

She swallowed and made up her mind. It would be somewhat harder for anyone to abduct her off a busy street than to winkle her out of her precarious hiding spot. A couple of gentlemen were coming along the path toward her, deep in conversation. As they passed, she stepped out on the path directly behind them, keeping so close that she was obliged to hear a very scabrous story concerning one man's father-in-law and what had happened when he chose to celebrate his birthday in a bawdy house. Before the end was reached, though, the street was reached, and she stepped away, walking fast down Ryder Street, with a sense of relief.

She was perspiring, in spite of the cool morning, and the pin thrust through her straw hat had loosened. She paused, took off the hat, and was dabbing her face with a handkerchief when a male voice spoke in her ear.

"So here ye are!" it said triumphantly. "Jaysus God!" This last was the result of her having whipped the eight-inch hatpin from its moorings and aimed it at his breast.

"Who the blood helly are you, and what do you mean by following me?" Minnie demanded, glaring at him. Then she saw his eyes lift, noticing something over her shoulder, and the words "two bodyguards" dropped into her mind like pebbles dropped in water. *Merde!*

"Two," she said flatly, and lowered the hatpin. "Mister O'Higgins, I presume? And . . . Mr. O'Higgins, as well?" she added, turning toward the other young man, who had come up behind her. He grinned at her and bowed extravagantly, sweeping off his cap.

"Raphael Thomas O'Higgins, me lady," he said. "Blood helly? Would that be a French expression, at all?"

"If you like," she said, still annoyed. "And you?" She swung back to face the first pursuer, who was also grinning from ear to ear.

"Michael Seamas O'Higgins, miss," he said, with a bob of the head. "Mick, to me friends, and me brother there is Rafe. Ye were expecting us, I see?"

"Hmph. How long have you been following me?"

"Since ye left the house, sure," Rafe said easily. "What was it spooked ye, would ye tell me? I thought we'd kept well back."

"To be honest, I don't know," she said. The rush of fright and flight was fading from her blood, and her annoyance with it. "I just suddenly had a . . . feeling. Just something at the back of my neck. But I didn't *know* someone was following me until I ran into the park and you"—nodding at Mick—"ran in after me."

The brothers O'Higgins exchanged a glance with lifted brows but seemed to take this at face value.

"Aye, then," Rafe said. "Well, we were to introjuice ourselves to ye at eleven o' the clock, and I hear the bells sayin' that's just what it is now . . . so, miss, is there anything we can be doin' for you today? Any errands to be run, parcels picked up, perhaps the little small quiet murder on the side . . . ?"

"How much is my father paying you?" she asked, beginning to be amused. "I doubt it extends to procuring murder."

"Oh, we come cheap," Mick assured her, straight-faced. "Though if it was to be anything of a fancy nature—beheading, say, or hiding multiple bodies—well, I won't say but what that might not run into money."

"That's all right," she assured him. "Should it come to that, I have a bit

of my own. And speaking of that"—the idea came to her as she re-pinned her hat—"I have several letters of credit, drawn on the Bankers on the Strand—you know the place? That's what you can do today: come with me to the bank and back again. I'll need cash in hand for one or two of my afternoon appointments."

4

REGIMENTAL BUSINESS

WINSTEAD TERRACE WAS A small row of discreetly fine townhouses that faced a similar terrace on the other side of a private park, its privacy protected by a tall fence of black iron and a locked gate.

Hal reached through the iron bars of the fence and carefully broke a twig from one of the small trees that pressed against it.

"What are you doing?" Harry demanded, stopping in mid-stride. "Picking a posy for your buttonhole? I don't think Grierson's much of a dandy."

"Nor am I," said Hal equably. "I wanted to see if this is what I thought it was, but it is."

"And what's that, pray?" Harry came back a step to look at the twig in Hal's hand. The foliage was cool on his fingers; it had rained a bit earlier and the leaves and flowers were still wet, water droplets sliding down his wrist, disappearing into the cloth of his frilled cuff.

He transferred the twig to his other hand and shook the water off, absently wiping his hand on his coat. He liked good linen and a well-fitting suit, but in fact he wasn't a dandy. It *was* necessary to impress Donald Grierson favorably, though, and to that end, he and Harry were both wearing semi-dress uniform, with a discreet but visible amount of gold lace.

"Cockspur," he said, showing Harry the two-inch thorns protruding from the twig. "It's a hawthorn of sorts."

"I thought hawthorns were hedges." Harry jerked his head toward the terrace, and Hal nodded, coming along.

"They can be. Or shrubs or trees. Interesting plant—the leaves are said to taste like bread and cheese, though I haven't tried."

Harry looked amused.

"I'll remember that, next time I'm in the country and not a pub in sight. Ready, are you?"

Hal might have felt annoyed at Harry's solicitousness, but his friend was—all too clearly—honestly worried for him. He drew breath and straightened his shoulders, admitting to himself that, *in* all honesty, he couldn't dismiss that worry as unfounded. He was getting better, though. He had to—there was the devil of a lot of work to be done if he had any hope of getting the regiment on its feet and ready to fight. And Major Grierson was going to help him do it.

"There's something else about hawthorn," he said, as they reached Grierson's door.

"What's that?" Harry was wearing his bird-dog look, alert and intent on the prey to be flushed, and Hal smiled privately to see it.

"Well, the green of the leaves symbolizes constancy, of course, but the flowers are said to—and I quote—'have the scent of a woman sexually aroused.'"

Harry's intent look switched instantly to the flowering twig in Hal's hand. Hal laughed, brushed the flowers under his own nose, then handed them to Harry, turning to lift the brass boar's-head knocker.

Good lord, it's true. The whiff of insinuating musk so distracted him that he scarcely noticed when the door opened. How the devil could something smell . . . slippery? He closed his fist involuntarily, with the very disconcerting feeling that he had touched his wife.

"My lord?" The servant who had opened the door was looking at him with a slightly puzzled frown.

"Oh," Hal said, snapping back to himself. "Yes. I am. I mean—"

"Major Grierson is expecting his lordship, I think?" Harry inserted himself between Hal and the inquiring face, which nodded obligingly and withdrew into the house, gesturing them to follow.

There were voices coming from the morning room to which they were escorted: a woman, and at least two men. Perhaps Grierson was married and his wife was accepting callers . . . ?

"Lord Melton!" Grierson himself—a big, bluff-looking, sandy-haired chap—rose from a settee and came to meet him, smiling. Hal's heart rose; he'd not met Grierson before, but his reputation was stellar. He'd served with a famous regiment of foot for years, fought at Dettingen, and was known as much for his organizational abilities as his courage. And organization was what the fledgling 46th needed, above all.

"So pleased to meet you," Grierson was saying. "Everyone's buzzing about this new regiment, and I want to hear all about it. Pansy, my dear, may I present his lordship, the Earl of Melton?" He half-turned, extending a hand to a small, darkly pretty woman of about his own age—which Hal estimated as thirty-five. "Lord Melton, my wife, Mrs. Grierson."

"Charmed, Mrs. Grierson."

Hal made a leg to Mrs. Grierson, who smiled at the attention—but his own attention was slightly distracted by Harry, behind him. Instead of advancing to be introduced and pay his own compliments, Harry had uttered a sort of throaty noise that might have been a growl in less-civil company.

Hal glanced in Harry's direction, saw what Harry was seeing, and felt as though he'd been punched in the stomach.

"We've met," said Reginald Twelvetrees, as Grierson turned to introduce him. Twelvetrees rose to his feet, cold-eyed.

"Indeed?" said Grierson, still smiling but now glancing warily between Hal and Twelvetrees. "I'd no idea. I trust you have no objection to Colonel Twelvetrees meeting with us, Lord Melton? And I trust that you, sir"—with a deferential nod to Twelvetrees—"have no objection to my inviting Colonel Lord Melton to join us?"

"Not in the slightest," said Twelvetrees, with a twitch in one cheek that was by no means a token smile. Still, he sounded as though he meant that "not in the slightest," and Hal began to feel a certain tightness in his chest.

"By all means," he said coolly, meeting Twelvetrees's stony gaze with one of his own. Reginald's eyes were the same color that Nathaniel's had been, a brown so dark as to seem black in some lights. Nathaniel's had been black as pitch, facing him in the dawn.

Mrs. Grierson excused herself and went out, saying that she would have refreshments brought, and the men settled, in the uneasy fashion of seabirds jealous of their rocky perches.

"Quite to my surprise, gentlemen, I find myself in the enviable position of being a valuable commodity," Grierson said, leaning affably forward. "As you may know, I fell ill in Prussia, was shipped home to recover, and fortunately did so. But it was a long convalescence, and by the time I was fit, my regiment had . . . well . . . I'm sure you know the general situation; I won't go into the particulars just now."

All three of his guests made small grunts of assent, with a few murmurs of decent sympathy. What had happened was that Grierson had been bloody lucky in falling ill when he did. There had been a truly scandalous mutiny a month after his removal to England, and when the mess had been cleaned up, half the surviving officers had been court-martialed, fifteen mutineers had been hanged, and the remnants dispersed to four other regiments. The original regiment had formally ceased to exist, and Grierson's commission with it.

The normal thing for a man in his position to do would be to buy a commission in another regiment. But Grierson was, as he bluntly put it, a valuable commodity. Not only was he a very capable administrator and

a good commander—he was popular, with other officers, with the War Office, and with the press.

Hal needed Grierson's expertise; he needed even more Grierson's connections. With Grierson on his staff, he could attract officers of a much higher caliber than he could do with money alone.

As to what Twelvetrees, colonel of a long established and very solid artillery regiment, might want with him, that was fairly obvious, too: he wanted Hal not to have Grierson.

"So, Lord Melton, tell me how things stand with you," Grierson said, once they'd got stuck into the wine and biscuits that Mrs. Grierson had sent in. "Who are your staff officers, to begin with?"

Hal set down his glass carefully and told him, in a calm voice, exactly who they were. Competent men, so far as he knew—but almost all of them quite young, with no experience of foreign campaigns.

"Of course," Harry put in helpfully, "that means that you would be quite senior in the regiment: have your pick of companies, postings, aides . . ."

"Just how many troops have you on your muster roll, Colonel?" Reginald didn't bother trying to sound neutral, and Grierson glanced at him. Not with disapproval, Hal saw, and his heart sped up a little.

"I cannot tell you exactly, sir," he said, with exquisite politeness. Sweat had begun to dampen his collar, though the room was cool. "We are conducting a major campaign of recruitment at the moment, and our numbers rise—substantially—each day." On a good day, they might get three new men—one of whom would not abscond with the bounty for signing— and from the smirk on Twelvetrees's face, Hal knew he was aware of this.

"Indeed," said Twelvetrees. "Untrained recruits. The Royal Artillery is at full strength presently. My company commanders have been with me for at least a decade."

Hal kept his temper, though he was beginning to feel slightly breathless from suppressed rage.

"In that case, Major Grierson may have less space in which to distinguish himself," he riposted smartly. "Whereas with us, sir . . ." He bowed to Grierson and felt momentarily giddy when he raised his head. "With us," he repeated more strongly, "you would have the satisfaction of helping to shape a fine regiment in . . . your own likeness, so to speak."

Harry chuckled in support, and Grierson smiled but politely. He'd also have the not-inconsiderable risk of failure and knew it.

Hal felt Harry stir uncomfortably next to him and took a deep breath, preparing to say something forceful about . . . about . . . The word had gone. Simply gone. He'd breathed in, and a trace of scent from the cockspur in Harry's buttonhole had touched his brain. He closed his eyes abruptly.

Major Grierson had luckily asked a question; Hal could hear Twelve-trees replying in a gruff, matter-of-fact way. Grierson said something else and Twelvetrees's voice relaxed a little, and quite suddenly it was Nathaniel's voice, and he opened his eyes and saw nothing of the cozy morning room, of the men there with him. He was cold, shaking with cold . . .

And his fingers were squeezing the cold pistol in his hand so hard the metal would leave marks on his palm. He'd fucked Esmé before he left to kill her lover. Waked her in the dark and taken her, and she'd wanted him—ferociously—or perhaps she had pretended it was Nathaniel in the dark. He knew it was the last time . . .

"Colonel?" A voice, a dim voice. "Lord Melton!"

"Hal?" Harry's voice, full of alarm. Harry, with him on the lawn, rain running down his face in a sunless dawn. He swallowed, tried to swallow, tried to breathe, but there was no air.

His eyes were open, but he couldn't see anything. The cold was spreading down the sides of his jaws, and he realized suddenly that . . .

He looked straight into Nathaniel's eyes and felt the bang and then it was . . .

HARRY HAD INSISTED on calling a carriage to take them home. Hal refused brusquely and strode—knees shaking, but he could walk, he *would* walk, dammit—away from Winstead Terrace.

He made it to the far side of the private garden—well away from the cockspur tree—where he stopped and gripped the cold black iron of the fence and carefully lowered himself to the pavement. His mouth tasted of brandy; Grierson had forced it down him, when he could breathe again.

"I've never bloody fainted in my *life*," he said. He was sitting, back against the fence, forehead on his knees. "Not even when they told me about Father."

"I know." Harry had sat down beside him. Hal thought briefly what flats they must look, two young soldiers got up in scarlet and gold lace, sitting on the pavement like a pair of beggars. He really didn't care.

"Actually," Hal said after a minute, "that's not true, is it? I passed out in the ham at tea last week, didn't I?"

"You just felt a bit queasy," Harry said stoutly. "Not eating for days, then two dozen sardines—enough to fell anyone."

"Two dozen?" Hal asked, and laughed despite everything. Not much of a laugh, but he turned his head and looked at Harry. Harry's face was creased with anxiety but relaxed a little when he saw Hal looking at him.

"At least that many. With mustard, too."

They sat a few moments, feeling easier. Neither of them wanted to say anything about what had just happened, and they didn't, but each could tell the other was thinking of it—how could they not?

"If it falls apart . . ." Harry began at last, then bent and looked at him searchingly. "You going to faint again?"

"No." Hal swallowed twice, then took a shallow breath—the only kind he could manage—and pushed himself to his feet, holding on to the iron fence. He had to let Harry know he could go, that he didn't have to try to carry on with this doomed enterprise, this fool's game. Though the thought of it made his throat close. He cleared it, hard, and repeated Harry's words: "If it does fall apart—"

Harry's hand on his arm stopped him. Harry's face was six inches from his own, the brown eyes clear and steady.

"Then we'll start again, old man," he said. "That's all. Come on; I need a drink, and so do you."

5

STRATEGY AND TACTICS

IT TOOK LESS THAN five minutes over the cake plate at Rumm's for Minnie to realize the depths of her father's treachery.

"Your style is very good, my dear," said Lady Buford. The chaperone was a thin, gray-haired lady with an aristocratically long nose and sharp gray eyes under heavy lids that had probably been languorously appealing in her youth. She gave a small, approving nod at the delicate white daisies embroidered on Minnie's pink linen jacket. "I had thought, with your portion, that we might set our sights on a London merchant, but with your personal attractions, it *might* be possible to aim a little higher."

"My . . . portion?"

"Yes, five thousand pounds is quite attractive—we'll have a good selection, I assure you. You could have your pick of army officers"—she made an elegantly dismissive gesture, then wrapped long, bony fingers around the handle of her teacup—"and there are a few that are *quite* appealing, I admit. But there's the perpetual absences to be considered . . . and postings in insalubrious spots, should your husband wish you to accompany

him. Now, if he's killed, there's a reasonable pension, but it's nothing to what a sound merchant might leave—and if he should be wounded to such an extent as to exclude him from service . . ." She took a long, considering sip, then shook her head.

"No. We can certainly do better than the army. Or the navy, God help us. Sailors tend to be somewhat . . . un. *Couth*," she said, leaning toward Minnie and pursing her wrinkled lips in a whisper.

"God help us," Minnie repeated in a pious tone, though her fist was knotted in the folds of the tablecloth. *You utter weasel!* she thought toward her absent father. *Establish a social life for me, eh?*

Despite her astonished annoyance, though, she had to admit to being somewhat impressed. Five thousand pounds?

If he actually meant it . . . the cynical part of her mind put in. But he likely did. It would be *just* like him. He'd see it as killing two birds with one stone: getting her access to likely sources of salable information and simultaneously marrying her off to one of them, with Lady Buford as his unwitting accomplice.

And he had, to be fair, told her that he wanted an Englishman for her. She just hadn't thought he'd meant *now*. Really, she had to admire her father's twisted genius; who but a marriage procuress would know more—and have less hesitation in revealing what she knew—about the intimate familial and financial details of wealthy men?

Taking a deep breath, she let go of the fistful of tablecloth and did her best to look interested, in a demure sort of way.

"We'll avoid the navy, then," she said. "Do you think . . . I hope I am not immodest in suggesting it, but after all, five thousand pounds . . . What about minor—very minor," she added hastily, "members of the peerage?"

Lady Buford blinked but not as though taken aback; merely reordering her mental index, Minnie thought.

"Well, there are impoverished knights and baronets by the score," she said. "And if you are set on a title . . . But really, my dear, I wouldn't recommend that avenue unless you will have independent means of your own. Your portion would be instantly swallowed in sustaining some crumbling manor and you yourself would molder inside it, never getting to London or having a new dress from one year's end to the next."

"To be sure. I, um, *do* possess a, er . . . small competence, shall we say?"

"Indeed." Lady Buford's wispy brows rose in interest. "How small?"

"A thousand a year," Minnie said, wildly exaggerating the income from her small private ventures, which totaled less than a tenth of that sum. Still, it hardly mattered, as she wasn't actually marrying any of these theoretical impoverished baronets; she only needed to enter the social circles they—they and their more interesting brethren—inhabited.

"Hmm." Lady Buford assumed an inward look and drank tea. After a few moments' contemplation, she set down the cup with decision.

"You speak good French, your father says?"

"Mais oui."

Lady Buford looked at her sharply, but Minnie kept a straight face.

"Well, then. We'll begin with Lady Jonas's Thursday *salon*. It's literary and intellectual, but she usually has a good mix of available gentlemen, including European—though your father *did* specify an Englishman. . . . Well, we'll see. Then perhaps a play on Saturday evening. . . . We'll have a box; it's important that you be seen—have you something appropriate to wear?"

"I don't know," Minnie said honestly. "I've never been to a play; what *is* appropriate?"

Half an hour, two pots of China tea, and a dozen tea cakes (with cream) later, she made her way out into the street, a scribbled list of engagements in her hand and her head spinning with tippets, panniers, mantuas, swags, fans—she had a nice fan, luckily—and other items necessary to the pursuit and bagging of a wealthy and influential husband.

"A gun would be simpler," she muttered, thrusting the list into her pocket. "And certainly less expensive."

"What kind of gun?" said Mick O'Higgins with interest, appearing out of a nearby doorway.

"Never mind," she said. "We're going to a milliner's."

"Oh, a milliner's, is it?" He bowed and offered her his arm. "Nay bother, then. Sure, the bird'll be dead before they put it on your hat."

A week later . . .

HER APPOINTMENT BOOK was a pleasure to look at, a glory to hold. Made in Florence, the leather cover was the color of rich chocolate, with a pressed gilt design of looping vines and a glorious, explosive-looking flower in the center. Her father had informed her that the Chinese called it *"Chu"* and that it was a symbol of happiness. He'd given her the book for her seventeenth birthday.

He'd given her another one, too, before she left Paris: a rough-cut notebook such as an artist might use for sketching notes—and sketches were just what decorated its pages, made by her own hand. And coded into the sketches were the appointments made for those clients whose names were never spoken aloud.

The first few pages were decoys; the first *aide-memoire* was on the fifth page (the appointment being for the fifth of the month): a sketch of trees

overhanging a path, with the legend *Vauxhall Gardens* underneath. There were footprints on the path, leading the way into the shadows—three of them clearly marked, and half of another. Half-three in Vauxhall Gardens, on the third of June. On the facing page, a sketch of a wrapped parcel, like a birthday present. To be received . . .

That was for tomorrow. She set the sketchbook aside and picked up the chuppointment book, where the less-private clients were listed—those merely wanting to buy or sell books. Eight ticked off since her arrival in London; she'd been very efficient.

She rubbed a thumb gently over the exuberant bloom on the cover. She'd never seen a real *chu* flower. Perhaps she might come across a botanist in London who would have such a plant; she'd love to know what they smelled like.

At the back of the appointment book, between the creamy blank pages and the soft leather cover, was the letter. She had written and rewritten it several times. Wanting to be sure, but knowing there could be no surety in this.

In the morning, she'd give it to one of the O'Higginses. She'd known them long enough now to be sure they'd carry out her errands without question—well, without a *lot* of questions. She sent a good many notes and letters in the course of business; there was no reason why this one should seem at all odd.

Mrs. Simpson, Parson's Green, Peterborough Road

Her fingers were damp; she put the letter back before the ink of the direction should smear and closed the book upon it.

From the Chu Diary

Monday, June 1

11:00—Mr. H. R. Wallace, to view *Philologus Hebraeus* (Johannes Leusden). Offer also *Histoire de la Guerre des Juifs Contre les Romains* (Joseph Flavius) and *De Sacrificiis Libri Duo Quorum Altero Explicantur Omnia Judæorum, Nonnulla Gentium Profanarum Sacrificia* (William Owtram)

1:00—Misses Emma and Pauline Jones, to discuss catalog of late father's library. In Swansea(!) How blood helly will I get those shipped?

2:00—fitting at Myers, peach silk suit

4:00—Lady Buford, tea here, then Mrs. Montague's salon
8:00—Drury Lane Theatre, *Mahomet the Imposter*

Tuesday, June 2

9:00—bath
10:00—hairdresser
1:00—Lady Buford, for Viscountess Baldo's luncheon
5:00—the Hon. Horace Walpole, to view Italian titles
      (arrange tea)

Wednesday, June 3

10:00—boating on Thames with Sir George Vance, Kt.,
      luncheon
3:30—Deer Park
7:00—Mrs. Annabelle Wrigley's rout

Note: Sir George young but boring; told L. Buford to cross him off. Met a promising gentleman named Hanksleigh at rout, knowledgeable about finance; seeing him for tea next week.

Note: Vauxhall Gardens charming (visit again next week)

Thursday, June 4

9:00—bath
10:00—body groomer (ouch)
11:00—hairdresser
1:00—measurements, Madame Alexander's, eau-de-nil ball
      gown
3:00—promenade in Hyde Park with Sir Robert Abdy, Bt.
8:00—supper party, Lady Wilford

Note: Lady Wilford's party well supplied. Two engagements for next week, and a promising conversation with the Marquess of Tewksbury about hocus-pocus in House of Lords.

Note: Also met Duke of Beaufort at supper, chatted briefly over asparagus mayonnaise. Asked me to ride with him in

Rotten Row next Tuesday. Declined on grounds that I have no horse, only to have him offer me one. Accepted. How hard can it be?

Friday, June 5

11:00—Baron Edgerly, to view French titles, elephant folio atlas
   1:30—Visit Mr. Smethurst, bookseller in Piccadilly, worm list
            of clients out of him if poss.
   4:30—Lady Buford, tea with Mrs. Randolph and her
            two daughters

Note: supper alone, thank God. Don't want to hear one more word spoken. Randolph girls complete *emmerdeuses*.

Note: reply from Mrs. Simpson. Monday, two o'clock.

Saturday, June 6

Beginning to attract clients desiring information rather than books. Father's work. Two this week. Said no to one, yes to Sir Roger Barrymore (request re character of man seeking to wed his daughter; met said man last week and could have told Sir Roger he's a wrong 'un on the spot, but will give him news next week to justify bill).

Sunday, June 7

Morning service, St. George's, Hanover Square, with Mr. Jaken (Exchange)—fond of organ music

4:00—tea, Lady Buford, review of progress
7:00—Evensong, St. Clement's, Mr. Hopworth, banker

6

UNEXPECTED INTRODUCTIONS

Monday, June 8

MINERVA RUBBED HER HANDS nervously on her petticoat to dry them, then poked for the dozenth time at her hair, though knowing it to be pinned up as securely as hair *could* be pinned; the skin of her face felt stretched, her eyebrows ludicrously arched. She glanced into the glass quickly—for the dozenth time—to assure herself that this was in fact not the case.

Would Mrs. Simpson come? She'd dithered about her mother all the way to London and for the two weeks since her arrival—and she hated dithering above all things. Make up one's mind and be done with it!

So she had, but for once, decision had not removed doubt. Maybe she should have gone to her mother's residence, appeared on the doorstep without warning. That had been her first impulse, and it was still strong. She'd finally decided instead to send a note—phrased with the utmost simplicity and the barest of facts—requesting the pleasure of Mrs. Simpson's company in her rooms in Great Ryder Street at two o'clock on Monday, the eighth of June.

She'd thought of sending a note asking permission to call upon Mrs. Simpson; that might have seemed more polite. But she feared the receipt of a rejection—or, still worse, silence—and so had issued an invitation instead. If her mother didn't come this afternoon, the doorstep option was still open. And by God, she would do it . . .

The note crackled in her pocket, and she pulled it out—again—unfolding it to read the message, written in a firm round hand—presumably Mrs. Simpson's—without salutation or signature, promise or rebuke.

Do you think this is wise? it said.

"Well, obviously not," she said aloud, cross, and shoved it back into her pocket. "What does that matter?"

The knock on the door nearly stopped her heart. She was here! She was early—it lacked a quarter hour of two o'clock—but perhaps Mrs. Simpson had been as eager as herself for the meeting, despite the cool reserve of the note.

The maid—Eliza, a solid middle-aged woman in a high state of starch, who had been engaged with the rooms—glanced at her and, at her nod,

went down the hall to answer the door. Minnie glanced in the looking glass again (*God, I look quite wild*), smoothed her embroidered overskirt, and assumed an aloof-but-cordial expression.

"Colonel Quarry, ma'am," said the maid, coming in and stepping aside to admit the visitor.

"Who?" said Minnie blankly. The tall gentleman who had appeared in the doorway had paused to look her over with interest; she lifted her chin and returned his regard.

He was wearing his scarlet uniform—infantry—and was quite handsome in a blunt sort of way. Dark and dashing—and well aware of it, she thought, concealing an inward smile. She knew how to handle this sort and allowed the smile to blossom.

"Your servant, ma'am," he said, with an answering flash of good teeth. He made her a very graceful leg, straightened, and said, "How old are you?"

"Nineteen," she said, adding two years without hesitation. "And you, sir?"

He blinked. "Twenty-one. Why?"

"I have an interest in numerology," she said, straight-faced. "Are you acquainted with the science?"

"Er . . . no." He was still eyeing her with interest, but the interest was of a different type now.

"What is your date of birth, sir?" she asked, sidling behind the small gilt desk and taking up a quill. "If you please?" she added politely.

"The twenty-third of April," he said, lips twitching slightly.

"So," she said, scratching briskly, "that is two plus three, which is five, plus four—April being the fourth month, of course," she informed him kindly. "Which makes nine, and then we add the digits of your year of birth, which makes . . . one plus seven plus two plus three? Yes, just so . . . totaling twenty-two. We then add both twos together and end with four."

"Apparently so," he agreed, coming round the desk to look over her shoulder at the paper, where she had written a large four, circling it. He emitted a noticeable amount of heat, standing so close. "What does this signify?"

She relaxed slightly against the tightness of her stays. Now she had him. Once they got curious, you could get them to tell you anything.

"Oh, the four is the most masculine of numbers," she assured him—quite truthfully. "It designates an individual of marked strength and stability. Dependable, and exceedingly trustworthy."

He'd put his shoulders back half an inch.

"You're very punctual," she said, giving him a sidelong look from beneath her lashes. "Healthy . . . strong . . . you notice details and are

very good in controlling complex affairs. And you're loyal—very loyal to those you care for." She gave him a small but admiring smile to go with this.

Fours were capable and persistent but not swift thinkers, and, once again, she was surprised at just how often the numbers turned out to be right.

"Indeed," he said, and cleared his throat, looking mildly embarrassed but undeniably pleased.

At this point, she heard the subdued ticking of the longcase clock behind her and a bolt of apprehension shot through her. She needed to get rid of him, and promptly.

"But I doubt that a desire to learn the science of numerology accounts for the pleasure of your visit, sir."

"Well." He looked her up and down in an effort at assessment, but she could have told him it was far too late for that. "Well . . . to be blunt, madam, I wish to employ you. In a matter of . . . some discretion."

That gave her another small jolt. So he knew who—or rather what—she was. Still, that wasn't really unusual. It was, after all, a business in which all connections were by word of mouth. And she was certainly known by now to at least three gentlemen in London who might move in the circles to which Colonel Quarry had access.

No point in beating round the bush or being coy; she was interested in him but more interested in his leaving. She gave him a small bow and looked inquiring. He nodded back and took a deep breath. *Some discretion, indeed . . .*

"The situation is this, madam: I have a good friend whose wife recently died in childbed."

"I'm sorry to hear it," Minnie said quite honestly. "How very tragic."

"Yes, it was." Quarry's face showed what he was thinking, and the trouble was clear in his eyes. "The more so, perhaps, in that my friend's wife had been . . . well . . . having an affair with a friend of his for some months prior."

"Oh, dear," Minnie murmured. "And—forgive me—was the child . . . ?"

"My friend doesn't know." Quarry grimaced but relaxed a little, indicating that the most difficult part of his business had been communicated. "Bad enough, you might say . . ."

"Oh, I would."

"But the further difficulty—well, without going into the reasons why, we . . . I . . . would like to engage you to find proof of that affair."

That confused her.

"Your friend—he isn't sure that she *was* having an affair?"

"No, he's positive," Quarry assured her. "There were letters. But—

well, I can't really explain why this is necessary, but he requires proof of the affair for a . . . a . . . legal reason, and he will not countenance the idea of letting anyone read his wife's letters, no matter that she is beyond the reach of public censure nor that the consequences to himself if the affair is *not* proved may be disastrous."

"I see." She eyed him with interest. Was there really a friend, or was this perhaps his own situation, thinly disguised? She thought not; he was clearly grieved and troubled but not flushed—not ashamed or angry in the least. And he hadn't the look of a married man. At all.

As though her invisible thought had struck him on the cheek like a flying moth, he looked sharply at her, meeting her eyes directly. No, not a married man. And not so grieved or troubled that a spark didn't show clearly in those deep-brown eyes. She looked modestly down for a moment, then up, resuming her businesslike manner.

"Well, then. Have you specific suggestions as to how the inquiry might proceed?"

He shrugged, a little embarrassed.

"Well . . . I thought . . . perhaps you could make the acquaintance of some of Esmé's—that was her name, Esmé Grey, Countess Melton—some of her friends. And . . . er . . . perhaps some of . . . *his* . . . particular friends. The, um, man who . . ."

"And the man's name?" Picking up the quill, she wrote *Countess Melton,* then looked up expectantly.

"Nathaniel Twelvetrees."

"Ah. Is he a soldier, too?"

"No," and here Quarry *did* blush, surprisingly. "A poet."

"I see," Minnie murmured, writing it down. "All right." She put down the quill and came out from behind the desk, passing him closely so that he was obliged to turn toward her—and toward the door. He smelled of bay rum and vetiver, though he didn't wear a wig or powder in his hair.

"I'm willing to undertake your inquiry, sir—though, of course, I can't guarantee results."

"No, no. Of course."

"Now, I have a prior engagement at two o'clock"—he glanced at the clock, as did she: four minutes to the hour—"but if you would perhaps make a list of the friends you think might be helpful and send it round? Once I've assessed the possibilities, I can inform you of my terms." She hesitated. "May I approach Mr. Twelvetrees? Very discreetly, of course," she assured him.

He made a grimace, half shock and half amusement.

"Afraid not, Miss Rennie. My friend shot him. I'll send the list," he promised, and, with a deep bow, left her.

The door had barely closed behind him before there was another knock. The maid popped out of the boudoir, where she had been discreetly lurking, and glided silently over the thick red Turkey carpet.

Minnie felt her stomach lurch and her throat tighten, as though she'd been dropped out of a high window and caught by the neck at the last moment.

Voices. Men's voices. Disconcerted, she hurried into the hall, to find the maid stolidly confronting a pair of what were not quite gentlemen.

"Madam is—" the maid was saying firmly, but one of the men spotted Minnie and brushed past the maid.

"Miss Rennie?" he inquired politely, and at her jerky nod bowed with surprising style for one dressed so plainly.

"We have come to escort you to Mrs. Simpson," he said. And, turning to the maid, "Be so kind as to fetch the lady's things, if you please."

The maid turned, wide-eyed, and Minnie nodded to her. Her arms prickled with gooseflesh and her face felt numb.

"Yes," she said. "If you please." And her fingers closed on the paper in her pocket, damp with handling.

Do you think this is wise?

THERE WAS A coach outside, waiting. Neither of the men spoke, but one opened the door for her; the other took her by the elbow and helped her politely up into the conveyance. Her heart was pounding and her head full of her father's warnings about dealing with unvouched-for strangers— these warnings accompanied by a number of vividly detailed accounts of things that had happened to incautious persons of his own acquaintance as a result of unwariness.

What if these men had nothing to do with her mother but knew who her father was? There *were* people who—

With phrases like *"And they only found her* head . . .*"* echoing in her mind, it was several moments before she could take notice of the two gentlemen, both of whom had entered the coach behind her and were now sitting on the squabs opposite, watching her like a pair of owls. Hungry owls.

She took a deep breath and pressed a hand to her middle, as though to ease her stays. Yes, the small dagger was still reassuringly tucked inside her placket; the way she was sweating, it would be quite rusted by the time she had to use it. *If,* she corrected herself. *If* she had to use it . . .

"Are you all right, madam?" one of the men asked, leaning forward. His voice cracked sharply on "madam," and she actually looked at him

properly for the first time. Sure enough, he was a beardless boy. Taller than his companion, and pretty well grown, but a lad nonetheless—and his guileless face showed nothing but concern.

"Yes," she said, and, swallowing, pulled a small fan from her sleeve and snapped it open. "Just . . . a little warm."

The older man—in his forties, slender and dark, with a cocked hat balanced on his knee—at once reached into his pocket and produced a flask: a lovely object made in chased silver, adorned with a sizable chrysoberyl, she saw with surprise.

"Try this," he said in a pleasant voice. "It is orange-flower water, with sugar, herbs, the juice of blood oranges, and just a touch of gin, for refreshment."

"Thank you." She repressed the *"drugged and raped"* murmuring in her brain and accepted the flask. She passed it unobtrusively under her nose, but there was no telltale scent of laudanum. In fact, it smelled divine and tasted even better.

Both of the men saw the expression on her face and smiled. Not with the smile of satisfied entrapment, but with genuine pleasure that she enjoyed their offering. She took a deep breath, another sip, and began to relax. She smiled back at them. On the other hand . . . her mother's address lay in Parson's Green, and she had just noticed that they were heading steadily in the opposite direction. Or at least she thought so . . .

"Where are we going?" she asked politely. They looked surprised, looked at each other, eyebrows raised, then back at her.

"Why . . . to see Mrs. Simpson," the older gentleman said. The boy nodded and bowed awkwardly to her.

"Mrs. Simpson," he murmured, blushing.

And that was all anyone said for the remainder of the journey. She occupied herself with sipping the refreshing orange drink and with surreptitious observation of her . . . not captors, presumably. Escorts?

The gentleman who had given her the flask spoke excellent English, but with a touch of foreign sibilance: Italian, perhaps, or Spanish?

The younger man—he didn't really seem a boy, in spite of smooth cheeks and cracking voice—had a strong face and, regardless of his blushing, an air of confidence about him. He was fair and yellow-eyed, yet that brief glimpse when the two had looked at her in question had shown her a faint, vanishing resemblance between the two of them. Father and son? Perhaps so.

She flipped quickly through the ledger she carried in her head, in search of any such pair among her father's clients—or enemies—but found no one who met the description of her escorts. She took a deep breath, another sip, and resolved to think of nothing until they arrived at their destination.

Half an hour later, the flask was nearly empty and the coach lurched to a stop in what she thought was possibly Southwark.

Their destination was a small inn standing in a street of shops dominated by Kettrick's Eel-Pye House, this being evidently a successful eating place, judging by the crowds of people and the strong scent of jellied eels. Her belly rumbled as she got down from the carriage, but the sound was lost in the noises of the street. The boy bowed and offered her his arm; she took it, and putting on her most blandly pleasant face, she went with him inside.

IT WAS SHADOWY inside, light coming through two narrow, curtained windows. She noticed the smell of the place—hyacinths, how odd—but nothing more. Everything was a blur; all she felt was the beating of her heart and the solidness of the boy's arm.

Then a hallway, then a door, and then . . .

A woman. Blue dress. Soft-brown hair looped up behind her ears. Eyes. Pale-green eyes. Not blue like her own.

Minnie stopped dead, not breathing. For the moment, she felt an odd disappointment; the woman looked nothing like the picture she had carried with her all her life. This one was tall and thin, almost lean, and while her face was arresting, it wasn't the face Minnie saw in her mirror.

"Minerva?" the woman said, in a voice little more than a whisper. She coughed, cleared her throat explosively, and, coming toward Minnie, said much louder, "Minerva? Is it truly you?"

"Well, yes," said Minnie, not sure quite what to do. *She must be; she knows my real name.* "That's my name. And you are . . . Mrs. . . . . Simpson?" Her own voice broke quite absurdly, the final syllable uttered like the squeak of a bat.

"Yes." The woman turned her head and gave the two who had brought her a brief nod. The boy vanished at once, but the older man touched the woman's shoulder gently and gave Minnie a smile before following suit, leaving Minnie and Mrs. Simpson frankly staring at each other.

Mrs. Simpson was dressed well but quietly. She pursed her lips, looked sidelong at Minnie, as though estimating the possibility that she might be armed, then sighed, her square shoulders slumping.

"I'm not your mother, child," she said quietly.

Quiet as they were, the words struck Minnie like fists, four solid blows in the pit of her stomach.

"Well, who the bloody hell *are* you?" she demanded, taking a step backward. Every cautionary word she'd ignored came flooding back in her father's voice.

"... *kidnapped* ... *sold to a brothel* ... *shipped off to the colonies* ... *murdered for sixpence* ..."

"I'm your aunt, my dear," Mrs. Simpson said. The nettle grasped, she had regained some of her starch. "Miriam Simpson. Your mother is my sister, Hélène."

"Hélène," Minnie repeated. The name struck a spark in her soul. She had that much, at least. *Hélène.* A Frenchwoman? She swallowed.

"Is she dead?" she asked, as steadily as she could. Mrs. Simpson pursed her lips again, unhappy, but shook her head.

"No," she said, with obvious reluctance. "She lives. But . . ."

Minnie wished she'd brought a pocket pistol instead of a knife. If she had, she'd fire a shot into the ceiling right this minute. Instead, she took a step forward, so that her eyes were no more than inches from the green ones that didn't look like hers.

"Take me to her. Right now," she said. "You can tell me the story on the way."

7

ANNUNCIATION

T HE COACH CROSSED OVER the cobbles of a bridge with a great clattering of hooves and wheels. The racket was as nothing to the noise inside Minnie's head.

"A nun," Minnie said, as they passed onto a dirt road and the noise decreased. She sounded as blank as she felt. "My mother . . . was a *nun?*"

Mrs. Simpson—her aunt, Aunt Simpson, Aunt Miriam . . . she must get used to thinking of her that way—took a deep breath and nodded. With that bit of news out of the way, she had regained some of her composure.

"Yes. A sister of the Order of Divine Mercy, in Paris. You know of them?"

Minnie shook her head. She had thought she was prepared to hear any-thing, but she hadn't been, by a long chalk.

"What—what do they look like?" It was the first thing to come into her head. "Black, gray, white . . . ?"

Mrs. Simpson relaxed a little, bracing her back against the blue cush-ions to counter the jolting of the coach.

"Their habit is white, with a gray veil. They are a contemplative order but not cloistered."

"What does that mean, contemplative?" Minnie burst out. "What are they *contemplating*? Not their vows of chastity, apparently."

Her aunt looked startled, but her mouth twitched a little.

"Apparently not," she said. "Their chief occupation is prayer. Contemplation of God's mercy and His divine nature."

The day was cool enough, but Minnie felt hot blood rise from her chest to her ears.

"I see. So she—my mother—had an encounter with the Holy Spirit during a particularly intense prayer, did she?" She'd meant it sarcastically, but perhaps . . . "Wait a moment. My father *is* my father, isn't he?"

Her aunt overlooked the gibe.

"You are the daughter of Raphael Wattiswade, I assure you," she said dryly, with a glance at Minnie's face.

One of the small knots of doubt in Minnie's chest loosened. The possibility of this all being a hoax—if nothing more sinister—receded. Very few people knew her father's real name. If this woman did, then perhaps . . .

She sat back, crossed her arms, and fixed Mrs. Simpson with a hard look.

"So. What happened? And where are we going?" she added belatedly.

"To your mother," her aunt said tersely. "As to what happened . . . it was a book."

"Of course it was." Minnie's confidence in the woman's story moved up another small notch. "What book?"

"A Book of Hours." Mrs. Simpson waved away an inquisitive wasp that had flown in through the window. "I said that the order's chief occupation is prayer. They have others. Some of the nuns are scribes; some are artists. Soeur Emmanuelle—that's the name Hélène took when she entered the convent—was both," she explained, seeing Minnie's momentary confusion. "The order produces very beautiful books—things of a religious nature, of course, Bibles, devotionals—and sells them in order to support the community."

"And my father learned about this?"

Her aunt shrugged. "It's no secret. The order's books are well known, as are their skills. I imagine Raphael had dealt with the convent before. He—"

"He's never dealt with them, so far as I know, or I would have heard of them."

"Do you think he would risk your finding out?" her aunt said bluntly. "Whatever his defects of character, I will say the man knows how to keep a secret. He severed all connection with the order, after . . ." Her mouth

pressed tight and she made a flicking gesture with one hand that had nothing to do with the wasp.

Minnie's teeth were clenched, but she managed to get a few words out. "Bloody tell me what happened!"

Her aunt looked at her searchingly, the frills on her cap trembling with the vibration of the coach, then shrugged.

"Bon," she said.

What had happened ("in brief," said Mrs. Simpson) was that Raphael Wattiswade had acquired a very rare Book of Hours, made more than a century before. It was beautiful but in poor condition. The cover could be repaired, its missing jewels replaced—but some of the illustrations had suffered badly from the effects of time and use.

"And so Raphael came to the abbess of the order—a woman he knew well, in the course of business—and asked whether one of their more talented scribes might be able to restore the illustrations. For a price, of course."

Normally the book would simply be taken away to the scriptorium for examination and work, but in this case, some pages had been completely obliterated. Raphael, however, had discovered several letters from the original owner, rhapsodizing to a friend about his new acquisition and giving detailed descriptions of the more important illustrations.

"And he couldn't just give the letters to the abbess?" Minnie asked skeptically. Not that she could think why her father would purposely set out to seduce a nun he'd never heard of or set eyes on . . .

Mrs. Simpson shook her head.

"I said the book was from a previous age? The letters were written in German, and a very archaic form of that barbarous language. No one in the order was able to translate it."

Given that and the fragile state of the book, Soeur Emmanuelle was allowed to travel to Raphael's workshop—"With a proper *chaperon,* to be sure," Mrs. Simpson added, with a fresh pressing of the lips.

"To be sure."

Her aunt gave a very Gallic shrug. "But things happen, don't they?"

"Evidently they *do*." She eyed Mrs. Simpson, who, she thought, seemed tolerably free with her father's Christian name.

"C'est vrai. And what happened, of course, was you."

There was no good response to that, and Minnie didn't try to find one.

"She was only nineteen," her aunt finally said, looking down at her clasped hands, and speaking in a voice so soft that Minnie hardly heard it over the rumble of the coach. And how old had her father been? she wondered. He was forty-five now . . . twenty-eight. Maybe twenty-seven, allowing for the length of a pregnancy.

"Bloody old enough to know better," Minnie muttered, but under her

breath. "I suppose she—my mother"—she forced herself to say the words, which now felt shocking in her mouth— "was obliged to leave the order? I mean, you can't be pregnant in a convent, surely."

"You might be surprised," her aunt observed cynically. "But in this case, you're right. They sent her away, to a sort of asylum in Rouen—a terrible place." A flush had begun to burn on Mrs. Simpson's high cheekbones. "I heard nothing of it until Raphael appeared at my door one night, very distraught, to tell me she was gone."

"What did you do?"

"We went and got her," her aunt said simply. "What else?"

"You said 'we.' Do you mean you and . . . my father?"

Her aunt blinked, shocked.

"No, of course not. My husband and myself." She breathed deep, clearly trying to calm herself. "It—she—it was most distressing."

Soeur Emmanuelle, torn from the community that had been her home since she entered the convent as a twelve-year-old novice, treated as an object of shame, having no knowledge or experience of pregnancy, without friend or family, and locked up in an establishment that sounded very like a prison, had been first hysterical, then had gradually withdrawn into a state of despair and finally of stone-like silence, sitting and staring all day at the blank wall, taking no notice even of food.

"She was skin and bones when I found her," Mrs. Simpson said, her voice shaking with remembered fury. "She didn't even know me!"

Soeur Emmanuelle had very gradually been brought back to a cognizance of the world—but not the world she had left.

"I don't know whether it was leaving her order—they were her family!—or the shock of being with child, but . . ." She shook her head, desolation draining the color from her face. "She lost her reason entirely. Took no notice of her state and believed herself to be back in the convent, going about her usual work."

They had humored her, given her a habit, provided her with paint and brushes, vellum and parchment, and she had shown some signs of being aware of her surroundings—would talk sometimes and knew her sister. But then the birth had come, inexorable.

"She had refused to think about it," Mrs. Simpson said, with a sigh. "But there you were . . . pink and slimy and loud." Soeur Emmanuelle, unable to cope with the situation, lost her tentative grip on sanity and reverted to her earlier state of blank detachment.

So I drove my own mother to insanity and destroyed her life. Her heart had risen into her throat, a hard, pulsing lump that hurt with each beat. Still, she had to speak.

"The shock, you said." She licked dry lips. "Was it just . . . me? I mean, was it rape, do you think?"

To Minnie's infinite relief, Mrs. Simpson looked aghast at the word.

"*Nom de Dieu!* No. No, certainly not." Her mouth twisted a little as she recovered from the brief shock. "Say what you will about Raphael, I'm sure he's never taken a woman who wasn't willing. Mind, he can make them willing in very short order."

Minnie didn't want to hear one word about willing women and her father.

"Where, exactly, are we going?" she asked in a firm voice. "Where *is* my mother?"

"In her own world, *ma chère*."

IT WAS A modest farm cottage, standing by itself at the edge of a broad, sunny field, though the house itself was sheltered by well-grown oaks and beeches. Perhaps a quarter mile farther on was a small village that boasted a surprisingly large stone church, with a tall spire.

"I wanted her to be close enough to hear the bells," Mrs. Simpson explained, nodding toward the distant church as their coach came to a halt outside the cottage. "They don't keep the hours of praise as a Catholic abbey would, of course, but she doesn't usually realize that, and the sound gives her comfort."

She looked at Minnie for a long moment, biting her lip, doubt plain in her eyes. Minnie touched her aunt's hand, as gently as she could, though the pulse beating in her ears nearly deafened her.

"I won't hurt her," she whispered in French. "I promise you."

The look of doubt didn't leave her aunt's eyes, but her face relaxed a little and she nodded to the groom outside, who opened the door and offered his arm to help her down.

An anchorite, her aunt had said; Sister Emmanuelle believed herself to be an anchorite. A hermitess, fixed in place, her only duty that of prayer. "She feels . . . secure, I think," Mrs. Simpson had said, though the creases in her brow showed the shadow of long worry. "Safe, you know?"

"Safe from the world?" Minnie had asked. Her aunt had given her a very direct look, and the creases in her brow grew deeper.

"Safe from everything," she had said. "And everyone."

And so Minerva now followed her aunt to the door, filled with a mixture of anxiety, astonishment, sorrow, and—unavoidably—hope.

She'd *heard* of anchorites, of course; they were mentioned frequently in religious histories—of saints, monasteries, persecutions, reformations—but at the moment the word conjured up only a ridiculous vision of St. Simeon Stylites, who had lived on top of a pillar for thirty years—and, when his niece was orphaned, generously set *her* up with her very own

pillar, next to his. After a few years of this life, the niece had reportedly climbed down and decamped with a man, much to the disapproval of the history's author.

The door of the cottage opened, revealing a large, cheerful-looking woman who greeted Miriam Simpson warmly and looked with pleasant inquiry at Minnie.

"This is Miss Rennie," Mrs. Simpson said, gesturing toward Minnie. "I've brought her to see my sister, Mrs. Budger."

Mrs. Budger's sparse gray brows rose toward her cap, but she made a brief bob in Minnie's direction.

"Your servant, mum," she said, and flapped her apron at a large calico cat. "Shoo, cat. The lady's none o' your business. He knows it's nearly time for Sister's tea," she explained. "Come in, ladies, the kettle's a-boiling already."

Minnie was in a fever of impatience, this interrupted by stabs of icy terror.

"Soeur Emmanuelle, she still calls herself," Mrs. Simpson had explained on the way. "She spends her days—and often her nights"—her wide brow had creased at the words—"in prayer, but she does have visitors. People who've heard of her, who come to ask her prayers for one thing or another.

"At first, I was afraid," she'd said, and turned to look out the coach window at a passing farm wagon, "that they'd upset her, telling her their troubles. But she seems . . . better when she's listened to someone."

"Does she . . . talk to them?" Minnie had asked. Her aunt had glanced at her, then paused for a few seconds too long before saying, "Sometimes," and turning toward the window again.

It doesn't matter, she told herself, clenching her fists in the folds of her skirt to avoid strangling Mrs. Budger, who was slowly, slowly puttering around the hearth, assembling a few slices of buttered bread, a wedge of cheese, and a mug on a tray, at the same time fetching down a chipped teapot and three more stone mugs, a dented tin tea caddy, and a small, sticky blue pot of honey. *It doesn't matter if she won't speak to me. It doesn't even matter if she can't hear me. I just want to see her!*

8

THE BOOK OF HOURS

IT WAS A TINY stone building with a thatch; Minnie thought it must once have been a lambing shed or something of the kind. The thought made her inhale, nostrils flaring—and she blinked in surprise. There was certainly a smell, but it wasn't the warm agricultural fug of animals; it was the faint tang of incense.

Mrs. Simpson glanced up at the sun, halfway down the sky.

"You won't have long," she said, grunting a little as she lifted the heavy bar from the door. "It's almost time for None—what she thinks is None. When she hears the bells, she won't do anything until the prayer is done, and often she's silent afterward."

"None?"

"The hours," Mrs. Simpson said, pushing the door open. "Hurry, if you want her to speak with you."

Minnie was bewildered, but she did certainly want her mother to speak with her. She nodded briefly and ducked under the lintel into a sort of glowing gloom.

The glow came from a single large candle set in a tall iron stand and from a brazier on the floor next to it. Fragrant smoke rose from both, drifting near the sooty beams of the low ceiling. A dim light suffused the room, seeming to gather around the figure of a woman dressed in white robes, kneeling at a crude prie-dieu.

The woman turned, startled at the sound of Minnie's entry, and froze at sight of her.

Minnie felt much the same but forced herself to walk forward, slowly. Instinctively, she held out a hand, like one does to a strange dog, presenting her knuckles to be sniffed.

The woman rose with a slow rustling of coarse cloth. She wasn't veiled, which surprised Minnie—her hair had been roughly cropped but had grown out somewhat; it curved just under her ears, cupping the angles of her jaw. Thick, smooth, the color of wheat in a summer field.

Mine, Minnie thought, with a thump of the heart, and stared into the woman's eyes. Mrs. Simpson had been right. *Mine, too . . .*

"Sister?" she said tentatively, in French. "Soeur Emmanuelle?"

The woman said nothing, but her eyes had gone quite round. They

traveled down Minnie's body and returned to her face, intent. She turned her head and addressed a crucifix that hung on the plastered wall behind her.

"*Est-ce une vision, Seigneur?*" she said, in the rusty voice of one who seldom speaks aloud. "Is this a vision, Lord?" She sounded uncertain, perhaps frightened. Minnie didn't hear a reply from Christ on the cross, but Sister Emmanuelle apparently did. She turned back to face Minnie, drawing herself upright, and crossed herself.

"Erm . . . *Comment ca va?*" Minnie asked, for lack of anything better. Sister Emmanuelle blinked but didn't reply. Perhaps that wasn't the right sort of thing for a vision to say.

"I hope I see you well," Minnie added politely.

Mother, she thought suddenly, with a pang as she saw the grubby hem of the rough habit, the streaks of food on breast and skirt. *Oh, Mother . . .*

There was a book on the prie-dieu. Swallowing the lump in her throat, she walked past her mother to look at it but glanced up and saw the crucifix—it was a rich one, she saw, polished ebony with mother-of-pearl edging. The corpus had been made by another, truer hand, though—the body of Christ glowed in the candlelight, contorted in the grip of a knotted chunk of some dark wood, rubbed smooth. His face was turned away, invisible, but the thorns were carved sharp and vivid, sharp enough to prick your finger if you touched them. The outflung arms were only half freed from the wood, but the sense of entrapment, of unendurable agony, struck Minnie like a blow to the chest.

"*Mon Dieu,*" she said aloud. She said it in shock, rather than by way of prayer, but vaguely heard the woman behind her let go a held breath. She heard the rustle of cloth and straw—she hadn't noticed when she came in, but the floor was covered in clean straw—and forced herself to stand quite still, heart beating in her ears, though she longed to turn and embrace Sister Emmanuelle, seize and carry her, drag her, bring her out into the world. After a long moment during which she could hear the woman's breathing, she felt a touch on her shoulder. She turned round slowly.

Her mother was close now, close enough that Minnie could smell her. Surprisingly, she smelled sweet—a tang of sweat, the smell of clothes worn too long without washing, but incense perfumed her hair, the cloth of her robe, and the hand that touched Minnie's cheek. Her flesh smelled warm and . . . pure.

"Are you an angel?" Emmanuelle asked suddenly. Doubt and fear had come into her face again, and she edged back a step. "Or a demon?"

So close, Minnie could see the lines in her face—crow's-feet, the gentle crease from nose to mouth—but the face itself was a blurred mirror of the one she saw in her looking glass. She took a breath and stepped closer.

"I'm an angel," she said firmly. She'd spoken in English, without thinking, and Emmanuelle's eyes flew wide in shock. She took an awkward step backward and sank to her knees.

"Oh, no! Don't do that!" Minnie cried, distressed. "I didn't mean it—I mean, *Je ne veux pas . . .* " She stooped to raise her mother to her feet, but Emmanuelle had clapped her hands to her eyes and wouldn't be moved, only swaying to and fro, making small whimpering noises.

Then Minnie realized that they weren't just noises. Her mother was whispering, "RaphaelRaphaelRaphael," over and over. Panicked, she seized her mother's wrists and pulled her hands away from her face.

"Stop! *Arrêtez!* Please stop!"

Her mother stopped, gasping for breath, looking up at her. *"Est-ce qu'il vous a envoyé? Raphael L'Archange? Êtes-vous l'un des siens?"* "Did he send you? Raphael the archangel? Are you his?" Her voice quivered, but she had calmed a little; she wasn't struggling, and Minnie cautiously let go.

"No, no one sent me," she said, as soothingly as possible. "I came on my own, to visit you." Groping for something else to say, she blurted, *"Je m'appelle Minerve."*

Emmanuelle's face went quite blank.

What is it? Does she know that name? Mrs. Simpson hadn't said whether her mother might know her name.

And then she realized that the bells of the distant church were ringing. Perhaps her mother hadn't even heard her speak.

Helpless, she watched as Emmanuelle got laboriously to her feet, stepping on the hem of her robe and staggering. Minnie made to take the woman's arm, but Emmanuelle regained her balance and went to the prie-dieu, quickly but with no sense of panic. Her face was composed, all her attention focused on the book lying on the prie-dieu.

Seeing it now, Minnie realized at last what her aunt had meant by "None" and "hours." The book was a small, elegant volume with an aged green cover, set with tiny rounded cabochon jewels. And as Emmanuelle opened it, Minnie saw inside the glow of beautiful paintings, pictures of angels speaking to the Virgin, to a man with a crown, to a crowd of people, to Christ on the cross . . .

It was a Book of Hours, a devotional volume meant for rich lay people, made during the last age, with the psalms and prayers intended to be said during the monastic hours of worship: Matins, Lauds, Prime, Terce, Sext, None, Vespers, and Compline. None was the ninth hour—the prayer said at three o'clock in the afternoon.

Emmanuelle's head was bent over the open book, and she was praying aloud, her voice soft but audible. Minnie hesitated, not sure whether she should leave . . . but no. She wasn't ready to say goodbye to her mother—

the more so as the woman probably wouldn't notice her trying to take her leave. Instead, she came quietly to the prie-dieu and knelt down beside Emmanuelle, in the straw.

She knelt close enough that the pink linen of her gown nearly brushed the white habit. It wasn't cold in the shed, not with the brazier going, but nonetheless she could feel her mother's warmth and, for just a moment, surrendered to the vain hope she had brought here—of being seen, accepted, wrapped in her mother's love.

She closed her eyes against the starting tears and listened to Emmanuelle's voice, soft and husky but sure. Minnie swallowed and opened her eyes, making an effort to follow the Latin.

"Deus, in adjutorium meum intende; Domine, ad adjuvandum me festina . . ." "O God, come to my assistance; O Lord, make haste to help me . . ."

As the recitation of the Office of None went on, Minnie joined timidly in the prayers she could read adequately. Her mother took no notice, but Emmanuelle's voice got stronger, her back straighter, as though she felt the support of her imagined community around her.

Minnie could see that the book was very old—at least a hundred years, maybe more—and then realized with a small shock that she had seen it before. Her father had sold it—or one very like it—to Mother Hildegarde, the abbess of le Couvent des Anges, a hospital order of nuns. Minnie had delivered it to the good mother herself a year or so before. How had it come here?

In spite of the rawness of her own emotions, she found a small sense of peace in the words, even when she didn't understand them all. Emmanuelle seemed to grow both quieter and stronger as she spoke, and when she had finished, she stayed motionless, gazing up at the crucifix, an expression of the greatest tenderness on her face.

Minnie was afraid to rise, not wanting to disturb the sense of peace in the room, but her knees couldn't stand much more kneeling on the stones of the floor, straw covering or no. She took a deep breath and eased herself up. The nun seemed not to have noticed, still deep in communion with Jesus.

Minnie tiptoed toward the door, which, she saw, was now open a crack. She could see a movement of something blue through the gap—undoubtedly Mrs. Simpson, come to remove her. She turned suddenly, on impulse, and went back quickly to the prie-dieu.

"Soeur Emmanuelle?" she said very softly, and gently, slowly, laid her hands on her mother's shoulders, fragile under the white cloth. She swallowed hard, so her voice wouldn't shake. "You are forgiven."

Then she lifted her hands and went quickly away, the glow of the straw a blur of light around her.

WELL PAST MIDNIGHT

I T WAS TIME.
Argus House had fourteen bedrooms, not counting the servants' quarters. So far, Hal had not been able to bring himself to sleep in any of them. Not his own. He hadn't lain there since the dawn when he'd risen from Esmé's warm body and gone out in the rain to face Nathaniel.

"On your bloody croquet lawn!" he said aloud, but under his breath. It was after midnight, and he didn't want to wake any inquisitive servant. "You pretentious nit!"

Not Esmé's chaste blue and white boudoir next door, either. He couldn't bring himself even to open the door, not sure whether her ghost might still linger in the scented air or whether the room would be a cold and empty shell. Afraid to find out, either way.

He was standing now at the head of the stairs, the long corridor of bedrooms lit at this late hour by only three of the dozen sconces, the colors of a half dozen Turkey rugs melting into shadow. He shook his head and, turning, went downstairs.

He generally didn't sleep at night, anyway. Went out occasionally and roamed the dark paths of Hyde Park, sometimes stopping briefly to share a fire with some of the vagrants who camped there. More often sat up reading in the library 'til the wax from melting candles pattered onto tables and floors and Nasonby or Wetters came silently in with scrapers and new candles, even though he'd ordered the footmen to go to bed.

Then he'd read stubbornly on by the new light—Tacitus, Marcus Aurelius, Cicero, Pliny, Julius Caesar—losing himself in distant battles and the thoughts of long-dead men. Their fellowship comforted him, and he'd fall asleep with the dawn, curled up on the blue settee or sprawled on the cool marble floor, his head cushioned on the white hearth rug.

Someone would come silently and cover him. He'd usually wake to find someone standing over him with a luncheon tray and would rise with aching limbs and a foggy mind that took 'til teatime to clear again.

"This will *not* do," he said aloud, pausing at the door to the library. Not tonight.

He didn't go into the library, though it was brightly lit in expectation of his presence. Instead, he reached into the bosom of his shirt and pulled out the note. He'd been carrying it ever since it had arrived at teatime,

reading it again and again—and now opened it to read again, as though the words might have altered or disappeared.

His Royal Highness the Prince of Wales is pleased to invite you to visit him to discuss your proposals regarding the re-commissioning of the 46th Regiment of Foot, a project which is of the deepest interest to him. It would perhaps be most convenient for you to attend the princess's garden fête at the White House on Sunday, 21 June. A formal invitation will be sent you this week; should the arrangement be agreeable to you, please reply in the usual fashion.

"Agreeable," he said aloud, and felt an unaccustomed tingle of excitement, as he had every time he'd read the note. "Agreeable, he says!"

Agreeable, indeed—if dangerous. The prince had a good deal of power, a good deal of influence in military circles, including with the secretary at war. But he wasn't the king. And king and prince most assuredly did *not* agree. The king and his heir had been estranged—if not actually at loggerheads—for several years, and to court the favor of the one was to invite coldness from the other.

Still . . . it might be possible to tread the narrow line between the two and emerge with the support of both. . . .

But he knew he was in no sort of shape to undertake that kind of finesse, exhausted as he was in mind and body.

Besides. It was time. He knew it. He cast a brief, regretful look into the library, then reached out and gently closed the door on his book-lined refuge.

The house was quiet, and his footsteps made no sound on the thick rugs as he came back—at last—to Esmé's room. He opened the door without hesitation and went in.

There was no light and he left the door open behind him, crossing the room to draw back the drapes from the big double window. A pale wash of moonlight fell over him and he went back and closed the door, silently. Then slid the bolt.

The room was cold, and clean. A faint smell of beeswax polish and fresh linen lingered. No trace of her perfume.

He made his way half blind to the dressing table in her closet and felt about in the darkness until he found the chunky crystal bottle. He felt the smooth ground-glass stopper grate softly as he took it out and dabbed a drop of her scent inside his wrist—just as he'd seen her do it, a hundred times and more.

It was a scent made just for her, and for the instant she lived again within it: complex and heady, spicy and bitter—cinnamon and myrrh, green oranges and oil of carnations. Leaving the bottle open, he walked

back into the bedroom and came slowly to the curtained white bed. Put back the drapes and sat down.

Everything in her chamber was white or blue; the room was filled with shadow. Even the Bible on her nightstand was covered in white leather. Only the glints of gold or silver in jewel box and candlestick caught the light of the moon.

Without the hiss and crackle of a fire or the melting of candles, the air lay quiet. He could hear his own heart, beating slow and heavy. There was only him. And her.

"Em," he said softly, eyes closed. "I'm sorry." And whispered, so low he barely heard the words, "I miss you. God, I miss you."

Finally, finally, he let grief take him and wept for her then, for a long time.

"Forgive me," he said.

And at last lay down upon her white bed and let sleep take him, too, to whatever dreams it would.

10

DOWN TO BUSINESS

OVER THE NEXT TWO weeks, Minnie threw herself determinedly into the pursuit of business. She tried not to think of Soeur Emmanuelle, but thought of her mother hovered near her, like an angel on her shoulder, and after a bit she accepted this. There was, after all, nothing she could do about it, and at least now she knew her mother was alive. Perhaps even content.

Between increasing business—of both kinds—and Lady Buford's determined social agenda, Minnie scarcely had a minute to herself. When she wasn't going to view a collection of moldy hymnals in a garret down by the Thames or accepting sealed documents from her father's mysterious client in the Vauxhall Gardens, she was dressing for a card party in Fulham. The O'Higginses, faithful Irish wolfhounds, either accompanied or trailed her to every destination, their visibility depending on her errand.

She was pleased, therefore, to be able to combine Colonel Quarry's commission with Lady Buford's husband-hunting. Rather to Minnie's surprise, the latter involved a great deal of socializing with females.

"To be desirable, it is necessary to be talked about, my dear," Lady Buford told her over a glass of iced negus at Largier's tea shop (Madame Largier was French and thought tea itself a distinctly second-class beverage). "But you must be talked about *in the right way*. You must not suggest any hint of scandal, and—just as important—you must not cause jealousy. Be sweet and unassuming, always admire your companions' frocks and dismiss your own, and do not bat your eyes at their sons or brothers, should such be present."

"I've never batted my eyes at anyone in my life!" Minnie said indignantly.

"It isn't a difficult technique to master," Lady Buford said dryly. "But I trust you take my point."

Minnie did, and as she had no intention whatever of attracting a potential husband, she was extremely popular with the young women of society. Which turned out to be an unexpectedly good thing, because most young women had no discretion whatever, very little judgment, and would tell you the most unspeakable things without batting a single one of their own eyes.

They hadn't the least hesitation in telling her all about Esmé Grey; the late Countess Melton was a prime subject of gossip. But it wasn't the sort of gossip Minnie had expected to hear.

After a week's gentle prodding, Minnie had formed the distinct impression that women in general had not really liked Esmé—most of them had been afraid of her or envious of her—but that most men very definitely *had* liked her; hence, the envy. That being so, the lack of any hint of scandal was surprising.

There was quite a bit of public sympathy for Esmé; she *was* dead, and the poor little baby, too. . . . It was a tragic story, and people did love tragedy, as long as it wasn't theirs.

And certainly there was a good deal of talk (in lowered tones) about Lord Melton having shot poor Mr. Twelvetrees, which threw the countess into such a state of shock that she had gone into labor too early and died—but, surprisingly, there was no indication that Esmé's affair with Nathaniel had been noticed.

There was a great deal of speculation as to Lord Melton's motive for assassinating Mr. Twelvetrees—but apparently Esmé had been more than discreet, and there was no talk at all about Mr. Twelvetrees having paid her attention or even having been seen alone with her on any occasion.

There *was* a whisper of gossip to the effect that Lord Melton had killed Nathaniel because of an intrigue over an Italian singer, but the general opinion was that it had been over a matter of business; Nathaniel had been a failed curate who then became a stockbroker ("though he wrote the most *divine* poetry, my dear!"), and there had been a rumor of consider-

able losses incurred by the Grey family, attributed to Nathaniel's incompetence.

But as she continued to poke and prod, Minnie discovered an increasing sentiment along the lines that Colonel Quarry had mentioned: people were beginning to whisper that Lord Melton had killed Nathaniel in a fit of madness. After all, the duke ("though I'm told we mustn't call him by his title; he won't have it—and if *that's* not proof of madness . . .") had appeared nowhere in public since the death of his wife.

Given that the countess's death had occurred only two months earlier, Minnie thought this reticence perhaps reasonable, even admirable.

But as Lady Buford had been present during one of these exchanges, Minnie took the opportunity in the carriage going home to ask her chaperone's opinion of the Duke—or not—of Pardloe's marriage.

Lady Buford pursed her lips and tapped her closed fan against them in a considering manner.

"Well, there was a great deal of scandal over the first duke's death—had you heard about that?"

Minnie shook her head, in hopes of hearing more than her father's précis had provided, but Lady Buford was one who could tell the difference between facts and gossip, and her account of the first duke's supposed Jacobite associations was even briefer than Minnie's father's had been.

"It was quixotic at best—do you know that word, my dear?"

"I do, yes. You're speaking of—the second duke, would he be—Harold? He repudiated his title, is that what you mean?"

Lady Buford gave a small sniff and put away her fan in her capacious sleeve.

"It's actually not possible to repudiate a title, unless the king should give one leave to do so. But he did decline to use it, which amused some people, disgusted others who thought it affectation, and quite shocked society in general. Still . . . he'd been married a year before the first duke died, so Esmé had wed him with the expectation that he'd eventually succeed to the title. She hadn't given any indication that she regretted his decision—or even that she'd noticed it. That girl knew the meaning of 'aloof,'" Lady Buford added with approval.

"Were they in love, do you think?" Minnie asked, with genuine interest.

"Yes, I do," Lady Buford said, without hesitation. "She was French, of course, and quite striking—exotic, you might say. And Harold Grey is certainly an odd—well, I shouldn't say that, perhaps I merely mean unusual—young man. Their peculiarities seemed to complement each other. And neither one of them gave a single thought to what anyone else said or thought about *them*."

Lady Buford's sharp eyes had softened slightly, looking into memory,

and she shook her head, making the ring-necked dove on her hat bob precariously.

"It really was a tragedy," she said, with evident regret.

And that, despite further discreet inquiry, was apparently that.

SHE MET COLONEL Quarry, by arrangement, at a concert of sacred music in St. Martin-in-the-Fields. There were enough people there that it was possible to sit inconspicuously in one of the galleries; she could see the back of Quarry's head at the far end of the gallery, bent in apparent rapt attention to the music being performed below.

She normally enjoyed music of any kind, but as the vibration of the organ's pipes ceased rumbling through the boards underfoot and a single high, pure voice rose from the silence in a *Magnificat,* she felt a sudden sense of acute sorrow, seeing in memory a room of shadows and candlelight, the dirty hem of a white habit, a bent head and a slender neck beneath a bell of hair as golden as clean straw.

Her throat was tight and she bent her own head, shielding her face from view with a spread fan; it was a warm day, and whenever the music paused, the air in the gallery whispered with the movement of fans. No one noticed.

At last it was over, and she stood with the others, lingering by the railing as people filed out in a buzz of conversation that rose above the last strains of the recessional.

Quarry came strolling toward her—exaggeratedly casual, but, after all, he likely wasn't used to intrigue, and if someone *did* notice, "intrigue" (in the vulgar meaning of the term) was exactly what they'd think it was.

"Miss Rennie!" he said, as though surprised by her presence, and swept her a bow. "Your most obedient servant, ma'am!"

"Why, Colonel Quarry!" she said, fluttering her fan coquettishly. "*What* a surprise! I'd no notion that you enjoyed sacred music."

"Can't stand it," he said amiably. "I'd have gone mad in another minute if they hadn't stopped that caterwauling. What the devil have you found out?"

She told him without preamble what her researches had discovered—or, rather, had *not* discovered.

"Damn," he said, then hunched his shoulders as two women going past gave him a shocked look.

"I mean," he said, lowering his voice, "my friend is quite certain that it actually happened. The, um . . ."

"Affair. Yes, you said he had letters proving it but that he wouldn't let

anyone read them. Reasonably enough." She wasn't sure why she was interested in this business, but there was something oddly fascinating about it. She ought just to give him a bill for her time and leave it at that, but . . .

"Do you know where he keeps these letters?" she asked.

"Why . . . I suppose they're in his father's library desk. He usually keeps correspondence there. Wh—" He stopped abruptly, looking hard at her. She shrugged a little.

"I told you what the talk is, about your friend's state of mind. And if the letters are the only proof that he had a reason—and an honorable one—for what he did . . ." She paused delicately. Quarry's face darkened, and she felt the shift in his body as his hands curled.

"Are you suggesting that—that I take the—I could never do such a thing! It's dishonorable, impossible! He's my friend, dammit!" He looked aside, swallowing, and unclenched his fists.

"For God's sake, if he found that I'd done such a thing, I . . . I think he'd—" He stopped, all too clearly envisioning the possible results of such a discovery. The blood was draining from his cheeks, and a wash of pale-blue light from a stained-glass window made him look suddenly corpse-like.

"I wasn't suggesting that, sir," Minnie said, as meekly as possible. "Not at all! Naturally a gentleman such as yourself, and a devoted friend, couldn't—wouldn't—*ever* do such a thing." *And if you did,* she thought, watching his face, *he'd know it the second he looked at you. You couldn't lie your way out of a children's tea party, poor sod.*

"But," she said, and glanced deliberately around, so that he could see they were now alone in the gallery, save for a group of women at the far side, leaning over the rail and waving to acquaintances in the nave below. *"But,"* she repeated in a low voice, "if the letters were simply to . . . be delivered anonymously to . . . ?" She paused and cocked a brow.

He swallowed again, audibly, and looked at her for a long moment.

"The secretary at war," he blurted, as though trying to get the words out before he thought better.

"I see," she said, relaxing inwardly. "Well. That does seem very . . . drastic. Perhaps I can think of some other avenue of inquiry. There must be some intimate friend of the late countess that I haven't yet discovered." She put a hand very lightly on his arm.

"Leave the matter with me for another few days, Colonel. I'm sure something useful will occur to one of us."

11

GARDEN PARTY

1 June, AD 1744
Paris

My Dearest,

*Having heard nothing to the contrary, I assume that all is well
with you. I've received a special Request, through a Friend;
an English Collector by the Name of Mr. Bloomer wishes to discuss
a special Commission. His Letter, with Details of his Requirements,
a List of Resources with which to meet those Requirements, and a
Note of acceptable Payment, will follow under separate Covers.*

Your most affectionate Father,
R. Rennie

"MR. BLOOMER" HAD SPECIFIED His Royal Highness
the Prince of Wales's residence at Kew for their meeting, on the
twenty-first of June—Midsummer's Day. Minnie's diary carried a sketch
of various flowers and fruits to mark the occasion; the White House (as it
was casually known) had notable gardens, and a private tea (*Admission by
Invitation Only*) was being held in said gardens by Princess Augusta, in
support of one of that lady's favorite charities.

It was a little *outré* for an unmarried young woman to go alone to such
an event, Minnie reflected, dressing for the occasion, but Mr. Bloomer
had specified that the agent do just that, sending a single ticket of invitation with his letter. Of course, he probably hadn't realized that the agent
would *be* a young woman.

It was a fine day out, and Minnie stepped down from the hansom at the
end of the long avenue that led along the riverbank and up to the—quite
large, if not quite palatial—house.

"I'll walk from here," she said to Rafe O'Higgins, who had accompanied her. "You can watch 'til I get in to the house, if you think you really
must." A number of colored parasols, broad-brimmed hats, and belled silk
skirts were swaying slowly along the walks that edged a huge reflecting
pool in the distance, like a parade of animated flowers—*very appropriate
to a garden party,* she thought, amused.

"I'll be picking ye up just here, then," Rafe said, ignoring her gibe. He pointed to a carved-stone horse tank that stood in a small lay-by. "Just here," he repeated, and looked at the sun. "It's just gone two—will ye be done with your business by four, d'ye think?"

"I've no idea," she said, standing on tiptoe to look as far as she could over the sea of green surrounding the house. Ornamental domes and shiny bits that might be glass or metal were visible through the trees, and she heard faint strains of music in the distance. She meant to explore the delights of Their Highnesses' royal residence and its gardens to the full, once she'd dealt with Mr. Bloomer.

Rafe rolled his eyes but good-naturedly.

"Aye, then. If ye're not here at four, I'll come back on the hour 'til I find you." He leaned down to address her nose to nose, hazel eyes boring into hers. "And if ye're not here by seven, I'm comin' in after you. Got that, have ye, Lady Bedelia?"

"Oh, piffle," she said, but in a genial manner. She'd bought a modest parasol of ruffled green silk and now unfurled it with a flourish, turning her back on him. "I'll see you anon."

"And's when anon, then?" he shouted behind her.

"Whenever I'm bloody ready!" she called back over her shoulder and strolled on, gently twirling.

The crowd was funneling in to a large central hall, where Princess Augusta—or so Minnie assumed the pretty, bejeweled woman with the big blue eyes and the incipient double chin to be—was greeting her guests, supported by several other gorgeously dressed ladies. Minnie casually faded into the crowd and bypassed the receiving line; no need to call attention to herself.

There were enormous refreshment tables at the back of the house, and she graciously accepted a glass of sherbet and an iced cake offered her by a servant; she nibbled as she wandered out into the gardens, with an eye to its design and the locations of various landmarks. She was to meet Mr. Bloomer at three o'clock, in the "first of the glasshouses." Wearing green.

Green she was, from head to toe: a pale-green muslin gown, with a jacket and overskirt in a printed French calico. And, of course, the parasol, which she erected again once outside the house.

It was clever of Mr. Bloomer to choose green, she thought; she was very visible among the much more common pinks and blues and whites the other women wore, though not so uncommon as to cause staring. Green didn't suit many complexions, but beyond that, green fabric tended to fade badly: Monsieur Vernet—an artist friend of her father's, quite obsessed with whales—had told her once that green was a fugitive color, a notion that delighted her.

Perhaps that was why trees changed the color of their leaves in autumn? The green slipped away somehow, leaving them to fade into a brownish death. But why, then, did they have that momentary blaze of red and yellow?

Such concerns were far from the plants surrounding her; it was midsummer, and everything was so verdant that, far from being conspicuous, had she stopped moving in the midst of all this burgeoning flora, she would have been almost invisible.

She found the glasshouses without difficulty. There were five of them, all in a row, glittering like diamonds in the afternoon sun, each one linked to its fellow by a short covered passageway. She was a bit early, but that shouldn't matter. She furled the parasol and joined the people passing in.

Inside, the air was heavy and damp, luscious with the smell of ripening fruit and heady blossom. She'd seen the king's Orangerie at Versailles once; this was much less impressive but much more appealing. Oranges and lemons and limes, plums, peaches and apricots, pears . . . and the intoxicating scent of citrus blossom floating over everything.

She sighed happily and drifted down the graveled pathways that led among the rows, murmuring apology or acknowledgment as she brushed someone in passing, never meeting anyone's eyes, and, finding herself momentarily alone beneath a canopy of quince trees, stopped to breathe the perfume of the solid yellow fruits overhead, the size of cricket balls.

A flash of red caught her own eye through the trees, and for an instant she thought it was an exotic bird, lured by the astonishing abundance of peculiar fruits. Then she heard male voices above the well-bred hum of the largely female guests, and a moment later her red bird stepped out into the wide graveled patch where the pathways intersected. A soldier, in fulldress uniform—a blaze of scarlet and gold, with shining black boots to the knee and a sword at his belt.

He wasn't tall; in fact, he was rather slight, with a fine-boned face seen in profile as he turned to say something to his companion. He stood very straight, though, shoulders square and head up, and there was something about him that reminded her of a bantam cock—something deeply fierce, innately proud, and completely unaware of its relative size. Ready to take on all comers, spurs first.

The thought entertained her so much that it was a moment before she noticed his interlocutor. The companion wasn't dressed as a soldier but was certainly very fine, too, in ocher velvet with a blue satin sash and some large medallion pinned to his chest—the Order of Something-or-Other, she supposed. He did, however, strongly resemble a frog, wide-lipped and pale, with rather big, staring eyes.

The sight of the two of them, rooster and frog, engaged in convivial

conversation, made her smile behind her fan, and she didn't notice the gentleman who had come up behind her until he spoke.

"Are you fond of opuntioid cacti . . . madam?"

"I might be, if I knew what they were," she replied, swinging round to see a youngish gentleman in a plum-colored suit gazing at her intently. He cleared his throat and cocked an eyebrow.

"Um . . . actually, I prefer succulents," she said, giving the agreed-upon countersign. She cleared her throat, as well, hoping she remembered the word. "Particularly the, um, euphorbias."

The question in his eyes vanished, replaced by amusement. He looked her up and down in a manner that might in other circumstances have been insulting. She flushed but held his gaze and raised her brows.

"Mr. Bloomer, I presume?"

"If you like," he said, smiling, and offered her his arm. "Do let me show you the euphorbias, Miss . . . ?"

A moment of panic: who should she be, or admit to being?

"Houghton," she said, seizing Rafe's mocking nickname. "Lady Bedelia Houghton."

"Of course you are," he said, straight-faced. "Charmed to make your acquaintance, Lady Bedelia."

He bowed slightly, she took his arm, and together they walked slowly into the wilderness.

They passed through minor jungles of philodendrons—but philodendrons that had never graced anything so plebeian as a parlor, with ragged leaves each half as large as Minnie herself, and a thing with great veined leaves the color of green ink and the look of watered silk.

"They're rather poisonous, philodendrons," Mr. Bloomer said, with a casual nod. "All of them. Did you know?"

"I shall make a note of it."

And then trees—ficus, Mr. Bloomer informed her (perhaps he hadn't chosen his *nom de guerre* at random, after all), with twisted stems and thick leaves and a sweet, musty smell, some of them with vines climbing their trunks with convulsive force, sturdy root-like hairs clinging to the thin bark.

And then, sure enough: the bloody euphorbias, in person.

She hadn't known things like that existed. Many of them didn't even look like proper plants, and some that did were strange perversions of the plant kingdom, with thick bare stems studded with cruel thorns, things that resembled lettuce—but a ruffled white lettuce with dark-red edgings that made it look as though someone had used it to mop up blood—

"They're rather poisonous, too, the euphorbias, but it's more the sap. Won't kill you, but you don't want to get it in your eyes."

"I'm sure I don't." Minnie took a better grip on her parasol, ready to

unfurl it in case any of the plants should take it into mind to spit at her; several of them looked as though they'd like nothing better.

"They call that one 'crown of thorns,'" Mr. Bloomer said, nodding at one particularly horrid thing with long black spikes sticking out in all directions. "Apt." He noticed her expression at this point and smiled, tilting his head toward the next house. "Come along; you'll like the next collection better."

"Oh," she said, in a small voice. Then, *"Oh!"* much louder. The new glasshouse was much bigger than the others, with a high, vaulted roof that filled the air with sun and lit the thousand—at least!—orchids that sprang from tables and spilled from trees in cascades of white and gold and purple and red and . . .

"Oh, my." She sighed in bliss, and Mr. Bloomer laughed.

They weren't alone in their appreciation. All of the glasshouses were popular—there had been a fair number of people exclaiming at the spiny, the grotesque, and the poisonous—but the orchid house was packed with guests, and the air was filled with a hum of amazement and delight.

Minnie inhaled as much as she could, sniffing. The air was scented with a variety of fragrances, enough to make her head swim.

"You don't want to smell that one." Mr. Bloomer, guiding her from one delight to the next, put out a shielding hand toward a large pot of rather dull green orchids with thick petals. "Rotting meat."

She took a cautious sniff and recoiled.

"And why on earth would an orchid want to smell like rotting meat?" she demanded.

He gave her a slightly queer look but smiled.

"Flowers put on the color and scent they require to attract the insects that pollinate them. Our friend the *Satyrium* there"—he nodded at the green things—"depends upon the services of carrion flies. Come, this one smells of coconut—have you ever smelt a coconut?"

They took their time in the orchid house—they could hardly do otherwise, given the slow-moving crowd—and despite Minnie's regret at leaving the exotic loveliness, she was relieved to pass into the last glasshouse in the row and find it nearly deserted. It was also cool, by contrast with the tropical heat created by so many bodies, and she breathed deep. The scents in here were subtle and modest by contrast, the plants small and ordinary-seeming, and quite suddenly she realized Mr. Bloomer's strategy.

The orchid house served as a sieve or barrier. Here they were quite alone, though standing in the open, where they could easily see anyone coming in time to alter their conversation to innocuous chat.

"To business, then?" she said, and Mr. Bloomer smiled again.

"Just so. You first or me?"

"You." It would be an exchange rather than a sale, but her half of the

bargain was concrete, and his was not. "Tell it to me," she said, focusing her concentration on his face—rather narrow but not displeasing; she could see humor in the creases near his mouth.

"You're quite sure you can remember?" he said dubiously.

"Certainly."

He drew breath, gave a short nod of his own, and began to talk.

Once more she took his arm, and they paced the aisles of the glasshouse, walking through patches of sun and shadow, while he told her various bits of information. She memorized these, repeating them back to him, now and again asking for clarification or repetition.

Most of the information had to do with financial matters, banking and the Exchange, the movement of money—between persons and between countries. A few tidbits of political gossip, but not many.

That surprised her; the information he was dealing *for* was all political in nature, and quite specific. Mr. Bloomer was hunting Jacobites. Particularly in England and Paris.

I can't think why, her father had remarked in the margin of his list. *It's true, Charles Stuart has come to Paris, but that's common Knowledge, and besides, everyone knows he'll never get anywhere; the Man's an Idiot. Still, you don't make Money by refusing to sell People what they want. . . .*

She was relieved when Mr. Bloomer finished. It hadn't been a long nor yet a complicated account, and she was sure that she had all the names and the necessary numbers securely fixed in mind.

"All right," she said, and took her own list—sealed—from the secret pocket sewn inside her jacket. She handed it over, making sure to meet his eyes as she did so. Her heart was beating fast and her palm was slightly moist, but he didn't appear suspicious.

Not that there was really anything wrong with what she'd done—she wasn't cheating Mr. Bloomer. Not exactly. Everything on her list was just as her father had specified . . . save that when she'd written it out fair, she'd left out James Fraser's name and the bits of information regarding his movements and interactions with Charles Stuart and his followers. She felt rather possessive, not to say protective, of Mr. Fraser.

Mr. Bloomer wasn't a fool; he opened the document and read it through, at least twice. Then he folded it up and smiled at her.

"Thank you, my dear. A pleasure to—"

He stopped suddenly and drew back a little. She turned to see what had struck him and saw the soldier, the bantam cock, coming in from the passage that led from the orchid house. He was alone, but his scarlet and gold made him glow like a tropical parrot as he stepped through a patch of sun.

"Someone you know?" she asked, low-voiced. *And someone you don't want to meet, I daresay.*

"Yes," Mr. Bloomer replied, and retired into the shadows of a tree fern.

"Will you do me a service, my dear? Go engage His Grace there in conversation for a few moments, while I take my leave."

He nodded encouragingly toward the advancing soldier, and as she took a hesitant step in that direction, he blew her a kiss and stepped round behind the tree fern.

There wasn't time to think what to say.

"Good afternoon," she said, smiling and bowing to the officer. "Isn't it pleasant in here, after all that crush?"

"Crush?" he said, looking faintly puzzled, and then his eyes cleared, focusing on her for the first time, and she realized that he hadn't actually seen her until she spoke to him.

"In the orchid house," she said, nodding toward the doorway he'd just come through. "I thought perhaps you'd come in here as I did, for refuge from the Turkish bath."

He was in fact sweating visibly in his heavy uniform, a bead of perspiration rolling down his temple. He wore his own hair—dark, she saw, in spite of the remnants of rice powder clinging to it. He seemed to realize that he'd been socially remiss, for he made her a deep bow, hand to his heart.

"Your servant, ma'am. I beg your pardon; I was . . ." Straightening, he trailed off with a vague gesture at the plants around them. "It is cooler here, is it not?"

Mr. Bloomer was still visible, near the door leading to the orchid house. He'd stopped, to her surprise, and she was somewhat displeased to realize that he was listening to her conversation—insipid as it was. She narrowed her eyes at him; he saw, and one corner of his long mouth turned up.

She moved closer to the soldier and touched his arm. He stiffened slightly, but there was no sign of repulsion on his face—quite the opposite, which was reassuring—and she said chattily, "Do you know what any of these plants are? Beyond orchids and roses, I'm afraid I'm a complete ignoramus."

"I know . . . some of them," he said. He hesitated for a moment, then said, "I actually came in here to see a particular flower that His Highness recommended to me just now."

"Oh, indeed?" she said, impressed. Her recollections of the frog in the ocher coat were undergoing a rapid readjustment, and she felt slightly faint at the thought that she'd been that close to the Prince of Wales. "Er . . . which flower was that, do you mind telling me?"

"Not at all. Pray let me show it to you. If I can find it." He smiled quite unexpectedly, bowed again, and gave her his arm, which she took with a small thrill, turning her back on the distant Mr. Bloomer.

"Go engage His Grace . . ." That's what he'd said: "His Grace." It had been a long time since she'd lived in London, and she'd rarely had occa-

sion to use English titles, but she was almost sure that you said "Your Grace" only to a duke.

She stole a quick sideways look at him; he wasn't tall but had a good six inches on her. Young, though . . . She'd always thought of dukes (when she thought of them at all) as gouty old men with paunches and dewlaps. This one couldn't be more than five-and-twenty. He was slender, though he still radiated that rooster-like fierceness, and he had a very striking face, but there were deep shadows under his eyes, and his cheeks had lines and hollows that made him seem older than she thought he probably was.

She felt suddenly sorry for him, and her hand squeezed his arm, quite without her meaning to do it.

He glanced down at her, surprised, and she snatched her hand back, diving into her pocket for a handkerchief that she pressed to her lips, feigning a coughing fit.

"Are you all right, madam?" he asked, concerned. "Shall I fetch you—" He turned to look toward the door that led back through the line of glass-houses, then turned back, courteously straight-faced. "I fear that were I to go and fetch you an ice, you'd be dead long before I returned. Shall I thump you on the back instead?"

"You shall not," she managed to say, and giving one or two small, lady-like hacks, dabbed her lips with the handkerchief and tucked it away. "Thank you, anyway."

"Not at all." He bowed but didn't offer her his arm again, instead nodding her to precede him toward a low table filled with an assortment of beautiful chinoiserie. *One more amazement,* she thought, seeing the array of delicate blue and white and gilded porcelain. Any one of these delicately painted bowls would cost a fortune, and here they were, filled with *dirt,* and used to display quite unremarkable flowers.

"These?" she said, turning to look at His Grace—ought she to ask his name? Offer hers?

"Yes," he said, though his voice now seemed hesitant, and she saw him very briefly clench his fists before advancing to the edge of the table. "They were brought from China—very . . . very rare."

She glanced at him, surprised at the catch in his voice.

"What are they, do you know?"

"They have a Chinese name . . . I don't recall it. I know a botanist, a Swedish fellow . . . he calls them chrysanthemum. *Chrystos*—gold, that is—and *anth, anthemon.* Means . . . flower."

She saw his throat bob above the edge of his leather stock as he swallowed and noticed with alarm that he was very pale.

"Sir?" she said, reaching tentatively for his arm. "Are you quite—are you well?"

"Yes, of course," he said, but his breath was coming fast, and the sweat was trickling down his neck. "I'm . . . I'll be . . . quite all ri—" He stopped suddenly, gasping, and leaned heavily on the table. The pots shifted a little and two of them chimed together, a high-pitched ringing that set her teeth on edge and made her skin jump.

"Perhaps you'd best sit down," she said, seizing him by the elbow and trying to lead him back a step, lest he fall face-first into hundreds of pounds of priceless porcelain and rare flowers. He stumbled back and sank to his knees in the gravel, clutching her arms, a heavy weight. She looked wildly about for help, but there was no one in the glasshouse. Mr. Bloomer had disappeared.

"I—" He choked, coughed, coughed harder, gulped air. His lips were slightly blue, which scared her. His eyes were open, but she thought he couldn't see; he let go of her and fumbled blindly at the skirts of his coat. "Need—"

"What is it? Is it in your pocket?" She stooped, pushed his hand away, groped through the folds of fabric, and felt something hard. There was a small pocket in the tail of his coat, and she thought for an instant that she hadn't expected it to be quite this way the first time she touched a man's buttocks, but she found her way into the pocket and extracted a blue enameled snuffbox.

"Is this what you want?" she asked dubiously, holding it out. Snuff seemed the very last thing likely to be helpful to a man in his state, surely. . . .

He took it from her, hands shaking, and tried to open the box. She took it back and opened it for him, only to find a tiny corked vial inside. With no idea what to do—she glanced wildly toward the entrance again, but no help appeared—she took the vial in hand, pulled the cork, and gasped, recoiling as the stinging fumes of ammonia rushed out.

She held the vial to his nose, and he gasped in turn, sneezed—all over her hand—then grabbed her hand and held the vial closer, taking one heroic breath before he dropped it.

He sat down heavily in the gravel, hunched over, and wheezed and snorted and gulped, as she surreptitiously wiped her hand on her petticoat.

"Sir . . . I'm going to go and find someone to help," she said, and made to do so, but his hand had shot out and grasped the fabric of her skirt. He shook his head, speechless, but after a moment got enough breath to say, "No. Be . . . all . . . right now."

She doubted that very much. Still, being conspicuous was the last thing she wanted, and he did seem, if not exactly *better*, at least less in danger of dying on the spot.

She nodded uncertainly, though she didn't think he saw her, and after looking about helplessly for a moment sat down gingerly on the rim of a raised bed full of what looked like pincushions, varying from things that would have fit in the palm of her hand (had they not been equipped with quite so many thorns) to ones much larger than her head. Her stays felt tight, and she tried to slow her breathing.

As her alarm subsided, she became aware of the distant chatter in the orchid house, which had just become noticeably louder and higher-pitched.

"Fred . . . rick," said the hunched form at her feet.

"What?" She bent over to look at him. He was still a bad color and breathing noisily, but he *was* breathing.

"Prince . . ." He flipped a hand toward the distant noise.

"Oh." She thought he meant that the Prince of Wales had come in to view the orchids, this causing the rising tide of excitement next door. In that case, she thought, they were probably safe from interruption for the present—no one would abandon His Royal Highness in order to look at pincushions and Chinese . . . whatever-they-weres.

His Grace had closed his eyes and appeared to be concentrating on breathing, which she thought a good thing. Moved by the desire to do something other than stare at the poor man, she rose and went over to the Chinese bowls.

All her attention had been for the porcelain, to start with, but now she examined the bowls' contents. Chrysanthemum, that's what he'd said. Most of the flowers were smallish, little tufty ball-like blossoms in cream or gold, with long stems and dark-green leaves. One was a pretty rusty color, though, and another bowl held a profusion of small purple blossoms. Then she saw a larger version, snowy white, and realized what she was looking at.

"*Oh!*" she said, quite loud. She glanced guiltily over her shoulder, then put out a hand and touched the flower very gently. There it was: the curved, symmetrical petals, tightly layered but airy, as though the flower floated above its leaves. It—they—had a noticeable fragrance, so close to. Nothing like the voluptuous, fleshy scents of the orchids; this was a delicate, bitter perfume—but perfume, nonetheless.

"Oh," she said again, more quietly, and breathed it in. It was clean and fresh and made her think of cold wind and pure skies and high mountains.

"*Chu,*" said the man sitting in the gravel behind her.

"Bless you," she said absently. "Are you feeling better?"

"The flowers. They're called *chu*. In Chinese. I apologize."

That made her turn round. He'd made it up onto one knee but was swaying a bit, plainly gathering his strength to try to rise. She reached

down and gripped his hand as solidly as she could. His fingers were cold, but his grip was firm. He looked surprised but nodded and, with a wheezing gasp, staggered to his feet, releasing her hand as he did so.

"I apologize," he said again, and inclined his head an inch. More than that and he might have fallen again, she thought, bracing herself uneasily to catch him if he did. "For discommoding you, madam."

"Not at all," she said politely. His eyes were rather unfocused, and she could hear his breath creaking in his chest. "Er . . . what the devil just happened to you? If you don't mind my asking."

He shook his head, then stopped abruptly, eyes closed.

"I—nothing. I shouldn't have come in here. Knew better."

"You're going to fall down again, I think," she said, and took him by the hand once more, guiding him to the raised bed, where she made him sit and sat beside him.

"You *should* have stayed at home," she said reprovingly, "if you knew you were ill."

"I'm not ill." He ran a trembling hand over the sweat on his face, which he then wiped carelessly on the skirts of his coat. "I—I just . . ."

She sighed and glanced at the doorway, then behind her. No other way out, and the chatter in the orchid house was still going strong.

"You just *what?*" she said. "I'm not dragging it out of you one word at a time. Tell me what's the matter with you, or I'm going in there and fetching His Highness out to look after you."

He gave her an astonished look, then started to laugh. And to wheeze. He stopped, fist to his mouth, and panted a bit, catching more breath.

"If you must know . . ." he said, and gulped air, "my father shot himself in the conservatory at our house. Three years ago . . . today. I . . . saw him. His body. Among the glass, all the plants, the—the light—" He looked up at the panes overhead, blinding with sun, then down at the gravel, patterned with the same light, and closed his eyes briefly. "It . . . disturbed me. I wouldn't have come—" He paused to cough. "Pardon me. I wouldn't have come here today, save that His Highness invited me, and I needed very much to meet him." His eyes, bloodshot and watering, met hers directly. They were blue, pale blue.

"In the unlikely event that you haven't heard the story: my father was accused of treason; he shot himself the night before they planned to arrest him."

"That's *very* terrible," Minnie said, appalled. Terrible in a number of ways—not least in the realization that this must be the Duke of Pardloe, the one her father had in mind as a potential . . . source. She avoided even thinking the word "victim."

"It was. He was *not* a traitor, as it happens, but there you are. The fam-

ily was disgraced, naturally. His regiment—the one he had raised, had built himself—was disbanded. I mean to raise it again." He spoke with a simple matter-of-factness and paused to mop his face with his hand again.

"Haven't you got a handkerchief? Here, have mine." She squirmed on the rough stones, digging for her pocket.

"Thank you." He wiped his face more thoroughly, coughed once, and shook his head. "I need support—patronage from high quarters—in that endeavor, and a friend managed an introduction to His Highness, who was kind enough to listen to me. I think he'll help," he added, in a meditative sort of way. Then he glanced at her and smiled ruefully. "Wouldn't help my cause to be found writhing on the ground like a worm directly after speaking to him, though, would it?"

"No, I can see that." She considered for a moment, then ventured a cautious question. "The *sal volatile*—" She gestured at the vial, fallen to the ground a few feet away. "Do you often feel faint? Or did you just . . . think you might need it today?"

His lips pressed tight at that, but he answered.

"Not often." He pushed himself to his feet. "I'm quite all right now. I'm sorry to have interrupted your day. Would you . . ." He hesitated, looking toward the orchid house. "Would you like me to present you to His Highness? Or to Princess Augusta, if you like; I know her."

"Oh. No, no, that's quite all right," Minnie said hastily, getting up, too. Regardless of her own desires, which didn't involve coming to the notice of royalty, she could see that the very last thing *he* wanted to do was to go anywhere near people, disheveled, shaken, and wheezing as he was. Still, he was pulling himself together before her eyes, firmness straightening his body. He coughed once more and shook his head doggedly, trying to rid himself of it.

"Your friend," he said, with the decisive air of one changing the subject, "do you know him well?"

"My fr—oh, the, um, gentleman I was talking to earlier?" Apparently Mr. Bloomer hadn't been quite fast enough in his disappearing act. "He isn't a friend. I met him by the euphorbias"—she gestured airily, as though she and the euphorbias were quite good chums—"and he began telling me about the plants, so we walked on together. I don't even know his name."

That made him look sharply at her, but it was, after all, the truth, and her look of innocence was apparently convincing.

"I see," he said, and it was obvious that he saw a good deal more than Minnie did. He thought for a moment, then made up his mind.

"I do know him," he said carefully, and wiped a hand under his nose. "And while I would not presume to tell you how to choose your friends, I don't think he's a good man with whom to associate. Should you meet

him again, I mean." He stopped, considering, but that was all he had to say on the subject of Mr. Bloomer. Minnie would have liked to know Bloomer's real name but didn't feel she could ask.

There was a short, awkward silence, in which they stared at each other, half-smiling and trying to think what to say next.

"I—" Minnie began.

"You—" he began.

The smiles became genuine.

"What?" she asked.

"I was going to say that I think the prince has likely left the orchids to their own devices by now. You ought to go along, before anyone comes in. You don't want to be seen alone in my company," he added, rather stiffly.

"I don't?"

"No, you don't," he said, his voice softer, regretful but still firm. "Not if you have any desire to be accepted in society. I meant what I said about my father and the family. I mean to change that, but for now . . ." Reaching out, he took her hands and drew her toward him, turning so they faced the entrance to the orchids. He was right; the conversation there had subsided to the mildly threatening hum of bumblebees.

"Thank you," he said, still more softly. "You're very kind."

There was a smudge of rice powder on his cheek; she stood a-tiptoe and wiped it off, showing him the white on her thumb.

He smiled, took her hand again, and, to her surprise, kissed the tip of her thumb.

"Go," he said, his voice very low, and let go her hand. She drew a deep breath and curtsied.

"I—all right. I'm . . . very happy to have made your acquaintance, Your Grace."

His face changed like lightning, startling her terribly. Just as fast, he got it—whatever "it" was—under control and was once more the civil king's officer. For that split second, though, he'd been pure rooster, an enraged cock ready to throw himself at an enemy.

"Don't call me that. Please," he added, and bowed formally. "I have not taken my father's title."

"I—yes, I see," she said, still shaken.

"I doubt it," he said quietly. "Goodbye."

He turned his back on her, took a few steps toward the Chinese bowls and their mysterious flowers, and stood still, gazing down at them.

Minnie seized her fallen fan and parasol, and fled.

12

WERY WENGEFUL

Dear Miss Rennie,

*May I beg the Honour of an Appointment with you at your
earliest convenience? I wish to propose a Commission that I think
very well suited to your considerable Talents.*

*Your Most Humble Servant,
Edward Twelvetrees*

MINNIE FROWNED AT THE note. It was commendably
brief but odd. This Twelvetrees spoke of her "talents" in a most
familiar sort of way; clearly he knew what those talents were—and yet he
gave no introduction, supplied no reference from one of her existing clients or connections. It made her uneasy.

Still, there was no sense of threat in the note, and she *was* in business.
No harm in seeing him, she supposed. She'd be under no obligation to
accept his commission if it, or he, seemed fishy.

She hesitated over whether to allow him to come to her rooms—but,
after all, he had sent the note here; plainly he knew where she lived. She
wrote back, offering to see him next day at three o'clock but making a
mental note to tell one of the O'Higginses to come a bit early and hide in
the boudoir, just in case.

"OH," SHE SAID, opening the door. "So that's it. I thought there was
something a trifle odd about your note."

"If you feel yourself offended, Miss Rennie, I willingly apologize." Mr.
Bloomer—alias Edward Twelvetrees, evidently—stepped in, not waiting
for invitation, and obliging her to take a step back. "But I imagine a
woman of your undoubted sense and experience might be willing to overlook a bit of professional subterfuge?"

He smiled at her, and, despite herself, she smiled back.

"I might," she said. "A professional, are you?"

"It takes one to know one," he said, with a small bow. "Shall we sit
down?"

She shrugged slightly and gave Eliza a nod, indicating that she might bring in a tray of refreshments.

Mr. Twelvetrees accepted a cup of tea and an almond biscuit but left the latter lying on his saucer and the former steaming away unstirred.

"I shan't waste your time, Miss Rennie," he said. "When I left you in the princess's glasshouse, I abandoned you—rather cavalierly, I'm afraid—to the company of His Grace, the Duke of Pardloe. Given the scandal attached to his family, I assumed at the time that you knew who he was, but from your manner when I observed you speaking with him, I revised this opinion. Was I right in thinking that you did *not* know him?"

"I didn't," Minnie said, keeping her composure. "But it was quite all right. We exchanged a few pleasantries, and I left." *Just how long were you watching us?* she wondered.

"Ah." He'd been watching her face intently but at this broke off his inspection long enough to add cream and sugar to his tea and stir it. "Well, then. The commission for which I wish to engage your services has to do with this gentleman."

"Indeed," she said politely, and picked up her own cup.

"I wish you to abstract certain letters from the duke's possession and deliver them to me."

She nearly dropped the cup but tightened her hold just in time.

"What letters?" she asked sharply. Now she knew what it was about his note that had struck her oddly. *Twelvetrees.* That was the name of the Countess of Melton's lover: Nathaniel Twelvetrees. All too plainly, this Edward was some relation.

And she heard in memory Colonel Quarry's words when she'd asked if she might speak with Nathaniel: *"Afraid not, Miss Rennie. My friend shot him."*

"Correspondence between the late Countess Melton and my brother Nathaniel Twelvetrees."

She sipped her tea, feeling Edward's gaze as hot on her skin as the breath from her cup. She set the cup down carefully and looked up. His face had an expression she'd seen on the faces of hawks fixing on their prey. But it wasn't she who was the prey here.

"That might be possible," she said coolly, though her heart had sped up noticeably. "Forgive me, though—are you sure such correspondence exists?"

He uttered a short laugh, quite without humor.

"It *did* exist, I'm sure of that."

"I'm sure you are," she said politely. "But if the correspondence is of the nature I surmise you mean—I *have* heard certain speculations—would the duke not have burned any such letters, following the death of his wife?"

Mr. Twelvetrees lifted one shoulder and let it fall, his eyes still fixed on hers.

"He might have done," he said. "And your immediate task would of course be to discover whether that is the case. But I have reason to believe that the correspondence still exists—and if it does, I want it, Miss Rennie. And I'll pay for it. Handsomely."

WHEN THE DOOR closed behind Edward Twelvetrees, she stood frozen for a moment, until she heard the door of her boudoir open, across the hall.

"Well, that's a rum cove," Rafe O'Higgins observed, with a nod toward the closed front door. Eliza, who had come in to take away the tray, inclined her head in sober agreement.

"Wengeful," she said. "Wery wengeful, 'e is. But oo'd blame him?"

Oo, indeed? Minnie thought, and suppressed the urge to laugh. Not from humor so much as from nerves.

"Aye, mibbe," Rafe said. He went to the window and, lifting the edge of the blue velvet curtain, looked carefully down into the street, where Edward Twelvetrees was presumably vanishing into the distance. "I'd say your man's inclined toward vengeance, sure. But what d'ye think he'd be after doing with these letters, if there are any?"

There was a brief silence, as all three of them contemplated the possibilities.

"Put 'em on broadsheets and sell 'em at a ha'penny a go?" Eliza suggested. "Could make a bit o' money out o' that, I s'pose."

"Make a lot more out of the duke," Rafe said, shaking his head. "Blackmail, aye? If the letters are juicy enough, I daresay His Grace would pay through the nose to keep just that from happening."

"I imagine so," Minnie said absently, though the echoes of her conversation with Colonel Quarry drowned out further suggestions.

"*. . . he requires proof of the affair for a . . . a . . . legal reason, and he will not countenance the idea of letting anyone read his wife's letters, no matter that she is beyond the reach of public censure nor that the consequences to himself if the affair is* not *proved may be disastrous.*"

What if numerology was less penetrating an art than usual and Harry Quarry wasn't a bluff, transparent four, after all? What if his care for Lord Melton was a charade? Twelvetrees had just openly engaged her to be his cat's-paw; what if Quarry had the same end in mind but was playing a double game?

If so . . . were the two men playing the *same* game? And if so, were they

in it together or working in opposition, whether known to each other or not?

She brought Quarry to mind, reliving their conversations and analyzing them, word for word, watching the emotions play out in memory across his broad, crudely handsome face.

No. One of the chief tenets of her family credo was "Trust no one," but one did have to make judgments. And she was as sure as it was possible to be that Harry Quarry's motive was what he had said: to protect his friend. And after all . . . Harry Quarry not only was convinced of the letters' existence but had a good notion of their location. True, he hadn't asked her to steal the letters, not explicitly, but had certainly done everything but.

She had promised Edward Twelvetrees nothing beyond an attempt to find out whether the letters *did* exist; if so, she'd said, then they could discuss further terms.

Well, then. The next step, at least, was clear.

"Rafe," she said, interrupting an argument between Rafe and Eliza as to whether Mr. Twelvetrees more resembled a ferret or an obelisk (she assumed they meant "basilisk" but didn't stop to find out), "I have a job for you and Mick."

13

THE LETTERS

M R. VAUXHALL GARDENS (alias Mr. Hosmer Thornapple, a wealthy broker on the Exchange, as Minnie had discovered by the simple expedient of having Mick O'Higgins follow him home) had proved to be not only an excellent client, with an insatiable appetite for Lithuanian illuminated manuscripts and Japanese erotica, but also a most valuable connection. Through him, she had acquired (besides a thin sheaf of sealed documents intended for her father's eyes) two fifteenth-century incunabula—one in excellent condition, the other needing some repair—and a tattered but originally beautiful small book by María Anna Águeda de San Ignacio, an abbess from New Spain, with handwritten annotations said to be in the hand of the nun herself.

Minnie hadn't enough Spanish to make out much of the content, but

it was the sort of small book that gave one pleasure simply to hold, and she had paused in her labors to do just that.

The sturdy sideboard in her parlor was stacked on one side with books and on the other with more books, these wrapped in soft cloth, then a layer of felt, one of lambswool, and then an outer skin of oiled silk, tied with tarred twine. Piles of packing materials were arrayed on the dining table, and several large wooden crates were wedged under it.

She trusted no one else to handle or pack the books for shipment back to Paris and was in consequence dust-stained and sweaty, in spite of the breeze from the open window. Just past Midsummer's Day, the weather had kept fine for a whole week, much to the astonishment of every Londoner she'd spoken to.

La Vida de la Alma. Close enough to the Latin to translate as "The Life of the Soul." It was soft-bound in a thin oxblood leather, worn by years—a lifetime?—of reading, with a stamped pattern of tiny pecten shells, each one edged with gilt. She touched one gently, feeling a great sense of peace. Books always had something to say, beyond the words inside, but it was rare to find one with so strong a character.

She opened it carefully; the paper inside was thin, and the ink had begun to fade with age but not to blur. The book had few illustrations, and those few, simple: a cross, the Lamb of God, the pecten shell, drawn larger—she'd seen that once or twice before, in Spanish manuscripts, but didn't know the significance. She must remember to ask her father. . . .

"Ah," she said, compressing her lips. "Father." She'd been trying not to think of him, not until she'd had time to sort out her emotions and consider what on earth she might say to him about her mother.

She'd thought of the woman called Sister Emmanuelle many times since leaving her in her hay-filled womb of light. The shock had faded, but the images of that meeting were printed on her mind as indelibly as the black ink inside this book. She still felt the sting of loss and the ache of sorrow—but the sense of peace from the book seemed somehow to shelter her, like a covering wing.

"Are you an angel?"

She sighed and set the book gently into its nest of cloth and felt. She'd have to talk to her father, yes. But what on earth would she say?

"Raphael . . ."

"If *you* have any answers," she said to the book and its author, "please pray for me. For us."

She wasn't crying, but her eyes were damp, and she wiped her face with the hem of her dusty apron. Before she could settle to work again, though, there came a knock at the door.

Eliza had gone out to do the shopping, so Minnie opened the door just as she was. Mick and Rafe O'Higgins stood shoulder to shoulder in

the hall, both of them smudged with soot and excited as terriers smelling a rat.

"We've got the letters, Bedelia!" Rafe said.

"*All* the letters!" Mick added, proudly holding up a leather bag.

"WE WAITED FOR the butler's day out," Mick explained, laying his booty out ceremoniously before her. "It's the butler what arranges for the sweeps to come in when needed, aye? So when we come to the door with our brooms and cloths—not to worry, we borrowed 'em, ye'll not have to pay—and said Mr. Sylvester had sent for us to attend to the library chimney . . ."

"Well, the housekeeper looked a bit squint-eyed," Rafe chimed in, "but she showed us along, and when we began to bang about and shout up the chimney and kick up soot, she left us quite to ourselves. And so . . ."

He swept a hand out over the table. All the letters, indeed. The bag had disgorged a small, flat wooden case, a leather folder, and a thin stack of letters, soberly bound with black grosgrain ribbon.

"Well done!" Minnie told them sincerely. She felt a flutter of excitement at sight of the letters, though a cautious excitement. The O'Higginses had, of course, brought away every letter they could find. They must have more than the countess's letters here, and she wondered for a brief moment whether some of the extras might be valuable . . . but dismissed the thought for now. As long as they'd found Esmé's . . .

"Did you get paid for sweeping the chimney?" she asked, out of curiosity.

"Sure and ye wound us, Lady Bedelia," Rafe said, clasping a battered hat over his heart and trying to look wounded. There was smut on his nose.

"O' course we did," Mick said, grinning. "T'wouldn't have been convincing, otherwise, would it?"

They were cock-a-hoop over their success, and it took nearly half a bottle of Madeira to celebrate said success enough for them to leave, but at last she closed the door upon them, rubbed at a smudge on the white doorjamb with her thumb, and walked slowly back to the table to see what she had.

She took the letters from their various wrappings and set them out in three neat piles. The letters from Esmé, Lady Melton, to her lover, Nathaniel Twelvetrees: those were the ones in the wooden box. The letters in the beribboned bundle were from said Nathaniel Twelvetrees to Esmé. And the leather folder held quite unexpected letters—from Harold, Lord Melton, to his wife.

Minnie had never felt the slightest reservation in reading someone else's letters. It was simply part of the work, and if she occasionally met someone in those pages whose voice struck her mind or heart, someone real—that was a bonus, something to treasure privately, with a sweet regret that she would never know the writer face-to-face.

Well, she'd certainly never know Esmé or Nathaniel face-to-face, she thought. As for Harold, Lord Melton—just looking at the untidy pile of crumpled, smoothed-out, ink-blotted sheets made the hairs prickle at the back of her neck.

Esmé first, she decided. Esmé was the center of it all. And it was Esmé's letters she'd been commissioned—more or less—to steal. A faint hint of perfume rose from the wooden box, something slightly bitter, fresh, and mysterious. Myrrh? Nutmeg? Dried lemon? Not sweet at all, she thought—nor, likely, was Esmé Grey.

Not all of the letters had dates, but she sorted them as best she could. All on the same stationery, an expensive linen rag paper, thick to the touch and pure white. The sentiments inscribed upon them were not pure at all.

Mon cher . . . Dois-je vous dire ce que je voudrais que vous me fassiez? "Shall I tell you what I want you to do to me?"

Minnie had read her way with interest through her father's entire stock of erotica when she was fourteen, accidentally discovering in the process that one didn't necessarily need a partner in order to experience the sensations so euphorically described therein. Esmé hadn't much literary style, but her imagination—surely some of it must *be* imagination?—was remarkable and expressed with a blunt freedom that made Minnie want to squirm, ever so slightly, in her seat.

Not that they were all like that. One was a simple two-line note making an assignation, another was a more thoughtful—and, surprisingly, a more intimate—letter describing Esmé's visit to—*oh, God,* Minnie thought, and wiped her hand on her skirt, as she'd begun to perspire—Princess Augusta and her fabulous garden.

Esmé had noted carelessly that she had no liking for the princess, whom she thought heavy in both body and mind, but that Melton had asked her to accept the invitation to tea in order to—and here Minnie translated Esmé's idiomatic French expression—"drench in melted butter" the vulgar woman and pave the way for Melton to discuss his military designs with the prince.

She then mentioned walking through the glass conservatories with the princess, paused to make comical, if offhandedly complimentary, comparisons between her lover's physical parts and various exotic plants—she mentioned the euphorbias, Minnie noted—and ended with a brief remark about the Chinese flowers called *chu.* She was attracted—Minnie snorted, reading this—by the "purity and stillness" of the blooms.

"*À les regarder, mon âme s'est apaisée,*" she had written. "It soothed my soul to look at them."

Minnie set the letter down, as gently as if it might break, and closed her eyes.

"You poor man," she whispered.

THERE WAS A DECANTER of wine on the sideboard. She poured a small glass, very carefully, and stood sipping it, looking at the desk and its burden of letters.

Someone real. She had to admit that Esmé Grey was definitely real. The impact of her personality was as palpable as though she'd reached out of the paper and stroked her correspondent's face. Teasing, erotic . . .

"Cruel," Minnie said aloud, though softly. To write to your lover and mention your husband?

"Hmph," she said.

And Esmé's partner in this criminal conversation? She glanced at the bundle of Nathaniel Twelvetrees's letters to his mistress. What bizarre quirk of mind had made Melton keep them? Was it guilt, a sort of hair shirt of the spirit?

And if so . . . guilt for killing Nathaniel Twelvetrees? Or guilt over Esmé's death? She wondered how quickly the one event had followed the other—had the shock of hearing of her lover's death brought on a miscarriage, or a fatally early labor, as gossip said?

Likely she'd never learn the answers to those questions, but while Melton had killed Nathaniel, he'd left the poet his voice; Nathaniel Twelvetrees could speak for himself.

She poured another glass of the wine—a heavy, aromatic Bordeaux; she felt she needed ballast—and unfolded the first of Nathaniel's letters.

For a poet, Nathaniel was a surprisingly pedestrian writer. His sentiments were expressed in sufficiently passionate language but a very common prose, and while he made a distinct effort to meet Esmé on her own ground, he was clearly not her match, in either imagination or expression.

Still, he was a poet, not a novelist; perhaps it wasn't fair to judge him by his prose style alone. In two of his letters, he mentioned an enclosure, a poem written in honor of his beloved. She checked the box: no poems. Maybe Melton had burned those—or Esmé had. Nathaniel's tone in presenting these literary gifts reminded Minnie very much of a naturalist's description she had read—of a type of male spider who brought his chosen mate an elaborately silk-wrapped parcel containing an insect and then leapt upon her whilst she was absorbed in unwrapping her snack, hastily achieving his purpose before she could finish eating and have him for dessert.

"She scared him," Minnie murmured to herself, with a sense of sympathy but one tempered with a mild contempt. "Poor worm." She was somewhat shocked to realize that contempt—and the more so to realize that Esmé had very likely felt the same.

Hence her invoking Melton's name in the letters to Twelvetrees? An attempt to sting him into greater ardor? She'd done it more than once; in fact—Minnie turned again to Esmé's letters—yes, she'd mentioned her husband, by name or reference, in every letter, even the two-line assignation: *My husband will be gone on his regimental business—come to me tomorrow in the oratory at four o'clock.*

"Huh," Minnie said, and sat back, eyeing the letters as she sipped her wine. They lay in stacks and single sheets and fans before her, with the as-yet-unread folder that held Melton's letters in the center. It looked not unlike a layout of the tarot—she'd had her own cards read several times in Paris, by an acquaintance of her father's named Jacques, who was practicing the art.

"Sometimes it's quite subtle," Jacques had said, shuffling the gaudy cards. "Especially the minor arcana. But then—sometimes it's obvious at first glance." This said, smiling, as he laid Death in front of her.

She had no opinion regarding the truth laid out in tarot cards, considering that to be no more than the reflection of the client's mind at the time of the reading. But she had definite opinions regarding letters, and she touched the two-line assignation thoughtfully.

Where had Esmé's letters come from? Would the Twelvetrees family have sent them to Lord Melton following Nathaniel's death? They might, she thought. What could be more painful to him? Though that argued both a subtlety of mind and a sense of refined cruelty that she saw no trace of in Nathaniel's letters and that she hadn't noticed in most English people.

Besides . . . what had made Melton challenge Twelvetrees in the first place? Surely Esmé hadn't confessed the affair to him. No . . . Colonel Quarry had said, or at least intimated, that Melton had found incriminating letters *written by his wife,* and that that was what . . .

She picked up the countess's pile again, frowning at the letters. Looking carefully, she could see that each one had a blot of ink or the occasional smear—one appeared to have had water spilled over the bottom edge. So . . . these were drafts of letters, later copied fair to be sent to Nathaniel? If so, though, why not throw the drafts into the fire? Why keep them and risk discovery?

"Or invite it," she said aloud, surprising herself. She sat up straight and read the letters through again, then set them down.

My husband will be gone . . . Every one. Every one of them noted Melton's absence—and his preoccupation with his nascent regiment.

Jacques was right; sometimes it was obvious.

Minnie shook her head, the wine fumes mingling with the dead countess's bitter perfume.

"*Pauvre chienne,*" she said softly. "Poor bitch."

14

NOTORIOUS BORES

IT WASN'T NECESSARY TO read Lord Melton's letters, but she couldn't possibly have stopped herself from doing so, and she picked one up as though it were a lit grenade that might go off in her hand.

It did. She read the five letters through without stopping. None were dated, and there was no way of telling the order in which they had been written; time had plainly been of no consequence to the writer—and yet it had meant everything. This was the voice of a man pushed off a cliff into the abyss of eternity and documenting his fall.

I will love you forever, I cannot do otherwise, but by God, Esmé, I will hate you forever, with all the power of my soul, and had I you before me and your long white neck in my hands I would strangle you like a fucking swan and fuck you as you died, you . . .

He might as well have picked up the inkwell and flung it at the page. The words were scrawled and blotted, big and black, and there were ragged holes torn through the paper where here and there he had stabbed the page with his quill.

She took a deep, gasping breath when she came to the end, feeling as though she hadn't breathed once in the reading. She didn't weep, but her hands were shaking, and the last letter slipped from her fingers and floated to the floor. Weighted with loss and a grief that didn't cut but clawed and, merciless, tore its prey to bloody ribbons.

SHE DIDN'T READ the letters again. It would have felt like a desecration. As it was, there was no need to read them over; she thought she would never forget a word of any of them.

She had to leave her rooms and walk for some time to regain any sense

of composure. Now and then she felt tears run down her cheeks and hastily blotted them before any passerby should notice and ask her trouble. She felt as though she'd wept for days or as though someone had beaten her. And yet it was nothing to do with her.

She felt one of the O'Higgins brothers following somewhere behind her, but he tactfully hung back. She walked from one end to the other of St. James's Park, and all the way around the lake, but finally sat down on a bench near a flotilla of swans, exhausted in mind and body both. Someone sat down on the other end of the bench—Mick, she saw, from the corner of her eye.

It was teatime; the bustle of the streets was dying down as people hurried home or dropped into a tavern or an ordinary to refresh themselves after a long day's labor. Mick coughed in a meaning manner.

"I'm not hungry," she said. "You go on, if you like."

"Now, Bedelia. Ye know fine I'm not goin' anywhere you don't go." He'd scooted along the bench and sat at her elbow, slouched and companionable. "Shall I be fetchin' ye a pie, now? Whatever's the trouble, it'll seem better on a full belly."

She wasn't hungry, but she *was* empty and, after a moment's indecision, gave in and let him buy her a meat pie from a pie man. The smell of it was so strong and good that she felt somewhat restored just from holding it. She nibbled the crust, felt the rich flood of juice and flavor in her mouth, and, closing her eyes, gave herself over to the pie.

"There, now." Mick, having long since finished his own pie, sat gazing benevolently at her. "Better, is it not?"

"Yes," she admitted. At least she could now think about the matter, rather than drowning in it. And while she hadn't been conscious of actually *thinking* at any time since leaving her rooms, evidently some back chamber of her mind had been turning things over.

Esmé and Nathaniel were dead. Harold, theoretical Duke of Pardloe, wasn't. That's what it came down to. She could do something about him. And she found that she was determined to do it.

"What, though?" she asked, having explained the matter to Mick in general terms. "I can't send those letters to the secretary at war—there's no way His Grace wouldn't find out, and I think it would kill him to know anyone had read them, let alone people who . . . who had any power over him, you know."

Mick pulled a face but allowed that this might be so.

"So what is it ye want to happen, Lady Bedelia?" he asked. "There's maybe another way of it?"

She drew a breath that went down to her shoes and let it out slowly.

"I suppose I want what Captain Quarry wants: to scotch the notion

that His Grace is insane and to get his regiment re-commissioned. I think I have to do both those things. But how?"

"And ye can't—or ye won't—do it with the letters. . . ." He eyed her sideways, to see if she might be convinced otherwise, but she shook her head at him.

"Get a false witness?" he suggested. "Bribe someone to say there was an affair betwixt the countess and the poet?"

Minnie shook her head dubiously.

"I'm not saying I couldn't find someone who would take a bribe," she said. "But not one who'd be believed. Most young women aren't good liars at all."

"No," he agreed. "You're one of a kind, so ye are." It was said with admiration, and she nodded briefly at the compliment but went on with her train of thought.

"The other thing is that it's easy enough to start a rumor, but once it's started, it's quite likely to take on a life of its own. You can't control it, I mean. If I got someone—man or woman—to say he or she knew about the affair, it wouldn't stop there. And because it wouldn't be the truth to start with, there's no telling where it might go. You don't set light to a fuse without knowing where it's laid," she added, raising a brow at him. "My father always told me that."

"A wise man, your father." Mick touched the brim of his hat in respect. "If it's not to be bribery and false witness, then . . . what might his honor, your da, recommend?"

"Well . . . forgery, most likely," she said with a shrug. "But I don't think writing a false version of those letters would be a great deal better than showing the originals, in terms of effect." She rubbed her thumb across her fingers, feeling the faint slick of lard from the piecrust. "Get me another pie, will you, Mick? Thinking is hungry work."

She finished the second pie and, thus fortified, reluctantly began to mentally revisit Esmé's letters. It was, after all, Countess Melton who was the *fons et origo* of all this misery.

Would you think it was worth it, I wonder? she thought toward the absent Esmé. Likely the woman had only wanted to make her husband jealous; she probably hadn't had the slightest intent of causing her husband to shoot one of his friends; most certainly she hadn't had any intent of dying, along with her child. That circumstance struck Minnie with a particular poignancy and, for some odd reason, made her think of her mother.

I don't suppose you intended anything that happened, either, she thought with compassion. *You certainly didn't intend me.* Still, she thought her mother's situation, while very regrettable, wasn't the theatrical tragedy that Esmé's had been. *I mean, we both survived.*

And speaking only for myself, she added, *I'm quite glad to be here. I'm reasonably sure that Father's pleased about that, too.*

A slight sound pulled her from her thoughts, and she perceived that Mick had adjusted his position, indicating silently that he thought it was getting late and best they begin walking back to Great Ryder Street.

He was right; the shadows of the huge trees had begun to edge across the path like a seeping stain of spilled tea. And the sounds had changed, too: the cawing laughter of the society women with their parasols had mostly vanished, replaced by the male voices of soldiers and businessmen and clerks, all heading for their tea with the single-mindedness of donkeys headed for their mangers.

She stood up and shook her skirts back into place, retrieved her hat and pinned it firmly to her hair. She nodded to Mick and indicated with a small movement of the hand that he should walk with her, rather than follow. She was wearing a decent but very demure blue gingham with a plain straw hat; she might easily pass for an upper housemaid walking out with an admirer, as long as they didn't meet anyone she knew—and that wasn't likely at this hour.

"This chap what his lordship shot," Mick said, after half a block. "They say he was a poet, was he?"

"So I'm told."

"Have ye maybe read any of his poems, like?"

She glanced at him, surprised.

"No. Why?"

"Well, it was just ye mentioned your da thinkin' highly o' forgery in some situations. I was wonderin' what ye might forge that would help, and it struck me—what if your man Twelvetrees had written a poem of an incriminatin' nature about the countess? Or, rather," he added, in case she was missing his point, which she wasn't, "what if ye were to write one for him?"

"It's a thought," she said slowly. "Perhaps quite a good thought, too— but let's turn it over for a bit, shall we?"

"Aye," said Mick, beginning to grow enthused. "Well, first off, o' course: what class of a forger might ye be, at all?"

"Not inspired," she admitted. "I mean, no hope of me doing a proper banknote. And I've really not done much in the way of true forgery, either—copying the writing of a real person, I mean. It's mostly writing a false letter but one that's meant for a person who doesn't know the sender. And only now and then, not often."

Mick emitted a low humming noise.

"Still, ye have got some of the man's letters to work from," he pointed out. "Could ye maybe trace a few words here and there and add in, between-like?"

"Maybe," she said dubiously. "But there's more to a good forgery than only the handwriting, you know. If it's going to a person who knows the sender, then the style needs to be a decent facsimile—has to resemble the real person's, I mean," she added quickly, seeing his lips start to shape "facsimile."

"And his style writing a poem could be different to what he'd do writing a letter?" Mick turned that over for a moment, considering.

"Yes. What if he was only known to write sonnets, I mean, and I wrote a sestina? Someone might smell a rat."

"I'll take yer word for it. Though I shouldn't think your man was in the habit of writin' love poems to the secretary at war, eh?"

"No," she said, a little tersely. "But if I wrote something shocking enough to justify his lordship shooting the man who wrote it, what are the chances that the secretary would show it to somebody else? Who might tell somebody else, and . . . and so on." She flipped a hand. "If it got to someone who could tell that Nathaniel Twelvetrees didn't write it, then what?"

Mick nodded soberly. "Then they'd maybe think your lordship did it himself, ye mean?"

"That's one possibility." On the other hand, the *other* possibility was undeniably fascinating.

They had reached Great Ryder Street and the scrubbed white steps that led up to her door. The scent of brewing tea floated up from the servants' areaway beside the steps, and her stomach curled in a pleasantly anticipatory fashion.

"It's a good idea, Mick," she said, and touched his hand lightly. "Thank you. I'll ask Lady Buford whether Nathaniel published any of his poetry. If I could read a bit of it, just to see . . ."

"Me money's on you, Lady Bedelia," Mick said, and, smiling at her, raised her hand and kissed it.

"NATHANIEL TWELVETREES?" Lady Buford was surprised and peered closely at Minnie through her quizzing glass. "I don't believe so. He was much given to declaiming his poetry at salons and I believe went so far as to give a theatrical reading at one point, but from what little I heard of his poetry—well, what little I hear of what people *said* of his poetry—I doubt that most printers would have considered it a promising financial venture."

She resumed watching the stage, this presently featuring a mediocre performance of "Charming Country Songs, by a Duette of Two Ladies," but tapped her closed fan now and then against her closed lips, an indication of continued thought.

"I believe," she said at once, when the next pause in the entertainment came, "that Nathaniel *did* have some of his poems privately printed. For the edification of his friends," she added, with a delicate lift of one strong gray brow. "Why do you ask?"

Fortunately the pause had given Minnie time enough to foresee that one, and she answered readily enough.

"Sir Robert Abdy was speaking of Mr. Twelvetrees at Lady Scroggs's rout the other night—rather scornfully," she added, with her own delicacy. "But as Sir Robert has his own pretensions in that line . . ."

Lady Buford laughed, a deep, engaging laugh that made people in the box next them turn round to look, and proceeded to say a few scornful— and deeply amusing—things of her own about Sir Robert.

But Minnie continued to think, through the appearances of a pair of Italian fire-eaters, a dancing pig (which disgraced itself onstage, to the delight of the audience), two purportedly Chinese gentlemen who sang a purportedly comic song, and several more acts of a similar ilk.

Privately printed. For the edification of his friends. There were at least two poems, written expressly for the edification of Esmé, Countess Melton. Where were they?

"I wonder," she said quite casually, as they began to make their way out through the throngs of theatergoers, "if Countess Melton was fond of poetry?"

Lady Buford was only half attending, being occupied in trying to catch the eye of an acquaintance on the far side of the theater, and replied absently, "Oh, I don't think so. Woman never read a book in her life, save the Bible."

"The Bible?" Minnie asked, incredulous. "I wouldn't have thought her a . . . a religious person." Lady Buford had succeeded in attracting the friend, who was wading forcefully toward them through the crowd, and spared a cynic smile for Minnie.

"She wasn't. But she did like to read the Bible and make fun of it to shock people. Only too easy to do, I'm afraid."

"SHE WOULDN'T HAVE thrown the poems away," Minnie argued to Rafe, who was disposed to be dubious. "They were *to* her, *about* her. No woman would throw away a poem that a man she cared for wrote about her—and most especially not a woman like Esmé."

"Has any man ever written you a love poem, Lady Bedelia?" he asked, teasing.

"No," she said primly, but felt herself blushing. A few men had done

just that—and she'd kept the poems, even though she didn't care all that much for the men who'd written them. Still . . .

"Mmm," Rafe conceded, with a waggle of his head. "But maybe your man Melton burnt them. *I* would, if some smellsmock had been sending my wife that class of thing."

"If he didn't burn the letters," Minnie said, "he wouldn't have burnt the poems, either. The poems couldn't possibly have contained anything worse."

Why didn't *he burn the letters?* she wondered, for at least the hundredth time. And to have kept *all* the letters—Esmé's, Nathaniel's . . . and his own.

Perhaps it was guilt, the need to suffer for what he'd done, obsessively reading them over. Perhaps it was confusion—some need or hope of making sense of what had happened, what they'd *all* done, in making this tragedy. He was the only one left to do it, after all.

Or . . . perhaps it was only that he still loved his wife and his friend, mourned them both, and couldn't bear to part with these last personal relics. His own letters were certainly filled with a heartbreaking grief, easily visible amongst the blots of rage.

"I think that she deliberately left the letters where her husband would find them," Minnie said slowly, watching a line of half-grown cygnets sailing after their mother. "But the poems . . . maybe those didn't have any pointed references to Lord Melton in them. If they were only about *her,* she might have kept them private, put them away somewhere safe, I mean."

"So?" Rafe was beginning to look wary. "We'll not get back in Argus House, ye know. Every servant in the place saw us last time."

"Ye-es." She stretched out a leg, considering her new calf-leather court shoes. "But I was wondering . . . might you have a . . . a sister, say, or perhaps a cousin, who wouldn't mind earning . . . say . . . five pounds?" Five pounds was half a year's pay for a house servant.

Rafe stopped dead and stared at her.

"Are ye wanting us to burgle the house or burn it down, for all love?"

"Nothing at all dangerous," she assured him, and batted her eyes, just once. "I just want you—or, rather, your female accomplice—to steal the countess's Bible."

IN THE END, stealing the book hadn't been necessary. Cousin Aoife, in her guise as a newly hired chambermaid, had simply gone through the Bible, this still resting chastely on the night table beside the countess's

bereft bed, removed from it a handful of folded papers, pocketed these, walked down the stairs and out to the privy behind the house, from whence she had modestly disappeared through a hole in the hedge, never to return.

"Anything ye can use, Lady Bedelia?" Mick and Rafe had both come up to her rooms the day after they'd delivered their prize and collected Aoife's wages.

"Yes." She hadn't slept at all the night before, and everything around her had a slightly dream-like quality, including the two Irishmen. She yawned, spreading her fan just in time, and blinked at them, then reached into her pocket and drew out a parchment cover, sealed with black wax and addressed to *Sir William Yonge, Secretary at War.*

"Can you ensure—and I do mean make *sure*—that Sir William will get this? I know," she said dryly, seeing Rafe make doe's eyes at her, "I wound you. Do it, though."

They laughed and went, leaving her to the silence of her room and the company of paper. Small barricades of books protected the table on which she'd made her magic, summoning the shade of her father with half a glass of Madeira, crossing herself and asking the blessing of her mother's prayers before picking up her quill.

Nathaniel Twelvetrees, bless his erotically inclined heart, had waxed lascivious in describing his mistress's charms. He had also, in one of the poems, mentioned various aspects of the place in which the lovers had disported themselves. He hadn't signed that one—but he had written *Yours forever, darling—Nathaniel* at the bottom of the other.

After some dithering, she had at last decided to take the risk in order to put the matter beyond doubt and, after filling two foolscap pages with practice attempts, had cut a fresh quill and written—in what she *thought* was a decent version of Nathaniel's hand and style—a title for his untitled poem: *Love's Constant Flowering: in Celebration of the Seventh of April.* And at the bottom—after a lot more practice: *Yours, in the flesh and in the spirit, darling Esmé—Nathaniel.*

If she was lucky, no one would ever think to investigate where Esmé, Countess Melton, had been on the seventh of April, but one of said countess's letters had made an assignation for that date, and the details of the place given in Nathaniel's poem matched what Minnie knew of the spot chosen for said assignation.

The poem made it clear, at least, that the Duke of Pardloe would have had more than adequate grounds for challenging Nathaniel Twelvetrees to a duel. And it certainly suggested that the countess had encouraged Twelvetrees's attentions, if not more—but it didn't disclose the true heart of the matter, let alone reveal Esmé's character or the painful intimacies of her husband.

So. Now it was done.

The letters—all of them—were still arrayed on the table in their tarot spread before her, silent witnesses.

"And what am I to do with *you*?" she said to them. She filled up her glass of wine and drank it slowly, contemplating.

The simplest thing—and by far the safest—was to burn them. Two considerations stopped her, though.

One. If the poem didn't work, the letters were the only evidence of the affair. In the last resort, she could give them to Harry Quarry and let him make what use he could—or would—of them.

Two. That final thought lingered in her mind, nibbled at her heart. *Why did he keep them?* Whether for guilt, grief, repentance, solace, or reminder—His Grace had kept them. They had value to him.

It was just past Midsummer's Day; the sun still hung in the sky, though it was past eight o'clock. She heard the bells of St. James's strike the hour and, draining her glass, made up her mind.

She'd have to put them back.

WHETHER IT WAS the influence of her mother's prayers or a benign intercession by Mother María Anna Águeda de San Ignacio, it was only three days following this rash decision that the opportunity to carry it out was put into Minnie's hands.

"Such news, my dear!" Lady Buford was quite flushed, from either heat or excitement, and fanned herself rapidly. "Earl Melton is holding a ball, in honor of his mother's birthday."

"What? I didn't know he had a mother. Er . . . I mean—"

Lady Buford laughed, growing noticeably pinker.

"Even that villain Diderot has a mother, my dear. But it's true that the dowager Countess of Melton is not strongly in evidence. She wisely decamped to France following her husband's suicide and has been living very quietly there ever since."

"But . . . she's coming back?"

"Oh, I doubt it extremely," Lady Buford said, and took out a rather worn lace handkerchief, with which she dabbed her forehead. "Is there tea, my dear? I find myself in dire need of a cup; summer air is so drying."

Eliza hadn't waited for a summons. Knowing Lady Buford's attitude toward tea, she had begun brewing a pot the moment Lady B's knock was heard at the door and now came trundling down the hall with a rattling tray.

Minnie waited with what patience she could summon for the necessary ceremony of pouring: the administration of three sugar lumps—Lady Bu-

ford had very few teeth left, and no wonder—a large dollop of cream, and the acquisition of exactly two ginger biscuits. Finally restored, Lady Buford patted her lips, stifled a soft belch, and sat up straight, ready for business.

"There's tremendous talk about it, of course," she said. "It's not even four months since the countess's death. And while I'm sure his mother is not planning to appear at this affair, choosing to celebrate her birthday is . . . audacious, but audacious without committing open scandal."

"I should think the . . . er . . . his lordship has had quite enough of that," Minnie murmured. "Um . . . what do you mean by 'audacious,' though?"

Lady Buford looked pleased; she enjoyed displaying her skills.

"Well. When someone—especially a man—does something unusual, you must always ask what it was they intended by the action. Whether or not that effect is achieved, the intent usually explains much.

"And in this instance," she said, plucking another biscuit delicately from the plate and dunking it into her tea to soften, "I think that his lordship means to put himself on display, in order to prove to society at large that he is not insane—whatever else he might be," she added thoughtfully.

Minnie wasn't so sure about Lord Melton's mental state but nodded obligingly.

"You see . . ." Lady Buford paused to nibble the edge of her softened biscuit, made an approving face, and swallowed. "You see, were he simply to host a rout or ball of the normal sort, he would seem light-minded and frivolous at best, cold and unfeeling at worst. He would also expose himself to considerable risk that no one would accept an invitation."

"But as it is?" Minnie prompted.

"Well, there's the factor of curiosity, which can never be overlooked." Lady Buford's rather pointed tongue darted out to capture a stray crumb, which was whisked out of sight. "But by making the occasion in honor of his mother, he more or less commands the loyalty of her friends—who are many—and also those who were friends of his late father but who couldn't openly support him. *And,*" she added, leaning forward portentously, "there are the Armstrongs."

"Who?" Minnie asked blankly. By this time she had quite an extensive social index of London but recognized no prominent person therein named Armstrong.

"The duke's mother is an Armstrong by birth," Lady Buford explained, "though her mother was English. But the Armstrongs are a very powerful Scottish family, from the Borders. And the rumor is that Lord Fairbairn—that's the duke's maternal grandfather, only a baron but very rich—is in London and will attend the . . . er . . . function."

Minnie was beginning to think tea inadequate to the occasion and rose to fetch the decanter of Madeira from the sideboard. Lady Buford made no demur.

"Of course you must go," Lady Buford said, having downed half a glassful at one gulp.

"Really?" Minnie was experiencing that sudden visceral emptiness that attends excitement, anticipation, and panic.

"Yes," Lady Buford said, with determination, and downed the rest, setting her glass down with a thump. "Almost all of your choicest prospects will be there, and there is nothing like competition to make a gentleman declare himself."

Now the sensation was one of unalloyed panic. What with one thing and another, Minnie had quite forgotten that she was meant to be husband-hunting. Just last week, she'd had two proposals, though luckily from fairly undistinguished suitors, and Lady Buford hadn't objected to her refusing them.

She finished her own Madeira and poured another for them both.

"All right," she said, feeling a slight spinning sensation. "What do you think I should wear?"

"Your very best, my dear." Lady Buford raised her refilled glass in a sort of toast. "Lord Fairbairn is a widower."

15

BURGLARY AND OTHER DIVERSIONS

THE *CARTE D'INVITATION* ARRIVED by messenger two days later, addressed to her simply as *Mademoiselle Wilhelmina Rennie*. Seeing her name—even a mistaken version of her assumed name—in black and white gave her a slight rippling sensation down the back. If she should be caught . . .

"Think about it, girl," said her father's logical voice, affectionate and slightly impatient. *"What if you are caught? Don't be afraid of unimagined possibilities; imagine the possibilities and then imagine what you'll do about them."*

Her father was, as usual, right. She wrote down every possibility she

could think of, from being refused admittance to Argus House, to being recognized at the ball by one of the clients she'd met this week, to being detected by a servant while returning the letters. And then she summoned the O'Higginses and told them what she wanted.

SHE'D COME LATE, smoothly inserting herself into a group of several giggly young women and their chaperones, avoiding the notice paid to guests who arrived singly and were announced to the crowd. The dancing had started; it was simple to find a place among the wallflowers, where she could watch without being seen.

She'd learned from Lady Buford the art of drawing men's eyes. She'd already known the art of avoiding them. Despite having worn her best— the soft river-green eau-de-nil gown—so long as she kept her head modestly lowered, hung about on the edge of a group, and didn't speak, she was unlikely to get a second glance.

Her eyes, though, knew just where to look. There were a number of soldiers in lavish uniform, but she saw Lord Melton instantly, as though there was no other man in the room. He stood by the enormous hearth, absorbed in conversation with a few other men; with no sense of surprise, she recognized Prince Frederick, bulging and amiable in puce satin, and Harry Quarry, fine in his own uniform. A small, fierce-looking man with an iron-gray wig and the features of a shrike stood at Melton's elbow— that must be Lord Fairbairn, she thought.

She sensed someone behind her and turned to see the Duke of Beaufort beaming down at her. He swept her a deep bow.

"Miss Rennie! Your most humble servant, I do assure you!"

"Charmed, as always, Your Grace." She batted her eyes at him over her fan. She'd known she was likely to meet people she knew—and she'd decided what to do about it. To wit, nothing special. She knew how to flirt and disengage, moving skillfully from one partner to another without causing offense. So she gave Sir Robert her hand, joined him for two dances, sent him for an ice, and disappeared to the ladies' retiring room for a quarter of an hour—long enough for him to have given up and sought another partner.

When she came back, moving cautiously, her eyes went at once to the hearth and discovered that Lord Melton and his companions had vanished. A group of bankers and stockbrokers, many of whom she knew, had replaced them by the fire, deep in financial conversation by the look of them.

She drifted inconspicuously around the room, watching, but Hal— *Lord Melton,* she corrected herself firmly—was nowhere to be found. Nor

was the prince, Harry Quarry, or the ferocious Scottish grandfather. Clearly the conversation had reached a stage where privacy was required.

Well enough. But she couldn't get on with her own job until the bloody man came back into sight. If he was having private discussions, chances were good that he was doing it in the library; she daren't risk walking in on him.

"Miss Rennie! What a vision you are! Come and dance with me, I insist!"

She smiled and raised her fan.

"Of course, Sir Robert. Charmed!"

It was more than half an hour before the men came back. The prince reappeared first, strolling to one of the refreshment tables with a look of pleased accomplishment on his face. Then Lord Fairbairn, who popped out of a door on the far side of the ballroom and stood against the wall, looking on with as amiable an expression as his forbidding features could manage.

And then Lord Melton and Harry emerged from the door that opened into the main hallway, chatting to each other with a casualness that failed entirely to cover their excitement. So, whatever Hal's business was with the prince, it had come to a successful conclusion.

Good. He'd stay here, then, celebrating.

She put down the half-finished glass of champagne and faded discreetly away in the direction of the retiring room.

She'd noted what she could—the locations of doors, mostly, and the quickest path should she need to get out fast. The library was down a side corridor, second door on the right.

The door stood open; the room warm and inviting, a good fire lit in the hearth and candles blazing, softly upholstered furniture in blue and pink against a wallpaper of wine-striped damask. She breathed deep, burped slightly, and felt the bubbles of champagne rise up the back of her nose, and, with a quick look up and down the hallway, stepped into the library and quietly closed the door behind her.

The desk was on the left side of the hearth, just as Mick had told her.

THE METAL WAS warm from being carried in her bosom, and her hands were trembling. She'd dropped the picks twice already.

"It's dead easy," Rafe had told her, handing over the two little brass instruments. "Just don't let yourself be hurried. Locks don't care for haste, and they'll defy and obstruct ye if ye try to rush them."

"Like women," Mick put in, grinning at her.

Under the O'Higginses' patient tutelage, she'd succeeded in unlocking

the drawer of her own desk with the picks, several times. She'd felt confident then, but it was a lot less easy to feel confident when you were committing burglary—well, reverse burglary, but that was even worse—in a duke's private library, with said duke and two hundred carousing witnesses no more than a stone's throw away.

Theoretically, this desk had the same type of lock. It was bigger, though, a solid brass plate with a beveled edge surrounding a keyhole that looked to her as big as a gun barrel at the moment. She took a deep breath, pushed the tension pick into the hole, and, as instructed, turned it to the left.

Then insert the feeler and pull it out gently, listening to the lock. The roar of the ballroom was muffled by the intervening walls, but music thrummed in her head, making it hard to hear. She sank to her knees, pressing her ear almost to the brass of the lock as she pulled out the pick. Nothing.

She'd been holding her breath and the blood was pounding in her ears, making it even harder to hear. She sat back on her heels, making herself breathe. Had she got it wrong?

Again. She put in the tension pick and turned it to the right. As slowly as she could, she slid the feeler in. She thought she felt something, but . . . She licked her lips and pulled the feeler gently out. Yes! A tiny ripple of sound as the pins dropped.

"Don't . . . bloody . . . rush," she whispered, and, wiping her hand on her skirt, took up the feeler again.

On the third try, she'd nearly got it—she could feel that there were five pins, and she had three, each making its soft little click—and then the doorknob turned behind her, with a much louder *click!*

She sprang to her feet with a stifled shriek, startling the footman who'd come in nearly as badly as he'd startled her. He said, "Oh!" and dropped the tray he was carrying, which struck the marble floor with a loud clang and spun like a top, clattering finally to a stop.

Minnie and the footman stared at each other, equally aghast.

"I—I beg your pardon, madam," he said, and squatted, fumbling with the tray. "I didn't know anyone was in here."

"That's . . . quite all right," she said, and paused to swallow. "I—I—felt a bit faint. Thought I'd just . . . sit . . . down for a moment. Out of—of the—the crowd."

Both picks were sticking out of the lock. She took a step backward and put one hand on the desk, to support herself. It wasn't pretense; her knees had gone to water, and cold sweat was chilling the back of her neck. But the footman couldn't see the lock, screened as it was by her eau-de-nil skirts.

"Oh. Of course, madam." With his tray now held to his chest like a

shield, the footman was regaining his composure. "May I bring you an ice? A glass of water?"

Jesus Lord, no!

But then she saw the small table at the far side of the hearth, flanked by two armchairs and holding a plate of savories, several glasses, and three or four decanters—one of these plainly filled with water.

"Oh," she said faintly, and gestured toward the table. "Perhaps . . . a little water?"

The instant he turned his back, she reached behind her and jerked the picks out of the lock. With trembling knees, she crossed the hearth and sank into one of the chairs, pushing the picks down beside the cushion, under cover of her skirts.

"Would you like me to fetch someone for you, ma'am?" The footman, having solicitously poured her water, was swiftly tidying away the decanters of spirit and what she now saw were used glasses onto his tray. Of course—this was where the duke had been having his meeting.

"No, no. Thank you. I'll be quite all right."

The footman glanced at her, then at the plate of savories, and, with a tiny shrug, left it on the table, bowed, and went out, pulling the door gently to behind him.

She sat quite still, forcing herself to breathe evenly. It was all right. Everything would be all right. She could smell the little savories—things wrapped in bacon, bits of anchovy and cheese. Her stomach rumbled; ought she to eat something, to steady her nerves, her hands?

No. She was still safe, but there was no time to waste. She wiped her hands on the arms of the chair, stood up, and marched back to the desk.

Tension pick. Right turn. Feeler to be sure of the pins. Probe. Raise the pins one by one, listening for each tiny metal *tink!* A pull. No. No, dammit! Try again.

Twice she had to get up, go drink water, and walk clockwise round the room—another of the O'Higginses' bits of advice—to calm herself before trying again.

But then . . . a sudden decisive metal *choonk* and it was done. Her hands were shaking so badly that she could barely get the three parcels out of her pockets, but get them she did. She yanked out the drawer and flung them in, then slammed the drawer with an exclamation of triumph.

"What the devil are you doing?" said a curious voice behind her. She shrieked and whirled round to find the Duke of Pardloe standing in the doorway and, behind him, Harry Quarry and another soldier.

"I say—" Harry began, plainly aghast.

"What's all this, then?" said the other man, peering curiously past Harry's shoulder.

"Don't trouble yourselves," the duke said, not looking back at them.

His eyes were fixed on hers, intent. "I'll take care of it." Without turning round, he grasped the edge of the door and pushed it shut in their staring faces.

For the first time, she heard the ticking of the little enamel clock on the mantelpiece and the hiss of the fire. She couldn't move.

He walked across the room to her, eyes still fixed on hers. The sweat on her body had chilled to snow and she shivered once, convulsively.

He took her carefully by the elbow and moved her to one side, then stood staring at the closed drawer and the picklocks sticking out of it, brassily accusing.

"What the *devil* have you been doing?" he said, and turned his head sharply to look at her. She barely heard him for the pounding of the blood in her ears.

"I—I—robbing you, Your Grace," she blurted. Finding that she could speak after all was a relief, and she gulped air. "So much must be obvious, surely?"

"Obvious," he repeated, with a faint tone of incredulity. "What on earth is there to steal in a library?"

This from a man whose shelves included at least half a dozen books worth a thousand pounds each; she could see them from here. Still, he had a point.

"The drawer was locked," she said. "Why would it be locked if there wasn't something valuable in it?"

He glanced instantly at the drawer and his face changed like lightning. *Oh, bloody* hell! she thought. *He'd forgotten the letters were there.* Or maybe not . . .

He turned on her then, and the air of slightly puzzled inquiry had vanished. He didn't seem to move but was suddenly much closer to her; she could smell the starch in his uniform and the faint odor of his sweat.

"Tell me who you are, 'Lady Bedelia,' " he said, "and exactly why you're here."

"I'm just a thief, Your Grace. I'm sorry." No chance of making it to the door, let alone out of the house.

"I don't believe that for an instant." He saw her glance and grasped her arm. "And you're not going anywhere until you tell me what you're here for."

She was light-headed with fear, but the faint implication that she *might* go somewhere seemed to offer at least the possibility that he wouldn't immediately summon a constable and have her arrested. On the other hand . . .

He wasn't waiting for her to make up her mind or a story. He tightened his grasp on her arm.

"Edward Twelvetrees," he said, and his voice was nearly a whisper, his face deadly white. "Did he send you?"

"No!" she said, but her heart nearly leapt out of her bodice at the name. He stared hard at her, then his eyes dropped, running the length of her shimmering green skirts.

"If I were to search you, madam—what would I find, I wonder?"

"An unclean handkerchief and a little bottle of scent," she said truthfully. Then added boldly, "If you want to search me, go ahead."

His nostrils flared a bit, and he pulled her aside.

"Stand there," he said shortly, then let go of her and yanked the picklocks from the drawer. He dipped a finger into the small pocket on his waistcoat and came out with a key, with which he unlocked the drawer and pulled it out.

Minnie's heart had changed its rhythm when he suggested searching her—no slower, but different—but now sped up to such a rate that she saw white spots at the corners of her eyes.

She hadn't put the letters back in their correct places; she couldn't— Mick hadn't taken notice. He'd know. She closed her eyes.

He said something under his breath, in . . . Latin?

She had to breathe and did so, with a gasp.

The hand was back, now gripping her shoulder.

"Open your eyes," he said, in a low, menacing voice, "and bloody look at me."

Her eyes popped open and met his, a winter blue, like ice. He was so angry that she could feel it vibrating through him like a struck tuning fork.

"What were you doing with my letters?"

"I—" Invention completely failed her, and she spoke the truth, hopelessly. "Putting them back."

He blinked. Looked at the open drawer, with the key still in the lock.

"You . . . er . . . you saw me," she said, and found enough saliva to swallow. "Saw me close the drawer, I mean. Er . . . didn't you?"

"I—" A small line had formed between his dark brows, deep as a paper cut. "I did." He let go of her shoulder and stood there, looking at her.

"How," he said carefully, "did you come to be in possession of my letters, may I ask?"

Her heart was still thundering in her ears, but some blood was coming back into her head. She swallowed again. Only the one possibility, wasn't there?

"Mr. Twelvetreees," she said. "He—he did ask me to steal the letters. I . . . wouldn't do it for him."

"You wouldn't," he repeated. One brow had risen slowly, and he was looking at her as though she were some exotic insect he'd found crawling

over his chrysanthemums. He cocked his head at the drawer in question. "Why not?"

"I liked you," she blurted. "When we . . . met at the princess's garden party."

"Indeed." A faint flush rose in his cheeks and the stiffness returned to his person.

"Yes." She met his eyes straight on. "I could tell that Mr. Twelvetrees *didn't* like you."

"That's putting it mildly," he said. "So you say he asked you to steal my letters—why did he think you would be the person to employ for such a venture? Do you steal things professionally?"

"Well, not often," she said, striving for composure. "It's more that we—I—discover information that may be of value. Just . . . inquiries here and there, you know. Gossip at parties, that sort of thing."

"We?" he repeated, both brows rising now. "Who are your confederates, may I ask?"

"Just my father and me," she said hastily, lest he recall the chimney sweeps. "It's . . . the family business, you might say."

"The family business," he repeated, with a faintly incredulous air. "Well . . . putting that aside, if you refused Edward Twelvetrees's commission, how did you come to be in possession of my letters, anyway?"

She commended her soul to a God she didn't quite believe in and threw her fate to the wind.

"Someone else must have stolen them for him," she said, with as much sincerity as possible. "But I had occasion to . . . be in his house, and I found them. I . . . recognized your name. I didn't read them," she added hastily. "Not once I saw that they were personal."

He'd gone white again. No doubt envisioning Edward Twelvetrees poring greedily over his most intimate wounds.

"But I—I knew what they must be, because of what Mr. Twelvetrees had told me. So I . . . took them back."

She was breathing a little more easily now. It was much easier to lie than to tell him the truth.

"You took them back," he said, and blinked, then looked hard at her. "And then you thought you'd come put them back in my house? Why?"

"I thought you . . . might want them," she said in a small voice, and felt her own cheeks flush. *Oh, God, he'll know I read them!*

"How very kind of you," he said dryly. "Why didn't you just send them to me anonymously, if your only intent was to return them?"

She took a small, unhappy breath and told him the truth, though she knew he wouldn't believe it.

"I didn't want you to be hurt. And you would be if you thought someone had read them."

"You what?" he said, incredulous.

"Shall I prove it?" she whispered, and her hand floated up without her actually willing it, to touch his face. "Your Grace?"

"What?" he said blankly. "Prove it?"

She couldn't think of anything at all to say so merely rose on her toes, hands on his shoulders, and kissed him. Softly. But she didn't stop, and her body moved toward his—and his toward hers—with the slow certainty of plants turning toward sun.

Moments later, she was kneeling on the hearth rug, fumbling madly under folds of eau-de-nil for the tapes of her petticoats, and Hal's—she was frightened and exhilarated to realize that she was thinking of him as Hal—uniform coat had struck the floor with a muffled crash of buttons, epaulets, and gold lace, and he was ripping at his waistcoat buttons, muttering to himself in Latin.

"What?" she said, catching the word "insane." "Who's insane?"

"Plainly you are," he said, stopping for a moment to stare at her. "Do you want to change your mind? Because you have roughly ten seconds to do so."

"It will take longer than that to get at my blasted bum roll!"

Muttering *"Irrumabo"* under his breath, he dropped to his knees, rummaged her petticoats, and seized the tie of her bum roll. Rather than untie it, he jerked it, broke the tie, slid the bum roll out of her clothes like a huge sausage, and flung it onto one of the wing chairs. Then he threw off his waistcoat and pushed her onto her back.

"What does *irrumabo* mean?" she said to the hanging crystals of the chandelier overhead.

"Me, too," he said, breathless. His hands were under her skirt, very cold on her bottom.

"You, too, *what*?" The middle part of him was between her thighs, very warm, even through the moleskin breeches.

"*I'm* insane," he said, as though this should be evident—and maybe it was, she thought.

"Oh," he added, looking up from the flies of his breeches, "*irrumabo* means 'fuck.'"

Three seconds later he was alarmingly hot and terrifyingly immediate and—

"Jesus Christ!" he said, and froze, looking down at her, his eyes huge with shock.

It hurt shockingly and she froze as well, taking shallow breaths. She felt his weight shift, knew he was about to leave her, and gripped his bottom to stop him. It was tight and solid and warm, an anchor against pain and terror.

"I said I'd prove it," she whispered, and pulled him in with all her

strength, arching her back. She let out a stifled shriek as he came the rest of the way, and he grabbed her and held her, keeping her from moving.

They lay face-to-face, staring at each other and gulping air like a pair of stranded fish. His heart was hammering so hard that she could feel it under the hand she had on his back.

He swallowed.

"You've proved it," he said at last. "Whatever it . . . What was it you wanted to prove again?"

Between the tightness of her stays and his weight, she hadn't enough breath to laugh, but she managed a small smile.

"That I didn't want to hurt you."

"Oh." His breathing was growing slower, deeper. *He isn't wheezing,* she thought.

"I didn't want—I didn't mean—to hurt you, either," he said softly. For an instant she saw him hesitate: should he pull away? But then decision settled on his features once more and he bent his head and kissed her. Slowly.

"It doesn't hurt that much," she assured him when he stopped.

"*Mendatrix.* That means 'liar.' Shall I—"

"No, you shan't," she said firmly. Over the first shock, her brain was now working again. "This is never going to happen again, so I mean to enjoy it—if such a thing is possible," she added, a little dubiously.

He didn't laugh, either, and his smile was only a trace—but it reached his eyes. The fire was hot on her skin.

"Yes, it is," he said. "Let me prove it."

Some little time later . . .

HE PUT OUT a hand to her and, dazed, she took it. His cold fingers closed tight on hers, and hers on his.

He took her to the back stairs, where he let go her hand—the stairs were too narrow to go side by side—and went down before her, glancing back now and then to be sure she hadn't disappeared or fallen. He looked as dazed as she felt.

Noise echoed up the wooden stairwell from the kitchens below—pots clanging, voices calling to and fro, the clash of crockery, a crash and subsequent cursing. The scent of roasting meat struck her in a gust of warm air, and she was suddenly ravenous.

He took her hand again and drew her away from the smell of food, through a plain, dim, unvarnished corridor into a larger one, with a canvas floor cloth that muffled their footsteps, into a broad corridor with a Tur-

key carpet in blue and gold and candles flickering in the bronze plates of reflectors that shed a bright, soft light over everything. Servants flitted past them like ghosts, carrying trays, jugs, garments, bottles, eyes averted.

It was like walking through a soundless dream: something between curiosity and nightmare, where you had no notion where you were going or what lay before you but were obliged to keep on walking.

He stopped abruptly and looked at her as though he'd found her walking through *his* dream—and perhaps it was, she thought, perhaps it was. He put a hand very lightly on her breast for an instant, fixing her in place, then vanished round a corner.

With him gone, her stunned senses began to awaken. She could hear music and voices, laughter. A strong smell of hot punch and wine; she'd drunk nothing save that first glass of champagne but now felt very drunk indeed. She opened and closed her fingers slowly, still feeling the grasp of his hand, hard and chilled.

Suddenly he was there again, and she felt his presence like a blow to her chest. He had her cape in his hand and swung it open, round her, enveloping her. As though it was part of the same movement, he took her in his arms and kissed her fiercely. Let go, panting, then did it again.

"You—" she said, but then stopped, having no idea what to say.

"I know," he said, as though he did, and with a hand under her elbow led her somewhere—she wasn't noticing anything anymore—and then there was a whoosh of cold, rainy night air and he was helping her up the step of a hansom cab.

"Where do you live?" he said, in an almost normal voice.

"Southwark," she said, sheer instinct preventing her from giving him her real address. "Bertram Street, Number Twenty-two," she added, inventing wildly.

He nodded. His face was white, his eyes dark in the night. The place between her legs burned and felt slippery. He swallowed and she saw his throat move, slick with rain and gleaming in the light from the lantern; he hadn't put on his neckcloth or his waistcoat, and his shirt was open under his scarlet coat.

He took her hand.

"I will call upon you tomorrow," he said. "To inquire after your welfare."

She didn't answer. He turned her hand over and kissed her palm. Then the door was shut and she was rattling alone over wet cobbles, her hand closed tight on the warmth of his breath.

She couldn't think. She felt wetness seep into her petticoats, with the slightly sticky feel of blood. The only thing floating through her mind was a remark of her father's. *"The English are notorious bores about virginity."*

16

IT WASN'T THAT HARD to disappear. The O'Higgins brothers were masters of the art, as they assured her.

"Leave it to us, sweetheart," Rafe said, taking the purse she handed him. "To a Londoner, the world beyond the end of his street is as furrin as the pope. All ye need do is keep away from the places folk are used to seein' ye."

She hadn't had much choice. She wasn't going anywhere near the Duke of Pardloe or his friend Quarry or the Twelvetrees brothers. But there was still business to be done before she could go back to Paris—books to be both sold and bought, shipments made and received—and a few bits of more-private business, as well.

So Minnie had written a note paying off Lady Buford and announcing her return to France and then stayed in Parson's Green with Aunt Simpson and her family for a month. She allowed the O'Higginses to do the more straightforward things and—with some reluctance—entrusted the more delicate acquisitions to Mr. Simpson and her cousin Joshua. There'd been two or three clients who had declined to meet with anyone save her, and though the temptation was considerable, the risk was too great, and she had simply not replied to those.

She had gone once with Aunt Simpson to the farm, to take leave of her mother. She hadn't been able to bring herself to go into Soeur Emmanuelle's chamber, though, and had only laid her head and hands against the cool wood of the door and wept silently.

But now it was all done. And she stood alone in the rain on the deck of the *Thunderbolt,* bobbing like a cork over the waves of the channel toward France. And her father.

THE LAST THING she would ever do, she vowed to herself, was to tell her father who it had been.

He knew who Pardloe was, what his family background had been, just how fragile his family's present grip on respectability. And thus Pardloe's vulnerability to blackmail.

Perhaps not outright blackmail . . . at least, she didn't want to believe her father engaged in that. He'd always told her to avoid it. Not on moral grounds—he had principles, her father, but not morals—but on the purely pragmatic grounds that it was dangerous.

"Most blackmailers are amateurs," he'd told her, handing her a small stack of letters to read—an educational exchange between a blackmailer and his victim, written in the late fifteenth century. "They don't know what it's decent to ask for, and they don't know how to quit, even if they wanted to. It doesn't take a victim long to realize that, and then . . . it's often death. For one or the other."

"In this instance"—he'd nodded at the crumbling brown-stained papers in her hand—"it was both of them. The woman being blackmailed invited the blackmailer to her home for dinner and poisoned him. But she used the wrong drug; it didn't kill him outright, but it worked fast enough for him to realize what she'd done, and he strangled her over the dessert."

No, he *probably* hadn't had any intent of blackmailing Pardloe himself.

At the same time, she was certainly intelligent enough to realize that the letters and documents her father dealt in were very often commissioned by or sold to persons who intended to *use* them for blackmail. She thought of Edward Twelvetrees and his brother and felt colder than the icy blast of the wind off the English Channel.

Were her father to realize that it was Pardloe who had debauched his daughter . . . *What on earth* would *he do?* she wondered.

He wouldn't scruple to kill Pardloe, if he could do it undetected, she was pretty sure of that. Though he *was* very pragmatic: he might just demand satisfaction of a financial nature as compensation for the loss of his daughter's virginity. That was a salable commodity, after all.

Or—the worst possibility of all—he might try to force the Duke of Pardloe to marry her.

That's what he'd wanted: to find her a rich English husband, preferably one well-placed in society.

"Over my dead body!" she said out loud, causing a passing deckhand to look at her strangely.

SHE'D REHEARSED IT on the journey back. How she'd tell her father—what she *wouldn't* tell him—what he might say, think, do . . . She had a speech composed—firm, calm, definite. She was prepared for him to shout, to rebuke, disown her, show her the door. She wasn't at all prepared for him to look at her standing in the doorway of the shop, gulp air, and burst into tears.

Flabbergasted, she said nothing and an instant later was being crushed in his arms.

"Are you all right?" He held her away from him, so he could look into her face, and swiped a sleeve across his own wet, anxious, gray-stubbled face. "Did the swine hurt you?"

She couldn't decide whether to say "What swine?" or "What are you talking about?" and instead settled on a dubious-sounding "No . . ."

He let go then and stepped back, reaching into his pocket for a handkerchief, which he handed her. She realized belatedly that she was sniffling and her own eyes were welling.

"I'm sorry," she said, all her speeches forgotten. "I didn't mean to. . . . to . . ." *But you did,* her heart reminded her. *You did mean it.* She swallowed that down with her tears and said instead, "I didn't mean to hurt you, Papa."

She hadn't called him that in years, and he made a sound as though someone had punched him in the belly.

"It's me that's sorry, girl," he said, his voice unsteady. "I let you go by yourself. I should never . . . I knew . . . Christ, I'll kill him!" Blood flooded his pale cheeks, and he slammed a fist on the counter.

"No, don't," she said, alarmed. "It was my fault. I—" *I what?*

He grabbed her by the shoulders and shook her, though not hard.

"Don't ever say that. It—whatever—however it happened, it wasn't your fault." His hands dropped away from her shoulders and he drew breath, panting as though he'd been running. "I—I—" He stopped and ran a trembling hand down his face, closing his eyes.

He took two more deep breaths, opened his eyes, and said, with some semblance of his normal calm, "Come and sit down, *ma chère.* I'll make us some tea."

She nodded and followed him, leaving her bag where it had fallen. The back room seemed at once completely familiar and quite strange, as though she had left it years ago rather than months. It smelled wrong, and she felt uneasy.

She sat down, though, and put her hands on the worn wooden tabletop. There was a spinning sensation in her head, and when she took a deep breath to try to stop it, the sense of seasickness came back, the smell of dust and ancient silk, stewed tea and the nervous sweat of many visitors curling into a greasy ball in her stomach.

"How . . . how did you find out?" she asked her father, in an effort to distract herself from the sense of clammy apprehension.

His back was to her, as he chiseled a chunk from the battered brick of tea and dropped it into the chipped Chinese pot with its blue peonies. He didn't turn around.

"How do you think?" he said evenly, and she thought suddenly of the spiders, the thousands of eyes, hanging motionless, watching . . .

"*Pardonnez-moi,*" she said, breathless, and, stumbling to her feet, blundered out into the corridor and to the alley door, where she threw up over the cobblestones outside.

She stayed outside for perhaps a quarter of an hour, letting the cold air in the shadows cool her face, letting the sounds of the city come back to her, the noise of the street a faint echo of normality. Then the bell of Sainte-Chapelle struck the hour, and all the others followed, the distant *bong* of Notre Dame de Paris telling Paris in a deep bronze voice that the hour was three o'clock.

"*It's almost time for None,*" her aunt had said. "*When she hears the bells, she won't do anything until the prayer is done, and often she's silent afterward.*"

"*None?*"

"*The hours,*" Mrs. Simpson had said, pushing the door open. "*Hurry, if you want her to speak with you.*"

She wiped her mouth on the hem of her skirt and went inside. Her father had finished making the tea; a fresh-poured cup sat by her place. She picked it up, took a mouthful of the steaming brew, swished it round her mouth, and spat it into the aspidistra.

"I saw my mother," she blurted.

He stared at her, so shocked that he didn't seem to breathe. After a long moment, he carefully unclenched his fists and laid his hands on the table, one atop the other.

"Where?" he said very quietly. His gaze was still fixed, intent on her face.

"In London," she said. "Did you know where she was—is?"

He'd started to think; she saw the thoughts flying behind his eyes. What did she know? Could he get away with lying? Then he blinked, took a breath, and let it out through his nose in a sigh of . . . decision, she thought.

"Yes," he said. "I . . . keep in touch with her sister. If you've met Emmanuelle, I imagine you've met Miriam, as well?" One of his unruly eyebrows went up, and she nodded.

"She said—said that you paid for her care. Have you seen her, though? Seen where they keep her, seen how she . . . is?" Emotion was rumbling through her like an approaching thunderstorm, and she had trouble keeping her voice steady.

"No," he said, and she saw he'd gone white to the lips, whether with anger or some other emotion, she couldn't tell. "I never saw her again, after she told me that she was with child." He swallowed, and his eyes went to his folded hands.

"I tried," he said, looking up as though she'd challenged him, even though she'd said nothing. "I went to the convent, spoke with the mother superior. She had me arrested." He laughed, shortly but not with humor. "Did you know that debauching a nun is a crime punishable by exposure in the pillory?"

"I imagine you bought your way out of it," she said, as nastily as she could.

"So would anyone capable of doing so, *ma chère*," he said, keeping his temper. "But I had to leave Paris. I hadn't met Miriam then, but I knew about her. I sent her word, and money, imploring her to find what they had done with Emmanuelle—to save her."

"She did."

"I know." He'd got hold of himself now and gave her a sharp look. "And if you've seen Emmanuelle, you know what her state is. She went mad when the child—"

"When *I* was born!" She slapped a hand on the table, and the cups chimed in their saucers. "Yes, I know. Do you bloody blame me for her—for what happened to her?"

"No," he said, with an obvious effort. "I don't."

"Good." She took a breath and blurted, "I'm pregnant."

He went dead white and she thought he might faint. She thought she might faint, too.

"No," he whispered. His eyes dropped to her middle, and a deep qualm there made her feel she might be sick again.

"No. I won't . . . I won't let such a thing happen to you!"

"You—" She wanted to strike him, might have done so had he not been on the other side of the table.

"Don't you dare tell me how I can get rid of it!" She swept the cup and saucer off the table, smashing them against the wall in a spray of Bohea. "I'd never do that—never, never, *never*!"

Her father took a deep breath and very consciously relaxed his posture. He was still white, and his eyes creased with emotion, but he had himself under control.

"That," he said softly, "is the last thing I would ever do. *Ma chère. Ma fille.*"

She saw that his eyes were full of tears and felt the blow in her heart. He'd come for her when she was born. Come for his child, cherished and kept her.

He saw her fists unclench and he took a step toward her, tentative, as though walking on ice. But she didn't recoil and didn't shout, and one more step and they were in each other's arms, both weeping. She'd so missed the smell of him, tobacco and black tea, ink and sweet wine.

"Papa . . ." she said, and then cried harder, because she'd never been

able to say "Mama" and never would, and this tiny, helpless thing she carried would never know a father. She'd never felt so sad—but at the same time comforted.

He'd cared. He'd come for her after she was born. He'd loved her. He always would—that was what he was saying now, murmuring into her hair, sniffing back the tears. He'd never let her be persecuted and abused as her mother was, never let harm come to her or to her child.

"I know," she said. Worn out, she rested her head on his chest, holding him as he held her. "I know."

17

RED WAX AND EVERYTHING

HAL STRODE OUT OF Sir William Yonge's office, boot heels brisk on the marble tiles and head held high. He nodded cordially to the soldier outside the door and made it down the stairs, along the hall, and out into the street, dignity intact. Harry was waiting across the street, anxious.

He saw Harry's face break into an enormous grin at sight of him, and then Harry threw back his head and howled like a wolf, to the startlement of Lord Pitt and two companions, who were coming along the pavement at the moment. Hal just managed to bow to them and then was across the street, hammering Harry's back and shoulders in joy. One-handed, because the other hand was clutching the precious certificate of commission to his bosom.

"God! We did it!"

"*You* did it!"

"No," Hal insisted, and shoved Harry in exhilaration. "Us. We did it. Look!" He waved the document, covered and sealed with red wax, under Harry's nose. "King's signature and everything! Shall I read it to you?"

"Yes, every word—but not out here." Harry gripped his elbow and hailed a passing cab. "Come on—we'll go to the Beefsteak; we can get a drink there."

Mr. Bodley, the club's steward, viewed them benignly as they tumbled into the club, calling for champagne and steak and more champagne, and within moments they were installed in the deserted dining room—it being

eleven o'clock in the morning—with a cold bottle to hand and steak ordered to follow.

". . . *commissioned this day by His Royal Majesty, by the grace of God, George the second* . . . oh, my God, I can't breathe . . . such a-a-*thing* . . ."

Hal laughed at that. His own chest had felt as though it were in a vise all the time he'd been in Sir William's office—but the vise had burst when he'd seen the certificate, with its unmistakable royal seal at the bottom, and now he breathed as freely as a newborn babe.

"Isn't it, though?" He could barely stand to have the certificate out of his hand and now reached out to trace the king's signature with a possessive forefinger. "I was sure when I went in there that it was all up, that Sir William would give me some cock-and-bull story for refusal, all the time eyeing me in that way people do when they think you're off your head and might just pick up an ax and brain them unexpectedly. Not that I haven't often felt that way," he added judiciously, and drained his glass. "Drink up, Harry!"

Harry did, coughed, and poured more.

"So what *did* happen? Was Yonge friendly, matter-of-fact . . . what did he *say*?"

Hal frowned, absently enjoying the fresh burst of dry bubbles on his tongue.

"Friendly enough . . . though I don't think I could tell quite *what* his manner was. Not nervous at all. And not that wary way politicals often are with me when they're thinking of Father."

Harry made a low noise in this throat, indicating complete understanding and sympathy—he'd been by Hal's side through his father's suicide and all the bloody mess that came afterward. Hal smiled at his friend and half-lifted his glass in silent acknowledgment.

"As to what he said, he greeted me very affably, asked me to sit, and offered me a currant biscuit."

Harry whistled.

"My God, you *are* honored. I hear he only gives biscuits to the king and the first minister. Though I imagine he'd give one to the queen, too, should she choose to visit his lair."

"I think the contingency is remote." Hal emptied the bottle and turned to call for another, but Mr. Bodley's tray was already at his elbow. "Oh, thank you, Mr. Bodley." He stifled a belch and realized that his head, while not swimming, was showing a slight disposition to float. "Do you think the steak will be long in coming?"

Mr. Bodley tilted his head from side to side in equivocation.

"A little time, my lord. But the cook has some wonderful small eel pies, just out of the oven—perhaps I could tempt you with a pair while you're waiting?"

Harry sniffed the fragrant air drifting in from the kitchen and closed his eyes in anticipatory bliss. The Beefsteak made their eel pies with the usual onion, butter, and parsley but also with nutmeg and dry sherry.

"Oh, God, yes."

Hal's mouth watered a bit at the thought—but the thought also brought a tightening of his body. Harry opened his eyes and looked surprised.

"What's the matter, old man?"

"Matter? Nothing." Mr. Bodley had freed the cork from its lead seal and now loosed it deftly with a soft burp and a hiss of rising bubbles. "Thank you, Mr. Bodley. Yes, eel pies by all means!"

"Eel pies," he repeated, as Mr. Bodley faded discreetly toward the kitchen. "The mention just reminded me of Kettrick's . . . and that young woman."

The thought of her—God damn it, why had he not even thought to make her tell him her real name? Lady Bedelia Houghton, for God's sake—caused its usual *frisson* of mixed emotions. Lust, curiosity, annoyance . . . longing? He didn't know if he'd put it that strongly, but he did have an intense desire to see her again, if only to find out what the devil she'd actually been doing. A desire now greatly intensified by his meeting with the secretary.

"Kettrick's?" Harry said, looking blank. "Kettrick's Eel-Pye House, you mean? And what young woman?"

Hal caught something in Harry's voice and gave his friend a sharp look.

"The girl I caught magicking the drawer of my desk, the night of the ball."

"Oh, that girl," Harry murmured, and buried his nose in his glass.

Hal looked harder at Harry. He hadn't told Harry everything—not by a long chalk, by God—but he *had* told him that he was satisfied with what she'd told him (actually, a long way from satisfied, but . . .) and that he'd sent her home in a coach and requested her address, which she'd given.

Only to discover that said address didn't exist, and when he'd tracked down the coach driver, an Irish rapscallion, the man had told him that the girl had professed to be starving—she was; he'd heard her stomach growling when he . . . oh, Jesus—and had asked him to put her down for a moment at Kettrick's. He had, and the girl had promptly walked through the house, out the back, and legged it down an alley, never to be seen again.

Which, Hal thought, was a sufficiently interesting story as to have stuck in Harry's mind. To say nothing of the fact that he'd several times mentioned the girl, as well as his efforts to find her, to Harry.

"Hmph," he said, drank more, and shook his head to clear it. "Well, regardless . . . there was a bit of cordial conversation, quite cordial, though all through it there was something . . . odd . . . in Sir William's manner.

Rather grave—that's why I thought he was working up to a refusal—but then . . . sympathetic."

"Really?" Harry's thick brows shot up. "Why, do you suppose?"

Hal shook his head again, baffled.

"I don't know. Only . . . at the end, when he'd given me the certificate and congratulated me, he shook my hand and held on to it for a moment, and . . . he gave me a brief word of condolence on my . . . my loss." He'd thought he had his emotions well in check, but the pang was sharp as ever and he was obliged to clear his throat.

"Only being decent, surely," Harry said gruffly. Hal saw, to his fascination, that the blood was rising up Harry's neck and into his cheeks.

"Yes," he said, and leaned back, casual, glass in hand, but an eye on Harry. "At the time, I was so elated that I wouldn't have cared if he'd told me that a crocodile had hold of my foot, but with more-sober thought . . ."

Harry hooted slightly at that but then settled into his glass, eyes on the tablecloth. The flush had spread to his nose, now faintly glowing.

"I wondered—actually, just now—whether perhaps it was some sort of oblique reference to that bloody petition. You know, the one Reginald Twelvetrees brought, claiming that I'd assassinated his brother while off my head."

"He—didn't actually mention the petition?"

Hal shook his head. "No."

The eel pies arrived at this moment, smoking and savory, and no more was said for a bit.

Hal wiped the last bit of juice out of the dish with a sop of bread, chewed blissfully, swallowed, then opened his eyes and gave Harry a straight look.

"What the devil do you know about that petition, Harry?"

He'd known Harry Quarry since Harry was two and himself five. Harry *could* lie, if given warning and enough time to prepare, but he couldn't lie to Hal and knew it.

Harry sighed, closed his eyes, and thought for a bit, then opened one eye cautiously. Hal raised both brows and laid his hands flat on the table, in demonstration of the fact that he wasn't about to either hit Harry or strangle him. Harry looked down and bit his lip.

"Harry," Hal said softly. "Whatever you did, I forgive you. Just bloody tell me, all right?"

Harry looked up, nodded, drew a deep breath, and did.

"Irrumabo," Hal said, more in astonishment than anger. "But you told her not to take the letters, you say. . . ."

"Yes. I swear I did, Hal." The flush had diffused and was beginning to fade. "I mean—I knew what you felt—about—"

"I believe you." Hal was feeling a bit flushed himself and looked away.

Mr. Bodley was approaching with fresh plates and silver, followed by one of the club's waiters, ceremoniously carrying a sizzling platter.

They sat quietly while the steak—accompanied by a heap of wild mushrooms, garnished with tiny boiled onions and glistening with butter—was served. Hal watched, smelled, made the appropriate noises of appreciation to Mr. Bodley, and asked for a bottle of good Bordeaux. All this was purely automatic, though; his mind was in the library, on the night of the ball.

"I didn't want you to be hurt." He could still see the look on her face when she'd said it, and he believed her now just as much as he had then, the firelight glowing in her eyes, on her skin, in the folds of her green dress. *"Shall I prove it?"*

And she had, after all, proved it. A violent shiver ran through him at the memory.

"Are you all right, old man?" Harry was looking at him anxiously, a forkful of steak halfway to his mouth.

"I—yes," he said abruptly. "But she wasn't stealing Esmé's—I mean—the letters from my desk; she was putting them *back*. I know she was; I saw her close the drawer before she saw me. So she didn't send them to Sir William, I'm sure of that."

Harry nodded slowly. "I . . . don't like to suggest such a thing," he said, looking unhappy. "I mean—I trusted her, foolish as that likely was. But could she . . . copies, perhaps? Because the way you describe Yonge's manner . . ."

Hal shook his head.

"I'd swear not. The way she . . . No. I'm sure not. If nothing else . . ." He hesitated, but it was, after all, Harry. He swallowed and went on, eyes fixed on his plate but his voice steady. "If Sir William had seen those letters, he couldn't have looked me in the face, let alone have behaved as he did. No. Something convinced him that I had cause to challenge Twelvetrees, I'm sure of that—but God alone knows what it was. Perhaps the—the girl—did find someone who . . . knew about the affair . . ." Blood burned in his cheeks, and the pattern on the fork was digging into his palm where he clutched it. "If someone of good character swore to it . . ."

Harry let out a breath, nodding.

"You're right. And—that *was* what I'd asked her to do. Er . . . ask about discreetly, I mean. Um . . . sorry."

Hal nodded but couldn't speak. He did forgive Harry, but the thought that someone—someone unknown to him—had known . . . He had a brief, vivid urge to seize a candle from the sconce and set his head on fire in order to obliterate the thought, but instead he closed his eyes and breathed deeply for a few moments. The tightness in his chest began to loosen.

Well. Nothing to be done about it now. And the regiment *was* all right. He felt a bit of his earlier euphoria return and opened his eyes. Yes, by God, it was. There was the certificate, red wax seal and all, right there on the linen cloth.

He unclenched the fork, made himself pick up the knife, and cut into his steak. Hot red juice ran out, and he saw in memory the small blood-stain on the white hearthrug. Heat washed over him as though he *had* set his hair on fire.

"One thing you could do, Harry—if you're of a mind . . ."

"Anything you like, old man."

"Help me find her."

Harry stopped, fork halfway to his half-opened mouth.

"Of course," he said slowly, and lowered the fork. "But—" But, his face said, they'd both been looking for the past three weeks. Miss Rennie had vanished as surely as though she'd gone up in smoke.

Hal suddenly laughed. Mr. Bodley had materialized with the Bordeaux, and a brimming glass sat by his elbow.

"Confusion to all Twelvetrees!" Harry said, hoisting his own glass. Hal returned the salute and drank deep. It was a gorgeous wine, deep, strong, and smelling of cherries and buttered toast. Another bottle of this—well, maybe two—and he might just feel able to deal with things.

"One thing my father always said to me, Harry: 'They can't beat you if you don't give up.' And"—he lifted the glass to his friend—"I don't."

Harry's face cleared and he gave Hal a lopsided smile, returning the toast. "No," he said. "God help us all, you don't."

18

TAKING FLIGHT

Amsterdam, Kalverstraat 18
January 3, 1745

MINNIE CAREFULLY BRUSHED powdered sugar off the ledger. The early queasiness of pregnancy had mostly passed, re-placed by the appetite of a ravening owl, according to her father.

"An *owl?*" she'd said, and he nodded, smiling. His shock had passed

along with her queasiness, and his face took on a rapt look sometimes when she caught him watching her.

"You look at food, *ma chère,* and turn your head to and then fro, as though you expect it to bolt, and then you swoop on it and—*gulp!*—it's gone."

"Bah," she said now, and looked to see if there were more *oliebollen* in the pottery jar, but, no, she'd finished them. Mortimer's antics had abated and he'd fallen into a stupor, as he usually did when she ate, but she was still hungry.

"Is dinner nearly ready?" she called downstairs to her father. In the usual Amsterdam style, the house was long and narrow, the shop on the ground floor, living quarters above, and the kitchen in the basement. A savory smell of roasting chicken had been creeping up the steps for the last hour, and she was famished, in spite of the *oliebollen.*

Instead of an answer, she heard the sound of her father's feet coming up the stairs, accompanied by a rattle of stoneware and pewter.

"It's not even noon," he said mildly, setting down a tray on the counter. "Dinner won't be ready for another hour at least. But I've brought you some coffee and rolls with honey."

"Honey?" She sniffed pleasurably. Even though the queasiness had mostly gone, the acute sensitivity to smells remained, and the strong aroma of coffee with fresh buttered rolls ravished her.

"That child is nearly as big as you are now," her father observed, with an eye to her protuberant belly. "*When* did you say it will be born?"

"In about three months," she said, reaching for a roll and ignoring the implication. "And the midwife says it will be just about double in size by then." She glanced down at Mortimer's bulge. "I don't actually think such a thing is possible, but that's what she says."

Her father laughed and, leaning across the counter, rested a hand lightly on the curve of his grandchild.

"*Comment ça va, mon petit?*" he said.

"What makes you think it's a boy?" she asked, though she didn't move away. It touched her when he spoke to the baby; he always did so with the greatest tenderness.

"Well, you call him—it—Mortimer," he pointed out, and with a gentle pat withdrew his hand. "I suppose that means *you* think he's a male."

"I was just taken by the advertisement on a bottle of English patent medicine: *Mortimer's Dissolving, Resolving, and Absolving Tonic—removes stains of any kind: physical, emotional, or moral.*"

That took him aback; he wasn't sure whether she was joking. She saved him by laughing herself and waved him away to the kitchen. She loved Sundays, when Hulda, the maid of all work, stayed at home with her family, leaving the two Snyders—Willem Snyder being her father's *nom de*

guerre in the Low Countries—to fend for themselves. Her father was a much better cook, and it was peaceful without Hulda's solicitous questions and repeated suggestions of "nice gentlemen" among the shop's clientele who might be willing to take on a young widow with a child, if Mr. Snyder was able to offer a sufficiently generous inducement. . . .

Frankly, she thought her father wouldn't be above it. But he wouldn't push her into anything, either. She thought he was actually loath to part with her—and Mortimer, no doubt.

She closed her eyes, savoring the contrast of bitter coffee followed by a bite of buttered roll dripping with honey. As though stimulated by the coffee, Mortimer suddenly stretched himself as far as possible, making her clutch her belly and gasp.

"You little bastard," she said to him, and paused to swallow the last of the honeyed bite. "Sorry. You're *not* a bastard." At least he wouldn't be, as far as he or the rest of the world knew. He'd be the posthumous child of . . . Well, she hadn't quite decided. For the moment he was the child of a Spanish captain of rifles named Mondragon, dead of fever in some conveniently obscure campaign, but she'd think of something better by the time Mortimer was old enough to ask questions.

Perhaps a German; there were enough small duchies and principalities among which to hide an irregular birth—though the Germans *were* annoyingly methodical about registering people. Italy—now, there was an unmethodical country for you, and it was warm. . . .

He wouldn't be an Englishman, though. She sighed and put a hand over the little foot poking inquisitively under her liver. Mortimer *could* be a girl, she supposed, but Minnie couldn't think of him as anything other than male. Because she couldn't think of him without thinking of his father.

Maybe she *would* marry. Eventually.

Time enough for such considerations. For the moment, there was an inconsistency in the accounts between September and October, and she took a fresh sheet of foolscap and picked up her quill, on the trail of an errant three guilders.

Half an hour later, the stray guilders finally captured and pinned firmly to their proper column, she stretched, groaned, and hoisted herself to her feet. Her belly, much given to odd noises of late, was gurgling in ominous fashion. If dinner wasn't ready yet, she was going to—

The bell over the door tinged briskly, and she looked up, surprised. The virtuous Protestants of Amsterdam would never think of going anywhere on Sunday but to church. The man standing in the doorway, though, was neither Dutch nor virtuous. He *was* wearing a British uniform.

"Your . . . Grace?" she said stupidly.

"Hal," he said. "My name's Hal." Then he caught full sight of her and turned as white as the spilled sugar on the counter. "Jesus Christ."

"It's not . . ." she began, sliding out from behind the counter, "what you think . . ." she ended faintly.

It didn't matter. He took an enormous breath and strode toward her. She dimly heard her father coming up the stairs but saw nothing but that bone-white face, caught between shock and determination.

He reached her, bent his knees, and picked her up.

"Jesus Christ!" he said again, this time in response to her weight, which was considerable. Clenching his teeth, he clutched her tightly and wove his way across the shop, staggering only slightly. He smelled wonderfully of bay leaves and leather.

The door stood open, with Harry Quarry holding it and a blast of cold winter air coming in. His solid, square face broke into an enormous grin as he met her eyes.

"Pleased to see you again, Miss Rennie. Hurry up, old man, somebody's coming."

"Minnie! Stop! You—" Her father's shout was cut off by the slam of the shop door, and a moment later she was dumped unceremoniously into a coach that stood waiting. Hal shot in after her, and Harry hung precariously off the coach's step, shouting at the driver, before swinging inside himself and slamming the door.

"Minnie!" Her father's shout reached her, faint but audible.

She tried to turn, to look out of the rear window, but couldn't manage it without actually standing up and rotating her entire body. Before she could even contemplate doing that, though, Hal had wriggled free of his blue military cloak and was tucking it round her. The warmth of his body surrounded her, and his face was no more than a few inches from hers, still white, the warmth of his breath on her cheek white, too, misting in the frigid air of the coach.

His hands were on her shoulders, steadying her against the jolting, and she thought he might kiss her, but a sudden lurch as the coach swung round a corner sent him staggering. He fell backward into the seat opposite, beside Harry Quarry, who was still grinning from ear to ear.

She took a deep breath and readjusted her skirts over her bulge.

"Where do you think you're taking me?"

He'd been staring at her intensely but evidently without actually seeing her, for her words made him jerk.

"What?"

"*Where* are you taking me?" she repeated, louder.

"I don't know," he said, and looked at Harry, beside him. "Where are we going?"

"Place on the Keizersgracht," Harry said with a shrug. "Called *De Gevulde Gans.*"

"The Stuffed Goose? You're taking me to a *pub?*" Her voice rose involuntarily.

"I'm taking you to be married," Hal said, frowning at her.

He was very pale, and a muscle near his mouth twitched—the only thing he couldn't control, she thought. Well, that, and her.

"I married a lady and she became a whore. I cannot complain if it should be the other way about this time."

"You think I'm a whore, do you?" She wasn't sure whether to be amused or insulted. Perhaps both.

"Do you normally sleep with your victims, madam?"

She gave him a long, level stare and folded her arms atop the rounded curve of her belly.

"I wasn't asleep, Your Grace, and if you had been, I think I would have noticed."

THE STUFFED GOOSE was a rather down-at-heel establishment, with a drunkard bundled in rags picturesquely huddled against the steps.

"Why did you pick *this* place?" she asked Harry, picking up her skirts to avoid a small heap of vomit on the stones and glancing at the grimy doorknob.

"The landlady's husband is a minister," he said reasonably, leaning to open the door for her. "And reputed not to be too fussed about things."

Things like a wedding license, she supposed. Though perhaps you didn't need one when getting married in a different country?

"Go in," said Hal impatiently, behind her. "It stinks out here."

"And you think it will be better inside?" she asked, pinching her nose in preparation. He was right, though: the breeze had shifted, and she caught the full impact of the drunkard's scent.

"Oh, God," she said, turned neatly on her heel, and threw up on the opposite side of the step.

"Oh, God," said Hal. "Never mind, I'll get you some gin. Now go inside, for God's sake." He pulled a large white handkerchief out of his sleeve, wiped her mouth briskly with it, and hustled her through the door.

Harry had already gone in and opened negotiations, in bad but serviceable Dutch, this augmented by a substantial purse, which he plonked on the bar with a loud clinking noise.

Hal, who apparently had no Dutch, interrupted Harry's conversation with the landlady behind the bar by removing a golden guinea from his pocket and tossing it onto the bar.

"Gin," he said.

Minnie had subsided onto a stool as soon as she entered and was curled over, eyes shut and Hal's handkerchief clutched in one hand, trying not to breathe. A moment later, though, the sharp, clean scent of juniper cut through the miasma of the pub and the hint of dead rat. She swallowed, made herself sit up, and took the cup of gin Hal handed her.

To her considerable surprise, it worked. The nausea subsided with the first sip, the desire to lie down on the floor faded, and within a few moments she felt relatively normal—or as normal as one might feel if six months' pregnant and on the verge of marrying Hal, she thought.

The minister, apparently rousted from bed and evidently suffering from an extreme form of *la grippe,* turned bleary eyes from Hal to Minnie, then back.

"You want to marry her?" The tone of incredulity seeped through the nasal congestion, slow and glutinous.

"Yes," said Hal. "Now, if you please."

The minister closed one eye and looked at him, then turned his head slowly to his wife, who tutted impatiently and said something rapid in Dutch, accompanied by a peremptory gesture. He hunched his shoulders against the tirade in a way indicating that such assaults were common. When she stopped speaking, he nodded in a resigned fashion, drew a sodden handkerchief from the pocket of his sagging breeches, and blew his nose.

Hal's hand tightened on Minnie's; he hadn't let go since they'd entered the pub, and she twitched, not quite pulling away. He looked down at her.

"Sorry," he said, and loosened—but didn't release—his grip.

"She's wis child," said the minister, in a reproachful tone.

"I know that," Hal said, tightening his hold once more. "Get on with it, please. At once."

"Why?" said Minnie, mildly provoked. "Do you have somewhere special you have to be?"

"No," he said, narrowing his eyes at her. "But I want the child to be legitimate, and I think you may give birth to it at any moment."

"I will *not,*" she said, offended. "You *know* I'm no more than six months gone!"

"You look like a—" Catching a glimpse of her eyes at this point, he shut his mouth abruptly, coughed, and turned his attention once more to the minister. "Do please continue, sir."

The man nodded, blew his nose again, and motioned to his wife, who bent to rummage beneath the bar, eventually emerging with a battered prayer book, its cover spotted with kronk rings.

Possessed of this talisman, the minister seemed to take heart and straightened up a little.

"You heb witnesses?" he asked Hal.

"Yes," said Hal, impatient. "He's—Harry? Dammit, he went out to pay the carriage. Stay here!" he commanded Minnie, and, dropping her hand, strode out.

The minister looked dubiously after him, then at Minnie. The end of his nose was moist and scarlet, and tiny veins empurpled his cheeks.

"You are willing to marry dis man?" he asked. "I see he is rich, but maybe better to take a poor man who will treat you well."

"Ze is zes maanden zwanger, idioot," said the minister's wife. "She's six months gone with child." *"Is dit die schurk die je zwanger heeft gemaakt?"* She removed the pipe from the corner of her mouth and gestured from the door to Minnie's belly: *"He's the no-good who got you pregnant?"* A hefty kick from the occupant made Minnie grunt and double over.

"Ja, is die schurk," she assured the woman, glancing over her shoulder to the door, where Hal's shadow in the window was visible, a larger shadow that must be Harry behind him.

The men entered with a blast of winter air and the woman exchanged a look with her husband. Both shrugged, and the minister opened the book and began thumbing through it in a helpless sort of way.

Harry smiled reassuringly at Minnie and patted her hand before lining up solidly beside Hal. Oddly enough, she did feel reassured. If a man like Harry was Hal's good friend, then perhaps—just perhaps—she wasn't wrong about him.

Not that it would make any difference at this point, she thought, feeling a strangely pleasant shiver run up her back. It felt as though she were about to jump off a cliff but feeling a great pair of wings unfurling at her back, even as she looked out into the wind.

"Mag ik uw volledige naam alstublieft?" "What are your names, please?" The landlady had pulled out a ratty register book—it might be the accounts for the pub, Minnie thought, looking at the stained pages. But the woman turned to a clean, blank page at the back of the book and dipped her quill, expectant.

Hal looked blank for a moment, then said firmly, "Harold Grey."

"Only two names?" Minnie said, surprised. "No titles?"

"No," he said. "It's not the Duke of Pardloe or even the Earl of Melton you're marrying. Just me. Sorry to disappoint you, if that's what you thought," he added, in a tone that actually sounded apologetic.

"Not at all," she said politely.

"My middle name's Patricius," he blurted. "Harold Patricius Gerard Bleeker Grey."

"Really?"

"Ik na gat niet allemaal opschrijven," the woman objected. "I'm not going to write all that."

"Bleeker—*dat is Nederlands,*" the minister said, in surprised approval. "Your family is Dutch?"

"My father's mother's mother," Hal said, equally surprised.

The woman shrugged and wrote down the words, repeating, "Harold . . . Bleeker . . . Grey," to herself. *"En u?"* she asked, looking up at Minnie.

Minnie would have thought her heart couldn't go any faster, but she was wrong. Loose as her stays were, she felt light-headed, and before she could gather enough breath to speak, Hal stepped in.

"She's called Wilhelmina Rennie," he told the woman.

"Actually, it's Minerva Wattiswade," she said, getting a solid breath. Hal looked down at her, frowning.

"Wattiswade? What's Wattiswade?"

"Not what," she said, with exaggerated patience. "Who. Me, in fact."

This appeared to be too much for Hal, who looked to Harry for help.

"She means her name isn't Rennie, old man. It's Wattiswade."

"Nobody's named Wattiswade," Hal objected, transferring the frown back to Minnie. "I'm not marrying you under an assumed name."

"I'm not bloody marrying *you* under an assumed name!" she said. "Gah!"

"What—"

"Your bloody baby kicked me in the liver!"

"Oh." Hal looked somewhat abashed. "You mean your name really *is* Wattiswade, then."

"Yes, I do."

He took a deep breath.

"All right. Wattiswade. Why—never mind. You'll tell me later why you've been calling yourself Rennie."

"No, I won't."

He glanced at her, brows raised high, and she could see him—for once—debating whether to say something. But then his eyes lost the look of a man talking to himself and focused on hers.

"All right," he said softly, and held out his hand to her, palm upward.

She took another breath, looked out into the void, and jumped.

"Cunnegunda," she said, and put her hand in his. "Minerva Cunnegunda Wattiswade."

He said nothing, but she could feel him vibrating slightly. She carefully didn't look at him. Harry seemed to be arguing about something with the woman—something to do with the need for a second witness, she thought, but she couldn't concentrate enough to make out the words. The smell of tobacco smoke and stale sweat was making her gorge rise again, and she swallowed hard, several times.

All right. They'd decided that Mrs. Ten Boom could be the second wit-

ness. Good. Mortimer turned a somersault, landing heavily. Perspiration had broken out on Minnie's temples, and her ears felt hot.

Suddenly she was possessed by the fear that her father would burst through the door at any moment. She wasn't afraid of his stopping this impromptu ceremony; she was quite sure Hal wouldn't let him—and that certainty steadied her. Still . . . she didn't want him here. This was hers alone.

"Hurry," she said to Hal, in a low voice. "Please, hurry."

"Get on with it," he said to the minister, in a voice that wasn't particularly loud but plainly expected to be obeyed. The Reverend Ten Boom blinked, coughed, and opened his book.

It was all in Dutch; she could have followed the words but didn't—what echoed in her ears were the never-spoken phrases from the letters.

Not Esmé's—his. Letters written to a dead wife, in passionate grief, in fury, in despair. He might as well have punctured his own wrist with the sharpened quill and written those words in blood. She looked up at him now, white as the winter sky, as though all the blood had run out of his body, leaving him drained.

But his eyes were a pale and piercing blue when he turned his dark-browed face toward her, and the fire in him was not quenched, by any means.

You didn't deserve him, she thought toward the absent Esmé and rested her free hand on her gently heaving stomach. *But you loved him. Don't fret; I'll take care of them both.*

AUTHOR'S NOTE

IF YOU NEVER READ MADELEINE L'ENGLE'S MARVEL-ous *A Wrinkle in Time* in your younger years, it's not too late. It's a wonderful story and I highly recommend it. If you *did* read it, though, you'll certainly remember this iconic line: *There is such a thing as a tesseract.*

In fact, there is such a thing as a tesseract, both as a geometrical and a scientific concept: Putting it crudely, it's a four-dimensional construct, in which the fourth dimension is time. And it's used as a fictional device to bring two separate space/time lines together, obviating the linear time between them. Much more convenient than a clunky old time machine.

Now, it's also a well-known fact that I stink at ages. I have only the vaguest general notion as to how old anyone in these stories is at any given point, I usually don't know when their birthdays are, and I don't really care. This drives both my copy editor and the more OCD-prone of my readers to distraction, and they Aren't Going to Be Happy about this, but really, there's no choice.

When I wrote *The Scottish Prisoner,* I randomly assigned ages to Hal's and Minnie's young sons, never thinking we'd see them again until they were adults (we have in fact seen all of them at one time or another as adults in *An Echo in the Bone* and in *Written in My Own Heart's Blood*).

Now . . . I *also* noted in *The Scottish Prisoner* that Jamie Fraser had met Minnie prior to her marriage, in Paris, and that they had known each other in the context of the Jacobite plots of that time. That's something of a plot point, has to do with both their characters and their subsequent actions, and so is important.

And I allowed Minnie to tell Lord John the circumstances of her marriage to his brother Hal. That's also important, as indicating something of the relationships between Minnie and Hal and just why he calls on her for help in intelligence matters at various points in later stories.

So—those two facts are important. How old the kids are *isn't* important.

But going back to tell more of Minnie and Hal's backstory, naturally I wanted to include Minnie's acquaintance with Jamie Fraser. Okay, that *had* to take place sometime in 1744, when the Frasers were in Paris, plotting away.

Minnie's pregnancy and the impending birth of her first son, Benjamin, had much to do with the marriage between Minnie and Hal and with her feelings about it. Ergo, Benjamin has to have been conceived sometime in 1744.

As the more nitpicking sort of reader will have instantly realized, if Benjamin *was* conceived in 1744 and born in 1745—as he has to have been—then he can't have been eight years old in 1760, when *The Scottish Prisoner* takes place. Only he was.

Obviously, the only way to reconcile Benjamin's age—as well as those of his brothers, Henry and Adam—is to draw the logical conclusion that a tesseract occurred somewhere between the writing of *The Scottish Prisoner* and "A Fugitive Green," and the boys will all be full-grown men next time we see them and it won't matter. Luckily, I have full confidence in the mental ability of my Very Intelligent Readers to grasp this concept and enjoy the story without further pointless fretting.

Whale Painters

At one point, while contemplating the subtle color of her eau-de-nil dress, Minnie refers mentally to her acquaintance with a Mr. Vernet, who is a whale painter.

Whale painting was actually a thing in the eighteenth century: There was great demand for the production of romantically watery, adventurous paintings, and thus there were specialists in that production. Claude Joseph Vernet was a real historical artist whose profession consisted mostly of painting seascapes, many including whales. As such, he would also be a great expert in the delineation of water and its many colors and thus in a position to tell Minnie about the concept of "a fugitive green"—i.e., green paint at that time was made with a pigment given to fading out eventually, unlike the more robust and permanent blues and grays.

And of course you all understand the metaphorical allusion of the title. (I actually included M. Vernet in order to make it clear to readers who don't speak French and don't necessarily stop to Google unknown terms while reading that eau-de-nil is, in fact, a shade of green.)

BESIEGED

Jamaica
Early May 1762

L ORD JOHN GREY DIPPED a finger gingerly into the little stone pot, withdrew it, glistening, and sniffed cautiously.

"Jesus!"

"Yes, me lord. That's what I said." His valet, Tom Byrd, face carefully averted, put the lid back on the pot. "Was you to rub yourself with *that* stuff, you'd be drawing flies in their hundreds, same as if you were summat that was dead. *Long* dead," he added, and muffled the pot in a napkin for additional protection.

"Well, in justice," Grey said dubiously, "I suppose the whale *is* long dead." He looked at the far wall of his office. There were a number of flies resting along the wainscoting, as usual, fat and black as currants against the white plaster. Sure enough, a couple of them had already risen into the air, circling lazily toward the pot of whale oil. "Where did you get that stuff?"

"The owner of the Moor's Head keeps a keg of it; he burns it in his lamps—cheaper nor even tallow candles, he says, let alone proper wax ones."

"Ah. I daresay." Given the usual smell of the Moor's Head on a busy night, nobody would notice the stink of whale oil above the symphony of other reeks.

"Easier to come by on Jamaica than bear grease, I reckon," Tom remarked, picking up the pot. "D'you want me to try it with the mint, me lord? It *might* help," he added, with a dubious wrinkle of the nose.

Tom had automatically picked up the oily rag that lived on the corner of Grey's desk and, with a dexterous flick, snapped a fat fly out of the air and into oblivion.

"Dead whale garnished with mint? That should cause my blood to be especially attractive to the more discriminating biting insects in Charles

Town—to say nothing of Canada." Jamaican flies were a nuisance but seldom carnivorous, and the sea breeze and muslin window screening kept most mosquitoes at bay. The swamps of coastal America, though . . . and the deep Canadian woods, his ultimate destination . . .

"No," Grey said reluctantly, scratching his neck at the mere thought of Canadian deer flies. "I can't attend Mr. Mullryne's celebration of his new plantation house basted in whale oil. Perhaps we can get bear grease in South Carolina. Meanwhile . . . sweet oil, perhaps?"

Tom shook his head decidedly.

"No, me lord. Azeel says sweet oil draws spiders. They come and lick it off your skin whilst you're asleep."

Lord John and his valet shuddered simultaneously, recollecting last week's experience with a banana spider—a creature with a leg span the size of a child's hand—that had burst unexpectedly out of a ripe banana, followed by what appeared at the time to be several hundred small offspring, at a garden party given by Grey to mark his departure from the island and to welcome the Honorable Mr. Houghton Braythwaite, his successor as governor.

"I thought he'd have an apoplexy on the spot," Grey said, lips twitching.

"Likely wishes he had."

Grey looked at Tom, Tom at Grey, and they burst into suffocated snorts of laughter at the memory of the Honorable Mr. Braythwaite's face on this occasion.

"Come, come," Lord John said, getting himself under control. "This will never do. Have you—"

The rumble of a carriage coming up the gravel drive of King's House interrupted him.

"Oh, God, is that him now?" Grey glanced guiltily round at the disarray of his office: A gaping half-packed portmanteau lolled in the corner, and the desk was strewn with scattered documents and the remnants of lunch, in no condition to be viewed by the man who would inherit it tomorrow. "Run out and distract him, will you, Tom? Take him to the receiving room and pour rum into him. I'll come and fetch him as soon as I've done . . . something . . . about this." He waved a hand at the debris, and Tom obligingly vanished.

Grey picked up the oiled rag and disposed of an unwary fly, then seized a plate scattered with bread crusts, blobs of custard, and fruit peelings and decanted this out of the window into the garden beneath. Thrusting the empty plate out of sight under the desk, he began hurriedly to gather papers into piles but was interrupted almost at once by the reappearance of Tom, looking excited.

"Me lord! It's General Stanley!"

"Who?" Grey said blankly. His mind, occupied with the details of imminent escape, refused to deal with anything that might interfere with said escape, but "Stanley" did ring a distant, small bell.

"Might be as he's your mother's husband, me lord?" Tom said, with a becoming diffidence.

"Oh . . . *that* General Stanley. Why didn't you say so?" John hastily grabbed his coat from its hook and shrugged into it, brushing crumbs off his waistcoat as he did so. "Show him in, by all means!"

John in fact liked his mother's third husband—she having been twice widowed when she acquired the general four years before—though any military intrusion at this point was something to be regarded warily.

Wariness was, as usual, justified. The General Stanley who eventually appeared was not the bluff, jaunty, self-confident man last seen in his mother's company. This General Stanley was hobbling with a stick, his right foot bound up in an immense bandage, and his face gray with pain, effort . . . and profound anxiety.

"General!" John seized him by the arm before he could fall over and guided him to the nearest chair, hastily removing a pile of maps from it. "Do sit down, please—Tom, would you . . . ?"

"Just here, me lord." Tom had dug Grey's flask out of the open traveling bag with commendable promptitude and now thrust it into General Stanley's hand.

The general accepted this without question and drank deeply.

"Dear Lord," he said, setting the flask on his knee and breathing heavily. "I thought I shouldn't make it from the landing." He took another drink, somewhat more slowly, eyes closed.

"More brandy, Tom, if you please?" Grey said, watching this. Tom gave the general an assessing look, not sure whether he might die before more brandy could be fetched, but decided to bet on the general's survival and disappeared in search of sustenance.

"God." The general looked a good deal short of human but distinctly better than he had. He nodded thanks to John and handed back the empty flask with a trembling hand. "The doctor says I mustn't drink wine—apparently it's bad for the gout—but I don't recall his mentioning brandy."

"Good," John said, glancing at the bandaged foot. "Did he say anything about rum?"

"Not a word."

"Excellent. I'm down to my last bottle of French brandy, but we've got quite a lot of rum."

"Bring the cask." The general was beginning to show a tinge of color and, at this point, began to be cognizant of his surroundings. "You were packing to leave?"

"I *am* packing to leave, yes," John said, the feeling of wariness develop-

ing small, prickling feet inside his stomach. "I'm meant to sail tonight, for Charles Town."

"Thank God. I was afraid I shouldn't make it in time." The general breathed audibly for a moment, then gathered himself. "It's your mother."

"*What's* my mother?" The wariness turned instantly to a flare of alarm. "What's happened to her?"

"Nothing, yet. Or at least I sincerely hope not." The general patted the air in a vague gesture of reassurance that failed singularly to reassure.

"Where the devil is she? And what in God's name is she up to now?" Grey spoke with more heat than filial respect, but panic made him edgy.

"She's in Havana," General Stanley said. "Minding your cousin Olivia."

This seemed like a moderately respectable thing for an elderly lady to be doing, and Grey relaxed slightly. But only slightly.

"Is she ill?" he asked.

"I hope not. She said in her last letter that there was an outbreak of some sort of ague in the city, but she herself was in good health."

"Fine." Tom had come back with the brandy bottle, and John poured himself a small glass. "I trust she's enjoying the weather." He raised an eyebrow at his stepfather, who sighed deeply and put his hands on his knees.

"I'm sure she is. The problem, my boy, is that the British Navy is on its way to lay siege to the city of Havana, and I really think it would be a good idea if your mother wasn't *in* the city when they get there."

FOR A MOMENT, John stood frozen, glass in hand, mouth open, and his brain so congested with questions that he was unable to articulate any of them. At last, he gulped the remains of his drink, coughed, and said mildly, "Oh, I see. How does my mother come to be in Havana to start with?"

The general leaned back and let out a long breath.

"It's all the fault of that Stubbs fellow."

"Stubbs . . . ?" It sounded vaguely familiar, but stunned as he was, Grey couldn't think why.

"You know, chap who married your cousin Olivia. Looks like a build-er's brick. What's his Christian name . . . Matthew? No, Malcolm, that's it. Malcolm Stubbs."

Grey reached for the brandy bottle, but Tom was already pouring a fresh glass, which he thrust into his employer's hand. He carefully avoided meeting Grey's eye.

"Malcolm Stubbs." Grey sipped brandy, to give himself time to think. "Yes, of course. I . . . take it that he's quite recovered, then?" On one

level, this was good news; Malcolm Stubbs had lost a foot and part of the adjoining leg to a cannonball at the Battle of Quebec, more than two years before. By good luck, Grey had fallen over him on the field and had the presence of mind to use his belt as a tourniquet, thus preventing Stubbs from bleeding to death. He vividly recalled the splintered bone protruding from the remnants of Malcolm's shin, and the hot, wet smell of blood and shit, steaming in the cold air. He took a deeper swallow of brandy.

"Yes, quite. Got an artificial foot, gets around quite well—even rides."

"Good for him," Grey said, rather shortly. There were a few other things he recalled about Malcolm Stubbs. "Is *he* in Havana?"

The general looked surprised.

"Yes, didn't I say? He's a diplomat of some kind now—sent to Havana last September."

"A diplomat," Grey repeated. "Well, well." Stubbs probably did diplomacy well—given his demonstrated skills at lying, deceit, and dishonor. . . .

"He wanted his wife and children to join him in Havana, once he had a suitable establishment, so—"

"Children? He had only the one son when I last saw him." *Only the one* legitimate *son,* he added silently.

"Two, now—Olivia gave birth to a daughter two years ago; lovely child called Charlotte."

"How nice." His memory of the birth of Olivia's first child, Cromwell, was nearly as horrifyingly vivid as his memories of the Battle of Quebec, if for somewhat different reasons. Both had involved blood and shit, though. "But Mother—"

"Your mother offered to accompany Olivia, to help with the children. Olivia's expecting again, and a long sea voyage . . ."

"Again?" Well, it wasn't as though Grey didn't *know* what Stubbs's attitude toward sex was . . . and at least the man was doing it with his wife. John kept his temper with some difficulty, but the general didn't notice, continuing with his explanations.

"You see, I was meant to be sailing to Savannah in the spring—now, I mean—to advise a Colonel Folliott, who's raising a local militia to assist the governor, and your mother was going to come with me. So it seemed reasonable that she go ahead with Olivia and help her to get settled, and I would arrange for her to join me when I came."

"Very sensible," John said. "That's Mother, then. And where does the British Navy come into it?"

"Admiral Holmes, me lord," Tom said, with a faint air of reproach. "He told you last week, when you had him to dinner. He said the Duke of Albemarle was a-coming to take Martinique away from the frogs and then see to Cuba."

"Oh. Ah."

Grey recalled the dinner, which had featured a remarkable dish that he had realized—too late—was the innards of pickled sea urchins, mixed with bits of raw fish and sea lettuce that had been cured with orange juice. In his desire to keep his guests—all recently arrived from London, and all lamenting the dearth of roast beef and potatoes in the Indies—from sharing his realization, he had called for lavish and repeated applications of a native palm liquor. This had been very effective; by the second glass, they wouldn't have known they were eating whale turds, should his adventurous cook have taken it into his head to serve that as a second course. Consequently, though, his own memories of the occasion were somewhat dim.

"He didn't say Albemarle was proposing to lay siege to the place, did he?"

"No, me lord, but that must've been his meaning, don't you think?"

"God knows," said John, who knew nothing about Cuba, Havana, or the Duke of Albemarle. "Or possibly you do, sir?" He turned politely to General Stanley, who was beginning to look better, under the influence of relief and brandy. The general nodded.

"I wouldn't," he admitted frankly, "save that I shared Albemarle's table aboard his flagship for six weeks. What I don't presently know about the harbor at Havana probably isn't worth knowing, but I take no credit for the acquisition of that knowledge."

The general had learned of Albemarle's expedition only the night before the fleet sailed, when a message from the War Office had reached him, ordering him aboard.

"At that point, of course, the ship would reach Cuba long before any message I could send to your mother, so I went aboard at once—*this*"—he glowered at his bandaged foot—"notwithstanding."

"Quite." John raised a hand in brief interruption and turned to his valet. "Tom,—run—and I do mean *run*—to Admiral Holmes's residence and ask him to call upon me as soon as is convenient. And by *convenient,* I mean—"

"Right now. Yes, me lord."

"Thank you, Tom."

Despite the brandy, Grey's brain had finally grasped the situation and was busy calculating what to do about it.

If the British Navy showed up in Havana Harbor and started shelling the place, it wasn't merely physical danger threatening the Stubbs family and Lady Stanley, also known as the Dowager Duchess of Pardloe. All of them would likely become immediate hostages of Spain.

"The moment we got within sight of Martinique and joined Monckton's forces there, I . . . er . . . requisitioned a small cutter to bring me here, as quickly as possible."

"Requisitioned, sir?" John said, smiling at the general's tone.

"Well, I stole it, to be perfectly frank," the general admitted. "I don't imagine they'd bring me to a court-martial, at my age . . . and I bloody don't care if they do." He sat upright, gray-stubbled chin outthrust and a glint in his eye. "All I care about is Benedicta."

WHAT THE GENERAL knew about the harbor at Havana was, generally speaking, that it was one of the finest deepwater harbors in the world, capable of accommodating a hundred ships of the line, and that it was guarded on either side by a large fortress: Morro Castle to the east, and La Punta on the west.

"La Punta's a working fortress, purely defensive; it overlooks the city, though of course one side faces the harbor. El Morro—that's what the Spaniards call it—is a bigger place and is the administrative headquarters of Don Juan de Prado, governor of the city. It's also where the main batteries controlling the harbor are located."

"With luck, I won't need to know that," John said, pouring rum into a glass of orange juice, "but I'll make a note of it, just in case."

Tom returned toward the end of the general's remarks, to report that Admiral Holmes was aware of the planned invasion but had no details concerning it, beyond the fact that Sir James Douglas, who was due to take command of the Jamaica squadron, had sent word that he wished to rendezvous with the squadron off Haiti, at the admiral's earliest convenience.

Through all of this discussion, Lord John had been making mental notes of anything that might conceivably be useful to him—and a parallel list of things here in Jamaica that might come in handy for an impromptu expedition to an island where he didn't speak the language. When he got up to pour more orange juice for the general, he asked Tom, in an undertone, to fetch Azeel from the kitchen.

"What did you mean, you stole the cutter?" John asked curiously, topping up the orange juice with rum.

"Well, that might be a slightly dramatic way to have put it," the general admitted. "The cutter normally attends the *Warburton,* and I do believe Captain Grace, who commands her, was intending to send Lieutenant Rimes off on an errand of his own. I nipped across to Albemarle's ship, though, and . . . er . . . preempted him."

"I see. Why—oh." He caught sight of Azeel, who had arrived but was waiting respectfully in the doorway to be summoned. "Do come in, my dear; I want you to meet someone."

Azeel entered but stopped short at sight of General Stanley, the look of

happy anticipation on her face turning at once to one of caution. She dropped a low curtsy to the general, modestly lowering her white-capped head.

"General, may I present Mrs. Sanchez, my housekeeper? Mrs. Sanchez, this is General Stanley, my stepfather."

"Oh!" she exclaimed in surprise, and then blushed—a lovely sight, as the color in her dark cheeks made her look like a black rose. "Your servant, sir!"

"Your most humble, madam." The general bowed as gallantly as possible while remaining seated. "You must forgive my not standing . . ." He gestured ruefully toward his bandaged foot.

She made a graceful gesture of dismissal and turned toward John.

"This is—your . . ." She groped for the word. "He is the next governor?"

"No, he's not my replacement," John said. "That's Mr. Braythwaite; you saw him at the garden party. No, the general has come to give me some disturbing news, I'm afraid. Do you think you could fetch your husband, Mrs. Sanchez? I wish to discuss the situation with you both."

She looked both astonished and concerned at this and studied him carefully to see if he meant it. He nodded, and she at once curtsied again and vanished, her sandal heels tapping on the tiles in agitation.

"Her husband?" General Stanley said, in some surprise.

"Yes. Rodrigo is . . . er . . . a sort of factotum."

"I see," said the general, who plainly didn't. "But if this Braythwaite is already on board, so to speak, won't he want to make his own domestic arrangements?"

"I imagine so. I, um, had had it in mind to take Azeel and Rodrigo with me to South Carolina. But they may be helpful to the present venture, if . . . er . . . if Rodrigo is sufficiently recovered."

"Has he been ill?" Worry creased the general's already-furrowed brow. "I hear the yellow jack comes to the West Indies at this season, but I hadn't thought Jamaica was badly affected."

"No, not ill, exactly. He had the misfortune to run afoul of a *houngan*— a sort of, um, African wizard, I believe—and was turned into a zombie."

"A what?" The look of worry was superseded by one of astonishment.

Grey drew a deep breath and took a long swallow of his drink, the sound of Rodrigo's own description echoing in his ears.

"Zombie are dead people, sah."

GENERAL STANLEY WAS still blinking in astonishment at Grey's brief description of the events that had culminated in his own appoint-

ment as military governor—Grey judiciously suppressing the facts that Azeel had commissioned an Obeah man to drive the previous governor mad and that Rodrigo had gone one step further and arranged to have the late Governor Warren killed and partially devoured—when the sound of footsteps echoed once again in the corridor. Two people this time: the clack of Azeel's sandals but now walking slowly, to accommodate the slightly limping gait of the booted person accompanying her.

Grey stood up as they came in, Azeel hovering protectively behind Rodrigo.

The young man stopped, taking a deep breath before bowing deeply to the gentlemen.

"Your . . . servant. Sah," he said to Grey, and then straightened, turned upon his axis, and repeated this process to the general, who watched him with a mixture of fascination and wariness.

Every time he saw Rodrigo, Grey's heart was torn between regret for what the young man had once been—and a cautious joy in the fact that some of that splendid young man seemed still to be present, intact, and might yet come back further.

He was still beautiful, in a way that made Grey's body tighten every time he saw that dark, finely carved head and the tall straight lines of his body. The lovely cat-like grace of him was gone, but he could walk again, almost normally, though one foot dragged a little.

It had taken weeks of careful nursing by Azeel—she was the only member of Grey's household who was not terrorized by Rodrigo's mere proximity—with help from Tom, who was afraid, too, but thought it wasn't becoming for an Englishman to admit it.

Rodrigo had been nothing more than a shell of himself when Grey had rescued him and Tom from the maroons who had kidnapped them, and no one had expected that he would survive. Zombies didn't. Drugged with zombie poison—Grey had little notion what was in the stuff, beyond the liver of some remarkably poisonous fish—and buried in a shallow grave, the person attacked by a *houngan* woke after some time to find himself apparently dead and buried.

Rising in a state of mental and physical disorientation, they numbly followed the orders of the *houngan,* until they died of starvation and the aftereffects of the drugs—or were killed. Zombies were (justifiably, Grey thought) viewed with horror by everyone, even by the people who had once loved them. Left without food, shelter, or kindness, they didn't last long.

But Grey had refused to abandon Rodrigo, and so had Azeel. She had brought him slowly, slowly back to humanity—and then had married him, to the extreme horror of everyone in King's Town.

"He's got back most of his speech," Grey explained to the general.

"But only Spanish, that being his first language. He only remembers a few scattered words of English. We"—he smiled at Azeel, who ducked her head shyly—"hope that will improve, too, given time. But for now . . . he tells his wife things in Spanish, and she translates them for me."

He explained the situation briefly to Azeel and Rodrigo—the young man could understand some English, if spoken slowly, but his wife filled in the missing bits for him.

"I would like you to go with me to Cuba," Grey said, looking from one to the other. "Rodrigo could go where I could not go, and hear and see things I couldn't. But . . . there might be some small danger, and if you choose not to go, I will give you enough money for passage to the colonies. If you do choose to come with me, I will take you from Cuba to America, and you will either remain in my employment or, if you prefer, I will find you a place there."

Man and wife exchanged a long look, and at last Rodrigo nodded.

"We . . . go," he said.

GREY HAD NEVER seen a black person turn white before. Azeel had gone the color of grimy old bones and was clutching Rodrigo's hand as though one or both of them were about to be dragged off by slavers.

"Are you given to seasickness, Mrs. Sanchez?" he asked, making his way to them through the confusion of the docks. She swallowed heavily but shook her head, unable to take her eyes off the *Otter*. Rodrigo was unable to take his eyes off her and was anxiously patting her hand. He turned to Grey, fumbling for English words.

"She . . . scare . . ." He looked helplessly back and forth between his wife and his employer. Then he nodded a bit, making up his mind, then looked at Grey while pointing to Azeel. He lowered his hand, indicating something—someone?—short. Then turned to the sea and flung his arm wide, gesturing to the horizon.

"*Africa,*" he said, turning back to Grey and putting his arm around his wife's shoulders. His face was solemn.

"Oh, Jesus," Grey said to Azeel. "You were brought from Africa as a child? Is that what he means?"

"Yes," she said, and swallowed again. "I was . . . very . . . small."

"Your parents? Were they . . ." His voice died in his throat. He'd seen a slave ship only once, and that at a distance. He would remember the smell for as long as he lived. And the body that had bobbed up suddenly beside his own ship, thrown overboard by the slaver. It might have been dead kelp or a blood-bleached scrap from a whaling ship, bobbing in the waves, emaciated, sexless, scarcely human. The color of old bones.

Azeel shook her head. Not in negation but in a vain refusal to think of dreadful things.

"Africa," she said softly. "They are dead. In Africa."

Africa. The sound of the word prickled over Grey's skin like a centipede, and he shook himself suddenly.

"It's all right," he said to her firmly. "You are free now." At least he hoped so.

He had managed her manumission a few months before, in recognition of her services during the slave rebellion during which the late Governor Warren had been killed by zombies. Or, rather, by men under the delusion that they were zombies. Grey doubted that this distinction had been appreciated by the governor.

Grey didn't know whether the girl had been Warren's personal property, and he didn't ask her. He'd taken advantage of his own doubt to tell Mr. Dawes, the governor's erstwhile secretary, that as there was no record of her provenance, they should assume that she was technically the property of His Majesty and should thus be omitted from the list of Governor Warren's belongings.

Mr. Dawes, an excellent secretary, had made a noise like a mildly consumptive sheep and lowered his eyes in acquiescence.

Grey had then dictated a brief letter of manumission, signed this as acting military governor of Jamaica (and thus His Majesty's agent), and had Mr. Dawes affix the most imposing seal in his collection—Grey thought it was the seal of the department of weights and measures, but it was done in red wax and looked very impressive.

"You have your paper with you?" he asked. Azeel nodded, obedient. But her eyes, large and black, lingered fearfully on the ship.

The master of the cutter, having been apprised of their presence, now popped up on deck and came down the gangplank to meet them.

"Lord John?" he asked respectfully, bowing. "Lieutenant Geoffrey Rimes, commander. Your servant, sir!"

Lieutenant Rimes looked about seventeen, very blond and small for his age. He was, however, wearing proper uniform and looked both cheerful and capable.

"Thank you, Lieutenant." Grey bowed. "I understand that you . . . er . . . obliged General Stanley by bringing him here. And that you are now willing to convey me and my party to Havana?"

Lieutenant Rimes pursed his lips in thought.

"Well, I suppose I can do that, my lord. I'm to rendezvous with the fleet here in Jamaica, but as they won't likely arrive for another two weeks, I think I can deliver you safe to Havana, then skip back here to make my meeting."

A small knot formed in Grey's stomach.

"You . . . mean to leave us in Havana?"

"Well, yes, my lord," he said cheerfully. "Unless you can manage your business within two days, I'll have to. Orders, you know." He pulled a commiserating face.

"I'm not really *meant* to be going to Havana, you know," the lieutenant said, leaning forward in a confidential manner and lowering his voice. "But I hadn't any orders to stay in Jamaica, either, if you know what I mean. As written, my orders just say I'm to rendezvous here with the fleet, after delivering the message to Admiral Holmes. As I've already done that . . . well, the navy's always willing to oblige the army—when it suits," he added honestly. "And I'm thinking it wouldn't do me any harm to have a look at Havana Harbor and be able to tell Admiral Pocock about it when he gets here. The Duke of Albemarle's in command of the expedition," he added, seeing Grey look blank. "But Admiral Pocock's in charge of the ships."

"To be sure."

Grey was thinking that Lieutenant Rimes was equally likely to rise to great heights in his service or to be court-martialed and hanged at Execution Dock, but he kept these thoughts to himself.

"Wait a moment," he said, calling the lieutenant's attention— momentarily distracted by the sight of Azeel Sanchez, brilliant as a macaw in a yellow skirt and sapphire-blue bodice—to himself.

"Do you mean that you intend actually to sail *into* Havana Harbor?"

"Oh, yes, my lord."

Grey cast a glance at the *Otter*'s unmistakable British colors, lifting gently in the tropical breeze.

"You will pardon my ignorance, I hope, Lieutenant Rimes—but are we not at war with Spain just now?"

"Certainly, my lord. That's where you come in."

"That's where *I* come in?" Grey felt a sort of cold, inexorable horror rising from the base of his spine. "In what capacity, may I ask?"

"Well, my lord, the thing is, I have to bring you into Havana Harbor; it's the only real anchorage on that coast. I mean, there are fishing villages and the like, but was I to land you in one of those places, you'd have to make your way overland to Havana, and it might take longer than you've got."

"I see . . ." said John, in a tone indicating quite the opposite. Mr. Rimes noticed this and smiled reassuringly.

"So, I'll bring you in under colors—they won't shell a cutter, I don't think, not until they see what's what—and deliver you as an official visitor of some sort. The general thought perhaps you might be bringing some message to the English consul there, but of course you'll know best about that."

"Oh, indeed." *It couldn't be patricide, could it?* he thought. Strangling a stepfather, particularly under the circumstances . . .

"It's all right, me lord," Tom put in helpfully. "I've brought your full-dress uniform. Just in case you might need it."

IN THE EVENT, the officer of the battery guarding the boom chain declined to allow Mr. Rimes to pass, but neither did he offer to sink him. There were a good many curious looks directed at the cutter, but Grey's party was allowed to come ashore. The officer's English was on a par with Grey's Spanish, but after a long conversation filled with vehement gesticulations, Rodrigo convinced him to provide transport into the city.

"What did you tell him?" Grey asked curiously, when at last they were allowed to pass through the battery guarding the west side of the harbor. An imposing fortress with a tall watchtower stood on a promontory in the distance, and he wondered whether this was Morro Castle or the other one.

Rodrigo shrugged and said something to Azeel, who answered.

"He didn't understand the word 'consul'—we don't, either," she added apologetically. "So Rodrigo said you have come to visit your mother, who is sick."

Rodrigo had been following her words with great concentration and here added something else, which she translated in turn.

"He says everybody has a mother, sir."

The address General Stanley had given was the Casa Hechevarria, in Calle Yoenis. When Grey and his fellow travelers were eventually delivered to the *casa* by a wagon driver whose normal cargo appeared to be untanned hides, the place proved to be a large, pleasant, yellow-plastered house with a walled garden and a beehive-like air of peaceful busyness about it. Grey could hear the murmur of voices and occasional laughter within, but none of the bees seemed inclined to answer the door.

After a wait of some five minutes had failed to produce anyone—let alone his mother or something comestible—Grey left his small, queasy party on the portico and ventured round the house. Splashing noises, sharp cries, and the reek of lye soap seemed to indicate that laundry was being done at no great distance. This impression was confirmed as he came round the corner of the house into a rear courtyard and was struck in the face by a thick cloud of hot, wet air, scented with dirty linen, woodsmoke, and fried plantains.

A number of women and children were working in the vicinity of a huge cauldron, this mounted on a sort of brick hearth with a fire beneath—this in turn being fed by two or three small, mostly naked children who

were poking sticks into it. Two women were stirring the mess in the cauldron with huge wooden forks, one of them bawling at the children in Spanish with what he assumed were dire warnings against being underfoot, not getting splashed with boiling water, and keeping well clear of the soap bucket.

The courtyard itself looked like Dante's Fifth Circle of Hell, with sullen gurglings from the cauldron and drifting wisps of steam and smoke giving the scene a sinister Stygian cast. More women were pinning up wet clothes on lines strung round the pillars supporting a sort of loggia, and still others were tending braziers and griddles in a corner, from which drifted the fragrant smells of food. Everyone was talking, all at once, in a Spanish punctuated by parrot-like shrieks of laughter. Knowing that his mother was much less likely to be interested in laundry than in food, he edged round the courtyard—totally ignored by everyone—toward the cooks.

He saw her at once; her back was turned to him, hair hanging casually down her back in a long, thick plait, and she was talking, waving her hands, to a coal-black woman who was squatting, barefooted, on the tiles of the courtyard, patting out some sort of dough onto a hot greased stone.

"That smells good," he said, walking up beside her. "What is it?"

"Cassava bread," she said, turning to him and raising an eyebrow. "And *platanos* and *ropa vieja*. That means 'old clothes,' and while the name is quite descriptive, it's actually very good. Are you hungry? Why do I bother asking?" she added before he could answer. "Naturally you are."

"Naturally," he said, and was, the last vestiges of seasickness vanishing in the scents of garlic and spice. "I didn't know you could speak Spanish, Mother."

"Well, I don't know about speaking, so much," she said, thumbing a straggle of graying blond hair out of her left eye, "but I gesture fluently. What are you doing here, John?"

He glanced round the courtyard; everyone was still at their work, but every eye was fixed on him, interested.

"Do any of your . . . um . . . associates here speak English? In a non-gestural sort of way?"

"A few of them speak a little, yes, and Jacinto, the butler, is pretty fluent. They won't understand you if you talk fast, though."

"I can do that," he said, lowering his voice a little. "In short, your husband sent me, and . . . but before I acquaint you with the situation,—I brought several people with me, servants, and—"

"Oh, did you bring Tom Byrd?" Her face blossomed into what could only be called a grin.

"Certainly. He, along with two . . . er . . . Well, I left them on the portico; I couldn't make anyone hear me at the door."

His mother said something in Spanish that he thought must be an in-delicate expletive, as it made the black woman blink and then grin herself.

"We *have* a porter, but he's rather given to drink," his mother said apologetically, and beckoned to one of the older girls hanging laundry. "Juanita! *Aquí,* if you please."

Juanita instantly abandoned her wet laundry and hastened over, drop-ping a perfunctory curtsy and staring at Grey in fascination.

"*Señora.*"

"*Es mi hijo,*" his mother said, pointing at him. "*Amigos de el . . .*" She twirled a forefinger, indicating circumnavigation, and pointed toward the front of the house, then jerked a thumb at a brazier over which an earth-enware pot was bubbling. "*Agua. Comida. Por favor?*"

"I'm deeply impressed," John said, as Juanita nodded, said something fast and indecipherable, and vanished, presumably to rescue Tom and the Sanchezes. "Is *comida* food, by any chance?"

"Very perceptive of you, my dear." His mother gestured to the black lady, pointed in turn to John and herself, stabbed a finger at various pots and skewers, then nodded at a door on the far side of the courtyard and took John by the arm. "*Gracias, Maricela.*"

She led him into a small, rather dark salon that smelled of citronella, candle wax, and the distinctively sewer-like aroma of small children.

"I don't suppose this is a diplomatic ambassage, is it?" she said, crossing the room to throw open a window. "I would have heard about that."

"I am for the moment incognito," he assured her. "And with any luck, we'll be out of here before anyone recognizes me. How fast can you orga-nize Olivia and the children for travel?"

She halted abruptly, hand on the windowsill, and stared at him.

"Oh," she said. Her expression had gone in an instant from surprise to calculation. "So it's come to that already, has it? Where's George?"

"WHAT DO YOU mean, has it come to that already?" Grey said, star-tled. He stared hard at his mother. "Do you *know* about the"—he glanced round and lowered his voice, though no one was in sight and the laughter and chittering from the patio continued unabated—"the invasion?"

Her eyes flew open wide.

"The *what?*" she said loudly, then glanced hastily over her shoulder toward the open door. "When?" she said, turning back and lowering her own voice.

"Well, now, more or less," Grey said. He got up and quietly closed the door. The racket from the patio diminished appreciably.

"General Stanley turned up on my doorstep in Jamaica a week ago, with the news that the British Navy was on its way to take Martinique and then—if all goes as planned—Cuba. He rather thought it would be a good idea for you and Olivia to leave before they get here."

"I quite agree with him." His mother closed her eyes and rubbed her hands hard over her face, then shook her head violently, as though dislodging bats, and opened her eyes again. "Where is he?" she asked, with some semblance of calm.

"Jamaica. He'd, um, managed to borrow a naval cutter while the navy was preparing to take Martinique and came ahead as fast as he could, in hopes of warning you in time."

"Yes, yes," she said impatiently, "very good of him. But why is he in Jamaica and not here?"

"Gout." And quite possibly a few other infirmities, but no point in worrying her. She looked sharply at him but didn't ask further.

"Poor George," she said, and bit her lip. "Well, then. Olivia and the children are in the country, staying with a Señora Valdez."

"How far in the country?" Grey was making hasty calculations. Three women, two children, three men . . . four, with Malcolm. Ah, Malcolm . . . "Is Malcolm with them?"

"Oh, no. I'm not sure where he *is*," she added dubiously. "He travels a good deal, and with Olivia gone, he often stays in Havana;—he has an office in La Punta—that's the fortress on the west side of the harbor. But he does sleep here now and then."

"Oh, does he?" Grey tried to keep the edge out of his voice, but his mother glanced at him sharply. He looked away. If she didn't know about Malcolm's proclivities, he wasn't going to tell her.

"I need to talk to him as quickly as possible," he said. "Meanwhile, we must fetch Olivia and the children back here, but without giving the impression that there's any sort of emergency. If you'll write a note that will accomplish that, I'll have Rodrigo and Azeel carry it—they can help Olivia to pack up and help mind the children on the way."

"Yes, of course."

There was a small secretaire, rustic in design, crouched in the shadows. He hadn't noticed it until his mother opened it and swiftly produced paper, quill, and inkwell. She uncorked the latter, found it dry, said something under her breath in Greek that sounded like a curse but probably wasn't, and, crossing the room quickly, removed a bunch of yellow flowers from a pottery vase and poured some of the water from it into the empty well.

She shook ink powder into the well and was stirring the mixture briskly with a bedraggled quill when something occurred belatedly to Grey.

"What did you mean, Mother, when you said, 'It's come to that already'? Because you didn't know about the invasion, did you?"

She glanced up at him sharply, ceasing to stir. Then she took a deep breath, like one marshaling her mental forces, visibly made a decision, and put down the quill and ink.

"No," she said, turning to him. "George had told me such a thing was being quietly discussed—but I left England with Olivia in September. War with Spain hadn't yet been declared, though anyone could have seen that it was coming. No," she repeated, and looked at him intently. "I meant the slave revolt."

John stared at his mother for the space of thirty seconds or so, then slowly sank onto a wooden pew that ran along the side of the room. He closed his eyes briefly, shook his head, and opened them.

"Is there anything to drink in this establishment, Mother?"

FED, WASHED, AND fortified with Spanish brandy, Grey left Tom to see to the unpacking and made his way on foot back through the city to the harbor, where the fortress of La Punta—smaller than El Morro (what was a *morro?* he wondered), but still impressive—guarded the western shore.

A few people glanced at him but with no more interest than he might attract in London, and upon reaching La Punta, he was surprised at the ease with which he was not only admitted but escorted promptly to the *oficina del Señor Stubbs.* Granted, the Spaniards had their own notions of military readiness, but this seemed quite lax for an island at war.

The soldier accompanying him rapped on a door, said something in Spanish, and, with a brief nod, left him.

Footsteps, and the door opened.

Malcolm Stubbs looked twenty years older than he had last time Grey had seen him. He was still broad-shouldered and thick-bodied, but he seemed to have softened and fallen in on himself, like a slightly decayed melon.

"Grey!" he said, his tired face brightening. "Wherever did you spring from?"

"Zeus's forehead, no doubt," Grey said. "Where have you come from, for that matter?" The skirts of Stubbs's coat were thick with red dust, and he smelled strongly of horse.

"Oh . . . here and there." Malcolm beat the dust perfunctorily from his coat and subsided into his chair with a groan. "Oh, God. Stick your head out and call for a servant, will you? I need a drink and some food before I perish."

Well, he did know the Spanish word for "beer" . . . Sticking his head out into the corridor as advised, he spotted two servant girls loitering by

the window at the far end, evidently talking to someone in the courtyard below, their conversation accompanied by a good deal of giggling.

Interrupting this colloquy with a brief "Hoy!" he said, *"Cerveza?"* in a tone of polite inquiry, following this with scooping motions toward his mouth.

"Sí, señor!" one of the girls said, with a hasty bob, adding something else in a questioning voice.

"Certainly," he said cordially. "Er . . . I mean, *sí!* Um . . . *gracias,*" he added, wondering what he had just agreed to. Both girls curtsied and vanished in a swirl of skirts, though, presumably to fetch something edible.

"What is *pulpo?*" he asked, returning to the office and sitting down opposite Malcolm.

"Octopus," Malcolm replied, emerging from the folds of a linen towel with which he'd been wiping dirt from his face. "Why?"

"Just wondered. Putting aside the usual inquiries about your health— are you all right, by the way?" he interrupted himself, looking down at what used to be Malcolm's right foot. The boot encircled a sort of cup or stirrup, made of stiff leather with wooden reinforcements on the sides. Both wood and leather were deeply stained from long use, but there was fresh bright blood on the stocking above.

"Oh, that." Malcolm glanced down indifferently. "It's all right. My horse broke down a few miles from the city, and I had to walk some way before I got another." Bending down with a grunt, he unbuckled the appurtenance and took it off—an action that Grey found oddly more disconcerting than sight of the stump itself.

The flesh was deeply ridged from the boot, and when Malcolm peeled the ragged stocking off, Grey saw that a wide ring of skin about the calf had been flayed raw. Malcolm hissed a little and closed his eyes, gently rubbing the end of the stump, the flesh there showing the pale blue of fresh bruising.

"Did I ever thank you, by the way?" Malcolm asked, opening his eyes.

"For what?" Grey said blankly.

"Not letting me bleed to death on that field in Quebec," Malcolm said dryly. "That slipped your mind, did it?"

Actually, it had. There had been a great many things happening on and off that field in Quebec, and the frantic moments of grappling to get his belt loose and jerked tight round Malcolm's spurting leg were just fragments—though vivid ones—of a fractured space where neither time nor thought existed; he'd been actually conscious that day of nothing beyond a sense of constant thunder—of the guns, of his heart, of the hooves of the Indians' horses, all one and pounding through his blood.

"You're welcome," he said politely. "As I say—putting the social courtesies to one side for the moment, I came to inform you that a rather large

British fleet is on its way to invade and capture the island. Am I correct, by the way, in my assumption that the local commander does not yet realize that war has been declared?"

Malcolm blinked. He stopped massaging his leg, straightened up, and said, "Yes. When?" His face had changed in an instant, from exhaustion and pain to alertness.

"I think you may have as long as two weeks, but it might be less." He gave Malcolm what details he had, as concisely as he could. Malcolm nodded, a line of concentration deepening between his brows.

"So I've come to remove you and your family," Grey finished. "And my mother, of course."

Malcolm glanced at him, one eyebrow raised.

"Me? You'll take Olivia and the children, of course—I'm very much obliged to you *and* General Stanley. But I'm staying."

"What? What the devil for?" John was conscious of a sudden surge of temper. "Besides a pending invasion, my mother tells me there's a bloody slave revolt in progress!"

"Well, yes," Malcolm said calmly. "That's mine."

Before Grey could sort out a coherent response to this statement, the door opened suddenly and a sweet-faced black girl with a yellow scarf round her head and an enormous battered tin tray in her hands sidled through it.

"*Señores,*" she said, curtsying despite the tray, and deposited it on the desk. "*Cerveza, vino rústico, y un poco comida: moros y cristianos*"—she unlidded one of the dishes, loosing a savory steam—"*maduros*"—that was fried plantains; Grey was familiar with those—"*y pulpo con tomates, aceitunas, y vinagre!*"

"*Muchas gracias, Inocencia,*" Malcolm said, in what sounded like a surprisingly good accent. "*Es suficiente.*" He waved a hand in dismissal, but instead of leaving, she came round the desk and knelt down, frowning at his mangled leg.

"*Está bien,*" Malcolm said. "*No te preocupes.*" He tried to turn away, but she put a hand on his knee, her face turned up to his, and said something rapid in Spanish, in a tone of scolding concern that made Grey raise his brows. It reminded him of the way Tom Byrd spoke to *him* when he was sick or injured—as though it were all his own fault, and he therefore ought to submit meekly to whatever frightful dose or treatment was being proposed—but there was a distinct note in the girl's voice that Tom Byrd's lacked entirely.

Malcolm shook his head and replied, his own manner dismissive but kindly, and laid his hand on the girl's yellow head for a moment. It *might* have been merely a friendly gesture, but it wasn't, and Grey stiffened.

The girl rose, shook her head reprovingly at Malcolm, and went out,

with a hint of flirtation in the sway of her skirts. Grey watched the door close behind her, then turned back to Malcolm, who had plucked an olive out of one dish and was sucking it.

"Inocencia, my arse," Grey said bluntly.

Malcolm's normal complexion being brick red, he didn't flush, but neither did he meet Grey's eye.

"Quite the usual sort of names they give girls, the Spanish," he said, discarding the olive pit and picking up a serving spoon. "You find young women called all kinds of things: Assumpción, Immaculata, Concepción . . ."

"Conception, indeed." This was said in a tone cold enough to make Malcolm's wide shoulders hunch a little, though he still wouldn't look at Grey.

"They call this *moros y cristianos*—that means 'moors and Christians'— the rice being Christians and the black beans Moors, d'you see?"

"Speaking of conception—and Quebec," Grey said, ignoring the food— though it smelled remarkably good, "your son by the Indian woman . . ."

Malcolm did glance at him then. He looked back at his plate, finished chewing, swallowed, and nodded, not looking at Grey.

"Yes. I did make inquiries—once I was mended. They told me the child had died."

That struck Grey in the pit of the stomach. He swallowed, tasting bile, and plucked a bit of something out of the dish of *pulpo* at random.

"I see. How . . . regrettable."

Malcolm nodded, wordless, and helped himself liberally to the octopus.

"Was it quite recent, this news?" The shock had gone through him like an ocean breaker. He remembered vividly the day when he had taken the infant—the child's mother having died of smallpox, he had bought the boy from his grandmother for a blanket, a pound of sugar, two golden guineas, and a small cask of rum—and carried him to the little French mission in Gareon. The boy had been warm and solid in his arms, looking up at him from round, unblinking dark eyes, as though trusting him.

"Oh. No. No, it was at least two years ago."

"Ah." Grey put the piece of whatever-it-was into his mouth and chewed slowly, the sense of shock fading into an immense relief—and then a growing anger.

Not a trusting man himself, he had given the priest money for the child's needs and told him this payment would continue—but only so long as the priest sent Grey a lock of the child's hair once a year, to prove his continued existence and presumed good health.

Malcolm Stubbs's natural hair was sandy and tightly curled as sheep's wool; when left to its own devices, it exploded from its owner's head like a ruptured mattress. Consequently, Malcolm usually kept his head polled

and wore a wig. He'd evidently been wearing one earlier but had taken it off and set it aside, and the inch of mad growth thus displayed strongly resembled the texture of the two small curls of dark cinnamon-colored hair that Grey had so far received from Canada, each one bound carefully with black thread and accompanied by a brief note of thanks and blessing from Father LeCarré—the latest, just before his departure for Jamaica.

The urge to bounce Malcolm's head off the desk and shove him face-down into the *pulpo* was strong, but Grey mastered it, chewing the bite of octopus—very flavorful, but in texture reminiscent of an artist's rubber—thoroughly before saying anything. He swallowed.

"Tell me about this slave revolt of yours, then."

MALCOLM DID LOOK at him now, considering. He nodded and reached, grunting, for the limp, bloodstained stocking hanging out of his artificial foot.

"We'll go up to the battlements," he said. "Not many of the servants speak any English—but that doesn't mean none of them understand it. And they do listen at doors."

Grey blinked as they emerged from the gloom of a stone stairwell into a pure and brilliant day, a blinding sky spinning with seagulls overhead. A stiff wind was coming off the water, and Grey removed his hat, tucking it under his arm lest it be carried away.

"I come up here several times a day," Malcolm said, raising his voice above the wind and the shrieks of the gulls. He had wisely left his own hat and wig below in his office. "To watch the ships." He nodded toward the expanse of the huge harbor, where several very large ships were anchored, these surrounded by coveys of smaller vessels, going to and from the shore.

"They're beautiful," Grey said, and they were. "But they're not doing anything, are they?" All sails were furled, all port lids closed. The ships lay at anchor, rocking slowly in the wind, masts and spars swaying stark and black against the blue of sea and sky.

"Yes," Malcolm said dryly. "Particularly beautiful when they're not doing anything. That's how I know the declaration of war hasn't yet been received; if it had, the decks would be black with men, and the sails would be reefed, not furled. And that's why I come up here morning, noon, and night," he added.

"Yes," Grey said slowly, "but . . . if in fact de Prado—that's the commander of the forces here?—if he *doesn't* know that war is declared,—why are these ships here already? I mean, plainly they're men of war, not merchantmen. Even I know that much."

Malcolm laughed, though without much humor.

"Yes, the cannons rather give it away, don't they? The Spanish have been expecting war to be declared for the last six months. General Hevia brought these ships in last November, and they've been lying in wait here ever since."

"Ah."

Malcolm gave him a raised brow.

"Ah, indeed. De Prado's expecting a declaration any day. That's why I sent Olivia and the children to the country. De Prado's staff all treat me with exquisite courtesy"—his mouth twitched a little—"but I can see them measuring me for leg-irons and a cell."

"Surely not, Malcolm," Grey said mildly. "You're a diplomat, not an enemy combatant. Presumably they'd either deport or detain you, but I can't see it coming to chains."

"Yes," Malcolm agreed, eyes fixed again on the ships, as though he feared they might have begun to move in the last few moments. "But if they find out about the revolt—and I really don't see how that can be avoided—I rather think that might alter their views on my claim to diplomatic immunity."

This was said with a sort of calm detachment that impressed Grey—reluctantly, but still. He glanced round to be sure they were not overheard.

There were a lot of soldiers up here but none close to them; the gray stone of the rooftop stretched away for a hundred yards in all directions. Grey could hear, faintly, shouts between an officer at the far end of the battlement and someone in the watchtower above. There was a small group of regulars—most of them black, Grey saw—stripped to the waist and sweating despite the wind, repairing a gap in the battlement with baskets of stones—and there were guards. Four guards at each corner of the battlements, stiffly upright, muskets shouldered. The fortress of La Punta was prepared.

A detachment of twelve men marched past, two by two, under the command of a young corporal shouting the Spanish equivalent of "Hup!" as they wheeled past the stubby watchtower. The corporal saluted smartly; Malcolm bowed and turned again to the vast expanse of the harbor. It was a clear day; John could just make out the great boom chain at the harbor mouth, a thin darkness in the water, like a snake.

"It was Inocencia who told me," Malcolm said abruptly, as the soldiers disappeared down a stairway at the far side of the rooftop. He cut his eyes at Grey, who said nothing. Malcolm turned his face back to the harbor and began to talk.

The revolt was planned among slaves from two of the large sugar plan-

tations near Havana. The original plan, according to Inocencia—whose cousin was a servant at Hacienda Mendez but was having an affair with one of the house slaves, whose brother was one of the ringleaders of the plot—had been to band together and kill the owners of the haciendas, loot the houses, which were very rich, and then escape through the countryside to the Golfo de Xaguas, on the other side of the island.

"Thinking that the soldiers wouldn't pursue them, being distracted by the imminent arrival of the English on this side, you see." Malcolm appeared quite unmoved by the putative murder of the plantation owners. "It wasn't a bad plan, if they chose their moment and waited 'til the English *did* arrive. There are dozens of small islands in the *golfo;* they might have hidden there indefinitely."

"But you discovered this plan, and rather than mentioning it to the *comandante* . . ."

Malcolm shrugged.

"Well, we are at war with the Spanish, are we not? Or if we weren't, it was obvious that we would be at any moment. I met with the two leaders of the revolt and, er, convinced them that there was a better way to achieve their ends."

"Alone? I mean—you went to meet these men by yourself?"

"Of course," Malcolm said simply. "I wouldn't have got near them had I come mob-handed. Didn't have a mob to hand, anyway," he added, turning to Grey with a self-conscious grin that suddenly took years off his careworn face.

"I met Inocencia's cousin at the edge of the Saavedra plantation, and she took me to a big tobacco shed," he went on, the grin fading. "It was almost nightfall, so darkish inside. Lots of shadows, and I couldn't tell how many men were there; it felt as though the whole place was moving and whispering, but likely that was just the drying leaves—they're quite big, did you know? A plant is almost the size of a man. They hang them up, up in the rafters, and they brush against each other with this dry sort of rustle, almost like they're tittering to themselves . . . put the wind up me, a bit."

Grey tried to imagine that meeting and, surprisingly, could envision it: Malcolm, artificial foot and all, limping alone into a dark shed to convince dangerous men to forgo their own murderous plans in favor of his. In Spanish.

"You aren't dead, so they listened to you," Grey said slowly. "What did you offer them?"

"Freedom," Malcolm said simply. "I mean,—the army goes about freeing slaves who enlist—why oughtn't the navy to be similarly enlightened?"

"I'm not so sure that a sailor's life is noticeably better than that of a

slave," Grey said dubiously. "In terms of food, they may be better off as they are."

"I don't mean they're to enlist, booby," Malcolm said. "But I'm sure I can persuade either Albemarle or Admiral Pocock that they should be freed in token of regard for their service. If they survive," he added thoughtfully.

Grey was beginning to think that Malcolm might actually be a decent diplomat. Still . . .

"Since you mention service—what, exactly, are you proposing that these men do?"

"Well, my first notion was that they might creep along the shoreline after dark and detach and sink the boom chain across the harbor mouth."

"A good notion," Grey said, still dubious, "but—"

"The batteries. Yes, exactly. I couldn't very well go down and ask to inspect the batteries, but . . ." He reached into his coat and withdrew a small brass telescope.

"Have a look," he said, passing this to Grey. "Wave it around a bit, so it doesn't look as though you're spying out the batteries particularly."

Grey took the telescope. His hands were chilled and the brass, warm from Malcolm's body, gave him an odd *frisson*.

He'd seen one of the batteries close to, on the way in; the one on the opposite side of the harbor was similarly equipped: six four-pounders and two mortars.

"It's not only that, of course," Grey said, handing back the telescope. "It's the—"

"Timing," Malcolm finished. "Yes. Even if the men could swim from down shore rather than come through the battery, it would have to be done with the British fleet actually in view, or the Spaniards would have time to raise the chain again." He shook his head regretfully. "No. What I'm thinking, though—and do say, if you have a better idea—is that we might be able to take El Morro."

"What?" Grey glanced across the channel at the towering hulk of Morro Castle. Set on a rocky promontory, it rose considerably higher than La Punta and commanded the entire channel, most of the harbor, and a good bit of the city, as well. "How, exactly?"

Malcolm bit his lip, not in concern but concentration. He nodded at the castle.

"I've been inside, several times. And I can make an occasion to go again. You'll go with me—it's a blessing that you should have come, John," he added, turning his head to Grey. "It makes things much easier."

"Does it, indeed?" Grey murmured. A faint uneasiness began to stir at the base of his spine. A seagull landed on the parapet near his elbow and gave him a beady yellow look, which didn't help.

"The governor's down with fever, at the moment, but he might be better tomorrow. I'll request a meeting to introduce you. While you're engaged with de Prado—or his lieutenant, if de Prado's still indisposed—I'll make an excuse, slip off, and manage to take note of the floor plan, entrances and exits, all that—" He broke off suddenly. "You did say two weeks?"

"About that. But there's no telling, is there? What if Martinique didn't surrender easily, or there was a typhoon as they left the island? It *could* be a month or more." Another thought struck him. "And then there are the volunteers from the American colonies. Lieutenant Rimes says a number of transports are meant to rendezvous with the fleet here."

Malcolm scratched his head. The clipped bronze curls rippled in the wind like shorn autumn grass.

What? John thought, quite shocked at the poetic image his errant brain had presented him with. He didn't even *like* Malcolm, let alone . . .

"I don't suppose the transports would come near the harbor until they'd joined the fleet," Malcolm pointed out. "But two weeks seems decent odds—and that's long enough to get Olivia and your mother safely off the island."

"Oh. Yes," John said, relieved at this apparent return to sanity. "I had Mother send a note to bring them back to—oh, damn. You did say you'd sent them away on purpose."

The seagull made a disapproving noise, defecated on the parapet, and launched itself into the air.

"I did, yes. I tried to persuade your mother to go with Olivia, but she insisted on staying. Said she's writing something and wanted to be left in peace for a few days." Malcolm turned his back on the harbor and stared contemplatively at the stones under his feet.

"*Adelante!*" A shout came from behind Grey; he turned at the sound of marching feet and clanking weaponry. Another detachment drilling. They clumped past, eyes fixed forward, but their corporal saluted Malcolm politely, including Grey with a brief nod and a sidelong glance.

Was it his imagination, or had the man's eyes lingered on his face?

"The thing is . . ." Malcolm said, waiting 'til the soldiers had receded into the distance. "I mean . . ." He coughed and fell silent.

Grey waited.

"I know you don't like me, John," Malcolm said abruptly. "Or respect me. I don't like myself all that much," he added, looking away. "But—will you help me?"

"I don't see that I have a choice," Grey said, leaving the question of liking alone. "But for what it's worth," he added formally, "I do respect you."

Malcolm's broad face lighted at this, but before he could say anything in reply, Grey became suddenly aware of a change around them. The men repairing the wall had leapt to their feet, gesturing and pointing, shouting excitedly.

Everyone was shouting, rushing toward the battlements overlooking the harbor. Caught in the crush, the two Englishmen pushed their way forward, far enough to see the ship. A small boat, a fast Spanish cutter, coming like the wind itself, its sails white as gull's wings, hurtling across the blue water toward them.

"Oh, Jesus," Grey said. "It's—is it?"

"Yes, it is. It must be." Malcolm grabbed him by the elbow and pulled him out of the crowd of excited Spaniards. "Come. Now!"

THE STAIRWELL WAS blind dark after the dazzle above, and Grey had to drag a hand along the rough stone wall to avoid falling. He *did* fall, slipping on one of the age-hollowed steps near the bottom, but was luckily saved by clutching Malcolm's sleeve.

"This way." There was more light below, bright flashes from the narrow windows at the ends of long corridors, dim flickering of lanterns on the walls, a strong smell of whale oil. Malcolm led the way down to his office, where he said something in rapid Spanish to the secretary, who rose, looking surprised, and went out. Malcolm closed the door and locked it.

"Now what?" Grey asked. His heart was beating fast, and he felt a sense of confusion: an alertness like that of impending battle, an absurd urge to flee, the urgent need to do *something* . . . but what? The first knuckle on his right hand was bleeding; he'd scraped it when he slipped on the stairs. He put it to his mouth in reflex, tasting silver blood and stone dust.

Malcolm was breathing harder than the brisk walk merited. He braced himself with both hands on the desk, looking down at the dark wood. Finally he nodded, shook himself like a dog, and straightened up.

"It's not as though I haven't been thinking about it," he said. "But I hadn't expected you to be here."

"Don't let me interfere with your plans," Grey said politely. Malcolm looked at him, startled, then laughed and seemed to settle into himself.

"Right," he said. "Well, there's the two things, aren't there? The slaves, and Olivia—and your mother, of course," he added hastily.

Grey thought he might himself have reversed those two items in order of importance, but, then, he didn't know just how dangerous the slaves might be. He nodded.

"Do you really think they'll arrest you?"

Malcolm lifted one heavy shoulder and let it fall.

"Yes, I do—but I don't know how long it might take them to get round to it. After all, I'm no particular threat, so far as they know." He went to the small window and peered out. Grey could hear shouting in the courtyard below, someone trying to create order in the midst of a rising gabble of Spanish voices.

"The thing is," Malcolm said, turning back from the window with a frown of concentration upon his face, "they'll know officially that war has been declared, as soon as the captain of that ship presents his letters to the governor. But do you think they know anything about the fleet?" He saw Grey's raised eyebrow and added hastily, "I mean,—the ship bringing the declaration—if that's truly what it is—they might have spotted the fleet or . . . —or heard word of it. In which case . . ."

Grey shook his head.

"It's a big ocean, Malcolm," he said. "And is there anything you'd do differently if the Spanish *did* know about the fleet?"

He was rather impatient with Malcolm's orderly exegesis. His own blood was up, and he needed to be moving.

"Actually, yes. Number one being, run—both of us. If they think the British are about to be on their doorstep, the second thing the Spanish will do—after putting both forts on full alert—is round up every British citizen in Havana, very likely starting with me. If they *don't* know that, we might still have a bit of time in hand."

Grey saw that Malcolm was needing to move, too; he'd begun to walk to and fro behind his desk, glancing out of the window each time he passed it. He was limping heavily; walking clearly hurt him, but he seemed oblivious to the pain.

"The Mendez slaves will be nervous—well, they are already—but they'll be bloody well stirred up by this news. I've got to go and talk to them, as quickly as possible. Reassure them, you know? If I don't, they may very well take the declaration of war as a signal to fall upon their owners and slaughter them on the spot—which, aside from being generally deplorable in terms of humanity, would be a complete waste of their value to us."

"Deplorable, yes." Grey felt a qualm at the thought of the inhabitants of Haciendas Mendez and Saavedra, sitting down peaceably to their suppers tonight, with no notion that they might be murdered at any moment by the servants bringing their food. It occurred to him—as perhaps it had to Malcolm—that the slaves of those two plantations were quite possibly not the only ones on the island of Cuba who might be inclined to take advantage of a British invasion to settle scores. But there wasn't much either Malcolm or he could do about that.

"You'd best go, then, at once. I'll see to the women and children."

Malcolm was rubbing a hand fiercely over his face, as though this might assist thought.

"Yes. You'll have to get them off the island before the fleet arrives. Here, take this." He pulled out a drawer and withdrew a small, fat leather pouch. "Spanish money—you'll attract less attention. Cojimar—I think that's your best bet."

"What and where is Cojimar?" Drums. There were drums now, beating a tattoo in the courtyard, and the clatter of boots and voices as men spilled out of the recesses of the fortress. How big was the force manning El Morro?

He didn't realize he'd spoken that last question aloud until Malcolm answered it, distracted.

"About seven hundred soldiers, maybe another three hundred supportives—oh, and the African laborers; perhaps another three hundred of them—they don't live in the fort, though." He met Grey's eyes and nodded, divining his next thought. "I don't know. They might join our men, they might not. If I had time . . ." He grimaced. "But I don't. Cojimar is—oh, wait." Turning, he seized the wig he'd taken off earlier from his desk and thrust it into Grey's hands.

"Disguise," he said, and smiled briefly. "You rather take the eye, John. Best if people don't notice you on the street." He snatched up the hat and crammed it on his own bare head, then unlocked the door and pulled it open, impatiently gesturing Grey ahead of him.

John went, asking over his shoulder, "Cojimar?"

"Fishing village." Malcolm was looking up and down the corridor. "It's east of Havana, maybe ten miles. If the fleet can't get into the harbor, it's the best anchorage for them. Small bay—oh, and a small fort, too. *El Castillo de Cojimar.* You'll want to keep clear of that."

"Yes, I'll do that," John said dryly. "I'll—" He'd been going to say that he'd send Tom Byrd with any news, but the words died in his throat. Malcolm would presumably be somewhere in the countryside, tending his slaves, by the time there was any news. That, or in captivity. Or—very possibly—dead.

"Malcolm," he said.

Malcolm turned his head sharply and saw John's face. He stopped dead for a moment, then nodded.

"Olivia," he said quietly. "Will you tell her—" He broke off and looked away.

"You know I will."

He put out a hand, and Malcolm grasped it, hard enough that the bones shifted. When they let go, his skinned knuckle burned, and he saw that there was blood from it on Malcolm's palm.

They spoke no more but went out into the corridor, walking fast.

THE WIG WOULD have been much too large, given Malcolm's round-headed resemblance to an oversize muskmelon, but Grey's own hair—yellow and noticeable, as Malcolm had so tactfully noted—was thick, and with it stuffed up inside the wig, the horsehair contrivance sat securely, if uncomfortably. He hoped that Malcolm didn't suffer from lice but forgot such minor concerns as he made his way through the throngs of people in the street outside La Punta.

There was an air of curiosity in the street; people glanced at the fortress as they passed, clearly sensing some disturbance from its daily routine. But the news had not yet spread; for that matter, Grey wondered whether the news had officially reached the office of the governor—or his sickbed, as the case might be. Neither he nor Malcolm had had any doubt; only the most urgent news would have got the cutter past the boom chain with such dispatch.

The guard at the fortress's street gate had given him no more than a casual glance before waving him through; as was the case in peacetime, there were nearly as many civilians as soldiers inside the fort, and there were plenty of fair-skinned, blue-eyed Spaniards. The cut of his suit was not in the Spanish style, but it was discreet and sober in color.

He was going to need a horse—that was the first thing. He could walk ten miles, but doing so in his court shoes would be both slow and painful—and making the round-trip of twenty miles on foot . . . He glanced up at the sky; it was well past noon. Granted, in this latitude, the sun wouldn't set before eight or nine o'clock, but . . .

"Why the devil didn't I ask Stubbs what the word for 'horse' is?" he muttered under his breath, threading his way through a district of fragrant market stalls filled with fruit—he recognized plantains, of course, and papayas, mangoes, coconuts, and pineapples, but there were odd dark-green things that he'd not seen before, with pebbly skins, and lighter-green objects that he thought *might* be custard apples—whatever they were, they smelled delicious. His stomach growled—despite the octopus, he was starving—but then his head snapped round as he smelled something of a distinctly different nature. Fresh manure.

IT WAS VERY LATE by the time he finally returned to Casa Hechevarria that night. A full moon sailed high overhead, and the air was thick with smoke and orange blossom and the smell of slowly roasting meat. He'd eaten in Cojimar easily enough, merely pointing at things in the tiny market square and offering what appeared to be the smaller coins in his

pouch, but Cojimar was no more than a sunstruck distant memory, and he was starving again.

He slid off the rented mule, wrapped the creature's reins over the railing in front of the house, and went to hammer on the door. His arrival had been noticed, though, and soft lantern light flooded out upon him as he came up the shallow wooden steps.

"Is that you, me lord?" Tom Byrd, bless him, stood framed in the open doorway, lantern in hand and round face creased with worry.

"What's left of me," Grey said. He cleared his throat, clogged with dust, spat into the flowering bush by the portico, and limped into the house. "Get someone to see to the mule, will you, Tom?"

"Right away, me lord. What's amiss with your foot, though?" Tom fixed an accusing gaze on Grey's right foot.

"Nothing." Grey made his way into the *sala,* dimly lit by a small candle before a holy picture of some sort—there were things with wings in it, which must be angels—and sat down with a sigh of relief. "The heel of my shoe came off whilst I was helping the mule out of a rocky ditch."

"He fell into a ditch with you, me lord?" Tom was deftly lighting more candles with a spill and now lifted this in order to examine Grey more closely. "I thought mules was meant to be sure-footed."

"There's nothing wrong with his feet, either," Grey assured him, leaning back and closing his eyes for a moment. The candlelight made red patterns on the insides of his eyelids. "I'd stopped for a piss, and he took the opportunity of my inattention to walk down into said ditch, which he did without the slightest difficulty, by the way. There were some of these things growing on the bushes there that he wanted to eat." Fumbling in his pocket, Grey produced three or four small, smooth green fruits.

"I tried to lure him out with a handful, but he was happy as he was, and eventually I was obliged to resort to force." Said force being applied by two young black women passing by, who had laughed at Grey's predicament but then resolved it, one of the women tugging at the reins and addressing the mule in what sounded like deeply pejorative terms while her friend prodded it sternly in the backside with a stick. Grey yawned hugely. At least he'd learned the word for mule—*mula,* which seemed very reasonable—along with a few other things that might come in handy.

"Is there any food, Tom?"

"Those are guavas, me lord," Tom said, nodding at the little fruits, which Grey placed on a side table. "You make jelly from 'em, but they maybe won't poison you if you eat 'em raw." He'd knelt and got Grey's shoes off in a matter of seconds, then stood and deftly plucked the battered wig off Grey's head, viewing it with an expression of deep disapproval. "I mean, if you can't wait while I go rouse the cook."

"Don't do that. It must be past midnight." Grey dubiously prodded one of the guavas, which seemed unripe—it was hard as a golf ball.

"Never mind, me lord, there'll be cold stuff in the larder," Tom assured him. "Oh—" he added, stopping at the door, wig dangling from one hand, "I forgot to say as Her Grace is gone."

"Her Gr—what? Where the devil has she gone?" Grey sat up straight, all thoughts of food, bed, and sore feet vanishing.

"A note came from a Señora Valdez late this morning, me lord, saying as how Mrs. Stubbs and her little girl was both ill with fever and asking would Her Grace please come. So she went," he added unnecessarily, and vanished, too. *"Chingado huevón!"* Grey said, standing up.

"What did you say, me lord?" Tom's voice came from somewhere down the hall.

"I don't know. Never mind. Get the food, please, Tom. And beer, if there is any."

A faint laugh, cut off by the muffled thump of a swinging door. He looked round the room, wanting to do something violent, but an ancient cat curled up on the back of a stuffed chair opened its great green eyes and glared at him out of the twilight, disconcerting him.

"Bloody hell," he muttered, and turned away. So, not only were Olivia and family *not* headed back to Havana, his mother had decamped—how long ago had she left? She couldn't have made it to the Valdez plantation before dark; she must be somewhere on the road—and as for Rodrigo and Azeel, God knew where *they* were. Had they even reached Olivia's rural hideaway yet?

He strode restlessly to and fro, the stone-tiled floor cool through his stockings. He had no idea in which direction the Valdez plantation lay; how far might it be from Cojimar?

Not that it mattered, if Olivia and her daughter were too ill to travel. A moment ago, his mind had been as exhausted as his body, empty of thought. Now he felt as though his head were filled with ants, all rushing in different directions, each with tremendous determination.

He could find a wagon. But how sick were they? He couldn't load desperately ill people into a wagon, drive them ten, twenty, thirty miles over rocky trails, and then decant them into a boat, which might take *how* long to reach a safe haven. . . . —What about food and water? The *peón*—that's what someone had called him, he had no idea what it meant—with whom he'd arranged to rent a small boat had promised water;—he could buy food, but—Jesus, how many people could he get aboard? Could he leave Rodrigo and Azeel, to be rescued later? No, he'd need them to talk with the boatman, and to help, if half his passengers were prostrate and heaving, needing to be tended. What if more of the party fell ill on the way?

What if the boatman succumbed to fever? What if his mother caught the fever and died at sea?

He could all too easily envision himself making landfall on some god-forsaken shore of the southern colonies with a boatload of his dead or dying family and servants . . .

"No!" he said aloud, clenching his fists. "No, that's bloody not going to happen."

"What's not going to happen?" Tom inquired, backing into the room with a small wheeled table, festooned with edibles. "There's a *lot* of beer, me lord. You could bathe in it, should the fancy take you."

"Don't tempt me." He closed his eyes briefly and took several deep breaths. "Thank you, Tom."

Plainly, he couldn't do anything tonight, and no matter what he did in the morning, he'd do it better if he had food and rest.

Hungry as he'd been half an hour before, his appetite seemed now to have deserted him. He sat down, though, and forced himself to eat. There were small patties of some kind of blood sausage, made with onions and rice, a hard cheese, the light, thin-crusted Cuban bread—he thought he'd heard someone call it a *flauta,* could that be right? Pickled vegetables of some kind. Beer. More beer.

Tom was hovering nearby, quiet but watchful.

"Go to bed, Tom. I'll be fine."

"That's good, me lord." Tom didn't bother trying to look as though he believed Grey; there was a deep crease between his valet's brows. "Is Captain Stubbs all right, me lord?"

Grey took a deep breath and another mouthful of beer.

"He was quite well when we parted this afternoon. As for tomorrow . . ." He hadn't meant to tell Tom anything *until* tomorrow; no point in destroying his sleep and peace of mind. But from the look on his young valet's face, it was much too late for any such kindly procrastinations.

"Sit down," Grey said. "Or, rather, get another cup and then sit down."

By the time he had finished explaining matters to Tom, nothing remained of his meal save crumbs.

"And Captain Stubbs means to make these slaves come into Havana and . . . do what?" Tom looked both horrified and curious.

"That, fortunately, is Captain Stubbs's concern. Did my mother say anything about the state of Olivia and her daughter? How ill they might actually be?"

Tom shook his head.

"No, me lord. But from the look on her face—Her Grace's face, I mean—the news must've been pretty bad. I'm sorry to say. She even left her story behind." Tom's face was grave in the flickering shadows. He'd lighted half a dozen thick candles, and despite muslin covering the win-

dows, clouds of tiny insects had filtered into the room like dust, their minuscule shadows frantic on the dim white walls.

The sight made Grey itch. He'd been ignoring insects all day and sported more than a dozen mosquito bites on neck and arms. A high, mocking *zeeeee!* sang past his ear, and he slapped at it in futile reflex. The gesture made Tom brighten.

"Oh!" he said. "Wait a bit, me lord, I've got summat for you."

He returned almost at once with a stoppered vial of blue glass, looking pleased with himself.

"Try that, me lord," he said, handing it to his employer. Grey pulled the stopper, and a delicious, rich scent floated out.

"Coconut oil," Tom said proudly. "The cook uses it, and she gave me some. I mixed the mint into it, for good measure, but she says the mosquitoes don't like the oil. Flies do," he added judiciously, "but most of them don't bite."

"Thank you, Tom." Grey had shucked his coat to eat; he rolled up his shirtsleeves and anointed himself, rubbing it into every inch of exposed skin. Something occurred to him.

"What did you mean, Tom? About my mother leaving her story behind—a book of some sort?"

"Well, I don't know as whether it *might* be a book," Tom said dubiously. "It's not one yet, but the servants say she writes some of it every day, so sooner or later . . ."

"She's *writing* a book?"

"So Dolores said, me lord. It's in there." He turned and lifted his chin toward the secretaire that Grey had seen his mother use—Christ, had it been only this morning?

Consumed by curiosity, Grey got up and opened the secretaire. Sure enough, there was a small stack of written pages, neatly bound with blue tape. The page on top was a title page—evidently she *did* mean it to be a book. It said, simply, *My Life*.

"A memoir?"

Tom shrugged.

"Dunno, me lord. None of the servants can read English, so they don't know."

Grey was torn between amusement, curiosity, and a certain unease. To the best of his knowledge, his mother had led a rather adventurous life— and he was well aware that his knowledge of that life was limited, by unspoken mutual consent. There were a lot of things he didn't want her to know about his own life; he could respect her secrets. Though, if she was writing them down . . .

He touched the manuscript lightly, then closed the lid of the secretaire. Food, beer, and the living, candlelit silence of the Casa Hechevarria had

quieted both his body and his mind. He could think of a thousand possibilities, but in fact, there was only one thing he could *do:* ride to the Valdez plantation as fast as he could and assess the situation when he got there.

Two weeks—about—before the British fleet arrived. Two weeks minus one. God willing, that would be enough time for him to sort things out.

"What did you say, Tom?"

Tom was piling up the empty dishes on the table but stopped to answer him.

"I said, that word you said—*huevón?*"

"Oh. Yes, I heard it from a young lady I met on the road from Cojimar. Do you know what it means?"

"Well, I know what Juanito *says* it means," Tom replied, striving for accuracy. "He says it means a chap what's lazy because his balls are too big to stir himself." Tom gave Grey a sidelong glance. "A lady said that to you, me lord?"

"She was speaking to the mule—or at least I hope she was speaking to the mule." Grey stretched himself, feeling the joints of his shoulders and arms pop, inviting the caress of sleep. "Go to bed, Tom. It will be a long day tomorrow, I'm afraid."

He paused on his way out, to look at the painting of the things with wings. They were angels, rendered crudely but with a simplicity that made them oddly moving. Four of them hovered protectively over an infant Christ, lying in his manger of straw, asleep. And where was Stubbs sleeping tonight? In a cold spring field, a dim tobacco shed?

"God bless you, Malcolm," he whispered, and went to seek his bed.

A MODEST COUGH woke him, well after dawn, to find Tom Byrd beside his bed, holding a tray containing breakfast, a steaming cup of the local equivalent of tea, and a note from his mother.

"Her Grace met Rodrigo and Azeel late last night," Tom informed him. "Them being on the way back posthaste to fetch her, and happen as how she stopped at the same inn where they were stopping to water their horses."

"She—my mother—isn't traveling by herself, surely?" At this stage of her life, he wouldn't put it past her, but . . .

"Oh, no, me lord," Tom assured him, with a slightly reproachful look. "She took Eleana and Fatima and three good lads by way of escort. Her Grace ain't afraid of things, but she's no ways reckless, as you might say."

Grey detected a certain emphasis on "she's" that he might have taken personally but chose to ignore in favor of reading his mother's message.

Dear John,

I trust Tom Byrd has told you that Olivia sent Word asking me to come to her at Hacienda Valdez. I met your two Servants at a Hovel somewhere on the Road, they being on their Way back with a similar but more detailed Message, this one written by the local Priest.

Padre Cespedes says that nearly everyone in the House is affected by the Illness, which he—having seen many Occurrences of Fever during his Years serving God near the Zapata Swamp—is sure is not a relapsing Fever, like the tertian Ague, but is almost certainly the Yellow Jack.

A small shock ran through him. "Fever" was a vague word, which might mean anything from a touch of the sun to malaria. Even "ague" might be a passing ill, easily shaken off. But "yellow jack" was stark and definite as a knife in the chest. Most of his army career had involved postings in northern climes; the closest he had come to the dread disease was the sight of ships—now and then—in Kingston Harbor, flying the yellow quarantine flag. But he'd seen the corpses being carried off those ships, too.

His hands had gone cold, and he wrapped one around the hot pottery cup while he read the rest.

Don't come here, unless I write to say so. There is one thing to be said for the Yellow Jack, which is that it is fearfully quick. All will likely be resolved—one Way or the Other—within a Week. That may leave enough Time in which to execute your original Intent. If not . . . not.

I think I will see you again, but should God will otherwise, tell Paul and Edgar, Hal and his family, that I love them, tell George— well, tell him that he knows my Heart and what I would say were we together. And for you, John . . . you are my dearest Son and I carry my Thought of you through all that lies before us.

Your Most Affectionate Mother

John swallowed several times before he could pick up the cup and drink from it. If she had ridden through the night, which seemed likely, she might be arriving at the plantation now. To meet . . .

Grey said something very obscene in German, under his breath. He put the cup back and swung out of bed, thrusting the letter at Tom; he couldn't speak coherently enough to transmit the contents.

He had to piss, and did so. This elemental act gave him some semblance

of control, and he shoved the utensil back under the bed and straightened up.

"Tom, go and ask where the nearest doctor is to be found. I'll dress myself."

Tom gave him a look, but not the look of profound doubt that might have been expected in response to his last statement. This was a very patient look, and one much older than Tom's years.

"Me lord . . ." he said, very gently, and set the letter on the chest of drawers. "If Her Grace wanted you to send a doctor, she'd've said so, don't you think?"

"My mother has very little faith in doctors." Neither did Grey, but, dammit, what *else* was he to do? "That doesn't mean one might not . . . help."

Tom looked at him for a long moment, then nodded soberly and went.

John could indeed dress himself, though his hands shook so much that he decided to forgo shaving. Malcolm's ghastly wig lay on the chest of drawers beside his mother's letter, looking like a dead animal. Ought he wear it?

Why? he wondered. He couldn't hide his Englishness from the doctor. He probably should send Jacinto to talk to the doctor, in any case. But he couldn't bloody stand to stay in the house, doing nothing. He picked up the now-lukewarm cup and drained the bitter contents. Christ, what *was* this stuff?

He rubbed more of Tom's coconut-oil concoction into his exposed skin, brushed his hair and bound it simply with a ribbon, then strode out to see what Tom had found out from the other servants.

They were on the patio, which seemed the center of the house. The usual cheerful racket was much subdued, though, and Ana-Maria crossed herself and bobbed a curtsy when she saw him.

"*Lo siento mucho, señor,*" she said. "*Su madre . . . su prima y los niños—*" She waved a graceful hand outward, encompassing his mother, Olivia, and the children, then again inward, this time indicating all the servants around her, and laid the hand on her heart, looking at him with a great compassion in her softly lined face. "*Tenemos dolor, señor.*"

He took her meaning clearly, if not every word, and bowed low to her, nodding to the other servants as he straightened.

"*Muchas gracias . . .*" Señora? Señorita? Was she married? He didn't know, so he just repeated, "*Muchas gracias,*" with more emphasis.

Tom wasn't among the servants; he'd likely gone to talk to Jacinto about doctors. John bowed again to the servants generally and turned toward the house.

There were voices toward the front of the house, speaking very rapid Spanish, with an occasional baffled word from Tom edging its way into the

conversation. Curious, John made his way past the *sala* and into the small vestibule, where he found Jacinto and Tom blocking the front door and heard a woman's voice outside, raised in agitation, saying his name.

"Necessito hablar con el Señor Grey! Ahorita!"

"What's going on?" He spoke sharply, and the two men turned toward him, allowing him a view of a yellow bandanna and the desperate face of Inocencia.

She seized the moment and pushed her way between the butler and Tom, snatched a crumpled note from her bosom, and thrust it into Grey's hand. Then she fell to her knees, clutching the hem of his coat.

"Por favor, señor!"

The note was limp with the sweat of her body, and the ink had blurred a little but was still clearly readable. There was neither salutation nor signature, and it was very short:

I'm nabbed, old cock. Your ball.

"WHAT DOES THIS MEAN, *señor?*" Jacinto had been reading the note over his shoulder, without the slightest attempt to pretend he wasn't. "This is . . . not English, is it?"

"It is," he assured the butler, carefully folding the note and putting it in his pocket. He felt as though someone had punched him in the chest, very hard, and he had trouble catching his breath.

It was English, all right—but English that no one but an Englishman would understand. And not even an Englishman like Tom—who was frowning at Inocencia in puzzlement—would know the meaning of that last, paralyzing sentence.

Your ball.

Grey swallowed, tasting the last bitterness of the breakfast drink, and made himself breathe deep. Then he stooped and raised Inocencia to her feet. She was gasping for breath, too, he saw, and there were tracks of dried tears on her cheeks.

"The consul has been arrested?" he asked. She looked helplessly from him to Jacinto, who coughed and translated what Grey had said. She nodded violently, biting her lower lip.

"Está en El Morro," she managed, gulping, and added something else that Grey couldn't follow. A quick back and forth, and Jacinto turned to Grey, his long old face very grave.

"This woman says that your friend was arrested at the city wall last night and has been taken to El Morro. That is where the *gobierno*—the government, excuse me—where they keep prisoners. This . . . lady"—he inclined his head, giving Inocencia the benefit of the doubt—"she saw

Señor Stubbs being taken to the governor's office soon after dawn, and so she waited nearby and followed when they took him down to—" He broke off to ask Inocencia a sharp question. She shook her head and said something in reply.

"He is not in the dungeon," Jacinto reported. "But he is locked in a room where they put gentlemen when it is necessary to contain them. She was able to come and talk to him through the door, once the guards had left, and he wrote this note and told her to hurry and bring it to you at once, before you left the city." Jacinto shot Grey a glance but then coughed and looked away. "He said you would know what to do."

Grey felt a black dizziness come over him and a prickle of rising hair on the back of his neck. His lips felt stiff.

"Did he, indeed."

"YOU CAN'T, ME LORD!" Tom stared at him, aghast.

"I'm very much afraid you're right, Tom," he said, striving for calm. "But I don't see that I have any choice but to try."

He thought Tom was going to be sick; the young valet's face was pale as the morning mist that blanketed the tiny garden where they'd gone for a bit of privacy. Grey was himself just as pleased that he hadn't had a chance to eat breakfast; he recalled Jamie Fraser telling him once, in inimitable Scottish fashion, that his "wame was clenched like a fist," a phrase that described his own present sensation to a T.

He'd have given a lot to have Fraser beside him on this occasion.

He'd have given almost as much to have Tom.

As it was, he was apparently going into battle supported by a stuttering ex-zombie, an African woman of unpredictable temper and known homicidal tendencies, and Malcolm Stubbs's concubine.

"It will be fine," he told Tom firmly. "Inocencia will provide an introduction to the ringleaders and establish my *bona fides.*" And if she failed to convince these men that Grey had any such qualifications, all of them would likely be for the chop within seconds: He'd seen machetes wielded with casually murderous ease yesterday—God, was it only yesterday?—by field hands on his way to Cojimar.

"And Rodrigo and Azeel will be there to help me speak to them," he added, with a little more confidence. To his surprise, when he had put the situation before them, the Sanchezes had shared a long marital look, then nodded soberly and said they would go.

"Rodrigo's a good 'un," Tom admitted reluctantly. "But he won't be no good to you in a fight, me lord." His own fists had been clenched

throughout the conversation, and it was clear that he had a higher opinion of his own abilities in that regard.

Actually, Grey thought, he might be right. Used as he was to Tom's constant presence, he hadn't taken conscious notice, but his valet was no longer the pie-faced seventeen-year-old who had bluffed his way into Grey's service. Tom had grown a few inches, and while not in Malcolm Stubbs's class in the matter of bulk, he'd definitely filled out. His shoulders were square and his freckled forearms nicely muscled. However . . .

"If it comes to that sort of fight, it wouldn't matter if I had an entire company of infantry with me," he said. He smiled at his valet with true affection. "And besides, Tom:—I cannot depend on anyone but you to see to things here. You must go with Jacinto to find a doctor—cost is no consideration; I'm leaving you with all of our English money, and there's enough gold there to buy half of Havana—and then take the man to the Valdez plantation, along with any medicines he thinks useful. I've written a note to my mother—" He reached into his bosom and withdrew a small folded square, sealed with smoky candle wax and stamped with his smiling half-moon signet. "See that she gets that."

"Yes, me lord." Tom glumly accepted the note and tucked it away.

"And then find someplace nearby to stay. Don't stay in the house; I don't want you to be exposed to the fever. But keep an eye on things: Visit twice a day, make sure the doctor does what he can, give Her Grace any assistance she'll let you give, and send back reports every day as to the state of things. I don't know when I'll get them"—*or if*— "but send them anyway."

Tom sighed but nodded.

Grey stopped, unable to think of anything else. The *casa* was well awake by now, and there was a muted sense of bustle in the distant patio, a rising scent of boiling beans and the sweetness of fried plantains. He hadn't told the house servants anything of his own unspeakable mission—they couldn't help, and to know anything at all of it would put both himself and them in danger. But they knew about the situation at Hacienda Valdez, and he'd heard the murmur of prayers and the clicking of rosary beads when he'd passed by the patio a few minutes ago. It was oddly comforting.

He reached out and clasped Tom's hand, squeezing.

"I trust you, Tom," he said softly.

Tom's Adam's apple bobbed in his throat. His deft, sturdy fingers turned and squeezed back.

"I know, me lord," he said. "You can."

FOUR DAYS LATER—it had taken more time than anticipated to find what was needed—Lord John Grey stood naked in the middle of a grove of mangoes, on a hill overlooking the hacienda of the Mendez family.

He'd seen the big house as they rode into the plantation, a sprawling establishment of rooms added over the years, odd wings sprouting from unexpected places, outbuildings scattered near it in an untidy constellation. *One of the complicated constellations,* he thought, looking down on it. *Cassiopeia, maybe, or Aquarius. One of the ones where you just take the ancient astronomer's word for what you're looking at.*

The windows in the main house had been lighted, with servants passing to and fro like shadows in the dusk, but he had been too far away to hear any of the noises of the place, and he was left with a queer sensation of having seen something ghostly that might suddenly be swallowed by the night.

In fact, it had been, in the sense that the hacienda was invisible from his present situation—and a good thing, too. His traveling clothes lay puddled on the leaf mold in which his bare feet were sunk, and small insects were treating his private parts with an unseemly familiarity. This caused him to rummage his pack first for the bottle of coconut–mint elixir and apply this lavishly before getting dressed.

Not for the first time—nor, he was sure, the last—he deeply regretted the absence of Tom Byrd. He *was* actually capable of dressing himself, though both he and Tom acted on the tacit assumption that he wasn't. But what he missed most at the moment was the sense of solemn ceremony that attended Tom's dressing him in full uniform. It was as though he assumed a different persona with scarlet coat and gold lace, Tom's respect giving him belief in his own authority, as though he put on not only uniform but armor and office.

He could bloody use that belief just now. He swore softly under his breath as he struggled into the moleskin breeches and brushed bits of leaf off each foot before pulling on his silk stockings and boots. It was a gamble, but he felt that the chances of these men taking him seriously, listening to him, and—above all—trusting him would be increased if he appeared not just as a stand-in for Malcolm Stubbs but as the incarnation of England, as it were: a true representative of the king. They had to trust that he could do what he said he would do for them, or it was all up. For the *hacendados*—and for him.

"Wouldn't do the bloody navy any good, either," he muttered, tying his neckcloth by feel.

Done at last, his traveling clothes bundled into the pack, he heaved a sigh of relief and stood still for a minute to gather himself, settle into the uniform.

He'd had no idea mango trees grew to such a size; this was an old

grove, the trees each more than a hundred feet in height, the leaves rising and falling gently on the evening breeze, making a sound like the sea overhead. Something slithered heavily in the fallen leaves near him and he froze. But the serpent—if that's what it was—continued on its way, untroubled by his presence.

Rodrigo, Azeel, and Inocencia were where he had left them, no more than a hundred yards away, but he felt entirely alone. His mind had gone blank, and he welcomed that respite. Windfalls of unripe fruit knocked down by a storm lay all around like pale-green cricket balls in the leaves, but the fruit still on the trees had gone yellow—he'd seen it in the twilight as they came up into the grove—and had begun to blush crimson. Now it was dark, and he only sensed the mangoes when he brushed a low-lying branch and felt the heavy swing of the fruit.

He was walking, not having made up his mind to do so nor remembering the taking of the first step, but walking, propelled into motion by a sense that it was time.

He came down through the grove and found Rodrigo and the girls on their feet, in murmured conversation with a tall, spare young woman—Inocencia's cousin, Alejandra, who would take them to the tobacco shed.

All of them turned to see him, and Alejandra's eyes widened, gleaming in the moonlight.

"*Hijo,*" she said in admiration.

"Thank you, madam," he said, and bowed to her. "Shall we go?"

HE'D IMAGINED IT vividly, from Malcolm's account. The bulk of the big tobacco barn, the dark, the whispering of the drying leaves overhead, the sense of waiting men . . . What Malcolm hadn't mentioned was the overpowering scent that lay in a cloud over the shed, a thick incense that reached out to grab him by the throat from thirty feet away. It wasn't unpleasant, by any means, but it was strong enough to make him breathe shallowly for a moment—and he needed all the breath he could get.

Cano. That was the name of the man he must convince. Cano was headman of the slaves of the Mendez plantation. There was a headman from Saavedra, too, named Hamid, but Alejandra said that it was Cano's opinion that counted most heavily among the slaves

"If he says yes, they all will do it," she had assured Grey.

There was a great deal more to the barn's atmosphere than the heavy scent of tobacco. He could smell the reek of constant sweat the instant he stepped inside—and the sharp, dark stink of angry men.

There was a single lantern burning, hung from a nail in one of the uprights supporting the high roof. It made a small pool of light, but the glow

of it diffused much farther, showing him the men massed in the shadows. No more than the curve of a skull, a shoulder, the gleam of light on black skin, the whites of staring eyes. Below the lantern stood two men, turned to meet him.

There was no question which was Cano. A tall black man, wearing only short, ragged breeches, though his companion (and most of the men in the shadows, as a sidelong glimpse confirmed) was dressed in both breeches and shirt and wore a spotted bandanna tied around his head.

No question why, either. Gray scars mottled Cano's back and arms like barnacle scars on an old whale—the marks of whips and knives. The man watched Grey approach and smiled.

Smiled to show that his front teeth were gone, but the canines remained, sharp and stained brown with tobacco.

"*Mucho gusto, señor,*" he said. His voice was light and mocking. Grey bowed, very correctly. Alejandra had come in behind him, and she made the introductions in soft, rapid Spanish. She was nervous; her hands were twisted in her apron and Grey could see sweat shining in the hollows under her eyes. Which was her lover? he wondered, this man or Hamid?

"*Mucho gusto,*" Grey said politely, when she had finished, and bowed to her. "Madam—will you be so good as to tell these gentlemen that I have brought with me two interpreters, so that we can be assured of understanding one another?"

At this cue, Rodrigo came in, Azeel a pace or two behind him. She looked as though she were wading into a pool filled with crocodiles, but Rodrigo's manner was cool and dignified. He wore his best black suit, with immaculate white linen that shone like a beacon in the grubby brown light of the barn.

There was a palpable ripple of interest—and a just-as-palpable hostility at sight of him. Grey felt it like a jab in the stomach. Christ, was he going to get Rodrigo killed, as well as himself?

And they don't even know what he is yet, he thought. He'd been told—often enough to believe it—that the fear of zombies was so great that sometimes even the rumor of it was enough that a crowd would fall upon the suspected person and beat them to death.

Well, best get on with it. He wasn't armed, save for the regimental dirk at his belt. Nothing was going to get them through this but words, so best start talking.

This he did, presenting his compliments (that got the breath of a laugh—encouraging . . .) and stating that he came as the friend and representative of Malcolm Stubbs, whom they knew. Nods of wary approval. He came (he said) also as the representative of the King of England, who intended to overthrow the Spaniards in Cuba and take the island.

This was pretty bold, and Azeel stammered a little as she said it for him,

but it went over quite well; it appeared that the crowd was quite united with the king in this desire.

"My friend, Señor Stubbs, has asked your help in this endeavor," Grey said, looking deliberately from one side of the barn to the other, speaking to all of them. "I have come to counsel with you and to decide how best to accomplish our desires, so that—"

"*Dónde está el Señor Malcolm?*" Cano interrupted him. "*Por qué él no está aquí?*"

That didn't need interpretation, but for the sake of protocol, he let Azeel translate it before replying that, alas, Señor Malcolm had been arrested and was imprisoned in Morro Castle. Hence he, John Grey, had come to carry out Señor Malcolm's plan.

A small rumble of doubt, a shuffling of bare feet in the dust.

"For your assistance in this matter, Señor Malcolm promised you your freedom. I promise this, too." He spoke as simply as he could, hoping that this would carry sincerity.

Exhalations, quiet murmurs. They were worried—and were more than right to be, he thought. The barn was hot, packed with so many men, and damp with their sweat and the exudations of the drying tobacco leaves. Sweat was seeping through his linen.

Suddenly the other man—Hamid, it must be—said something abrupt and jerked his chin at Grey. The man was bearded, and it occurred to Grey that perhaps he was a Mussulman.

"This gentleman wants to know how you will accomplish the things you speak of," Azeel said, glancing at Grey. "You are only one man. Do you have soldiers, weapons?"

Grey wondered what the views of the Prophet were with regard to zombies . . . because it was clear that he was going to have to use Rodrigo.

Rodrigo himself stood close beside his wife, his face calm and unmoving, despite the weight of eyes upon him, but Grey saw him straighten a little and take a deep breath.

"Tell Señor Hamid"—and Grey bowed to the bearded man—"that I am indeed one man . . . but I am an Englishman. And I am a man of my word. To show that this is true, I have brought my servant, Rodrigo Sanchez, who will tell them why they may believe me and trust what I say."

Heart thumping audibly in his ears, Grey stepped back and inclined his head toward Rodrigo. He saw Rodrigo squeeze Azeel's hand lightly, and drop it, before he moved forward.

Unhurried, composed, civilized in a way that these men had never known, Rodrigo picked up a wooden bucket standing near the wall, carried this to a central spot in the light of the lantern, turned it upside down, set it on the floor, and sat down. Very slowly, Azeel moved to stand behind him, her eyes fixed on the men in the shadows.

Rodrigo began to speak, his voice deep, soft but carrying. There was an audible intake of massed breath from the audience, and a ripple of horror moved through the barn. Azeel turned to Grey.

"My husband, he says . . ." Azeel's voice trembled, and she stopped to clear her throat. Then she straightened and, putting her hand on her husband's shoulder, spoke clearly.

"He says this: *'I have been dead. I died in the hands of a* houngan, *and I woke in my grave, smelling the rot of my own body. I could not move—how should I move? I was dead. And then, years later, I felt the air on my face and a hand on my arm. The* houngan *pulled me from my grave and told me that I was indeed dead. But that now I was a zombie.'*"

Grey felt the ripple of horror that moved through the room, and heard the intake of massed breath, the shocked murmur that had broken out at this. But Azeel put both hands on Rodrigo's shoulders and glared over his head, turning her eyes from one side of the room to the other.

"I tell you—listen!" she said violently. *"Escuchen!"*

Grey saw Cano jerk back a little, whether from affront or shock, he couldn't tell. But the man gave an explosive snort and over the murmuring in the shed said loudly, *"Háblanos!"* The murmurs stopped abruptly, and Azeel turned her head to look at Cano, the light of the lantern gleaming on her skin, in her eyes.

"Háblame," she said softly to Rodrigo. *"Sólo a mí. Háblame."* Speak to me. Only to me.

Rodrigo's hand rose slowly and rested on hers. He raised his chin and went on, Azeel translating softly for Grey as he spoke:

"I was dead, and a zombie, in the power of an evil man, in the power of hell. But this man—" He moved his head a little, indicating Grey. "This man, he came for me. He came alone, into the high mountains, and he walked into the cavern of Damballa, the great serpent—"

At this, exclamations and agitations broke out in such a confusion of noise that Rodrigo was obliged to stop speaking. This he did and went on sitting there, unmoved as a statue.

God, he's beautiful. The thought sparked for a moment in Grey's mind and then vanished as Rodrigo raised a slow hand, palm out. He waited, and the noise died away in a smother of shushings.

"In the cavern of serpents, this man walked—alone—through the dark and through demons. He turned the *houngan's* magic back upon himself, and then he came out of the cave and he took me back. By his own power, he raised me from death."

There was a moment's silence, as Azeel's soft words vanished among the hidden leaves and the dark bodies. Then Rodrigo nodded, once, and said simply, *"Es verdad."*

It's true.

Utter silence for a long moment, and then a murmur, another. Wonder. Doubt. Amazement. Grey thought the language had changed; they weren't all speaking Spanish but some other language—or perhaps languages. African tongues. He caught the word *"houngan,"* and Cano was looking sharply at him, eyes narrowed.

Then the bearded man spoke gruffly to Grey in English, jerking his chin at Rodrigo.

"Tell your zombie to go outside."

Grey exchanged a quick look with Rodrigo, who nodded very slightly and stood up.

"If you will oblige me, Señor Sanchez?" Grey bowed, gesturing toward the door. Rodrigo returned the bow, moving very slowly, and walked with equal slowness to the big open door. Grey thought he might be exaggerating the stiffness of his gait, but perhaps he was imagining that.

Had it worked? *"Your zombie,"* the man had said. Did they believe that he had rescued Rodrigo from the *houngan*, from death, or did they think that he was himself some sort of English *houngan* who controlled Rodrigo and had compelled him to make that speech? Because if so . . .

Rodrigo's black form merged with the night and disappeared. There was a noticeable relaxation of the atmosphere, as though every man there had released a sigh of relief.

Cano and the bearded man exchanged a long look, and after a moment, Hamid nodded reluctantly.

Cano turned to Grey and said something in Spanish. Azeel, who had gone nearly as stiff as her husband as he walked away, pulled her eyes away from the open door and translated Cano's question.

"So, then. How shall we do this thing?"

Grey let out a long, long breath.

Simple as the concept was, it took no little time to explain. Some of the slaves had seen a cannon—all of them had heard one fire, though only in the far distance, when the cannons of the two fortresses were fired on holidays or to salute a ship coming in—but almost none of them had any notion of the operation of a gun.

A space on the floor was swept free of tracks and trampled tobacco leaves and another lantern was brought. The men gathered close. Grey drew the outline of a gun in the reddish dirt with a stick, talking slowly and simply as he explained the loading and firing of a cannon, and repeatedly pointed out the touchhole.

"Here is where they put fire. The powder"—he prodded the barrel—"explodes"—a murmur of confusion, explanations from those who had seen this thing—"and BOOM!" Everyone looked stunned for an instant, then broke into laughter. When the repetitions of "BOOM!" had died down, he pointed again at the touchhole.

"Fire," he said, and waited expectantly.

"Fire!" several voices said happily.

"*Exactamente,*" he said, and, smiling at them, reached into his pocket. "Look."

"*Miren,*" Azeel said, but it was unnecessary. Every eye was fixed on the six-inch metal spike in Grey's hand. He had a large bag of them in his pack, of different sizes, as he'd had to take whatever he could find from the various ironmongers and ship chandlers of Havana, but from what Inocencia and Azeel had been able to tell him of the guns in Morro Castle, he thought they would suffice.

He squatted above his drawing and mimed pushing the spike into the touchhole. Then he pulled a small hammer from his other pocket and pounded the spike vigorously into the dirt.

"No fire," he said, looking up.

"*Bueno!*" said several voices, and there was much murmuring and nudging.

He took a deep breath of the thick, intoxicant air. So far, so good. His heart was thumping audibly in his ears and seemed to be going much faster than usual.

It took much longer to explain the map. Only a few had seen a map or chart before, and it was very difficult for some of them to make the mental connection between lines on a piece of paper and the positions of corridors, doors, rooms, cannon batteries, and powder stores in El Castillo de los Tres Reyes Magos del Morro. They *had* all seen the fortress itself, at least: when they were taken from ships onto the dock, on their way to the slave markets in the city.

Sweat was running down Grey's back under his uniform coat, his body throbbing with the effects of moist heat and mental tension, and he took the coat off, to avoid fainting.

Finally, a consensus of sorts was achieved. Inocencia very bravely said that she would go into the fortress with the men and help to show them where the guns were. This was met with a moment's silence, and then Hamid nodded at her and raised a brow at Cano, who, after a moment's hesitation, also nodded, and a murmur of approval rustled through the men.

Nearly done. He resisted the urge to give in to relief, though. The last item on his agenda might spike his personal guns—or get him killed. He rolled up the crude maps that Inocencia had drawn and handed them ceremoniously to Cano. Then he withdrew from his pack another rolled paper—this one blank—a capped inkwell, and a quill.

His head was not so much spinning as it was floating, and he had some difficulty in fixing his eyes on things. He made an effort, though, and spoke firmly to Cano.

"I will write here that you are performing a great service for the King of England and that I say you should receive your freedom for doing this thing. I am a . . . *God, let me get this right . . . un hombre de gracia,* and I will sign my name." *Hombre de gracia* was as close as Azeel could come to the notion of "nobleman."

He waited, watching their faces, while Azeel translated this. Wary, curious, some—the younger ones—with a touch of hope that stabbed at his heart.

"You must then put down your names. If you do not . . . have letters . . . you can tell me your name, I will write it, and you can make a mark to say it is yours."

Instant alarm, much looking to and fro, the shine and flicker of eyes in the dark, agitation, a gabble of voices. He raised a hand and waited patiently. It took several minutes, but at last they calmed enough for him to speak again.

"I will go with you into the castle, too," he pointed out. "What if I am killed? Then I will not be there to tell the king you should have your freedom. But this will tell him." He tapped a finger on the blank sheet.

"What if some of you become lost in the city after we leave the castle? If you go later to the chief of the English sailors and say to him that you have done this great thing and now you must be free, how will he believe you?" He tapped again.

"This will speak for you. You will tell the English chief your name, and he will see it on this paper and know what you say is true."

". . . *es verdad."* Azeel looked as though she, too, was about to faint from the strain, the heat, and—no doubt—the fear of the situation, but her voice was loud and firm.

Cano and Hamid had drawn together, were engaged in a low-voiced debate. Sweat was dripping from the tail of Grey's hair; he could feel it hitting the small of his back through his shirt with the regularity of grains of sand—slow grains of sand, he thought wryly, very slow—in an hourglass.

At last they settled things between them, though, and Cano took several steps forward, to face Grey himself. He spoke, looking intently into Grey's face from a distance of no more than a foot; Grey could smell the man's breath, hot with tobacco and with a hint of rot from his teeth.

"He says," Azeel said, and stopped to work a little saliva into her mouth, "he says that they will do it. But you must make three papers—one for you, one for him, and one for Hamid, because if you are killed and have the only paper, what good is it?"

"Very reasonable," Grey said gravely. "Yes, I will do that."

The sense of relief ran through his limbs like warm water. But he wasn't quite done yet.

"One thing," Grey said, and took a breath. Too deep a breath; it made him dizzy, and he took another, shallower.

Cano inclined his head, listening.

"The people in the haciendas—the Mendez family, the Saavedras—I know what your intention was, and we will say no more of that. But you must assure me that these people will not be harmed, will not be killed."

"*. . . Ellos no serán asesinados.*" Azeel's voice was soft now, remote, as though she was reading the terms of a contract. Which, Grey reflected, it was, in all justice.

Cano's nostrils flared at that, and there was a low sound—not quite a growl—from the men in the shadows. The sound of it made Grey's scalp contract.

The man nodded, as though to himself, then turned to look into the shadows, first to one side and then the other, deliberate, as a barrister might look to see the temper of a jury. Then he turned back to Grey and nodded again.

"*No los mataremos,*" he said.

"We will not kill them," Azeel whispered.

Grey's heart had stopped thumping and now seemed to be beating with unusual slowness. The thought of fresh, clean air steadied his mind.

Without thinking about it, he spat into his palm, as soldiers and farmers did, and held out his hand. Cano's face went quite blank for an instant but then he nodded, made a small "huh" under his breath, spat in his hand, and clasped Grey's.

He had an army.

TOO LATE. That was his first thought when he heard the firing of artillery in the distance as they approached the city. The British fleet had arrived, and the siege of Havana was begun. A moment's heavy breathing, though, and the panic passed. It didn't matter, he realized, and a wave of relief went over him.

Ever since Malcolm had first sprung this plan on him, the matter of timing had been in his mind: the notion that the slaves' raid must happen just before the arrival of the fleet. But Malcolm's reference had been with respect to his original plan, having the slaves sabotage the boom chain, to allow the fleet into the harbor.

That truly wouldn't have worked, unless the fleet was in sight when the chain was sunk; any delay and the Spaniards would have it raised again. But the spiking of the fortress's guns . . . that would be helpful at any time.

Granted, he thought, tilting his head to try to gauge the direction of the firing, it would certainly be more dangerous to carry out such a mis-

sion with the fortress's gun crews in place. On the other hand, said gun crews would be focused entirely on their business. It was very likely that the gun crews would be taken completely unaware. For the first few moments.

It was going to be a bloody business, on both sides. He didn't like the thought but didn't shy away from it. It was war, and he was—once again—a soldier.

Still, his mind was uneasy. He had no doubt of the slaves' ferocity or their will, but to pit completely untrained, lightly armed men against practiced soldiers in close combat . . .

Wait. Perhaps a night attack—could that be managed? He reined his mule in to a walk, the better to think it out.

With the British Navy on their doorstep, the guns of El Morro would never sleep—but neither would they necessarily be manned at full strength during the night watches. He'd seen enough, during his brief excursion to Cojimar, to convince him that the small harbor there was the only possible base for an attack on Morro Castle. What were the distances?

General Stanley had referred repeatedly to an intended siege of Havana. Clearly the navy knew about the boom chain, and, just as clearly, an effective siege must be mounted from the ground, not from ships. So—

"*Señor!*" A shout from the line of wagons broke his train of thought, but he tucked the notion safely away for further analysis. He didn't want the slaves to be butchered, if it could be helped; still less did he want to suffer the same fate.

THEY WERE WELL in sight of the city wall of Havana now. In one way, the fleet's arrival was fortuitous: A city under siege needed food, above all things. Faced with the problem of getting a hundred slaves past the city guard, Hamid had suggested loading the plantation wagons with anything that came to hand and letting each wagon be accompanied by a half dozen men, there presumably to do the unloading and delivery. Between the two plantations, they could muster ten wagons—with driver and assistant, that was eighty men. The rest could easily slip in by ones and twos.

A decent plan, but what, Grey had asked, about the plantations' owners, their servants? It would take time to load wagons, and their departure couldn't be easily concealed. An alarm would be raised, surely?

No, no, he was assured. The wagons were kept in barns near the fields. The loading would happen by night; they would be gone before daylight. And, Cano added, through Azeel, the female slaves who worked in the house could be relied upon to create distractions, as necessary. The thought

made him grin his empty black grin, wolf teeth flashing yellow in the lantern light.

It had worked, insofar as no one had come shouting out of the hacienda, demanding to know what was going on as the wagons rumbled out by moonlight. Now, what might happen when the owners and overseers discovered that a hundred able-bodied slaves were missing . . .

But whatever distractions the women had devised had evidently been effective. No one had pursued them.

He stopped the wagons just out of sight of the city gate, had a hasty check-round with the various teams, reassuring the men and making sure everyone knew where and when they were to meet—and that all the machetes were carefully concealed. Even though he had packed away his uniform and was once more in mufti—complete with Malcolm's wig—he thought it better not to come into Havana with the wagons. He would go back to the Casa Hechevarria with Rodrigo and Azeel and find out from Jacinto what the news of the invasion was; Inocencia would try to speak with Malcolm in Morro Castle and, in the process, discover anything in the present situation that might be of strategic value.

"*Muchas gracias,* my dear," he told her, and bowed low over her hand. "Azeel, please tell her that we could not even contemplate this venture without her courage and help. The entire British Navy is in her debt."

Inocencia's lips made a smile, and she bobbed her head in response, but Grey could see that she was trembling with exhaustion, and her brilliant eyes were sunk in her face. Tears quivered on her lashes.

"It will be all right," he said, taking her hand. "We will succeed—and we will rescue Señor Stubbs. I promise you."

She swallowed and nodded, wiping her face on the edge of her filthy apron. Her mouth twitched, as though she meant to say something, but she changed her mind and, pulling her hand free, dropped him a curtsy, turned, and hurried away, lost at once in the crowd of women in the market, all pushing and shouting in an effort to procure food.

"She is afraid," Azeel said quietly, behind him.

She's not the only one . . . He'd felt a coldness at the bone ever since he walked into the tobacco shed, and it hadn't gone away, though the day was bright and sunny. There was a small flame of excitement at the prospect of action, though, and it was normal for the nerves to be raw—

There was a sharp report from the direction of El Morro, echoed at once by another, and he was suddenly on the Plains of Abraham in Quebec, the cannon firing from the walls, and the army waiting, waiting there on the open ground, waiting in the agony of delay . . .

He shook himself like a dog and felt better.

"It will be all right," he said again, firmly, and turned in to the Calle Yoenis.

HE COULD TELL, at once that something had happened. There was no singing, no chatter from the patio, no one working in the garden. He did hear muted voices, and food was being cooked—but there was no spice in the air. Only the slightly soapy smell of long-boiled beans and scorched eggs.

He walked rapidly through the empty front rooms, and his heart stopped as he heard a baby's high-pitched squall.

"Olivia?" he called. The muted voices paused, though the baby's mewling continued.

"John?" His mother stepped out of the *sala,* peering into the murk of the unlighted corridor. She was disheveled, her hair in a half-unraveled plait, and she had a tiny baby in her arms.

"Mother." He hurried to her, his heart suddenly feeling as though it had come loose in his chest. She took a step toward him that brought her face into the bar of sunlight from a window, and one look told him.

"Jesus," he said under his breath, and reached out to embrace her, draw her close, as though he could fix her in space, prevent her talking, put off knowing for one minute more. She was shaking.

"Olivia?" he said quietly into her hair, and felt her nod. The baby had stopped fussing but was moving between them, odd, small, tentative proddings.

"Yes," his mother said, and drew a long, quivering breath. He let go of her and she stepped back in order to look him in the face. "Yes, and poor little Ch-Charlotte, too." She bit her lip briefly and straightened herself.

"The yellow fever has two stages," she said, and lifted the child to her shoulder. It had a head like a small cantaloupe, and Grey was reminded shockingly of its father. "If you survive the first stage—it lasts several days—then sometimes you recover. If not, there's a lull in the fever—a day or two when the—the person seems to be improving, but then . . . it comes back."

She closed her eyes for a moment, and he wondered when she had last slept. She looked at once a thousand years old and ageless, like a stone.

"Olivia," she said, and opened her eyes, patting the child's tiny back, "recovered, or seemed to. Then she went into labor, and—" She lifted the baby slightly in illustration. "But the next day . . . it came back. She was dead in—in hours. It took Charlotte a day later . . . she was . . . so small. So fragile."

"I am so sorry," Grey said softly. He had been fond of his cousin, but his mother had raised Olivia from the age of ten, when his cousin had lost her own parents. A thought came to him.

"Cromwell?" he asked, afraid to hear but needing to know. He'd deliv-

ered Olivia's son, very much by accident, but as a result had always felt close to the boy.

His mother gave him a watery smile.

"He's fine. The fever never touched him, thank God. Nor this little one." She cupped a hand behind the infant's fuzzy skull. "Her name is Seraphina. Olivia had time . . . to hold her, at least, and give her a name. We christened her at once, in case . . ."

"Give her to me, Mother," he said, and took the child from her arms. "You need to go and sit down, and you need something to eat."

"I'm not—" she began automatically, and he interrupted her.

"I don't care. Go sit down. I'll go and blow up the cook."

She tried to give him a smile, and the twitch of her lips reminded him with a jolt of Inocencia. And everything else. His own mourning would have to wait.

IF YOU *HAD* TO attack a fortress at night, on foot and lightly armed, doing it with black men was distinctly an advantage, Grey thought. The barely risen moon was a crescent, a thread of light against the dark sky. Cano's men had removed their shirts and, dressed only in rough canvas breeches, they were no more than shadows, flowing barefoot and silent through the empty marketplace.

Cano himself materialized suddenly behind Grey's shoulder, announced by a waft of foul breath.

"*Ahorita?*" he whispered. Now?

Grey shook his head. Malcolm's wig was wadded up in his pocket and he had assumed instead an infantryman's cap—a contrivance of steel plates, punctured and laced together, to be worn under a uniform hat—this covered with a black knitted cap. He felt as though his head were melting, but it would turn the blade of a sword—or a machete.

"Inocencia," he murmured, and Cano grunted in reply and faded back into the night. The girl wasn't yet late; the church bells had only just rung midnight.

Like any self-respecting fortress, El Castillo de los Tres Reyes Magos del Morro—the Castle of the Three Magi of the Hump, as Azeel had kindly translated its full name for him—the hump being the big black rock at the opening of the harbor—had only one way in and one way out. It also had steeply sloping walls on all sides, to deter both climbers and cannonballs.

True, there were small penetrations on the water side, used for the disposal of garbage or inconvenient bodies, or for the arrival of provisions or the secret deliverance of a guest or prisoner held incognito. Those were of

no use in the present venture, though, as the only possible approach was by boat.

One bell bonged the quarter hour. Two for the half hour. Grey had just pulled his head covering off in order to avoid fainting when there was a stir in the darkness nearby.

"*Señor?*" said a soft, low voice by his elbow. "*Es listo. Venga!*"

"*Bueno,*" he whispered back. "*Señor Cano?*"

"*Aquí.*" Cano *was aquí,* so quickly that Grey realized the man must have been standing no more than a few feet away.

"*Venga,* then." Grey moved his head toward the fortress, then paused to put on his two caps. By the time he had managed this, they were all there, a breathing mass like a herd of cattle, eyes shining now and then in an errant gleam of light.

He took Inocencia by the arm, to prevent her being lost or trampled, and they walked quietly into the small stone guardhouse that shielded the castle's entrance, for all the world like a bride and groom walking sedately into church, followed by a horde of machete-wielding wedding guests.

This absurd fancy disappeared directly as they stepped into the torchlit room. There were four guards, one slumped over a table, the others on the floor. Inocencia shuddered under his hand, and, glancing at her in the flickering light, he saw that her dark dress was torn at the shoulder, and her lip was bleeding. She had drugged the guards' wine, but evidently it hadn't acted fast enough.

"*Bueno,*" he whispered to her, and squeezed her arm. She didn't smile but nodded, swallowed hard, and gestured toward the door on the other side of the guards' room.

This was the entrance to the fortress proper, portcullis and all, and his heart began to beat in his ears as they passed beneath its teeth with no sound but the shuffle of feet and the occasional clink from the bags of metal spikes.

He had gone over and over the maps of the floors, knew where the batteries were—though not which ones were manned at the moment. Inocencia led them into a broad corridor half-lit by torches, with doors on either side. She jerked her chin upward—a stairway at the end.

Up. He could hear the panting of the men behind him—even barefoot they made a lot of noise; surely they would be heard.

They were. A surprised-looking guard stood at the head of the stair, his musket still on his shoulder. Grey rushed him and knocked him down; the men behind him knocked *him* down and trampled him in their eagerness. There was a gurgle and the smell of blood, and something wet soaked through the knee of his breeches.

Up again, no longer in the lead, following the rush of men. He had lost Inocencia but saw her up ahead, being pulled along by Hamid and an-

other of the Mussulman slaves, heads covered with dark bandannas. Another stair, pushing and shoving, grunting bodies hot for a fight.

The next guard had his musket out and fired on them. Shouts from the guard, though he was quickly borne down. Shouts from beyond him and a draft of cold air—the first battery, on the rooftop.

"*Primero!*" Grey bellowed, and a gang of slaves rushed the first cannon. He didn't wait to see how they fared; he was already plunging down a stairwell at the far end of the roof, shouting, "*Segundo!*" at the top of his voice, then pawing and shoving through a clot of slaves and cannon crew that had poured after him and collided, struggling in the narrow space at the foot of the stair.

He shouted, "*Tres! Tres!*" but he couldn't be heard. The air was thick with shrieks and curses and the reek of blood and sweat and fury.

He pushed out of the scrum and pressed himself against a wall, panting for breath. They were gone now, out of anyone's control. He heard the dull *bong* of hammer on iron, though—at least one man had remembered their purpose . . . then the ring and clash of others, striking through the riot. Yes!

Suddenly the Mussulman who had accompanied Hamid burst out of the crowd, Inocencia clutched by the arm. He hurled her at Grey like a bag of wheat and he caught her in much the same way, grunting at the impact.

"*Jesús, Maria, Jesús, Maria,*" she was gasping, over and over. She was splattered with blood, blotches showing wet on the black of her dress, and her eyes showed white all around.

"Are you hurt? Er . . . *dolor?*" he shouted in her ear. She stared at him, dazed.

He must get her out. She'd done all she promised.

"*Venga!*" he shouted in her ear, and jerked her after him, back toward the stair.

"No!" she panted, setting her heels. "*Allí!*" He didn't know that word, but she was dragging him toward the far end of the corridor. This meant leapfrogging squirming bodies on the floor, but he followed her without demur, throwing his body between her and a cannoneer armed with a ramrod. It hit him in the shoulder, numbing his arm, but didn't knock him down. Someone had dropped a bag of spikes, spilling them on the floor, and he nearly fell as these rolled under his feet, clinking on the stones.

They had almost reached the momentary sanctuary of the stairhead when something hit him on the head and he collapsed to his knees. His vision had gone black and his ears were ringing, but through it he could hear Inocencia shrieking at the top of her voice, calling his name.

He struggled blindly, trying to reach the wall so he could get up, but another blow came in from the right. It was a machete—he heard the blade rip the air an instant before the dull *thunk* of metal rang through his head.

Shock and nausea rocked him back against the wall, but he had a hand on the dagger at his waist. He scrabbled it free and, crouching as low as he could, flung himself round on his knees, slashing. He hit someone. The impact jarred the knife from his hand, but his vision was coming back and he found the dagger again, through flashing black and white lights.

Another scream from Inocencia, this one pure terror. He stumbled to his feet, dagger in hand. A scarred back just before him . . . Cano brought down his machete with murderous force and Inocencia dropped to the floor, blood spraying from her head. Without a second's hesitation, Grey thrust the dagger up beneath the man's ribs, as hard as he could.

Cano stiffened, dropped his machete, clattering. He swayed, and fell, but Grey was already by Inocencia's side, scooping her into his arms.

"Fucking bloody hell, oh, bloody hell, please, God . . ." He staggered with her into the stairwell and leaned against the wall for a moment, fighting for breath. She stirred, saying something he couldn't hear for the ringing in his ears.

"No . . ." He shook his head, meaning that he didn't understand, and she flung out a hand, pointing down, emphatically, down, *down!*

"All right." He took a tighter hold and caromed down the narrow stair, slipping and crashing into the stones, then finding his footing once again. He could hear the battle still raging above—but also heard through the fading buzz in his ears the clash of steel and hammers.

He tried to exit at the next landing, but she was having none of it and urged him down, still down. The spots were thickening at the corners of his eyes again, and he smelled damp and seaweed, the brackish scent of low tide.

"Jesus Christ, where are we?" he gasped. He had to set her down but tried to support her with one arm.

"Malcolm," she gasped. "Malcolm," and pointed to a crooked passageway that curved away to the right.

It was like the sort of nightmare that involves endless repetition of something insane, he thought. The last such nightmare hadn't smelled like a dead octopus, though . . .

"Aquí!" She squirmed suddenly and he lost his grip on her. She staggered and crashed into a door that looked as though it had been left outdoors for a century or two. *Still pretty solid,* he thought dimly.

"God, do you mean I have to break it *down?*"

She ignored him, swaying as she fumbled in her skirts. Her face, her

hair, and her shoulder were drenched with blood, and her hands shook so hard that she dropped the keys as soon as she found them. They landed in a clash of metal, drops of blood blooming on the stones around them.

John fumbled in his sleeve for a handkerchief, in some hope of stopping the bleeding, and there ensued an awkward struggle, him trying to tie the cloth around her head, she bending and snatching vainly at the keys, falling every time she bent over.

Grey finally said something in German and grabbed the keys himself. He thrust the handkerchief into Inocencia's twitching fingers and stabbed at the door.

"Quién es?" said Malcolm's voice, quite loudly, near his ear.

"Es mi, querida!" Inocencia collapsed against the door, palms plastered to the wood, and left streaks of blood as she slid slowly down it. Grey dropped the keys, fell to his knees, and grabbed his handkerchief out of her limp hand. He found Malcolm's wig in his pocket, wadded it, and bound it as tightly to her head as he could. There was a long slash through her scalp, and her left ear was hanging by a thread, but he thought dimly that it wasn't that bad—if she didn't bleed to death.

She was gray as a storm cloud and gasping heavily, but her eyes were open, fixed on the door.

Malcolm had been shouting for the last few minutes, pounding on the door 'til it shook. Grey stood up and kicked it several times. The pounding and shouting stopped for a moment.

"Malcolm?" Grey said, bending to look for the keys. "Bloody get dressed. We're leaving as soon as I get this damn door open."

BY THE TIME they reached the main level of the fortress, most of the noise above had ceased. Grey could still hear shouts and the sounds of an occasional scuffle;—there was a lot of muffled Spanish that had an official tone—the officers of the fortress marshaling men, assessing damage, starting the clearing up.

He'd told the slaves: *"Spike the guns, and run. Don't wait about for your companions or for anything else. Make your way into the city and hide. When you think it's safe, go to Cojimar, where the British ships are. Ask for General Stanley or the admiral. Tell them my name."*

He'd given a letter of explanation, and the document signed by the slaves, to Tom Byrd, with instructions to find General Stanley. He hoped Tom had made it to the siege lines without being shot—but he'd sent Tom because of his face. No one could doubt he was an Englishman, at whatever distance.

The night outside was quiet. He breathed the clean sea air and felt the touch of it soft on his face. Then he touched Malcolm's arm—Malcolm was carrying the girl—and pointed toward Calle Yoenis.

"We'll go to my mother's house," he said. "I'll tell you everything when we get there."

SOME LITTLE TIME later, too restless to sit, he limped from the *sala* into the garden and leaned against a flowering quince tree. His ears still rang with the sounds of steel, and he closed his eyes, seeking silence.

Maricela had assured him that Inocencia would live. She herself had stitched the ear back on and applied a *pulpa* of several herbs whose names Grey didn't recognize. Malcolm was still with her. Grey hadn't had the strength to tell Malcolm that he was now a widower rather than an adulterer. The night would vanish, all too soon, but for the moment, time had no meaning. Nothing need be done.

He couldn't know the extent of the slaves' success—but they *had* been successful. Even in the brief frantic interstices of the fighting, he'd seen a dozen guns spiked, and heard the ring of hammers above as he'd half-fallen down the stairs with Inocencia. As he and Malcolm had made their way out of the fortress with her, he'd heard Spanish shouting from the rooftop, furious and thick with curses.

He stood among the fragrant bushes for what seemed a long time, feeling his heart beat, content simply to be breathing. He stirred, though, at the sounds of the garden gate opening and low voices.

"Tom?" he came out from under his sheltering quince, to find both Tom and Rodrigo—both of whom were amazingly, if flatteringly, delighted to see him.

"We thought you was done for, sure, me lord," Tom said for the third or fourth time, following Grey into the kitchen. "You sure you're all right, are you?"

The tone of accusing doubt in this question was so familiar that Grey felt tears come to his eyes. He blinked them away, though, assured Tom that he was somewhat banged about but essentially undamaged.

"Gracias a Dios," Rodrigo said, with such heartfelt sincerity that Grey looked at him in surprise. He said something else in Spanish that Grey didn't understand; John shook his head, then stopped abruptly, wincing.

Tom looked at Rodrigo, who made a small helpless gesture at his inability to be understood and nodded at Tom, who took a deep breath and looked at his employer searchingly.

"What?" Grey said, somewhat disturbed by their solemn attitudes.

"Well, me lord," Tom squared his shoulders, "it's just what Rodrigo told me this afternoon—after you left." He glanced at Rodrigo, who nodded again.

"See, he's been a-wanting to tell you, ever since you come back from the plantations, but he didn't want his wife or Inocencia to hear it. But he got Jacinto to come translate for him, so he could tell me."

"Tell you what?" Grey was discovering the stirrings of hunger and was rummaging through the larder, pulling out sausages and cheese and a jar of some kind of fruit preserve.

"Well, he told me about what happened when you talked to the slaves in the tobacco shed and when the one man told him to leave because he's a zombie." Tom looked protectively at Rodrigo; he'd quite lost any sense of fear about it.

"So he didn't want to stay too near—he says sometimes people gets very upset about him—and he walked down toward the plantation house."

Approaching the house, Rodrigo had come upon the woman Alejandra—Inocencia's cousin, the one who had revealed the slave revolt, in hopes that Inocencia's English lover might be able to do something before anything dreadful could happen.

"She was worried, you could see, Rodrigo says, and talked a lot about her lover—that's Hamid, what he says you met—and how she didn't want him or the others to die, and they would if . . . well, anyway, they got summat close to the big hacienda, and she stopped sudden."

Alejandra had stood there in the darkness, her white dress seeming to float in the air beside Rodrigo like a ghost. He stood with her, quiet, waiting to see what she would say next. But she hadn't spoken, only stood frozen for what seemed a long time but probably wasn't, the night wind rising and stirring her skirts.

"Then she took his arm and said they should go back, and they did. But . . ." Tom coughed, his round face troubled, and looked at Rodrigo again.

"Rodrigo said Azeel told him on the way back to Havana what happened in the shed. What you said to that man, Cano, and what he said to you—about the people what owned the plantation."

"Yes?" Grey paused in the act of buttering a chunk of bread.

Rodrigo said something quiet, and Tom nodded.

"He said something didn't seem right while they were looking at the house. There were servants going in and out, but it just didn't feel right to him. And when he heard what this Cano said to you—"

"*No los mataremos,*" Grey said, suddenly uneasy. "'We will not kill them'?"

Rodrigo nodded, and Tom cleared his throat.

"You can't kill somebody what's already dead, can you, me lord?"

"Already . . . no. No, you can't mean that the slaves had already . . . No." But a worm of doubt was taking up residence in his stomach, and he put the bread down.

"The . . . wind," Rodrigo said, with his usual agonizing pause to find an English word. *"Muerto."*

He lifted his hand, a beautiful, slender hand, and drew his knuckles gently beneath his nose.

"I . . . know . . . the smell . . . of death."

COULD IT BE TRUE? Grey was too exhausted to feel more than a distant sense of cold horror at the notion, but he couldn't dismiss it. Cano had not struck him as a patient man. He could easily imagine that the slave had grown frustrated when Malcolm didn't appear soon enough and had decided to carry out his original plan. But then when Grey *did* come—Christ, he must have arrived on the heels of the . . . the massacre . . .

He remembered his sight of the hacienda: lights burning inside but so quiet. No sense of movement within; only the silent passage of the house-slaves outside. And the stink of anger in the tobacco shed. He shuddered.

He took his leave of Tom and Rodrigo but, too tired and shocked to sleep, then sought refuge in the *sala,* which seemed always to have light. One of the kitchen maids, undoubtedly roused by Tom, came in with a pitcher of wine and a plate of cheese; she smiled sleepily at him, murmured, *"Buenas noches, señor,"* and stumbled back toward her bed.

He couldn't eat, or even sit down, and after a moment's hesitation went out again, into the deserted patio. He stood there for some time, looking up into the black velvet sky. What time was it? The moon had set and surely dawn could not be far off, but there was no trace of light save the distant stars.

What should he do? Was there anything he *could* do? He thought not. There was no way of telling whether Rodrigo was right—and even if he was (a small, cold feeling at the back of Grey's neck was inclined to believe it) . . . there was nothing to be done, no one to tell who could investigate, let alone try to find the murderers, if murderers they were.

The city lay suspended between the Spanish and the British invaders; there was no telling when the siege would be successful—though he thought it would. The spiking of El Morro's guns would help, but the navy must be informed, so as to take advantage of it.

Come dawn, he would try to leave the city with his mother and the children and his servants. He thought it could be managed easily enough; he had brought as much gold from Jamaica as he could, and there was more than enough left to bribe their way past the guard at the city gate.

What then? Exhausted as he was, he wasn't even thinking, just watching dimly as the future unrolled in small, disjointed pictures: a carriage for his mother and the children and Azeel, himself on the stubborn white mule, two more animals for Tom and Rodrigo.

The slaves' contract . . . if any of them had survived . . . freedom . . . the general could see to that . . .

Malcolm and the girl . . . he wondered dimly for a moment about Inocencia; why had Cano tried to kill her . . . ?

Because she saw him try to kill you, fathead, some dim, dispassionate watcher in his skull observed. *And he had to kill you, for fear you'd find out what they'd done at Hacienda Mendez . . .*

Freedom . . . even if they'd? . . . but Cano was dead, and Grey would never know who was guilty of what.

"Not my place . . ." he murmured and shut his eyes.

His hand touched the breast of his shirt and found it stiff with dried blood. He'd left his uniform coat in the kitchen . . . perhaps one of the women could clean it. He'd need to wear it again, to approach the British lines in Cojimar . . . Cojimar . . . a brief vision of white graveled sand, sunlight, fishing boats . . . the tiny white stone fort, like a doll's house . . . find General Stanley.

Thought of the general drew his fragmented thoughts together, a magnet in a scatter of loose iron filings. Someone to depend on . . . a man to share the burden . . . he wanted that, above all things.

"Oh, God," he whispered, and moths touched his face, gentle in the dark.

HE WAS GROWING COLD. He went back inside to the *sala* and found his mother sitting there. She had taken the manuscript from the secretaire; it sat on the small table beside her, her hand resting on it and a distant look in her eyes. He didn't think she'd noticed him come in.

"Your . . . manuscript," John said awkwardly. His mother came back abruptly from wherever she had been, her eyes alert but calm.

"Oh," she said. "You read it?"

"No, no," he said, embarrassed. "I . . . I only wondered . . . why are you writing your memoirs? I mean, that *is* what it is, isn't it?"

"Yes, it is," she said, looking faintly amused. "It would have been quite all right if you'd read it—you may read it whenever you like, in fact, though perhaps it would be better to wait until I've finished. If I do."

He felt a small sense of relaxation at this. His mother was both honest and blunt by nature, and the older she got the less she cared for anyone's opinion save her own—but she did have a very deep degree of emotional

perception. She was reasonably sure that whatever she'd written wouldn't embarrass him seriously.

"Ah," he said. "I wondered whether perhaps you meant it for publication. Many"—he choked off the words "old people" just in time, replacing them with—"people who've led interesting lives choose to, er, share their adventures in print."

That made her laugh. It was no more than a low, soft laugh, but nonetheless it brought tears to her eyes, and he thought it was because he'd inadvertently cracked the shell she'd built over the course of the last weeks and let her own feelings bubble back to the surface. The thought made him happy, but he looked down to hide it, pulled a clean handkerchief from his sleeve, and handed it to her without comment.

"Thank you, dear," she said, and, having dabbed her eyes, shook her head.

"Persons who have truly interesting lives *never* write about them, John—or not with an eye to publication, at least. The ability to keep their own counsel is one of the things that makes them interesting and is also what causes other truly interesting people to confide in them."

"I assure you, Mother," he said dryly, "you are undoubtedly the most interesting woman I've ever met."

She snorted briefly and gave him a direct look.

"I suppose that's why you haven't yet married, is it?"

"I didn't think a wife needed to be interesting," he replied, with some honesty. "Most of the ones I know certainly aren't."

"How true," she said briefly. "Is there any wine in the house, John? I've got rather fond of Spanish wine since I've been here."

"*Sangria* do you? One of the maids brought me a pitcher of it, but I hadn't drunk any yet." He got up and fetched the pitcher—a beautiful smooth stoneware thing the color of mulberries—and brought it with a pair of glasses to the table between their chairs.

"That will be perfect," she said, and leaned forward with a sigh, massaging her temples. "Oh, God. I go about all day, feeling that none of it is real, that everything is just as I left it, and then suddenly—" She broke off and dropped her hands, her features drawn with pain and tiredness. "Suddenly it's real again."

She glanced at the secretaire as she said this, and John caught a hint of something in her voice. He poured the wine carefully, not to let the sliced lemons and oranges floating in it fall out into the glasses, and didn't speak until he'd put the pitcher down and taken his seat again.

"When you write it down . . ." he said. "Does that make it—whatever it is—real again? Or does the act of putting it into words make it unreal? You know, something . . . separate from yourself." What had happened at El Morro had taken place mere hours before, and yet it seemed like years.

But the scent of blood and guns hung about him like a shroud, and his muscles still twitched with the memory of desperate exertion.

His own words brought back to him the letters he had written now and then. The phantoms, as he thought of them: letters he'd written to Jamie Fraser—honest, conversational, heartfelt, and very real. No less real because he'd burned them all.

His mother looked at him in surprise, then took a meditative sip of the cool spiced wine.

"Both," she said at last. "It's completely real to me as I write it—and should I go back to read it again later, it's real again." She paused for a moment, thinking. "I can live in it," she said softly. She finished her wine—the glasses were small, the sort of cup called a shot glass because the heavy base made it possible to slam it on the table with a loud report at the conclusion of a toast—and carefully poured more.

"But when it's done, and I leave it . . ." She sipped again, the scent of red wine and oranges softening the smells of travel and sickness in her clothes. "It . . . seems somehow to separate itself from me. I can set it—whatever it was, whatever it is—aside in my mind then, just as I set aside the page."

"How very useful," John murmured, half to himself, thinking that he must try that. The wine was dissolving his own sense of sorrow and exhaustion—if only temporarily. The room grew peaceful around them, candlelight warm on the plastered walls, the wings of angels.

"But as to why—" His mother refilled his glass, and hers again.

"It's a duty. The book—should it be a book—I'll have it printed and bound, but privately. It's for you and the other boys, for the children—for Cromwell and Seraphina," she added softly, and her lips quivered for an instant.

"Mother," he said quietly, and laid his hand on hers. She bent her head and put her free hand on his, and he saw how the tendrils of her hair, still thick, once blond like his own but mostly silver now, escaped from their plait and curled on her neck.

"A duty," she said, holding his hand between her own. "The duty of a survivor. Not everyone lives to be old, but if you do, I think you owe it to those who didn't. To tell the stories of those who shared your journey . . . for as long as they could."

She closed her eyes and two tears ran down her cheeks.

He put his arm around her and drew her head down on his shoulder, and they sat silently together, waiting for the light to come back.

AUTHOR'S NOTE

Whale Oil

WHALE OIL VERSUS SPERMACETI. NOW, SEE, I ACTU-ally read the entirety of the infamous "list of whales" chapter of *Moby-Dick* and thought it was hilarious. But I admit that I was (at one point in a highly checkered career) a marine biologist, so I may have been slightly more aligned with Melville's frame of reference than is the casual modern reader, who might be inclined to think of whale oil as being the same thing as spermaceti (assuming the CMR to be sufficiently widely read as to have encountered "spermaceti" in print at all).

In fact, though, these are two completely different (though equally combustible) substances. Whale oil is rendered from the flensed blubber of slaughtered whales. In other words, it's the liquefied body fat of something that feeds mostly on small crustaceans. Body chemistry being what it is, an organism that stores energy in body fat also tends to store iffy chemicals it encounters in the same depository.

Your own body, for instance, stores excess hormones in your body fat, as well as various toxic or otherwise dubious compounds like PBCs, strontium, and insecticides.

The point here is that dead crustaceans are rather pungent. Think of the last time you left a package of thawed frozen shrimp in your refrigerator for a week. These aromatic compounds are stored in the body fat of things that eat the organism that makes them.

I first encountered this phenomenon when I had a postdoctoral appointment in which my principal job was dissecting gannets. These are big diving seabirds (related to boobies) that feed largely on squid. Their body fat smells like rotting squid, especially when you put it in a drying oven in order to desiccate it. So if you're burning whale oil in your lamps (it was cheap, as Tom Byrd notes), your establishment is probably going to smell like week-old krill. And, being fat, it makes smoke when you burn it.

Spermaceti, by contrast, is not body fat as such—though it is oily

and very burnable. It's an oil that is secreted and stored in the head case (basically, a storage compartment for this oil) of a sperm whale. The appearance of this liquid—white, thickish, slippery—is why they're called sperm whales; that's what the old whalers thought the gunk was, though plainly occurring in the wrong place. . . . However, the point here is that spermaceti was also very popular as lamp fuel and general lubricant—because it didn't stink. It's very clean-burning and almost odorless. But it's much more limited in availability, as only sperm whales make it, and thus much more expensive than whale oil.

So, what's the sperm whale using this substance for? Nobody knows, though speculation is that it's part of the whale's sensory system, perhaps acting as an echolocation device, assisting the whale to locate things like giant squid (a major component of its diet, and I'm profoundly grateful that I will likely never be called upon to dissect and analyze the body tissues of a sperm whale) in the black abyssal depths.

Ambassadors, Consuls, and British Diplomats

An ambassador is an appointed office in the British diplomatic service, and very formal. An ambassador may receive official tenders from the foreign power to which he is appointed—declarations of war, statements of intent, official notices of concern, etc.—and by and large acts as the delegated (nonmilitary) authority of the British government within his own territory (they didn't have female ambassadors in the eighteenth century; it was always "his").

A consul is a much less formal office, though also appointed by the government. A consul's duties are to look after the welfare of British citizens in the country to which he's appointed. He would assist with things like permits to do business, small trade agreements, the relief of British citizens who have run into trouble in the foreign country, and so on. He does not have full diplomatic powers but is generally regarded as part of the diplomatic service.

Now, Britain didn't have a real ambassador to Cuba until sometime in the late 1800s. They did have consuls appointed for some time prior to the appointment of a real ambassador, though, and Malcolm Stubbs would have been one of these.

The Siege of Havana

The thing about a siege is that it's usually rather a long-drawn-out affair. The 1762 siege of Havana (there was more than one, so we

need to specify the date) took several weeks, beginning with the arrival of the Duke of Albemarle's fleet (under the command of Admiral George Pocock—a real person, and, no, I have No Idea whether he might have anything to do with anyone else we've met recently . . .) on June 6 and ending on August 14, when the British entered the conquered city.

It was a fairly traditional siege, in that the British were obliged to throw up breastworks from which to fire. That's the traditional way of referring to the act of erecting or digging barriers to shelter the besieging forces—and in some instances, it *is* pretty rapid. Others, not so much.

At Havana, the rock of the promontory on which the fortress of El Morro sat was impervious to digging and prevented a head-on advance. The British (or, rather, the American volunteers from Connecticut and New Hampshire—though, mark me, these men *were* still Englishmen at the time) had to blast trenches through the hard coral rock to approach from the sides and erect wooden breastworks above the trenches to cover the advance. This was naturally a tedious business, made worse by mosquitoes and yellow fever (which killed an enormous number of both besiegers and inhabitants of the city).

If you want an account of the actual siege, there are plenty of them available online, some with considerable detail. However . . . this particular story is not really *about* the siege (let alone how many ships of the line and how many men took part—21 ships of the line, 24 lesser warships, and 168 other vessels, mainly transports, carrying 14,000 seamen and marines, 3,000 hired sailors, and 12,862 regulars, if you *do* care) but about Lord John and his personal sense of honor and responsibility.

That being so, I've chosen to shorten the duration of the siege considerably rather than find a way for Lord John to spend an extra six weeks doing nothing.

Now, I will note that while the slave revolt at the Mendez and Saavedra plantations is a fictional one, there *were* several slave revolts on Cuba during the second half of the eighteenth century, and such an event would not be improbable in the least.

Likewise, while I found no account of the guns of El Morro being spiked, it *is* true that the siege was finally ended by a naval bombardment of the fortress—taking advantage of the sudden silence of most of the guns of the castle's battery.

And there is a historical note that ninety slaves were given their freedom after the battle, *in return for their services during the siege.*

ACKNOWLEDGMENTS

I'd like to acknowledge . . .

The invaluable suggestions regarding the French bits of dialogue contributed by Bev LaFrance (France), Gilbert Sureau (French Canada), and a number of other nice people whose names I unfortunately didn't write down at the time . . .

The assistance of Maria Syzbek in the delicate matter of Polish vulgarities (any errors in grammar, spelling, or accent marks are entirely mine) and of Douglas Watkins in the technical descriptions of small-plane maneuvers (also the valuable suggestion of the malfunction that brought Jerry's Spitfire down) . . .

The help of several people in researching aspects of Jewish history, law, and custom for "Virgins": Elle Druskin (author of *To Catch a Cop*), Sarah Meyer (registered midwife), Carol Krenz, Celia K. and her Reb mom, and especially Darlene Marshall (author of *Castaway Dreams*). I'm indebted also to Rabbi Joseph Telushkin's very helpful book *Jewish Literacy*. Any errors are mine . . .

Eve Ackermann and Elle Druskin for helpful notes and references regarding Sephardic wedding traditions and rituals . . .

Catherine MacGregor and her Francophone associates, especially Madame Claire Fluet, for unblushing help with the lascivious French bits . . .

Selina Walker and Cass DiBello for kind assistance with eighteenth-century London geography . . .

Simcha Meijer, for help with the Dutch language bits, and to a number of helpful Dutch readers on Facebook, for suggestions as to appropriate powdered sugar pastries for a pregnant lady . . .

And a number of kind Cuban Facebook readers, for helpful observations and suggestions on the color of Cuban dirt, the appearance of Cuban bread, traditional Cuban food, and the correct spelling of "inocencia" . . .

And the wonderful Penguin Random House team who have, as usual, killed themselves to produce a wonderful book: My editor, Jennifer Hershey, for insight and helpful suggestions, Anne Speyer, who did most of the heavy lifting on this one, Erin Kane for useful Spanish suggestions, our heroically prompt and always astute copy editor, Kathy Lord, and—as always—Virginia Norey, for the beautiful design of the book.

ABOUT THE AUTHOR

DIANA GABALDON is the #1 *New York Times* bestselling author of the wildly popular Outlander novels—*Outlander, Dragonfly in Amber, Voyager, Drums of Autumn, The Fiery Cross, A Breath of Snow and Ashes* (for which she won a Quill Award and the Corine International Book Prize), *An Echo in the Bone,* and *Written in My Own Heart's Blood*—as well as the related Lord John Grey books *Lord John and the Private Matter, Lord John and the Brotherhood of the Blade, Lord John and the Hand of Devils,* and *The Scottish Prisoner;* two works of nonfiction, *The Outlandish Companion, Volumes 1* and *2;* the Outlander graphic novel *The Exile;* and *The Official Outlander Coloring Book.* She lives in Scottsdale, Arizona, with her husband.

dianagabaldon.com
Facebook.com/AuthorDianaGabaldon
Twitter: @Writer_DG

To inquire about booking Diana Gabaldon for a speaking engagement, please contact the Penguin Random House Speakers Bureau at speaker@penguinrandomhouse.com.